THE ADAPTED MIND

THE ADAPTED MIND

Evolutionary Psychology and the Generation of Culture

Edited by
Jerome H. Barkow
Leda Cosmides
John Tooby

New York   Oxford
OXFORD UNIVERSITY PRESS

Oxford University Press

Oxford   New York   Toronto
Delhi   Bombay   Calcutta   Madras   Karachi
Kuala Lumpur   Singapore   Hong Kong   Tokyo
Nairobi   Dar es Salaam   Cape Town
Melbourne   Auckland   Madrid

and associated companies in
Berlin   Ibadan

First published in 1992 by Oxford University Press, Inc.,
198 Madison Avenue, New York, NY l00l6

First issued as an Oxford University Press Paperback, 1995

Library of Congress Cataloging-in-Publication Data
The adapted mind: evolutionary psychology and the generation of
culture / edited by Jerome H. Barkow, Leda Cosmides, John Tooby.
p. cm. Includes bibliographical references and index.
ISBN 0-19-506023-7
ISBN 0-19-510107-3 (PBK)
1. Genetic psychology.   2. Cognition and culture.   3. Behavior
evolution.   4. Sociobiology.   I. Barkow, Jerome H.   II. Cosmides,
Leda.   III. Tooby, John.
BF711.A33   1992   155.7—dc20   91-25307

9 8 7
Printed in the United States of America
on acid-free paper.

ACKNOWLEDGMENTS

Jerome H. Barkow thanks his wife Irma and his children Philip and Sarah for their patience and support during the never-ending preparation of this volume. Much thanks, too, to the Dalhousie colleagues who listened to ideas and made suggestions, and to Dalhousie University's Research Development Fund which provided some financial support.

This volume was prepared, in part, while Leda Cosmides and John Tooby were Fellows at the Center for Advanced Study in the Behavioral Sciences in Stanford, California, and they are deeply indebted to the Center's staff for their many kindnesses. They are grateful for the Center's financial support during this period, as well as that provided by the Gordon P. Getty Trust, the Harry Frank Guggenheim Foundation, and NSF Grant BNS87-00864 to the Center. In addition, they would like to thank NSF Grant BNS85-11685 to Roger Shepard and NSF Grant BNS9157-449 to John Tooby for research support.

The research and thinking of Cosmides and Tooby about the convergence between evolutionary biology, psychology, and the social sciences have been shaped and sustained by many friends and colleagues. Outstanding among these have been Donald Brown, David Buss, Martin Daly, Irven DeVore, Paul Ekman, Gerd Gigerenzer, Steven Pinker, Paul Romer, Paul Rozin, Roger Shepard, Dan Sperber, John Staddon, Valerie Stone, Donald Symons, George Williams, and Margo Wilson. Cosmides and Tooby are particularly grateful for the institutional shelter as well as the intellectual comradeship provided by Irven DeVore of Harvard University and Roger Shepard of Stanford University. The final stages of editing were completed at the Zentrum für interdisziplinäre Forschung (ZiF) of the University of Bielefeld, and Cosmides and Tooby would like to thank Peter Weingart and the ZiF for their support and assistance.

The editors owe a special debt of gratitude to Joan Bossert of Oxford University Press, whose encouragement and enthusiasm for this project have been extraordinary. We also thank Louise Page, Melodie Wertelet, and Constance Devanthéry-Lewis who, together with Joan, have unflaggingly attended to all aspects of the production of this book. Thanks also to Larry and Michelle Sugiyama for their invaluable assistance in the final hours of preparation.

Finally, the editors would like to express their heartfelt gratitude to the contributors to the volume, who labored far above and beyond the call of duty. The unusual plan for this volume and its chapters, involving the attempt to integrate levels of explanation within a single structured format (one that began with selection pressures, then moved to mechanisms and to cultural output) placed a heavy burden on the contributors. Their willingness to work with the editors, often through an extended series of drafts, in order to accommodate this conceptual organization is deeply appreciated.

Contents

VIII. NEW THEORETICAL APPROACHES TO CULTURAL PHENOMENA

CONTRIBUTORS

Jerome H. Barkow
Department of Anthropology
Dalhousie University
Halifax, Nova Scotia B3H 1T2
Canada

Paul Bloom
Department of Psychology
University of Arizona
Tucson, AZ 85721

Michael J. Boulton
Psychology Section
School of Health and Community
    Studies
Sheffield City Polytechnic
Sheffield S10 2TN
United Kingdom

David M. Buss
Department of Psychology
University of Michigan
Ann Arbor, MI 48109

Leda Cosmides
Department of Psychology
University of California
Santa Barbara, CA 93106

Martin Daly
Department of Psychology
McMaster University
Hamilton, Ontario L8S 4K1
Canada

Marion Eals
Department of Psychology
York University
North York, Ontario M3J 1P3
Canada

Bruce J. Ellis
Department of Psychology
University of Michigan
Ann Arbor, MI 48109

Anna T. C. Feistner
Jersey Wildlife Preservation Trust
Trinity, Jersey
Channel Islands

Anne Fernald
Department of Psychology
Stanford University
Stanford, CA 94305

Judith H. Heerwagen
Departments of Architecture and
    Psychosocial Nursing
University of Washington
Seattle, WA 98195

Stephen Kaplan
Department of Psychology
University of Michigan
Ann Arbor, MI 48109

Alan T. Lloyd
Fox Run Hospital
St. Clairsville, OH 43950

Janet Mann
Department of Psychology
Georgetown University
Washington, DC 20057

W. C. McGrew
Scottish Primate Research Group
Department of Psychology
University of Stirling
Stirling FK9 4LA
Scotland

Randolph M. Nesse
Department of Psychiatry
University of Michigan Medical
    Center
Ann Arbor, MI 48109

Gordon H. Orians
Department of Zoology and
    Institute for Environmental
    Studies
University of Washington
Seattle, WA 98195

Steven Pinker
Department of Brain and Cognitive
    Sciences
Massachusetts Institute of
    Technology
Cambridge, MA 02139

Margie Profet
Divison of Biochemistry and
    Molecular Biology
University of California
Berkeley, CA 94720

Roger N. Shepard
Department of Psychology
Stanford University
Stanford, CA 94305

Irwin Silverman
Department of Psychology
York University
North York, Ontario M3J 1P3
Canada

Peter K. Smith
Department of Psychology
University of Sheffield
Sheffield S10 2TN
United Kingdom

Donald Symons
Department of Anthropology
University of California
Santa Barbara, CA 93106

John Tooby
Department of Anthropology
University of California
Santa Barbara, CA 93106

Margo Wilson
Department of Psychology
McMaster University
Hamilton, Ontario L8S 4K1
Canada

THE ADAPTED MIND

Introduction: Evolutionary Psychology and Conceptual Integration

LEDA COSMIDES, JOHN TOOBY, AND JEROME H. BARKOW

The Adapted Mind is an edited volume of original, commissioned papers centered on the complex, evolved psychological mechanisms that generate human behavior and culture. It has two goals: The first is to introduce the newly crystallizing field of evolutionary psychology to a wider scientific audience. Evolutionary psychology is simply psychology that is informed by the additional knowledge that evolutionary biology has to offer, in the expectation that understanding the process that designed the human mind will advance the discovery of its architecture. It unites modern evolutionary biology with the cognitive revolution in a way that has the potential to draw together all of the disparate branches of psychology into a single organized system of knowledge. The chapters that follow, for example, span topics from perception, language, and reasoning to sex, pregnancy sickness, and play. The second goal of this volume is to clarify how this new field, by focusing on the evolved information-processing mechanisms that comprise the human mind, supplies the necessary connection between evolutionary biology and the complex, irreducible social and cultural phenomena studied by anthropologists, sociologists, economists, and historians.

Culture is not causeless and disembodied. It is generated in rich and intricate ways by information-processing mechanisms situated in human minds. These mechanisms are, in turn, the elaborately sculpted product of the evolutionary process. Therefore, to understand the relationship between biology and culture one must first understand the architecture of our evolved psychology (Barkow, 1973, 1980a, 1989a; Tooby & Cosmides, 1989). Past attempts to leapfrog the psychological—to apply evolutionary biology directly to human social life—have for this reason not always been successful. Evolutionary psychology constitutes the missing causal link needed to reconcile these often warring perspectives (Cosmides & Tooby, 1987).

With evolutionary psychology in place, cross-connecting biology to the social sciences, it is now possible to provide conceptually integrated analyses of specific questions: analyses that move step by step, integrating evolutionary biology with psychology, and psychology with social and cultural phenomena (Barkow, 1989a; Tooby & Cosmides, 1989). Each chapter in this volume is a case study of the difficult task of integrating across these disciplinary boundaries. Although it has been said that the first expressions of new and better approaches often look worse than the latest and most elaborated expressions of older and more deficient ones, we think these chapters are illuminating contributions to the human sciences that stand up well against prevailing approaches. Nevertheless, readers should bear in mind that none of these chapters are meant to be the last word "from biology" or "from psychology"; they are not intended to definitively settle issues. They are better thought of as "first words," intended to open new lines of investigation and to illustrate the potential inherent in this new outlook.

CONCEPTUAL INTEGRATION IN THE BEHAVIORAL AND SOCIAL SCIENCES

Conceptual integration—also known as *vertical integration*[1]—refers to the principle that the various disciplines within the behavioral and social sciences should make themselves mutually consistent, and consistent with what is known in the natural sciences as well (Barkow, 1980b, 1982, 1989a; Tooby & Cosmides, this volume). The natural sciences are already mutually consistent: the laws of chemistry are compatible with the laws of physics, even though they are not reducible to them. Similarly, the theory of natural selection cannot, even in principle, be expressed solely in terms of the laws of physics and chemistry, yet it is compatible with those laws. A conceptually integrated theory is one framed so that it is compatible with data and theory from other relevant fields. Chemists do not propose theories that violate the elementary physics principle of the conservation of energy: Instead, they use the principle to make sound inferences about chemical processes. A compatibility principle is so taken for granted in the natural sciences that it is rarely articulated, although generally applied; the natural sciences are understood to be continuous.

Such is not the case in the behavioral and social sciences. Evolutionary biology, psychology, psychiatry, anthropology, sociology, history, and economics largely live in inglorious isolation from one another: Unlike the natural sciences, training in one of these fields does not regularly entail a shared understanding of the fundamentals of the others. In these fields, paying attention to conceptual integration and multidisciplinary compatibility, while not entirely unknown, is unusual (Campbell, 1975; Hinde, 1987; Symons, 1979). As a result, one finds evolutionary biologists positing cognitive processes that could not possibly solve the adaptive problem under consideration, psychologists proposing psychological mechanisms that could never have evolved, and anthropologists making implicit assumptions about the human mind that are known to be false. The behavioral and social sciences borrowed the idea of hypothesis testing and quantitative methodology from the natural sciences, but unfortunately not the idea of conceptual integration (Barkow, 1989a; Tooby & Cosmides, this volume).

Yet to propose a psychological concept that is incompatible with evolutionary biology is as problematic as proposing a chemical reaction that violates the laws of physics. A social science theory that is incompatible with known psychology is as dubious as a neurophysiological theory that requires an impossible biochemistry. Nevertheless, theories in the behavioral and social sciences are rarely evaluated on the grounds of conceptual integration and multidisciplinary, multilevel compatibility.

With *The Adapted Mind,* we hope to provide a preliminary sketch of what a conceptually integrated approach to the behavioral and social sciences might look like. Contributors were asked to link evolutionary biology to psychology and psychology to culture—a process that naturally entails consistency across fields.

The central premise of *The Adapted Mind* is that there is a universal human nature, but that this universality exists primarily at the level of evolved psychological mechanisms, not of expressed cultural behaviors. On this view, cultural variability is not a challenge to claims of universality, but rather data that can give one insight into the structure of the psychological mechanisms that helped generate it. A second premise is that these evolved psychological mechanisms are adaptations, constructed by natural selection over evolutionary time. A third assumption made by most of the contributors is that the evolved structure of the human mind is adapted to the way of life of Pleistocene hunter-gatherers, and not necessarily to our modern circumstances.

What we think of as all of human history—from, say, the rise of the Shang, Minoan, Egyptian, Indian, and Sumerian civilizations—and everything we take for granted as normal parts of life—agriculture, pastoralism, governments, police, sanitation, medical care, education, armies, transportation, and so on—are all the novel products of the last few thousand years. In contrast to this, our ancestors spent the last two million years as Pleistocene hunter-gatherers, and, of course, several hundred million years before that as one kind of forager or another. These relative spans are important because they establish which set of environments and conditions defined the adaptive problems the mind was shaped to cope with: Pleistocene conditions, rather than modern conditions. This conclusion stems from the fact that the evolution of complex design is a slow process when contrasted with historical time. Complex, functionally integrated designs like the vertebrate eye are built up slowly, change by change, subject to the constraint that each new design feature must solve a problem that affects reproduction better than the previous design. The few thousand years since the scattered appearance of agriculture is only a small stretch in evolutionary terms, less than 1% of the two million years our ancestors spent as Pleistocene hunter-gatherers. For this reason, it is unlikely that new complex designs—ones requiring the coordinated assembly of many novel, functionally integrated features—could evolve in so few generations (Tooby & Cosmides, 1990a, 1990b). Therefore, it is improbable that our species evolved complex adaptations even to agriculture, let alone to postindustrial society. Moreover, the available evidence strongly supports this view of a single, universal panhuman design, stemming from our long-enduring existence as hunter-gatherers. If selection had constructed complex new adaptations rapidly over historical time, then populations that have been agricultural for several thousand years would differ sharply in their evolved architecture from populations that until recently practiced hunting and gathering. They do not (Barkow, 1980a, 1989a, 1990).

Accordingly, the most reasonable default assumption is that the interesting, complex functional design features of the human mind evolved in response to the demands of a hunting and gathering way of life. Specifically, this means that in relating the design of mechanisms of the mind to the task demands posed by the world, "the world" means the Pleistocene world of hunter-gatherers. That is, in considering issues of functionality, behavioral scientists need to be familiar with how foraging people lived. We cannot rely on intuitions honed by our everyday experiences in the modern world. Finally, it is important to recognize that behavior generated by mechanisms that are adaptations to an ancient way of life will not necessarily be adaptive in the modern world. Thus, our concern in this volume is with adaptations—mechanisms that evolved by natural selection—and not with modern day adaptiveness (Symons, this volume; see also Barkow, 1989a, 1989b).

Aside from the two opening, orienting chapters and the concluding one, each chapter of *The Adapted Mind* focuses on an adaptive problem that our hunter-gatherer ancestors would have faced: a problem that affected reproduction, however distally, such as finding mates, parenting, choosing an appropriate habitat, cooperating, communicating, foraging, or recovering information through vision. We asked each contributor to consider three questions: (1) What selection pressures are most relevant to understanding the adaptive problem under consideration?; (2) What psychological mechanisms have evolved to solve that adaptive problem?; and (3) What is the relationship between the structure of these psychological mechanisms and human culture? We chose these three questions because there are interesting causal relationships between selection pressures and psychological mechanisms on the one hand, and between psychological mechanisms and cultural forms on the other.

There is now a rich literature in evolutionary biology and paleoanthropology that allows one to develop useful models of selection pressures, and there have been for many decades in anthropology, sociology and other social sciences rich descriptions of social and cultural phenomena. Using the above three questions, *The Adapted Mind* is intended to supply the missing middle: the psychological mechanisms that come between theories of selection pressures on the one hand and fully realized sociocultural behavior on the other. By concentrating on evolved mechanisms, this collection represents a departure from both traditional anthropology and various evolutionarily inspired theories of culture and behavior. Although both of these fields recognize that culture and cultural change depend critically upon the transmission and generation of information, they have frequently ignored what should be the causal core of their field: the study of the evolved information-processing mechanisms that allow humans to absorb, generate, modify, and transmit culture—the psychological mechanisms that take cultural information as input and generate behavior as output (Barkow, 1978, 1989a; Tooby & Cosmides, 1989). Our goal in this collection is to focus on these mechanisms in order to see where a more precise understanding of their structure will lead.

Because an evolutionary perspective suggests that there will be a close functional mesh between adaptive problems and the design features of the mechanisms that evolved to solve them, each chapter of *The Adapted Mind* focuses on an adaptive problem, and each discusses what kind of psychological mechanisms one might expect natural selection to have produced to solve that problem. Evidence from the literatures of psychology, anthropology, and evolutionary biology was brought to bear on these hypotheses whenever possible. Many of the authors also addressed a few of the implications that the psychological mechanisms they studied might have for culture. The relationship between psychology and culture can be complex, and in some cases the psychological mechanisms are not yet sufficiently well-understood to make any meaningful statement. Nevertheless, in the interests of conceptual integration, the contributors to *The Adapted Mind* have tried, insofar as it has been possible, to bring data from cross-cultural studies to bear on their psychological hypotheses, to point out when the psychological mechanisms discussed can be expected to cause variation or uniformity in practices, preferences, or modes of reasoning across cultures, or to discuss what implications the psychological mechanisms concerned might have for various theories of cultural change.

BASIC CONCEPTS IN EVOLUTIONARY PSYCHOLOGY AND BIOLOGY

The organization of *The Adapted Mind* is unusual: Few works in psychology or the social sciences are organized around adaptive problems. The decision to do so was theoretically motivated. The first two chapters, "The Psychological Foundations of Culture," by Tooby and Cosmides, and "On the Use and Misuse of Darwinism in the Study of Human Behavior," by Symons, as well as the last chapter, "Beneath New Culture Is Old Psychology," by Barkow, present the theoretical program that animates this volume (see also Barkow, 1989a, 1990; Brown, 1991; Cosmides & Tooby, 1987; Daly & Wilson, 1988; Sperber, 1985; Symons, 1979; Tooby & Cosmides, 1989, 1990b). But because this volume is aimed at a broad social science audience, each discipline of which is familiar with different concepts and terms, it may prove helpful to begin with a brief orientation to what the contributors to this volume mean when they use terms such as *mind, selection, adaptive problem,* and *evolutionary psychology.*

Evolutionary psychology is psychology informed by the fact that the inherited architecture of the human mind is the product of the evolutionary process. It is a conceptually integrated approach in which theories of selection pressures are used to generate hypotheses about the design features of the human mind, and in which our knowledge of psychological and behavioral phenomena can be organized and augmented by placing them in their functional context. Evolutionary psychologists expect to find a functional mesh between adaptive problems and the structure of the mechanisms that evolved to solve them. Moreover, every psychological theory—even the most doctrinairely "anti-nativist"—carries with it implicit or explicit evolutionary hypotheses. By making these hypotheses explicit, one can evaluate whether psychological theories are consistent with evolutionary biology and paleoanthropology and, if not, investigate which field needs to make changes.

There are various languages within psychology for describing the structure of a psychological mechanism, and many evolutionary psychologists take advantage of the new descriptive precision made possible by cognitive science. Any system that processes information can be described in at least two different, mutually compatible and complementary ways. If asked to describe the behavior of a computer, for example, one could characterize the ways in which its physical components interact—how electrons flow through circuits on chips. Alternatively, one could characterize the programs that the system runs—what kind of information the computer takes as input, what rules or algorithms it uses to transform that information, what kinds of data structures (representations) those rules operate on, what kinds of output it generates. Naturally, programs run by virtue of the physical machine in which they are embodied, but an information-processing description neither reduces to nor can replace a physical description, and vice versa. Consider the text-editing program "Wordstar." Even though it can run on a variety of different hardware architectures, it always has the same functional design—the same key strokes will delete a line, move a block of text, or print out your file. It processes information in the same way no matter what kind of hardware it is running on. Without an information-processing description of Wordstar, you will not know how to use it or what it does, even if you are intimately acquainted with the hardware in which it is embodied. A physical description cannot tell one what the computer was designed to do; an information-processing description cannot tell one the physical processes by virtue of which the programs are run.

In psychology, it has become common to describe a brain as a system that processes information—a computer made out of organic compounds rather than silicon chips. The brain takes sensorily derived information from the environment as input, performs complex transformations on that information, and produces either data structures (representations) or behavior as output. Consequently, it, too, can be described in two mutually compatible and complementary ways. A neuroscience description characterizes the ways in which its physical components interact; a cognitive, or information-processing, description characterizes the "programs" that govern its operation. In cognitive psychology, the term *mind* is used to refer to an information-processing description of the functioning of the brain, and not in any colloquial sense. Behavioral descriptions can be illuminating, but manifest behavior is so variable that descriptions that capture and explain this variability inevitably require an explication of the psychological mechanisms and environmental conditions that generate it (see Symons, this volume).

An account of the evolution of the mind is an account of how and why the information-processing organization of the nervous system came to have the functional properties that it does. Information-processing language—the language of cognitive psychology—is simply a way of getting specific about what, exactly, a psychological mechanism does. In this volume, most psychological mechanisms are described in information-processing terms, either explicitly or implicitly. Research in some areas of psychology is so new that it is too early to develop hypotheses about the exact nature of the rules and representations involved. Nevertheless, the contributors have focused on the kinds of questions that will allow such hypotheses to be developed, questions such as: What kinds of information are available in the environment for a psychological mechanism designed for habitat selection, or mate selection, or parenting to use? Is there evidence that this information *is* used? If so, how is it evaluated? What kinds of affective reactions does it generate? How do people reason about this information? What information do they find memorable? What kinds of information are easy to learn? What kinds of decision rules guide human behavior? What kinds of cross-cultural patterns will these mechanisms produce? What kinds of information will they cause to be socially transmitted?

One doesn't have to look far to find minds that are profoundly different from our own: The information-processing mechanisms that collectively comprise the human mind differ in many ways from those that comprise the mind of an alligator or a bee or a sparrow or a wolf. The minds of these different species have different *design features:* different perceptual processes, different ways of categorizing the world, different preferences, different rules of inference, different memory systems, different learning mechanisms, and so on. These differences in psychological design cause differences in behavior: Upon perceiving a rattlesnake, a coyote might run from it, but another rattlesnake might try to mate with it.

Darwin provided a naturalistic explanation for the design features of organisms, including the properties of the minds of animals, not excepting humans. He wanted to explain how complex functional design could emerge in species spontaneously, without the intervention of an intelligent artificer, such as a divine creator. Darwin's explanation—natural selection—provides an elegant causal account of the relationship between adaptive problems and the design features of organisms. An *adaptive problem* is a problem whose solution can affect reproduction, however distally. Avoiding predation, choosing nutritious foods, finding a mate, and communicating with others are examples of adaptive problems that our hominid ancestors would have faced.

The logic of his argument seems inescapable. Imagine that a new design feature arises in one or a few members of a species, entirely by chance mutation. It could be anything—a more sensitive retina, a new digestive enzyme, a new learning mechanism. Let's say that this new design feature solves an adaptive problem better than designs that already exist in that species: The more sensitive retina allows one to see predators faster, the new digestive enzyme allows one to extract more nutrients from one's food, the new learning mechanism allows one to find food more efficiently. By so doing, the new design feature causes individuals who have it to produce more offspring, on average, than individuals who have alternative designs. If offspring can inherit the new design feature from their parents, then it will increase in frequency in the population. Individuals who have the new design will tend to have more offspring than those who lack it, those of their offspring who inherit the new design will have more offspring, and so on, until, after enough generations, every member of the species will have the new design feature. Eventually, the more sensitive retina, the better digestive enzyme, the more reliable learning mechanism will become universal in that species, typically found in every member of it.

Darwin called this process *natural selection.* The organism's interaction with the environment—with "nature"—sets up a feedback process whereby nature "selects" one design over another, depending on how well it solves an adaptive problem (a problem that affects reproduction).

Natural selection can generate complex designs that are *functionally organized*—organized so that they can solve an adaptive problem—because the criterion for the selection of each design feature is functional: A design feature will spread only if it solves an adaptive problem better than existing alternatives. Over time, this causal feedback process can create designs that solve adaptive problems well—designs that "fit" the environment in which the species evolved. Random processes, such as mutation and drift, cannot, by themselves, produce complex designs that are functionally organized because the probability that all the right design features will come together simply by chance is vanishingly small. By definition, random processes contain no mechanism for choosing one design over another based on its functionality. Evolution by natural selection is the only presently validated explanation for the accumulation of functional design features across generations.

The emerging field of evolutionary psychology attempts to take advantage of Darwin's crucial insight that there should be a functional mesh between the design features of organisms and the adaptive problems that they had to solve in the enviroment in which they evolved. By understanding the selection pressures that our hominid ancestors faced—by understanding what kind of adaptive problems they had to solve—one should be able to gain some insight into the design of the information-processing mechanisms that evolved to solve these problems. In doing so, one can begin to understand the processes that underlie cultural phenomena as well.

COMPLEMENTARY APPROACHES TO FUNCTIONAL ANALYSIS

The most common approach that evolutionarily oriented behavioral scientists have taken is to start with a known phenotypic phenomenon, such as pregnancy sickness, language, or color vision, and to try to understand what its adaptive function was—why that design was selected for rather than alternative ones. To do this, one must show that it is well designed for solving a specific adaptive problem, and that it is not more parsimoniously explained as a by-product of a design that evolved to solve some other adaptive problem (Williams, 1966; Symons, this volume). This is a difficult enterprise, but a necessary one: Until one understands a mechanism's adaptive function, one does not have a fully satisfying, conceptually integrated account of why it exists and what it does. More critically, asking functional questions and placing the phenomenon in a functional context often prompts important new insights about its organization, opening up new lines of investigation and bringing to light previously unobserved aspects and dimensions of the phenomenon. A number of contributions to *The Adapted Mind* take this approach (e.g., Boulton & Smith, Nesse & Lloyd, Profet, Pinker & Bloom, and Shepard). Going from a known psychological phenomenon to a theory of adaptive function is the most common form of conceptual integration between evolutionary biology and psychology.

With equal validity, however, one can take the analysis in the opposite direction as well (see Figure I.1). One can use theories of adaptive function to help one discover psychological mechanisms that were previously unknown. When one is trying to discover the structure of an information-processing system as complex as the human brain, knowing what its components were "designed" to do is like being given an aerial map of a territory one is about to explore by foot. If one knows what adaptive functions the human mind was designed to accomplish, one can make many educated guesses about what design features it should have, and can then design experiments to test for them. This can allow one to discover new, previously unsuspected, psychological mechanisms.

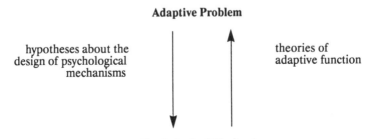

Figure I.1 The consideration of adaptive function can inform research into human behavior and psychological architecture in a variety of ways. The two most direct paths are schematized here. First, knowledge of the adaptive problems and ancestral conditions that human hunter-gatherers faced can lead to new hypotheses about the design of psychological mechanisms that evolved to solve them. Such heuristic analyses can supply crucial guidance in the design of experiments to discover previously unknown psychological mechanisms—investigations that researchers who neglect functional analysis would not have thought to conduct. Secondly, researchers can start with a known psychological phenomenon, and begin to investigate its adaptive function, if any, by placing it in the context of hunter-gatherer life and known selection pressures. The discovery of the functional significance of a psychological phenomenon is not only worthwhile in its own right, but clarifies the organization of the phenomenon, and prompts the discovery of new associated phenomena.

Empirically minded researchers, distrustful of "theory" (by which they often mean facts or principles drawn from unfamiliar fields), frequently ask why they should bother thinking about evolutionary biology: Why not just investigate the mind and behavior, and simply report what is found? The answer is that understanding function makes an important and sometimes pivotal contribution to understanding design in systems that are otherwise bewildering in their complexity. This point is illustrated by a story from the engineering community about the utility of knowing something's function. Reportedly, at a conference, an engineering professor carried a relatively simple circuit around to the various participants, asking them each to guess what its function was. Despite many guesses, none of the assembled engineers was able to figure it out. Finally, on the last day of the conference, the professor went up to the podium and asked the audience members to sketch the design of a circuit that would be able to perform a function that he then named. Everyone was able to do this rapidly, and when they were finished they were surprised to see that they had just drawn a picture of the same circuit that he had been showing them, the circuit whose function they had been unable to guess.[2] Behavioral scientists have been nearly defeated by the complexity of the behavior they confront. Guidance as to function vastly simplifies the problem of organizing the data in a way that illuminates the structure of the mind.

Our hominid ancestors had to be able to solve a large number of complex adaptive problems, and do so with special efficiency. By combining data from paleontology and hunter-gatherer studies with principles drawn from evolutionary biology, one can develop a task analysis that defines the nature of the adaptive information-processing problem to be solved. David Marr (1982) called this kind of task analysis a *computational theory*. Once one understands the nature of the problem, one can then generate very specific, empirically testable hypotheses about the structure of the information-processing mechanisms that evolved to solve it. A number of contributors to *The Adapted Mind* adopted this research strategy (e.g., Buss, Cosmides & Tooby, Mann, Silverman & Eals). One virtue of this approach is that it is immune to the usual (but often vacuous) accusation of post hoc storytelling: The researcher has predicted in advance the properties of the mechanism.

Using an evolutionarily derived task analysis to generate hypotheses about the structure of our cognitive processes can lead one to look for mechanisms that would otherwise have been overlooked. Silverman and Eals's chapter on spatial cognition is a good example. Research on spatial cognition has been proceeding for 100 years without the benefit of an evolutionary perspective, and the only kinds of mechanisms discovered were ones that produced a male performance advantage. But by asking what kind of spatial cognition a Pleistocene woman would have needed to be good at solving the adaptive problem of foraging for plant foods, Silverman and Eals were able to discover a new class of mechanisms involved in spatial cognition, which produce a 60% female advantage.

Psychologists should be interested in evolutionary biology for the same reason that hikers should be interested in an aerial map of an unfamiliar territory that they plan to explore on foot. If they look at the map, they are much less likely to lose their way.

THE HARVEST OF CONCEPTUAL INTEGRATION

Conceptual integration has been such a powerful force in the natural sciences not only because it allows scientists to winnow out improbable hypotheses or build aesthetically pleasing bridges between disciplines, but because it has been crucial to the discovery of new knowledge. For example, the atomic theory allowed chemists to see thermodynamics in a new way: The atomic theory was connected to Newtonian mechanics through the kinetic theory of heat, and thermodynamics was recast as statistical mechanics. When quantum theory was subsequently developed in physics, statistical mechanics was modified in such a way that it could explain not only the thermal and mechanical properties of matter, but its magnetic and electrical properties as well (Holton, 1973). The emergence of Mendelian genetics at the turn of the century solved a major puzzle in Darwinian theory. By showing that pre-Mendelian theories of blending inheritance were false—i.e., that tall and short plants need not produce medium offspring, that red and white flowers need not produce pink flowers, and so on—Mendelian genetics showed that natural selection could, in fact, create new species, a proposition that theories of blending inheritance had called into question. Subsequently, the combination of Mendelian genetics, Darwinian theory, and newly developed approaches to statistics led to the Modern Synthesis, which in turn made possible a family of new sciences, from population genetics to behavioral ecology.

Conceptual integration generates this powerful growth in knowledge because it allows investigators to use knowledge developed in other disciplines to solve problems in their own. The causal links between fields create anchor points that allow one to bridge theoretical or methodological gaps that one's own field may not be able to span. This can happen in the behavioral and social sciences, just as it has happened in the natural sciences. Evidence about cultural variation can help cognitive scientists decide between competing models of universal cognitive processes; evidence about the structure of memory and attention can help cultural anthropologists understand why some myths and ideas spread quickly and easily while others do not (e.g., Mandler et al., 1980; Sperber, 1985, 1990); evidence from evolutionary biology can help social psychologists generate new hypotheses about the design features of the information-processing mechanisms that govern social behavior; evidence about cognitive adaptations can tell evolutionary biologists something about the selection pressures that were present during hominid evolution; evidence from paleoanthropology and hunter-gatherer studies can tell developmental psychologists what kind of environment our developmental mechanisms were designed to operate in; and so on.

At present, crossing such boundaries is often met with xenophobia, packaged in the form of such familiar accusations as "intellectual imperialism" or "reductionism." But by calling for conceptual integration in the behavioral and social sciences we are neither calling for reductionism nor for the conquest and assimilation of one field by another. Theories of selection pressures are not theories of psychology; they are theories about some of the causal forces that produced our psychology. And theories of psychology are not theories of culture; they are theories about some of the causal mechanisms that shape cultural forms (Barkow, 1973, 1978, 1989a; Daly & Wilson, 1988; Sperber, 1985, 1990; Tooby & Cosmides, 1989, this volume). In fact, not only do the principles of one field not reduce to those of another, but by tracing the relationships between fields, additional principles often appear.

Instead, conceptual integration simply involves learning to accept with grace the irreplaceable intellectual gifts offered by other fields. To do this, one must accept the tenet of mutual consistency among disciplines, with its allied recognition that there are causal links between them. Compatibility is a misleadingly modest requirement, however, for it is an absolute one. Consequently, accepting these gifts is not always easy, because other fields may indeed bring the unwelcome news that favored theories have problems that require reformulation. Inattention to the compatibility requirement has led to many conceptual wrong turns in the social sciences (Barkow, 1989a; Tooby & Cosmides, this volume) as well as in evolutionary biology (Symons, this volume; Tooby & Cosmides 1990b). But fortunately errors can be avoided in the future by scrutinizing hypotheses in each field in the light of what is known in other fields. Investigators planning to apply such an approach will need to develop simultaneous expertise in at least two "adjacent" fields. Toward this end we hope that training in the behavioral and social sciences will move away from its present fragmented and insular form and that students will be actively encouraged to gain a basic familiarity with relevant findings in allied disciplines.

In the final analysis, it is not unaided empiricism that has made the natural sciences so powerful, but empiricism wedded to the power of inference. Every field has holes and gaps. But when there are causal links that join fields, the holes that exist in one discipline can sometimes be filled by knowledge developed in another. What the natural sciences have discovered is that this is a process with positive feedback: The more that is known—the more that can be simultaneously brought to bear on a question—the more that can be deduced, explained, and even observed. If we, as behavioral and social scientists, change our customs and accept what mutual enrichment we can offer one another, we can be illuminated by the same engine of discovery that has made the natural sciences such a signal human achievement.

NOTES

1. The idea that two statements cannot contradict each other and both be true was old when Aristotle formalized it, and it is only a small step from that to the commonplace idea that claims from different scientific disciplines should not contradict each other either, without at least one of them being suspected of being in error. Such a notion would seem too obvious to discuss were it not for the bold claims of autonomy made for the social sciences, accompanied by the institutionalized neglect of neighboring disciplines (Barkow, 1989c). It is, perhaps, one of the astonishing features of intellectual life in our century that cross-disciplinary consistency should be treated as a radical claim in need of defense, rather than as a routine tool of inference (Tooby & Cosmides, this volume). In any case, the central idea is simply one of consistency or compatibility across sciences, and conceptual integration and vertical integration are simply different names for this principle.

The adjective *vertical* in *vertical integration* (Barkow, 1980b, 1982, 1989a) emphasizes, alongside the notion of mutual compatibility, the notion that certain disciplines exist in a structured relationship with each other, such as physics to chemistry, and chemistry to biology. Each field "lower" in such a structure deals with principles that govern more inclusive sets of phenomena. For example, the laws of physics apply to chemical phenomena, and the principles of physics and chemistry apply to biological phenomena, but not the reverse. By the same token, however, each field "higher" up in the structure requires additional principles special to its more restricted domain (e.g., living things, humans) that are not easily reduced to the principles found in the other fields (e.g., natural selection is not derivable from chemistry).

We will generally use the term "conceptual integration" to avoid the connotation that vertical relationships between disciplines imply some epistemological or status hierarchy among sciences. For example, Lord Kelvin's criticism of Darwinism was based on Kelvin's erroneous calculation of the age of the earth. This case demonstrates that when physics and biology conflict, it is certainly possible that physics is in error. Moreover, the array of modern disciplines (from geochemistry to astrophysics to paleodemography to neuropharmacology) makes heterarchical relationships often seem more natural than any vertical ordering. Sciences should learn from and strive for consistency with every other field, from those existing in a clearly vertical relationship, such as chemistry is to physics, to those standing in more complex relationships, such as paleontology to psychology.

2. Our thanks to Jim Stellar for passing on to us this parable about the usefulness of functional approaches.

REFERENCES

Barkow, J. H. (1973). Darwinian psychological anthropology: A biosocial approach. *Current Anthropology, 14*(4), 373–388.

Barkow, J. H. (1978). Culture and sociobiology. *American Anthropologist, 80*(1), 5–20.

Barkow, J. H. (1980a). Biological evolution of culturally patterned behavior. In J. S. Lockard (Ed.), *The evolution of human social behavior*, (pp. 227–296). New York: Elsevier.

Barkow, J. H. (1980b). Sociobiology: Is this the new theory of human nature? In A. Montagu (Ed.), *Sociobiology examined*, (pp. 171–192). New York and London: Oxford University Press.

Barkow, J. H. (1982). Begged questions in behavior and evolution. In G. Davey (Ed.), *Animal models of human behavior: Conceptual, evolutionary, and neurological perspectives.* Chichester: John Wiley & Son.

Barkow, J. H. (1989a). *Darwin, sex, and status: Biological approaches to mind and culture.* Toronto: University of Toronto Press.

Barkow, J. H. (1989b). The elastic between genes and culture. *Ethology and Sociobiology, 10,* 111–129.

Barkow, J. H. (1989c). Broad training for social scientists [Letter to the editor]. *Science, 243*(4894):992.

Barkow, J. H. (1990). Beyond the DP/DSS controvery. *Ethology and Sociobiology, 11,* 341–351.

Brown, D. E. (1991). *Human universals.* New York: McGraw-Hill.

Campbell, D. T. (1975). On the conflicts between biological and social evolution and between psychology and moral tradition. *American Psychologist, 30,* 1103–1126.

Cosmides, L., & Tooby, J. (1987). From evolution to behavior: Evolutionary psychology as the missing link. In J. Dupre (Ed.), *The latest on the best: Essays on evolution and optimality*, (pp. 227–306). Cambridge, MA: The MIT Press.

Daly, M., & Wilson, M. (1988). *Homicide.* New York: Aldine.

Hinde, R. A. (1987). *Individuals, relationships, and culture: Links between ethology and the social sciences.* Cambridge: Cambridge University Press.

Holton, G. (1973). *Introduction to concepts and theories in physical science* (2nd ed.). (Revised by S. G. Brush.) Reading, MA: Addison-Wesley.

Mandler, J. M., Scribner, S., Cole, M., & DeForest, M. (1980). Cross-cultural invariance in story recall. *Child Development, 51,* 19–26.

Marr, D. (1982). *Vision: A computational investigation into the human representation and processing of visual information.* San Francisco: Freeman.

Sperber, D. (1985). Anthropology and psychology: Towards an epidemiology of representations. *Man (N.S.), 20,* 73–89.

Sperber, D. (1990). The epidemiology of beliefs. In C. Fraser & G. Geskell (Eds.), *Psychological studies of widespread beliefs.* New York: Oxford University Press.

Symons, D. (1979). *The evolution of human sexuality.* New York: Oxford University Press.

Tooby, J. & Cosmides, L. (1989). Evolutionary psychology and the generation of culture, part I. Theoretical considerations. *Ethology & Sociobiology, 10,* 29–49.

Tooby, J., & Cosmides, L. (1990a). On the universality of human nature and the uniqueness of the individual: The role of genetics and adaptation. *Journal of Personality, 58,* 17–67.

Tooby, J., & Cosmides, L. (1990b). The past explains the present: Emotional adaptations and the structure of ancestral environments. *Ethology and Sociobiology, 11,* 375–424.

Williams, G. C. (1966). *Adaptation and natural selection: A critique of some current evolutionary thought.* Princeton, NJ: Princeton University Press.

I

THE EVOLUTIONARY AND PSYCHOLOGICAL FOUNDATIONS OF THE SOCIAL SCIENCES

1

The Psychological Foundations of Culture

JOHN TOOBY AND LEDA COSMIDES

INTRODUCTION: THE UNITY OF SCIENCE

One of the strengths of scientific inquiry is that it can progress with any mixture of empiricism, intuition, and formal theory that suits the convenience of the investigator. Many sciences develop for a time as exercises in description and empirical generalization. Only later do they acquire reasoned connections within themselves and with other branches of knowledge. Many things were scientifically known of human anatomy and the motions of the planets before they were scientifically explained.

—GEORGE WILLIAMS,
Adaptation and Natural Selection

Disciplines such as astronomy, chemistry, physics, geology, and biology have developed a robust combination of logical coherence, causal description, explanatory power, and testability, and have become examples of how reliable and deeply satisfying human knowledge can become. Their extraordinary florescence throughout this century has resulted in far more than just individual progress within each field. These disciplines are becoming integrated into an increasingly seamless system of interconnected knowledge and remain nominally separated more out of educational convenience and institutional inertia than because of any genuine ruptures in the underlying unity of the achieved knowledge. In fact, this development is only an acceleration of the process of conceptual unification that has been building in science since the Renaissance. For example, Galileo and Newton broke down the then rigid (and now forgotten) division between the celestial and the terrestrial—two domains that formerly had been considered metaphysically separate—showing that the same processes and principles applied to both. Lyell broke down the distinction between the static present and the formative past, between the creative processes operating in the present and the geological processes that had operated across deep time to sculpt the earth. Maxwell uncovered the elegant principles that unified the many disparate electrical and magnetic phenomena into a single system.

And, one by one, the many gulfs separating life from nonlife were bridged and then closed: Harvey and others found that the macrostructure of the body turned out to operate according to comprehensible mechanical principles. Wöhler's synthesis of urea showed that the chemistries of the living and the nonliving were not forever separated by the occult operation of special vitalistic forces. In Wöhler's wake, the unraveling of the molecular biology of the gene and its regulation of cellular processes has shown how many of the immensely complex and functionally intricate mechanisms that constitute life are realized in molecular machinery: the *élan vital* turned out to be nothing other than this microscopic functional intricacy. Most critically, Darwin showed how even the intricately articulated functional organization of living systems (then only observable at the macroscopic level) could be explained as the product of intelligible natural causes operating over the expanse of deep time. In so doing, he conceptually united the living and the nonliving into a single system of principled causation, and the entire diversity of plant, animal, and microbial species into a single tree of descent. Darwin took an equally radical step toward uniting the mental and physical worlds, by showing how the mental world—whatever it might be composed of—arguably owed its complex organization to the same process of natural selection that explained the physical organization of living things. Psychology became united with the biological and hence evolutionary sciences.

The rise of computers and, in their wake, modern cognitive science, completed the conceptual unification of the mental and physical worlds by showing how physical systems can embody information and meaning. The design and construction of artificial computational systems is only a few decades old, but already such systems can parallel in a modest way cognitive processes—such as reason, memory, knowledge, skill, judgment, choice, purpose, problem-solving, foresight, and language—that had supposedly made mind a metaphysical realm forever separated from the physical realm, and humans metaphysically disconnected from the causal network that linked together the rest of the universe. These intellectual advances transported the living, the mental, and the human—three domains that had previously been disconnected from the body of science and mystified because of this disconnection—into the scientifically analyzable landscape of causation.

One useful way to organize this knowledge is as a principled history of the universe. Starting with some characterizable initial condition (like the Big Bang), each successive state of the system is described, along with the principles that govern the transitions from state to state. To the extent that our scientific model is well-developed, we should be able to account for the types of entities that emerge (pulsars, tectonic plates, ribosomes, vision, incest avoidance) and their distribution and location in the causal matrix. Such a history—in its broadest outlines—is well on its way to being constructed, from an initial quantum state, to the formation and distribution of particles during the early expansion, to the cooling and formation of atoms, the formation of galaxies, stellar evolution, the synthesis of heavier nuclei, and, of parochial interest to us, the local history of the solar system. This includes the formation of the sun and planets; the geochemistry of prebiotic earth; the generation of complex organic compounds; the emergence of the initial ancestral reproducing chemical system; the evolution of the genetic code and prokaryotic design; the emergence of eukaryotic sexual organisms, multicellular plants, animals, and fungi; and the rest of the history of life on earth.

In this vast landscape of causation, it is now possible to locate "Man's place in nature" to use Huxley's famous phrase and, therefore, to understand for the first time what humankind is and why we have the characteristics that we do. From this vantage point, humans are self-reproducing chemical systems, multicellular heterotrophic mobile organisms (animals), appearing very late in the history of life as somewhat modified versions of earlier primate designs. Our developmental programs, as well as the physiological and psychological mechanisms that they reliably construct, are the natural product of this evolutionary history. Human minds, human behavior, human artifacts, and human culture are all biological phenomena—aspects of the phenotypes of humans and their relationships with one another.

The rich complexity of each individual is produced by a cognitive architecture, embodied in a physiological system, which interacts with the social and nonsocial world that surrounds it. Thus humans, like every other natural system, are embedded in the contingencies of a larger principled history, and explaining any particular fact about them requires the joint analysis of all the principles and contingencies involved. To break this seamless matrix of causation—to attempt to dismember the individual into "biological" versus "nonbiological" aspects—is to embrace and perpetuate an ancient dualism endemic to the Western cultural tradition: material/spiritual, body/ mind, physical/mental, natural/human, animal/human, biological/social, biological/ cultural. This dualistic view expresses only a premodern version of biology, whose intellectual warrant has vanished.

This expansive new landscape of knowledge has not always been welcome, and many have found it uncongenial in one respect or another. The intellectual worlds we built and grew attached to over the last 3,000 years were laid out before much was known about the nature of the living, the mental, and the human. As a result, these intellectual worlds are, in many important respects, inconsistent with this new unified scientific view and, hence, are in need of fundamental reformulation. These established intellectual traditions and long-standing habits of mind seem, to many, to be more nourishing, more comfortable and, therefore, more valuable than the alternative prospect of new and unfamiliar scientific knowledge. To pick a single example, the shift from a universe designed to embody a moral and spiritual order to a universe that is undesigned and is structured only by a causal order engendered an immeasurably greater cultural dislocation than that which occurred when Copernicus identified the sun rather than the earth as the center of the planetary orbits. Consequently, the demystifications that have taken place since 1859 have been painful and have precipitated considerable resistance to accepting these discoveries and their implications. With the appearance of Darwinism, the full scope of the emerging unified account was, for the first time, apparent. Therefore, much of the opposition has specifically revolved around evolution and its application to humans. Gladstone, for example, in a debate with Huxley, captured in his choice of language the widely shared, visceral sense of revulsion caused by the claim "that natural selection and the survival of the fittest, all in the physical order, exhibit to us the great arcanum of creation, the sun and the center of life, so that mind and spirit are dethroned from their old supremacy, are no longer sovereign by right, but may find somewhere by charity a place assigned them, as appendages, perhaps only as excrescences, of the material creation" (Gladstone, quoted in Gould, 1988, p. 14). The dislocations in world view stemming from this process of conceptual unification led to a growing demand for, and production of, conceptual devices and rationales to divorce the natural sciences from the human social and inner landscape, to blunt the implications of monism and Darwinism, and to restore a comfortable distance between the human sciences and the world of natural causation. To many scholarly communities, conceptual unification became an enemy, and the relevance of other fields a menace to their freedom to interpret human reality in any way they chose.

Thus, despite some important exceptions, the social sciences have largely kept themselves isolated from this crystalizing process of scientific integration. Although social scientists imitated many of the outward forms and practices of natural scientists (quantitative measurement, controlled observation, mathematical models, experimentation, etc.), they have tended to neglect or even reject the central principle that valid scientific knowledge—whether from the same or different fields—should be mutually consistent (see Cosmides, Tooby, & Barkow, this volume). It is this principle that makes different fields relevant to each other, and part of the same larger system of knowledge. In consequence, this insularity is not just an accident. For many scholars, it has been a conscious, deeply held, and strongly articulated position, advanced and defended since the inception of the social sciences, particularly in anthropology and sociology. Durkheim, for example, in his *Rules of the Sociological Method*, argued at length that social phenomena formed an autonomous system and could be only explained by other social phenomena (1895/1962). The founders of American anthropology, from Kroeber and Boas to Murdock and Lowie, were equally united on this point. For Lowie, "the principles of psychology are as incapable of accounting for the phenomena of culture as is gravitation to account for architectural styles," and "culture is a thing *sui generis* which can be explained only in terms of itself. . . . *Omnis cultura ex cultura*" (1917/1966, p. 25–26; p. 66). Murdock, in his influential essay "The science of culture," summed up the conventional view that culture is "independent of the laws of biology and psychology" (1932, p. 200).

Remarkably, while the rest of the sciences have been weaving themselves together through accelerating discoveries of their mutual relevance, this doctrine of intellectual isolationism, which has been the reigning view in the social sciences, has only become more extreme with time. With passionate fidelity, reasoned connections with other branches of knowledge are dismissed as ignorant attempts at crude reductionism, and many leading social scientists now openly call for abandoning the scientific enterprise instead. For example, Clifford Geertz advocates abandoning the ground of principled causal analysis entirely in favor of treating social phenomena as "texts" to be interpreted just as one might interpret literature: We should "turn from trying to explain social phenomena by weaving them into grand textures of cause and effect to trying to explain them by placing them into local frames of awareness" (1983, p. 6). Similarly, Edmund Leach rejects scientific explanation as the focus of anthropology: "Social anthropology is not, and should not aim to be, a 'science' in the natural science sense. If anything it is a form of art . . . . Social anthropologists should not see themselves as seekers after objective truth . . . ." (Leach, 1982, p. 52). These positions have a growing following, but less, one suspects, because they have provided new illumination than because they offer new tools to extricate scholars from the unwelcome encroachments of more scientific approaches. They also free scholars from all of the arduous tasks inherent in the attempt to produce scientifically valid knowledge: to make it consistent with other knowledge and to subject it to critical rejection on the basis of empirical disproof, logical inconsistency, and incoherence. In any case, even advocates of such avenues of retreat do not appear to be fully serious about them because few are actually willing to accept what is necessarily entailed by such a stance: Those who jettison the epistemological standards of science are no longer in a position to use their intellectual product to make any claims about what is true of the world or to dispute the others' claims about what is true.

Not only have the social sciences been unusual in their self-conscious stance of intellectual autarky but, significantly, they have also been relatively unsuccessful as sciences. Although they were founded in the 18th and 19th centuries amid every expectation that they would soon produce intellectual discoveries, grand "laws," and validated theories to rival those of the rest of science, such success has remained elusive. The recent wave of antiscientific sentiment spreading through the social sciences draws much of its appeal from this endemic failure. This disconnection from the rest of science has left a hole in the fabric of our organized knowledge of the world where the human sciences should be. After more than a century, the social sciences are still adrift, with an enormous mass of half-digested observations, a not inconsiderable body of empirical generalizations, and a contradictory stew of ungrounded, middle-level theories expressed in a babel of incommensurate technical lexicons. This is accompanied by a growing malaise, so that the single largest trend is toward rejecting the scientific enterprise as it applies to humans.

We suggest that this lack of progress, this "failure to thrive," has been caused by the failure of the social sciences to explore or accept their logical connections to the rest of the body of science—that is, to causally locate their objects of study inside the larger network of scientific knowledge. Instead of the scientific enterprise, what should be jettisoned is what we will call the Standard Social Science Model (SSSM): The consensus view of the nature of social and cultural phenomena that has served for a century as the intellectual framework for the organization of psychology and the social sciences and the intellectual justification for their claims of autonomy from the rest of science. Progress has been severely limited because the Standard Social Science Model mischaracterizes important avenues of causation, induces researchers to study complexly chaotic and unordered phenomena, and misdirects study away from areas where rich principled phenomena are to be found. In place of the Standard Social Science Model, there is emerging a new framework that we will call the Integrated Causal Model. This alternative framework makes progress possible by accepting and exploiting the natural connections that exist among all the branches of science, using them to construct careful analyses of the causal interplay among all the factors that bear on a phenomenon. In this alternative framework, nothing is autonomous and all the components of the model must mesh.

In this chapter, we argue the following points:

1. There is a set of assumptions and inferences about humans, their minds, and their collective interaction—the Standard Social Science Model—that has provided the conceptual foundations of the social sciences for nearly a century and has served as the intellectual warrant for the isolationism of the social sciences.

2. Although certain assumptions of this model are true, it suffers from a series of major defects that make it a profoundly misleading framework. These defects have been responsible for the chronic difficulties encountered by the social sciences.

3. Advances in recent decades in a number of different disciplines, including evolutionary biology, cognitive science, behavioral ecology, psychology, hunter-gatherer studies, social anthropology, biological anthropology, primatology, and neurobiology have made clear for the first time the nature of the phenomena studied by social scientists and the connections of those phenomena to the principles and findings in the rest of science. This allows a new model to be

constructed—the Integrated Causal Model—to replace the Standard Social Science Model.

4. Briefly, the ICM connects the social sciences to the rest of science by recognizing that:

   a. the human mind consists of a set of evolved information-processing mechanisms instantiated in the human nervous system;
   b. these mechanisms, and the developmental programs that produce them, are adaptations, produced by natural selection over evolutionary time in ancestral environments;
   c. many of these mechanisms are functionally specialized to produce behavior that solves particular adaptive problems, such as mate selection, language acquisition, family relations, and cooperation;
   d. to be functionally specialized, many of these mechanisms must be richly structured in a content-specific way;
   e. content-specific information-processing mechanisms generate some of the particular content of human culture, including certain behaviors, artifacts, and linguistically transmitted representations;
   f. the cultural content generated by these and other mechanisms is then present to be adopted or modified by psychological mechanisms situated in other members of the population;
   g. this sets up epidemiological and historical population-level processes; and
   h. these processes are located in particular ecological, economic, demographic, and intergroup social contexts or environments.

On this view, culture is the manufactured product of evolved psychological mechanisms situated in individuals living in groups. Culture and human social behavior is complexly variable, but not because the human mind is a social product, a blank slate, or an externally programmed general-purpose computer, lacking a richly defined evolved structure. Instead, human culture and social behavior is richly variable because it is generated by an incredibly intricate, contingent set of functional programs that use and process information from the world, including information that is provided both intentionally and unintentionally by other human beings.

THE STANDARD SOCIAL SCIENCE MODEL

The Central Logic of the Standard Social Science Model

But one would be strangely mistaken about our thought if, from the foregoing, he drew the conclusion that sociology, according to us, must, or even can, make an abstraction of man and his faculties. It is clear, on the contrary, that the general characteristics of human nature participate in the work of elaboration from which social life results. But they are not the cause of it, nor do they give it its special form; they only make it possible. Collective representations, emotions, and tendencies are caused not by certain states of the consciousnesses of individuals but by the conditions in which the social group, in its totality, is placed. Such actions can, of course materialize only if the individual natures are not resistant to them; *but these individual natures are merely the indeterminate material that the social factor molds*

and transforms. Their contribution consists exclusively in very general attitudes, in vague and consequently plastic predispositions which, by themselves, if other agents did not intervene, could not take on the definite and complex forms which characterize social phenomena.
—DURKHEIM, 1895/1962,
pp. 105–106, emphasis added.

Humans everywhere show striking patterns of local within-group similarity in their behavior and thought, accompanied by profound intergroup differences. The Standard Social Science Model (SSSM or Standard Model) draws its enduring persuasive power by starting with these and a few other facts, rooted in direct experience and common knowledge. It then focuses on one salient causal and temporal sequence: how individuals change over their development from "unformed" infants into complexly competent adult members of their local social group, and how they do so in response to their local human environment. The central precepts of the SSSM are direct and seemingly inescapable conclusions drawn from these facts (D. E. Brown, 1991), and the same reasoning appears in author after author, from perhaps its most famous early expression in Durkheim (1895/1962), to its fully conventional modern adherents (with updated conceptual ornamentation) such as Geertz (1973).

The considerations that motivate the Standard Social Science Model are as follows:

Step 1. The existence of rapid historical change and the multitude of spontaneous human "cross-fostering experiments" effectively disposes of the racialist notion that human intergroup behavioral differences of any significance are attributable to genetic differences between groups. Infants everywhere are born the same and have the same developmental potential, evolved psychology, or biological endowment—a principle traditionally known as *the psychic unity of humankind.* The subsequent growth of knowledge over this century in genetics and human development has given strong empirical support to the conclusion that infants from all groups have essentially the same basic human design and potential. Human genetic variation, which is now directly detectable with modern electrophoretic techniques, is overwhelmingly sequestered into functionally superficial biochemical differences, leaving our complex functional design universal and species-typical (Tooby & Cosmides, 1990a). Also, the bulk of the variation that does exist is overwhelmingly inter-individual and within-population, and not between "races" or populations. By the nature of its known distribution, then, genetic variation cannot explain why many behaviors are shared within groups, but not between groups. That is, genetic variation does not explain why human groups dramatically differ from each other in thought and behavior. (Significantly, this is the only feature of the SSSM that is correct as it stands and that is incorporated unmodified into the Integrated Causal Model. Why it turns out to be true, however, depends on the existence of complex evolved psychological and physiological adaptations—something explicitly or implicitly denied by adherents of the SSSM.)

Step 2. Although infants are everywhere the same, adults everywhere differ profoundly in their behavioral and mental organization.

These first two steps, just by themselves, have led to the following widely accepted deduction: Because, it is reasoned, a "constant" (the human biological endowment observable in infants) cannot explain a "variable" (intergroup differences in complex adult mental or social organization) the SSSM concludes that "human nature" (the evolved structure of the human mind) cannot be the cause of the mental organization of adult humans, their social systems, their culture, historical change, and so on.

Step 3. Even more transparently, these complexly organized adult behaviors are absent from infants. Infants do not emerge speaking, and they appear to lack virtually every recognizable adult competency. Whatever "innate" equipment infants are born with has traditionally been interpreted as being highly rudimentary, such as an unorganized set of crude urges or drives, plus the ability to learn—certainly nothing resembling adult mental organization. Because adult mental organization (patterned behavior, knowledge, socially constructed realities, and so on) is clearly absent from the infant, infants must "acquire" it from some source outside themselves in the course of development.

Step 4. That source is obvious: This mental organization is manifestly present in the social world in the form of the behavior and the public representations of other members of the local group. Thus, the stuff of mental organization is categorizable according to its source: (1) the "innate" (or inborn or genetically determined, etc.), which is supplied "biologically" and is what you see in the infant, and (2) the social (or cultural or learned or acquired or environmental), which contains everything complexly organized and which is supplied by the social environment (with a few exceptions supplied by the physical environment and nonsocial learning). "[C]ultural phenomena . . . are in no respect hereditary but are characteristically and without exception acquired" (Murdock, 1932, p. 200). This line of reasoning is usually supported by another traditional argument, the deprivation thought experiment: "Undirected by culture patterns—organized systems of significant symbols—man's behavior would be virtually ungovernable, a mere chaos of pointless acts and exploding emotions, his experience virtually shapeless" (Geertz, 1973, p. 46). Humans raised without a social or cultural environment would be "mental basket cases" with "few useful instincts, fewer recognizable sentiments, and no intellect" (Geertz, 1973, p. 49). Because, it is reasoned, an effect disappears when its cause is withdrawn, this thought experiment is believed to establish that the social world is the cause of the mental organization of adults.

Step 5. The causal arrow in this process has a clear directionality, which is directly observable in the individual's development. The cultural and social elements that mold the individual precede the individual and are external to the individual. The mind did not create them; they created the mind. They are "given," and the individual "finds them already current in the community when he is born" (Geertz, 1973, p. 45). Thus, the individual is the creation of the social world and, it appears to follow, the social world cannot be the creation of "the individual." If you are reading this chapter, you learned English and did not create it. Nor did you choose to learn English (assuming you are a native speaker) any more than any effect chooses its cause; this action of the social world on the individual is compulsory and automatic—"coercive," to use Durkheim's phrase. Adult mental organization is socially determined. Moreover, by looking at social processes in the vast modern societies and nation-states, it is obvious that the "power asymmetry" between "the individual" and the social world is huge in the determination of outcomes and that the reciprocal impact of the individual on the social world is negligible. The causal flow is overwhelmingly or entirely in one direction. The individual is the acted upon (the effect or the outcome) and the sociocultural world is the actor (the cause or the prior state that determines the subsequent state).

Step 6. Accordingly, what complexly organizes and richly shapes the substance of human life—what is interesting and distinctive and, therefore, worth studying—is the variable pool of stuff that is usually referred to as "culture." Sometimes called "extra-somatic" or "extragenetic" (e.g., Geertz, 1973) to emphasize its nonbiological origins and nature, this stuff is variously described as behavior, traditions, knowledge, significant symbols, social facts, control programs, semiotic systems, information, social organization, social relations, economic relations, intentional worlds, or socially constructed realities. However different these characterizations may appear to be in some respects, those who espouse them are united in affirming that this substance—whatever its character—is (in Durkheim's phrase) "external to the individual." Even so psychological a phenomenon as thinking becomes external: "Human thought is basically both social and public—. . . its natural habitat is the house yard, the marketplace, and the town square. Thinking consists not of 'happenings in the head' (though happenings there and elsewhere are necessary for it to occur) but of a traffic in what have been called, by G.H. Mead and others, significant symbols—words for the most part . . . ." (Geertz, 1973, p. 45). "The individual" contributes only the infant's impoverished drives, unformed tendencies, and capacity to be socialized.

These first six steps constitute the SSSM's account of the causal process whereby what is assumed to be an initially formless infant is transformed into a fully human (i.e., fully cultural) being. The next important element in the SSSM is its approach to answering the question, "If culture creates the individual, what then creates culture?"

Before describing the SSSM's answer to this question, however, we need to make an important aspect of the question explicit: Human life is complexly and richly ordered. Human life is not (solely) noise, chaos, or random effect (contra Macbeth). Although the substance of human life, like human speech, is various and contingent, it is still, like human speech, intricately patterned. Many attempt to capture this perception with the phrase that human cultures (e.g., human symbol systems) are "meaningful." Human conduct does not resemble white noise. In a way that is analogous to William Paley's argument from design in his *Natural Theology*, one must ask: If there is complex and meaningful organization in human sociocultural life, what is the creator or artificer of it? Entropy, perturbation, error, noise, interaction with other systems, and so on, are always operating to influence culture (and everything else), so clearly not everything in culture is orderly. Equally, if these processes were all that were operating, complex order would never appear and would quickly degrade even if it did. Just as finding a watch on the heath, already complexly organized, requires that one posit a watchmaker (Paley, 1828), finding out that human life is complexly ordered necessitates the search for the artificer or source of this order (see Dawkins, 1986, for an exceptionally lucid general analysis of the problem of explaining complex order, its importance as a question, and the extremely narrow envelope of coherent answers). So, the question is not so much, What are the forces that act on and influence human culture and human affairs? but rather, What is the generator of complex and significant *organization* in human affairs?

Step 7. The advocates of the Standard Social Science Model are united on what the artificer is not and where it is not: It is not in "the individual"—in human nature or evolved psychology—which, they assume, consists of nothing more than what the infant comes equipped with, bawling and mewling, in its apparently unimpressive initial performances. Because the directional flow of the organization is from the outer world inward into "the individual," the direction toward which one looks for the source of the organization is likewise clear: outward into the social world. As Durkheim says, "[w]hen the individual has been eliminated, society alone remains" (1895/1962, p. 102).

Step 8. The SSSM maintains that the generator of complex and meaningful organization in human life is some set of emergent processes whose determinants are realized at the group level. The sociocultural level is a distinct, autonomous, and self-caused realm: "Culture is a thing *sui generis* which can be explained only in terms of itself. . . . *Omnis cultura ex cultura*" (Lowie, 1917/1966, p. 25–26). For Alfred Kroeber, "the only antecedents of historical phenomena are historical phenomena" (Kroeber, 1917). Durkheim was equally emphatic: "The determining cause of a social fact should be sought among the social facts preceding it and not among the states of individual consciousness"; that is, phenomena at the sociocultural level are mostly or entirely caused by other phenomena at the sociocultural level (Durkheim, 1895/1962, p. 110). It must be emphasized that this claim is not merely the obvious point that social phenomena (such as tulip bulb mania, the contagious trajectory of deconstructionist fashions, or the principles of supply and demand) cannot be understood simply by pointing inside the head of a single individual. It is, instead, a claim about the *generator* of the rich organization everywhere apparent in human life. What is generated even includes individual adult psychological phenomena, which are themselves simply additional social constructions. For Durkheim (and for most anthropologists today), even emotions such as "sexual jealousy" and "paternal love" are the products of the social order and have to be explained "by the conditions in which the social group, in its totality, is placed." As Geertz argues, "Our ideas, our values, our acts, even our emotions, are, like our nervous system itself, cultural products—products manufactured, indeed, out of tendencies, capacities, and dispositions with which we were born, but manufactured nonetheless" (1973, p. 50). Similarly, Shweder describes "cultural psychology" as "the study of the way cultural traditions and social practices regulate, express, transform, and permute the human psyche, resulting less in psychic unity for humankind than in ethnic divergences in mind, self and emotion" (Shweder, 1990, p. 1).

Step 9. Correspondingly, the SSSM denies that "human nature"—the evolved architecture of the human mind—can play any notable role as a generator of significant organization in human life (although it is acknowledged to be a necessary condition for it). In so doing, it removes from the concept of human nature all substantive content, and relegates the architecture of the human mind to the delimited role of embodying "the capacity for culture." Human nature is "merely the indeterminate material that the social factor molds and transforms. [This] contribution consists exclusively in very general attitudes, in vague and consequently plastic predispositions which, by themselves, if other agents did not intervene, could not take on the definite and complex forms which characterize social phenomena" (Durkheim, 1895/1962, p. 106). As Hatch comments, the "view that the Boasians had struggled to foster within the social sciences since almost the turn of the century" was that the human mind is "almost infinitely malleable" (1973, p. 236). Socialization is the process of externally supplied "conceptual structures molding formless talents" (Geertz, 1973, p. 50).

Social scientists who paid any attention to neuroscience, ethology, and cognitive psychology were increasingly, if uneasily, aware of the evidence that the nervous system was complex and not well characterized by the image of the "blank slate." Nonetheless, aside from paying some lip service to the notion that *tabula rasa* empiricism was untenable, this changed nothing important in the SSSM. The blank slate was traded in for blank cognitive procedures.[2] The mind could be seen as complex, but its procedures were still assumed to be content-free. As long as environmental input could enter and modify the system, as it clearly could, environmental input was presumed to orchestrate the system, giving it its functional organization. It doesn't matter if the clay of the human mind has some initial shape (tendencies, dispositions), so long as it is soft enough to be pounded by the external forces into any new shape required. Thus, for Geertz, who is attracted to the language if not the actual substance of cognitive science, the mind is not a slate, blank or otherwise (he dismisses this as a straw man position "which no one of any seriousness holds" or perhaps ever held [Geertz, 1984, p. 268]), but it is instead the tabula rasa's fully modern equivalent, a general-purpose computer. Such a computer doesn't come pre-equipped with its own programs, but instead—and this is the essential point—it obtains the programs that tell it what to do from the outside, from "culture." Thus, the human mind is a computer that is "desperately dependent upon such extragenetic, outside-the-skin control mechanisms" or "programs" "for the governing of behavior" (Geertz, 1973, p. 44).

This eliminates the concept of human nature or its alternative expression, the evolved psychological architecture, as useful or informative concepts. As Geertz puts it, "[t]he rise of the scientific concept of culture amounted to . . . the overthrow of the view of human nature dominant in the Enlightenment . . .", that is, that "[man] was wholly of a piece with nature and shared in the general uniformity of composition which natural science . . . had discovered there" with "a human nature as regularly organized, as thoroughly invariant, and as marvelously simple as Newton's universe" (Geertz, 1973, p. 34). Instead, the view entailed in the modern "scientific concept of culture" is that "humanity is as various in its essence as in its expression" (Geertz, 1973, p. 37). Geertz does not mean, of course, that infants vary due to genetic differences, but that all significant aspects of adult mental organization are supplied culturally. As deeply as one can go into the mind, people here are different from people there, leading to "the decline of the uniformitarian view of man" (Geertz, 1973, p. 35).

The conclusion that human nature is an empty vessel, waiting to be filled by social processes, removed it as a legitimate and worthwhile object of study. Why study paper when what is interesting is the writing on it and, perhaps even more important, the author (the perennially elusive generative social processes)? Since there could be no content, per se, to the concept of human nature, anything claimed to be present in human nature was merely an ethnocentric projection of the scholar making the claim. Thus, attempts to explore and characterize human nature became suspect. Such efforts were (and are) viewed as simply crude attempts to serve ideological ends, to manufacture propaganda, or to define one way of being as better and more natural than others.

Step 10. In the SSSM, the role of psychology is clear. Psychology is the discipline that studies the process of socialization and the set of mechanisms that comprise what anthropologists call "the capacity for culture" (Spuhler, 1959). Thus, the central concept in psychology is learning. The prerequisite that a psychological theory must meet to participate in the SSSM is that any evolved component, process, or mechanism must be equipotential, content-free, content-independent, general-purpose, domain-general, and so on (the technical terms vary with movement and era). In short, these mechanisms must be constructed in such a way that they can absorb any kind of cultural message or environmental input equally well. Moreover, their structures must themselves impose no particular substantive content on culture. As Rindos (1986, p.

315) puts it, "the specifics that we learn are in no sense predetermined by our genes." Learning is thus the window through which the culturally manufactured pre-existing complex organization outside of the individual manages to climb inside the individual. Although this approach deprives psychological mechanisms of any possibility of being the generators of significant organization in human affairs, psychologists get something very appealing in exchange. Psychology is the social science that can hope for general laws to rival those of the natural sciences: general laws of learning, or (more recently) of cognitive functioning. The relationship of psychology to biology is also laid out in advance by the SSSM: In human evolution, natural selection removed "genetically determined" systems of behavior and replaced them with general-purpose learning mechanisms or content-independent cognitive processes. Supposedly, these more general systems were favored by evolution because they did not constrain human behavior to be maladaptively inflexible (e.g., Geertz, 1973; Harris, 1979; Montagu, 1964). Neurobiology is the account of how these general mechanisms are instantiated in our nervous system.

Consequently, the concepts of learning, socialization, general-purpose (or content-independent) cognitive mechanisms, and environmentalism have (under various names and permutations) dominated scientific psychology for at least the last 60 years. Skinnerian behaviorism, of course, was one of the most institutionally successful manifestations of the SSSM's program for psychology, but its antimentalism and doctrinaire scientism made it uncongenial to those who wanted an account of their internal experience. More importantly, its emphasis on individual histories of reinforcement limited the avenues through which culture could have its effect. It proved an easy target when cognitive science provided precise ways of characterizing and investigating the mental as a system that processes information, a characterization that seemed to offer easier avenues for cultural transmission than laboriously organized schedules of reinforcement. Although cognitive psychologists threw out behaviorism's cumbersome antimentalism, they uncritically adopted behaviorism's equipotentiality assumption. In mainstream cognitive psychology, it is *assumed* that the machine is free of content-specialized processes and that it consists primarily of general-purpose mechanisms. Psychologists justify this assumption by an appeal to parsimony: It is "unscientific" to multiply hypothesized mechanisms in the head. The goal, as in physics, is for as few principles as possible to account for as much as possible. Consequently, viewing the mind as a collection of specialized mechanisms that perform specific tasks appears to be a messy approach, one not worth pursuing. Anthropologists and sociologists easily accommodated themselves to these theoretical changes in psychology: Humans went from being viewed as relatively simple equipotential learning systems to very much more complex equipotential information-processing systems, general-purpose computers, or symbol manipulators (see, e.g., Sahlins, 1976a, 1976b).

Within psychology there are, of course, important research communities that fall outside of the SSSM and that have remained more strongly connected to the rest of science, such as physiological psychology, perception, psychophysics, (physiological) motivation, psycholinguistics, much of comparative psychology, and a few other areas. Moreover, to explain how organisms remain alive and reproduce (and to make some minimal attempt to account for the focused substance of human life), psychologists have found it necessary to posit a few content-oriented mechanisms: hunger, thirst, sexual motivation, and so on. Nevertheless, the tendency has been to keep these elements restricted to as small a class as possible and to view them as external to the important central learning or cognitive processes. They are incorporated as, for example, reinforcers operating by drive reduction. Cognitive psychologists have, for the most part, labored to keep any such content-influenced elements extrinsic to the primary cognitive machinery. Indeed, they have usually avoided addressing how functional action—such as mate choice, food choice, or effort calculation—takes place at all. The principles of concept formation, of reasoning, of remembering, and so forth, have traditionally been viewed as uninfected prior to experience with any content, their procedures lacking features designed for dealing with particular types of content. Modular or domain-specific cognitive psychologists, in dissenting from this view, are abandoning the assumptions of the Standard Social Science Model.

Of course, readers should recognize that by so briefly sketching large expanses of intellectual history and by so minimally characterizing entire research communities, we are doing violence to the specific reality of, and genuine differences among hundreds of carefully developed intellectual systems. We have had to leave out the qualifications and complexities by which positions are softened, pluralisms espoused, critical distinctions lost, and, for that matter, lip service paid. This is inevitable in attempting so synoptic a view. In what is surely a graver defect, we have had to omit discussion of the many important dissident subcommunities in sociology, anthropology, economics, and other disciplines, which have sloughed off or never adopted the Standard Social Science Model. In any case, we simply hope that this sketch captures a few things that are true and important, to compensate for the unavoidable simplifying distortions and omissions. Most obviously, there are no pure types in the world, and scholars are quoted not to characterize the full richness of their individual views, which usually undergo considerable evolution over their intellectual development anyway, but rather to illustrate instances of a larger intellectual system. It is the larger intellectual system we are criticizing, and not the multitude of worthwhile research efforts that have gone on inside its structure. We think the roof of the Standard Social Science Model has collapsed, so to speak, because the overall architectural plan is unsound, not because the bricks and other building materials are defective. The detailed research efforts of hundreds of scientists have produced critically important knowledge that has transformed our understanding of the world. In this criticism, we are looking for an architectural design for the social sciences that is worthy of the intelligence and labor of those whose research goes on within their compass.

The Standard Social Science Model's Treatment of Culture

This logic has critically shaped how nearly every issue has been approached and debated in the social sciences. What we are concerned with here, however, is the impact of the Standard Social Science Model on the development of modern conceptions of culture, its causal role in human life, and its relationship to psychology. Briefly, standard views of culture are organized according to the following propositions (see also D. E. Brown, 1991, p. 146; Tooby & Cosmides, 1989a):

1. Particular human groups are properly characterized typologically as having "a" culture, which consists of widely distributed or nearly group-universal behavioral practices, beliefs, ideational systems, systems of significant symbols, or informational substance of some kind. Cultures are more or less bounded entities, although cultural elements may diffuse across boundaries.

2. These common elements are maintained and transmitted "by the group," an entity that has cross-generational continuity.

3. The existence of separate streams of this informational substance, culture, transmitted from generation to generation, is the explanation for human within-group similarities and between-group differences. In fact, all between-group differences in thought and behavior are referred to as cultural differences and all within-group similarities are regarded as the expressions of a particular culture. Since these similarities are considered to be "cultural," they are, either implicitly or explicitly, considered to be the consequence of informational substance inherited jointly from the preceding generation by all who display the similarity.

4. Unless other factors intervene, the culture (like the gene pool) is accurately replicated from generation to generation.

5. This process is maintained through learning, a well-understood and unitary process, that acts to make the child like the adult of her culture.

6. This process of learning can be seen, from the point of view of the group, as a group-organized process called socialization, imposed by the group on the child.

7. The individual is the more or less passive recipient of her culture and is the product of that culture.

8. What is organized and contentful in the minds of individuals comes from culture and is socially constructed. The evolved mechanisms of the human mind are themselves content-independent and content-free and, therefore, whatever content exists in human minds originally derives from the social or (sometimes) nonsocial environment.

9. The features of a particular culture are the result of emergent group-level processes, whose determinants arise at the group level and whose outcome is not given specific shape or content by human biology, human nature, or any inherited psychological design. These emergent processes, operating at the sociocultural level, are the ultimate generator of the significant organization, both mental and social, that is found in human affairs.

10. In discussing culture, one can safely neglect a consideration of psychology as anything other than the nondescript "black box" of learning, which provides the capacity for culture. Learning is a sufficiently specified and powerful explanation for how any behavior acquires its distinct structure and must be the explanation for any aspect of organized human life that varies from individual to individual and from group to group.

11. Evolved, "biological," or "innate" aspects of human behavior or psychological organization are negligible, having been superseded by the capacity for culture. The evolution of the capacity for culture has led to a flexibility in human behavior that belies any significant "instinctual" or innate component (e.g., Geertz, 1973; Montagu, 1968, p. 11; Sahlins, 1976a & b), which, if it existed, would have to reveal itself as robotlike rigid behavioral universals. To the extent that there may be any complex biological textures to individual psychology, these are nevertheless organized and given form and direction by culture and, hence, do not impart any substantial character or content to culture.

On the Reasonableness of the Standard Social Science Model

There are, of course, many important elements of truth in the tenets of the SSSM, both in its core logic and in its treatment of culture. The SSSM would not have become as decisively influential if it did not have a strong surface validity, anchored in important realities. For example: It is true that infants are everywhere the same. Genetic differences are superficial. There is within-group similarity of behavior and there are between-group differences, and these persist across generations, but also change over historical time. Highly organized socially communicated information exists outside of any particular individual at any one time (in the cognitive mechanisms of other individuals), and over time this information can be internalized by the specific individual in question. And so on.

Nevertheless, the Standard Social Science Model contains a series of major defects that act to make it, as a framework for the social sciences, deeply misleading. As a result, it has had the effect of stunting the social sciences, making them seem falsely autonomous from the rest of science (i.e., from the "natural sciences") and precluding work on answering questions that need to be answered if the social sciences are to make meaningful progress as sciences. After a century, it is time to reconsider this model in the light of the new knowledge and new understanding that has been achieved in evolutionary biology, development, and cognitive science since it was first formulated. These defects cluster into several major categories, but we will limit our discussion to the following three:

1. The central logic of the SSSM rests on naive and erroneous concepts drawn from outmoded theories of development. For example, the fact that some aspect of adult mental organization is absent at birth has no bearing on whether it is part of our evolved architecture. Just as teeth or breasts are absent at birth, and yet appear through maturation, evolved psychological mechanisms or modules (complex structures that are functionally organized for processing information) could develop at any point in the life cycle. For this reason, the many features of adult mental organization absent at birth need not be attributed to exposure to transmitted culture, but may come about through a large number of causal avenues not considered in traditional analyses.

2. More generally, the SSSM rests on a faulty analysis of nature-nurture issues, stemming from a failure to appreciate the role that the evolutionary process plays in organizing the relationship between our species-universal genetic endowment, our evolved developmental processes, and the recurring features of developmental environments. To pick one misunderstanding out of a multitude, the idea that the phenotype can be partitioned dichotomously into genetically determined and environmentally determined traits is deeply ill-formed, as is the notion that traits can be arrayed along a spectrum according to the degree that they are genetically versus environmentally caused. The critique of the SSSM that has been emerging from the cognitive and evolutionary communities is not that traditional accounts have underestimated the importance of biological factors relative to environmental factors in human life. Instead, the target is the whole framework that assumes that "biological factors" and "environmental factors" refer to mutually exclusive sets of causes that exist in some kind of explanatory zero-sum relationship, so that the more one explains "biologically" the less there is to explain "socially" or "environmentally." On the contrary,

as we will discuss, environmentalist claims necessarily require the existence of a rich, evolved cognitive architecture.

3. The Standard Social Science Model requires an impossible psychology. Results out of cognitive psychology, evolutionary biology, artificial intelligence, developmental psychology, linguistics, and philosophy converge on the same conclusion: A psychological architecture that consisted of nothing but equipotential, general-purpose, content-independent, or content-free mechanisms could not successfully perform the tasks the human mind is known to perform or solve the adaptive problems humans evolved to solve—from seeing, to learning a language, to recognizing an emotional expression, to selecting a mate, to the many disparate activities aggregated under the term "learning culture" (Cosmides & Tooby, 1987; Tooby & Cosmides, 1989a). It cannot account for the behavior observed, and it is not a type of design that could have evolved.

The alternative view is that the human psychological architecture contains many evolved mechanisms that are specialized for solving evolutionarily long-enduring adaptive problems and that these mechanisms have content-specialized representational formats, procedures, cues, and so on. These richly content-sensitive evolved mechanisms tend to impose certain types of content and conceptual organization on human mental life and, hence, strongly shape the nature of human social life and what is culturally transmitted across generations. Indeed, a post-Standard Model psychology is rapidly coalescing, giving a rapidly expanding empirical foundation to this new framework. In fact, historically, most of the data already gathered by psychologists supports such a view. It required a strongly canalized interpretative apparatus to reconcile the raw data of psychology with the central theoretical tenets of SSSM psychology.

Before examining in detail what is wrong with the SSSM and why the recognition of these defects leads to the formulation of a new model with greater explanatory power, it is necessary first to alleviate the fears of what would happen if one "falls off the edge" of the intellectual world created by the SSSM. These fears have dominated how alternative approaches to the SSSM have been treated in the past and, unless addressed, will prevent alternatives from being fairly evaluated now. Moreover, the Standard Model has become so well-internalized and has so strongly shaped how we now experience and interpret social science phenomena that it will be difficult to free ourselves of the preconceptions that the Standard Model imposes until its Procrustean operations on psychology and anthropology are examined.

THE WORLD BUILT BY THE STANDARD SOCIAL SCIENCE MODEL

The Moral Authority of the Standard Social Science Model
The Case Against Nativism

The overwhelming success of the Standard Social Science Model is attributable to many factors of which, arguably, the most significant has been its widespread moral appeal. Over the course of the century, its strong stand against explaining differences between races, classes, sexes, or individuals by hypothesizing underlying biological differences has been an important element in combating a multitude of searing horrors and oppressions, from the extermination of ethnic groups and the forced sterilization of the poor to restrictive immigration laws and legally institutionalized sex and race discrimination. The depth of these tragedies and the importance of the issues involved have imbued the SSSM and its central precept, "environmentalism," with an imposing moral stature. Consequently, the positions of individual scholars with respect to the SSSM have been taken to imply allegiances with respect to the larger social and moral conflicts around the world. Thus, to support the SSSM was to oppose racism and sexism and to challenge the SSSM was, intentionally or not, to lend support to racism, sexism, and, more generally (an SSSM way of defining the problem), "biological determinism." If biological ideas could be used to further such ends, then ideas that minimized the relevance of biology to human affairs, such as the tenets of the SSSM, could only be to the good.

In this process, all approaches explicitly involving nativist elements of whatever sort became suspect. In consequence, fundamentally divergent—even opposing—programs and claims have become enduringly conflated in the minds of 20th-century social scientists. Most significant was the failure to distinguish adaptationist evolutionary biology from behavior genetics. Although the adaptationist inquiry into our universal, inherited, species-typical design is quite distinct from the behavior genetics question about which differences between individuals or sets of individuals are caused by differences in their genes, the panspecific nativism typical of adaptationist evolutionary biology and the idiotypic nativism of behavior genetics became confused with each other (Tooby and Cosmides, 1990a). Obviously, claims about a complexly organized, universal human nature, by their very character, cannot participate in racist explanations. Indeed, they contradict the central premises of racialist approaches. Yet, despite this fact, adaptationist approaches and behavior genetics remain inextricably intertwined in the minds of the majority of social scientists.

The second strong moral appeal of the Standard Social Science Model derives from its emphasis on human malleability and the hope it, therefore, gave for social melioration or social revolution. The claim of John B. Watson, the founder of behaviorism, exemplifies this optimism about the power of scientifically directed socialization (as well as the usual implicit conflation of idiotypic and panspecific nativism):

> Give me a dozen healthy infants, well-formed, and my own specified world to bring them up in and I'll guarantee to take any one at random and train him to become any type of specialist I might select—doctor, lawyer, artist, merchant-chief, and yes, even beggar-man and thief, regardless of his talents, penchants, tendencies, abilities, vocations, and race of his ancestors (Watson, 1925, p. 82).

As D. E. Brown (1991, p. 61) comments, "In hindsight it is clear that this famous statement about the influence of the environment on individual differences is entirely compatible with the most extreme of the faculty or modular views of the human mind—in which it comprises numerous innate and highly specific mechanisms." But this thought experiment was interpreted by the social science mainsteam, Watson included, as demonstrating that "people are the products of their societies or cultures." Therefore, "change society or culture and you change people. . . . Intelligent, scientific socialization can make us whatever we want to be" (D. E. Brown, 1991, p. 61). Moreover, "[t]he equation of an arch environmentalism (including cultural relativism) with optimism about the practical application of social science to the problems of society remains a force to the present" (D. E. Brown, 1991, p. 62). More critically, the belief that the mind is "almost infinitely malleable" (or, in more modern terms, is a general-purpose computer) means that humans are not condemned to the status quo, and need not inevitably fight wars, or have social classes, or manifest sex-differentiated roles, or live in families, and so on.

If the "happy" ability of the mind to "quite readily take any shape that is presented" (Benedict, 1934/1959, p. 278) is the ameliorator's ideal because it is believed to be logically necessary to allow social change, then dissent from the SSSM tends to be framed as claims about "constraints" or limits on this malleability. This, in turn, is taken to imply a possible intractability to social problems—the stronger the biological forces are, the more we may be constrained to suffer from certain inevitable expressions of human darkness. Thus, the debate on the role of biology in human life has been consistently framed as being between optimistic environmentalists who plan for human betterment and sorrowful, but realistic nativists who lament the unwelcome inevitability of such things as aggression (e.g., Ardrey, 1966; Lorenz, 1966), or who (possibly even gleefully) defend the status quo as inevitable and natural (e.g., Goldberg, 1973, on patriarchy). These nativists are, in turn, "debunked" by the tireless opponents of "biological determinism" (e.g., Chorover, 1979; Lewontin, Rose, & Kamin, 1984; Montagu, 1968, 1978), who place each new biological intrusion onto social science territory in the context of the bitter lessons of the century. (Environmentalist holocausts are, of course, edited out of this chronology.)

This morality play, seemingly bound forever to the wheel of intellectual life, has been through innumerable incarnations, playing itself out in different arenas in different times (rationalism versus empiricism, heredity versus environment, instinct versus learning, nature versus nurture, human universals versus cultural relativism, human nature versus human culture, innate behavior versus acquired behavior, Chomsky versus Piaget, biological determinism versus social determinism, essentialism versus social construction, modularity versus domain-generality, and so on). It is perennial because it is inherent in how the issues have been defined in the SSSM itself, which even governs how the dissidents frame the nature of their dissent. Accordingly, the language of constraint and limitation is usually adopted by biologically oriented behavioral scientists themselves in describing the significance of their own work. Thus we even have titles such as *The Tangled Wing: Biological Constraints on the Human Spirit* (Konner, 1982), *Biological Boundaries of Learning* (Seligman and Hager, 1972), "Constraints on Learning" (Shettleworth, 1972), and "Structural Constraints on Cognitive Development" (Gelman, 1990a). Biologically oriented social and behavioral scientists often see themselves as defining limits on the possible. Environmentalists see themselves as expanding the borders of the possible. As we will see, this framing is profoundly misleading.

Wrong Diagnosis, Wrong Cure

Driven by these fears to an attitude that Daly and Wilson (1988) have termed "biophobia," the social science community lays out implicit and sometimes explicit ground rules in its epistemological hierarchy: The tough-minded and moral stance is to be skeptical of panspecific "nativist" claims; that is, of accounts that refer in any way to the participation of evolved psychological mechanisms together with environmental variables in producing outcomes, no matter how logically inescapable or empirically well-supported they may be. They are thought to be explanations of last resort and, because the tough-minded and skeptical can generate particularistic alternative accounts for any result at will, this last resort is rarely ever actually arrived at. For the same reason, it is deemed to be the moral stance to be correspondingly credulous of "environmentalist" accounts, no matter how vague, absurd, incoherent, or empirically contradicted they may be. These protocols have become second nature (so to speak) to nearly everyone in the social science community. This hierarchy is driven by the fear of falling off the edge of the Standard Social Science Model, into unknown regions where monsters such as "biological" or "genetic determinism" live.

What, in fact, is an environmentalist account? There are two brands of environmentalism: coherent environmentalism and incoherent environmentalism, which correspond approximately to environmentalism as defended and environmentalism as practiced. As Daly and Wilson (1988, p. 8) comment, "[a]ll social theorists, including the staunchest antinativists, seek to describe human nature at some cross-culturally general level of abstraction" and would be "distressed should their theories . . . prove applicable to Americans but not to Papuans." Both Skinner (1957) and Chomsky (1975)—opponents in a paradigmatic case of an environmentalist-nativist debate—posit the existence of universal evolved psychological mechanisms, or what cognitive psychologists have called "innate mechanisms." As Symons (1987) points out, most of what passes for the nature-nurture debate is not about the need to posit evolved mechanisms in theories. Everyone capable of reasoning logically about the problem accepts the necessity of this. As Symons makes clear, what the debate often seems to be about is how general or content-specific the mechanisms are: Skinner proposes conditioning mechanisms that apply to all situations, while Chomsky proposes specialized mechanisms particularly designed for language. Consequently, coherent environmentalists acknowledge that they are positing the existence of evolved developmental or psychological mechanisms and are willing to describe (1) the explicit structure of these mechanisms, and (2) what environmental variables they interact with to produce given outcomes. By this standard, of course, Chomsky is an environmentalist, as was Skinner, as are we, along with most other evolutionary psychologists and evolutionarily informed behavioral scientists. Equally, all coherent behavioral scientists of whatever orientation must be nativists in this same sense, and no coherent and fully specified hypothesis about behavior can avoid making nativist claims about the involvement of evolved structure.

Incoherent environmentalists, on the other hand, are those who propose theories of how environments regulate behavior or even psychological phenomena without describing or even mentioning the evolved mechanisms their theories would require to be complete or coherent. In practice, communities whose rules of discourse are governed by incoherent environmentalism consider any such trend toward explicitness to be introducing vague and speculative variables and—more to the point—to be in bad taste as well. The simple act of providing a complete model is to invoke evolved design and, hence, to court being called a genetic or biological determinist. Given that all coherent (fully specified) models of psychological processes necessarily entail an explication of how environmental variables relate to the inherited architecture or developmental machinery, this attitude has the effect of portraying psychologists who are clear about all causal steps as more soft-minded and speculative than those who remain vague about the crucial elements necessary to make their theories coherent (e.g., Cheng & Holyoak, 1989). (This criticism is, of course, fully symmetrical: Incoherent nativists are those scholars who talk about how evolution structures behavior without attempting to describe the structure of the evolved adaptations that link evolution, environment, and behavior in adaptively patterned ways; for discussion, see Symons, this volume; Tooby & Cosmides, 1990b.)

The problem with an epistemological hierarchy that encourages incoherence and discourages coherence (aside from the fully sufficient objection that it has introduced major distortions into the body of scientific knowledge) is that it is completely unnecessary, even on its own terms. Not only is the cure killing the patient—social science—but also the diagnosis is wrong and the patient is not menaced by the suspected malady. In the first place, as discussed, models of a robust, universal human nature by their very character cannot participate in racist explanations of intergroup differences. This is not just a definitional trick of defining human nature as whatever is universal. There are strong reasons to believe that selection usually tends to make complex adaptations universal or nearly universal, and so humans must share a complex, species-typical and species-specific architecture of adaptations, however much variation there might be in minor, superficial, nonfunctional traits. As long-lived sexual reproducers, complex adaptations would be destroyed by the random processes of sexual recombination every generation if the genes that underlie our complex adaptations varied from individual to individual. Selection in combination with sexual recombination tends to enforce uniformity in adaptations, whether physiological or psychological, especially in long-lived species with an open population structure, such as humans (Tooby & Cosmides, 1990b). Empirically, of course, the fact that any given page out of *Gray's Anatomy* describes in precise anatomical detail individual humans from around the world demonstrates the pronounced monomorphism present in complex human physiological adaptations. Although we cannot yet directly "see" psychological adaptations (except as described neuroanatomically), no less could be true of them. Human nature is everywhere the same.

The Malleability of Psychological Architecture versus the Volatility of Behavioral Outcomes

If the fear that leaving the Standard Social Science Model will lead to racist doctrines is unfounded, what of the issue of human malleability? Does a biologically informed approach necessarily imply an intractability of undesired social and behavioral outcomes and an inevitability of the status quo? After all, isn't the basic thrust of biologically informed accounts against malleability and in favor of constraints and limits on human aspirations?

No. The central premise of an opposition between the mind as an inflexible biological product and the mind as a malleable social product is ill-formed: The notion that inherited psychological structure constrains is the notion that without it we would be even more flexible or malleable or environmentally responsive than we are. This is not only false but absurd. Without this evolved structure, we would have no competences or contingent environmental responsiveness whatsoever. Evolved mechanisms do not prevent, constrain, or limit the system from doing things it otherwise would do in their absence. The system could not respond to "the environment" (that is, to selected parts of the environment in an organized way) without the presence of mechanisms designed to create that connection. Our evolved cognitive adaptations—our inherited psychological mechanisms—are the means by which things are affirmatively accomplished. It is an absurd model that proposes that the potentially unfettered human mind operates by flailing around and is only given structure and direction by the "limits" and "constraints" built in by "biology." Instead, any time the mind generates any behavior at all, it does so by virtue of specific generative programs in the head, in conjunction with the environmental inputs with which they are presented. Evolved structure does not constrain; it creates or enables (Cosmides & Tooby, 1987).

Given that we are all discussing universal human design and if, as Symons argues, all coherent behavioral scientists accept the reality of evolved mechanisms, then the modern nature-nurture debate is really about something else: the character of those evolved mechanisms (Symons, 1987). Does the mind consist of a few, general-purpose mechanisms, like operant conditioning, social learning, and trial-and-error induction, or does it also include a large number of specialized mechanisms, such as a language acquisition device (Chomsky, 1975; Pinker, 1984; Pinker & Bloom, this volume), mate preference mechanisms (Buss, 1989, this volume; Ellis, this volume; Symons, 1979), sexual jealousy mechanisms (Daly, Wilson, & Weghorst, 1982; Wilson & Daly, this volume), mother-infant emotion communication signals (Fernald, this volume), social contract algorithms (Cosmides, 1989; Cosmides & Tooby, 1989, this volume; Gigerenzer & Hug, in press), and so on? This is the point of separation between the Standard Social Science Model and the Integrated Causal Model, and it is the main focus of this volume.

How, then, does the issue of the number and specificity of evolved mental mechanisms bear on the issue of the inevitability of undesired behavioral outcomes? As we will discuss and review later, the same answer applies: General mechanisms turn out to be very weak and cannot unassisted perform at least most and perhaps all of the tasks humans routinely perform and need to perform. Our ability to perform most of the environmentally engaged, richly contingent activities that we do depends on the guiding presence of a large number of highly specialized psychological mechanisms (Cosmides & Tooby, 1987; Rozin, 1976; Symons, 1987; Tooby & Cosmides, 1990b). Far from constraining, specialized mechanisms enable competences and actions that would not be possible were they absent from the architecture. This rich array of cognitive specializations can be likened to a computer program with millions of lines of code and hundreds or thousands of functionally specialized subroutines. It is because of, and not despite, this specificity of inherent structure that the output of computational systems is so sensitively contingent on environmental inputs. It is just this sensitive contingency to subtleties of environmental variation that make a narrow intractability of outcomes unlikely.

The image of clay, and terms such as "malleability," "flexibility," and "plasticity" confuse two separate issues: (1) the detailed articulation of human evolved psychological design (i.e., what is the evolved design of our developmental programs and of the mechanisms they reliably construct), and (2) the fixity or intractability of expressed outcomes (what must people do, regardless of circumstance). The first question asks what evolved organization exists in the mind, while the second asks what events will inevitably occur in the world. Neither "biology," "evolution," "society," or "the environment" directly impose behavioral outcomes, without an immensely long and intricate intervening chain of causation involving interactions with an entire configuration of other causal elements. Each link of such a chain offers a possible point of intervention to change the final outcome. For this reason, computer programs present a far better model of the situation: The computer does nothing without them, they frequently involve superbly complex contingent branching and looping alternatives, they can (and the procedures in the human mind certainly do) take as input environmental variables that create cascading changes in subsequent computational events and final outcomes, and the entire system may respond dramatically and dynamically to direct intervention (for example, the alteration of even a single instruction) at any of a great number of locations in the program.

Moreover, we know in advance that the human psychological system is immensely flexible as to outcome: Everything that every individual has ever done in all of human history and prehistory establishes the minimum boundary of the possible. The maximum, if any, is completely unknown. Given the fact that we are almost entirely ignorant of the computational specifics of the hundreds or thousands of mechanisms that comprise the human mind, it is far beyond the present competence of anyone living to say what are and are not achievable outcomes for human beings.

It is nevertheless very likely to be the case that we will find adaptive specializations in the human mind that evolved to make, under certain circumstances, choices or decisions that are (by most standards) ethically unacceptable and often lead to consensually undesirable outcomes (e.g., male sexual proprietariness, Wilson & Daly, this volume; discriminative parental solicitude, Daly & Wilson, 1981; Mann, this volume). If one is concerned about something like family violence, however, knowing the details of the mechanisms involved will prove crucial in taking any kind of constructive or ameliatory action. "Solutions" that ignore causation can solve nothing.

In any case, the analysis of the morality or practicality of intervention to prevent undesired outcomes—"ontogenetic engineering" (Daly, Wilson & Weghorst, 1982)—is beyond the scope of this discussion. Our point here is simply that leaving behind the SSSM does not entail accepting the inevitability of any specific outcome, nor does it entail the defense of any particular aspect of the status quo. Instead, for those genuinely concerned with such questions, it offers the only realistic hope of understanding enough about human nature to eventually make possible successful intervention to bring about humane outcomes. Moreover, a program of social melioration carried out in ignorance of human complex design is something like letting a blindfolded individual loose in an operating room with a scalpel—there is likely to be more blood than healing. To cure, one needs to understand; lamenting disease or denouncing the researchers who study its properties has never yet saved a life. At present, we are decades away from having a good model of the human mind, and this is attributable in no small measure to a misguided antinativism that has, for many, turned from being a moral stance into a tired way of defending a stagnated and sterile intellectual status quo. There are, of course, no guarantees, and it is at least logically possible that understanding our complex array of evolved mechanisms will offer no way to improve the human condition. But, if that is the case, it will be the first time in history that major sets of new discoveries turned out to be useless.[3]

The Empirical Authority of the Standard Social Science Model
The Division of Labor: Content-Independent Psychology

One major consequence of the adoption of the Standard Social Science Model has been the assignment of a division of labor among the social sciences. It gave each field its particular mission, stamped each of them with its distinctive character, and thereby prevented them from making much progress beyond the accumulation of particularistic knowledge. Anthropology, as well as sociology and history, study both the impor-

tant and variable content of human life (the signal) and the more vaguely defined processes and contingent events that generated it (the artificer or author of the signal).
Psychology studies the medium on which this socially generated content is inscribed,
the process of inscription, and the mechanisms that enable the inscription to take
place. (The SSSM also assigns to psychology and to psychological anthropology the
task of cataloging, at the individual level, the particularistic psychological phenomena
that are created by the action of each culture on individuals; e.g., what do American
college sophomores get anxious about?).

In advance of any data, the Standard Model defined for psychology the general
character of the mechanisms that it was supposed to find (general-purpose, content-
independent ones), its most important focus (learning), and how it would interpret the
data it found (no matter what the outcome, the origin of content was to be located
externally—for example, in the unknowably complex unobserved prior history of the
individual—and not "internally" in the mind of the organism). Psychologists certainly
were not forced by the character of their data into these types of conclusions (e.g., Breland & Breland, 1961). Instead, they had to carefully design their experiments so as to
exclude evolutionarily organized responses to biologically significant stimuli by eliminating such stimuli from their protocols (e.g., by using stimulus-impoverished Skinner boxes or the currently widespread practice of eliminating "emotionally charged"
stimuli from cognitive experiments). This was done in the name of good experimental
design and with the intention of eliminating contaminating "noise" from the exploration of the content-independent mechanisms that were thought to exist.

The Division of Labor: Particularistic, Content-Specific Anthropology

Even more than psychology, anthropology was shaped by the assumptions inherent in
the SSSM's division of labor: A content-independent (or content-free) psychology
symbiotically requires a content-supplying anthropology to provide the agent—culture—that transforms malleable generalized potential into specifically realized human
beings. So anthropology's mission was to study the particular (Geertz, 1973, p. 52).
Consequently, anthropology became the custodian of the key explanatory concept in
the paradigm, "culture." Belief in culture, as a substance passed across generations
causing the richly defined particularity of adult mental and social organization defines
one's membership in the modern social science community. The invocation of culture
became the universal glue and explanatory variable that held social science explanations together: Why do parents take care of their children? It is part of the social role
their culture assigns to them. Why are Syrian husbands jealous? Their culture tied their
status to their wife's honor. Why are people sometimes aggressive? They learn to be
because their culture socializes them to be violent. Why are there more murders in
America than in Switzerland? Americans have a more individualistic culture. Why do
women want to look younger? Youthful appearance is valued in our culture. And so
on.

Although using culture as an all-purpose explanation is a stance that is difficult or
impossible to falsify, it is correspondingly easy to "confirm." If one doubts that the
causal agent for a particular act is transmitted culture, one can nearly always find similar prior acts (or attitudes, or values, or representations) by others, so a source of the
contagion can always be identified. Culture is the protean agent that causes everything
that needs explaining in the social sciences, apart from those few things that can be

explained by content-general psychological laws, a few drives, and whatever superorganic processes (e.g., history, social conflict, economics) that are used to explain the particularities of a specific culture. Psychologists, then, need not explain the origin of complexly specific local patterns of behavior. They can be confident that anthropologists have done this job and have tracked, captured, defined, and analyzed the causal processes responsible for explaining why men are often sexually jealous or why women often prefer to look youthful.

In defining culture as the central concept of anthropology, the SSSM precluded the development of the range of alternative anthropologies that would have resulted if, say, human nature, economic and subsistence activity, ecological adaptation, human universals, the organization of incentives inside groups, institutional propagation, species-typical psychology, or a host of other reasonable possibilities had been selected instead. More critically, because of the way in which the SSSM frames the relationship between culture and the human mind, anthropology's emphasis on relativity and explanatory particularism becomes inescapable, by the following logic: If the psyche is general-purpose, then all organized content comes from the outside, from culture. Therefore, if something is contentful, then it must be cultural; if it is cultural, then—by the nature of what it is to be cultural—it is plastically variable; if it is plastically variable, then there can be no firm general laws about it. Ergo, there can be no general principles about the content of human life (only the contentless laws of learning). The conclusion is present in the premises. The relativity of human behavior, far from being the critical empirical discovery of anthropology (Geertz, 1973, 1984), is something imposed *a priori* on the field by the assumptions of the SSSM, because its premises define a program that is incapable of finding anything else. Relativity is no more "there" to be found in the data of anthropologists than a content-independent architecture is "there" to be found in the data of psychologists. These conclusions are present in the principles by which these fields approach their tasks and organize their data, and so are not "findings" or "discoveries" at all.

The consequences of this reasoned arrival at particularism reverberate throughout the social sciences, imparting to them their characteristic flavor, as compared with the natural sciences. This flavor is not complexity, contingency, or historicity: Sciences from geology to astronomy to meteorology to evolutionary biology have these in full measure. It is, instead, that social science theories are usually provisional, indeterminate, tentative, indefinite, enmeshed in an endlessly qualified explanatory particularism, for which the usual explanation is that human life is much more complex than mere Schrödinger equations or planetary ecosystems. Because culture was held to be the proximate (and probably the ultimate) cause of the substance and rich organization of human life, the consensus was, naturally, that documenting its variability and particularity deserved to be the primary focus of anthropological study (e.g., Geertz, 1973). This single proposition alone has proven to be a major contributor to the failure of the social sciences (Tooby & Cosmides, 1989a). Mainstream sociocultural anthropology has arrived at a situation resembling some nightmarish short story Borges might have written, where scientists are condemned by their unexamined assumptions to study the nature of mirrors only by cataloging and investigating everything that mirrors can reflect. It is an endless process that never makes progress, that never reaches closure, that generates endless debate between those who have seen different reflected images, and whose enduring product is voluminous descriptions of particular phenomena.

The Empirical Disproof of a Universal Human Nature

The view that the essence of human nature lies in its variousness and the corresponding rejection of a complex, universal human nature is not advanced by anthropologists simply as an assertion. Instead, it is presented as a dramatic and empirically well-supported scientific discovery (Geertz, 1984) and is derived from a particular method through which the limits of human nature are explored and defined. This method, a logical process of elimination, "confirmed" that the notion of human nature was empirically almost vacuous. Since infants are everywhere the same, then anything that varies in adults can only (it was reasoned) be cultural and, hence, socially inherited and, hence, socially manufactured. The method depends on accepting the premise that behavior can only be accounted for in these two ways: (1) as something "biological," or inborn, which is, therefore, inflexibly rigid regardless of environment and (because of the psychic unity of humankind) everywhere the same, or (2) as sociocultural, which includes everything that varies, at a minimum, and perhaps many things that happen by accident to be universal as well.

Whenever it is suggested that something is "innate" or "biological," the SSSM-oriented anthropologist or sociologist riffles through the ethnographic literature to find a report of a culture where the behavior (or whatever) varies (for a classic example, see Mead's 1949 *Male and Female*). Upon finding an instance of reported variation (or inventing one through strained interpretation; see again, Mead, 1949), the item is moved from the category of "innate," "biological," "genetically determined," or "hardwired" to the category of "learned," "cultural," or "socially constructed." Durkheim succinctly runs through the process, discussing why sexual jealousy, filial piety, and paternal love must be social constructions, despite claims to the contrary: "History, however, shows that these inclinations, far from being inherent in human nature, are often totally lacking. Or they may present such variations in different human societies that the residue obtained after eliminating all these differences—which alone can be considered of psychological origin—is reduced to something vague and rudimentary and far removed from the facts that need explanation" (Durkheim, 1895/1962, p. 106). Because almost everything human is variable in one respect or another, nearly everything has been subtracted from the "biologically determined" column and moved to the "socially determined" column. The leftover residue of "human nature," after this process of subtraction has been completed, is weak tea indeed, compared to the rich and engaging list of those dimensions of life where humans vary. No wonder Geertz (1973) finds such watered-down universals no more fundamental or essential to humans than the behaviors in which humans vary. Psychologists have, by and large, accepted the professional testimony of anthropologists and have, as part of their standard intellectual furniture, the confidence that other cultures violate virtually every universal claim about the content of human life. (D. E. Brown [1991] offers a pivotal examination of the history and logic of anthropological approaches to human universals, cultural variation, and biology, and this entire discussion is informed by his work.)

Discovering Regularities Depends on Selecting Appropriate Frames of Reference

Because of the moral appeal of antinativism, the process of discrediting claims about a universal human nature has been strongly motivated. Anthropologists, by each new claim of discovered variability, felt they were expanding the boundaries of their discipline (and, as they thought, of human possibility itself) and liberating the social sciences from biologically deterministic accounts of how we are inflexibly constrained to live as we do. This has elevated particularism and the celebration of variability to central values inside of anthropology, strongly asserted and fiercely defended.

The most scientifically damaging aspect of this dynamic has not been the consequent rhetorical emphasis most anthropologists have placed on the unusual (Bloch, 1977; Goldschmidt, 1960; Symons, 1979; see, especially, D. E. Brown, 1991). As Bloch (1977, p. 285) says, it is the "professional malpractice of anthropologists to exaggerate the exotic character of other cultures." Nor is the most damaging aspect of this dynamic the professionally cultivated credulousness about claims of wonders in remote parts of the world, which has led anthropologists routinely to embrace, perpetuate, and defend not only gross errors (see Freeman, 1983, on Mead and Samoa; Suggs, 1971, on Linton and the Marquesas) but also obvious hoaxes (e.g., Casteneda's UCLA dissertation on Don Juan; or the gentle "Tasaday," which were manufactured by officials of the Marcos regime).

The most scientifically damaging aspect of this value system has been that it leads anthropologists to actively reject conceptual frameworks that identify meaningful dimensions of cross-cultural uniformity in favor of alternative vantage points from which cultures appear maximally differentiated. Distinctions can easily be found and endlessly multiplied, and it is an easy task to work backward from some particular difference to find a framework from which the difference matters (e.g., while "mothers" may exist both there and here, motherhood here is completely different from motherhood there because mothers there are not even conceptualized as being blood kin, but rather as the wife of one's father, etc., etc.). The failure to view such variation as always profoundly differentiating is taken to imply the lack of a sophisticated and professional appreciation of the rich details of ethnographic reality.

But whether something is variable or constant is not just "out in the world"; it is also a function of the system of categorization and description that is chosen and applied. The distance from Paris to Mars is complexly variable, so is the location of Paris "constant" and "inflexible" or is it "variable?" In geography, as in the social sciences, one can get whichever answer one wants simply by choosing one frame of reference over another. The order that has been uncovered in physics, for example, depends on the careful selection of those particular systems of description and measure that allow invariances to appear. These regularities would all disappear if physicists used contingently relative definitions and measures, such as their own heartbeat to count units of time (the speed of light would slow down every time the measurer got excited—"relativity" indeed).

Other sciences select frameworks by how much regularity these frameworks allow them to uncover. In contrast, most anthropologists are disposed to select their frameworks so as to bring out the maximum in particularity, contingency, and variability (e.g., how are the people they study unique?). Certainly one of the most rewarded of talents inside anthropology is the literary ability to express the humanly familiar and intelligible as the exotic (see, e.g., Geertz's description of a raid by the authorities on a cock fight in Bali; Geertz, 1973; see D. E. Brown, 1991, for a lucid dissection of the role of universals in this example, and Barkow, 1989, on how Balinese cock fighting illustrates the conventional psychology of prestige). Anthropologists' attraction to frameworks that highlight particularistic distinctions and relationships has nearly pre-cluded the accumulation of genuine knowledge about our universal design and renders anthropologists' "empirically" grounded dismissal of the role of biology a matter of convention and conjuring rather than a matter of fact.

Beneath Variable Behavior Lie Universal Mechanisms

The social science tradition of categorizing everything that varies as "nonbiological" fails to identify much that is "biological." This is because anthropologists have chosen ill-suited frames of reference (such as those based on surface "behavior" or reflective "meaning") that accentuate variability and obscure the underlying level of universal evolved architecture. There may be good reasons to doubt that the "behavior" of mar-riage is a cross-cultural universal or that the articulated "meaning" of gender is the same across all cultures, but there is every reason to think that every human (of a given sex) comes equipped with the same basic evolved design (Tooby & Cosmides, 1990a). The critical question is not, for example, whether every human male in every culture engages in jealous behaviors or whether mental representations attached to situations of extra-pair mating are the same in every culture; instead, the most illuminating ques-tion is whether every human male comes endowed with developmental programs that are designed to assemble (either conditionally or regardless of normal environmental variation) evolutionarily designed sexual jealousy mechanisms that are then present to be activated by appropriate cues. To discern and rescue this underlying universal design out of the booming, buzzing confusion of observable human phenomena requires selecting appropriate analytical tools and frames of reference.

Genetics, for example, had enormous difficulty making progress as a science until geneticists developed the distinction between genotype and phenotype—between the inherited basis of a trait and its observable expression. We believe a similar distinction will prove necessary to the development of an integrated social science. We will refer to this as the distinction between the *evolved* (as in evolved mechanisms, evolved psy-chology, evolved architecture, etc.) and the *manifest* (as in manifest psychology, man-ifest behavior, etc.). One observes variable manifest psychologies or behaviors between individuals and across cultures and views them as the product of a common, under-lying evolved psychology, operating under different circumstances. The mapping between the evolved architecture and manifest behavior operates according to prin-ciples of expression that are specified in the evolved developmental mechanisms and the psychological mechanisms they reliably construct; manifest expressions may differ between individuals when different environmental inputs are operated on by the same procedures to produce different manifest outputs (Cosmides & Tooby, 1987; Tooby & Cosmides, 1989b).

For example, some individuals speak English while others do not, yet everyone passes through a life stage when the same species-typical language acquisition device is activated (Pinker & Bloom, this volume). In fact, if an individual survives a child-hood of aberrant social isolation she may never acquire a language and may be inca-pable of speaking; yet, she will have had the same species-typical language acquisition device as everyone else. So what at the behavioral level appears variable ("speaks English," "speaks Kikuyu"; or, even, "speaks a language," "does not speak any lan-guage") fractionates into variable environmental inputs and a uniform underlying design, interacting to produce the observed patterns of manifest variation. The fog enveloping most social science debates would blow away if all hypotheses were completely spelled out, through analyzing each situation into environmental conditions, evolved architecture, and how their interaction produces the manifest outcome.

Standard Model partisans have confidently rested their empirical case on what now appears to be uncertain ground: that manifest universality across cultures is the observation that evolutionary hypotheses about human nature require and that, on the other hand, cross-cultural variability establishes that the behavior in question is uncontaminated by "biology" and is, instead, solely the product of "culture" or "social processes." The recognition that a universal evolved psychology will produce variable manifest behavior given different environmental conditions exposes this argument as a complete non sequitur. From a perspective that describes the whole integrated system of causation, the distinction between the biologically determined and the nonbiologically determined can be seen to be a nondistinction.

In its place, the relevant distinction can be drawn between what Mayr (among others) called open and closed behavior programs (Mayr, 1976). This terminology distinguishes mechanisms that are open to factors that commonly vary in the organism's natural environment and, hence, commonly vary in their manifest expression from those that are closed to the influence of such factors and are, consequently, uniform in their manifest expression. The human language acquisition device is an open behavior program whose constructed product, adult competency in the local language, varies depending on the language community in which the individual is raised. Certain facial emotional displays that manifest themselves uniformly cross-culturally (Ekman, 1973) may be examples of closed behavior programs. The Standard Social Science Model's method of sorting behavior by its cross-cultural uniformity or variability of expression into "biologically determined" and "socially determined" categories in reality sorts behaviors into those generated by closed behavior programs, and those generated by open behavior programs. In neither case can the analysis of the "determination" of behavior be made independent of "biology," that is, independent of understanding the participation of the evolved architecture. For this reason, the whole incoherent opposition between socially determined (or culturally determined) phenomena and biologically determined (or genetically determined) phenomena should be consigned to the dustbin of history, along with the search for a biology-free social science.

The Search for the Artificer

If psychology studies the content-independent laws of mind and anthropology studies the content-supplying inheritances of particular cultures, one still needs to find the content-determining processes that manufacture individual cultures and social systems. The Standard Social Science Model breaks the social sciences into schools (materialist, structural-functionalist, symbolic, Marxist, postmodernist, etc.) that are largely distinguished by how each attempts to affirmatively characterize the artificer, which they generally agree is an emergent group-level process of some kind. It is far beyond the scope of this chapter to review and critique these attempts to discover somewhere in the social system what is in effect a generative computational system. Nevertheless, it is important to recognize that the net effect of the central logic of the Standard Social Science Model has been to direct the quest for the ultimate cause or generator of significant mental and social organization outward away from the rich computational architecture of the human mind. It is there where sufficiently powerful ordering processes—ones capable of explaining the phenomena—are primarily to be found. As will be discussed later, it is there where the actual generators of organization are principally (though not exclusively) located and could be productively investigated. And understanding this architecture is an indispensible ingredient in modeling or understanding whatever super-individual processes exist.

This is not to say that there aren't many important phenomena and processes operating at the population level, which, for example, modify the nature and distribution of representations (for non-SSSM analyses, see, e.g., Sperber, 1985, 1990, on epidemiological approaches to cultural transmission; see also Boyd & Richerson, 1985; Campbell, 1965, 1975; Durham, 1991; and others, who examine analogs to natural selection operating at the cultural level). But because the traditional SSSM efforts to characterize these generative processes make them entirely external to "the individual" as well as independent of species-typical psychology, these accounts tend to share a certain ineradicable indefiniteness of location and substance. The SSSM attempt to abstract social processes away from the matrix of interacting psychological architectures necessarily fails because the detailed structure of psychological mechanisms is inextricably bound up in how these social processes operate. One might say that what mostly remains, once you have removed from the human world everything internal to individuals, is the air between them. This vagueness of ontology and causal mechanism makes it difficult to situate these hypothetical generative processes with respect to our knowledge of the rest of the natural world (Sperber, 1986). Moreover, attempting to locate in these population-level processes the primary generator of significant organization has caused these processes to be fundamentally misunderstood, mystified, and imbued with such unwarranted properties as autonomy, teleology, functionality, organism-like integration, intelligence, intentionality, emotions, need-responsiveness, and even consciousness (see, e.g., Durkheim, 1895/1962; Harris, 1979; Kroeber, 1915; Marx, 1867/1909; Merton, 1949; Parsons, 1949; Radcliffe-Brown, 1952; see Harris, 1968, for review and discussion).

Of course, the social system is not like a person or an organism or a mind, self-ordering due to its own functionally integrated mechanisms. It is more like an ecosystem or an economy whose relationships are structured by feedback processes driven by the dynamic properties of its component parts. In this case, the component parts of the population are individual humans, so any social dynamics must be anchored in models of the human psychological architecture. In contrast, the customary insistence on the autonomy (or analytic separability) of the superorganic level is why there have been so few successful or convincing causal models of population-level social processes, including models of culture and social organization (apart from those originating in microeconomics or in analogies drawn from population biology, which do not usually take SSSM-style approaches; see, e.g., Boyd & Richerson, 1985; Schelling, 1978).

Rejecting the design of individuals as central to the dynamics is fatal to these models, because superorganic (that is, population-level) processes are not just "out there," external to the individual. Instead, human superindividual interactions depend intimately on the representations and other regulatory elements present in the head of every individual involved and, therefore, on the systems of computation inside each head. These govern what is selected from "out there," how this is represented, what procedures act on the representations, and what behaviors result that others can then observe and interact with in a population dynamic fashion (Sperber, 1985, 1986, 1990;

Tooby & Cosmides, 1989a). These psychological mechanisms are primarily where there is sufficient anti-entropic computational power to explain the rich patterning of human life. The design of the human psychological architecture structures the nature of the social interactions humans can enter into, as well as the selectively contagious transmission of representations between individuals. Only after the description of the evolved human psychological architecture has been restored as the centerpiece of social theory can the secondary anti-entropic effects of population-level social dynamics be fully assessed and confidently analyzed. Hence, the study of population-level social and cultural dynamics requires a sophisticated model of human psychology to undergird it (see Barkow, 1989, this volume; Cosmides & Tooby, 1989, this volume; Daly, 1982; Sperber, 1985, 1990; Tooby & Cosmides, 1989a).

The Division of Labor: The Social Sciences versus the Natural Sciences

The single most far-reaching consequence of the Standard Social Science Model has been to intellectually divorce the social sciences from the natural sciences, with the result that they cannot speak to each other about much of substance. Where this divorce has been achieved can be precisely located within the model. Because biology and evolved psychology are internal to the individual and because culture—the author of social and mental organization—is seen as external to the individual, the causal arrow from outside to inside logically insulates the social sciences from the rest of the natural sciences, making them autonomous and the natural sciences substantively irrelevant. This set of propositions is the locus of the primary break between the social and the natural sciences. Although there has been a causal flow across four billion years of evolutionary time, its ability to causally shape the human present is impermeably dammed at the boundaries of the individual—in fact, well within the individual, for evolution is thought to provide nothing beyond an account of the origins of the drives, if any, and of the general-purpose, content-free learning or computational equipment that together comprise the SSSM's minimalist model of human nature.

Thus, whatever their empirical success may be, the claims made by (to pick some obvious examples) ethologists, sociobiologists, behavioral ecologists, and evolutionary psychologists about the evolutionary patterning of human behavior can be simply dismissed out of hand as wrong, without requiring specific examination, because causality does not flow outward from the individual or from psychology, but inward from the social world (Sahlins, 1976a). Or, as Durkheim put it nearly a century before, "every time that a social phenomenon is directly explained by a psychological phenomenon, we may be sure that the explanation is false" (Durkheim, 1895/1962, p. 103). Organic evolution manufactured the biological substratum, the human capacity for culture—"the breadth and indeterminateness of [man's] inherent capacities" (Geertz, 1973, p. 45)—but otherwise reaches a dead end in its causal flow and its power to explain.

Finally, it would be a mistake to think that the Standard Social Science Model reflects the views solely of social scientists and is usually resisted by biologists and other natural scientists. It is instead considered the common sense and common decency of our age. More particularly, it is a very useful doctrine for biologists themselves to hold. Many of them vigorously defend its orthodoxies, adding their professional imprimatur to the social scientists' brief for the primacy of culture or social forces over "biology" (see, e.g., Gould, 1977a, 1977b; Lewontin, Rose & Kamin, 1984). This does not happen simply because some are drawn to the formidable moral authority of the Standard

Model and the mantle it confers. Even for those of a genuinely scientific temperament, fascinated with biological phenomena for their own sake, such a doctrine is a godsend. The Standard Model guarantees them, *a priori*, that their work cannot have implications that violate socially sanctified beliefs about human affairs because the Standard Model assures them that biology is intrinsically disconnected from the human social order. The Standard Model therefore frees those in the biological sciences to pursue their research in peace, without having to fear that they might accidentally stumble into or run afoul of highly charged social or political issues. It offers them safe conduct across the politicized minefield of modern academic life. This division of labor is, therefore, popular: Natural scientists deal with the nonhuman world and the "physical" side of human life, while social scientists are the custodians of human minds, human behavior, and, indeed, the entire human mental, moral, political, social, and cultural world. Thus, both social scientists and natural scientists have been enlisted in what has become a common enterprise: the resurrection of a barely disguised and archaic physical/mental, matter/spirit, nature/human dualism, in place of an integrated scientific monism.

THE EVOLUTIONARY CONTRIBUTION TO INTEGRATED EXPLANATION

Rediscovering the Relevance of Evolutionary Biology

If the adoption of the Standard Social Science Model has not led to a great deal of natural science-like progress, that is surely not a good argument against it. Its convenience has no bearing on whether it is true. What, then, are the reasons for believing it is false? There are a number of major problems that independently lead to the rejection of the SSSM—some emerging from evolutionary biology, some from cognitive science, and some from their integration. We will discuss three of these problems, arguing that (1) the Standard Social Science Model's analysis of developmental or "nature-nurture" issues is erroneous; (2) the general-purpose, content-free psychology central to the SSSM could not have been produced by the evolutionary process and, therefore, is not a viable candidate model of human psychology; and (3) a psychology of this kind cannot explain how people solve a whole array of tasks they are known to routinely perform.

For advocates of the Standard Social Science Model, evolution is ignored because it is irrelevant: The explanatory power of evolution ends with the emergence of the content-free computational equipment that purportedly constitutes human nature. This equipment does not impose any form on the social world, but instead acquires all of its content from the social world. The supposed erasure of content-sensitive and content-imparting structure from hominid psychological architecture during our evolution is what justifies the wall of separation between the natural and the social sciences. If this view were correct, then evolution would indeed be effectively irrelevant to the social sciences and the phenomena they study.

In contrast, proponents of the Integrated Causal Model accept that, in addition to whatever content-independent mechanisms our psychological architecture may contain, it also contains content-specific devices, including those computationally responsible for the generation and regulation of human cultural and social phenomena. These content-specific mechanisms are adaptations (as are content-independent mechanisms), and evolved to solve long-enduring adaptive problems characteristic of our hunter-gatherer past. Because of their design, their operation continually imparts evolutionarily patterned content to modern human life. If this view is correct, then the specifics of evolutionary biology have a central significance for understanding human thought and action. Evolutionary processes are the "architect" that assembled, detail by detail, our evolved psychological and physiological architecture. The distinctive characteristics of these processes are inscribed in the organizational specifics of these designs. Consequently, an understanding of the principles that govern evolution is an indispensable ally in the enterprise of understanding human nature and an invaluable tool in the discovery and mapping of the species-typical collection of information-processing mechanisms that together comprise the human mind. The complex designs of these mechanisms are the main causal channels through which the natural sciences connect to and shape the substance of the "social" sciences.

Thus, the relevance of evolutionary biology does not in the least depend on our being "just like" other species, which we obviously are not (Tooby & DeVore, 1987). Each species has its own distinctive properties stemming from its own unique evolutionary history. Evolutionary biology is fundamentally relevant to the study of human behavior and thought because our species is the product of naturalistic terrestrial processes—evolutionary processes—and not of divine creation or extraterrestrial intervention. However unusual our properties may be from a zoological point of view— and we have every reason to believe humans followed a unique evolutionary trajectory (Tooby & DeVore, 1987)—we need an account of how they were produced in the natural world of causation over evolutionary history (Boyd & Richerson, 1985). Such accounts are constructed from (1) the principles that govern the evolutionary process (such as natural selection and drift), (2) the designs of ancestral hominid species, and (3) the particular ancestral environments and contingent historical events hominids encountered during their evolutionary history.

Reproduction, Feedback, and the Construction of Organic Design

> [T]hese elaborately constructed forms, so different from each other, and dependent on each other in so complex a manner, have all been produced by laws acting around us. These laws, taken in the largest sense, being . . . Reproduction; Inheritance which is almost implied by reproduction; Variability . . . and as a consequence . . . Natural Selection. . . .
>
> —CHARLES DARWIN,
> *The Origin of Species*

While physicists tend to start their causal history with the Big Bang, biologists usually select a different, later event: the emergence of the first living organism. Life (or the instances we have so far observed) is a phenomenon that originated on earth three to four billion years ago through the formation of the first living organism by contingent physical and chemical processes. What is life? What defining properties qualified some ancient physical system as the first living organism? From a Darwinian perspective, it is the reproduction by systems of new and similarly reproducing systems that is the defining property of life. An organism is a self-reproducing machine. All of the other properties of living organisms—cellular structure, ATP, polypeptides, the use of carbon's ability to form indefinitely large chains, DNA as a regulatory element—are incidental rather than essential, and the logic of Darwinism would apply equally to self-reproducing robots, to self-reproducing plasma vortices in the sun, or to anything else that reproduces with the potential for inheritable change (mutation). From reproduction, the defining property of life, the entire elegant deductive structure of Darwinism follows (Dawkins, 1976; Tooby & Cosmides, 1990b; Williams, 1985).

Very simple proto-bacteria emerged early in terrestrial history, as chemical machines that constructed additional chemical machines like themselves. Because reproduction means the construction of offspring designs like the parent machines, one could imagine this leading to an endless chain of proliferating systems identical to the original parent. This is not what happened, of course, because mutations or random modifications are sometimes introduced into offspring designs by accident, with far-reaching consequences. Most random modifications introduce changes into the organism's organization that interfere with the complex sequence of actions necessary for self-reproduction. But a small proportion of random modifications happen to cause an enhancement in the average ability of the design to cause its own reproduction. In the short run, the frequency of those variants whose design promotes their own reproduction increases, and the frequency of those variants whose design causes them to produce fewer (or no) offspring decreases. Consequently, one of two outcomes usually ensues: (1) the frequency of a design will drop to zero—i.e., go extinct (a case of negative feedback); or (2) a design will outreproduce and thereby replace all alternative designs in the population (a case of positive feedback). After such an event, the population of reproducing machines is different from its ancestors because it is equipped with a new and more functionally organized design or architecture. Thus, the fact that alternative design features give rise to reproductive performance differences creates the system of positive and negative feedback called natural selection. Natural selection guides the incorporation of design modifications over generations according to their consequences on their own reproduction.

Over the long run, down chains of descent, this cycle of chance modification and reproductive feedback leads to the systematic accretion within architectures of design features that promote or formerly promoted their own propagation. Even more importantly, the reproductive fates of the inherited properties that coexist in the same organism are linked together: What propagates one design feature tends to propagate others (not perfectly, but sufficiently for a coherent design to emerge; see Cosmides & Tooby, 1981). This means that traits will be selected to work together to produce the same outcomes and to enhance each other's functionality. Frequently, then, these accumulating properties will sequentially fit themselves together into increasingly functionally elaborated machines for reproduction, composed of constituent mechanisms—called adaptations—that solve problems that are either necessary for reproduction or increase its likelihood (Darwin, 1859; Dawkins, 1986; R. Thornhill, 1991; Tooby & Cosmides, 1990b; Williams, 1966, 1985). As if by the handiwork of an invisible and nonforesightful engineer, element after element is added to a design over generations, making it a more functional system for propagation under the conditions prevailing at the time each new element was added. At present, there is no extant alternative theory for how organisms acquired complex functional organization over the course of their evolution (Dawkins, 1986).

What is most compelling about Darwin's approach is that the process of natural selection is an inevitable by-product of reproduction, inheritance, and mutation. Like water running downhill, over generations organisms tend to flow into new functional designs better organized for effective propagation in the environmental context in which they evolved. There is, however, another method, besides selection, by which mutations can become incorporated into species-typical design: chance. The sheer impact of many random accidents may cumulatively propel a useless mutation upward in frequency until it crowds out all alternative design features from the population. Its presence in the architecture is not explained by the (nonexistent) functional consequences it has on reproduction, and so it will not be coordinated with the rest of the organism's architecture in a functional way.

But despite the fact that chance plays some role in evolution, organisms are not primarily chance agglomerations of stray properties. To the extent that a feature has a significant effect on reproduction, selection will act on it. For this reason, important and consequential aspects of organismic architectures are shaped by selection. By the same token, those modifications that are so minor that their consequences are negligible on reproduction are invisible to selection and, therefore, are not organized by it. Thus, chance properties drift through the standard designs of species in a random way, unable to account for complex organized design and, correspondingly, are usually peripheralized into those aspects that do not make a significant impact on the functional operation of the system (Tooby & Cosmides, 1990a, 1990b).

In short, evolution (or descent with modification, to use Darwin's phrase) takes place due to the action of both chance and natural selection, causing descendants to diverge in characteristics from their ancestors, down chains of descent. Over evolutionary time, this appears as a succession of designs, each a modification of the one preceding it. Generation by generation, step by step, the designs of all of the diverse organisms alive today—from redwoods and manta rays to humans and yeast—were permuted out of the original, very simple, single-celled ancestor through an immensely long sequence of successive modifications. Each modification spread through the species either because it caused its own propagation, or by accident. Through this analytic framework, living things in general and each species in particular can be located in the principled causal history of the universe. Moreover, the specific design features of a species' architecture can also be causally located in this history: they exist either because of chance incorporation or because they contributed to the functional operation of the architecture. The theory of evolution by natural selection vastly expanded the range of things that could be accounted for, so that not only physical phenomena such as stars, mountain ranges, impact crater's, and alluvial fans could be causally located and explained but also things like whales, eyes, leaves, nervous systems, emotional expressions, and the language faculty.

The modern Darwinian theory of evolution consists of the logically derivable set of causal principles that necessarily govern the dynamics of reproducing systems, accounting for the kinds of properties that they cumulatively acquire over successive generations. The explicit identification of this core logic has allowed the biological community to develop an increasingly comprehensive set of principles about what kinds of modifications can and do become incorporated into the designs of reproducing systems down their chains of descent, and what kinds of modifications do not (Dawkins, 1976, 1982, 1986; Hamilton, 1964, 1972; Maynard Smith, 1964, 1982; Williams, 1966). This set of principles has been tested, validated, and enriched by a comprehensive engagement with the empirical reality of the biological world, from functional and comparative anatomy, to biogeography, to genetics, to immunology, to embryology, to behavioral ecology, and so on. Just as the fields of electrical and mechanical engineering summarize our knowledge of principles that govern the design of human-built machines, the field of evolutionary biology summarizes our knowledge of the engineering principles that govern the design of organisms, which can be thought of as machines built by the evolutionary process (for good summaries, see Daly & Wilson 1984a; Dawkins, 1976, 1982, 1986; Krebs & Davies, 1987). Modern evolutionary biology constitutes, in effect, an "organism design theory." Its principles can be used both to evaluate the plausibility of the psychology posited by the Standard Social Science Model and to guide the construction of a better successor psychology.

The Peculiar Nature of Biological Functionality

In certain narrowly delimited ways, then, the spontaneous process of evolution parallels the intentional construction of functional machines by human action. But, whereas machines built by human engineers are designed to serve a diverse array of ends, the causal process of natural selection builds organic machines that are "designed" to serve only one very specialized end: the propagation into subsequent generations of the inherited design features that comprise the organic machine itself.

Because design features are embodied in organisms, they can, generally speaking, propagate themselves in only two ways: (1) by solving problems that will increase the probability that the organism they are situated in will produce offspring, or (2) by solving problems that will increase the probability that the organism's kin will produce offspring (Hamilton, 1964; Williams & Williams, 1957). An individual's relatives, by virtue of having descended from a common ancestor, have an increased likelihood of having the same design feature as compared to other conspecifics, so their increased reproduction will tend to increase the frequency of the design feature. Accordingly, design features that promote both direct reproduction and kin reproduction, and that make efficient trade-offs between the two, will replace those that do not. To put this in standard biological terminology, design features are selected for to the extent that they promote their inclusive fitness (Hamilton, 1964). For clarity, we will tend to call this propagation or design-propagation.

Selection, then, is the only known account for the natural occurrence of complexly organized functionality in the inherited design of undomesticated organisms. Moreover, selection can only account for functionality of a very narrow kind: approximately, design features organized to promote the reproduction of an individual and his or her relatives (Dawkins, 1986; Williams, 1966). Fortunately for the modern theory of evolution, the only naturally occurring complex functionality that has ever been documented in plants, animals, or other organisms is functionality of just this kind, along with its derivatives and by-products. Mammals evolved adaptations to produce milk to feed their own young, infectious micro-organisms mimic human biochemistry to escape immune surveillance so they can survive and reproduce, and plants produce oxygen as a waste product of feeding themselves through photosynthesis, and not for the pleasure of watching humans breathe or forests burn. Of course, human breeders artificially intervene, and one could easily imagine, as an alternative to a Darwinian world, a benevolent deity or extraterrestrial being creating the properties of living things in order to serve human convenience rather than the organisms' own reproduction. Wild horses could be born with saddle-shaped humps, luggage racks, and a spontaneous willingness to be ridden; chronic bacterial infections could jolt humans with caffeine every morning 45 seconds before they need to get up. Similarly, the non-living world could be full of intricate functional arrangements not created by humans, such as mountains that naturally mimic hotels down to the details of closet hangers, electric wiring, and television sets. But this is not the world we live in. We live in a world of biological functionality and its derivatives, traceable originally to the operation of natural selection on reproducing systems (Darwin, 1859; Dawkins, 1976, 1986). To be able to understand the world of biological phenomena, one must be able to recognize this peculiar functional organization and distinguish it from the products of chance.

Of course, the fact that living things are machines organized to reproduce themselves and their kin does not mean that evolutionary functional analysis focuses narrowly on such issues as copulation or pregnancy (things intuitively associated with reproduction) over, say, taste preferences, vision, emotional expression, social categorization, coalition formation, or object recognition. A life history of successfully achieved reproduction (including kin reproduction) requires accomplishing the entire tributary network of preconditions for and facilitations of reproduction in complex ecological and social environments. Of course, this includes all of the information-gathering, inference, and decision-making that these tasks entail. For this reason, humans display a diverse range of adaptations designed to perform a wide and structured variety of subsidiary tasks, from solicitation of assistance from one's parents, to language acquisition, to modeling the spatial distribution of local objects, to reading the body language of an antagonist.

Finally, the behavior of individual organisms is caused by the structure of their adaptations and the environmental input to them; it is not independently governed by the principle of individual fitness maximization. Individual organisms are best thought of as adaptation-executers rather than as fitness-maximizers. Natural selection cannot directly "see" an individual organism in a specific situation and cause behavior to be adaptively tailored to the functional requirements imposed by that situation. To understand the role of selection in behavior, one must follow out all steps in the chain: Selection acting over evolutionary time has constructed the mechanisms we have inherited in the present, and it is this set of mechanisms that regulates our behavior—not natural selection directly. These mechanisms situated in particular individuals frequently—but by no means always—bring about a functional coordination between the adaptive demands of particular situations and associated behavioral responses.

Thus, the biological concept of functionality differs from the folk notion of functionality as goal-seeking behavior. Although some of our evolved psychological mechanisms probably operate through goal-seeking, surely none of them has fitness-maximization as a mentally represented goal (see Symons, this volume). Those goal-seeking mechanisms that do exist most likely embody proximate goals, such as "stay warm" or "protect your infant," rather than ultimate goals, such as "maximize your fitness" or "have as many offspring as possible." Indeed, goals of the latter kind are probably impossible to instantiate in any computational system (Symons, 1987, 1989, this volume; see also Barkow, 1989; Cosmides & Tooby, 1987, 1992; Daly & Wilson, 1988; Irons, 1983, p. 200; Tooby & Cosmides, 1990b. For somewhat contrary views, see, e.g., Alexander, 1979, 1987 and Turke, 1990).

For this reason, an adaptationist approach does not properly involve explaining or interpreting *individual* behavior in specific situations as "attempts" to increase fitness (Symons, 1989, this volume; Tooby & Cosmides, 1990b). To make the distinction between these alternative views of evolutionary explanation clear—humans as fitness-maximizers (fitness-teleology) versus humans as adaptation-executors (adaptationism)—a brief example will serve. Fitness teleologists may observe a situation and ask something like, "How is Susan increasing her fitness by salting her eggs?" An adaptationist would ask, instead, "What is the nature of the evolved human salt preference mechanisms—if any—that are generating the observed behavior and how did the structure of these mechanisms mesh with the physiological requirements for salt and the opportunities to procure salt in the Pleistocene?" So, in viewing cases of behavior, the adaptationist question is not, "How does this or that action contribute to this particular individual's reproduction?" Instead, the adaptationist questions are, "What is the underlying panhuman psychological architecture that leads to this behavior in certain specified circumstances?" and "What are the design features of this architecture—if any—that regulate the relevant behavior in such a way that it would have constituted functional solutions to the adaptive problems that regularly occurred in the Pleistocene?

What Adaptations Look Like

For the reasons outlined above, the species-typical organization of the psychology and physiology of modern humans necessarily has an evolutionary explanation and an evolutionarily patterned architecture. This is not a vague speculation or an overreaching attempt to subsume one discipline inside another, but constitutes as solid a fact as any in modern science. In fact, this conclusion should be a welcome one because it is the doorway through which a very rich body of additional knowledge—evolutionary biology—can be brought to bear on the study of psychological architecture. At its core, the discovery of the design of human psychology and physiology is a problem in reverse engineering: We have working exemplars of the design in front of us, but we need to organize our sea of observations about these exemplars into a map of the causal structure that accounts for the behavior of the system. Psychology has never been limited by a lack of observations. Fortunately, the knowledge that humans are the product of evolution supplies us with a powerful set of tools—the concepts of evolutionary functional analysis—for organizing these observations into useful categories so that the underlying systems of order can be discerned.

The most illuminating level of description for organizing observations about living species is usually in terms of their adaptations (and associated evolutionary categories). This system of description has some warrant on being considered a privileged frame of reference because the complex functional organization that exists in the design of organisms was injected into them through the construction of adaptations by natural selection. Adaptations are the accumulated output of selection, and selection is the single significant anti-entropic or ordering force orchestrating functional organic design (Dawkins, 1986). So if one is interested in uncovering intelligible organization in our species-typical psychological architecture, discovering and describing its adaptations is the place to begin.

To understand what complex adaptations "look like," it will help to begin concretely with a standard example, the vertebrate eye and its associated neural circuitry. (For its role in understanding adaptations, see Pinker & Bloom, this volume; for a discussion of color vision, see Shepard, this volume.) The eye consists of an exquisitely organized arrangement of cells, structures, and processes, such as (1) a transparent protective outer coating, the cornea; (2) an opening, the pupil, through which light enters; (3) an iris, which is a muscle that surrounds the pupil and constricts or dilates the aperture, regulating the amount of light entering the eye; (4) a lens, which is both transparent and flexible, and whose curvature and thickness can be adjusted to bring objects of varying distances into focus; (5) the retina, a light-sensitive surface that lies in the focal plane of the lens: this multilayered neural tissue lining the inside back of the eyeball is, in effect, a piece of the brain that migrated to the eye during fetal development; (6) classes of specialized cells (rods and cones) in the retina that transform sampled properties of ambient light through selective photochemical reactions into electrochemical impulses; (7) the activation by these electrochemical impulses of neighboring bipolar cells, which, in turn, feed signals into neighboring ganglion cells, whose axons converge to form the optic nerve; (8) the optic nerve, which carries these signals out of the eye and to the lateral geniculate bodies in the brain; (9) the routing of these signals to the visual cortex, into a series of retinotopic maps and other neural circuits, where they are further analyzed by a formidable array of information-processing mechanisms that also constitute crucial parts of the visual system.

The dynamic regulatory coordination present in the operation of the eye is also striking: The variable aperture modulates the amount of light entering the eye in coordination with ambient illumination; the eyes are stereoscopically coordinated with each other so that their lines of vision converge on the same object or point of interest; the thickness and curvature of the lens is modulated so that light from the object being viewed is focused on the retina and not in front of it or behind it; and so on. Through a more detailed description, this list could easily be extended to include many thousands of specialized features that contribute to the functionality of the system (tear ducts, eyelids, edge detectors, muscle systems, specific photochemical reactions, and so on) through the orchestrated arrangement of hundreds of millions of cells. This is even more true if we were to go beyond a taxonomically generalized description of the vertebrate eye and relate specific design features in particular vertebrates to the particular environments and visual tasks they faced. Frogs, for example, have retinal "bug detectors"; the rabbit retina has a variety of specialized devices, including a "hawk detector" (Marr, 1982, p. 32), and so on. It is important to appreciate that this organization is not just macroscopic, but extends down to the organized local relationships that subsets of cells maintain with each other, which perform such computations as edge detection and bug detection—and beyond, down into the specific architecture of the constituent cells themselves. Thus, rods and cones have a distinctive design and layout that includes specialized organelles that adjust the size and shape of the photoreactive regions; they have membranes that fold back on themselves to form sacs localizing the photoreceptive pigments; they have specialized chemistry so that light induces these pigments to undergo chemical changes that ultimately result in a change of membrane potential; they are arranged so that this change of membrane potential effects the release of neurotransmitters to neighboring bipolar glion cells, and so on.

Thus, the eye is an extraordinarily complex arrangement of specialized features that does something very useful for the organism. Moreover, this structure was originally absent from the ancestral design of the original single-celled founding organism, so the appearance of eyes in modern organisms must be explained as a succession of modifications across generations away from this initial state. It is easy to see how selection, through retaining those accidental modifications that improved performance, could start with an initial accidentally light-sensitive nerve ending or regulatory cell and transform it, through a large enough succession of increasingly complex functional forms, into the superlatively crafted modern eye (see, e.g., Dawkins, 1986). In fact, eyes (light-receptive organs) have evolved independently over 40 times in the history of animal life from eyeless ancestral forms (Mayr, 1982).

Of course, there are certainly many nonselectionist processes in evolution by which descendants are modified away from ancestral forms—drift, macromutation, hitchhiking, developmental by-product, and so on. But selection is the only process that directs change by retaining variants that are more functional. Thus, selection is the only causal process that has a systematic tendency to propel the system in the direction of increasingly functional arrangements, instead of into the immeasurably larger array of nonfunctional arrangements that the system could move to at each of the innumerable choice points in the evolution of designs. In contrast, nonselectional processes can produce functional outcomes only by chance because a new modification's degree of functionality plays no role in determining whether nonselectional processes will cause it to be retained or eliminated. For this reason, evolutionary processes other than selection are properly classified as "chance processes" with respect to the evolution of adaptive complexity. It would be a coincidence of miraculous degree if a series of these function-blind events, brought about by drift, by-products, hitchhiking, and so on, just happened to throw together a structure as complexly and interdependently functional as an eye (Dawkins, 1986; Pinker & Bloom, this volume). For this reason, nonselectionist mechanisms of evolutionary change cannot be seen as providing any reasonable alternative explanation for the eye or for any other complex adaptation. Complex functional organization is the signature of selection.

The eye is by no means a unique case. Immunologists, for example, have traced out a similar, immensely articulated architecture of complexly interrelated defenses (the blood monocytes, histiocytes, free macrophages, T-lymphocytes, B-lymphocytes, spleen, thymus, and so on). In fact, virtually every organ that has been examined so far betrays a complex functionality unmatched as of yet by any system engineered by humans. More than a century of research and observation confirms that selection builds into organisms a complex functional organization of an eye-like precision and quality.

Still, although many social and biological scientists are willing to concede that the body is full of the most intricately functional machinery, heavily organized by natural selection, they remain skeptical that the same is true of the mind. Moreover, partisans of the Standard Social Science Model insist on the Cartesian distinction between the material world of anatomy and physiology and the mental world of psychology, vigorously resisting attempts to see them as different descriptions of the same integrated system, subject to the same organizing principles. Arguments by Chomsky and others that our psychological architecture should contain "mental organs" for the same reasons that the rest of the body contains physical organs (i.e., that different tasks require functionally different solutions) have yet to convince the majority of psychologists outside of perception and language (Chomsky, 1975; Marshall, 1981).

Thus, Lewontin is expressing a thoroughly orthodox SSSM skepticism toward the idea that the human psychological architecture is functionally organized when he suggests that "[h]uman cognition may have developed as the purely epiphenomenal consequence of the major increase in brain size, which, in fact, may have been selected for quite other reasons" (Lewontin, 1990, p. 244). At least as numerous are those researchers who detect in human thought and behavior something more than the sheer accident that Lewontin sees, and yet who ask: Aren't psychological (or neurophysiological) mechanisms expected to be less well-engineered than physiological organs? High degrees of functionality are all very well for eyes, intestines, and immune systems, but what about the constituent structures of the human psychological architecture? Are there at least any examples of well-engineered psychological adaptations that might parallel physiological adaptations?

What is most ironic about this question is that perhaps the single most uncontroversial example of an adaptation—an example that is conceded to be well-engineered by even the most exercised of the anti-adaptationists—*is* a psychological adaptation: the eye. As Epicharmus pointed out two and half millennia ago, "Only mind has sight and hearing; all things else are deaf and blind." The eye and the rest of the visual system perform no mechanical or chemical service for the body; it is an information-processing adaptation. This information-processing device is designed to take light incident on a two-dimensional body surface and—through applying information-processing procedures to this two-dimensional array—construct cognitive models of the local three-dimensional world, including what objects are present, their shapes, their locations, their orientations, their trajectories, their colors, the textures of their surfaces, as well as face recognition, emotional expression recognition, and so on. Indeed, for those committed to a Cartesian world view, one could think of the eye as a tube that traverses metaphysical realms, one end of which obtrudes into the physical realm, the other into the mental. For modern monists, however, these two realms are simply alternative descriptions of the same thing, convenient for different analytic purposes. The "mental" consists of ordered relationships in physical systems that embody properties typically running under labels such as "information," "meaning," or regulation. From this point of view, there is no Cartesian tube: both ends of the visual system are physical and both are mental.

Because psychologists as a community have been resistant to adaptationist thinking, it was an enormous (although nonaccidental) stroke of good fortune that the visual system extrudes a "physical end" to the surface of the body and that this "physical end" bears a remarkable resemblance to the camera, a functional machine designed by humans. Selection has shaped the physical structure of the eye so that it reflects and exploits the properties of light, the geometry of the three-dimensional world, the refracting properties of lenses, and so on; the camera has a similar structure because it was designed by human engineers to reflect and exploit these same properties. These parallels between camera and eye were clues that were so obvious and so leading that it became a reasonable enterprise to investigate the visual system from a functionalist perspective. Researchers started with the physical end and followed the "Cartesian tube" upward and inward, so to speak, into the mind. In so doing, they have discovered increasingly complex and specialized computational machinery: edge detectors, motion detectors, shape detectors, depth boundary detectors, bug detectors (in the third neural layer of the retina of frogs), stereoscopic disparity analyzers, color constancy mechanisms, face recognition systems, and on and on.

Hundreds of vision researchers, working over decades, have been mapping this exquisitely structured information-processing adaptation, whose evolutionary function is scene analysis—the reconstruction of models of real-world conditions from a two-dimensional visual array. As more and more functional subcomponents are explored, and as artificial intelligence researchers try to duplicate vision in computational systems attached to electronic cameras, four things have become clear (Marr, 1982; Poggio, Torre, & Koch, 1985). The first is that the magnitude of the computational problem posed by scene analysis is immensely greater than anyone had suspected prior to trying to duplicate it. Even something so seemingly simple as perceiving the same object as having the same color at different times of the day turns out to require intensely specialized and complex computational machinery because the spectral distribution of light reflected by the object changes widely with changes in natural illumination (Shepard, this volume). The second conclusion is that as a psychological adaptation (or set of adaptations, depending on whether one is a lumper or splitter), our visual system is very well-engineered, capable of recovering far more sophisticated information from two-dimensional light arrays than the best of the artificially engineered systems developed so far. The third is that successful vision requires specialized neural circuits or computational machinery designed particularly for solving the adaptive problem of scene analysis (Marr, 1982). And the fourth is that scene analysis is an unsolvable computational problem unless the design features of this specialized machinery "assume" that objects and events in the world manifest many specific regularities (Shepard, 1981, 1984, 1987a; Marr, 1982; Poggio et al., 1985). These four lessons—complexity of the adaptive information-processing problem, well-engineered problem-solving machinery as the evolved solution, specialization of the problem-solving machinery to fit the particular nature of the problem, and the requirement that the machinery embody "innate knowledge" about the problem-relevant parts of the world—recur throughout the study of the computational equipment that constitutes human psychology (Cosmides & Tooby, 1987, 1992; Tooby & Cosmides, 1989a, 1990b; on language, see Chomsky, 1975; Pinker, 1989; on vision, see Marr, 1982; Poggio et al., 1985).

These discoveries of superlative "engineering" in the visual system have been paralleled in the study of the other sense organs, which are simply the recognizable transducing ends of an intricate mass of psychological adaptations that consist of increasingly more complex and integrative layers of specialized neural processing. For a variety of reasons, the information-processing adaptations involved in perception have been the only psychological mechanisms that have been studied for decades and in depth from a functionalist perspective. Contributing factors include the fact that their functionality is obvious to all sensate humans and their scientific study was rescued from the metaphysical doubt that hangs over other psychological phenomena because their associated physical transducing structures provided a "materialist" place to begin. Arguably the most important factor, however, was that these were the only mechanisms for which psychologists had any good standards of what counted as biologically successful problem-solving. Unacquainted with evolutionary biology, few psychologists know that there are standards for successful problem-solving in other realms as well, such as social behavior. Unless one knows what counts as a biologically successful outcome, one simply cannot recognize or investigate complex functional design or assess the extent to which a design is well-engineered.

Consequently, at present it is difficult to assess how well psychological adaptations measure up against the intricacy and functionality of other adaptations. We can only judge on the basis of the restricted set that have already been studied extensively from a functionalist perspective—the perceptual mechanisms. Of course, because the paradigmatic example of a well-engineered adaptation, used for over 130 years in biology, is a psychological adaptation, we know that in at least some cases our evolved information-processing machinery incorporates complex functional design of the highest order. Indeed, when the eye does appear in debates over Darwinism, it is usually used by anti-Darwinians, who insist that the eye is far too perfect a mechanism to have been constructed by natural selection. In general, whenever information-processing mechanisms have been studied from an evolutionary functional perspective—human vision and audition, echolocation in bats, dead-reckoning in desert ants, and so on—the results have indicated that the brain/mind contains psychological adaptations at least as intricately functional as anything to be found in the rest of the body.

One could perhaps argue that perceptual mechanisms are exceptional cases because they are evolutionarily older than those psychological adaptations that are distinctively human-specific, and so have had more time to be refined. There are many reasons to suspect this is not the case. But, even if it were, it would only suggest that purely human-specific adaptive problems, such as extensive tool use or extensive reciprocation, would have problem-solving adaptations less exquisite than vision, hearing, maternal care, threat-perception, motivational arbitration, mate selection, foraging, emotional communication, and many other problems that have been with us for tens of millions of years. It would not mean that we have no adaptations to human-specific problems at all. One reason the case of language is so illuminating is that it speaks to exactly this issue of the potential for complex functionality in human-specific adaptations. The language faculty is the only human information-processing system outside of perception that has been studied extensively with clear standards of what counts as functional performance, and the facts of psycholinguistics weigh in heavily against the hypothesis that human-specific adaptations have had insufficient time to evolve the same highly elaborated, intricately interdependent functionality characteristic of perceptual mechanisms (Pinker & Bloom, this volume). The language faculty has the same hallmarks of overwhelmingly functional complex design that the visual system does, and yet we know it is a recent and human-specific adaptation that evolved after the hominids split off from the (rest of the) great apes (Pinker & Bloom, this volume). The claim that language competence is a simple and poorly engineered adaptation cannot be taken seriously, given the total amount of time, engineering, and genius that has gone into the still unsuccessful effort to produce artificial systems that can remotely approach—let alone equal—human speech perception, comprehension, acquisition, and production.

Finally, behavioral scientists should be aware that functional organic machines look very different from the kinds of systems human engineers produce using planning and foresight. Human engineers can start with a clean drawing board, designing systems from scratch to perform a function cleanly, using materials selected particularly for the task at hand. Evolving lineages are more like the proverbial ship that is always at sea. The ship can never go into dry dock for a major overhaul; whatever improvements are made must be implemented plank by plank, so that the ship does not sink during its modification. In evolution, successive designs are always constructed out of modifications of whatever preexisting structures are there—structures linked (at least in the short run) through complex developmental couplings. Yet these short-run limitations do not prevent the emergence of superlatively organized psychological and physiological adaptations that exhibit functionality of the highest known order—higher, in fact, than human engineers have been able to contrive in most cases. This is because the evolutionary process continues to operate over large numbers of individuals and over enormous stretches of time, with selection relentlessly hill-climbing. To anthropomorphize, selection achieves its results through "tinkering," saving large numbers of frequently small and independent improvements cumulatively over vast expanses of time (Jacob, 1977). Thus, chains of successive modifications may be very large indeed to arrive at an increasingly sophisticated "solution" or problem-solving mechanism. The fact that alternative modifications are randomly generated—and that selection at any one time is limited to choosing among this finite set of actual alternatives—means that the evolutionary process might by chance "overlook" or "walk by" a specific solution that would have been obvious to a human engineer, simply because the correct mutations did not happen to occur. The fact that evolution is not a process that works by "intelligence" cuts both ways, however. Precisely because modifications are randomly generated, adaptive design solutions are not precluded by the finite intelligence of any engineer. Consequently, evolution can contrive subtle solutions that only a superhuman, omniscient engineer could have intentionally designed.

So, although organisms are functionally designed machines, they look very different from the machines that humans build. For this reason, the science of understanding living organization is very different from physics or chemistry, where parsimony makes sense as a theoretical criterion. The study of organisms is more like reverse engineering, where one may be dealing with a large array of very different components whose heterogeneous organization is explained by the way in which they interact to produce a functional outcome. Evolution, the constructor of living organization, has no privileged tendency to build into designs principles of operation that are simple and general. Evolution operates by chance—which builds nothing systematic into organisms—and by selection—which cumulatively adds modifications, regardless of whether they add complexity. Thus, psychologists are not likely to find a few satisfying general principles like Maxwell's equations that unify all psychological phenomena, but instead a complex pluralism of mechanisms. Satisfying general principles will instead be found at the next level up, in the principles of evolutionary functionalism that explain the organization of these mechanisms. At an engineering or mechanism level, knowledge will have to be constructed mechanism by mechanism, with the organization of the properties of each mechanism made intelligible by knowing the specific evolved function of that mechanism. Thus, the computational mechanisms that generate maternal love, grammar acquisition, mate selection, kin-directed assistance, or reciprocation can be expected to parallel Ramachadran's characterization of perception as,

> essentially a "bag of tricks;" that through millions of years of trial and error, the visual system evolved numerous short-cuts, rules-of-thumb and heuristics which were adopted not for their aesthetic appeal or mathematical elegance but simply because they *worked* (hence the "utilitarian" theory). This is a familiar idea in biology but for some reason it seems to have escaped the notice of psychologists, who seem to forget that the brain is a biological organ just like the pancreas, the liver, or any other specialized organ (Ramachadran, 1990, p. 24).

Adaptations, By-products, and Random Effects

The most fundamental analytic tool for recognizing an adaptation is its definition. Stripped of complications and qualifications, an adaptation is (1) a system of inherited and reliably developing properties that recurs among members of a species that (2) became incorporated into the species' standard design because during the period of their incorporation, (3) they were coordinated with a set of statistically recurrent structural properties outside the adaptation (either in the environment or in the other parts of the organism), (4) in such a way that the causal interaction of the two (in the context of the rest of the properties of the organism) produced functional outcomes that were ultimately tributary to propagation with sufficient frequency (i.e., it solved an adaptive problem for the organism). (For a more extensive definition of the concept of *adaptation*, see Tooby & Cosmides, 1990b). Adaptations are mechanisms or systems of properties crafted by natural selection to solve the specific problems posed by the regularities of the physical, chemical, developmental, ecological, demographic, social, and informational environments encountered by ancestral populations during the course of a species' or population's evolution (for other discussions of adaptation, see Pinker & Bloom, this volume; Symons, 1989, this volume; R. Thornhill, 1991; Tooby & Cosmides, 1990a; Williams, 1966, 1985; see Dawkins, 1986, for his discussion of adaptations under the name *adaptive complexity*).

Thus, chance and selection, the two components of the evolutionary process, explain different types of design properties in organisms, and all aspects of design must be attributed to one of these two forces. Complex functional organization is the product and signature of selection. Reciprocally, the species-typical properties of organisms attributable to chance will be no more important, organized, or functional than can be attributed to chance. The conspicuously distinctive cumulative impacts of chance and selection allow the development of rigorous standards of evidence for recognizing and establishing the existence of adaptations and distinguishing them from the non-adaptive aspects of organisms caused by the nonselectionist mechanisms of evolutionary change (Pinker & Bloom, this volume; Symons, this volume; R. Thornhill, 1991; Tooby & Cosmides, 1990b; Williams, 1966, 1985). Complex adaptations are usually species-typical (Tooby & Cosmides, 1990a); moreover, they are so well-organized and such good engineering solutions to adaptive problems that a chance coordination between problem and solution is effectively ruled out as a plausible explanation. Adaptations are recognizable by "evidence of special design" (Williams, 1966); that is, by recognizing certain features of the evolved species-typical design of an organism "as components of some special problem-solving machinery" (Williams, 1985, p. 1) that solve an evolutionarily long-standing problem. Standards for recognizing special design include factors such as economy, efficiency, complexity, precision, specialization, and reliability, which, like a key fitting a lock, render the design too good a solution to a defined adaptive problem to be coincidence (Williams, 1966). Like most other methods of empirical hypothesis testing, the demonstration that something is an adaptation is always, at the core, a probability assessment concerning how likely a situation is to have arisen by chance. The lens, pupil, iris, optic nerve, retina, visual cortex, and so on, are too well coordinated both with each other and with environmental factors—such as the properties of light and the reflectant properties of surfaces—to have arisen by chance.

In addition to adaptations, the evolutionary process commonly produces two other outcomes visible in the designs of organisms: (1) concomitants or by-products of adaptations (recently nicknamed "spandrels"; Gould & Lewontin, 1979); and (2) random effects. The design features that comprise adaptations became incorporated into the standard design because they promoted their own frequency and are, therefore, recognizable by their organized and functional relationships to the rest of the design and to the structure of the world. In contrast, concomitants of adaptations are those properties of the phenotype that do not contribute to functional design per se, but that happen to be coupled to properties that are, and so were dragged along into the organism's design because of selection on the design features to which they are linked. They may appear organized, but they are not *functionally* organized.

The explanation for any specific concomitant or spandrel is, therefore, the identification of the adaptation or adaptations to which it is coupled, together with the reason why it is coupled. For example, bones are adaptations, but the fact that they are white is an incidental by-product. Bones were selected to include calcium because it conferred hardness and rigidity to the structure (and was dietarily available), and it simply happens that alkaline earth metals appear white in many compounds, including the insoluble calcium salts that are a constituent of bone. From the point of view of functional design, by-products are the result of "chance," in the sense that nothing in the process of how they came to be incorporated into a design other than sheer coincidence would cause them to be coordinated solutions to any adaptive problem. For this reason, by-products are expected not to contribute to the solution of adaptive problems more often or more effectively than chance could explain. Finally, of course, entropic effects of many types act to introduce functional disorder into the design of organisms. They are recognizable by the lack of coordination that they produce within the architecture or between it and the environment, as well as by the fact that they frequently vary between individuals. Classes of entropic processes include mutation, evolutionarily unprecedented environmental change, individual exposure to unusual circumstances, and developmental accidents. Of course, one can decompose organisms into properties (or holistic relations) according to any of an infinite set of alternative systems. But, unless one applies a categorization system designed to capture their functional designs or adaptations, organisms will seem to be nothing but spandrels, chemistry, and entropy.

Recognizing Psychological Adaptations: Evolutionary, Cognitive, Neural, and Behavioral Levels of Analysis

Capturing Invariance in Functional Organization: Behavioral, Cognitive, and Neuroscience Descriptions

If the psychological architectures of organisms are infused with complex functional organization, this is not always easy to see. Precisely because functional organization may be very complex, and embedded in an even more bewildering array of variable and intricate by-products, it may appear to the unaided intellect to be indistinguishable from chaos or poor design. Unless one knows what to look for—unless, at the very least, one knows what counts as functional—one cannot recognize complex functionality even when one sees its operation.

Sciences prosper when researchers discover the level of analysis appropriate for describing and investigating their particular subject: when researchers discover the level where invariance emerges, the level of underlying order. What is confusion, noise, or random variation at one level resolves itself into systematic patterns upon the discovery of the level of analysis suited to the phenomena under study.

How, then, should the psychological architectures of organisms be described so as to capture a level of underlying functional order? Three different languages for describing psychological phenomena are commonly used: the behavioral, the cognitive, and the neurobiological. Each language has strengths and weaknesses for different scientific purposes. For the purpose of discovering, analyzing, and describing the functional organization of our evolved psychological architecture, we propose that the information-processing language of cognitive science is the most useful.

In the first place, this is because the evolutionary function of the brain is the adaptive regulation of behavior and physiology on the basis of information derived from the body and from the environment. Alternative design features are selected for on the basis of how well they solve adaptive problems—problems whose solution affects reproduction. How an organism processes information can have an enormous impact on its reproduction. It is, therefore, meaningful to ask what kind of cognitive design features would have constituted good solutions to adaptive information-processing problems that persisted over many generations. Evolutionary biology and hunter-gatherer studies supply definitions of the recurrent adaptive problems humans faced during their evolution, and cognitive psychology describes the information-processing mechanisms that evolved to solve them. By combining insights from these two fields, the functional organization of the mind can be brought into sharp relief.

Second, adaptations are usually species-typical. Consequently, to capture evolved functional organization, one needs a language that can describe what is invariant across individuals and generations. This process of description is key: By choosing the wrong descriptive categories, everything about an organism can seem variable and transitory to the extent that "plasticity" or "behavioral variability" can seem the single dominant property of an organism. In contrast, well-chosen categories can bring out the hidden organization that reappears from individual to individual and that, consequently, allows psychological phenomena to be described both economically and precisely.

Purely behavioral categories are seldom able to capture meaningful species-typical uniformity: Are humans "aggressive" or "peaceful," "pair-bonding" or "polygynous," "rational" or "irrational?" With much justice, Geertz, echoing Kroeber, dismissed large and vague behavioral universals, such as marriage and religion, as "fake" (1973, p. 101). Human phenomena accurately described and categorized solely in terms of behavioral outcomes appear endlessly variable; they seem to manifest a kaleidoscopic welter of erratic and volatile phenomena, which makes any underlying uniformity—let alone functional design—difficult to see. Exceptions, such as reflexes and fixed action patterns, occur in the very few cases where the mapping between stimulus and behavior is simple and immediate. Behavioral characterizations of anything much more complicated rapidly become so watered down with exceptions that, at best, one ends up with vague portrayals employing terms such as "capacity," "predisposition," "urge," "potential," and so on—things too murky to be helpful either in describing adaptations or in predicting behavior.

Perhaps more important, however, is that behavior is not a phenomenon *sui generis*. It is the product of mechanisms that process information. Mechanisms that produce behavior can be usefully studied on a variety of different descriptive and explanatory levels. Neuroscientists describe the brain on a physical level—as the interaction of neurons, hormones, neurotransmitters, and other organic aspects. In contrast, cognitive psychologists study the brain as an information-processing system—that is, as a collection of programs that process information—without reference to the exact neurophysiological processes that perform these tasks. A cognitive description specifies what kinds of information the mechanism takes as input, what procedures it uses to transform that information, what kinds of data structures (representations) those procedures operate on, and what kinds of representations or behaviors it generates as output. The study of cognition is the study of how humans and other animals process information.

To understand subsequent arguments clearly, it is important to keep in mind exactly what we mean by the cognitive or information-processing level. Like all words, "cognitive" is used to mean many different things. For example, some researchers use it in a narrow sense, to refer to so-called "higher mental" processes, such as reasoning, as distinct from other psychological processes, such as "emotion" or "motivation"; that is, to refer to a concept that corresponds more or less to the folk notion of reasoning while in a calm frame of mind. In contrast, we are using the word *cognitive* in a different and more standard sense. In this chapter, we use terms such as *cognitive* and *information-processing* to refer to a language or level of analysis that can be used to precisely describe any psychological process: Reasoning, emotion, motivation, and motor control can all be described in cognitive terms, whether the processes that give rise to them are conscious or unconscious, simple or complex. In cognitive science, the term *mind* refers to an information-processing description of the functioning of an organism's brain and that is the sense in which we use it. (For a more detailed discussion of the nature of cognitive explanations, see Block, 1980; Fodor, 1981; or Pylyshyn, 1984.)

For example, ethologists have traditionally studied very simple cognitive programs. A newborn herring gull, for instance, has a cognitive program that defines a red dot on the end of a beak as salient information from the environment, and that causes the chick to peck at the red dot upon perceiving it. Its mother has a cognitive program that defines pecking at her red dot as salient information from her environment, and that causes her to regurgitate food into the newborn's mouth when she perceives its pecks. These simple programs adaptively regulate how herring gulls feed their young. (If there is a flaw anywhere in these programs—i.e., if the mother or chick fails to recognize the signal or to respond appropriately—the chick starves. If the flaw has a genetic basis, it will not be passed on to future generations. By such feedback, natural selection shapes the design of cognitive programs.)

These descriptions of the herring gull's cognitive programs are entirely in terms of the functional relationship among different pieces of information; they describe two simple information-processing systems. Moreover, precise descriptions of these cognitive programs can capture the way in which information is used to generate adaptive behavior. Of course, these programs are embodied in the herring gull's neurobiological "hardware." Knowledge of this hardware, however, is not necessary for understanding the programs as information-processing systems. Presumably, one could build a silicon-based robot, using chemical processes completely different from those present in the gull's brain, that would produce the same behavioral output (pecking at red dot) in response to the same informational input (seeing red dot). The robot's cognitive programs would maintain the same functional relationships among pieces of information and would, therefore, be, in an important sense, identical to the cognitive programs of the herring gull. But the physical processes that implement these programs in the robot would be totally different.

Although all information-processing mechanisms operate by virtue of the physical processes that implement them, cognitive descriptions and physicalist ones are not equivalent, but complementary. They cannot be reduced to each other. For this reason, the information-processing descriptions of cognitive science are not merely metaphors in which brains are compared to computers. Their status as an independent level of psychological explanation can be established by considering the fact that the same information-processing relationships can be embodied in many different physical arrangements. The text-editing program Wordstar, for example, can run on machines with many different kinds of physical architectures, but it always has the same functional design at an information-processing level—the same key strokes will move the cursor, delete a word, or move a block of text. And the robot "gull" will still peck at a red dot, even though its programs are embodied in silicon chips rather than in neurons. These relationships can be described independently of their physical instantiation in any particular computer or organism, and can be described with precision. Thus, an information-processing program, whether in an organism or in a computer, is a set of invariant relationships between informational inputs and "behavioral" outputs. Moreover, from the point of view of the adaptive regulation of behavior, it is the cognitive system of relationships that counts. Given that the correct information-processing steps are carried out, selection pressures on psychological mechanisms are "blind" to the specific physical implementation of their information-processing structure (except insofar as different physical implementations may vary in factors such as metabolic cost). Because the primary function of the brain is the adaptive regulation of behavior and physiology in response to information, natural selection retains neural mechanisms on the basis of their ability to create functionally organized relationships between information and behavior (e.g., the sight of a predator activates inference procedures that cause the organism to hide or flee) or between information and physiology (e.g., the sight of a predator increases the organism's heart rate in preparation for flight). The mechanism is selected to create the correct information-behavior or information-physiology relationship and, so long as the physical implementation produces this relationship, its particular form is free to vary according to other factors. Indeed, at certain points in development, injury to the human brain can sometimes be "repaired" in the sense that different neurons re-create the same ordered relationship between information and behavior that the damaged ones had implemented prior to the injury (Flohr, 1988). When "rewiring" of this kind occurs, the information-processing relationship is preserved, not its physical instantiation.

In short, it is primarily the information-processing structure of the human psychological architecture that has been functionally organized by natural selection, and the neurophysiology has been organized insofar as it physically realizes this cognitive organization. Because the function of the brain is informational in nature, its richly organized functional structure is only visible when its properties are described in cognitive terms. Much of great interest can be learned by investigating the brain in neurobiological terms, but its adaptive dimension will remain invisible unless and until its mechanisms are described in a language that is capable of expressing its informational functions.

For these reasons, the invariant functional organization of complex psychological adaptations is more likely to be captured by cognitive descriptions than by neuroscience ones.[4] Just as mathematics is an indispensable language for describing many scientific phenomena, information-processing language is a precise descriptive vehicle for capturing how complex systems functionally interact with complex environments. What mathematics is for physics, cognitive descriptions can be for a science of psychology and behavior.

The use of information-processing language is not enough, however. Alone, it is no more useful for discovering invariances in functional organization than any other descriptive language. Unless one knows what counts as functional, one cannot rec-ognize complex functional design even when one sees its operation. Is friendship functional? Is anger? Is joining a group? Is pregnancy sickness? Unless one knows what adaptive problems a species encountered during its evolutionary history and what would have counted as solutions to these problems, these questions are unanswerable. To discover invariances in the functional organization of the human mind, the language and methods of cognitive science must be used in concert with principles drawn from evolutionary biology.

Where Evolutionary Biology and Cognitive Psychology Meet

Conceptual systems, models, and theories function as organs of perception: They allow new kinds of evidence and new relationships to be perceived (Popper, 1972). As Einstein remarked, "it is the theory which decides what we can observe" (Heisenberg, 1971, p. 63). The tools of evolutionary functional analysis function as an organ of perception, bringing the blurry world of human psychological and behavioral phenomena into sharp focus and allowing one to discern the formerly obscured level of our richly organized species-typical functional architecture.

Theories about selection pressures operating in ancestral environments place important constraints on—and often define—what can count as an adaptive function. Indeed, many theories of adaptive function define what would count as adaptive *information-processing*. Consider, for example, Hamilton's rule, which describes the selection pressures operating on mechanisms that generate behaviors that have a reproductive impact on an organism and its kin (Hamilton, 1964). The rule defines (in part) what counts as biologically successful outcomes in these kinds of situations. These outcomes often cannot be reached unless specific information is obtained and processed by the organism.

In the simplest case of two individuals, a mechanism that produces acts of assistance has an evolutionary advantage over alternative mechanisms if it reliably causes individual i to help relative j whenever $C_i < r_{i,j}B_j$. In this equation, C_i = cost to i of rendering an act of assistance to j, measured in terms of foregone reproduction, B_j = benefit to j of receiving that act of assistance, measured in terms of enhanced reproduction, and $r_{i,j}$ = the probability that a randomly sampled gene will be present at the same locus in the relative due to joint inheritance from a common ancestor.

Other things being equal, the more closely psychological mechanisms reliably produce behavior that conforms to Hamilton's rule, the more strongly they will be selected for. Under many ecological conditions, this selection pressure defines an information-processing problem that organisms will be selected to evolve mechanisms to solve.

Using this description of an adaptive problem as a starting point, one can immediately begin to define the cognitive subtasks that would have to be addressed by any set of mechanisms capable of producing behavior that conforms to this rule. What information-processing mechanisms evolved to reliably identify relatives, for example? What criteria and procedures do they embody—for example, do these mechanisms define an individual as a sibling if that individual was nursed by the same female who nursed you? What kind of information is processed to estimate $r_{i,j}$, the degree of relatedness? Under ancestral conditions, did siblings and cousins co-reside, such that one might expect the evolution of mechanisms that discriminate between the two? After all, $r_{i, full\ sib} = 4r_{i,\ first\ cousin}$. What kind of mechanisms would be capable of estimating the magnitudes of the consequences of specific actions on one's own and on others' reproduction? How are these various pieces of information combined to produce behavior that conforms to Hamilton's rule? And so on.

This example highlights several points. First and most important, it shows how knowledge drawn from evolutionary biology can be used to discover functional organization in our psychological architecture that was previously unknown. Hamilton's rule is not intuitively obvious; no one would look for psychological mechanisms that are well-designed for producing behavior that conforms to this rule unless they had already heard of it. After Hamilton's rule had been formulated, behavioral ecologists began to discover psychological mechanisms that embodied it in nonhuman animals (Krebs & Davies, 1984). Unguided empiricism is unlikely to uncover a mechanism that is well-designed to solve this kind of problem.

Second, this example illustrates that one can easily use the definition of an adaptive problem to generate hypotheses about the design features of information-processing mechanisms, even when these mechanisms are designed to produce social behavior. It allows one to break the adaptive problem down into cognitive subtasks, such as kin recognition and cost/benefit estimation, in the same way that knowing that the adaptive function of the visual system is scene analysis allows one to identify subtasks such as depth perception and color constancy.

Third, the example shows how knowing the ancestral conditions under which a species evolved can suggest hypotheses about design features of the cognitive adaptations that solve the problem. For example, co-residence is a reliable cue of sib-hood in some species, but other cues would have to be picked up and processed in a species in which siblings and cousins co-reside.

Fourth, Hamilton's rule provides one with a standard of good design for this particular problem. Such standards are an essential tool for cognitive scientists because they allow them to identify whether a hypothesized mechanism is capable of solving the adaptive problem in question and to decide whether that mechanism would do a better job under ancestral conditions than alternative designs. This allows one to apply the powerful methods of learnability analysis outside of psycholinguistics, to adaptive problems involving social behavior (see pp. 73–77, on evolutionary functional analysis).

Fifth, this example illustrates how insights from evolutionary biology can bring functional organization into clear focus at the cognitive level, but not at the neurobiological level. Hamilton's rule immediately suggests hypotheses about the functional organization of mechanisms described in information-processing terms, but it tells one very little about the neurobiology that implements these mechanisms—it cannot be straightforwardly related to hypotheses about brain chemistry or neuroanatomy. Once one knows the properties of the cognitive mechanisms that solve this adaptive problem, however, it should be far easier to discover the structure of the neural mechanisms that implement them.

The intellectual payoff of coupling theories of adaptive function to the methods and descriptive language of cognitive science is potentially enormous. By homing in on the right categories—ultimately adaptationist categories—an immensely intricate, functionally organized, species-typical architecture can appear, with perhaps some additional thin films of frequency-dependent or population-specific design as well (e.g., McCracken, 1971). Just as one can now flip open *Gray's Anatomy* to any page and find an intricately detailed depiction of some part of our evolved species-typical morphology, we anticipate that in 50 or 100 years one will be able to pick up an equivalent reference work for psychology and find in it detailed information-processing descriptions of the multitude of evolved species-typical adaptations of the human mind, including how they are mapped onto the corresponding neuroanatomy and how they are constructed by developmental programs.

The Impact of Recurrent Environmental and Organismic Structure on the Design of Adaptations

Organisms transact the business of propagation in specific environments, and the persistent characteristics of those environments determine the dangers, opportunities, and elements the organism has to use and to cope with in its process of propagation. Consequently, the structure of the environment causes corresponding adaptive organization to accumulate in the design of the organism (Shepard, 1987a; Tooby & Cosmides, 1990b). For example, the design of eyes reflects the properties of light, objects, and surfaces; the design of milk reflects the dietary requirements of infants (and what was dietarily available to mothers); the design of claws reflects things such as the properties of prey animals, the strength of predator limbs, and the task of capture and dismemberment. This functional organization in the organism—its set of adaptations—is designed to exploit the enduring properties of the environment in which it evolved (termed its environment of evolutionary adaptedness, or EEA) and to solve the recurring problems posed by that environment. Adaptations evolve so that they mesh with the recurring structural features of the environment in such a way that reproduction is promoted in the organism or its kin. Like a key in a lock, adaptations and particular features of the world fit together tightly, to promote functional ends.

Moreover, from the point of view of any specific design feature or adaptation, the rest of the encompassing organism itself constitutes an enduring environmental structure as well. New adaptations or design features will be selected for on the basis of how well they productively coordinate with the persistent characteristics of this internal environment, as well as with the external environment. This is why adaptations evolve to fit together with each other within the organism so well. Thus, the adaptive mesh between tendon, muscle, and bone is no different in principle than the adaptive mesh between foraging mechanisms and the ecological distribution of food and cues reliably correlated with its presence (Real, 1991). Obviously, therefore, adaptations may solve endogenous adaptive problems and may improve over evolutionary time without necessarily being driven by or connected to any change in the external environment.

Long-term, across-generation recurrence of conditions—external, internal, or their interaction—is central to the evolution of adaptations, and it is easy to see why. Transient conditions that disappear after a single or a few generations may lead to some temporary change in the frequency of designs, but the associated selection pressures will disappear or reverse as often as conditions do. Therefore, it is only those conditions that recur, statistically accumulating across many generations, that lead to the construction of complex adaptations. As a corollary, anything that is recurrently true (as a net statistical or structural matter) across large numbers of generations could potentially come to be exploited by an evolving adaptation to solve a problem or to improve performance. For this reason, a major part of adaptationist analysis involves sifting for these environmental or organismic regularities or invariances. For example, mental states, such as behavioral intentions and emotions, cannot be directly observed. But if there is a reliable correlation over evolutionary time between the movement of human facial muscles and emotional state or behavioral intentions, then specialized mechanisms can evolve that infer a person's mental state from the movement of that person's facial muscles (Ekman, 1973, 1984; Fridlund, in press). Indeed, evidence drawn from cognitive neuroscience indicates that we do have mechanisms specialized for "reading" facial expressions of emotion (Etcoff, 1983, 1986).

To begin with, a cognitive adaptation can, through exploiting the world's subtle statistical structure, go far beyond the information it is given, and reconstruct from fragmentary cues highly accurate models of local conditions by exploiting these relationships (e.g., self-propelled motion is correlated with the presence of an animal; a sharp discontinuity in reflected light intensity is correlated with the presence of an edge). This evolutionary Kantian position has already been richly vindicated in the fields of perception and psychophysics (see, e.g., Marr, 1982; Shepard, 1981, 1984, 1987a, this volume), where the representations that our evolved computational systems construct go far beyond what is "logically" warranted solely by the sensory information itself, usually settling on single preferred interpretations. Our minds can do this reliably and validly because this fragmentary information is operated on by evolved procedures that were selected precisely because they reflect the subtle relationships enduringly present in the world (e.g., shading cues that are correlated with shape and depth, time-location relationships that are correlated with the most probable kinematic trajectories followed by natural objects). These mechanisms supply a privileged organization to the available sense data so that the interaction of the two generates interpretations that usually correspond to actual conditions in the external world. In the absence of specialized mechanisms that assume and rely on certain relationships being characteristic of the world, recovering accurate models of the external world from sense data would be an insoluble computational problem (Marr, 1982; Poggio et al., 1985).

Parallel ideas form the centerpiece of Chomskyan psycholinguistics: Children must be equipped with specialized mechanisms ("mental organs") that are functionally organized to exploit certain grammatical universals of human language. Otherwise, language learning would be an unsolvable computational problem for the child (Chomsky 1957, 1959, 1975, 1980; Pinker 1979, 1982, 1984, 1989; Wexler & Culicover, 1980). The discovery and exploratory description of such universal subtle relationships present in the "world" of human language is a primary activity of modern linguists and psycholinguists. Proposed mechanisms for language learning that do not include specialized procedures that exploit these relationships have been repeatedly shown to be inadequate (e.g., Pinker 1989, 1991; Pinker & Prince, 1988). As in perception, adaptations for grammar acquisition must mesh with the enduring structure of the world. But in this case, the recurrent structure to be meshed with is created by the species-typical design of other (adult) human minds, which produce grammars that manifest certain relationships and not others.

Due to common evolutionary ancestry, the living world of plants and animals is structured into species and other more inclusive units that share large sets of properties in common: Wolves resemble other wolves, mammals other mammals, and so on. Living things occur in so-called natural kinds. This is another enduring set of relationships in the world that our minds evolved to exploit. Ethnobiologists and cognitive anthropologists such as Atran and Berlin have shown that the principles humans spontaneously use in categorizing plants and animals reflect certain aspects of this enduring structure, and are the same cross-culturally as well (Atran, 1990; Berlin, Breedlove, & Raven, 1973).

In the last decade, the field of cognitive development has been revolutionized by the discovery that the principles of inference that infants and children bring to the tasks of learning are organized to reflect the particular recurrent structure of specific problem domains, such as object construal and motion, the differences between artifacts and living kinds, physical causality, and so on (see, e.g., Carey & Gelman, 1991). These evolved, domain-specific cognitive specializations have been shown to be specialized according to topic and to develop in the absence of explicit instruction.

For example, contrary to the Piagetian notion that infants must "learn" the object concept, recent research has shown that (at least) as early as 10 weeks—an age at which the visual system has only just matured—infants already have a sensorily-integrated concept of objects as entities that are continuous in space and time, solid (two objects cannot occupy the same place at the same time), rigid, bounded, cohesive, and move as a unit (e.g., Spelke, 1988, 1990, 1991). Indeed, when infants of this age are shown trick displays that violate any of these assumptions, they indicate surprise—one could almost say in such cases that the object concept embodied in their evolved mechanisms causes them to "disbelieve" the evidence of their senses (Leslie, 1988). By 27 weeks, infants already analyze the motion of inanimate objects into submovements and use this parsing to distinguish causal from noncausal relationships (Leslie, 1988; Leslie & Keeble, 1987). Needless to say, these are all relationships that accurately reflect the evolutionarily long-enduring structure of the world.

A. Brown (1990) has shown that early causal principles such as "no action at a distance" guide learning about tool use in children as young as 18 months; these children categorize tools for use according to functional properties (e.g., has a hooked end for pulling) over nonfunctional properties (e.g., color). In contrast, the same children have great difficulty learning how to use a tool when its mechanism of action appears to violate one of their concepts about physical causality—concepts that mirror certain aspects of Newtonian mechanics.

Very young children also make sharp distinctions between the animate and inanimate worlds. Throughout our evolutionary history, being an animal has been reliably—if imperfectly—correlated with self-generated motion, whereas inanimate objects rarely move unless acted upon by an outside force. Recent research suggests that young children use this cue to distinguish the animate from the inanimate worlds, and make very different inferences about the two (Gelman, 1990b; Premack, 1990). More generally, experiments by Keil (1989) and others indicate that the kind of inferences children spontaneously make about "natural kinds," such as animals, plants, and substances, differ sharply from those they are willing to make about human-made artifacts. Natural kinds are viewed as having invisible "essences" that bear a causal relation to their perceptual attributes, whereas artifacts are defined by how their perceptual attributes subserve their (intended) function. In an important series of experiments, Gelman and Markman (1986, 1987; Markman, 1989) found that natural kinds were a powerful organizer of inference in young children. In general, being a member of a natural kind carries more inferential weight than being perceptually similar. In addition, children give more weight to natural kind membership when reasoning about traits that actually *are* more likely to vary as a function of membership in a natural kind, such as breathing, than when reasoning about traits that are more likely to vary as a function of perceptual similarity, such as weight or visibility at night (for summary, see Markman, 1989).

These principles apply far beyond these few simple cases. The world is full of long-enduring structure, and the mind appears to be full of corresponding mechanisms that use these structural features to solve a diverse array of adaptive problems: geometrical and physical relationships that shape the probability of various trajectories (Shepard, 1984), biomechanically possible and impossible motions (Shiffrar & Freyd, 1990), momentum effects on trajectories (Freyd, 1987), correlations between the ingestion of plant toxins and teratogenesis (Profet, 1988, this volume), privileged relationships between the gravitational field and the orientation of objects in the world (Triesman, 1977), and on and on. It is only for expository convenience that we have mostly focused on mechanisms bearing on categorization and inference ("knowledge"), rather than on motivation, emotion, and decision making ("value"). The structure of the world is reflected in the nature of behavior-regulating systems as well because the long-term statistical structure of the world systematically creates relationships between choices and adaptive consequences. (For a discussion of how emotional adaptations reflect the relationship between decisions and the detailed structure of ancestral conditions, see Tooby & Cosmides, 1990b.) Mind/world relationships extend all the way from the ease with which people acquire fears of spiders and snakes (Marks, 1987; Seligman, 1971), to the more subtle impact that aesthetic factors have on habitat choice and wayfinding (Kaplan, this volume; Orians & Heerwagen, this volume), to the relative unwillingness of adults to have sex with people with whom they co-resided for long periods during childhood (McCabe, 1983; Parker & Parker, 1986; Pastner, 1986; Shepher, 1983; Westermarck, 1891; Wolf, 1966, 1968; Wolf & Huang, 1980; N. W. Thornhill, 1991), to the intensity with which parents and children may come to love each other (Bowlby, 1969), to the often violent passions humans exhibit when they discover the existence of spousal infidelity (Daly & Wilson, 1988; Wilson & Daly, this volume).

For those who study psychological adaptations, the long-enduring structure of the world provides a deeply illuminating source of knowledge about the evolved architecture of the mind. As Shepard has so eloquently put it, there has been the evolution of a mesh between the principles of the mind and the regularities of the world, such that our minds reflect many properties of the world (Shepard, 1987a). Many statistical and structural relationships that endured across human evolution were "detected" by natural selection, which designed corresponding computational machinery that is specialized to use these regularities to generate knowledge and decisions that would have been adaptive in the EEA. Because the enduring structure of ancestral environments *caused* the design of psychological adaptations, the careful empirical investigation of the structure of environments, from a perspective that focuses on adaptive problems and outcomes, can provide powerful guidance in the exploration of the mind. The long-term structure of the ancestral world is worth knowing, worth studying, and worth relating to psychology. This realization vastly widens the scope of information that can be brought to bear on questions in psychology: Evolutionary biology, paleo-anthropology, hunter-gatherer studies, behavioral ecology, botany, medicine, nutrition, and many other fields can be mined for information that suggests specific hypotheses, guides one toward productive experimentation, and informs one about the broad array of functionally specialized mechanisms that are likely to be present. The stuff of the mind is the stuff of the world, and so the investigation of the rich structure of the world provides a clearly observable and empirically tractable—if not royal—road into the hidden countries of the mind.

THE CENTRAL ELEMENTS OF EVOLUTIONARY FUNCTIONAL ANALYSIS

Approaching the coordination between the structure of the ancestral world and the design features of adaptations with an engineering sensibility is what gives empirical specificity and inferential power to evolutionary functional analysis. The following are five structured components that can be fit together in such an analysis.

1. *An adaptive target*: a description of what counts as a biologically successful outcome in a given situation. Out of the infinite set of potential behavioral sequences, which small subset would count as a solution to the adaptive problem? Here, one wants to know which behavioral outcomes will have the property of enhancing the propagation of the psychological designs that gave rise to them. For example, out of all the substances in the world, which should the organism eat and which should it avoid? With whom should the organism mate? How much parental care should it devote to each offspring? When should the organism join a coalition? What inferences should be drawn on the basis of the retinal display about the location of various surfaces? In defining an adaptive target, the goal is to ascertain whether the proposed behavioral outcome, in combination with all the other activities and outcomes produced by the organism, will enhance design propagation under ancestral conditions.

2. *Background conditions*: a description of the recurrent structure of the ancestral world that is relevant to the adaptive problem. One wants to know what features of the ancestral world were sufficiently stable to support the evolution of a design that could produce an adaptive target. This could be a part of the external environment, another part of the standard design of the organism, or a combination of the two. This includes the information available to solve the problem, the environmental and endogenous obstacles to solving the problem, and so on. So, for example, the regular spatial orientation of human eyes with respect to each other, the face, and the ground constitute background conditions for the evolution of face recognition mechanisms in infants. Often, but not always, the ancestral world will be similar to the modern world (e.g., the properties of light and the laws of optics have not changed). However, one needs to know something about hunter-gatherer studies and paleoanthropology to know when ancestral conditions germane to the adaptive problem diverge from modern conditions. Of course, when there is a difference between the two, ancestral conditions are the applicable ones for the purpose of analyzing the functional design of an adaptation because they are the cause of that design. Modern environments are relevant to the analysis of the ontogeny of mechanisms and their calibration. It is important to keep in mind that a mechanism that was capable of producing an adaptive target under ancestral conditions may not be capable of doing so under modern ones. Our visual system fails to maintain color constancy under sodium vapor lamps in modern parking lots (Shepard, this volume), and attempting to understand color constancy mechanisms under such unnatural illumination would have been a major impediment to progress.

3. *A design*: a description of the articulated organization of recurrent features in the organism that together comprise the adaptation or suspected adaptation. A design

description of the eye, for example, would include a specification of its species-typical parts and the manner in which they interact to produce an adaptive target.

The design—or even the existence—of a proposed information-processing mechanism is frequently unknown. Indeed, an appropriate functional description of a design is often what one is trying to discover. When this is the case, this step in an evolutionary functional analysis would be the construction of a hypothesis about the existence and design features of a psychological adaptation. This might include what environmental cues the mechanism monitors, what information it draws from other mechanisms, how it categorizes and represents this information, what procedures or decision rules transform the informational input, what kinds of representations or behaviors it produces as output, which mechanisms use its output for further processing, how its output is used by other mechanisms to generate behavior, and so on. The more causally explicit one can make the design description at the cognitive level, the better. Eventually, one hopes to have a description of the neurobiological implementation of the adaptation as well.

4. *A performance examination*: a description of what happens when the proposed adaptation mechanistically interacts with the world. What range of outcomes does the design actually produce? Like putting a new aircraft prototype in a wind tunnel, what one is looking for is a good causal or "engineering" analysis of how the proposed design actually performs under conditions that are representative of situations our ancestors routinely faced, and how it performs under present conditions as well. For a proposed language acquisition device, for example, one wants to model how its information-processing procedures perform when they encounter normal linguistic environments, in order to see whether the interaction of procedures and environment assembles an increasingly elaborated computational system capable of producing intelligible and grammatical sentences. Similarly, one wants to model how psychological mechanisms in women or men interact with their social and informational environments to produce mating preferences. We want to emphasize that we are looking here for a mechanistic or causal description of how the system generates output given input. Statements like, "the human child learns its culture through imitation and generalization" are not models of how input generates output. They are too unspecified to qualify as hypotheses or explanations; we should have ceased treating them as such a long time ago.

5. *A performance evaluation*: a description or analysis of how well (or how poorly) the design, under circumstances paralleling ancestral conditions, managed to produce the adaptive target (the set of biologically successful outcomes). The better the mechanism performs, the more likely it is that one has identified an adaptation.

It is just as important, however, to see whether the proposed mechanism produces the behaviors one actually observes from real organisms under modern conditions. If it does, this suggests that the researcher is converging on a correct description of the design of the mechanisms involved, whether they are producing behavior that is currently adaptive or not. The Westermarck incest avoidance mechanism, for example, passes both tests. It produces adaptive outcomes under ancestral (and many modern) conditions (e.g., distaste for sex between siblings who co-resided as children), and it also explains the nonadaptive outcomes that are observed under certain modern conditions (e.g., distaste for sex between kibbutz crèche mates who co-resided as children [Shepher, 1983]; distaste for sex with spouses who were adopted into one's family at a young age and with whom one was raised [Wolf & Huang, 1980]).

In short, an evolutionary functional analysis consists of asking a series of engineering questions: Would the proposed design have interacted with properties of the ancestral world to produce target adaptive outcomes? Does the proposed design interact with properties of the modern world to produce outcomes that one actually observes in real organisms, whether these outcomes are adaptive or not? Is there an alternative design that is better able to generate adaptive targets under ancestral conditions? If so, then are there any background conditions that one has overlooked that would have prevented the alternative design from evolving? And so on.

Natural selection is the process that shapes biological form to match function, and this link between form and function has been a critically illuminating relationship in thousands of applications. Ever since Harvey's question about why there were valves in the veins led him to discover the circulation of the blood, functional questions about organismic design have been a powerful engine for the discovery of new knowledge (Mayr, 1983). Those even distantly connected to organismic biology have become aware of the spectacular functionalist revolution that has transformed the field over the last 30 years, placing adaptationism on a new and far more rigorous foundation (Hamilton, 1964; Maynard Smith, 1982; see, especially, Williams, 1966). The reason why Lewontin and Gould's accusation (famous among social scientists) that adaptationism consists of post hoc storytelling has so resoundingly failed to impress practicing evolutionary biologists is that they saw on a daily basis that adaptationism was anything but post hoc (Gould & Lewontin, 1979; for discussion, see Pinker & Bloom, this volume). Simply put, an explanation for a fact by a theory cannot be post hoc if the fact was unknown until after it was predicted by the theory and if the reason the fact is known at all is because of the theory. Functionalist analysis in biology has motivated thousands of predictions about new and critical phenomena, whose subsequent discovery confirmed the productivity of the emerging paradigm. Lewontin and Gould's critique has primarily impressed those outside of evolutionary and organismic biology who have not been exposed on a professional basis to the flood of new findings that were both generated and economically organized by the newly emerging functionalist principles.

When they are linked together, the five components outlined above not only provide a framework for the explanation of facts that are already known; they also form a powerful heuristic system for the generation of new knowledge. Depending on which questions you need answered and what information you already have, you can put these relationships to a number of richly productive alternative uses. For example, if you are trying to discover the structure of unknown psychological mechanisms, you first need to integrate steps 1 and 2 together into a definition of an adaptive problem (what Marr called a computational theory or task analysis; Marr, 1982). You need to determine things such as what information was routinely available in the environment and in the organism to solve the problem (step 2), and what outcomes constituted a successful solution to the problem (step 1). From this, you can begin to develop hypotheses about the nature of the information-processing mechanisms that might have evolved to solve the problem, and then empirically test for their presence. (For a discussion of this approach, see Marr, 1982, and Cosmides & Tooby, 1987. For some applications of this approach to specific psychological problems: on vision, see Marr, 1982; on mechanisms specialized for reasoning about social exchange, see Cosmides, 1989, Cosmides & Tooby, 1989, this volume, and Gigerenzer & Hug, in press; on mechanisms regulating parental solicitude, see Mann, this volume, and Daly & Wilson, 1988). In short, by using steps 1 and 2, one can create a hypothesis about function that leads to the discovery of form. This use of the elements of evolutionary functional analysis guides the researcher step by step from a definition of an adaptive problem to the discovery and mapping of the mechanisms that solve it.

An alternative starting point is step 3: a well-specified candidate hypothesis about the structure of an information-processing mechanism. So, for example, you might hypothesize that operant conditioning explains the acquisition of natural language grammars. To proceed with an evolutionary functional analysis, you would then need to develop a description of the relevant environmental features (step 2) and define what counts as a successful outcome (step 1). You would then proceed to steps 4 and 5—performance examination and evaluation. If your hypothesis about design is correct, then the step 4 performance examination will reveal that the design's interaction with the relevant environment features is at least *capable* of producing a successful outcome. The performance evaluation of step 5 will allow you to determine whether the design hypothesized in step 3 is better at producing adaptive outcomes than alternative designs.

We will refer to the application of steps 4 and 5 as the *solvability criterion*. To be correct, a cognitive adaptation must be capable of solving the proposed problem or generating behavior that we know humans routinely perform and of doing so given the relevant background conditions. Although this may seem like an obvious step, theories in psychology are rarely evaluated in this way, which has allowed entire research communities to labor under the impression that, say, associationism or imitation constitute effective explanations of the phenomena they studied. Such tests of computational performance—or *learnability analyses* as they are called when applied to learning tasks—were pioneered in psycholinguistics by Pinker and colleagues (1979, 1984, 1989, 1991; Pinker & Prince, 1988; Wexler & Culicover, 1980) in order to evaluate which theories of language acquisition could actually account for the fact that children learn the language of their local community. By using this method one can, in fact, rule out entire classes of theories as inadequate, without having to empirically test each one of an inexhaustible set of trivial variants. Because there are an infinite number of alternative theories, empirical falsification is not by itself a practical research strategy; it must be combined with other sources of valid inference if one is to be able to draw larger and more interesting conclusions. For psychologists, the analysis of computational performance is one way of doing this.

Yet another approach to evolutionary functional analysis begins with noting the existence of a complexly articulated and recurrent phenotypic pattern—for example, eyes, teeth, pregnancy sickness, or sexual jealousy—and investigating whether it might be the expression of an adaptation (Williams, 1966, p. 10). In such cases, one is following the logic in yet another direction: Given a known phenotypic structure (step 3), one dissects the environment (step 2) and the requirements for reproduction (step 1), to find out whether they compose a well-defined adaptive problem for which the reliable outcomes of the design (step 4) constitute a well-engineered solution (step 5). Profet's proposal that pregnancy sickness constitutes an adaptation to limit maternal ingestion of teratogens during the most vulnerable phases of embryogenesis is an excellent application of this approach (Profet, this volume). It should be stressed that this is the only type of functionalist analysis to which Gould and Lewontin's accusation of post hoc storytelling could possibly apply, even in principle, since it is the only one that works backward from known facts about phenotypic design. Yet, even here, the critique could only apply if all facts about the environment, the other parts of the organism, and the structure believed to be an adaptation were known in advance. In practice, this is never the case. This form-to-function approach is just as productive as the others because it leads to the prediction and organization of previously unknown facts, usually about additional design features of the organism as well as about the recurrent structure of the world. For example, the study of the visual system has profited immensely from the fact that scientists knew that the eye exists and that the visual system's function is to perform scene analysis given data transduced by the eye. Indeed, the functionalist approach to the study of vision has generated one of the most sophisticated and least ad hoc bodies of knowledge in psychology. As Mayr put it, summarizing the historical record in response to accusations that adaptationist research was simply post hoc storytelling, "The adaptationist question, 'What is the function of a given structure or organ?' has been for centuries the basis for every advance in physiology" (1983, p. 32). Adaptationist principles can provide equally powerful guidance for research in psychology as well.

Even if every aspect of a mechanism were already known, examining the detailed transactions between selected features of the environment and selected parts of the mechanism would clarify many features of its functional organization, such as which aspects of the design perform the work (e.g., which aspects of pregnancy sickness cause the mother to avoid ingesting teratogens) and which are functionless or even harmful side effects (such as calorie reduction during the first trimester). Naturally, the form-to-function approach does include the risk of answering the post hoc "why" question that Gould and Lewontin so disdain; that is, of explaining why already known features of biological designs came to be as they are. But even physics and geology run the "risk" of addressing such Kiplingesque post hoc questions as why Mercury has an orbit that deviates from the predictions of Newtonian mechanics, why Asia has the Himalayas, or why the universe has its present set of four interactions, temporal asymmetry, background radiation, and particle distribution. In science, this is usually called "explanation."

TOWARD A POST-STANDARD MODEL VIEW OF DEVELOPMENT

Development from an Adaptationist Perspective

The recognition that organisms are integrated collections of problem-solving mechanisms organized to propagate their designs brings with it an adaptationist framing of development. An adaptation is, by its nature, an improbably good organization of elements and so will not often spontaneously come into existence merely by chance. Instead, for adaptations to exist, they must be specifically constructed from the materials present in evolutionarily normal environments. Accordingly, the developmental programs and machinery responsible for assembling an adaptation correctly are also adaptations. As adaptations, they themselves have complex structures that assume and require recurrent features of the world, and that interact with this recurrent structure to produce biologically functional targeted outcomes.

Hence, the primary function or target of developmental adaptations (which include the genes) is to reconstruct in offspring the evolved functional organization that was present in their parents, which is predominantly species-typical design. The genes and the mechanisms of genetic transmission are, of course, adaptations to the problem of faithfully replicating into the offspring critical information necessary to reconstruct this design. The genes come embedded in a matrix of cellular and developmental machinery constituting an additional set of adaptations that use the genetic structure as regulatory elements to institute and to guide embryogenesis and subsequent development along species-standard pathways. For this reason, it is useful to think of the genes together with the developmental machinery as one integrated suite of adaptations—the developmental programs—and to distinguish the minor idiosyncratic features of an individual's genes and zygotic machinery from the recurrent or species-typical dimensions that have endured long enough to have been organized by natural selection. The latter specify the species-standard physiological and psychological architecture visible in all humans raised in normal environments, whereas the former specify the usually minor perturbations within that architecture (Tooby & Cosmides, 1990a).

Why do we so often connect complex adaptations or evolved architectures with concepts such as *species-typical, human universal, species-standard, recurrent,* and so on? This is because when humans are described from the point of view of their complex adaptations, differences tend to disappear, and a universal architecture stands out in stark relief. This is both empirically the case (nearly everyone has two eyes, two hands, the same sets of organs, and so on) and theoretically expected to be the case if organisms are primarily collections of complex adaptations. The logic is straightforward (Tooby & Cosmides, 1990a; see also Tooby, 1982):

1. A species is a group of organisms with a common history of interbreeding and a continuing ability to interbreed to form offspring who can typically reproduce at least as well as their parents.
2. To survive and reproduce in a complex world, organisms need complex problem-solving machinery (complex adaptations).
3. Complex adaptations are intricate machines that require complex "blueprints" at the genetic level. This means that they require coordinated gene expression, involving hundreds or thousands of genes to regulate their development.
4. Sexual reproduction automatically breaks apart existing sets of genes and randomly generates in the offspring new, never before existing combinations of genes at those loci that vary from individual to individual.
5. If genes differed from individual to individual in ways that significantly impacted the developed design of the component parts of complex adaptations, then existing genetic combinations whose developed expressions had fit together into complex adaptations would be pulled apart by sexual recombination. Equally, new combinations would be thrown randomly together, resulting in phenotypes whose parts were functionally incompatible. This is because parts in any complex machine are functionally interdependent: If you tried to build a new car engine out of a mixture of parts from a Honda and a Toyota, the parts would not fit together. To build a new engine whose component parts fit together, you would have to salvage parts from two "parents" that were of the same make and model.
6. Because sexual recombination is a random process, it is improbable that all of the genes necessary for a complex adaptation would be together in the same

individual if the genes coding for the components of complex adaptations varied substantially between individuals.

7. Therefore, it follows that humans, and other complex, long-lived, outbreeding organisms, must be very nearly uniform in those genes that underlie our complex adaptations.

8. By the same token, sexually reproducing populations of organisms freely tolerate genetic variation to the extent that this variation does not impact the complex adaptive organization shared across individuals. To return to our car engine example, the color of the parts is functionally irrelevant to the operation of the car and so can vary arbitrarily and superficially among cars of the same make and model; but the shapes of the parts are critical to functional performance and so cannot vary if the "offspring" design is to function successfully.

These constraints on variation apply with equal force to psychological adaptations: Even relatively simple cognitive programs, "mental organs," or neurological structures must contain a large number of interdependent processing steps, limiting the nature of the variation that can exist without violating the functional integrity of the psychological adaptation. The psychic unity of humankind—that is, a universal and uniform human nature—is necessarily imposed to the extent and along those dimensions that our psychologies are collections of complex adaptations. Therefore, it is *selection* interacting with sexual recombination that tends to impose near uniformity at the functional level in complex adaptive designs (as well as in whatever is developmentally coupled to complex functional structure). It is selection that is responsible for what we have been calling our universal evolved psychological and physiological architecture.

There is no small irony in the fact that Standard Social Science Model hostility to adaptationist approaches is often justified through the accusation that adaptationist approaches purportedly attribute important differences between individuals, races, and classes to genetic differences. In actuality, adaptationist approaches offer the explanation for why the psychic unity of humankind is genuine and not just an ideological fiction; for why it applies in a privileged way to the most significant, global, functional, and complexly organized dimensions of our architecture; and for why the differences among humans that are caused by the genetic variability that geneticists have found are so overwhelmingly peripheralized into architecturally minor and functionally superficial properties. If the anti-adaptationists were correct (e.g., Gould & Lewontin, 1979) and our evolved architectures were not predominantly sets of complex adaptations or properties developmentally coupled to them, then selection would not act to impose cross-individual uniformity, and individuals would be free to vary in important ways and to any degree from other humans due to genetic differences. If the world were, in fact, governed by nonselectionist forces, then the psychic unity of humankind would simply be a fiction.

Modern geneticists, through innovative molecular genetic techniques, have certainly discovered within humans and other species large reservoirs of genetic variability (Hubby & Lewontin, 1966; Lewontin & Hubby, 1966; see reviews in Ayala, 1976, and Nevo, 1978). But it is only an adaptationist analysis that predicts and explains why the impact of this variability is so often limited in its scope to micro-level biochemical variation, instead of introducing substantial individuating design differences. The study of the operation of selection on complex mechanisms makes it difficult to see how more than a tiny fraction of this variation could be constitutive of complex psychological or physiological adaptations.[5]

Thus, human design resolves itself into two primary tiers: First, an encompassing functional superstructure of virtually universal, complexly articulated, adaptively organized developmental, physiological, and psychological mechanisms, resting on a universally shared genetic basis; and, second, low level biochemical variation creating usually slight individuating perturbations in this universal design due to the existence of a reservoir of genetic variability in the species. There may also be some thin films of population-specific or frequency-dependent adaptive variation on this intricate universal structure (see, e.g., Durham, 1991; McCracken, 1971), but for a number of reasons these will be very small in magnitude next to the complex structure of a universal human nature (for discussion, see Tooby & Cosmides, 1990a, 1990b). The primary function of developmental adaptations is to reconstruct in each new individual this complex, functional architecture, and the primary focus of adaptationists is the study of this universal structure.

The fact that humans in ordinary environments reliably develop a clearly recognizable species-typical architecture should in no way be taken to imply that any developed feature of any human is immutable or impervious to modification or elimination by sufficiently ingenious ontogenetic intervention. Nothing about humans could possibly be immune from developmental intervention, simply because we are physical systems open to contact and manipulation by the rest of the world; we are not something made unalterable by inexorable supernatural predestination. People frightened of the myth that biology is destiny can be reassured (just as others may be alarmed) by the fact that there are no limits to what could be done, especially by evolutionarily novel measures: Deliver the right quanta to the right ribosomes or other locations at the right times and anyone or anything could be successively modified into a watermelon or an elephant. In contrast, Standard Social Science Model advocates, such as Gould, tend to equate evolved biological design with immutability without any logical or empirical warrant. As Gould expresses his rather magical belief, "If we are programmed to be what we are, then these traits are ineluctable. We may, at best, channel them, but we cannot change them either by will, education, or culture" (Gould, 1977c, p. 238).

In actuality, the very openness of development to intervention poses a critical set of adaptive problems for developmental adaptations. Their primary function is to successfully reconstruct each functionally necessary detail of our species-typical architecture, including the tens or hundreds of thousands of specific components and arrangements that endow us with a lens, a retina, an optic nerve, language, maternal attachment, emotions, retinotopic maps, ten fingers, a skeleton, color constancy, lungs, a representational system embodying the implicit theory that others have minds, an ability to cooperate, spatial cognition, and so on. Each of these adaptations constitutes a very narrow target of improbably good functional organization. Because the world is full of potential disruptions, there is the perennial threat that the developmental process may be perturbed away from the narrow targets that define mechanistic workability, producing some different and nonfunctional outcome. Developmental adaptations are, therefore, intensely selected to evolve machinery that defends the developmental process against disruption (Waddington, 1962). Profet (this volume) provides an elegant analysis of a psychological adaptation designed to defend against just such threats to adaptive development, protecting embryogenesis from the potentially disruptive plant toxins in the mother's diet through modifying her dietary decisions during pregnancy. More generally, developmental programs are often designed to respond to environmentally or genetically introduced disorder through feedback-driven compensation that redirects development back toward the successful construction of adaptations. Thus, developmental processes have been selected to defend themselves against the ordinary kinds of environmental and genetic variability that were characteristic of the environment of evolutionary adaptedness, although not, of course, against evolutionarily novel or unusual manipulations.

Of course, unlike human-built machines that have a static architecture until they break down, organisms are systematically transformed by developmental adaptations over their life histories from zygote to senescence. Thus, the task facing developmental adaptations is not to assemble a machine of fixed design, but rather to assemble and modify the set of expressed adaptations according to a moving target of age, sex, and circumstance-dependent design specifications. For example, adaptive problems are often specific to a particular life stage, and so the organism must be developmentally timed to have the necessary adaptations for that stage, regardless of whether, as a side effect, they happen to appear before or persist after they are needed (e.g., the placenta, fetal hemoglobin, the sucking reflex, the ability to digest milk, the fear of strangers, ovulation, the ability to be sexually aroused, milk production, and so on).

Hence, the Standard Model assumption—critical to its logic—that the mental organization present in adults but absent from newborns must be "acquired" from the social world has no conceptual foundation and is, in many cases, known to be empirically false. In the worldview of the SSSM, biological construction goes on in the uterus, but at birth the child is "biologically complete" except for growth; at this point, it is surrendered into the sole custody of social forces, which do the remainder of the construction of the individual. This, of course, reflects folk biology, captured in the two dictionary definitions of *innate* as "present from birth" and as "intrinsic." Social constructivist arguments frequently take the form that because thus-and-such is absent at birth, or doesn't appear until after age seven, or until after puberty, it is obviously "learned" or "socially constructed." As a result, a common, but generally irrelevant feature of "nativist" versus "environmentalist" debates is over what is "present from birth." This confuses (among other things) the question of whether something is *expressed* at the time of birth with whether there exists in the individual evolved developmental mechanisms that may activate and organize the expression of an adaptation at some point in the life cycle. Developmental processes continue to bring additional adaptations on line (as well as remove them) at least until adulthood, and there is an increasing amount of evidence to suggest that age-driven adaptive changes in psychological architecture continue throughout adulthood (see, e.g., Daly & Wilson, 1988). Thus, just as teeth and breasts are absent at birth and develop later in an individual's life history, perceptual organization, domain-specific reasoning mechanisms, the language acquisition device, motivational organization, and many other intricate psychological adaptations mature and are elaborated in age-specific fashions that are not simply the product of the accumulation of "experience." Consequently, psychological adaptations may be developmentally timed to appear, disappear, or change operation to mesh with the changing demands of different age-specific tasks, such as parenting, emotional decoding of the mother's voice, language acquisition, species-appropriate song learning, and so on (Daly & Wilson, 1988; Fernald, this volume; Marler, 1991; Newport, 1990).

Equally, although most human psychological and physiological adaptations appear to be sexually monomorphic, some are obviously sexually differentiated to address those adaptive problems whose task demands were recurrently disparate for females and males over evolutionary time (e.g., Buss, 1987, 1989, 1991, this volume; Daly & Wilson, 1988; Ellis, this volume; Silverman & Eals, this volume; Symons, 1979; Wilson & Daly, this volume). For any particular gender difference, many psychologists are interested in whether it was caused (1) by sexually monomorphic psychologies encountering differential treatment by the social world, or (2) by sexually differentiated developmental mechanisms encountering treatment from the social world, whether that treatment was uniform or differential. As interesting as this question may be, however, the fact that an expressed gender difference may first appear after birth, or even late in life, is evidence neither for nor against either of these views.

For these reasons, one needs to distinguish an organism's evolved design or species-typical architecture from its "initial state" (Carey, 1985a); that is, its state at whatever point in development one chooses to define as "initial" (birth, conception, fetus prior to gonadal or neural sexual differentiation, puberty, or whatever). Not all features of evolved human design are or can be present at any one time in any one individual. Thus, the genetically universal may be developmentally expressed as different maturational designs in the infant, the child, the adolescent, and the adult; in females and males; or in individuals who encounter different circumstances. Pregnancy sickness is arguably a feature of our evolved universal design, but it does not appear in males, children, or women who have never become pregnant; it is only present in sexually mature women while they are pregnant. Thus, when we use terms such as "evolved design," "evolved architecture," or even "species-typical," "species-standard," "universal," and "panhuman," we are not making claims about every human phenotype all or even some of the time; instead, we are referring to the existence of evolutionarily organized developmental adaptations, whether they are activated or latent. Adaptations are not necessarily expressed in every individual. They only need to have been expressed often enough in our evolutionary history to have been targets of selection, and, hence, to have been organized by selection so that they reliably develop under appropriate circumstances. For this reason, adaptations and adaptive architecture can be discussed and described at (at least) two levels: (1) the level of reliably achieved and expressed organization (as, for example, in the realized structure of the eye), and (2) at the level of the developmental programs that construct such organization. To avoid cumbersome expressions, we do not usually bother to terminologically distinguish successfully assembled expressed adaptive architecture from the more fundamental developmental adaptations that construct them. Context usually makes obvious which is being discussed.

Selection Regulates How Environments Shape Organisms

Many social and biological scientists have labored under the false impression that only certain things are under the "control," or "influence," or "determination" of the genes or of biology. According to this view, evolutionary approaches are only applicable to those traits under such "genetic control," and the greater the environmental influence or control, the smaller the domain of things for which evolutionary analyses properly apply (e.g., Sahlins, 1976a; Gould, 1977a, 1977b, 1977c; note, especially, Gould's Standard Model contrast of "genetic control" with the "purely cultural"). In this dualistic conception, the genes are "biological" and evolved, while "the environment"—including the social environment—is nonbiological and nonevolved. In consequence, the environment is held to be something that not only can attenuate, nullify, or even reverse "genetic forces" but may break the causal chain entirely, liberating human affairs from the causal patterning of evolution. For proponents of the SSSM, it is self-evident that the causal forces of evolution and "biology" are located solely inside the organism and are expressed in an unadulterated form only at birth, if then. In contrast, the causal forces of the environment are seen as external to the organism, as having their own independent causal history, and as having no particular reason to act on the organism in such a way as to preserve or elaborate the organism's initial biological organization. In short, the environment is conceptualized as obviously nonbiological in character. Development is consequently portrayed as a process in which the newborn organism—usually seen as a passive clay-like object with some initial biologically given form—is pounded or sculpted by the active and nonbiological environment according to its accidents, structure, or agenda. It follows from this view that biology can only express itself in human life if it is unalterable or at least rigid enough to resist the pounding forces of the environment—a bombardment that begins at birth. One might think of the stubbornly biological aspects of human life as the hardened part of the clay, while the more plastic parts are easily shaped by the environment and quickly lose their initial biological form. Consequently, even if advocates of the SSSM do not want to dichotomize traits cleanly into two sets (e.g., hardened versus wet clay), they could array them by this criterion as more or less biologically determined; that is, as more or less environmentally influenced.

Despite its tenacity in the social sciences at large, this Standard Model view of development has been abandoned by many cognitive scientists and by biologists because it rests on a series of fallacies and misconceptions. To begin with, despite the routine use of such dualistic concepts and terms by large numbers of researchers throughout the social and biological sciences, there is nothing in the real world that actually corresponds to such concepts as "genetic determination" or "environmental determination." There is nothing in the logic of development to justify the idea that traits can be divided into genetically versus environmentally controlled sets or arrayed along a spectrum that reflects the relative influence of genes versus environment. And, most critically, the image of "the environment" as a "nonbiological" causal influence that diminishes the "initial" evolved organization of humans rests on the failure to appreciate the role that the evolutionary process plays in organizing the relationship between our species-universal genetic endowment, our evolved developmental processes, and the recurring features of developmental environments.

In the first place, every feature of every phenotype is fully and equally codetermined by the interaction of the organism's genes (embedded in its initial package of zygotic cellular machinery) and its ontogenetic environments—meaning everything else that impinges on it. By changing either the genes or the environment any outcome can be changed, so the interaction of the two is always part of every complete explanation of any human phenomenon. As with all interactions, the product simply cannot be sensibly analyzed into separate genetically determined and environmentally determined components or degrees of influence. For this reason, *everything*, from the most delicate nuance of Richard Strauss's last performance of Beethoven's Fifth Symphony to the presence of calcium salts in his bones at birth, is totally and to exactly the same extent genetically and environmentally codetermined. "Biology" cannot be segregated off into some traits and not others.

Nevertheless, one could understand and acknowledge that all human phenomena are generated by gene-environment interactions, yet believe that the existence and participation of the environment in such interactions insulates human phenomena from interesting evolutionary patterning. After all, if only our genes evolved, whereas the form of the environment is generated by other processes (such as geology, cultural transmission, epidemiology, and meteorology) then the gene-environment interaction seems to blunt the organizing effects evolution might otherwise have on human life. Although this view seems quite reasonable, a close examination of how natural selection actually adaptively organizes gene-environment interactions over time leads to a very different conclusion, which might be summed up by the counterintuitive claim that "the environment" is just as much the product of evolution as are the genes.

To understand why this is so, one needs to distinguish "the environment" in the sense of the real total state of the entire universe—which, of course, is not caused by the genes or the developmental mechanisms of any individual—from "the environment" in the sense of those particular aspects of the world that are rendered developmentally relevant by the evolved design of an organism's developmental adaptations. It is this *developmentally relevant environment*—the environment as interacted with by the organism—that, in a meaningful sense, can be said to be the product of evolution, evolving in tandem with the organism's organized response to it. The confusion of these two quite distinct senses of "environment" has obscured the fact that the recurrent organization of the environment contributes a biological inheritance parallel to that of the genes, which acts co-equally with them to evolutionarily organize the organism throughout its life.

The assumption that only the genes are evolved reflects a widespread misconception about the way natural selection acts. Genes are the so-called units of selection, which are inherited, selected, or eliminated, and so they are indeed something that evolves. But every time one gene is selected over another, one design for a developmental program is selected over another as well; by virtue of its structure, this developmental program interacts with some aspects of the environment rather than others, rendering certain environmental features causally relevant to development. So, step by step, as natural selection constructs the species' gene set (chosen from the available mutations), it constructs in tandem the species' developmentally relevant environment (selected from the set of all properties of the world). Thus, *both the genes and the developmentally relevant environment are the product of evolution.*

Even more crucially, by selecting one developmental program over another, the evolutionary process is also selecting the mechanisms that determine how the organism will respond to environmental input, including environmental input that varies. A developmental mechanism, by virtue of its physical design, embodies a specification for how each possible state of the developmental environment is to be responded to, if encountered. This is a central but little understood point: There is nothing "in" the environment that by itself organizes or explains the development, psychology, morphology, or behavior of any organism. "The" environment affects different organisms in different ways. We find the smell of dung repellent; dung flies are attracted to it. Temperature at incubation determines the sex of an alligator, but not of a human

(Bull, 1983). A honeybee larva that is fed Royal Jelly will become a queen bee rather than a sterile worker, but Royal Jelly will not have this effect on a human baby. Many bats navigate by sound echoes that humans cannot even hear. Rats have an elaborate sense of smell, which their food choice mechanisms use, but their navigation mechanisms ignore smell cues entirely in favor of geometric cues (Gallistel, 1990). Indeed, this last example shows that the developmentally relevant environment is not just organism-specific, it is mechanism-specific. In other words, the actual relationship between environmental conditions and developmental outcomes *is created by the design of the developmental procedures that exist in the organism* and, within the limit of the physically possible, mechanisms could be designed into the system to create a causal relationship between any imaginable environmental input and any imaginable output. In principle, genetic engineers could build honeybee larvae that develop into workers if, and only if, they are exposed to recitations of Allen Ginsberg's "Howl."

Aside from physical necessity, then, it is the evolved design of the organism that decides what organized consequences the environment can have on it. The rules that govern how environments impact the developing organism have themselves evolved and have been shaped by selection. Consequently, the evolutionary process determines how the environment shapes the organism. Over evolutionary time, genetic variation in developmental programs (with selective retention of advantageous variants), explores sampled properties out of the total environment potentially available to be interacted with. This process discovers which recurrent features are useful in the task of organizing and calibrating psychological adaptations and which recurrent features are unreliable or disruptive. It renders the latter irrelevant to development.

A natural response is to claim that although the genes are highly stable, replicated with few mutations from generation to generation, the environment is volatile, rendering any developmental process coordinating the two ineffectual. Once again, however, our intuitions are not a privileged perspective from which one can declare the world to be either stable or variable. Whether the world is "stable" or "variable" depends on the categorization system used or, to put it another way, on which parts of the world are selected to be processed by a mechanism.

Consider, for example, the following thought experiment. Imagine that an identical pool shot is set up every generation on a rather odd pool table. Three of the four cushions wobble continuously and unpredictably, but one happens to be stable. The "genes" determine the exact direction the cue ball is hit each time, while the "environment" (i.e., the angle of the cushions when struck) determines how the shot will be reflected back. Whether a particular shot successfully sinks the target ball in a pocket (i.e., whether it achieves the adaptive target) is determined by the interaction of the direction of the shot and the orientation of the cushion at the time the ball hits it (i.e., the interaction of genes and environment determine the outcome). Assume also that there is variation in the direction of the shot (i.e., in the "genes") and that successful shots cause genes to be retained.

Over the long run, feedback-driven selection will come to determine which direction the ball is hit. In determining this direction, it will also end up selecting the stable cushion for the bank shot, and not the wildly oscillating ones. It will end up directing the shot at exactly that spot along the stable cushion from which the shots are stably successful.

Similarly, selection will design developmental adaptations that respond to those aspects of the world that have a relatively stable recurrent structure, such that the mesh between the two will reliably produce design-propagating outcomes. Just as selection has acted on genetic systems to keep mutations to tolerable levels, selection has acted to "choose" the more stable parts of the environment to render developmentally relevant, such that these aspects of the environment stably mesh with developmental programs to produce reliably developing adaptive architectures.

The Standard Model framing says that the world pre-exists and is not caused by the organism, so that the world's effect on the organism will have no particular tendency to organize the developing organism according to any evolved or adaptive pattern. Equally, the pool-table cushion pre-existed each shot and was not created by them, and the laws of physics determined how each shot would be reflected back. So, in a static SSSM analysis, it is self-evident that the outcome is the mixture of two factors, one "biological" and one "nonbiological," with the nonbiological diluting, obliterating, or even reversing the biological. In contrast, an evolutionary analysis points out that the shot, through its careful targeting, picked out the particular cushion hit and the exact location hit. Over time, the selective retention of successful shots will organize the effect that the pre-existing environment had on the trajectories of the shots and the outcome of the game. The pre-existing structure of the world was exploited to impose an organization on the outcome that it would not otherwise have had.

In this same fashion, the evolutionary process explores and sifts the environment for aspects that will usefully organize the developing organism. The evolutionary process puts to work sources of organization and information anywhere they are unearthed, whether in the genes or in the environment, in a mother's smile or in a companion's expression of surprise. Selection has crafted the design of the developmental programs so that organisms tap into these reservoirs of information or hook themselves to environmental forces that help to construct them. Thus, the genomes of organisms have evolved to "store" organization and information that is necessary or helpful for development in the structure of the world itself. For example, for a developing child, the information in the minds of other humans, properly used, is a very useful source of information to use in their development, as are the linguistic patterns encountered in the local language community and patterns in local social behavior. Natural selection has intricately orchestrated developmental mechanisms so that things in the developmentally relevant world have been assigned an appropriate causal role, from gravity, plants, and three-dimensionality, to language, mothers, and social groups. Evolution shapes the relationship between the genes and the environment such that they both participate in a coordinated way in the construction and calibration of adaptations. Thus, evolutionarily patterned structure is coming in from the environment, just as much as it is coming out from the genes.

Accordingly, "biology" is not some substance that is segregated or localized inside the initial state of the organism at birth, circumscribing the domain to which evolutionary analyses apply. It is also in the organization of the developmentally relevant world itself, when viewed from the perspective imposed by the evolved developmental mechanisms of the organism. Thus, nothing the organism interacts with in the world is nonbiological to it, and so for humans cultural forces are biological, social forces are biological, physical forces are biological, and so on. The social and cultural are not alternatives to the biological. They are aspects of evolved human biology and, hence, they are the kinds of things to which evolutionary analysis can properly be applied.

Social scientists need to recognize that humans have evolved to expect, rely on, and take advantage of the richly structured participation of the environment—including the human social and cultural environment—in the task of adaptive development. Our developmental and psychological programs evolved to invite the social and cultural worlds in, but only the parts that tended, on balance, to have adaptively useful effects. Programs governing psychological development impose conceptual frameworks on the cultural and social worlds; choose which parts of the environment are monitored; choose how observations and interactions are categorized, represented, and interrelated; decide what entities to pursue interactions with; and, most importantly, determine what algorithms or relationships will organize environmental input into developmental change or psychological output. Consequently, the study of developmental adaptations is a central branch of evolutionary psychology. Understanding these adaptations will make visible the subtly stable structure of the developmentally relevant world and illuminate the evolutionary patterning in how human beings respond to smiles, to language, and to the cultural knowledge in others' minds. Each human, by expressing his or her species-typical architecture, contributes to the environmental regularities that others inhabit and rely on for their development.

For these reasons, it is a complete misconception to think that an adaptationist perspective denies or in the least minimizes the role of the environment in human development, psychology, behavior, or social life. Environmentalists have been completely correct about the importance of environmental input in the explanation of human behavior. Humans more richly and complexly engage the variable features of the environment than any other species we know of. It is this perception that has maintained environmentalism as the predominant viewpoint in the social sciences, despite its crippling inadequacies as an analytic framework. The terms "culture," "socialization," "intelligence," and "learning" are labels for poorly understood families of processes that reflect this complex and overwhelming human engagement with environmental inputs. Any viable theory of the evolved architecture of humans must reflect this reality and must be environmentalist in this sense. As discussed, the incoherence of Standard Model environmentalism stems from (1) the widespread failure to recognize that environmental responsiveness requires a complex evolved design (expressible as either a set of developmental adaptations or as a reliably developing psychological architecture; (2) the refusal to investigate or specify the nature of this architecture or these programs; and (3) the failure to recognize that the regulatory structure of these programs specifies the relationship between environmental input and behavioral, developmental, or psychological output.

For social scientists, of course, this recognition requires a radical change in practice: Every "environmentalist" explanation about the influence of a given part of the environment on humans will—if it is to be considered coherent—need to be accompanied by a specific "nativist" hypothesis about the evolved developmental and psychological mechanisms that forge the relationship between the environmental input and the hypothesized psychological output. All "environmentalist" theories necessarily depend upon and invoke "nativist" theories, rendering environmentalism and nativism interdependent doctrines, rather than opposed ones. For post-Standard Model researchers, these incoherent traditional dichotomies (genetic/environmental, biological/social, nativist/environmental) are being abandoned, as is reflected, for example, in the title of a recent article, "Learning by instinct" (J. Gould & Marler, 1987).

The Impact of the Recurrent Structure of Human Life and Human Culture on the Design of Psychological Adaptations

The evolved mesh between the information-processing design of human psychological adaptations, their developmentally relevant environments, and the stably recurring structure of humans and their environments is pivotal to understanding how an evolutionary psychological approach to culture differs from that of the Standard Social Science Model. For traditional anthropologists, cultures vary from place to place, and there is nothing privileged about a conceptual framework that categorizes human thought and action so as to capture underlying patterns of cross-cultural uniformity, as against the infinite class of perspectives by which human thought and behavior appear everywhere different (Geertz, 1973, 1983, 1984; see D. E. Brown, 1991, for a critique of this view). Nevertheless, from the "point of view" of natural selection, such uniformities—however subtle and unimportant to professionally neutral anthropological minds—are indeed privileged, and for a very simple reason. However variable cultures and habitats may have been during human evolution, selection would have sifted human social and cultural life (as well as everything else) for obvious or subtle statistical and structural regularities, building psychological adaptations that exploited some subset of these regularities to solve adaptive problems. (As we will discuss, one of the problems that had to be solved using regularities was the problem of learning "culture" itself.)

Thus, Geertz's starting point, that humans have evolved to use culture, is obviously true (although not in the slavish sense he envisions). But the next step in his logic—that humans don't have general cultures, only particular ones, and so evolved to realize themselves only through cultural particularity—is the error of naive realism. No instance of anything is intrinsically (much less exclusively) either "general" or "particular"—these are simply different levels at which any system of categorization encounters the same world. When you meet Roger Shepard you are, at one and the same time, meeting both a particular (and distinctive) individual and a manifestation of humanity in general, embodying innumerable species-typical characteristics. So it is with cultures. Selection operated across ancestral hominid populations according to what were, in effect, systems of categorization, screening cross-cultural variability for any recurrent relationships that were relevant to the solution of adaptive problems. To be thoroughly metaphorical, natural selection scrutinized the structure of human cultural and social environments, searching for regularities that could be used to engineer into our evolved architecture effective techniques for adaptive problem-solving. Thus, the issue is: During the Pleistocene, were there any statistical and structural uniformities to human life from culture to culture and habitat to habitat, from any perspective—no matter how subtle or abstract or unobservable—that could have been used by species-typical problem-solving machinery for the adaptive regulation of behavior and physiology? Geertz sees (modern) cultures as irredeemably particularized, confidently dismissing talk of meaningful human universals as nearly vacuous. Did natural selection "see" the human world the same way?

The answer is obvious, once the question is asked. Anthropological orthodoxy to the contrary, human life is full of structure that recurs from culture to culture, just as the rest of the world is. (Or, if one prefers, there are innumerable frames of reference within which meaningful cross-cultural uniformities appear, and many of these statistical uniformities and structural regularities could potentially have been used to solve adaptive problems.) Exactly which regularities are, in fact, part of the developmentally relevant environment that is used by our universal architectures is a matter to be empirically determined on a mechanism by mechanism, case by case basis. Such statistical and structural regularities concerning humans and human social life are an immensely and indefinitely large class (D. E. Brown, 1991): adults have children; humans have a species-typical body form; humans have characteristic emotions; humans move through a life history cued by observable body changes; humans come in two sexes; they eat food and are motivated to seek it when they lack it; humans are born and eventually die; they are related through sexual reproduction and through chains of descent; they turn their eyes toward objects and events that tend to be informative about adaptively consequential issues; they often compete, contend, or fight over limited social or subsistence resources; they express fear and avoidance of dangers; they preferentially associate with mates, children, and other kin; they create and maintain enduring, mutually beneficial individuated relationships with nonrelatives; they speak; they create and participate in coalitions; they desire, plan, deceive, love, gaze, envy, get ill, have sex, play, can be injured, are satiated; and on and on. Our immensely elaborate species-typical physiological and psychological architectures not only constitute regularities in themselves but they impose within and across cultures all kinds of regularities on human life, as do the common features of the environments we inhabit (see D. E. Brown, 1991, for an important exploration of the kinds and significance of human universals).

Human developmental mechanisms have been born into one cultural environment or another hundreds of billions of times, so the only truly long-term cumulatively directional effects of selection on human design would have been left by the statistical commonality that existed across cultures and habitats. Consequently, the sustained impact of these cross-culturally recurrent relationships sculpted the problem-solving mechanisms of the human mind to expect and exploit the common structure of human cultures and human life; that is, natural selection constructed adaptations specialized to mesh with the detailed structural regularities common to our ancestral cultural environments. For this reason, not only does natural selection privilege frames of reference that reveal patterns of universality in human life but our evolved psychological architecture does also. Embedded in the programming structure of our minds are, in effect, a set of assumptions about the nature of the human world we will meet during our lives. So (speaking metaphorically) we arrive in the world not only expecting, Geertzian fashion, to meet some particular culture about whose specifically differentiated peculiarities we can know nothing in advance. We also arrive expecting to meet, at one and the same time, and in one and the same embodiment, the general human culture as well—that is, recognizably human life manifesting a wide array of forms and relations common across cultures during our evolution (or at least some set out of the superset). Thus, human architectures are "pre-equipped" (that is, reliably develop) specialized mechanisms that "know" many things about humans, social relations, emotions and facial expressions, the meaning of situations to others, the underlying organization of contingent social actions such as threats and exchanges, language, motivation, and so on.

To take only one example, humans everywhere include as part of their standard conceptual equipment the idea that the behavior of others is guided by invisible internal entities, such as "beliefs" and "desires"—reflecting what Dennett calls "the intentional stance" (1987). Of course, this way of thinking seems so natural to us that it is difficult to see that there is anything to explain: It is tempting to think that beliefs and desires are "real" and that, therefore, humans everywhere simply learn to see the world as it really is. Side-stepping the complex question of whether this panhuman folk psychology is an accurate way of capturing "real" human psychology (i.e., whether it is a true or a complete description), we simply want to point out that things such as beliefs and desires are inherently unobservable hidden variables used to explain observations that could be explained by any of an infinite set of alternative theories (in fact, psychologists have come up with many such theories). Therefore, a belief in beliefs and desires cannot be justified by observations alone, so the fact that it is conventional among humans to "theorize" about others in this fashion is not inexorably mandated by their experience or otherwise required by the structure of the external world. For the same set of nonmandated ideas to have emerged everywhere on earth, our developmental programs or cognitive architectures must impose this way of interpreting the world of other humans on us.

In fact, an intensive research effort in the field of cognitive development has recently provided substantial support for the hypothesis that our evolved psychological architecture includes procedures that cause very young children to reliably develop a belief-desire folk psychology—a so-called "theory of mind" (e.g, Astington, Harris, & Olson, 1988; Leslie, 1987, 1988; Perner, 1991; Wellman, 1990; Wimmer & Perner, 1983). Developmental psychologists have been finding that even 2- and 3-year-olds make different inferences about "mental entities" (dreams, thoughts, desires, beliefs) than about "physical entities." Moreover, children typically "explain" behavior as the confluence of beliefs and desires (e.g., Why has Mary gone to the water fountain? *Because* she has a *desire* for water (i.e., she is thirsty) and she *believes* that water can be found at the water fountain). Such inferences appear to be generated by a domain-specific cognitive system that is sometimes called a "theory of mind" module (Leslie, 1987). This module consists of specialized computational machinery that allows one to represent the notion that "agents" can have "attitudes" toward "propositions" (thus, "Mary" can "believe" that "X," "Mary" can "think" that "X," and so on). Between the ages of 3 and 5 this domain-specific inferential system develops in a characteristic pattern that has been replicated cross-culturally in North America, Europe, China (Flavell, Zhang, Zou, Dong & Qui, 1983), Japan (Gardner, Harris, Ohmoto & Hamazaki, 1988), and a hunter-gatherer group in Camaroon (Avis & Harris, in press). Moreover, there is now evidence suggesting that the neurological basis of this system can be selectively damaged; indeed, autism is suspected to be caused by a selective neurological impairment of the "theory of mind" module (Baron-Cohen, Leslie, & Frith, 1985; Leslie, 1987, 1988; Leslie & Thaiss, 1990).

This research indicates that a panhuman "theory of mind" module structures the folk psychology that people develop. People in different cultures may elaborate their folk psychologies in different ways, but the computational machinery that guides the development of their folk notions will be the same, and some of the notions developed will be the same as well. Humans come into the world with the tendency to organize their understanding of the actions of others in terms of beliefs and desires, just as they organize patterns in their two-dimensional retinal array under the assumption that the world is three-dimensional and that objects are permanent, bounded, and solid.

Thus, not only do evolved mechanisms assume certain things will tend to be true of human life but these specialized procedures, representational formats, cues, and categorization systems impose—out of an infinite set of potential alternatives—a detailed organization on experience that is shared by all normal members of our species. There is certainly cultural and individual variability in the exact forms of adult mental organization that emerge through development, but these are all expressions of what might be called a single human metaculture. All humans tend to impose on the world a common encompassing conceptual organization, made possible by universal mechanisms operating on the recurrent features of human life. This is a central reality of human life and is necessary to explain how humans can communicate with each other, learn the culture they are born into, understand the meaning of others' acts, imitate each other, adopt the cultural practices of others, and operate in a coordinated way with others in the social world they inhabit. By *metaculture,* we mean the system of universally recurring relationships established and constituted by (1) our universal evolved species-typical psychological and physiological architectures, (2) the interaction of these architectures with each other in populations, (3) their interaction with the developmentally relevant recurrent structure of human natural and cultural environments, and (4) their patterned standard impact on human phenomena.

Social scientists have traditionally considered there to be a tension or explanatory competition between human universals and transmitted cultural variability: the more of one, the less of the other (D. E. Brown, 1991). However, careful causal analysis of the information-processing tasks required to learn transmitted culture leads to what is very nearly the opposite conclusion. In fact, it is only the existence of this common metacultural structure, which includes universal mechanisms specialized to mesh with the social world, that makes the transmission of variable cultural forms possible.

To make this clear, consider the question of how it is possible for pre-linguistic children to deduce the meanings of the words they hear when they are in the process of learning their local language for the first time. The child's task of discovering the meanings of words involves isolating, out of an infinite set of possible meanings, the actual meanings intended by other speakers (e.g., Carey, 1982, 1985a; Quine, 1960). Children can infer the meanings of messages in the local, but unknown language only because they, like cryptographers, have a priori statistical knowledge about likely messages, given the situational context. To solve the problem of referential ambiguity, the child's procedures for semantic analysis must depend on the fact that our universal evolved psychological architectures impose on the world enough standard and recurrent interpretations between speaker and listener to make the deduction of a core lexicon possible. Since the infant is new to the culture and ignorant of it, these shared interpretations cannot be supplied by the culture itself, but must be supplied by the human universal metaculture the infant or child shares with adults by virtue of their common humanity. (In contrast, the Standard Model's initially content-free general process child mind would share no common interpretations with local adults and could rely on no necessary imposition of common event construals by both speaker and listener.) Thus, the system for assigning correct semantic meanings to culturally arbitrary signs necessarily relies on the presence of species-typical cognitive adaptations and on the nonarbitrariness of meaning systems that inhabit these cognitive adaptations. These mechanisms reliably identify evolutionarily recurrent situations (such as threat, play, or eating) in such a way that the participants have similar construals of the situation and responses to it, including things likely to be said about it.

For example, children who are just learning their local language interpret novel words using Markman's "whole object assumption" and her "taxonomic assumption." The whole object assumption causes them to interpret the novel word "cup" as referring to a whole cup, and not to its handle, the porcelain it is made of, a cup on a saucer, a cup of tea (and so on); the "taxonomic assumption" causes them to interpret "cup" as referring to all objects of the same type, and not to the particular cup being pointed to at that moment (Markman, 1989; Markman & Hutchinson, 1984). Of course, the operation of these assumptions depends, in turn, on interpretations generated by the kinds of domain-specific inferential systems discussed earlier, which define what entities and relations count as whole objects, animals, plants, people, natural kinds, artifacts, taxonomic categories, and so on (Carey & Gelman, 1991). Still other domain-specific reasoning procedures may privilege certain interpretations of social relations. Thus, social contract algorithms have both intrinsic definitions for the terms used by their procedures and cues for recognizing which elements in recurrent situations correspond to those terms (Cosmides, 1989; Cosmides & Tooby, 1989, this volume). Consequently, these evolved reasoning specializations may sometimes function as nuclei around which semantic inference is conducted. Emotional expressions also function as metacultural cues that assign standardized meanings to the contingent elements of situations (see Fernald, this volume; Tooby & Cosmides, 1990b). For example, if someone reacts with fear, others interpret this as a reaction to danger and attempt to identify in the situation what the dangerous entity is, re-evaluating various stimuli. They may scan the local environment, organizing their search by a categorization system that privileges some things (e.g., snakes) over others (e.g., flowers) (Cook, Hodes, & Lang, 1986).

Thus, we have the surprising result that it is the shared species-typical mechanisms and common metacultural framings that make it possible for a child to learn what is culturally variable: in this case, the meanings of words in the local language. This argument, in fact, generalizes beyond language: The variable features of culture can be learned solely because of the existence of an encompassing universal human metaculture. The ability to imitate the relevant parts of others' actions (Meltzoff, 1988), the ability to reconstruct the representations in their minds, the ability to interpret the conduct of others correctly, and the ability to coordinate one's behavior with others all depend on the existence of human metaculture. Sperber and Wilson (1986) have written at length about how, for successful communication to be possible, both sender and receiver must share a great many assumptions about the world. The less they mutually assume, the more difficult it is to communicate until, in the limiting case, they cannot communicate at all. The child arrives in the culture free of any knowledge about its particularities, and so the only way the child initially can be communicated with is through what is mutually manifest between the child and the adults by virtue of their common humanity (e.g., Fernald, this volume). The same is true, as Sperber (1982) concisely points out, of ethnographers: The best refutation of cultural relativity is the activity of anthropologists themselves, who could not understand or live within other human groups unless the inhabitants of those groups shared assumptions that were, in fact, very similar to those of the ethnographer. Like fish unaware of the existence of water, interpretativists swim from culture to culture interpreting through universal human metaculture. Metaculture informs their every thought, but they have not yet noticed its existence.

So the beginning of this section, in which we discussed how natural selection sifted cultural variability throughout the Pleistocene for uniformities, gave only a one-sided analysis of how, despite cultural variability, universals still existed. It is even more important to realize that contentful human universals make possible the very existence of transmitted cultural variability (what is usually called "culture"), which would otherwise be impossible. Therefore, the development of increasing cultural variation throughout the Pleistocene was made possible by the evolution of psychological specializations that exploited the regularities of human metaculture in order to learn the variable features of culture. To return to a position William James stated a century ago, to behave flexibly, humans must have more "instincts" than other animals, not fewer.

THE TRANSITION TO POST-STANDARD MODEL PSYCHOLOGY

The Decline of Standard Model Psychology

The progression from Standard Model psychology to post-Standard Model psychology was driven largely by the emergence of new and more rigorous standards that psychological theories are now expected to meet. As the field grew more sophisticated, various communities of psychologists began insisting on *causal accounts* of how hypothesized Standard Model mechanisms produced their effects: What are the networks of cause and effect that, step by step, lead from input to output? In the social sciences, no model of the human psychological architecture seemed impossible when its proponents didn't have to specify by what methods it generated human behavior. The cognitive revolution, with its emphasis on formal analysis, made clear that theories needed to be made causally explicit to be meaningful, and it supplied psychologists with a far more precise language and set of tools for analyzing and investigating complexly contingent, information-responsive systems. When examined from this perspective, most traditional theories turned out to be both incomplete and incapable of accounting for large classes of observed phenomena. Indeed, most no longer seemed to qualify as hypotheses at all. For example, "learning" ceased to be seen as an explanation for behavior, but instead was recognized as a label for a loosely defined class of phenomena generated by as yet unknown procedures. For modern psychologists the key question became: What is the explicit description of these procedures?

Over the last three decades, the hard work of discovering procedures that could actually account in detail for observed behavior and competences has led to the widespread conclusion that our evolved psychological architecture must include a large set of mechanisms of a very different character than Standard Model psychologists had envisioned. The most fundamental shift from Standard Model to post-Standard Model psychology has been the abandonment of the axiom that evolved psychological mechanisms must be largely—or exclusively—general-purpose and free of any contentful structure not put there by experience (e.g., Carey & Gelman, 1991; Chomsky, 1975; Cosmides & Tooby, 1987, 1992; Gallistel, 1990; Gigerenzer, 1991b; Gigerenzer & Murray, 1987; Herrnstein, 1977; Pinker, 1984; Rozin, 1976; Rozin & Schull, 1988; Shepard, 1984, 1987a; Symons, 1987). Many psychologists have been forced by their data to conclude that both human and nonhuman minds contain—in addition to whatever general-purpose machinery they may have—a large array of mechanisms that are (to list some of the terms most frequently used) functionally specialized, content-dependent, content-sensitive, domain-specific, context-sensitive, special-purpose, adaptively specialized, and so on. Mechanisms that are functionally specialized have been called (with some differences in exact definition) adaptive specializations by

Rozin (1976), modules by Fodor (1983), and cognitive competences or mental organs by Chomsky (1975, 1980).

Consequently, the core of the debate is not really about whether the reliably developing design of the mind evolved—the answer to that question can only be yes. The debate is, instead, over whether our evolved psychological architecture is predominantly domain-general (Symons, 1987). Did the human mind evolve to resemble a single general-purpose computer with few or no intrinsic content-dependent programs (e.g., Gould, 1979)? Or does its evolved architecture more closely resemble an intricate network of functionally dedicated computers, each activated by different classes of content or problem, with some more general-purpose computers embedded in the architecture as well (e.g., Chomsky, 1975, 1980; Cosmides & Tooby, 1987; Gallistel, 1990; Gazzaniga, 1985; Rozin, 1976; Symons, 1987)? In other words, does the human mind come equipped with any procedures, representational formats, or content-primitives that evolved especially to deal with faces, mothers, language, sex, food, infants, tools, siblings, friendship, and the rest of human metaculture and the world?

Solvability and the Formal Analysis of Natural Competences

Thirty years ago, Noam Chomsky inaugurated a new era in the behavioral sciences when he began to explore psychological questions by analyzing the capacities of well-specified computational systems (Chomsky 1957, 1959). His approach was distinctive. To evaluate existing psychological theories, he first made their underlying assumptions about computational mechanisms explicit. He then tested the ability of these computational mechanisms to solve real, natural problems that humans were known to be able to solve. In his first application of this method, he attempted to evaluate the adequacy of behaviorist accounts of language, particularly as presented in Skinner's then recently published book, *Verbal Behavior* (1957). When Chomsky examined the behaviorist account of language in the light of these criteria, he found that it suffered from a series of difficulties that precluded it from being a persuasive explanation for human linguistic competence.

Chomsky's research program brought the serious deficiencies of the Standard Model into plain view because it combined two key ingredients: (1) the study of tasks related to a natural, complex, real-world competence that humans were known to have, and (2) the use of formal solvability analyses to explore the actual computational capacities of mechanisms hypothesized to generate explicitly defined outcomes. A theory about the design of a mechanism cannot be correct if, under the relevant conditions, that design cannot solve the problem or generate the performance that the theory claims it can; this can be determined using a solvability analysis, as outlined in pp. 73–77.

Language was a pivotal choice for a test of domain-general accounts of behavior because language—particularly syntax—involved complex but clearly specifiable patterns of behavior that humans were already known to be able to produce under natural conditions without elaborate experimental manipulations. Within this domain, one could precisely and unambiguously define criteria for recognizing what behavioral patterns humans could and did routinely produce (grammatical versus ungrammatical sentences). Therefore, one could define what output any mechanism hypothesized to account for these behavioral patterns had to produce as well. In contrast, no one could tell whether associationist mechanisms or general-purpose symbol-processing mechanisms could account for phenomena such as "religion," "marriage," or "politics" because no one had an unambiguous empirical definition of human performance in these spheres.

By specifying what counts as the production of grammatical utterances or the acquisition of the grammar of a human language, psycholinguists working within the Chomskyan research tradition have been using solvability analyses to show that a task routinely mastered by four-year-old children is too richly structured to be accounted for by any known general-purpose mechanism operating in real time (Chomsky, 1975, 1980; Pinker, 1979, 1984, 1989, 1991; Pinker & Bloom, this volume; Wexler & Culicover, 1980). Despite three decades of intensive efforts by Standard Model psychologists to get general-purpose cognitive machinery to learn grammar, their theories have fared no better than did their behaviorist predecessors. To take a recent example, through careful solvability analyses, Pinker and Prince were able to show that newly proposed domain-general connectionist and associationist models were computationally insufficient to solve even so narrow a problem as the acquisition of the past tense in English (Pinker, 1991; Pinker & Prince, 1988). These mechanisms failed precisely because they lacked computational machinery specialized for the acquisition of grammar.

Thirty years of such findings have forced many cognitive psychologists, against their inclination, to accept domain-specific hypotheses about language learning—to conclude that humans have as part of their evolved design a language acquisition device (LAD), which incorporates content-dependent procedures that reflect in some form "universal grammar" (Chomsky, 1975, 1980; Pinker, 1979, 1984, 1989, 1991; Wexler & Culicover, 1980). In this view, the architecture of the human mind contains content-specialized mechanisms that have evolved to exploit the subtle cross-culturally recurring features of the grammars of human language communities—one facet of human metaculture (Pinker & Bloom, this volume).

The introduction of solvability analyses and the increasing demand for well-specified information-processing models have exposed the deficiencies of Standard Model theories in other areas of psychology as well (see, e.g., Carey & Gelman, 1991; Cosmides, 1989; Cosmides & Tooby, 1989, this volume; Gelman & Markman, 1986, 1987; Keil, 1989; Leslie, 1987; Markman, 1989). Standard Model theories are usually so underspecified that one cannot make their underlying assumptions about computational mechanisms procedurally explicit. To the extent that they can be evaluated, however, when they are faced with real world tasks that humans routinely solve, they consistently perform poorly or not at all.

In fact, the large-scale theoretical claims of Standard Model psychology never had a strong empirical base. Limited empirical support could be produced for Standard Model domain-general theories, but only so long as research was confined to the investigation of experimenter-invented, laboratory limited, arbitrary tasks. The occasional matches between domain-general theories and data sets have been chronically weak and experimentally fragile. These restricted empirical successes depended on carefully picked experimental venues, such as pigeons isolated from conspecifics pecking for food in stimulus-depauperated environments or humans learning lists of nonsense syllables. Standard Model theories of mechanisms have maintained themselves as empirically credible primarily through pretheoretical decisions concerning what kinds of experiments were considered meaningful and through assumptions imposed a priori on the class of hypotheses that would be entertained. For humans and nonhumans alike, exposure to biologically significant stimuli and natural tasks elicits complexly patterned performances that Standard Model theories are unable to predict or explain. So, to keep behavioral phenomena in line with theory, Standard Model psychologists had to keep humans and other species outside of ecologically valid circumstances, away from any biologically significant stimuli, and test them on artificial problems that subjects would not have had to solve in their environment of evolutionary adapt- edness (for discussion, see Beach, 1950; Breland & Breland, 1961; Herrnstein, 1977; Lockard, 1971). Although these weaknesses have now mostly been abandoned in the study of other species, they unfortunately remain endemic in many areas of human psychology.

Once animal behavior researchers let the pigeon out of its barren artificial cage, a rich flock of behavioral phenomena appeared, and questions inevitably arose about the mechanisms that guide the animal to do all the different things it needs to do in natural environments to survive and reproduce. Thus, ethology (or behavioral ecol- ogy, sociobiology, or animal behavior) played an important corrective role by provid- ing examples of the tasks organisms solve and the complex performances they exhibit in more natural conditions (Daly & Wilson, 1984b; Krebs & Davies, 1984; Lorenz, 1965; Rozin & Schull, 1988; Tinbergen, 1951; Wilson, 1975). These fields carefully documented functionally interpretable behaviors that lie far outside anything that Standard Model psychology and a short list of drives could explain. Researchers inves- tigating the now well-known selection pressure expressed by Hamilton's rule (see pp. 67–68) documented an enormous array of kin-directed assistance in nonhuman ani- mals—behaviors completely undreamed of in Standard Model psychology (Hamil- ton, 1964; Williams & Williams, 1957; for review, see Krebs & Davies, 1984). Infant macaques become emotionally attached to immobile cloth figures even though they nurse ("are reinforced") on another structure entirely (Harlow, Harlow, & Suomi, 1971). There are reports from an entire range of species—from langurs to lions to rodents—of newly resident males killing the unweaned infants of their predecessors, thereby accelerating ovulation in their new mates (Hrdy, 1977; Hausfater & Hrdy, 1984). Ring doves may expend considerable effort to monitor the sexual behavior of their mates (Erickson & Zenone, 1976). There was the discovery of the complex pat- terns of food reciprocation in vampire bats—phenomena difficult to account for using traditional notions of general-purpose cognition, conditioning, and drive reduction (e.g., Wilkinson 1988, 1990). From echolocation to parental care, to celestial naviga- tion, to courtship, to coalitional action in chimpanzees, to seasonal migration, to decoying predators away from nests, to communication in bees, to "friendship" and dominance in baboons, nonhuman behavior is full of tasks and organized behaviors that do not remotely fit into Standard Model psychology. This burgeoning body of phenomena caused many animal behavior researchers to break away from the narrow experimental paradigms and narrow questions of the Standard Social Science Model.

In human psychology, the observational basis for Standard Model theories was equally circumscribed, but escape from its narrow experimental paradigms has been more difficult than in nonhuman psychology. Standard Model psychologists had no salient reason for suspecting that different psychological mechanisms would be acti- vated by different kinds of tasks. Human activities appeared to be so variable—both between cultures and among individuals within a culture—that the notion that some tasks and problems might be more "natural" than others did not seem conspicuously sensible. Although most psychologists were faintly aware that hominids lived for millions of years as hunter-gatherers or foragers, they did not realize that this had theoretical implications for their work. More to the point, however, the logic of the Standard Social Science Model informed them that humans were more or less blank slates for which no task was more natural than any other. Until the emergence of a community of Chomskyan psycholinguists, mainstream psychology had been overwhelmingly dominated by general-purpose learning and cognitive theories. In consequence, the same processes were assumed to account for learning and action in all domains of human activity, from suckling at the breast to incest avoidance, language learning, and alliance negotiations among Dani warriors.

By questioning the assumption that all tasks were created equal, Chomsky exposed how narrowly chosen Standard Model research topics had actually been and how over-reaching the extrapolation had been from these topics to the rest of human thought and action. The rise of Chomskyan psycholinguistics constituted a decisive turning point in the development of human psychology because it introduced the subversive idea that some tasks might awaken associated competences that were more "natural" than others: more functionally specialized, more complex, more reliably developing, more species-differentiated, and, therefore, more worthy of detailed exploration (Marr & Nishihara, 1978).

The Rise of Domain-Specific Psychology

The Chomskyan revolution in the study of language slowly began to legitimize the exploration of models of our evolved psychological architecture that did not assume a priori that all tasks are solved by the same set of content-independent processes. In diverse subcommunities, the gradually expanding freedom to consider domain-specific hypotheses alongside more orthodox ones has led to their increasing acceptance. Performance in virtually every kind of experimental situation is sensitive to the content and context of the task, and domain-specific hypotheses tend to organize, account for, and predict this performance better than their Standard Model predecessors. Although social and behavioral scientists outside of cognitive, comparative, and physiological psychology still routinely assume a domain-general human mind, within the community of psychologists who rigorously study mechanisms this view is in retreat and disarray. Standard Model psychology has been able to persist only in those research communities that avoid formal analysis entirely or that avoid using it to study performance on ecologically valid, natural tasks.[6]

Thus, researchers who ask hard questions about how organisms actually solve problems and who focus on the real performance of organisms on natural tasks have had to abandon the idea that the mind is free of content-specialized machinery. Researchers who study color vision, visual scene analysis, speech perception, conceptual development in children, mental imagery, psychophysics, locomotion, language acquisition, motor control, anticipatory motion computation, face recognition, biomechanical motion perception, emotion recognition, social cognition, reasoning, and the perception and representation of motion, for example, cannot account for the psychological phenomena they study by positing computational mechanisms that are solely domain-general and content-independent (see, e.g., Bizzi, Mussa-Ivaldi, & Giszter, 1991; Carey & Gelman, 1991; Etcoff, 1986; Freyd, 1987; Kosslyn, 1980; Liberman & Mattingley, 1985, 1989; Lindblom, 1986, 1988; Maloney & Wandell, 1985; Marr, 1982; Pinker, 1984, 1989; Poggio et al., 1985; Proffitt & Gilden, 1989; Shepard, 1981,

1984, 1987a; Shiffrar & Freyd, 1990; Spelke, 1988, 1990). In fact, the reality has always been that every field of psychology bristles with observations of content-dependent phenomena. Freedom from the axiom that all psychological phenomena must be explained by content-independent machinery has allowed psychologists to move ahead to explore—and to view as meaningful—the rich content-sensitive effects that permeate psychological phenomena (e.g., Astington et al., 1988; A. Brown, 1990; Carey, 1985b; Carey & Gelman, 1991; Cosmides, 1989; Cosmides & Tooby, this volume; Gelman & Markman, 1986, 1987; Gigerenzer & Hug, in press; Gigerenzer & Murray, 1987; Keil, 1989; Manktelow & Over, 1991). Formerly, these omnipresent content effects were considered an embarrassment to be explained away or else dismissed as noise. Now they are considered to be primary data about the structure of the mind.

Outside of cognitive psychology, the emergence of post-Standard Model approaches derived their impetus from branches of evolutionary biology. In the 1950s and 1960s, the successful application of evolutionary approaches to animal behavior in ethology and its successor disciplines provided evidence of domain-specific mechanisms that was difficult to ignore (e.g., attachment, emotion, phobias, mating, foraging, navigation). This trend was accelerated by the rapid advances in evolutionary biology over the last three decades, which made the previously clouded connection between evolution and behavior somewhat clearer. These advances included (1) more coherent approaches to the nature-nurture issue, (2) a more rigorous foundation for the theory of natural selection (Williams, 1966), (3) formal analyses of what behaviors would be favored by selection in a variety of newly explored domains (e.g., Charnov, 1976; Hamilton, 1964; Maynard Smith, 1982; Stephens & Krebs, 1986; Trivers, 1971, 1972, 1974; Williams, 1966), and (4) a cascade of successful applications of these theories to animal behavior (Alexander, 1974; Daly & Wilson, 1984b; Krebs & Davies, 1984; Wilson, 1975).

Just as in the case of nonhuman behavior, evolutionarily informed studies of human choice, motivation, emotion, and action also bristle with documented phenomena that cannot be accounted for with content-independent architectures and a short list of drives, rewards, or reinforcers (the chapters in this volume are a small sampling of such cases). For example, ever since the Harlows demolished the myth that an infant's love for its mother was a conditioned response to food rewards, the rich collection of co-adapted mechanisms in the mother and infant has been a productive focus of psychological investigation (e.g., Bowlby, 1969). Profet (this volume) identifies a maternal psychological adaptation for the protection of the fetus during embryogenesis. Fernald (this volume) explores the communicative adaptations mothers have to the infant's perceptual limitations. Moreover, cross-cultural regularities in fall-rise patterns of maternal fundamental frequency provide an elegant illustration that a child and adult initially communicate by virtue of what they share through their common human metaculture. Communication through such human universals is a precondition for the child's acquisition of the culturally specific. Facial expressions of emotion represent another evolved modality through which humans communicate situation-construals, and the cross-culturally stable features of emotional expression provide another critical foundation for human metaculture. Ekman and his colleagues have established one of the earliest and most sophisticated traditions of evolutionary psychological research, and these studies of emotional expression represent a major achievement in modern psychology (e.g., Ekman, 1973, 1982, 1984; Ekman & Friesen, 1975; Ekman, Levenson, & Friesen, 1983). Etcoff (1986) has marshalled substantial neuropsychological evidence that humans have mechanisms specialized for the identification of emotional expression—an adaptation to an important, stable feature of ancestral social and cultural environments.

Indeed, ever since Darwin (1871, 1872), emotions have been seen as the product of the evolutionary process and usually, although not always, as functional adaptations (e.g., Arnold, 1960, 1968; Chance, 1980; Daly, et al., 1982; Eibl-Eibesfeldt, 1975; Ekman, 1982; Frijda, 1986; Hamburg, 1968; Izard, 1977; Otte, 1974; Plutchik, 1980; Tomkins, 1962, 1963; Tooby & Cosmides, 1990b; and many others). Functional or not, the emotions collectively provide a dense and pervasive network of domain-specific phenomena that have consistently resisted assimilation into any Standard Model theory. However, in contrast to their Standard Model reputation as crude and indiscriminate responses, on close scrutiny each specific emotion appears to be an intricately structured information-sensitive regulatory adaptation. In fact, the emotions appear to be designed to solve a certain category of regulatory problem that inevitably emerges in a mind full of disparate, functionally specialized mechanisms—the problem of coordinating the menagerie of mechanisms with each other and with the situation being faced (Tooby, 1985; Tooby & Cosmides, 1990b; Nesse, 1990).

Daly and Wilson have been exploring the evolved complexity and functional subtlety of the human motivational system. They have produced a substantial body of findings supporting specific hypotheses they derived from a broad array of adaptationist theories (Daly & Wilson, 1981, 1982, 1984b, 1987a, 1987b, 1988; Daly, et al., 1982; Wilson & Daly, 1985, 1987, this volume). Their particular interest has been the evolved motivational systems that regulate parental care, spousal relations, sexual jealousy, sexual proprietariness, and risk-taking. By using behavioral phenomena such as violence and homicide as dependent measures, they have been able to investigate many aspects of these evolved motivational systems—including how their operation is affected by factors such as gender, age, kinship, reproductive value, number of children, and other situational variables (see also Mann, this volume). Similarly, in the area of human mate choice and sexuality, the work of Symons, Buss, and many others shows that the construct of a "sex drive" is completely inadequate to cope with the structured richness of the situational factors processed by the differentiated sexual psychologies of men and women across cultures (e.g., Buss, 1987, 1989, 1991, this volume, in prep.; Ellis, this volume; Sadalla, Kenrick, & Vershure, 1987; Symons, 1979; Townsend, 1987). These studies indicate that existing theories of motivation will have to be replaced with theories positing a far more elaborate motivational architecture, equipped with an extensive set of evolved information-processing algorithms that are contingently sensitive to a long list of situational contents and contexts.

Thus, the examination of even a small sampling of non-Standard Model behavioral studies by a handful of researchers such as Bowlby, Daly and Wilson, Ekman, Fernald, Marks, Buss, and Symons leads to the conclusion that the human mind contains evolved emotional and motivational mechanisms that are specifically targeted to address adaptive problems involved in parenting, emotional communication with infants and adults, kinship, mate choice, sexual attraction, aggression, the avoidance of danger, mate guarding, effort allocation in child care, and so on. That is, humans have psychological adaptations that contain contentful structure specifically "about" their mothers, "about" their children, "about" the sexual behavior of their mates, "about" those identified by cues as kin, "about" how much to care for a sick child, and so on, and these contents are not derived exclusively from either a short list of drives or from culturally variable, socially learned "values."

In short, the central tenets of Standard Model psychology are contradicted by results from a large and rapidly growing body of research on humans and nonhumans from the cognitive community, from the evolutionary community, from the behavioral ecology community, and from other research communities as well (for example, much of psychobiology, comparative psychology, and neuroscience). Content-independent mechanisms simply cannot generate or explain the richly patterned behaviors and knowledge structures that appear when one's research focus is widened beyond arbitrary laboratory tasks to include the complex performances orchestrated by natural competences on real world tasks. Moreover, unlike most Standard Model theories and results, these kinds of studies and hypotheses withstand cross-cultural scrutiny and indicate that a great deal of the substance of social life attributed to "culture" around the world is in fact caused by the operation of contingently responsive domain-specific mechanisms. These converging results are accumulating into a strikingly different picture than that provided by the Standard Social Science Model. They indicate that a universal, evolved psychological architecture that is filled with contingently responsive mechanisms infuses distinctively human patterns into the life of every culture.

The Frame Problem and the Weakness of Content-Independent, Domain-General Mechanisms
From Flexibility to Adaptive Flexibility

In the passage from Standard Model to post-Standard Model psychology it seems fair to say that the greatest reversal lay in how content-independence and domain-generality came to be regarded. Many modern researchers recognize that content-independent, general problem-solvers are inherently weak in comparison to content-specialized mechanisms. From a traditional point of view, however, it seemed sensible to regard generality as an enhancement of the capacity of a system: The system is not prevented from assuming certain states or kept from doing what is adaptive (or desirable) by a "rigid" or "biased" architecture. Generality of application seems like such an obvious virtue and content-independence seems like such an obvious road to flexible behavior, what could possibly be wrong with them? As Marvin Harris puts this line of reasoning, "Selection in the main has acted against genetically imposed limitations on human cultural repertoires" (1979, p. 136). Why rigidly prevent the system from engaging in certain behaviors on those occasions when they would be advantageous? Moreover, why not have an "unbiased" architecture in which the actual structure of local circumstances impresses a true picture of itself in a free, objective, and unconstrained way? In this view, content-specificity in evolved psychological design is imbued with all the legendary attributes of "biology"—rigidity, inflexibility, and constraint. It is viewed as preventing the system from achieving advantageous states that would otherwise naturally come about.

So what, after all, is so wrong with domain-general systems? Why do cognitive psychologists and artificial intelligence researchers consistently find them too weak to solve virtually any complex real world task? Why isn't "flexibility" in the form of content-independence a virtue? The answers to these questions emerge from one clarification and from two basic facts. The two facts have already been touched on many times: (1) possibilities are infinite; and (2) desirable outcomes—by any usual human, evolutionary, or problem-solving standard—are a very small subset of all possibilities.

The clarification concerns the kind of plasticity and flexibility that are implicitly being referred to. Literally, plasticity, or flexibility, is the simple capacity to vary in some dimension. The more dimensions of possible variation, the greater the "plasticity." Hence, a lump of clay is very plastic with respect to shape (although not with respect to substance, density, and so on). Similarly, there is an infinite number of ways that humans and other animals could potentially act. The difficulty lies in the fact that the overwhelming majority of behavioral sequences would be lethal in a few hours, days, or weeks. The set of behaviors that leads even to temporary individual survival— let alone to reproduction or design-propagation—constitutes an extremely miniscule subset of all possible behavioral sequences. Thus, the property of freely varying behavior in all dimensions independent of conditions is not advantageous: It is evolutionarily and individually ruinous.

Accordingly, to be endowed with broad behavioral plasticity unconnected to adaptive targets or environmental conditions is an evolutionary death sentence, guaranteeing that the design that generates it will be removed from the population. Designs that produce "plasticity" can be retained by selection only if they have features that guide behavior into the infinitesimally small regions of relatively successful performance with sufficient frequency. In reality, terms such as flexibility or plasticity are implicitly used to mean something very different from the simple "capacity to vary." They are implicitly used to mean the capacity to adjust behavior (or morphology) as a coordinated response to the specifics of local conditions so that the new behavior is particularly appropriate to or successful in the specific circumstances faced.

This narrowly specialized form of flexibility requires three components: (1) a set of mechanisms that define an adaptive target (such as finding food, finding home, or finding a mate); (2) a set of mechanisms that can compute or otherwise determine what responses are most likely to achieve the adaptive target in each specific set of circumstances that one is likely to encounter; and (3) the ability to implement the specific response once it is determined. Plasticity in the "lump of clay/capacity to vary" sense refers only to the third component: If an organism has correctly computed what it is advantageous to do, then (and only then) is it disadvantageous to be inflexibly prevented from implementing those changes by some fixed element of the system.

In fact, plasticity (e.g., variability) tends to be injurious everywhere in the architecture except where it is guided by well-designed regulatory mechanisms that improve outcomes or at least do no harm. It would be particularly damaging if these regulatory mechanisms were themselves capriciously "plastic," instead of rigidly retaining those computational methods that produce advantageous responses to changing conditions. Thus, plasticity is only advantageous for those specific features of the organism that are governed by procedures that can compute the specific changes or responses that will be, on average, more successful than a fixed phenotype. Adaptive flexibility requires a "guidance system" (Cosmides & Tooby, 1987; Tooby, 1985).

The most important conclusion to be derived from this line of reasoning is that adaptive flexibility can only evolve when the mechanisms that make it possible are embedded within a co-evolved guidance system. Consequently, the expansion of behavioral and cognitive flexibility over evolutionary time depended acutely on how well-designed these computational guidance systems became. There is nothing in the ability to vary per se that naturally leads systems to gravitate toward producing successful performances. It is the guidance system itself that is doing the bulk of the interesting regulation of outcomes, with the "potential to vary" component explaining very little about the situation. Thus, Gould's (1977a, 1979) faith in the explanatory power of the SSSM concept of generalized human "biological potential" depends either on (1) an unjustified teleological panglossianism (e.g., unguided processes, such as accidental brain growth, just happen to "work out for the best," giving humans the desire to care for their children, to defend themselves when attacked, to cooperate; the ability to recognize faces, to find food, to speak a language . . .), or (2) the unacknowledged existence of co-evolved cognitive adaptations that guide behavioral plasticity toward the achievement of adaptive targets.

It is the necessary existence of these co-evolved guidance systems that has, for the most part, escaped the attention of Standard Model advocates. In fact, the SSSM ediface is built on the conflation of two distinct notions of flexibility: (1) flexibility as the absence of any limits on responses, and (2) flexibility as the production of contextually appropriate responses. Advocates of the SSSM imagine that flexibility in the first sense—an absence of limits on variation—is easy to computationally arrange (just remove all "constraints"). But they also assume this is the same as—or will automatically produce—flexibility in the second sense: adaptive, successful, or contextually appropriate behavior. Post-Standard Model psychology rests on the recognition that flexibility in this second sense is not something that is teleologically inevitable once constraints are removed, but is, instead, something very improbable and difficult to achieve, requiring elaborate functionally organized machinery.

The Weakness of Content-Independent Architectures

If the doors of perception were cleansed everything would appear to man as it is, infinite.
—WILLIAM BLAKE

If plasticity by itself is not only useless but injurious, the issue then becomes, what kind of guidance systems can propel computational systems sufficiently often toward the small scattered islands of successful outcomes in the endless expanse of alternative possibilities? Attempts over the last three decades to answer this question have led directly to two related concepts, called by artificial intelligence researchers and other cognitive scientists *combinatorial explosion* and the *frame problem*.

Combinatorial explosion is the term for the fact that with each new degree of freedom added to a system, or with each new dimension of potential variation added, or with each new successive choice in a chain of decisions, the total number of alternative possibilities faced by a computational system grows with devastating rapidity. For example, if you are limited to emitting only one out of 100 alternative behaviors every successive minute (surely a gross underestimate: raise arm, close hand, toss book, extend foot, say "Havel," etc.), after the second minute you have 10,000 different behavioral sequences from which to choose, a million by the third minute, a trillion by six minutes, and 10^{120} possible alternative sequences after only one hour—a truly unimaginable number. Every hour, each human is surrounded by a new and endless expanse of behavioral possibility. Which leads to the best outcome? Or, leaving aside optimality as a hopelessly utopian luxury in an era of diminished expectations, which sequences are nonfatal? The system could not possibly compute the anticipated outcome of each alternative and compare the results, and so must be precluding without complete consideration the overwhelming majority of branching pathways. What are the principles that allow us to act better than randomly?

Combinatorial explosion attacks any system that deals with alternatives, which means any system that is flexible in response or has decisions to make. The more flexible the system, the greater the problem. Even worse, knowledge acquisition is impossible for a computational system equipped only with the limited information it can gain through its senses; this is because the number of alternative states of affairs in the world that are consistent with its sense data is infinite. For example, if cognitive mechanisms are attempting to infer the meaning of an unknown word, there is an infinite set of potential meanings. If perceptual mechanisms are trying to construct a three dimensional model of the local world from a visual array, there is an infinite number of different ways to do it that are all consistent with the array. For any finite sample of sentences encountered, there exists an infinite number of alternative grammars that could have generated them. If one is making a decision about how to forage there is, practically speaking, an infinite number of possibilities. Moreover, random choice is not a general solution to the problem because for most adaptive or humanly defined problems the islands of success are infinitesimal next to the illimitable seascapes of failure. And for biological systems, success and failure are not arbitrary. The causal world imposes a nonarbitrary distinction between detecting in one's visual array the faint outline of a partly camouflaged stalking predator and not detecting it because of alternative interpretive procedures. Nonpropagating designs are removed from the population, whether they believe in naive realism or that everything is an arbitrary social construction.

The inexhaustible range of possibilities latent in behavior, categorization, interpretation, decision, and so on, is a not just an abstract philosophical point. It is an implacable reality facing every problem-solving computational system. Each prelinguistic child trying to learn her own language or to induce new knowledge about the world is faced with this problem; so is every artificial intelligence system. In artificial intelligence research, it is called the "frame problem" (Boden, 1977); in linguistics, this problem is called the "poverty of the stimuli" (Chomsky, 1975); in semantics, it is called the problem of "referential ambiguity" (Gleitman & Wanner, 1982); in developmental psychology, it is called the "need for constraints on induction" (Carey, 1985a); in perception, they say that the stimulus array "underdetermines" the interpretation. Any design for an organism that cannot generate appropriate decisions, inferences, or perceptions because it is lost in an ocean of erroneous possibilities will not propagate, and will be removed from the population in the next generation. As selection pressures, combinatorial explosion and the frame problem are at least as merciless as starvation, predation, and disease.

With this as background, the converging results from artificial intelligence, perception, cognitive development, linguistics, philosophy, and evolutionary biology about the weaknesses of domain-general content-independent mechanisms are not difficult to fathom. One source of difficulty can be sketched out quickly. If a computational system, living or electronic, does not initially know the solution to the problem it faces then its procedures must operate to find a solution. What methods do content-independent systems bring to problem-solving? To describe a system as domain-general or content-independent is to say not what it is but only what it lacks: It lacks any specific a priori knowledge about the recurrent structure of particular situations or problem-domains, either in declarative or procedural form, that might guide the system to a solution quickly. It lacks procedures that are specialized to detect and deal with particular kinds of problems, situations, relationships, or contents in ways that differ from any other kind of problem, situation, relationship, or content. By definition, a domain-general system takes a "one size fits all" approach.

To understand the importance of this, consider the definition of an adaptation. An *adaptation* is a reliably developing structure in the organism, which, because it meshes with the recurrent structure of the world, causes the solution to an adaptive problem. It is easy to see how a specific structure, like a bug detector in a frog's retina, in interaction with bug trajectories in the local environment, solves a feeding problem for the frog. It is easy to see how the Westermarck sexual disinterest mechanism combines with the co-residence cue to diminish the probability of sex between close relatives (Shepher, 1983; Wolf & Huang, 1980). When the class of situations that a mechanism is designed to solve is more narrowly defined, then (1) the situations will have more recurrent features in common, and therefore (2) the mechanism can "know" more in advance about any particular situation that is a member of this class. As a result, (3) the mechanism's components can embody a greater variety of problem-solving strategies. This is because mechanisms work by meshing with the features of situations and, by definition, narrowly defined situations have more features in common. Our depth perception mechanism has this property, for example: It works well because it combines the output of many small modules, each sensitive to a different cue correlated with depth. In addition, (4) the narrower the class, the more likely it is that a good, simple solution exists—a solution that does not require the simultaneous presence of many common features. The frog can have a simple "bug detector" precisely because insects share features with one another that are not shared by many members of more inclusive classes, such as "animals" or "objects."

In contrast, the more general a problem-solving technique is, the larger the range of situations across which the procedure must successfully apply itself. When the class of situations that a mechanism must operate over is more broadly defined, then (1) the situations will have fewer recurrent features in common, therefore (2) the mechanism can "know" less in advance about any particular situation that is a member of this class. Because (3) broadly defined situations have so few features in common for a mechanism to mesh with, there exist fewer strategies capable of solving the problem.

This result is logically inevitable. Every kind of problem-solving strategy that applies to a more inclusive class also applies to every subset within it; but not every strategy that applies to a narrowly defined class will apply to the larger classes that contain it (e.g., all insects are objects, but not all objects are insects). By identifying smaller and smaller problem domains on the basis of an increasing set of recurrent similarities, more and more problem-solving strategies can be brought to bear on that set. Conversely, by widening the problem domain that a mechanism must address, strategies that worked correctly on only a subset of problems must be abandoned or subtracted from the repertoire because they give incorrect answers on the newly included problems in the enlarged domain. As problem domains get larger and more broadly defined, a smaller and smaller set of residual strategies is left that remains applicable to the increasingly diverse set of problems. At the limit of perfect generality, a problem-solving system can know nothing except that which is always true of every situation in any conceivable universe and, therefore, can apply no techniques except those that are applicable to all imaginable situations. In short, it has abandoned virtually anything that could lead it to a solution.

This weakness of domain-general architectures arises not because all relatively general problem-solving techniques are useless; indeed, many are very useful—the ability to reject propositions because they are contradicted, the ability to associate, and the ability to recalibrate based on the consequences of actions, for example. The weakness arises because content-sensitivity and specialization are eliminated from the architecture. By definition, a content-independent architecture does not distinguish between different problem-domains or content classes; therefore, it is restricted to employing only general principles of problem-solving that can apply to all problems.

In contrast, a content-dependent domain-specific architecture does identify situations as members of specific problem domains and content classes. Because of this, it can maintain a repertoire of specialized problem-solving techniques that are only activated when they encounter the delimited domains to which they are applicable (e.g., snakes, sex with kin, grammar, falling in love, faces). At the same time, a pluralistic architecture can simultaneously activate every other problem-solving technique appropriate to the larger and more inclusive classes that contain the problem encountered (for faces: face recognition, object recognition, association formation, and so on). Thus, a domain-specific architecture can deploy every general problem-solving technique at the disposal of a domain-general architecture and a multitude of more specific ones as well. This sensible approach to organizing a problem-solving architecture is exactly what is ruled out by SSSM advocates of a content-independent mind whose procedures operate uniformly over every problem or domain.

To put it in adaptationist terms, what does the work of adaptive problem-solving for organisms is (1) the recurrent structure of the world relevant to the problem, in interaction with (2) the recurrent structure of the adaptation. The more broadly defined the problem domain is (1) the less recurrent structure can be supplied by the world (because more diverse situations have less recurrent structure in common), and (2) the less recurrent structure can be supplied by the adaptation in the form of problem-solving procedures that are solution-promoting across a diverse class of situations. The erosion of both sets of problem-solving structures—those in the adaptation and those in the world—increasingly incapacitates the system. This can sometimes be compensated for, but only through a correspondingly costly increase in the amount of computation used in the attempt to solve the problem. The less the system knows about the problem or the world to begin with, the more possibilities it must contend with. Permutations being what they are, alternatives increase exponentially as generality increases and combinatorial explosion rapidly cripples the system. A mechanism unaided by domain-specific rules of relevance, specialized procedures, "preferred" hypotheses, and so on could not solve any biological problem of routine complexity in the amount of time the organism has to solve it, and usually could not solve it at all.

It is the perennial hope of SSSM advocates within the psychological community that some new technology or architecture (wax impressions, telephone switching, digital computers, symbol-processing, recursive programming languages, holograms, non-von Neumann architectures, parallel-distributed processing—a new candidate emerges every decade or so), will free them to return to empiricism, associationism, domain-generality and content-independence (where the SSSM tells them they should go). Nevertheless, the functional necessity of content-specificity emerges in every technology because it is a logical inevitability. Most recently, researchers are establishing this all over again with connectionism (e.g., Jacobs, Jordan, & Barto, 1990; Miller &

Todd, 1990; Pinker & Prince, 1988; Todd & Miller, 1991a, 1991b). Combinatorial explosion and the frame problem are obstacles that can only be overcome by endowing computational architectures with contentful structure. This is because the world itself provides no framework that can decide among the infinite number of potential category dimensions, the infinite number of relations, and the infinite number of potential hypotheses that could be used to analyze it.

The Necessity of Frames

Artificial intelligence research is particularly illuminating about these issues because explicitness is demanded in the act of implementing as programs specific hypotheses about how problems can be solved. By the program's operation, one can tell a great deal about the adequacy of the hypothesis. Moreover, artificial intelligence researchers became interested in getting computers and robots to perform real world tasks, where, just as in evolutionary biology, action is taken in a real, structured, and consequential environment. As a result, artificial intelligence researchers can tell unambiguously whether the decisions the system makes are a success or a failure. The range of problems studied in artificial intelligence widened beyond cognitive psychology's more traditional, philosophy-derived concerns, to include problems such as the regulation of purposive action in a three-dimensional world.

To their great surprise, artificial intelligence researchers found that it was very difficult to discover methods that would solve problems that humans find easy, such as seeing, moving objects or even tying shoelaces. To get their programs to handle even absurdly simplified tasks (such as moving a few blocks around), they were forced to build in substantial "innate knowledge" of the world. As a practical matter, this "knowledge" was either in the form of (1) content-dependent procedures matched closely to the structural features of the task domain within which they were designed to operate, or (2) representations (data structures) that accurately reflected the task domain (i.e., "knowledge of the world"). To move an object, make the simplest induction, or solve a straightforward problem, the computer needed a sophisticated model of the domain in question, embodied either in procedures or representations. Artificial intelligence research demonstrated in a concrete, empirical form, the long-standing philosophical objections to the tabula rasa (e.g., Hume, 1977/1748; Kant, 1966/1781; Popper, 1972; Quine, 1960, 1969). These demonstrations have the added advantage of bracketing just how much "innate" structure is necessary to allow learning to occur.

Artificial intelligence researchers call the specific contentful structures that problem-solving systems need to be endowed with *frames*. For this reason, the consistent inability of systems without sufficiently rich and specialized frames to solve real problems is called the *frame problem* (e.g., Boden, 1977; F. M. Brown, 1987; Fodor, 1983). A frame provides a "world-view": It carves the world into defined categories of entities and properties, defines how these categories are related to each other, suggests operations that might be performed, defines what goal is to be achieved, provides methods for interpreting observations in terms of the problem space and other knowledge, provides criteria to discriminate success from failure, suggests what information is lacking and how to get it, and so on. For example, one might apply a spatial/object frame to a situation. In such a frame, the local world is carved into empty space and objects, which are cohesive, have boundaries defined by surfaces, and move as a unit. They have locations and orientations with respect to each other. They have trajectories and, if solid, cannot pass through one another (unless they change the shape of the object passed through), and so on. In such a framing, humans are simply objects like any other and are not expected to pass through other solid objects. Alternatively, one might have a coalitional framing (present, for example, in a football game or a war), in which humans are a relevant and differentiated entity and are construed as animate goal-seeking systems that are members of one of two mutually exclusive social sets; the members of each set are expected to coordinate their behavior with each other to reach goals that cannot be mutually realized for both sets; the goal of each set is to thwart the purposes of the other, and so on. In our own work, we have attempted to sketch out some of the framing necessary for humans to engage in social exchange (Cosmides, 1989; Cosmides & Tooby, 1989, this volume). Very general mechanisms have frames as well: In the formal logic of the propositional calculus, the problem-space is defined syntactically in terms of sets of propositions, truth values, and rules of inference such as *modens ponens* and *modus tollens*. In this frame, the content of the propositions is irrelevant to the operation of the rules of inference.

The solution to the frame problem and combinatorial explosion is always the same whether one is talking about an evolved organism or an artificial intelligence system. When the information available from the world is not sufficient to allow learning to occur or the problem to be solved, it must be supplied from somewhere else. Because the world cannot supply to the system what the system needs first in order to learn about the world, the essential kernals of content-specific framing must be supplied initially by the architecture. For an artificial intelligence system, a programmer can supply it. For organisms, however, it can only be supplied through the process of natural selection, which creates reliably developing architectures that come equipped with the right frames and frame-builders necessary to solve the adaptive problems the species faced during its evolutionary history.

Because of their survival into the present, we know for a fact that living species can reliably solve an enormous array of problems necessary to consistently reproduce across thousands of generations in natural environments. Moreover, the signal lesson of modern evolutionary biology is that this adaptive behavior requires the solution of many information-processing problems that are highly complex—far more complex than is commonly supposed (Cosmides & Tooby, 1987, 1989). If one bothers to analyze virtually any adaptive problem human hunter-gatherers solve, it turns out to require an incredible amount of evolved specialization (see, e.g, Cosmides & Tooby, 1989, this volume). Given (1) the complexity of the world, (2) the complexity of the total array of adaptive tasks faced by living organisms, and (3) the sensitive frame-dependence of problem-solving abilities, the psychological architecture of any real species must be permeated with domain-specific structure to cause reliable reproduction in natural environments. Current research in cognitive psychology and artificial intelligence indicates that Standard Model theories are far too frame-impoverished to solve even artificially simplified computational problems (e.g., identifying and picking up soda cans in the MIT artificial intelligence laboratory), let alone the complex information-processing problems regularly imposed by selective forces operating over evolutionary time.

Our minds are always automatically applying a rich variety of frames to guide us through the world. Implicitly, these frames appear to us to be part of the world. For precisely this reason, we have difficulty appreciating the magnitude, or even the existence of, the frame problem. Just as the effortlessness of seeing led artificial intelligence researchers to underestimate the complexity of the visual system, the automatic and effortless way in which our minds frame the world blinds us to the computational complexity of the mechanisms responsible. When anthropologists go to other cultures, the experience of variation awakens them to things they had previously taken for granted in their own culture. Similarly, biologists and artificial intelligence researchers are "anthropologists" who travel to places where minds are far stranger than anywhere any ethnographer has ever gone. We cannot understand what it is to be human until we learn to appreciate how truly different nonhuman minds can be, and our best points of comparison are the minds of other species and electronic minds. Such comparisons awaken us to an entire class of problems and issues that would escape us if we were to remain "ethnocentrically" focused on humans, imprisoned by mistaking our mentally imposed frames for an exhaustive demarcation of reality.

When we examine electronic minds that truly have no frames and then try to give them even a few of our own real world capacities, we are made forcefully aware of the existence of the immensely intricate set of panhuman frames that humans depend on to function in the world, to communicate with one another, and to acquire additional frames through social inference from others (i.e., to "learn culture"). Geertz's (1973) studies in Bali acquainted him with some of the culturally variable frames that differ between Bali and the United States, but, as his writings make clear, they left him oblivious to the encompassing, panhuman frames within which these variable elements were embedded (D. E. Brown, 1991). If he had widened his scope to include other animal species, he would have been made strongly aware of this dense level of universal and human-specific metacultural frames—a level that should interest every anthropologist because it permeates and structures every aspect of human life. Indeed, if Geertz had widened his scope still further to include electronic minds as points of comparison, he might have come to realize the sheer magnitude of what must be supplied by evolution to our psychological architectures for us to be recognizably human. Perhaps he might also have come to recognize that he and the "simple" Balinese fighting cocks he watched even shared many frames lacking from artificial intelligence systems (about things such as space, motion, vision, looming threats, pain, hunger, and, perhaps, conflict, rivalry, and status changes after fights). Biology, cognitive psychology, and artificial intelligence research comprise a new form of ethnography, which is revealing the previously invisible wealth of evolved frames and specialized framebuilders that our evolved psychological architecture comes equipped with.

The Evolvability Criterion and Standard Model Architectures

In a solvability analysis, the researcher asks whether a proposed architecture is capable of generating a behavior that we know humans (or the relevant species) regularly engage in, whether adaptive or not. But one can also evaluate a proposed architecture by asking how it would fare in solving the actual adaptive problems a species is known to have regularly confronted and solved during its evolutionary history. Because nonhuman and human minds—i.e., the computational systems responsible for regulating behavior—were produced by the evolutionary process operating over vast expanses of time, tenable hypotheses about their design must be drawn from the class of designs that evolution could plausibly—or at least possibly—have produced. To be adequate, proposed designs must be able to account for the solution of the broad array of distinct problems inherent in reliable reproduction over thousands of generations under ancestral conditions. In short, a candidate design must satisfy the *evolvability criterion*. In essence, designs that are more plausible according to criteria drawn from evolutionary biology are to be preferred over designs that are less plausible.

Some rules for evaluating hypotheses by the evolvability criterion are as follows:

1. Obviously, at a minimum, a candidate architecture must be able to perform all of the tasks and subtasks necessary for it to reproduce. It can have no properties that preclude, or make improbable, its own reproduction across multiple generations in natural environments. Just by itself, this is a difficult criterion to meet. No known Standard Model psychological architecture can solve all or even very many of the problems posed by reproduction in natural environments.

2. Given that human minds evolved out of prehuman primate minds, a hypothesis should not entail that an architecture that is substantially inferior at promoting its own propagation (its inclusive fitness) replace an architecture that was better designed to promote fitness under ancestral conditions. There is now known to be an entire range of competences and specialized design features that enhance propagation in a large array of other species. A candidate architecture should be at least roughly comparable to them in their ability to solve the classes of adaptive problems humans and other primate species mutually faced. For this reason, it is not sufficient to incorporate into a general-purpose system a few drives that account for why the organism does not die of thirst or hunger in a few days. Even though some psychological architectures of this kind might conceivably manage their own reproduction under artificially protected circumstances, they contain nothing that would solve other obvious propagation-promoting tasks that have called forth adaptive specializations in innumerable other species. Thus, to be plausible, a proposed human architecture should cause individuals to help relatives more or less appropriately, to defend sexual access to their mates, to forage in a relatively efficient way, and so on. The SSSM view that human evolution was a process of erasing "instincts" violates the evolvability criterion unless it can be shown that for each putatively "erased" adaptive specialization, the general-purpose mechanism that is proposed to have replaced it would have solved the adaptive problem better (Tooby, 1985; Tooby & DeVore, 1987). To our knowledge, no general mechanism operating under natural circumstances has ever been demonstrated to be superior to an existing adaptive specialization.

3. A candidate architecture should not require the world to be other than it really is. For example, models of grammar acquisition that assume that adults standardly correct their children's grammatical errors do not meet this condition (Pinker, 1989). Nor do socialization models that require children to be taught where their own interests lie by individuals with conflicting interests—for many domains, this class even includes the child's own parents and siblings (e.g., Hamilton, 1964; Trivers, 1974). An architecture that was completely open to manipulation by others, without any tendency whatsoever to modify or resist exploitive or damaging social input would be strongly selected against. For this reason, cognitive architectures that are passive vehicles for arbitrary semiotic systems are not plausible products of the evolutionary process.

4. In a related vein, a candidate theory should not invoke hypotheses that require assumptions about the coordinated actions of others (or any part of the environment) unless it explains how such coordination reliably came about during Pleistocene hunter-gatherer life. For example, if the model proposes that people acquire certain adaptive skills or information from others through, say, imitation or conversation,

that model needs to explain how these others reliably obtained the (correct) information and where the information originated. If the blind lead the blind, there is no advantage to imitation. Consequently, acceptable models should not employ shell games, such as the venerable "adaptive knowledge comes from the social world."

5. A candidate model must not propose the existence of complex capacities in the human psychological architecture unless these capacities solve or solved adaptive (design-propagative) problems for the individual. That is, social scientists should be extremely uneasy about positing an improbably complex structure in the system with the capacity to serve nonbiological functional ends, unless that capacity is a by-product of functionality that evolved to serve adaptive ends. Selection builds adaptive functional organization; chance almost never builds complex functional organization. So positing complex designs that serve the larger social good, or that complexly manipulate symbolic codes to spin webs of meaning, or that cause one to maximize monetary profit, all violate the evolvability criterion unless it can be shown that these are side effects of what would have been adaptive functional organization in the Pleistocene. Similarly, one should not posit the existence of complex functional designs that evolved to solve adaptive problems that emerged only very recently. Complex functionality requires time to evolve and, therefore, can arise only in response to longstanding adaptive problems (Dawkins, 1982, 1986; Tooby & Cosmides, 1990a).

Over the course of their evolution, humans regularly needed to recognize objects, avoid predators, avoid incest, avoid teratogens when pregnant, repair nutritional deficiencies by dietary modification, judge distance, identify plant foods, capture animals, acquire grammar, attend to alarm cries, detect when their children needed assistance, be motivated to make that assistance, avoid contagious disease, acquire a lexicon, be motivated to nurse, select conspecifics as mates, select mates of the opposite sex, select mates of high reproductive value, induce potential mates to choose them, choose productive activities, balance when walking, avoid being bitten by venomous snakes, understand and make tools, avoid needlessly enraging others, interpret social situations correctly, help relatives, decide which foraging efforts have repaid the energy expenditure, perform anticipatory motion computation, inhibit one's mate from conceiving children by another, deter aggression, maintain friendships, navigate, recognize faces, recognize emotions, cooperate, and make effective trade-offs among many of these activities, along with a host of other tasks. To be a viable hypothesis about human psychological architecture, the design proposed must be able to meet both solvability and evolvability criteria: It must be able to solve the problems that we observe modern humans routinely solving and it must solve all the problems that were necessary for humans to survive and reproduce in ancestral environments. No existing version of Standard Model psychology can remotely begin to explain how humans perform these tasks.

Over the course of this chapter, we have touched on how domain-specific mechanisms are empirically better supported than domain-general mechanisms, on why domain-general mechanisms cannot give rise to routinely observable behavioral performances, on why domain-specific architectures are usually more functional than domain-general architectures, and, especially, on why it is implausible or impossible for predominantly content-independent, domain-general computational systems to perform the tasks necessary for survival and reproduction in natural environments. The main arguments that we have reviewed here (and elsewhere; see Cosmides & Tooby, 1987, 1992; Tooby & Cosmides, 1990b) are as follows:

1. In order to perform tasks successfully more often than chance, the architecture must be able to discriminate successful performance from unsuccessful performance. Because a domain-general architecture by definition has no built-in content-specific rules for judging what counts as error and success on different tasks, it must have a general rule. Unfortunately, there is no useable general cue or criterion for success or failure that can apply across domains. What counts as good performance for one task (e.g., depth perception) is completely different from what counts as good performance for other tasks (e.g., incest avoidance, immune regulation, avoiding contagion, imitating, eating). The only unifying element in discriminating success from failure is whether an act promotes fitness (design-propagation). But the relative fitness contribution of a given decision cannot be used as a criterion for learning or making choices because it is inherently unobservable by the individual (for discussion, see Cosmides & Tooby, 1987, 1992; Tooby & Cosmides, 1990b). Consequently, our evolved psychological architecture needs substantial built-in content-specific structure to discriminate adaptive success from failure. There needs to be at least as many different domain-specific psychological adaptations as there are evolutionarily recurrent functional tasks with different criteria for success.

2. As discussed at length, domain-general, content-independent mechanisms are inefficient, handicapped, or inert compared to systems that also include specialized techniques for solving particular families of adaptive problems. A specialized mechanism can make use of the enduring relationships present in the problem-domain or in the related features of the world by reflecting these content-specific relationships in its problem-solving structure. Such mechanisms will be far more efficient than general-purpose mechanisms, which must expend time, energy, and risk learning these relationships through "trial and possibly fatal error" (Shepard, 1987a).

3. Many problems that humans routinely solve are simply not solvable by any known general problem-solving strategy, as demonstrated by formal solvability analyses on language acquisition (e.g., Pinker, 1979, 1984, 1989, 1991; Pinker & Prince, 1988). We think that the class of such problems is large and, as discussed above, includes at a minimum all motivational problems.

4. Different adaptive problems are often incommensurate. They cannot, in principle, be solved by the same mechanism (Chomsky, 1980). To take a simple example, the factors that make a food nutritious are different from those that make a human a good mate or a savannah a good habitat. As Sherry and Schacter point out, "functional incompatibility exists when an adaptation that serves one function cannot, because of its specialized nature, effectively serve other functions. The specific properties of the adaptation that make it effective as a solution to one problem also render it incompatible with the demands of other problems" (1987, p. 439).

5. Many adaptive courses of action can be neither deduced nor learned by general criteria alone because they depend on statistical relationships that are unobservable to the relevant individual. For a content-independent system to learn a relationship, all parts of the relationship must be perceptually detectable. This is frequently not the case. Natural selection can "observe" relationships that exist between a sensory cue, a decision rule, and a fitness outcome that is inherently unobservable to the individual making the decision (Tooby & Cosmides, 1990b), as in the case of pregnancy sickness (Profet, 1988, this volume) or the Westermarck incest avoidance mechanism (Shepher, 1983; Wolf & Huang, 1980). This is because natural selection does not work by inference or computation: It took the real problem, "ran the experiment," and

retained those designs whose information-processing procedures led over thousands of generations to the best outcome. Natural selection, through incorporating content-specific decision rules, allows the organism to behave as if it could see and be guided by relationships that are perceptually undetectable and, hence, inherently unlearnable by any general-purpose system.

6. As discussed, the more generally framed problems are, the more computational systems suffer from combinatorial explosion, in which proliferating alternatives choke decision and learning procedures, bringing the system to a halt. If it were true that, as Rindos (1986, p. 315) puts the central tenet of the Standard Social Science Model, "the specifics that we learn are in no sense predetermined by our genes," then we could learn nothing at all.

7. Everything a domain-general system can do can be done as well or better by a system that also permits domain-specific mechanisms because selection can incorporate any successful domain-general strategies into an architecture without displacing its existing repertoire of domain-specific problem-solvers.

Without further belaboring the point, there is a host of other reasons why content-free, general-purpose systems could not evolve, could not manage their own reproduction, and would be grossly inefficient and easily outcompeted if they did. Equally important, these arguments apply not simply to the extreme limiting case of a completely content-free, domain-general architecture but to all Standard Model architectures, as conventionally presented. The single criterion that any proposed human psychological architecture must solve all the problems necessary to cause reliable reproduction under natural conditions is decisive. When taken seriously and considered carefully, it leads to the conclusion that the human psychological architecture must be far more frame-rich and permeated with content-specific structure than most researchers (including ourselves) had ever suspected.

The Content-Specific Road to Adaptive Flexibility

The ability to adjust behavior flexibly and appropriately to meet the shifting demands of immediate circumstances would, of course, be favored by selection, other things being equal. What organism would not be better off if it could solve a broader array of problems? Moreover, the psychologies of different species do differ in the breadth of situations to which they can respond appropriately, with humans acting flexibly to a degree that is zoologically unprecedented. Humans engage in elaborate improvised behaviors, from composing symphonies to piloting aircraft to rice cultivation, which collectively indicate a generality of achieved problem-solving that is truly breathtaking. Although many human acts do not successfully solve adaptive problems, enough do that the human population has increased a thousandfold in only a few thousand years. If general-purpose mechanisms are so weak, how can the variability of observed human behavior be reconciled with its level of functionality?

As discussed above, there is little in content-independent, domain-general strategies of problem-solving that by themselves can account for functional behavior, whether it is flexible or not. In contrast, specialized mechanisms can be very successful and powerful problem-solvers, but they achieve this at the price of addressing a narrower range of problems than a more general mechanism. If these were the only two alternatives, organisms would be limited to being narrow successes or broad failures, and the human case of broad adaptive flexibility could not be accounted for.

The solution to the paradox of how to create an architecture that is at the same time both powerful and more general is to bundle larger numbers of specialized mechanisms together so that in aggregate, rather than individually, they address a larger range of problems. Breadth is achieved not by abandoning domain-specific techniques but by adding more of them to the system. By adding together a face recognition module, a spatial relations module, a rigid object mechanics module, a tool-use module, a fear module, a social-exchange module, an emotion-perception module, a kin-oriented motivation module, an effort allocation and recalibration module, a child-care module, a social-inference module, a sexual-attraction module, a semantic-inference module, a friendship module, a grammar acquisition module, a communication-pragmatics module, a theory of mind module, and so on, an architecture gains a breadth of competences that allows it to solve a wider and wider array of problems, coming to resemble, more and more, a human mind. The more a system initially "knows" about the world and its persistent characteristics, and the more evolutionarily proven "skills" it starts out with, the more it can learn, the more problems it can solve, the more it can accomplish. In sharp contrast to the Standard Model, which views an absence of content-specific structure as a precondition for richly flexible behavior, the analysis of what computational systems actually need to succeed suggests the opposite: that the human capacity for adaptive flexibility and powerful problem-solving is so great precisely because of the number and the domain-specificity of the mechanisms we have. Again, this converges on William James's argument that humans have more "instincts" than other animals, not fewer (James, 1892; Symons, 1987).

Moreover, there are many reasons to think that the number of function-general mechanisms and function-specific mechanisms in an architecture are not inversely related in a zero-sum relationship, but are positively related (Rozin, 1976). Content-specialized mechanisms dissect situations, thereby creating a problem space rich with relevant relationships that content-independent mechanisms can exploit (e.g., Cosmides & Tooby, under review; Gigerenzer, Hoffrage & Kleinbölting, 1991; Shepard, 1987b). Thus, the more alternative content-specialized mechanisms an architecture contains, the more easily domain-general mechanisms can be applied to the problem spaces they create without being paralyzed by combinatorial explosion. Although domain-general mechanisms may be weak in isolation, they can valuably broaden the problem-solving range of an architecture if they are embedded in a matrix of adaptive specializations that can act as a guidance system (Rozin, 1976). For example, humans have powerful specialized social inference mechanisms that reflect the contentful structure of human metaculture, allowing humans to evaluate and interpret others' behaviors. This provides a foundation for the human-specific ability to imitate others (Galef, 1988; Meltzoff, 1988), greatly increasing the range of situations to which they can respond appropriately.

Therefore, what is special about the human mind is not that it gave up "instinct" in order to become flexible, but that it proliferated "instincts"—that is, content-specific problem-solving specializations—which allowed an expanding role for psychological mechanisms that are (relatively) more function-general. These are presently lumped into categories with unilluminating labels such as "the capacity for culture," "intelligence," "learning," and "rationality." It is time for the social sciences to turn from a nearly exclusive focus on these embedded, more function-general mechanisms to a wider view that includes the crucial, and largely neglected, superstructure of evolved functional specializations. Equally, we need to explore how the two classes of mechanisms are interwoven so that their combined interactive product is the zoologically unique yet evolutionarily patterned breadth of functional behaviors.

EVOLUTIONARY PSYCHOLOGY AND THE GENERATION OF CULTURE

The Pluralistic Analysis of Human Culture and Mental Organization

> Malinowski maintained that cultural facts are partly to be explained in psychological terms. This view has often been met with skepticism or even scorn, as if it were an easily exposed naive fallacy. What I find fallacious are the arguments usually leveled against this view. What I find naive is the belief that human mental abilities make culture possible and yet do not in any way determine its content and organization.
>
> —DAN SPERBER (1985, p. 73)

A large and rapidly growing body of research from a diversity of disciplines has shown that the content-independent psychology that provides the foundation for the Standard Social Science Model is an impossible psychology. It could not have evolved; it requires an incoherent developmental biology; it cannot account for the observed problem-solving abilities of humans or the functional dimension of human behavior; it cannot explain the recurrent patterns and characteristic contents of human mental life and behavior; it has repeatedly been empirically falsified; and it cannot even explain how humans learn their culture or their language. With the failure of Standard Model psychology, and the emergence of a domain-specific psychology, the remaining logic of the Standard Social Science Model also collapses.

In this chapter, we have limited ourselves to analyzing some of the defects of the Standard Social Science Model, concentrating on the untenability of the psychology that forms the foundation for its theory of culture. Along the way, we have touched on only a handful of the changes that an evolutionary psychological approach would introduce into the theoretical foundations of the social sciences. These and the remarks that follow should not, however, be mistaken for a substantive discussion of what a new theory of culture that was based on modern biology, psychology, and anthropology would look like. Still less should they be mistaken for a presentation of the mutually consistent conceptual framework (what we have been calling the Integrated Causal Model) that emerges when the various biological, behavioral, and social science fields are even partially integrated and reconciled. Because our argument has been narrowly focused on psychology, we have been unable to review or discuss the many critical contributions that have been made to this embryonic synthesis from evolutionary biology, anthropology, neurobiology, sociology, and many other fields. These must be taken up elsewhere. In particular, readers should be aware that the ideas underlying the Integrated Causal Model are not original with us: They are the collaborative product of hundreds of individual scholars working in a diverse array of fields over the last several decades.[7] Indeed, the collaborative dimension of this new framework is key. The eclipse of the Standard Model and the gradual emergence of its replacement has resulted from researchers exploring the natural causal connections that integrate separate fields (see, e.g., Barkow, 1989, on the importance of making psychology consistent with biology and anthropology consistent with psychology). The research program we and others are advocating is one of integration and consistency, not of psychological or biological reductionism. (See Atran, 1990; Daly & Wilson, 1988; and Symons, 1979, for examples of how such integrative approaches can be applied to specific problems.)

What does the rise of domain-specific psychology mean for theories of culture? By themselves, psychological theories do not and cannot constitute theories of culture. They only provide the foundations for theories of culture. Humans live and evolved in interacting networks that exhibit complex population-level dynamics, and so theories and analyses of population-level processes are necessary components for any full understanding of human phenomena. Nevertheless, increasing knowledge about our evolved psychological architecture places increasing constraints on admissible theories of culture. Although our knowledge is still very rudimentary, it is already clear that future theories of culture will differ significantly in a series of ways from Standard Social Science Model theories. Most fundamentally, if each human embodies an evolved psychological architecture that comes richly equipped with content-imparting mechanisms, then the traditional concept of culture itself must be completely rethought.

Culture has been the central concept of the Standard Social Science Model. According to its tenets, culture is a unitary entity that expresses itself in a trinity of aspects. (1) It is conceived as being some kind of contingently variable informational substance that is transmitted by one generation to another within a group: Culture is what is socially learned. (2) Because the individual mind is considered to be primarily a social product formed out of the rudimentary infant mind, all or nearly all adult mental organization and content is assumed to be cultural in derivation and substance: Culture is what is contentful and organized in human mental life and behavior. (3) Humans everywhere show striking patterns of local within-group similarity in their behavior and thought, accompanied by significant intergroup differences. The existence of separate streams of transmitted informational substance is held to be the explanation for these group patterns: Cultures are these sets of similarities, and intergroup or cross-location differences are called cultural differences.

In the absence of a content-free psychology, however, this trinity breaks into separate pieces because these three sets of phenomena can no longer be equated. We have already sketched out why the human mind must be permeated with content and organization that does not originate in the social world. This breaks apart any simple equivalence between the first two meanings of "culture." Nevertheless, even for those who admit that the mind has some content that is not socially supplied, the distribution of human within-group similarities and between-group differences remains the most persuasive element in the Standard Model analysis. These salient differences are taken to confirm that the socially learned supplies most of the rich substance of human life (see "The Standard Social Science Model", pp. 24–34). Because Standard Model advocates believe that a constant—our universal evolved architecture—cannot explain what varies, they can see no explanation for "cultural differences" other than differences in transmitted information.

Although this conclusion seems compelling, a simple thought experiment illustrates why it is unfounded. Imagine that extraterrestrials replaced each human being on earth with a state-of-the-art compact disk juke box that has thousands of songs in its repertoire. Each juke box is identical. Moreover, each is equipped with a clock, an automated navigational device that measures its latitude and longitude, and a circuit that selects what song it will play on the basis of its location, the time, and the date.

What our extraterrestrials would observe would be the same kind of pattern of within-group similarities and between-group differences observable among humans: In Rio, every juke box would be playing the same song, which would be different from the song that every juke box was playing in Beijing, and so on, around the world. Each juke box's "behavior" would be clearly and complexly patterned because each had been equipped with the same large repertoire of songs. Moreover, each juke box's behavior would change over time, because the song it plays is a function of the date and time, as well as of its location. Juke boxes that were moved from location to location would appear to adopt the local songs, sequences, and "fashions." Yet the generation of this distinctive, culture-like pattern involves no social learning or transmission whatsoever. This pattern is brought about because, like humans, the juke boxes (1) share a universal, highly organized, architecture, that (2) is designed to respond to inputs from the local situation (e.g., date, time, and location).

All humans share a universal, highly organized architecture that is richly endowed with contentful mechanisms, and these mechanisms are designed to respond to thousands of inputs from local situations. As a result, humans in groups can be expected to express, in response to local conditions, a variety of organized within-group similarities that are not caused by social learning or transmission. Of course, these generated within-group similarities will simultaneously lead to systematic differences between groups facing different conditions. To take a single example, differences in attitudes toward sharing between hunter-gatherer groups may be evoked by ecological variables (for discussion, see Cosmides & Tooby, this volume).

Thus, complex shared patterns that differ from group to group may be evoked by circumstances or may be produced by differential transmission. For this reason, the general concept of "culture" in the Standard Model sense is a conflation of *evoked culture* and *transmitted culture* (as well as of metaculture and other components). Given that the mind contains many mechanisms, we expect that both transmitted and evoked factors will play complementary roles in the generation of differentiated local cultures. The operation of a richly responsive psychology, plus the ability to socially "learn," can jointly explain far more about "culture" and cultural change than either can alone. For example, when members of a group face new and challenging circumstances (drought, war, migration, abundance), this may activate a common set of functionally organized domain-specific mechanisms, evoking a new set of attitudes and goals. The newly evoked psychological states will make certain new ideas appealing, causing them to spread by transmission, and certain old ideas unappealing, causing them to be discarded. In contrast, the Standard Model "do what your parents did" concept of culture is not a principle that can explain much about why cultural elements change, where new ones come from, why they spread, or why certain complex patterns (e.g., pastoralist commonalities) recur in widely separated cultures. Of course, many anthropologists implicitly recognize these points, but they need to make the links between the cultural processes they study and the underlying evolved content-organizing psychology they are assuming explicit. For example, economic and ecological anthropology, to be coherent, necessarily assume underlying content-specialized psychological mechanisms that forge relationships between environmental and economic variables and human thought and action.

It is especially important for post-Standard Model researchers to recognize that the environmental factors that cause contentful mental and behavioral organization to be expressed are not necessarily the processes that constructed that organization. In the case of the juke box, it would be a mistake to attribute the organized content manifest in the music to the environmental stimuli (i.e., the location, date, and time) that caused one song to be played rather than another. The stimuli did not compose the music; they merely caused it to be expressed. Similarly, our psychological architectures come equipped with evolved contentful organization, which can remain latent or become activated depending on circumstances and which may vary in its expression according to procedures embodying any degree of complexity. Because our psychological architecture is complexly responsive, the Standard Model practice of equating the variable with the learned is a simple non sequitur. The claim that some phenomena are "socially constructed" only means that the social environment provided some of the inputs used by the psychological mechanisms of the individuals involved.

In short, observations of patterns of similarities and differences do not establish that the substance of human life is created by social learning. In any specific case, we need to map our evolved psychological architecture to know which elements (if any) are provided by transmission, which by the rest of the environment, which by the architecture, and how all these elements causally interact to produce the phenomenon in question. Admittedly, the juke box thought experiment is an unrealistically extreme case in which a complex, functionally organized, content-sensitive architecture internalizes no transmitted informational input other than an environmental trigger. But this case is simply the mirror image of the SSSM's extreme view of the human mind as a content-free architecture where everything is provided by the internalization of transmitted input. Our central point is that in any particular domain of human activity, the programming that gives our architecture its ability to contingently respond to the environment may or may not be designed to take transmitted representations as input. If it does, it may mix in content derived from its own structure and process the resulting representations in complex and transformative ways. The trinity of cultural phenomena can no longer be equated with one another. Our complex content-specific psychological architecture participates in the often distinct processes of generating mental content, generating local similarities and between-group differences, and generating what is "transmitted." Indeed, it also participates in the complex process of internalizing what others are "transmitting."

Inferential Reconstruction and Cultural Epidemiology

The Standard Social Science Model has been very effective in promulgating the unity of the trinity. The socially learned, the set of within-group commonalities and between-group differences, and the contentful organization of human mental and social life have been so thoroughly conflated that it is difficult to speak about human phenomena without using the word *culture*. For this reason, we will use *culture* to refer to any mental, behavioral, or material commonalities shared across individuals, from those that are shared across the entire species down to the limiting case of those shared only by a dyad, regardless of why these commonalities exist. When the causes of the commonality can be identified, we will use a qualifier, such as "evoked."

So things that are cultural in the sense of being organized, contentful, and shared among individuals may be explained in a number of different ways. Within-group commonalities may have been evoked by common circumstances impacting universal architectures. An even larger proportion of organized, contentful, and shared phenomena may be explained as the expression of our universal psychological and physiological architectures in interaction with the recurrent structure of the social or non-social world—what we earlier called metaculture. Metaculture includes a huge range of psychological and behavioral phenomena that under Standard Model analyses have been invisible or misclassified (D. E. Brown, 1991). Because the Standard Model attributed everything that was contentful and recurrent to some form of social learning, it misinterpreted phenomena such as anger upon deliberate injury, grief at a loss, the belief that others have minds, treating species as natural kinds, social cognition about reciprocation, or the search for food when hungry as socially manufactured products.

Nevertheless, after the evoked and the metacultural have been excluded, there still remains a large residual category of representations or regulatory elements that reappear in chains from individual to individual—"culture" in the classic sense. In giving up the Standard Social Science Model, we are not abandoning the classic concept of culture. Instead, we are attempting to explain what evolved psychological mechanisms cause it to exist. That way we can get a clearer causal understanding of how psychological mechanisms and populational processes shape its content, and thereby restrict its explanatory role in social theory to the phenomena that it actually causes.

This subset of cultural phenomena is restricted to (1) those representations or regulatory elements that exist originally in at least one mind that (2) come to exist in other minds because (3) observation and interaction between the source and the observer cause inferential mechanisms in the observer to recreate the representations or regulatory elements in his or her own psychological architecture. In this case, the representations and elements inferred are contingent: They could be otherwise and, in other human minds, they commonly are otherwise. Rather than calling this class of representations "transmitted" culture, we prefer terms such as *reconstructed culture*, *adopted culture*, or *epidemiological culture*. The use of the word "transmission" implies that the primary causal process is located in the individuals from whom the representations are derived. In contrast, an evolutionary psychological perspective emphasizes the primacy of the psychological mechanisms in the learner that, given observations of the social world, inferentially reconstruct some of the representations existing in the minds of the observed. Other people are usually just going about their business as they are observed, and are not necessarily intentionally "transmitting" anything.

More precisely, an observer (who, for expository simplicity, we will call the "child") witnesses some finite sample of behavior by others (e.g., public representations, such as utterances or other communicative acts; people going about their affairs; people responding to the child's behavior). The task of the mechanisms in the child is to (1) reconstruct within themselves on the basis of these observations a set of representations or regulatory elements that (2) are similar enough to those present in the humans she lives among so that (3) the behavior her mechanisms generate can be adaptively coordinated with other people and her habitat. Thus, the problem of learning "culture" lies in deducing the hidden representations and regulatory elements embedded in others' minds that are responsible for generating their behavior. To the extent that the child's mechanisms make mistakes—and mistakes are endemic—and reconstruct the wrong underlying representations and regulatory elements, she will not be able to predict other people's behavior, interpret their transactions with one another in the world, imitate them, communicate with them, cooperate with them, help them, or even anticipate or avoid their hostile and exploitive actions.

Why did ancestral hominid foragers evolve mechanisms that allowed them to reconstruct the representations present in the minds of those around them? Leaving aside the question of their costs and limitations, the advantage of such mechanisms is straightforward. Information about adaptive courses of action in local conditions is difficult and costly to obtain by individual experience alone. Those who have preceded an individual in a habitat and social environment have built up in their minds a rich store of useful information. The existence of such information in other minds selected for specialized psychological adaptations that were able to use social observations to reconstruct some of this information within one's own mind (e.g., Boyd & Richerson, 1985; Tooby & DeVore, 1987). By such inferential reconstruction, one individual was able to profit from deducing what another already knew. When such inferential reconstruction becomes common enough in a group, and some representations begin to be stably re-created in sequential chains of individuals across generations, then the structure of events begins to warrant being called "cultural."

As discussed earlier, this task of reconstruction would be unsolvable if the child did not come equipped with a rich battery of domain-specific inferential mechanisms, a faculty of social cognition, a large set of frames about humans and the world drawn from the common stock of human metaculture, and other specialized psychological adaptations designed to solve the problems involved in this task (see, e.g., Boyer, 1990; Sperber, 1985, 1990; Sperber & Wilson, 1986; Tooby & Cosmides, 1989a). Consequently, epidemiological culture is also shaped by the details of our evolved psychological organization. Thus, there is no radical discontinuity inherent in the evolution of "culture" that removes humans into an autonomous realm. Mechanisms designed for such inferential reconstruction evolved within a pre-existing complex psychological architecture and depended on this encompassing array of content-structuring mechanisms to successfully interpret observations, reconstruct representations, modify behavior, and so on. Solving these inferential problems is not computationally trivial, and other species, with a few possible minor exceptions, are not equipped to perform this task to any significant degree (Galef, 1988).

Moreover, outside of contexts of competition, knowledge is not usually devalued by being shared. Consequently, to the substantial extent that individuals in a hunter-gatherer group had interests in common and had already evolved mechanisms for inferential reconstruction, selection would have favored the evolution of mechanisms that facilitated others' inferences about one's own knowledge (as, for example, by communicating or teaching). The mutual sharing of valuable knowledge and discoveries has a dramatic effect on the usefulness of mechanisms that attempt to adaptively adjust behavior to match local conditions (Boyd & Richerson, 1985; Tooby & DeVore, 1987). Because of combinatorial explosion, knowledge of successful local techniques is precious and hard to discover, but relatively cheap to share (once again, ignoring the cost of the psychological mechanisms that facilitate or perform such sequential reconstruction). Within limits, this creates economies of scale: The greater the number of individuals who participate in the system of knowledge sharing, (1) the larger the available pool of knowledge will be, (2) the more each individual can derive from the pool, (3) the more advantageous reconstructive adaptations will be, and (4) the more it would pay to evolve knowledge-dependent mechanisms that could exploit this set of local representations to improvise solutions to local problems. This collaborative information-driven approach to the adaptive regulation of behavior can be thought of as the "cognitive niche" (Tooby & DeVore, 1987). The mutual benefit of such knowledge sharing led to the co-evolution of sets of adaptations, such as language, elaborated communicative emotional displays, and pedagogy that coordinate specialized processes of inferential reconstruction with the specialized production of behaviors designed to facilitate such reconstruction (e.g., Ekman, 1984; Freyd, 1983; Fridlund, in press; Pinker & Bloom, this volume; Premack, in prep.; Sperber & Wilson, 1986).

Because reconstructive inferences are often erroneous and what others "know" is often of dubious quality or irrelevant, such inferential processes could not have evolved without adaptations that assessed to some degree the value of such reconstructed knowledge and how it fits in with knowledge derived from other sources. If a representation is easy to successfully reconstruct and is evaluated positively, then it will tend to spread through inter-individual chains of inference, becoming widely shared. If it is difficult to reconstruct or evaluated as not valuable, it will have only a restricted distribution or will disappear (Sperber, 1985, 1990; Sperber & Wilson, 1986). This evaluation process gives sequentially reconstructed culture its well-known, if partial, parallels to natural selection acting on genes; that is, the selective retention and accumulation of favored variants over time (e.g., Barkow, 1989; Boyd & Richerson, 1985; Campbell, 1965, 1975; Cavalli-Sforza & Feldman, 1981; Dawkins, 1976; Durham, 1991; Lumsden & Wilson, 1981). Moreover, these psychological mechanisms endow sequentially reconstructed culture with its epidemiological character, as a dynamically changing distribution of elements among individuals living in populations over time. As Sperber says, "Cultural phenomena are ecological patterns of psychological phenomena. They do not pertain to an autonomous level of reality, as anti-reductionists would have it, nor do they merely belong to psychology as reductionists would have it" (1985, p. 76).

The more widely shared an element is, the more people are inclined to call it "cultural," but there is no natural dividing point along a continuum of something shared between two individuals to something shared through inferential reconstruction by the entire human species (Sperber, 1985, 1990). The Standard Model practice of framing "cultures" as sets of representations homogeneously shared by nearly all members of discrete and bounded "groups" does not capture the richness of the ecological distribution of these psychological elements, which cross-cut each other in a bewildering variety of fractal patterns. Language boundaries do not correspond to subsistence practice boundaries, which do not correspond to political boundaries or to the distribution of rituals (for discussion, see Campbell & LeVine, 1972). Within groups, representations occur with all kinds of different frequencies, from beliefs passed across generations by unique dyads, such as shamanistic knowledge or mother-daughter advice, to beliefs shared by most or all members of the group.

The belief that sequentially reconstructed representations exist primarily in bounded cells called "cultures" derives primarily from the distribution of language boundaries, which do happen to be distributed more or less in this fashion. As Pinker and Bloom (this volume) point out, communication protocols can be arbitrary, but must be shared between sender and receiver to be functional. The benefit of learning an arbitrary linguistic element is proportional to how widely it is distributed among a set of interacting individuals. Therefore, well-designed language acquisition mechanisms distributed among a local set of individuals will tend to converge on a relatively homogeneous set of elements: It is useful for all local individuals to know the same local language. Although there are other reasons why reconstructed elements may show sharp coordinated boundaries (e.g., ethnocentrism, common inheritance, sharp habitat boundaries, geographical barriers, and so on), most classes of representations or regulatory elements dynamically distribute themselves according to very different patterns, and it is probably more accurate to think of humanity as a single interacting population tied together by sequences of reconstructive inference than as a collection of discrete groups with separate bounded "cultures."

Finally, the reconstruction of regulatory elements and representations in a psychological architecture should not be thought of as a homogeneous process. Given that our minds have a large set of domain-specific mechanisms, it seems likely that different mechanisms would be selected to exploit social observations in different ways and have quite distinct procedures for acquiring, interpreting, and using information derived from the social world. Certainly, the language acquisition device appears to have its own special properties (e.g., Pinker, 1989), and many other domains appear to follow their own special rules (Carey & Gelman, 1991). It seems unlikely in the extreme that the different modules underlying mate preferences (Symons, 1979; Buss, in prep.), food preferences (e.g., Galef, 1990), display rules for emotional expression (Ekman, 1982), fears (Cook et al., 1986) and so on, process social observations according to a single unitary process. Moreover, to the extent that there may exist a large, potentially interacting store of representations in the mind (see, e.g., Fodor 1983), nothing in the psychological architecture necessarily segregates off representations derived through "epidemiological culture" from representations and regulatory elements derived from other sources.

This brief sketch suggests a few of the features future theories of culture may incorporate, once the Standard Model concept of learning is discarded. These are organized by two themes. First, what is presently attributed to "culture" will come to be pluralistically explained as metaculture, evoked culture, epidemiological culture, and individual mental contents that are internally generated and not derived through inferential reconstruction (see table below). Second, with the fall of content-independent learning, the socially constructed wall that separates psychology and anthropology (as well as other fields) will disappear. The heterogeneous mechanisms comprising our evolved psychological architecture participate inextricably in all cultural and social phenomena and, because they are content-specialized, they impart some contentful patterning to them. Indeed, models of psychological mechanisms, such as social exchange, maternal attachment, sexual attraction, sexual jealousy, the categorization of living kinds, and so on, are the building blocks out of which future theories of culture will, in part, be built (Sperber, 1990; Tooby & Cosmides, 1989a). By no means do

Table 1.1 Decomposing the Traditional Concept of Culture

Metaculture	Evoked culture	Epidemiological culture
Mechanisms functionally organized to use cross-cultural regularities in the social and nonsocial environment give rise to panhuman mental contents and organization.	Alternative, functionally organized, domain-specific mechanisms are triggered by local circumstances; leads to within-group similarities and between-group differences.	Observer's inferential mechanisms construct representations similar to those present in others; domain-specific mechanisms influence which representations spread through a population easily and which do not.

we deny or minimize the existence of emergent phenomena, such as institutions, or the fact that population-level processes alter the epidemiological distribution of cultural contents over time. The point is simply that cultural and social phenomena can never be fully divorced from the structure of the human psychological architecture or understood without reference to its design.

The Twilight of Learning as a Social Science Explanation

Advocates of the Standard Social Science Model have believed for nearly a century that they have a solid explanation for how the social world inserts organization into the psychology of the developing individual. They maintain that structure enters from the social (and physical) world by the process of "learning"—individuals "learn" their language, they "learn" their culture, they "learn" to walk, and so on. All problems—whether they are long-enduring adaptive problems or evolutionarily unprecedented problems—are solved by "learning." In the intellectual communities dominated by the SSSM, learning has been thought to be a powerful explanation for how certain things come about, an explanation that is taken to refer to a well-understood and well-specified general process that someone (i.e., the psychological community) has documented. For this reason, "learning," and such common companion concepts as "culture," "rationality," and "intelligence," is frequently invoked as an alternative explanation to so-called "biological" explanations (e.g., sexual jealousy did not evolve, it is learned from culture; one doesn't need to explain how humans engage in social exchange: They simply used their "reason" or "intelligence").

Of course, as most cognitive scientists know (and all should), "learning"—like "culture," "rationality," and "intelligence"—is not an explanation for anything, but is rather a phenomenon that itself requires explanation (Cosmides & Tooby, 1987; Tooby & Cosmides, 1990b). In fact, the concept of "learning" has, for the social sciences, served the same function that the concept of "protoplasm" did for so long in biology. For decades, biologists could see that living things were very different from nonliving things, in that a host of very useful things happened inside of living things that did not occur inside of the nonliving (growth, the manufacture of complex chemicals, the assembly of useful structures, tissue differentiation, energy production, and so on). They had no idea what causal sequences brought these useful results about. They reified this unknown functionality, imagining it to be a real substance, and called it "protoplasm," believing it to be the stuff that life was made of. It was a name given to a mystery, which was then used as an explanation for the functional results that remained in genuine need of explanation. Of course, the concept of protoplasm eventually disappeared when molecular biologists began to determine the actual causal sequences by which the functional business of life was transacted. "Protoplasm" turned out to be a heterogeneous collection of incredibly intricate functionally organized structures and processes—a set of evolved adaptations, in the form of microscopic molecular machinery such as mitochondria, chloroplasts, the Krebs cycle, DNA transcription, RNA translation, and so on.

Similarly, human minds do a host of singularly useful things, by which they coordinate themselves with things in the world: They develop skill in the local community's language; upon exposure to events they change behavior in impressively functional ways; they reconstruct in themselves knowledge derived from others; they adopt the practices of others around them; and so on. Psychologists did not know what causal sequences brought these useful results about. They reified this unknown functionality, imagining it to be a unitary process, and called it "learning." "Learning" is a name given to the unknown agent imagined to cause a large and heterogeneous set of functional outcomes. This name was (and is) then used as an explanation for results that remained in genuine need of explanation. We expect that the concept of learning will eventually disappear as cognitive psychologists and other researchers make progress in determining the actual causal sequences by which the functional business of the mind is transacted. Under closer inspection, "learning" is turning out to be a diverse set of processes caused by a series of incredibly intricate, functionally organized cognitive adaptations, implemented in neurobiological machinery (see, e.g., Carey & Gelman, 1991; Gallistel, 1990; Pinker 1989, 1991; Real, 1991). With slight qualifications about the exact contexts of usage, similar things could be said for "culture," "intelligence," and "rationality." The replacement of the concept of protoplasm with a real understanding of the vast, hidden, underlying worlds of molecular causality has transformed our understanding of the world in completely unexpected ways, and we can only anticipate that the same will happen when "learning" is replaced with knowledge.

NOTES

1. Philosophers and historians of science sometimes use the phrase "unity of science" as a term of art to refer to an axiomatized reductionistic approach to science. We are not using it in this sense, but rather in its common sense meaning of mutual consistency and relevance.

2. For a very illuminating discussion of how various "tools" (from wax tablets to general-purpose computers to methods of statistical inference) have served as metaphors for the structure of the human mind, see "From tools to theories: A heuristic of discovery in cognitive psychology" by Gerd Gigerenzer (1991a).

3. Nevertheless, given the sorry history of the social sciences, in which every new research program has loosed a deluge of half-baked nostrums and public policy prescriptions on generally unconsenting victims, an important caution is in order. The human mind is the most complex phenomenon humans have encountered and research into it is in its infancy. It will be a long time before scientific knowledge of the aggregation of mechanisms that comprise the human psychological architecture is reliable enough and comprehensive enough to provide the basis for confident guidance in matters of social concern.

4. Finding invariances in cognitive architecture should, in turn, help neuroscientists in their search for the neural mechanisms that implement them. In neuroscience (as everywhere else), researchers practicing unguided empiricism rapidly become lost in a forest of complex phenomena without knowing how to group results so that the larger scale functional systems can be recognized. The evolved functional organization of cognitive programs offers an independently discoverable, intelligible and privileged system for ordering and relating neuroscientific phenomena: The brain itself evolved to solve adaptive problems, and its particular systems of organization were selected for because they physically carried out information-processing procedures that led to the adaptive regulation of behavior and physiology.

5. In fact, the actual distribution and character of genetic variation fits well with theories that explain it as mutations, selectively neutral variants (Nei, 1987), quantitative variation and, especially, as the product of parasite-driven frequency-dependent selection for biochemical individuality (see e.g., Clarke, 1979; Hamilton & Zuk, 1982; Tooby, 1982). Briefly, the more biochemically individualized people in a community are, the more difficult it is for disease micro-organisms adapted to one individual's biochemistry to contagiously infect neighboring humans. Thus, so long as this variation doesn't disrupt the higher level uniformity of functional

integration in complex adaptations, selection will favor the maintenance of individualizing low level protein variability in the tissues attacked by parasites, generating a large reservoir of a restricted kind of genetic variability (Tooby, 1982; Tooby & Cosmides, 1990a).

6. Researchers in phenomena-oriented fields, such as personality and social psychology, usually avoid formal analysis, so one is never sure what kind of computational mechanisms are hypothesized to generate performance. And, although many who do mainstream cognitive research on memory, problem-solving, and decision-making make formal models, these are devised to predict performance on artificial, evolutionarily unprecedented tasks, such as chess-playing or cryptarithmetic. Naturally, the phenomena-oriented researchers reject the latter's formal analyses as sterile and irrelevant to their interests, whereas those who do rigorous analyses of artificial tasks regard the phenomena-oriented researchers as wooly headed and unscientific. But neither community considers the possibility that the assumptions of the SSSM might be the problem.

7. See, for example, Alexander, 1979; Atran, 1990; Barkow, 1973, 1978, 1989; Berlin & Kay, 1969; Boyer, 1990; Bowlby, 1969; Boyd & Richerson, 1985; D. E. Brown, 1991; Buss, 1989, 1991; Campbell, 1965, 1975; Cavalli-Sforza & Feldman, 1981; Chagnon, 1988, Chagnon & Irons, 1979; Cheney & Seyfarth, 1990; Chomsky, 1959, 1975, 1980; Cloak, 1975; Clutton-Brock & Harvey, 1979; Crawford & Anderson, 1989; Crawford, Smith, & Krebs, 1987; Daly & Wilson, 1981, 1984a, 1987a, 1988; Dawkins, 1976, 1982, 1986; Dennett, 1987; Dickemann, 1981; Durham, 1991; Eibl-Eibesfeldt, 1975; Ekman, 1982; Fodor, 1983; Fox, 1971; Freeman, 1983; Freyd, 1987; Fridlund, in press; Galef, 1990; Gallistel, 1990; Garcia, 1990; Gazzaniga, 1985; Gelman, 1990a; Ghiselin, 1973; Glantz & Pearce, 1989; Hamilton, 1964; Hinde, 1987; Hrdy, 1977; Irons, 1979; Jackendoff 1992; Kaplan & Hill, 1985; Keil, 1989; Konner, 1982; Krebs & Davies, 1984; Laughlin & d'Aquili, 1974; Lee & DeVore, 1968, 1976; Leslie, 1987, 1988; Lockard, 1971; Lorenz, 1965; Lumsden & Wilson, 1981; Marr, 1982; Marshall, 1981; Maynard Smith, 1982; Nesse, 1991; Pinker, 1989; Real, 1991; Rozin, 1976; Rozin & Schull, 1988; Seligman & Hager, 1972; Shepard, 1981, 1984, 1987a; Shepher, 1983; Sherry & Schacter, 1987; Spelke, 1990; Sperber, 1974, 1982, 1985, 1986, 1990; Sperber & Wilson, 1986; Staddon, 1988; Symons, 1979, 1987, 1989; N. W. Thornhill, 1991; R. Thornhill, 1991; Tiger, 1969; Tinbergen, 1951; Trivers, 1971, 1972; Van den Berghe, 1981; de Waal, 1982; Williams, 1966, 1985; Wilson, 1971, 1975, 1978; Wilson & Daly, 1987; Wolf & Huang, 1980; Wrangham, 1987.

ACKNOWLEDGMENTS

We are deeply indebted to Jerry Barkow, Don Brown, David Buss, Nap Chagnon, Martin Daly, Lorraine Daston, Irven DeVore, Gerd Gigerenzer, Steve Pinker, Roger Shepard, Dan Sperber, Don Symons, Valerie Stone, and Margo Wilson for many enlightening discussions of the issues addressed in this chapter. We are especially grateful to Jerry Barkow, Lorraine Daston, and Steve Pinker for their incisive comments on prior drafts. This chapter was prepared, in part, while the authors were Fellows at the Center for Advanced Study in the Behavioral Sciences. We are grateful for the Center's support, as well as that provided by the Gordon P. Getty Trust, the Harry Frank Guggenheim Foundation, NSF Grant BNS87–00864 to the Center, NSF Grant BNS9157–449 to John Tooby, and the McDonnell Foundation. We would also like to thank Peter Weingart and the Zentrum fur interdisziplinaire Forschung (ZiF) of the University of Bielefeld, Germany, for its support during the latter phases of this project. We gratefully acknowledge the many late course corrections supplied by Rob Boyd, Bill Durham, Peter Hejl, Hans Kummer, Sandy Mitchell, Peter Molnar, Pete Richerson, Neven Sesardic, Joan Silk, Nancy Thornhill, Peter Weingart, and the rest of the ZiF community. Most importantly, we would like to thank Ranoula, Gerd, Thalia, Aubri, and Nancy, without whom this paper might never have been completed.

REFERENCES

Alexander, R. D. (1974). The evolution of social behavior. *Annual Review of Ecology & Systematics, 5,* 325–383.

Alexander, R. D. (1979). *Darwinsim and human affairs.* Seattle: University of Washington Press.

Alexander, R. D. (1987). *The biology of moral systems.* Hawthorne, NY: Aldine.

Ardrey, R. (1966). *The territorial imperative.* New York: Atheneum.

Arnold, M. B. (1960). *Emotion and personality.* New York: Columbia University Press.

Arnold, M. B. (1968). *The nature of emotion.* London: Penguin Books.

Astington, J. W., Harris, P. L., & Olson, D. R. (Eds.). (1988). *Developing theories of mind.* Cambridge, UK: Cambridge University Press.

Atran, S. (1990). *The cognitive foundations of natural history.* New York: Cambridge University Press.

Avis, J., & Harris, P. L. (in press). Belief-desire reasoning among Baka children: Evidence for a universal conception of mind. *Child Development.*

Ayala, F. (1976). *Molecular evolution.* Sunderland, MA: Sinauer.

Barkow, J. H. (1973). Darwinian psychological anthropology: A biosocial approach. *Current Anthropology, 14(4),* 373–388.

Barkow, J. H. (1978). Culture and sociobiology. *American Anthropologist, 80(1),* 5–20.

Barkow, J. H. (1989). *Darwin, sex, and status: Biological approaches to mind and culture.* Toronto: University of Toronto Press.

Baron-Cohen, S., Leslie, A., & Frith, U. (1985). Does the autistic child have a "theory of mind"? *Cognition, 21,* 37–46.

Beach, F. A. (1950). The snark was a boojum. *American Psychologist, 5,* 115–124.

Benedict, R. (1934/1959). Anthropology and the abnormal. In M. Mead (Ed.), *An anthropologist at work: Writings of Ruth Benedict.* Boston: Houghton-Mifflin, pp. 262–283.

Berlin, B., Breedlove, D., & Raven, P. (1973). General principles of classification and nomenclature in folk biology. *American Anthropologist, 75,* 214–242.

Berlin, B., & Kay, P. (1969). *Basic color terms: Their universality and evolution.* Berkeley: University of California Press.

Bizzi, E., Mussa-Ivaldi, F. A., & Giszter, S. (1991). Computations underlying the execution of movement: A biological perspective. *Science, 253,* 287–291.

Bloch, M. (1977). The past and the present in the present. *Man, 12,* 278–292.

Block, N. (1980). What is functionalism? In N. Block (Ed.), *Readings in philosophy of psychology.* Cambridge, MA: Harvard University Press.

Boden, M. (1977). *Artificial intelligence and natural man.* New York: Basic Books.

Bowlby, J. (1969). *Attachment.* New York: Basic Books.

Boyer, P. (1990). *Tradition as truth and communication: Cognitive description of traditional discourse.* New York: Cambridge University Press.

Boyd, R., & Richerson, P. J. (1985). *Culture and the evolutionary process.* Chicago: University of Chicago Press.

Breland, K., & Breland, M. (1961). The misbehavior of organisms. *American Psychologist, 16,* 681–684.

Brown, A. (1990). Domain-specific principles affect learning and transfer in children. *Cognitive Science, 14,* 107–133.

Brown, D. E. (1991). *Human universals.* New York: McGraw-Hill.

Brown, F. M. (1987). *The frame problem in artificial intelligence.* Los Altos, CA: Morgan Kaufman.

Bull, J. J. (1983). *The evolution of sex determining mechanisms.* Menlo Park, CA: Benjamin/Cummings.

Buss, D. M. (1987). Sex differences in human mate selection criteria: An evolutionary perspective. In *Sociobiology and psychology,* C. B. Crawford, M. F. Smith, & D. L. Krebs (Eds.). Hillsdale, NJ: Erlbaum.

Buss, D. M. (1989). Sex differences in human mate preferences: Evolutionary hypotheses tested in 37 cultures. *Behavioral and Brain Sciences, 12,* 1–49.

Buss, D. M. (1991). Evolutionary personality psychology. *Annual Review of Psychology 42:* 459–491.

Buss, D. M. (in prep.). *Sexual strategies.* New York: Basic Books.

Campbell, D. T. (1965). Variation and selective retention in sociocultural evolution. In R. W. Mack, G. I. Blanksten, & H. R. Barringer (Eds.), *Social change in underdeveloped areas: A reinterpretation of evolutionary theory.* Cambridge, MA: Schenkman.

Campbell, D. T. (1975). On the conflicts between biological and social evolution and between psychology and moral tradition. *American Psychologist, 30,* 1103–1126.

Campbell, D. T., & Levine, R. A. (1972). *Ethnocentrism: Theories of conflict, ethnic attitudes, and group behavior.* New York: Wiley.

Carey, S. (1982). Semantic development: The state of the art. In E. Wanner & L. Gleitman (Eds.), *Language acquisition: State of the art.* London: Cambridge University Press.

Carey, S. (1985a). Constraints on semantic development. In J. Mehler & R. Fox (Eds.), *Neonate cognition.* Hillsdale, NJ: Erlbarum, pp. 381–398.

Carey, S. (1985b). *Conceptual change in childhood.* Cambridge, MA: MIT Press.

Carey, S., & Gelman, R. (Eds.). (1991). *The epigenesis of mind: Essays on biology and cognition.* Hillsdale, NJ: Erlbaum.

Cavalli-Sforza, L. L., & Feldman, M. W. (1981). *Cultural transmission and evolution: A quantitative approach.* Princeton: Princeton University Press.

Chagnon, N. (1988). Life histories, blood revenge, and warfare in a tribal population. *Science, 239,* 985–992.

Chagnon, N., & Irons, W. (Eds.). (1979). *Evolutionary biology and human social behavior: An Anthropological perspective.* North Scituate, MA: Duxbury Press.

Chance, M.R.A. (1980). An ethological assessment of emotion. In R. Plutchik & H. Kellerman (Eds.), *Emotion: Theory, research, and experience.* New York: Academic Press, pp. 81–111.

Charnov, E. L. (1976). Optimal foraging: The marginal value theorem. *Theoretical Population Biology, 9,* 129–136.

Cheney, D. L., & Seyfarth, R. (1990). *How monkeys see the world.* Chicago: University of Chicago Press.

Cheng, P., & Holyoak, K. (1989). On the natural selection of reasoning theories. *Cognition, 33,* 285–313.

Chomsky, N. (1957). *Syntactic structures.* The Hague: Mouton & Co.

Chomsky, N. (1959). Review of Skinner's "Verbal Behavior." *Language 35,* 26–58.

Chomsky, N. (1975). *Reflections on language.* New York: Random House.

Chomsky, N. (1980). *Rules and representations.* New York: Columbia University Press.

Chorover, S. (1979). *From genesis to genocide.* Cambridge, MA: MIT Press.

Clarke, B. (1979). The evolution of genetic diversity. *Proceedings of the Royal Society, London, B, 205,* 453–474.

Cloak, F. T. (1975). Is a cultural ethology possible? *Human Ecology, 3,* 161–182.

Clutton-Brock, T. H., & Harvey, P. H. (1979). Comparison and adaptation. *Proceedings of the Royal Society, London, B, 205,* 547–565.

Cook, E. W., III, Hodes, R. L., & Lang, P. J. (1986). Preparedness and phobia: Effects of stimulus content on human visceral conditioning. *Journal of Abnormal Psychology, 95,* 195–207.

Cosmides, L. (1989). The logic of social exchange: Has natural selection shaped how humans reason? Studies with the Wason selection task. *Cognition, 31,* 187–276.

Cosmides, L., & Tooby, J. (1981). Cytoplasmic inheritance and intragenomic conflict. *Journal of Theoretical Biology, 89,* 83–129.

Cosmides, L., & Tooby, J. (1987). From evolution to behavior: Evolutionary psychology as the missing link. In J. Dupre (Ed.), *The latest on the best: Essays on evolution and optimality.* Cambridge, MA: MIT Press, pp. 277–306.

Cosmides, L., & Tooby, J. (1989). Evolutionary psychology and the generation of culture. Part II: A computational theory of social exchange. *Ethology & Sociobiology, 10,* 51–97.

Cosmides, L., & Tooby, J. (1992). From evolution to adaptations to behavior: Toward an integrated evolutionary psychology. In Roderick Wong (Ed.), *Biological perspectives on motivated and cognitive activities.* Norwood, NJ: Ablex.

Cosmides, L., & Tooby, J. (under review). Are humans good intuitive statisticians after all? Rethinking some conclusions of the literature on judgment under uncertainty.

Crawford, C. B., Smith, M. F., & Krebs, D. L. (Eds.). (1987). *Sociobiology and psychology.* Hillsdale, NJ: Erlbaum.

Crawford, C. B., & Anderson, J. L. (1989). Sociobiology: An environmentalist discipline? *American Psychologist 44(12):*1449–1459.

Daly, M. (1982). Some caveats about cultural transmission models. *Human Ecology, 10,* 401–408.

Daly, M., & Wilson, M. (1981). Abuse and neglect of children in evolutionary perspective. In R. D. Alexander & D. W. Tinkle (Eds.), *Natural selection and social behavior.* New York: Chiron.

Daly, M., & Wilson, M. (1982). Homicide and kinship. *American Anthropologist, 84:*372–378.

Daly, M., & Wilson, M..(1984a). *Sex, evolution and behavior.* Second Edition. Boston: Willard Grant.

Daly, M., & Wilson, M. (1984b). A sociobiological analysis of human infanticide. In: S. Hrdy & G. Hausfater (Eds.). *Infanticide: Comparative and evolutionary perspectives.* New York: Aldine, pp. 487–502.

Daly, M., & Wilson, M. (1987a). Evolutionary psychology and family violence. In C. B. Crawford, M. F. Smith & D. L. Krebs (Eds.), *Sociobiology and psychology.* Hillsdale, NJ: Erlbaum.

Daly, M., & Wilson, M. (1987b). The Darwinian psychology of discriminative parental solicitude. *Nebraska Symposium on Motivation. 35,* 91–144.

Daly, M., & Wilson, M. (1988). *Homicide.* New York: Aldine.

Daly, M., Wilson, M., & Weghorst, S. J. (1982). Male sexual jealousy. *Ethology and Sociobiology, 3,* 11–27.

Darwin, C. (1859). *On the origin of species.* London: Murray.

Darwin, C. (1871). *The descent of man and selection in relation to sex.* London: Murray.

Darwin, C. (1872). *The expression of emotion in man and animals.* London: Murray.

Dawkins, R. (1976). *The selfish gene.* New York: Oxford University Press.

Dawkins, R. (1982). *The extended phenotype.* San Francisco: W. H. Freeman.

Dawkins, R. (1986). *The blind watchmaker.* New York: Norton.

Dennett, D. C. (1987). *The intentional stance.* Cambridge, MA: MIT Press.

Dickemann, M. (1981). Paternal confidence and dowry competition: A biocultural analysis of purdah. In R. D. Alexander & D. W. Tinkle (Eds.), *Natural selection and social behavior.* New York: Chiron, pp. 417–438.

Durham, W. (1991). *Coevolution: Genes, culture, and human diversity.* Stanford: Stanford University Press.

Durkheim, E. (1895/1962). *The rules of the sociological method.* Glencoe, IL: Free Press.

Eibl-Eibesfeldt, I. (1975). *Ethology: The biology of behavior.* Second edition. New York: Holt, Rinehart, & Winston.

Ekman, P. (1973). Cross-cultural studies of facial expression. In P. Ekman (Ed.), *Darwin and facial expression: A century of research in review.* New York: Academic Press, pp. 169–222.

Ekman, P. (Ed.) (1982). *Emotion in the human face.* Second Edition. Cambridge, UK: Cambridge University Press.

Ekman, P. (1984). Expression and the nature of emotion. In P. Ekman & K. Scherer (Eds.), *Approaches to emotion.* Hillsdale, NJ: Erlbaum, pp. 319–343.

Ekman, P., & Friesen, W. V. (1975). *Unmasking the face.* New York: Prentice Hall.

Ekman, P., Levenson, R. W., & Friesen, W. V. (1983). Autonomic nervous system activity distinguishes among emotions. *Science, 221,* 1208–1210.

Erickson, C. J., & Zenone, P. G. (1976). Courtship differences in male ring doves: Avoidance of cuckoldry? *Science, 192,* 1353–1354.

Etcoff, N. (1983). Hemispheric differences in the perception of emotion in faces. Doctoral dissertation, Boston University.

Etcoff, N. (1986). The neuropsychology of emotional expression. In G. Goldstein & R. E. Tarter (Eds.), *Advances in clinical neuropsychology, Vol. 3.* New York: Plenum.

Flavell, J. H., Zhang, X-D, Zou, H., Dong, Q., & Qui, S. (1983). A comparison of the appearance-reality distinction in the People's Republic of China and the United States. *Cognitive Psychology, 15,* 459–466.

Flohr, H. (Ed.). (1988). *Post-lesion neural plasticity.* Berlin: Springer-Verlag.

Fodor, J. A. (1981). The mind-body problem. *Scientific American, 244,* 124–133.

Fodor, J. A. (1983). *The modularity of mind.* Cambridge, MA: MIT Press.

Fox, R. (1971). The cultural animal. In J. F. Eisenberg & W. S. Dillion (Eds.), *Man and beast: Comparative social behavior.* Washington, DC: Smithsonian Institution Press, pp. 273–296.

Freeman, D. (1983). *Margaret Mead and Samoa: The making and unmaking of an anthropological myth.* Cambridge, MA: Harvard University Press.

Freyd, J. J. (1983). Shareability: The social psychology of epistemology. *Cognitive Science, 7,* 191–210.

Freyd, J. J. (1987). Dynamic mental representations. *Psychological Review 94:* 427–438.

Frijda, N. H. (1986). *The emotions.* London: Cambridge University Press.

Fridlund, A. J. (in press). Evolution and facial action in reflex, social motive, and paralanguage. In P. K. Ackles, J. R. Jennings, & M.G.H. Coles (Eds.), *Advances in psychophysiology* (Vol. 4). London: Jessica Kingsley, Ltd.

Galef, B. G., Jr. (1988). Imitation in animals: History, definition, and interpretation of data from the psychological laboratory. In T. R. Zentall & B. G. Galef (Eds.), *Social learning: Psychological and biological perspectives.* Hillsdale, NJ: Erlbaum, pp. 3–28.

Galef, B. G. Jr. (1990). An adaptationist perspective on social learning, social feeding, and social foraging in Norway rats. In D. Dewsbury (Ed.), *Contemporary issues in comparative psychology.* Sunderland, MA: Sinauer.

Gallistel, C. R. (1990). *The organization of learning.* Cambridge, MA: MIT Press.

Garcia, J. (1990). Learning without memory. *Journal of Cognitive Neuroscience, 2,* 287–305.

Gardner, D., Harris, P. L., Ohmoto, M., & Hamazaki, T. (1988). Japanese children's understanding of the distinction between real and apparent emotion. *International Journal of Behavioral Development, 11,* 203–218.

Gazzaniga, M. S. (1985). *The social brain.* New York: Basic Books.

Geertz, C. (1973). *The interpretation of cultures.* New York: Basic Books.

Geertz, C. (1983). *Local knowledge: Further essays in interpretive anthropology.* New York: Basic Books.

Geertz, C. (1984). Anti anti-relativism. *American Anthropologist, 86,* 263–278.

Gelman, R. (1990a). Structural constraints on cognitive development: Introduction to a special issue of *Cognitive Science. Cognitive Science, 14,* 3–9.

Gelman, R. (1990b). First principles organize attention to and learning about relevant data: Number and the animate-inanimate distinction as examples. *Cognitive Science, 14,* 79–106.

Gelman, S., & Markman, E. (1986). Categories and induction in young children. *Cognition, 23,* 183–208.

Gelman, S., & Markman, E. (1987). Young children's inductions from natural kinds: The role of categories and appearances. *Child Development, 58,* 1532–1540.

Ghiselin, M. T. (1973). Darwin and evolutionary psychology. *Science, 179,* 964–968.

Gigerenzer, G. (1991a). From tools to theories: A heuristic of discovery in cognitive psychology. *Psychological Review, 98,* 254–267.

Gigerenzer, G. (1991b). How to make cognitive illusions disappear: Beyond "heuristics and biases." *European Review of Social Psychology, 2,* 83–115.

Gigerenzer, G., Hoffrage, U., & Kleinbölting, H. (1991). Probabilistic mental models: A Brunswikian theory of confidence. *Psychological Review. 98(4):* 506–528.

Gigerenzer, G., & Hug, K. (in press). Domain-specific reasoning: Social contracts, cheating and perspective change. *Cognition.*

Gigerenzer, G., & Murray, D. (1987). *Cognition as intuitive statistics.* Hillsdale, NJ: Erlbaum.

Glantz, K., & Pearce, J. K. (1989). *Exiles from Eden.* New York: Norton.

Gleitman, L. R., & Wanner, E. (1982). Language acquisition: The state of the state of the art. In E. Wanner & L. R. Gleitman (Eds.), *Language acquisition: The state of the art.* Cambridge, UK: Cambridge University Press, pp. 3–48.

Goldberg, S. (1973). *The inevitability of patriarchy.* New York: Morrow.

Goldschmidt, W. (1960). Culture and human behavior. In A.F.C. Wallace (Ed.), *Men and cultures: Selected papers of the Fifth International Congress of Anthropological and Ethnological Sciences (1956).* Philadelphia: University of Pennsylvania Press, pp. 98–104.

Gould, J. L., & Marler, P. (1987). Learning by instinct. *Scientific American, 256,* 74–85.

Gould, S. J. (1977a). Biological potentiality vs. biological determinism. In *Ever since Darwin: Reflections in natural history.* New York: Norton, pp. 251–259.

Gould, S. J. (1977b). So cleverly kind an animal. In *Ever since Darwin: Reflections in natural history.* New York: Norton, pp. 260–267.

Gould, S. J. (1977c). The nonscience of human nature. In *Ever since Darwin: Reflections in natural history.* New York: Norton, pp. 237–242.

Gould, S. J. (1979). Panselectionist pitfalls in Parker & Gibson's model of the evolution of intelligence. *Behavioral and Brain Sciences, 2,* 385–386.

Gould, S. J. (1988). This view of life. *Natural History, 9,* 14.

Gould, S. J., & Lewontin, R. C. (1979). The spandrels of San Marco and the Panglossian program: A critique of the adaptationist programme. *Proceedings of the Royal Society of London 250,* 281–288.

Hamburg, D. A. (1968). Emotions in the perspective of human evolution. In S. L. Washburn & P. C. Jay (Eds.), *Perspectives on human evolution.* New York: Holt, pp. 246–257.

Hamilton, W. D. (1964). The genetical evolution of social behavior. *Journal of Theoretical Biology, 7,* 1–52.

Hamilton, W. D. (1972). Altruism and related phenomena, mainly in social insects. *Annual Review of Ecology and Systematics, 3,* 193–232.

Hamilton, W. D., & Zuk, M. (1982). Heritable true fitness and bright birds: A role for parasites? *Science, 218,* 384–387.

Harlow, H. F., Harlow, M. K., & Suomi, S. J. (1971). From thought to therapy: Lessons from a primate laboratory. *American Scientist, 59,* 538–549.

Harris, M. (1968). *The rise of anthropological theory.* New York: Crowell.

Harris, M. (1979). *Cultural materialism: The struggle for a science of culture.* New York: Random House.

Hatch, E. (1973). *Theories of man and culture.* New York: Columbia University Press.

Hausfater, G., & Hrdy, S. B. (Eds.). (1984). *Infanticide: Comparative and evolutionary perspectives.* New York: Aldine.

Heisenberg, W. (1971). *Physics and beyond: Encounters and conversations.* New York: Harper & Row.

Herrnstein, R. J. (1977). The evolution of behaviorism. *American Psychologist 32,* 593–603.

Hinde, R. A. (1987). *Individuals, relationships, and culture: Links between ethology and the social sciences.* Cambridge, UK: Cambridge University Press.

Hrdy, S. B. (1977). *The langurs of Abu.* Cambridge, MA: Harvard University Press.

Hubby, J. L., & Lewontin, R. C. (1966). A molecular approach to the study of genic heterozygosity in natural populations: I. The number of alleles at different loci in *Drosophila pseudoobscura. Genetics, 54,* 577–594.

Hume, D. (1977/1748). *An enquiry concerning human understanding.* (E. Steinberg, Ed.). Indianapolis: Hackett.

Irons, W. (1979). Natural selection, adaptation, and human social behavior. In: Napoleon Chagnon and William Irons, (Eds.), *Evolutionary biology and human social behavior.* North Scituate, MA: Duxbury Press, pp. 4–39.

Irons, W. (1983). Human female reproductive strategies. In S. K. Wasser (Ed.), *Social behavior of female vertebrates.* New York: Academic Press.

Izard, C. E. (1977). *Human emotions.* New York: Plenum.

Jackendoff, R. (1992). *Languages of the mind.* Cambridge, MA: Bradford Books/MIT Press.

Jacob, F. (1977). Evolution and tinkering. *Science 196:* 1161–1166.

Jacobs, R. A., Jordan, M. I., & Barto, A. G. (1990). Task decomposition through competition in a modular connectionist architecture: The what and where vision tasks. *COINS Technical Report 90-27,* Dept. of Computer & Information Science, University of Massachusetts, Amherst, MA 01003.

James, W. (1892). *The principles of psychology.* London: Macmillan.

Kant, I. (1966/1781). *Critique of pure reason.* New York: Anchor Books.

Kaplan, H., & Hill, K. (1985). Food-sharing among Ache foragers: Tests of explanatory hypotheses. *Current Anthropology, 26,* 223–246.

Keil, F. C. (1989). *Concepts, kinds, and cognitive development.* Cambridge, MA: MIT Press.

Konner, M. (1982). *The tangled wing: Biological constraints on the human spirit.* New York: Holt, Rinehart, & Winston.

Kosslyn, S. M. (1980). *Image and mind.* Cambridge, MA: Harvard University Press.

Krebs, J. R., & Davies, N. B. (1984). *Behavioural ecology: An evolutionary approach.* Second Edition. Sunderland, MA: Sinauer.

Krebs, J. R., & Davies, N. B. (1987). *An introduction to behavioural ecology.* Oxford: Blackwell Sceintific Publications.

Kroeber, A. (1915). The eighteen professions. *American Anthropologist, 17,* 283–289.

Kroeber, A. (1917). The superorganic. *American Anthropologist, 19,* 163–213.

Laughlin, C. D., & d'Aquili, E. G. (1974). *Biogenetic structuralism.* New York: Columbia University Press.

Leach, E. (1982). *Social anthropology.* New York: Oxford University Press.

Lee, R. B., & DeVore, I. (Eds.). (1968). *Man the hunter.* Chicago: Aldine-Atherton.

Lee, R. B., & DeVore, I. (Eds.). (1976). *Kalahari hunter-gatherers.* Cambridge, MA: Harvard University Press.

Leslie, A. M. (1987). Pretense and representation: The origins of "theory of mind." *Psychological Review, 94,* 412–426.

Leslie, A. M. (1988). The necessity of illusion: Perception and thought in infancy. In L. Weiskrantz, (Ed.), *Thought without language.* Oxford: Clarendon Press, pp. 185–210.

Leslie, A. M., & Keeble, S. (1987). Do six-month-old infants perceive causality? *Cognition, 25,* 265–288.

Leslie, A. M., & Thaiss, L. (1990). Domain specificity in conceptual development: Evidence from autism. Paper presented at conference on "Cultural knowledge and domain specificity," Ann Arbor, Michigan.

Lewontin, R. C. (1990). Evolution of cognition. In D. Osherson & E. E. Smith, (Eds.), *Thinking: an invitation to cognitive science. Vol. 3.* Cambridge, MA: MIT Press, pp. 229–246.

Lewontin, R. C., & Hubby, J. L. (1966). A molecular approach to the study of genic heterozygosity in natural populations: II. Amount of variation and degree of heterozygosity in natural populations of *Drosophila pseudoobscura. Genetics, 54,* 595–609.

Lewontin, R. C., Rose, S., & Kamin, L. (1984). *Not in our genes.* New York: Pantheon.

Liberman, A., & Mattingley, I. (1985). The motor theory of speech preception revised. *Cognition, 51,* 1–36.

Liberman, A., & Mattingley, I. (1989). A specialization for speech perception. *Science, 243,* 489–496.

Lindblom, B. (1986). Phonetic universals in vowel systems. In J. J. Ohala & J. J. Jaeger (Eds.), *Experimental phonology.* New York: Academic Press.

Lindblom, B. (1988). Role of phonetic content in phonology. In W. Dressler (Ed.), *Proceedings of the Sixth International Phonology Meeting and Third International Morphology Meeting.* Vienna: University of Vienna.

Lockard, R. (1971). Reflections on the fall of comparative psychology: Is there a message for us all? *American Psychologist, 26,* 22–32.

Lorenz, K. (1965). *Evolution and the modification of behavior.* Chicago: University of Chicago Press.

Lorenz, K. (1966). *On aggression.* London: Methuen.

Lowie, R. H. (1917/1966). *Culture and ethnology.* New York: Basic Books.

Lumsden, C., & Wilson, E. O. (1981). *Genes, mind, and culture.* Cambridge, MA: Harvard University Press.

Maloney, L. T., & Wandell, B. A. (1985). Color constancy: A method for recovering surface spectral reflectance. *Journal of the Optical Society of America A, 3,* 29–33.

Manktelow, K. I., & Over, D. (1991). Social roles and utilities in reasoning with deontic conditionals. *Cognition, 39,* 85–105.

Markman, E. M. (1989). *Categorization and naming in children: Problems of induction.* Cambridge, MA: MIT Press.

Markman, E. M., & Hutchinson, J. E. (1984). Children's sensitivity to constraints on word meaning: Taxonomic vs. thematic relations. *Cognitive Psychology, 16,* 1–27.

Marks, I. M. (1987). *Fears, phobias, and rituals.* New York: Oxford University Press.

Marler, P. (1976). Social organization, communication, and graded signals: The chimpanzee and the gorilla. In P.P.G. Bateson & R. A. Hinde (Eds.), *Growing points in ethology.* Cambridge, UK: Cambridge University Press.

Marler, P. (1991). The instinct to learn. In S. Carey & R. Gelman (Eds.), *The epigenesis of mind.* Hillsdale, NJ: Erlbaum, pp. 37–66.

Marr, D. (1982). *Vision: A computational investigation into the human representation and processing of visual information.* San Francisco: Freeman.

Marr, D., & Nishihara, H. K. (1978). Visual information-processing: Artificial intelligence and the sensorium of sight. *Technology Review,* October, 28–49.

Marshall, J. C. (1981). Cognition and the crossroads. *Nature, 289,* 613–614.

Marx, K. (1867/1909). *Capital.* E. Unterman, trans. Chicago: C. H. Kerr.

Maynard Smith, J. (1964). Group selection and kin selection. *Nature, 20,* 1145–1147.

Maynard Smith, J. (1982). *Evolution and the theory of games.* Cambridge, UK: Cambridge University Press.

Mayr, E. (1976). Behavior programs and evolutionary strategies. In E. Mayr (Ed.), *Evolution and the diversity of life: Selected essays.* Cambridge, MA: Harvard University Press, pp 694–711.

Mayr, E. (1982). *The growth of biological thought.* Cambridge, MA: Harvard University Press.

Mayr, E. (1983). How to carry out the adaptationist program. *The American Naturalist, 121,* 324–334.

McCabe, S. (1983). FBD marriage: Further support for the Westermarck hypothesis of the incest taboo? *American Anthropologist, 85,* 50–69.

McCracken, R. (1971). Lactase deficiency: An example of dietary evolution. *Current Anthropology, 12,* 479–517.

Mead, M. (1949). *Male and female.* New York: Morrow.

Meltzoff, A. N. (1988). The human infant as *Homo imitans.* In T. R. Zentall & B. G. Galef, Jr., *Social learning: Psychological and biological perspectives.* Hillsdale, NJ: Erlbaum, pp. 319–341.

Merton, R. (1949). *Social theory and social structure.* Glencoe, IL: Free Press.

Miller, G. F., & Todd, P. M. (1990). Exploring adaptive agency I: Theory and methods for simulating the evolution of learning. In D. S. Touretskz, J. L. Elman, T. J. Sejnowski, & G. E. Hinton (Eds.), *Proceedings of the 1990 Connectionist Models Summer School.* San Mateo, CA: Morgan Kauffman, pp. 65–80.

Montagu, M.F.A. (Ed.). (1964). *Culture: Man's adaptive dimension.* Chicago: University of Chicago Press.

Montagu, M.F.A. (1968). *Man and aggression.* New York: Oxford University Press.

Montagu, M.F.A. (Ed.). (1978). *Learning nonaggression.* New York: Oxford University Press.

Murdock, G. P. (1932). The science of culture. *American Anthropologist, 34,* 200–215.

Nei, M. (1987). *Molecular evolutionary genetics.* New York: Columbia University Press.

Nesse, R. M. (1990). Evolutionary explanations of emotions. *Human Nature, 1,* 261–290.

Nevo, E. (1978). Genetic variation in natural populations: Patterns and theory. *Theoretical Population Biology, 13,* 121–177.

Newport, E. (1990). Maturational constraints on language learning. *Cognitive Science, 14,* 11–28.

Otte, D. (1974). Effects and functions in the evolution of signaling systems. *Annual Review of Ecology and Systematics, 5,* 385–417.

Paley, W. (1828). *Natural theology.* Second Edition. Oxford: J. Vincent.

Parker, H., & Parker, S. (1986). Father-daughter sexual abuse: An emerging perspective. *American Journal of Orthopsychiatry, 56,* 531–549.

Parsons, T. (1949). *The structure of social action.* New York: Free Press.

Pastner, C. M. (1986). The Westermarck hypothesis and first cousin marriage: The cultural modification of negative imprinting. *Journal of Anthropological Research, 24,* 573–586.

Perner, J. (1991). *Understanding the representational mind.* Cambridge, MA: MIT Press.

Pinker, S. (1979). Formal models of language learning. *Cognition 7:*217–283.

Pinker, S. (1982). A theory of the acquisition of lexical interpretive grammars. In J. Bresnan (Ed.), *The mental representation of grammatical relations.* Cambridge, MA: MIT Press.

Pinker, S. (1984). *Language learnability and language development.* Cambridge, MA: Harvard University Press.

Pinker, S. (1989). *Learnability and cognition: The acquisition of argument structure.* Cambridge, MA: MIT Press.

Pinker, S. (1991). Rules of language. *Science 253:*530–535.

Pinker, S., & Bloom, P. (1990). Natural language and natural selection. *Behavioral and Brain Sciences 13,* 707–784.

Pinker, S., & Prince, A. (1988). On language and connectionism: Analysis of a parallel distributed processing model of language acquisiton. *Cognition, 28,* 73–193.

Plutchik, R. (1980). *Emotion: A psychoevolutionary synthesis.* New York: Harper & Row.

Poggio, T., Torre, V., & Koch, C. (1985). Computational vision and regularization theory. *Nature, 317,* 314–319.

Popper, K. R. (1972). *Objective knowledge: An evolutionary approach.* London: Oxford University Press.

Premack, D. (1990). The infant's theory of self-propelled objects. *Cognition, 36,* 1–16.

Premack, D. (in prep.) *Theory of mind.*

Profet, M. (1988). The evolution of pregnancy sickness as protection to the embryo against Pleistocene teratogens. *Evolutionary Theory, 8,* 177–190.

Proffitt, D. R., & Gilden, D. L. (1989). Understanding natural dynamics. *Journal of Experimental Psychology: Human Perception and Performance, 15,* 384–393.

Pylyshyn, Z. W. (1984). *Computation and cognition: Toward a foundation for cognitive science.* Cambridge, MA: MIT Press.

Quine, W.V.O. (1960). *Word and object.* Cambridge, MA: MIT Press.

Quine, W.V.O. (1969). *Ontological relativity and other essays.* New York: Columbia University Press.

Radcliffe-Brown, A. R. (1952). *Structure and function in primitive society.* Glencoe, IL: Free Press.

Ramachadran, V. S. (1990). Visual perception in people and machines. In: A. Blake & T. Troscianko (Eds.), *AI and the eye.* New York: Wiley.

Real, L. A. (1991). Animal choice behavior and the evolution of cognitive architecture. *Science, 253,* 980–986.

Rindos, D. (1986). The evolution of the capacity for culture. *Current Anthropology 27,* 315–332.

Rozin, P. (1976). The evolution of intelligence and access to the cognitive unconscious. In J. M. Sprague & A. N. Epstein (Eds.), *Progress in psychobiology and physiological psychology.* New York: Academic Press.

Rozin, P., & Schull, J. (1988). The adaptive-evolutionary point of view in experimental psychology. In R. C. Atkinson, R. J. Herrnstein, G. Lindzey, and R. D. Luce (Eds.), *Stevens's handbook of experimental psychology.* New York: Wiley.

Sahlins, M. D. (1976a). *The use and abuse of biology: An anthropological critique of sociobiology.* Ann Arbor: University of Michigan Press.

Sahlins, M. D. (1976b). *Culture and practical reason.* Ann Arbor: University of Michigan Press.

Sadalla, E. K., Kenrick, D. T., & Vershure, B. (1987). Dominance and heterosexual attraction. *Journal of Personality and Social Psychology, 52,* 730–738.

Sapolsky, R. (1983). Endocrine aspects of social instability in the olive baboon *(Papio anubis). American Journal of Primatology, 5,* 365–379.

Schelling, T. C. (1978). *Micromotives and macrobehavior.* New York: Norton.

Seligman, M.E.P. (1971). Phobias and preparedness. *Behavior therapy, 2,* 307–320.

Seligman, M.E.P., & Hager, J. L. (1972). *Biological boundaries of learning.* New York: Meredith.

Shepard, R. N. (1981). Psychophysical complementarity. In M. Kubovy & J. R. Pomerantz, *Perceptual organization.* Hillsdale, NJ: Erlbaum.

Shepard, R. N. (1984). Ecological constraints on internal representation: Resonant kinematics of perceiving, imagining, thinking, and dreaming. *Psychological Review, 91,* 417–447.

Shepard, R. N. (1987a). Evolution of a mesh between principles of the mind and regularities of the world. In J. Dupre (Ed.), *The latest on the best: Essays on evolution and optimality.* Cambridge, MA: MIT Press.

Shepard, R. N. (1987b). Towards a universal law of generalization for psychological science. *Science, 237,* 1317–1323.

Shepher, J. (1983). *Incest: A Biosocial Approach.* New York: Academic Press.

Sherry, D. F., & Schacter, D. L. (1987). The evolution of multiple memory systems. *Psychological Review, 94,* 439–454.

Shettleworth, S. J. (1972). Constraints on learning. In D. S. Lehrman, R. A. Hinde, & E. Shaw (Eds.), *Advances in the study of behavior, Vol. 4.* New York: Academic Press.

Shiffrar, M., & Freyd, J. J. (1990). Apparent motion of the human body. *Psychological Science, 1,* 257–264.

Shweder, R. (1990). Cultural psychology: What is it? In J. Stigler, R. Shweder, & G. Herdt (Eds.), *Cultural psychology.* Cambridge, UK: Cambridge University Press.

Skinner, B F. (1957). *Verbal behavior.* New York: Appleton.

Spelke, E. S. (1988). The origins of physical knowledge. In L. Weiskrantz (Ed.), *Thought without language.* Oxford: Clarendon Press, pp. 168–184.

Spelke, E. S. (1990). Principles of object perception. *Cognitive Science, 14,* 29–56.

Spelke, E. S. (1991). Physical knowledge in infancy: Reflections on Piaget's theory. In S. Carey & R. Gelman (Eds.), *The epigenesis of mind.* Hillsdale, NJ: Erlbaum, pp. 133–169.

Sperber, D. (1974). *Rethinking symbolism.* Cambridge, UK: Cambridge University Press.

Sperber, D. (1982). *On anthropological knowledge.* Cambridge, UK: Cambridge University Press.

Sperber, D. (1985). Anthropology and psychology: Towards an epidemiology of representations. *Man, 20,* 73–89.

Sperber, D. (1986). Issues in the ontology of culture. In R. Marcus, G. Dorn, & P. Weingartner (Eds.), *Logic, methodology and philosophy of science VII. Proceedings of the Seventh International Congress of Logic, Methodology and Philosophy of Science, Salzburg 1983.* Amsterdam: North Holland, pp. 557–571.

Sperber, D. (1990). The epidemiology of beliefs. In C. Fraser and G. Gaskell (Eds.), *The social psychological study of widespread beliefs.* Oxford: Clarendon Press.

Sperber, D., & Wilson, D. (1986). *Relevance: Communication and cognition.* Cambridge, MA: Harvard University Press.

Spuhler, J. N. (1959). *The evolution of man's capacity for culture.* Detroit: Wayne State University Press.

Staddon. J.E.R. (1988). Learning as inference. In R. C. Bolles & M. D. Beecher (Eds.), *Evolution and learning.* Hillsdale, NJ: Erlbaum.

Stephens, D. W., & Krebs, J. R. (1986). *Foraging theory.* Princeton, NJ: Princeton University Press.

Suggs, R. C. (1971). Sex and personality in the Marquesas: A discussion of the Linton-Kardiner Report. In D. S. Marshall & R. C. Suggs (Eds.), *Human sexual behavior.* New York: Basic Books, pp. 163–186.

Symons, D. (1979). *The evolution of human sexuality.* New York: Oxford University Press.

Symons, D. (1987). If we're all Darwinians, what's the fuss about? In C. B. Crawford, M. F. Smith, & D. L. Krebs (Eds.), *Sociobiology and psychology.* Hillsdale, NJ: Erlbaum, pp. 121–146.

Symons, D. (1989). A critique of Darwinian anthropology. *Ethology and Sociobiology, 10,* 131–144.

Thornhill, N. W. (1991). An evolutionary analysis of rules regulating human imbreeding and marriage. *Behavioral and Brain Sciences, 14,* 247–293.

Thornhill, R. (1991). The study of adaptation. In M. Bekoff & D. Jamieson (Eds.), *Interpretation and explanation in the study of behavior.* Boulder, CO: Westview Press.

Tiger, L. (1969). *Men in groups.* New York: Random House.

Tinbergen, N. (1951). *The study of instinct.* New York: Oxford University Press.

Todd, P. M., & Miller, G. F. (1991a). Exploring adaptive agency II: Simulating the evolution of associative learning. In J. A. Meyer & S. W. Wilson (Eds.), *From animals to animats: Proceedings of the First International Conference of Simulation of Adaptive Behavior.* Cambridge, MA: MIT Press, pp. 306-315.

Todd, P. M., & Miller, C. F. (1991b). Exploring adaptive agency III: Simulating the evolution of habituation and sensitization. In H. P. Schwefel & R. Manner (Eds.), *Parallel problem solving from nature.* Berlin: Springer-Verlag, pp. 307-313.

Tomkins, S. S. (1962). *Affect, imagery, consciousness. Vol. I.* New York: Springer.

Tomkins, S. S. (1963). *Affect, imagery, consciousness. Vol. II.* New York: Springer.

Tooby, J. (1982). Pathogens, polymorphism and the evolution of sex. *Journal of Theoretical Biology, 97,* 557–576.

Tooby, J. (1985). The emergence of evolutionary psychology. In D. Pines (Ed.), *Emerging syntheses in science.* Santa Fe: Santa Fe Institute.

Tooby, J., & Cosmides, L. (1989a). Evolutionary psychology and the generation of culture, Part I. Theoretical considerations. *Ethology & Sociobiology, 10,* 29–49.

Tooby, J., & Cosmides, L. (1989b). The innate versus the manifest: How universal does universal have to be? *Behavioral and Brain Sciences, 12,* 36–37.

Tooby, J., & Cosmides, L. (1990a). On the universality of human nature and the uniqueness of the individual: The role of genetics and adaptation. *Journal of Personality, 58,* 17–67.

Tooby, J., & Cosmides, L. (1990b). The past explains the present: Emotional adaptations and the structure of ancestral environments. *Ethology and Sociobiology, 11,* 375–424.

Tooby, J., & DeVore, I. (1987). The reconstruction of hominid behavioral evolution through strategic modeling. In W. G. Kinzey (Ed.), *The evolution of human behavior: Primate models.* Albany: SUNY Press.

Townsend, J. M. (1987). Sex differences in sexuality among medical students: Effects of increasing socioeconomic status. *Archives of Sexual Behavior, 16,* 425–441.

Treisman, M. (1977). Motion sickness: An evolutionary hypothesis. *Science, 197,* 493–495.

Trivers, R. L. (1971). The evolution of reciprocal altruism. *Quarterly Review of Biology, 46,* 35–57.

Trivers, R. L. (1972). Parental investment and sexual selection. In B. Campbell (Ed.), *Sexual selection and the descent of man 1871–1971.* Chicago: Aldine.

Trivers, R. L. (1974). Parent-offspring conflict. *American Zoologist, 14,* 249–264.

Turke, P. W. (1990). Which humans behave adaptively, and why does it matter? *Ethology and Sociobiology, 11,* 305–339.

Van den Berghe, P. (1981). *The ethnic phenomenon.* New York: Elsevier.

de Waal, F. (1982). *Chimpanzee politics: Power and sex among apes.* New York: Harper.

Waddington, C. H. (1962). *New patterns in genetics and development.* New York: Comumbia University Press.

Watson, J. B. (1925). *Behaviorism.* New York: Norton.

Wellman, H. M. (1990). *The child's theory of mind.* Cambridge, MA: MIT Press.

Westermarck, E. A. (1891). *The history of human marriage.* New York: Macmillan.

Wexler, K., & Culicover, P. (1980). *Formal principles of language acquisition.* Cambridge, MA: MIT Press.

Wilkinson, G. S. (1988). Reciprocal altruism in bats and other mammals. *Ethology and Sociobiology, 9:*85–100.

Wilkinson, G. S. (1990). Food sharing in vampire bats. *Scientific American.* February, 76–82.

Williams, G. C. (1966). *Adaptation and natural selection: A critique of some current evolutionary thought.* Princeton, NJ: Princeton University Press.

Williams, G. C. (1985). A defense of reductionism in evolutionary biology. *Oxford surveys in evolutionary biology, 2,* 1–27.

Williams, G. C., & Williams, D. C. (1957). Natural selection of individually harmful social adaptations among sibs with special reference to social insects. *Evolution 17:*249–253.

Wilson, E. O. (1971). *The insect societies.* Cambridge, MA: Harvard University Press.

Wilson, E. O. (1975). *Sociobiology: The new synthesis.* Cambridge, MA: Harvard University Press.

Wilson, E. O. (1978). *On human nature.* Cambridge, MA: Harvard University Press.

Wilson, M., & Daly, M. (1985). Competitiveness, risk taking, and violence: the young male syndrome. *Ethology and Sociobiology. 6:*59–73.

Wilson, M., & Daly, M. (1987). Risk of maltreatment of children living with step-parents. In R. Gelles & J. Lancaster (Eds.), *Child abuse and neglect: Biosocial dimensions.* New York: Aldine.

Wimmer, H., & Perner, J. (1983). Beliefs about beliefs: Representation and constraining function of wrong beliefs in young children's understanding of deception. *Cognition, 13,* 103–128.

Wolf, A. (1966). Childhood association, sexual attraction and the incest taboo. *American Anthropologist, 68,* 883–898.

Wolf, A. P. (1968). Adopt a daughter-in-law, marry a sister: A Chinese solution to the problem of the incest taboo. *American Anthropologist, 70,* 864–874.

Wolf, A. P., & Huang, C. (1980). *Marriage and adoption in China 1845–1945.* Stanford: Stanford University Press.

Wrangham, R. W. (1987). The significance of African apes for reconstructing human social evolution. In W. G. Kinzey (Ed.), *The evolution of human behavior: Primate models.* Albany: SUNY Press, pp. 51–71.

2

On the Use and Misuse of Darwinism in the Study of Human Behavior

DONALD SYMONS

A biological explanation should invoke no factors other than the laws of physical science, natural selection, and the contingencies of history.

GEORGE C. WILLIAMS

Darwin's theory of evolution by natural selection answered one of the great existential questions: "Why are people?" (Dawkins, 1976). But once we know why people are, a second question immediately suggests itself: "What of it?" (Medawar, 1982). Had such tough-minded thinkers of the seventeenth and eighteenth centuries as Thomas Hobbes and Samuel Johnson been apprised of Darwin's discovery, says Medawar, they might well "have demanded to know what great and illuminating new truth about mankind followed from our realisation of his having evolved" (p. 191). This essay is a meditation on the question, "What of it?"

One of the major aims of this essay is to critically analyze the following hypothesis, which many scholars believe to be entailed by the proposition that human beings are the products of natural selection: Human behavior per se can be expected to be adaptive (i.e., reproduction-maximizing), and hence a science of human behavior can be based on analyses of the reproductive consequences of human action. My critique of this hypothesis perhaps can be introduced most easily by way of an example.

Because I wrote a book about the evolution of human sexuality (Symons, 1979), I am sometimes invited to lecture on this topic. During such lectures, I present various hypotheses about the psychological mechanisms that underpin human sexual behavior and about the selective forces that shaped these mechanisms. For example, I claim that, other things being equal, men tend to be more strongly sexually attracted to women with whom they have never had sexual relations than they are to women with whom they regularly have sexual relations. This phenomenon results, I argue, not from a generalized tendency to become bored by familiarity, but from the operation of a specialized psychological mechanism. Now, there is nothing in the laws of physical science that can account for the existence of this mechanism; nor does it somehow follow as an inevitable consequence of biological law or sexual reproduction; nor does such a mechanism exist universally in male animals (indeed, many scientists, including some evolutionists, implicitly deny its existence in human males). This mechanism, I argue, was produced by natural selection during the course of human evolutionary history because opportunities sometimes existed for males to sire offspring at little "cost" (a trivial amount of sperm, a few moments of their time) by copulating with new females. Also, opportunities often existed for males of high status or exceptional competitive abilities to acquire multiple mates. In such circumstances, even if only a tiny fraction of sexual impulses could be consummated, over the course of thousands of generations males with a roving eye, a taste for partner variety, and the ability to discriminate low- from high-risk opportunities produced more offspring, on the average, than did males with different psychological characteristics.

After the lecture I am often asked some version of the following question: "Since polygamy is illegal, and modern women generally practice contraception, a roving eye currently doesn't seem to make much reproductive sense; so why do men still have it?" My usual answer is this: Natural selection takes hundreds or thousands of generations to fashion any *complex* adaptation. The brain/mind mechanisms that constitute human nature were shaped by selection over vast periods of time in environments different in many important respects from our own, and it is to these ancient environments that human nature is adapted. Because modern contraceptive technology has existed for an evolutionarily insignificant amount of time, we have no adaptations specifically designed to deal with it, and a roving eye may be less adaptive (i.e., less reproduction-maximizing) today than it once was, or perhaps maladaptive (even in the ancestral environments in which human nature evolved, no doubt some males met early deaths as a direct consequence of their roving eyes). Indeed, if all women began practicing perfectly effective contraceptive techniques, chose to be impregnated only by their husbands, never married polygynously, and never married divorced men, over the course of several thousand years selection might eliminate the human male's roving eye.

Although the questioner generally seems satisfied with this answer, the subsequent discussion almost invariably reveals that a fundamental conceptual error, which motivated the question in the first place, remains unresolved. That error—which is almost as pervasive among behavioral scientists as it is among laymen—is to assume that if life's machinery was designed by natural selection—i.e., by nonrandom differential reproduction—then this machinery ought to incarnate or instantiate some sort of generalized reproductive striving. In other words, the error is to confuse the general process that produces adaptations with the adaptations themselves (Tooby & Cosmides, 1989a). The human brain/mind is an integrated bundle of complex mechanisms *(adaptations).* Each mechanism was designed by natural selection in past environments to promote the survival of the genes that directed its construction *by serving some specific function*—i.e., by performing some specific task, such as regulating blood pressure, perceiving edges, or detecting cheaters in social exchanges. No mechanism could possibly serve the general function of promoting gene survival because there simply is no general, universally effective way of doing so (Cosmides & Tooby, 1987). What works in one species may not work in another; what works in the infant of the species may not work in the adult; what works in the female of the species may not work in the male; what works in a given species at one time may not work at another time; what works in solving one kind of biological problem may not work in solving another. And, in every case, "what works" is determined in the crucible of evolutionary time.

One source of confusion is that we are used to thinking of human behavior as being uniquely flexible and responsive to environmental variation. Human behavior is flexible, of course, but this flexibility is of means, not ends, and the basic experiential goals that motivate human behavior are both inflexible and specific. For example, assume that we, along with many other primates, possess a specialized gustatory mechanism underpinning the sensation of sweetness. This mechanism was shaped by natural selection in ancestral populations because a sugar-producing fruit is most nutritious when its sugar content is highest, hence individuals who detected and liked sugar produced, on the average, more progeny than did individuals who could not detect sugar or who actually preferred the taste of green or overripe or rotten fruit. Since human behavior is so flexible, we have been able to develop virtually an infinite number of ways of obtaining sugar; but the goal of eating sugar remains the same—to experience the sensation of sweetness.

In modern industrial societies, where refined sugar is abundantly available, the human sweet tooth may be dysfunctional, but sugar still tastes sweet, and the goal of experiencing sweetness still motivates behavior. That's how we're made. We can decide to avoid refined sugar, but we can't decide to experience a sensation other than sweetness when sugar is on our tongues. If we decide to forgo the pleasure of sweetness in order to reduce the risk of tooth decay, this conscious decision will be in the service of other specific goals (probably related to material cost, physical attractiveness, and pain).

In summary, although human behavior is uniquely flexible, the goal of this behavior is the achievement of specific experiences—such as sweetness, being warm, and having high status. Our flexibility of means and our inflexible experiential ends are underpinned by an array of psychological mechanisms that is universal among *Homo sapiens* (sex and age differences excepted) and finite (Tooby & Cosmides, 1990). By contrast, the behaviors that these mechanisms produce are not universal (species-typical movement patterns such as walking, suckling, and frowning excepted) and are infinitely variable. We possess the particular array of psychological mechanisms that we do, not because there is anything inherent in physical or biological law that makes them inevitable, but because in ancestral populations selection favored certain variants of each mechanism over the available alternatives, and genes specifically for each psychological mechanism became established in the human gene pool. (There are, however, no genes specifically for human behaviors—species-typical movement patterns excepted.) In short, human beings, like all other organisms, have been designed by selection to strive for *specific* goals, not the *general* goal of reproduction-maximizing: There can be no such thing as a generalized reproduction-maximizer mechanism because there is no general, universally effective way to maximize reproduction. Thus, there is no reason to expect whatever psychological mechanisms underpin our unique flexibility of behavior to maximize reproductive success in evolutionarily novel environments: These mechanisms merely make possible novel behavioral means to the same old specific ends.

My primary goal in this essay is to convince the reader that because Darwinism is a theory of adaptation it illuminates human behavior only insofar as it illuminates the adaptations that constitute the machinery of behavior. In other words, the link between Darwinian theory and behavior is psychology (Barkow, 1984, Cosmides & Tooby, 1987). I proceed as follows: First, I describe the adaptationist program in biology. Second, I try to show how this program can be applied fruitfully to human psychological adaptations, using the perception of sexual attractiveness as an example. Third, I illustrate how social scientists, whose goal is to illuminate phenomena that are not themselves adaptations, can use evolutionary psychology to guide their research. And fourth, I argue that no approach to human behavior can be simultaneously psychologically agnostic and genuinely Darwinian.

THE ADAPTATIONIST PROGRAM

The goal of the adaptationist program in biology is "to recognize certain of [the organism's] features as components of some special problem-solving machinery" (Williams, 1985, p. 1). As Mayr (1983) notes, the study of adaptation long antedates Darwin: "The adaptationist question, 'What is the function of a given structure or organ?' has been for centuries the basis for every advance in physiology" (p. 328). But it was Darwin who provided the first and only scientifically coherent account of the origin and maintenance of adaptations: evolution by natural selection (Darwin, 1859; Dawkins, 1986). To most evolutionary biologists, to say that a structural, behavioral, physiological, or psychological trait is an adaptation is to say that the trait was designed by selection to serve a specific function (Thornhill, 1990; Williams, 1966). Since one can arbitrarily partition phenotypes in an infinite number of ways—the overwhelming majority of which will be useless for any sort of biological analysis—the identification of adaptations is central to the study of the phenomena of life: By identifying adaptations one carves the phenotype at its natural, functional joints. "The study of adaptation is not an optional preoccupation with fascinating fragments of natural history, it is the core of biological study" (Pittendrigh, 1958, p. 395).

To claim that a trait is an adaptation is thus to make a claim about the *past:* "The one thing about which modern authors are unanimous is that adaptation is not teleological, but refers to something produced in the past by natural selection" (Mayr, 1983, p. 324). It is also to make a claim about *design:* Natural selection is not mere differential reproduction, it is "differential reproduction *in consequence of . . . design differences*" (Burian, 1983, p. 307). When one claims that a feature of an organism is an adaptation, "one is claining not only that the feature was brought about by differential reproduction among alternative forms, but also that the relative advantage of the feature vis-à-vis its alternatives played a significant causal role in its production" (Burian, 1983, p. 294). Finally, given modern understandings of the genetical basis of reproduction, to claim that a trait is an adaptation is to make a certain kind of claim about *genes.* There has been much confusion on this point because the question, "Are there genes for trait X?" can be a question about any one of three entirely different—yet often confused—subjects: ontogeny, heritability, and adaptation. When people ask whether there are genes for a given trait, they generally have some version of the adaptationist question in mind, and yet the answer they are given usually assumes that they were asking the ontogenetic or the heritability question.

The ontogenetic question is, Did genes play a role in the development of trait X? The answer is always yes, no matter what trait X is—adaptation or pathology, idiosyncrasy or species-typical organ—since every part of every organism emerges only via interactions among genes, gene products, and myriad environmental phenomena. And because the answer to the ontogenetic question is always yes, it is uninformative.

Second is the heritability question: Is any of the population variance in trait X caused by genetic variance? This rather specialized question is asked primarily by population and behavior geneticists, personality psychologists, medical researchers, and plant and animal breeders. For example, if trait X is milk production in a particular population of cows, and if the answer is no (i.e., if the heritability is zero), then the dairy farmer knows that attempts to increase milk production via selective breeding would be futile. The heritability question can be asked about any phenotypic trait, whether or not the trait is an adaptation, and the answer will reveal little about the evolutionary past. In particular, a yes answer to this question is not evidence that trait X is an adaptation, and a no answer is not evidence for the opposite conclusion. The ability to lactate obviously is an adapation, whether or not any of its aspects are currently heritable. In fact, since selection tends to use up genetic variation, strong and consistent selection pressures favoring trait X usually drive its heritability to zero.

Third is the adaptationist question: Was trait X per se designed by selection to serve some function; i.e., is it an adaptation? When people ask whether there are genes for trait X, this is the question whose answer they are usually seeking. Small wonder, then, that they are perplexed when the useless, inevitably-affirmative answer to the ontogenetic question is palmed off on them, or they are told that the answer depends on the outcome of heritability studies! Evidence that there are genes for trait X in the adaptationist sense (i.e., evidence that trait X is an adaptation) comes from the study of phenotopic design. Design can be demonstrated by comparative studies of evolutionary convergences and divergences (Curio, 1973; Mayr, 1983) and by "engineering" analyses, in which it is recognized in the precision, economy, efficiency, constancy, and complexity with which effects are achieved (Curio, 1973; Dawkins, 1986; Thornhill, 1990; Williams, 1966, 1985).

The adaptationist program applied to the study of the human brain/mind has come to be called "evolutionary psychology" (see Barkow, 1980, 1984, 1986; Cosmides & Tooby, 1987; Daly & Wilson, 1984, 1986, 1988; Symons, 1987b, 1989; Tooby, 1985; Tooby & Cosmides, 1989a, 1989b). In the following section, I try to illustrate the usefulness of the evolutionary psychological approach by considering briefly a single topic, the perception of sexual attractiveness.

THE EVOLUTIONARY PSYCHOLOGY OF SEXUAL ATTRACTIVENESS

Since Darwin's theory of evolution by natural selection is the only viable explanation of the origin and maintenance of the machinery of life (Dawkins, 1982, 1986), it is a powerful heuristic tool for investigating the phenomena of life: It "provides a guide and prevents certain kinds of errors, raises suspicions of certain explanations or observations, suggests lines of research to be followed, and provides a sound criterion for recognizing significant observations on natural phenomena" (Lloyd, 1979, p. 18). Selectional thinking (i.e., thinking informed and inspired by Darwin's theory) can provide a guide to the study of human sexuality, not because it typically inspires startling or counterintuitive hypotheses (although some day it may do so), but because it leads one to expect that an array of specialized (specifically "sexual") psychological mechanisms underpins human sexual behavior. To explain why this is so, and why this expectation differs so profoundly from the psychological assumptions implicit in most social science accounts of human sexual behavior, requires an introductory digression.

Every theory of human behavior implies a human psychology. This includes theories that attribute human behavior to "culture": If human beings have culture, while rocks, tree frogs, and lemurs don't, it must be because human beings have a different psychological makeup from that of rocks, tree frogs, and lemurs. And every psychological theory implies a human nature; that is, every psychological theory implies that the brain/mind comprises mechanisms that are typical of *Homo sapiens* as a species, in the same sense that having arms rather than wings is typical of *Homo sapiens*. The phrase "human nature" thus does not imply the existence of a mysterious essence or Platonic ideal but rather the possibility of evaluating scientifically alternative hypotheses about brain/mind mechanisms: The brain/mind either is or is not sexually dimorphic; it either does or does not contain mechanisms specialized for detecting and preferring landscapes of certain sorts (Orians & Heerwagen, this volume), and so forth.

Historically, there have been two basic conceptions of human nature: the empiricist conception, in which the brain is thought to comprise only a few domain-general, unspecialized mechanisms; and the nativist conception, in which the brain is thought to comprise many, domain-specific, specialized mechanisms. It is no accident that Darwinians typically favor some version of the latter: The adaptationist program entails thinking in terms of *function,* and the brain clearly has many varied functions; that is, it has been designed by natural selection to solve many different kinds of problems. Each kind of problem is likely to require its own distinctive kind of solution. It is no more probable that some sort of general-purpose brain/mind mechanism could solve all the behavioral problems an organism faces (find food, choose a mate, select a habitat, etc.) than it is that some sort of general-purpose organ could perform all physiological functions (pump blood, digest food, nourish an embryo, etc.) or that some sort of general-purpose kitchen device could perform all food processing tasks (broil, freeze, chop, etc.). There is no such thing as a "general problem solver" because there is no such thing as a general problem.

Now consider in this light the question of the brain/mind mechanisms that are responsible for human perceptions of sexual attractiveness. Anthropologists (indeed, most social scientists) typically attribute such perceptions to "culture." In doing so, they surely imply that this perception is merely one by-product of some—necessarily general-purpose—"capacity-for-culture" mechanisms, which underpin the myriad phenomena that are attributed to culture (kinship systems, pottery styles, religious beliefs, etc.). In essence, to attribute the perception of sexual attractiveness to culture is to deny, implicitly, that human beings possess *specialized* psychological mechanisms underpinning this perception; that is, it is to deny that this perception *exists* as an independent psychological phenomenon with its own distinctive rules and principles. Similarly, to attribute the perception of sexual attractiveness to "learning" is to imply that this perception is merely one by-product of some sort of general-purpose "learning mechanism."

Selectional thinking and comparative data on nonhuman animals should arouse deep suspicions of "cultural" and "learning" explanations of the perception of sexual attractiveness. Most Darwinists expect that human beings will be found to possess domain-specific, specialized mechanisms underlying this perception. The adaptationist rationale for this expectation applies not only to human beings but to all complex, sexually reproducing animals: Finding a mate represents a crucial adaptive problem, and all objects in the environment are not equally valuable as potential mates, just as all objects in the environment are not equally valuable as potential food. Thus, selection has produced brain/mind adaptations specialized to detect conspecifics of the opposite sex who evidence high "mate value," just as selection has produced adaptations specialized to detect objects that evidence high nutritional value.

There is nothing inherent in eucalyptus leaves that makes them tasty or not tasty: Like beauty, tastiness is in the adaptations of the beholder. Eucalyptus leaves are tasty to koalas but not to human beings because the two species have different feeding adaptations. Similarly, the psychological mechanisms that underpin the perception of sexual attractiveness among koalas must differ from the mechanisms that underpin koala food choice because, historically, there would have been no reproductive payoff for koalas in choosing to mate with eucalyptus leaves. Of course, in a given species many of the same psychological adaptations are likely to be involved in both food and mate choice: basic mechanisms of visual perception, for example. But the central evaluative mechanisms *must be* different because the qualities that make for nutritious food are not the qualities that make for a good mate (though we may link the two with such metaphors as "sweet"). In short, just as specialized, distinctively sexual anatomy exists below the neck, so the Darwinist expects it to exist above the neck.

If the perception of sexual attractiveness were the product of some sort of generalized "capacity-for-culture" mechanisms, then standards of sexual attractiveness would vary capriciously cross-culturally and would be impossible to predict in advance for a heretofore unknown people. But if the evolutionary psychological argument outlined above is approximately correct, and specialized adaptations are responsible for human perceptions of sexual attractiveness, then fundamental cross-cultural regularities in standards of sexual attractiveness would exist, cross-cultural variation in these standards would be explicable largely in terms of universal, specialized psychological mechanisms operating on varied inputs, and standards of sexual attractiveness among a heretofore unknown people would be predictable in advance with a reasonable degree of accuracy.

Available evidence overwhelmingly favors the evolutionary psychological expectation (Daly & Wilson, 1983, Symons, 1979, 1987a). For example, human males universally seem to be maximally sexually attracted, other things being equal, to certain physical correlates of female nubility, i.e., to certain physical characteristics indicative of a human female who has recently begun fertile menstrual cycles and who has not yet borne a child. In the human environment of evolutionary adaptedness (EEA)— i.e., the Pleistocene environment in which the overwhelming majority of human evolution occurred—such a woman probably would have been between 15 and 18 years of age. In modern contracepting societies women can maintain a relatively youthful appearance far longer than was possible in the EEA, hence in these societies the effect of age on women's sexual attractiveness is much less marked than it was in the past or is today among band-level peoples (whose environments in some respects resemble the EEA). Nevertheless, I'll hazard the prediction that no society will be found in which most men perceive 38-year-old women as more sexually attractive than 18-year-old women (see Symons, 1979, 1987a).

As a second example, available evidence indicates that women are maximally sexually attracted, other things being equal, to men who exhibit signs of high status (Symons, 1979; Daly & Wilson, 1983). I predict that this preference will be found universally and that it will be as characteristic of high-status women as it is of low-status women. The particular correlates or indexes of male status do, of course, vary; what is invariant is the psychological adaptation that specifies the rule "prefer signs of high status."

These predictions about male and female sexual preferences have two sources: First, the hypothesized psychological mechanisms that inform the predictions make excellent adaptive sense; and, second, the existence of such psychological mechanisms is strongly implied by the available data on sexual attractiveness (Symons, 1979; also see Buss, 1989; Daly & Wilson, 1983; Ellis, this volume; Symons, 1987a, 1987b; Townsend, 1987, 1989). Although selectional thinking is an important source of inspiration for the evolutionary psychologist, nature always gets the last word. For example, it would seem to make excellent adaptive sense for human males to be able to detect female ovulation and to find ovulating females most sexually attractive; but the preponderance of evidence is that no such adaptation exists (Doty, 1985).

Because the goal of evolutionary psychology, as noted above, is to discover and describe adaptations, evolutionary psychological hypotheses are necessarily hypotheses about the past, including the nature of the EEA (Tooby & DeVore, 1987), about phenotypic design, and about genes, in the adaptationist sense. The hypothesis that human males evolved specialized female-nubility-preferring psychological mechanisms thus entails the following assumptions: (a) During the evolutionary past, heritable variation existed among ancestral males in tendencies to be sexually attracted to certain physical correlates of female nubility; (b) males who preferred nubile females outreproduced, on the average, males with different sexual preferences, specifically because of the former's preference for nubility; (c) selection designed at least one psychological mechanism specifically for nubility-preferring; and (d) genes for nubility-preferring thus became established in human gene pools.

Since the brain/mind mechanisms that collectively constitute human nature were designed by natural selection in the EEA, they must be described solely in terms of phenomena that existed in the EEA; but the kinds of data that can be used to evaluate evolutionary psychological hypotheses are potentially limitless, and evolutionarily novel phenomena can be just as informative as phenomena that existed in the EEA, or more so (Tooby & Cosmides, 1989b). As Daly and Wilson (1986) note, "You're starting to get a real handle on how the hunger mechanism works . . . when you know how to get a stuffed rat to eat or a starved one to abstain" (p. 189). Human activities in modern industrial societies can be looked upon as unplanned experiments on human nature, analogous to the planned experiments on nonhuman animal nature with which ethologists supplement their field studies. The sugar, salt, and fat served in fast food restaurants tell us at least as much about the nature and evolution of the mechanisms that underpin our appetite as hunter/gatherer menus do. In evaluating the hypothesis that human males evolved specialized female-nubility-preferring mechanisms, here are some of the kinds of data that might prove to be relevant: observations of human behavior in public places, literary works (particularly the classics, which have passed the tests of time and translation), questionnaire results, the ethnographic record, measurement of the strength of penile erection in response to photographs of women of various ages, analyses of the effects of cosmetics, observations in brothels, the effects of specific brain lesions on sexual preferences, skin magazines, and discoveries in neurophysiology.

But surely the most compelling tests of this hypothesis, at least at present, would be cross-cultural studies designed specifically for that purpose. Cultural anthropologists are uniquely well situated to conduct research on psychological universals, since cross-cultural comparisons are the cornerstone of anthropology. Furthermore, anthropological research among band-level peoples (supplemented by the findings of archaeology and paleoanthropology) provides insight into the human EEA not obtainable from any other source. Cross-cultural data also can inspire evolutionary psychological hypotheses. For example, it was empirical evidence in the ethnographic record (rather than selectional thinking) that led van den Berghe and Frost (1986) to hypothesize that human males evolved a specialized psychological mechanism to prefer females with a skin tone somewhat lighter than the female average.

Now it might be argued that the human male's taste for nubile females is hardly front page news, that the existence of multibillion dollar industries producing cosmetics designed to make post-nubile women look younger is common knowledge, and that insofar as evolutionary psychologists emphasize such phenomena they merely belabor the obvious. There are two rejoinders to this argument. First, there has been essentially no systematic psychological research on variation in female sexual attractiveness with age. Are human males really most strongly sexually attracted to nubile females? If so, by what specific cues is nubility detected? Systematic changes in women's skin texture with age are obvious candidates, but are there perhaps also subtle changes in facial proportions that men unconsciously detect and respond to? Is the age of maximal female sexual attractiveness in some degree a function of the age or status of the male who is doing the evaluating? Academic psychology is silent on such matters. The human visual system "obviously" includes mechanisms designed to maintain size, color, and shape constancies, yet the obviousness of perceptual constancies quite properly has not prevented psychologists from investigating them. To the extent that evolutionary psychology focuses attention and inspires research on neglected but important topics, such as the perception of sexual attractiveness, it is valuable indeed.

Second, whenever a social scientist attributes something like the human male's sexual attraction to nubile females to cultural conditioning, social learning, socialization, etc., and whenever a social scientist begins a discussion of such matters with the phrase "In our culture . . ." (gratuitously implying that things are different elsewhere), he implicitly rejects a nativist conception of human nature and embraces an empiricist conception. As long as social scientists continue to make assumptions about human nature that have essentially no chance of being correct, even seemingly mundane findings of evolutionary psychology will be scientifically significant.

EVOLUTIONARY PSYCHOLOGY AND SOCIAL SCIENCE

Even though the goal of most social scientific research is to illuminate phenomena that are not themselves adaptations, social scientists can sometimes use evolutionary psychology as a guide. For example, Chagnon's (1988) investigation of kin term manipulation among the Yanomamo, an Amazonian tribal people, was inspired by his basic evolutionary psychological assumption that selection has designed each human being to have some goals that are achievable only at the expense of other human beings. This assumption implies that competition must always have permeated human affairs (Alexander, 1975), which, in turn, led Chagnon to suspect that among a people like the Yanomamo, where kinship classification has profound social ramifications and where there often exists enough ambiguity about the "correct" classification to provide scope for strategic manipulation, people are unlikely to be mere passive acceptors of their kinship system, as anthropologists have usually assumed. Rather, people are likely to be active critics and manipulators of their kinship system. And this is precisely what Chagnon found to be the case among the Yanomamo; for example, a Yanomamo man may begin calling a particular young woman in his village by a kin term that transfers her from a category in which she was ineligible to marry his son to a category in which she is a potential daughter-in-law. If this reclassification becomes generally accepted, it will inevitably harm some people's interests, and fights often erupt over such attempts to manipulate the kinship system (also see Chagnon, 1968, p. 65).

Now, Chagnon did not claim that his data point to the existence of a heretofore undreamed of psychological mechanism specialized for kin term manipulation (i.e., he did not argue that kin term manipulation per se is an adaptation). In fact, he did not claim that these data shed light on human psychology at all (although surely they do bolster his basic assumptions about human nature). Rather, he realized that traditional anthropological views of kinship and kin term usage are informed by a conception of a human nature that is, to a Darwinian, highly implausible. Although he did not form a specific hypothesis about psychological mechanisms, Chagnon's general selection-minded assumptions about human nature inspired both suspicions of traditional anthropological views of kinship and the collection of data on kin term manipulation (which few other anthropologists had thought to collect).

As this example illustrates, evolutionary psychology is useful to social scientists, at least at present, not because it typically inspires startling or counterintuitive hypotheses (as soon as one understands the social ramifications of kin classification among the Yanomamo, one immediately empathizes with the temptation to manipulate the system), but because it leads to the questioning of the basic assumptions about human nature that are implicit in traditional social scientific research and theory. From evolutionary biologist John Maynard Smith's (1982) vantage point outside the social sciences, it may seem trivial to demonstrate "that human behaviour is influenced by kinship, and that kinship has something to do with genetic relationship, because surely, despite some very odd remarks by anthropologists, that is uncontroversial?" (p. 3). From a vantage point within the social sciences, however, these "odd remarks" are, alas, the very warp and weft of accepted theory: Kinship often *is* thought to be independent of genetic relationship, hence the importance of an evolutionary psychological perspective on human nature.

DARWINIAN SOCIAL SCIENCE

In contrast to social scientists who use evolutionary psychology as a guide to research, some evolution-minded scholars seek to construct a psychologically agnostic science of human behavior based on the hypothesis that human beings are reproduction (or inclusive fitness) maximizers. This approach has been variously called human sociobiology, human behavioral ecology, evolutionary biological anthropology, and, by me, Darwinian anthropology (Symons, 1989). I'll refer to it here as "Darwinian social science" (since not all of its practitioners are anthropologists) and use the abbreviation DSS to mean both Darwinian social science and Darwinian social scientist. In referring to someone as a DSS I mean to imply only that he wears a DSS hat in the specific context under discussion, not in all his research. (And it should not be inferred that all research classified as human sociobiology, human behavioral ecology, or evolutionary biological anthropology is DSS.) For examples of DSS see Chagnon and Irons (1979) and Betzig, Borgerhoff Mulder and Turke (1988). In this section I describe DSS; in subsequent sections I critically analyze it. I will attempt to show that DSS is not genuinely Darwinian and that the reproductive data DSSes have collected rarely shed light on human nature or the selective forces that shaped that nature.

The central DSS hypothesis is that "evolved behavioral tendencies" cause human "behavior to assume the form that maximizes inclusive fitness" (Irons, 1979, p. 33). Turke and Betzig (1985) put it this way: "Modern Darwinian theory predicts that human behavior will be adaptive, that is, designed to promote maximum reproductive success (RS) through available descendent and nondescendent relatives" (p. 79). The research of Crook and Crook (1988) is a typical example of DSS. Crook and Crook collected reproductive data on Tibetans who did and who did not marry polyandrously and concluded, from an analysis of these data, that polyandry is adaptive (i.e., fitness-promoting) in certain highly unusual environmental conditions that some Tibetans have encountered in recent times. Now, polyandry, like all human activities, results from the operation of some array of brain/mind mechanisms; but polyandry is an *adaptation* only if at least one of these mechanisms was designed by selection specifically to produce it. In other words, it is an adaptation only if at least one psychological mechanism owes its form to the greater reproductive success of individuals who married polyandrously, in certain circumstances, in ancestral populations. If no such specialized mechanism exists, polyandry is not an adaptation, even though it may currently be adaptive—i.e., fitness-promoting—in certain modern environments. Since the specific environmental features to which Tibetan polyandry is adapted, according to Crook and Crook, include agricultural estates, animal husbandry, primogeniture, monasticism, artistocrats, landlords, governments, and taxation, none of which existed in the human EEA, there is no reason to suppose that polyandry is an adaptation.

The research of Crook and Crook differs from Chagnon's investigation of Yanomamo kin term manipulation in the following way: Chagnon's goal was to illuminate an important anthropological problem, the nature of kin term usage. He realized that traditional anthropological views of this matter are based on assumptions about human psychology that are almost certainly wrong. Thus, he became suspicious of these traditional views and was inspired to collect novel data on kin term manipulation. The result was a major contribution to the study of kinship, a subject of interest to anthropologists and to many other people. By contrast, Crook and Crook did not use evolutionary psychology to question previous views of polyandry. There is no reason to suppose that their assumptions about human nature differ from the assumptions made by other students of polyandry; i.e., there is no reason to suppose that Crook and Crook's reproductive data will—or should—prompt other students of polyandry to question or to modify previously held assumptions about human psychology. Indeed, there is no particular reason to suppose that Crook and Crook's reproductive data have any bearing on the anthropology of polyandry; they simply argued that polyandrous marriages are fitness-promoting in certain highly unusual circumstances. Crook and Crook believe that the scientific significance of their reproductive data lies in the fact that these data test the following prediction: "The central prediction made in a Darwinian perspective," they write, "is that humans are endeavouring consciously or unconsciously to optimize their reproductive success. It is then a matter of research to discover whether individuals marrying in contrasting ways in different contexts are in fact showing behaviour that does promote their genetic fitness" (p. 98).

To understand the scientific significance of this or of any prediction, one needs to know what theory, hypothesis, or assumption would be called into question by the prediction's disconfirmation. Since Darwin's theory of adaptation through natural selection is "the only workable theory we have to explain the organized complexity of life" (Dawkins, 1982, p. 35), there is no known scientific alternative to the theory that human nature is the product of natural selection. This theory is not on trial, and it obviously would not have been called into question—even in the slightest degree—by the disconfirmation of Crook and Crook's prediction about the adaptiveness of Tibetan polyandry. Williams (1985) points out that "predictions are tested to check on the truth of an understanding of the phenomena investigated. This understanding always includes special assumptions in addition to those of the major theory, and it is these that are confirmed or modified according to the outcome of an investigation" (p. 18). To evaluate the scientific significance of DSS research, such as that of Crook and Crook (1988), it is therefore necessary to discover and to evaluate the "special assumptions" (i.e., assumptions in addition to the theory that human nature is the product of natural selection) that underlie the DSS prediction that human behavior will be adaptive.

ADAPTATION AND ADAPTIVENESS

In describing the DSS enterprise, Irons (1979) writes, "The statement that a particular form of behavior is adaptive to a particular environment is a statement about its effect on survival and reproduction and nothing more" (p. 38; also see Dunbar, 1988). But the statement that a particular form of behavior is an *adaptation* to a particular environment does not imply the current existence of beneficial effects on survival and reproduction; it implies that during the course of evolutionary history selection produced *that particular form of behavior* because that form served a specific function more efficiently than available alternative forms did (Thornhill, 1990, Williams, 1966). The focus of DSS research thus is not adaptation but rather what Betzig (1989) calls "adaptiveness."

Studies of adaptiveness are not unique to DSS. They have been pursued for three reasons: First, to shed light on adaptations. For example, Lack (1954) collected evidence that songbirds producing clutches that are smaller or larger than average typically fledge fewer offspring than do conspecifics producing average-size clutches, and he argued that these data support the hypothesis that producing the average-size clutch is an adaptation. (Even though Lack did not describe specific physiological mechanisms, his hypothesis was that the songbird species in question possesses at least one physiological mechanism *specialized* to produce the average-size clutch.)

Although studies of adaptiveness may sometimes play a role in the adaptationist program, it would be a non sequitur to conclude that measurements of differential reproduction illuminate adaptations from the premise that adaptations are produced by differential reproduction. For a number of reasons, correlating individual variation in the expression of a trait with reproductive success is normally an ineffective or ambiguous way to study adaptation. For example, one reason is that such correlations generally are superfluous. That the lens of the vertebrate eye is adapted to focusing light on the retina is unambiguously manifested in the eye's design. Comparing the fertility of living individuals whose lenses focused light behind, on, and in front of their retinas would be pointless (such comparisons could not undermine the evidence of design no matter what results were obtained). A second reason is that the expression of an adaptation designed to cope with fitness-threatening exigencies might often correlate *negatively* with reproductive success. If, say, fever in mammals functions to combat pathogen infection, then individuals with the highest fevers might typically have fewer offspring than individuals without fevers (who were not infected in the first place) and individuals with lower fevers (who were infected less severely or who happened to have more efficient immune systems). A third reason is that an adaptation that is genetically fixed in population (i.e., with a heritability of zero) might vary capriciously with environmental variation, and, hence, its expression probably would not be correlated with reproductive success. Fourth, correlations too small to detect may have great selective significance over evolutionary time. Fifth, a given trait may covary with reproductive success merely because both are correlated with a third variable. Sixth, a given trait may promote fitness—and hence correlate positively with reproductive success—because it currently produces some effect other than its evolved function. Seventh, the hypothesis that trait X is an adaptation does not imply that trait X is currently adaptive (e.g., the human sweet tooth is an adaptation whether or not it is currently adaptive). As Hailman (1980) remarks, "Correlation of individual variation with reproductive success . . . is a mental briar patch that scratches all who enter" (p. 189).

The measurement of reproductive differentials has always been a problematical tool in the adaptationist program because such measurement, in and of itself, "simply does not measure degree of fitness or adaptedness in any of the senses relevant to evolutionary theory—all of which have something to do with *systematic or designed* fit with the environment, with causally mediated propensities to reproductive success" (Burian, 1983, p. 299; also see Thornhill, 1990). The hypothesis that human males evolved specialized female-nubility-preferring psychological mechanisms, for example, does not imply any of the following: (a) that selection is currently favoring a sexual preference for nubile females; (b) that variation in the phenotypic expression of this preference is correlated with male reproductive succes; and (c) that comparing the reproductive success of males who are and who are not sexually attracted to nubile females (other things being equal) would illuminate either male psychology or the evolutionary processes that produced that psychology.

The second reason for studying adaptiveness is that such investigations may sometimes shed light on the selective forces that maintain an adaptation when these differ from the forces that produced it. For example, the mottled pattern of black-headed gulls' eggs strongly implies a camouflage function. Since experiments have shown that parent birds sit better on normal than on artificially solid-colored eggs, however, the mottled pattern may be maintained by selection not only because mottled eggs are more cryptic but also because they provide a more effective sitting stimulus for parents (Tinbergen, 1967). (This does not imply that a function of mottling is to provide an effective sitting stimulus: That parents prefer mottled to solid-colored eggs reveals something about the design of black headed gulls' brains, not their eggs.)

Third, Caro and Borgerhoff Mulder (1987) argue that studying the adaptiveness of current human activities might make it possible to predict the future course of evolution. To accurately predict that selection will favor currently adaptive activity X, however, one would have to know that the tendency to perform X is heritable; that selection pressures on the X-producing psychological and physiological mechanisms in other contexts will not counter or simply wash out any tendency for selection to favor the performance of X specifically; and that the relevant environmental conditions will persist for an evolutionarily significant span of time. The likelihood of knowing these things strikes me as being sufficiently remote to rule out a prognostic rationale for studying the adaptiveness of human activities.

DSSes study adaptiveness for yet another reason: They assume that such investigations provide the raw material for building a science of human behavior. This assumption, however, is a conceptual muddle. DSSes invoke such notions as "Darwinism," "modern evolutionary theory," "natural selection," "sexual selection," "kin selection," and "inclusive fitness" to justify their inquiries into the adaptivenss of human behavior, but evolutionary fact and theory can be logically coupled to the study of human behavior only via adaptations (Barkow, 1984; Cosmides & Tooby, 1987). Except for such species-typical behavioral patterns as walking, running, smiling, and crying, human behavior per se was not designed by selection; rather, most human behavior is the product of the interaction of myriad psychological mechanisms, and it is these mechanism that were designed by selection. To the extent that DSSes' descriptions of human behavior are agnostic with respect to psychological mechanisms, they are adaptation-free, past-free, and gene-free, and, therefore, not logically coupled to evolutionary fact and theory.

Dawkins (1982) elaborates this point in the following anecdote about a seminar he attended in which a DSS interpreted human polyandry in terms of kin selection:

> Though fascinated by the information he presented, I tried to warn him of some difficulties in his hypothesis. I pointed out that the theory of kin selection is fundamentally a genetic theory, and that kin-selected adaptations to local conditions had to come about through the replacement of alleles by other alleles, over generations. Had his polyandrous tribes been living, I asked, under their current peculiar conditions for long enough—enough generations—for the necessary genetic replacement to have taken place? Was there, indeed, any reason to believe that variations in human mating systems are under genetic control at all?
>
> The speaker, supported by many of his anthropological colleagues in the seminar, objected to my dragging genes into the discussion. He was not talking about genes, he said, but about a social behavior pattern.... I tried to persuade him that it was *he* who had 'dragged genes in' to the discussion although, to be sure, he had not mentioned the word gene in his talk.... You cannot talk about kin selection, or any other form of Darwinian selection, *without* dragging genes in, whether you do so explicitly or not. By even speculating about kin selection as an explanation of differences in tribal mating systems, my anthropologist friend was implicitly dragging genes into the discussion. It is a pity he did not make it *explicit,* because he would then have realized what formidable difficulties lay in the path of his kin selection hypothesis: either his polyandrous tribes had to have been living, in partial genetic isolation, under their peculiar conditions for a large number of centuries, or natural selection had to have favoured the universal occurrrence of genes programming some complex 'conditional strategy.' (pp. 27–28)

According to Irons (1979), the DSS focus on human behavior rather than psychology has a distinctively Darwinian justification: "It is actual behavior which influences reproductive success directly" (p. 9). This argument, however, should be stood on its head: Darwin's theory of natural selection is a theory of adaptation, a *historical* account of the origin and maintenance of phenotypic design. In the study of adaptation, the key issue is whether differential reproductive success historically influenced the design of a given trait, not whether the trait currently influences differential reproductive success. In short, nothing in the theory of evolution by natural selection justifies an adaptation-agnostic science of adaptiveness. As J. Tooby (personal communication, 1989) notes, the study of adaptivenss merely draws metaphorical inspiration from Darwinism, whereas the study of adaptation *is* Darwinian. The tendency of DSSes to focus on adaptiveness stems in part, I believe, from their assumption that the social science Rosetta stone, the key to deciphering the deep structure of human affairs, is reproduction.

REPRODUCTION MINDEDNESS

Not only did Darwin's theory of natural selection account for adaptation and function, it gave them more precise meanings. Adaptations are designed to promote reproduction in specific environments—in modern terms, to promote the survival of genes. "Reproduction mindedness" thus can be useful to students of adaptation. The function of the peacock's tail, for example, would never be discovered by someone who assumed that the tail was designed to promote the survival of peacocks (rather than to promote the survival of peacock genes).

Reproduction mindedness, however, is no conceptual magic carpet, capable of soaring over questions of adaptation. Consider, for example, Alexander's (1987) reproduction-minded discussion of human male sexuality. Alexander argues that "for men much of sexual activity has had as a main (ultimate) significance the initiating of pregnancies" (p. 216). "This argument," he says, "does not predict, for example, that men will *never* wish to use contraceptives or have vasectomies (both evolutionary novelties), but it does predict that most males will be reluctant to do the first and *exceedingly* reluctant to do the second" (p. 218). He continues:

> It will probably be necessary to defend part of the above argument further. Thus, some men will undoubtedly assert that they are *more* interested in sex if there is no possibility of pregnancy, or that they are only interested in sex when contraceptives are used. Although it is not the principal theme of this essay, what is required is a review of the interplay of proximate and ultimate mechanisms. A man behaving so as to avoid pregnancies, and who derives from an evolutionary background of avoiding pregnancies, should be expected to favor copulation with women who are for age or other reasons incapable of pregnancy. A man derived from an evolutionary process in which securing of pregnancies typically was favored, may be expected to be most interested sexually in women most likely to become pregnant and near the height of the reproductive probability curve. This means that men should usually be expected to anticipate the greatest sexual pleasure with young, healthy, intelligent women who show promise of providing superior parental care. (p. 218)

Whether or not Alexander's predictions about male responses to contraceptives and vasectomies are correct, his justification for these predictions is at best incomplete. Neither the fact that men derive from an evolutionary process in which securing pregnancies was favored nor the fact that men typically fancy young, healthy women justifies the prediction that most men will be reluctant to use contraceptives. As discussed above, the psychological mechanisms that underpin the human male's perception of female sexual attractiveness were designed by natural selection in the EEA to assess *specific correlates* of mate value; for example, the human male's predilections for young and healthy women presumably reflect the operation of specialized psychological mechanisms: in ancestral populations, men with heritable tendencies to be sexually atttracted to certain physical correlates of youth and health typically outreproduced men whose tastes ran to physical correlates of old age and disease. (Since an intelligent mate may often have been a better mate, Alexander's assumption that men evolved a sexual taste for intelligent women is a psychological hypothesis worth investigating.) To be logically justified, a prediction about male sexual responses to evolutionary novelties, such as modern contraceptive technology, must *necessarily* be based on specific knowledge or assumptions about the psychological machinery of sexual attraction. In the absence of such psychological knowledge or assumptions, information about the probable reproductive consequences of sexual intercourse with a given woman (e.g., information about her use of oral contraceptives) provides no basis whatever for predicting her sexual attractiveness. Merely knowing that male psychology is the product of natural selection is not enough.

PREDICTION AND KNOWLEDGE

The obsession with falsifiable prediction and quantification that is endemic in the social sciences may in part account for the emphasis in DSS on the testing of predictions about differential reproduction. Yet the history of the social sciences stands as a monument to the proposition that testing falsifiable predictions and acquiring scientifically significant knowledge are not synonymous. All predictions that have some minimal empirical content and are not excessively vague are falsifiable, but falsifiability per se is no guarantee of scientific significance. The key question with respect to any prediction is this: If it is disconfirmed, what theory, hypothesis, or assumption is thereby called into question? The more precisely a student of adaptation characterizes phenotypic design the less ambiguous his predictions will be and the more likely it is that testing these predictions will illuminate adaptation. Similarly, the more precise an evolution-minded social scientist's assumptions are about human nature, the more closely coupled his predictions will be to evolutionary fact and theory, and the more likely it is that testing these predictions will be scientifically significant.

To illustrate this point, I will contrast two predictions made by Betzig (1988) in her evolution–inspired ethnographic research on Ifaluk, an atoll in the Western Caroline Islands. Her first prediction is that Ifaluk chiefs will profit materially from their high status. This prediction, I believe, is based on definite, albeit implicit, assumptions about phenotypic design (human nature). Despite evidence in the ethnographic record that in many pre-state societies high-ranking individuals control the distribution of resources and skim off the fat for themselves and their relatives, Betzig argues that two theoretical positions—substantivism and Marxism—hold that in pre-state societies there is always plenty of everything, and, hence, those in power distribute resources in a disinterested, nonexploitive fashion. Evolutionary psychological assumptions made Betzig suspicious of these theories. She investigated the matter and found that "the material advantage in sharing food on Ifaluk is not on the subordinates' side. . . . Nor, in this respect, is access to these life sustaining resources equal. . . . Rather, consistent with the Darwinian prediction, chiefs appear to gain productively by their positions" (Betzig, 1988, p. 55).

Had Betzig's "Darwinian prediction" been disconfirmed, certain assumptions about human nature might have been called into question. Conversely, the confirmation of her prediction may undermine the—admittedly vague—assumptions about human nature that have led some social scientists to romanticize pre-state peoples and to grossly exaggerate cultural variability. As Betzig notes, in essence she has demonstrated that the relationship between power and access to material resources is the same on Ifaluk as it is in the societies most of use are familiar with (presumably because human nature is the same on Ifaluk as it is elsewhere). Furthermore, whether Ifaluk chiefs do or do not obtain a disproportionate share of material resources is potentially relevant to long-standing anthropological questions about the relationship between political power and resource acquisition, questions also of interest to other social scientists, historians, and people in general. In this respect, Betzig's research is analogous to that of Chagnon (1988), discussed above, on Yanomamo kin term manipulation.[1]

Now consider a second Betzig prediction: high-ranking Ifaluk men and their male successors will have higher fertility than low-ranking men. What assumption would be called into question if this prediction were to be disconfirmed? It is extremely unlikely that any assumption about the psychology of status or the evolution of that psychology would be affected. That human beings everywhere pursue status—indeed, that "status" is universally a meaningful psychological category—implies that in ancestral populations high-status individuals typically outreproduced low-status individuals (a proposition supported by many studies of nonhuman animals [Wilson, 1975] and band-level human societies [Symons,1979]). There is no known or suspected alternative explanation for the existence of the human status motive. Indeed, Vining (1986) notes that the correlations between status and reproductive success in modern Western societies is typically *inverse,* yet few, if any, of the many commentators on Vining's article seem to believe that such data can test any significant hypothesis about the psychology of status striving or the selective forces that produced that psychology. Furthermore, since no competing social science theory predicts anything at all about the relationship between status or material resources and reproductive success, the outcome of a test of Betzig's second prediction cannot contribute to the resolution of any theoretical issue in the social sciences. Nor is it likely that assumptions about human nature implicit in other social science theories would be called into question by the confirmation of Betzig's prediction.

But Betzig's reproductive prediction most likely was not intended to test assumptions about human nature or the selective forces that produced that nature, nor was it intended to undermine rival social science theories. Rather, it was probably intended to test the central DSS hypothesis that human behavior is designed to maximize inclusive fitness. Betzig did confirm her reproductive prediction; does this confirmation support the DSS hypothesis? Not necessarily. The question "Are human beings inclusive fitness maximizers?" can be met only with another question: "Compared with what?" One approach (which Betzig did not pursue) is to compare the behavior of ethnographic subjects with an imaginary social engineer's ideal design for inclusive fitness-maximizing behavior. Given the particular circumstances in which the ethnographic subjects find themselves, and given the range of options open to them, how closely does their behavior approximate the engineering ideal? Kitcher (1985) uses this approach to analyze several of DSS's most famous ethnographic examples, and he concludes that the ethnographic subjects' behavior does not closely approximate the social engineering ideal.

Kitcher's criterion may, however, be too stringent: Probably no DSSes actually believe that human behavior achieves some unconstrained ideal of fitness maximization. For one thing, DSSes are aware of the many factors that can prevent natural selection from achieving optimal phenotypic design. For another, most DSSes probably would admit that since human behavior per se was not designed by selection, it might for that reason alone fail to achieve the social engineering ideal. And finally, DSSes know that current human environments differ in many respects from the EEA. Alexander (1979) sums up the DSS hypothesis this way: "I have not suggested that culture precisely tracks the interests of the genes—obviously this is not true—but that, in historical terms, it does so much more closely than we might have imagined" (p. 142).

DSS's central hypothesis thus appears to be something like the following: Modern Darwinian theory predicts that human behavior will be surprisingly adaptive. The special assumptions that this hypothesis seems to entail (in addition to the uncontested assumption that human nature is the product of natural selection) are that (a) patterns perceivable in mechanism-agnostic descriptions of human behavior approximate fitness-maximizing designs to a surprising degree; (b) evolutionarily novel environments cause human behavior to deviate from the path of fitness maximization less than one might have imagined; and (c) the accumulated measurements of the reproductive consequences of human behavior will lead to a deep understanding of human affairs.

The central DSS hypothesis—and the special assumptions it comprises—can be evaluated by considering the following questions: (a) Is the hypothesis really derived from Darwinian theory? (b) Is it accurate? (c) Are tests of its accuracy scientifically significant? With respect to the first question, since Darwin's is a theory of adaptation, not adaptiveness, the DSS hypothesis is not clearly derived from the theory of evolution by natural selection. Even if DSSes had time machines and could conduct research in the EEA, there would be no Darwinian justification for an adaptation-free study of adaptiveness. The problems become acute, however, when human behavior is observed, as it always is, in evolutionarily novel environments. In fact, since the adaptations that underpin human behavior were designed by selection to function in specific environments, there is a principled Darwinian argument for assuming that behavior in evolutionarily novel environments will often be *mal*adaptive. But no Darwinian argument justifies the assumption that such behavior will be more adaptive than one might have imagined. In addition, the hypothesis entails comparing the adaptiveness of human behavior with people's expectations about adaptiveness, and it is unclear how a prediction about people's expectations could be derived from Darwinian theory.

The next question is whether the DSS hypothesis is accurate. Strictly speaking, we can't really be sure because we don't know how adaptive people generally imagine human behavior to be (we don't know, for example, whether people are likely to be surprised by Betzig's reproductive data). No rival social science theory predicts anything at all about adaptiveness per se, and it is probably a subject to which most people have never given a moment's thought. In the EEA, human behavior presumably was as adaptive as that of any other animal species in its EEA, although whether it was *surprisingly* adaptive is anybody's guess. As I have argued elsewhere (Symons, 1987c), however, in modern industrial societies human behavior is so poorly designed to maximize fitness, given the available reproductive opportunities, that it is hard to imagine anyone being surprised by how adaptive it is. Consider, for example, the reproductive opportunities available in the United States today to a healthy white woman who wanted to maximize her fitness: She could bear a baby every year or two, from nubility to menopause, and give it up for adoption, confident that it would be cherished by a middle-class family. A reasonably young male member of the *Forbes'* 400 could use his fortune to construct a reproductive paradise in which women and their children could live in modest affluence and security for life (as long as paternity was verified). Such a man might well sire two orders of magnitude more offspring than the current average for male members of the *Forbes'* 400. In a world in which people actually wanted to maximize inclusive fitness, opportunities to make deposits in sperm banks would be immensely competitive, a subject of endless public scrutiny and debate, with the possibility of reverse embezzlement by male sperm bank officers an ever-present problem. It is difficult to picture clearly a modern industrial society in which people strive to maximize inclusive fitness because such a society would have so little in common with our own.

Finally, there is the question of the scientific significance of tests of the DSS hypothesis: If certain human behaviors were shown to be surprisingly adaptive, some of the demonstrations of adaptiveness might inspire novel hypotheses about the nature or evolution of the human brain/mind (Betzig, 1989; Thornhill, 1990); but it is difficult to see how a science of human behavior could be based on the demonstrations themselves. In fact, the possibility of a mechanism-agnostic science of human behavior of any sort—Darwinian or otherwise—is by no means a matter of universal agreement. It is often argued that the history of the social sciences (which are characteristically mechanism-agnostic) is a history of failures to establish reliable and useful bodies of knowledge and theory (Lindblom & Cohen, 1979, Murdock, 1972, Rosenberg, 1980, Ziman, 1978). "A good rule of thumb to keep in mind," quips Searle (1984), "is that anything that calls itself 'science' [e.g., Christian Science, military science, library science, social science] probably isn't" (p.11).

In summary, the hypothesis that human behavior is surprisingly adaptive does not derive from Darwinian theory and is almost certainly wrong in modern industrial environments. This hypothesis is presumably falsifiable, but that does not necessarily imply that testing it is scientifically significant. The accumulated measurements of the reproductive consequences of human behavior are unlikely to lead to a deep understanding of human affairs because such measurements merely demonstrate that various activities either are or aren't adaptive; they do not constitute new knowledge about human nature or the evolutionary processes that produced that nature (Tooby & Cosmides, 1989b).

CONCLUSION

The adaptationist program, whose goal is to recognize certain features of organisms as components of some special problem-solving machinery (Williams, 1985), has been the foundation of biology for centuries. Darwin's theory of evolution by natural selection added to this program a scientifically coherent account of the origin and maintenance of adaptations as well as a powerful heuristic tool: Darwinism provides a guide to research on phenotypic design, prevents certain kinds of errors, and raises suspicions of certain explanations and theories. The subject matter of the adaptationist program is structural, physiological, behavioral, or psychological phenotypic features that have been shaped by selection to serve some function. Although adaptations are produced by differential reproduction, studying differential reproduction does not necessarily illuminate adaptations. As Williams (1966) remarks, "Measuring reproductive success focuses attention on the rather trivial problem of the degree to which an organism actually achieves reproductive survival. The central biological problem is not survival as such, but design for survival" (p. 159). Current effects of phenotypic features—including beneficial effects on reproduction—thus are relevant to the adaptationist program only insofar as the study of such effects illuminates adaptations.

Evolutionary psychology is the application of the adaptationist program to the study of the human brain/mind. Evolutionary psychologists assume that the brain/mind has many functions—i.e., that it has been designed by selection to solve many different kinds of problems, each of which is likely to require its own distinctive kind of solution—and, therefore, that the brain/mind comprises many domain-specific, specialized mechanisms. For example, selectional thinking leads to the expectation that human perceptions of sexual attractiveness are underpinned by many specialized mechanisms (which operate according to their own distinctive rules and principles) rather than by some sort of generalized "learning" or "capacity-for-culture" mechanisms.

Cultural anthropologists are uniquely well situated to investigate evolutionary psychological hypotheses because the cornerstone of anthropology is cross-cultural comparison and because anthropological reseach among band-level peoples provides insight into the human environment of evolutionary adaptedness. Social scientists can use evolutionary psychology as a guide to research. For example, evolutionary psychological assumptions made Betzig (1988) suspicious of Marxist and substantivist claims about the relationship between political power and resource acquisition in pre-state societies. She investigated the matter on Ifaluk, and her findings are significant both with respect to questions of human nature and with respect to long-standing issues in the social sciences. On the other hand, her psychologically agnostic demonstration that high-ranking Ifaluk men and their male successors have higher fertility than low-ranking men does not seem to me to be scientifically significant because these data do not bear on questions of human nature or the selective forces that shaped that nature, nor do they have any apparent relevance for issues in the social sciences.

In sum, Darwin's theory of natural selection sheds light on human behavior only insofar as it sheds light on the adaptations that constitute the machinery of behavior: A science of human behavior cannot be simultaneously psychologically agnostic and genuinely Darwinian. Although great and illuminating new truths about human beings may not follow automatically from the realization that we evolved, Darwinism can be a guide and source of inspiration for students of human nature.

ACKNOWLEDGMENTS

I thank Jerome Barkow, Donald Brown, Napoleon Chagnon, Leda Cosmides, Bruce Ellis, Kalman Glantz, Yonie Harris, Sarah Hrdy, Randy Thornhill, John Tooby, and Richard Wrangham for their careful and thoughtful readings of various drafts of this essay and for their many insightful comments; I'm particularly grateful to Margie Profet, who made it shorter, sweeter, and more comprehensible—even to me. Thanks too to Alexander Robertson and John Tooby for "personal communications."

NOTE

1. There are two potential objections to Betzig's interpretation of these data. First, it is an open question whether those who consider themselves to be substantivists and Marxists would agree that their theoretical positions imply that Ifaluk chiefs can be expected *not* to skim off the fat for themselves and their relatives (A. Robertson, personal communication, 1989). Nevertheless, it seems clear to me that there is a widespread belief that people in non-state societies are nicer and less self-interested than people in state societies, and it is to this belief that Betzig's data speak. Second, Alexander (1988) argues that Betzig does not really demonstrate that Ifaluk chiefs exploit their subordinates: Chiefs may acquire material resources from those beneath them as

fair compensation for services rendered, just as a highly paid CEO may be worth her salary. This cogent argument does not materially affect the moral I wish to draw from Betzig's research: Betzig's prediction implies a cynical view of human nature—which she takes to be the Darwinian view—in which people in pre-state societies, like people in state societies, often *will* exploit one another when the opportunity arises and the cost is not prohibitive (whether or not they are actually doing so in this particular case).

REFERENCES

Alexander, R. D. (1975). The search for a general theory of behavior. *Behavioral Science, 20,* 77–100.

Alexander, R. D. (1979). *Darwinism and human affairs.* Seattle, WA: University of Washington Press.

Alexander, R. D. (1987). *The biology of moral systems.* New York: Aldine de Gruyter.

Alexander, R. D. (1988). Evolutionary approaches to human behavior: What does the future hold? In L. Betzig, M. Borgerhoff Mulder, & P. Turke (Eds.), *Human reproductive behaviour: A Darwinian perspective* (pp. 317–341). New York: Cambridge University Press.

Barkow, J. H. (1980). Sociobiology: Is this the new theory of human nature? In A. Montagu (Ed.), *Sociobiology examined* (pp. 171–197.) New York: Oxford University Press.

Barkow, J. H. (1984). The distance between genes and culture. *Journal of Anthropological Research, 40,* 367–379.

Barkow, J. H. (1986). Central problems of sociobiology. *Behavioral and Brain Sciences, 9,* 188.

Betzig, L. (1988). Redistribution: Equity or exploitation? In L. Betzig, M. Borgerhoff Mulder, & P. Turke (Eds.), *Human reproductive behaviour: A Darwinian perspective (pp. 49–63).* New York: Cambridge University Press.

Betzig, L. (1989). Rethinking human ethology: A response to some recent critiques. *Ethology and Sociobiology, 10,* 315–324.

Betzig, L., Borgerhoff Mulder, M., & Turke, P. (Eds.). (1988). *Human reproductive behaviour: A Darwinian perspective.* New York: Cambridge University Press.

Burian, R. M. (1983). "Adaptation." In M. Grene (Ed.), *Dimensions of Darwinism* pp. 287–314). New York: Cambridge University Press.

Buss, D. M. (1989). Sex differences in human mate preferences: Evolutionary hypotheses tested in 37 cultures. *Behavioral and Brain Sciences, 12,* 1–49.

Caro, T. M., & Borgerhoff Mulder, M. (1987). The problem of adaptation in the study of human behavior. *Ethology and Sociobiology, 8,* 61–72.

Chagnon, N. A. (1968). *Yanomamo: The fierce people.* New York: Holt, Rinehart and Winston.

Chagnon, N. A. (1988). Male Yanomamo manipulations of kinship classifications of female kin for reproductive advantage. In L. Betzig, M. Borgerhoff Mulder, & P. Turke (Eds.), *Human reproductive behaviour: A Darwinian perspective (pp. 23–48).* New York: Cambridge University Press.

Chagnon, N. A., & Irons, W. (Eds.). (1979). *Evolutionary biology and human social behavior: An anthropological perspective.* North Scituate, MA: Duxbury Press.

Cosmides, L., & Tooby, J. (1987). From evolution to behavior: Evolutionary psychology as the missing link. In J. Dupre (Ed.), *The latest on the best: Essays on evolution and optimality* (pp. 277–306). Cambridge, MA: MIT Press.

Crook, J. H., & Crook, S. J. (1988). Tibetan polyandry: Problems of adaptation and fitness. In L. Betzig, M. Borgerhoff Mulder, & P. Turke (Eds.), *Human reproductive behaviour: A Darwinian perspecitve.* (pp. 97–114). New York: Cambridge University Press.

Curio, E. (1973). Towards a methodology of teleonomy. *Experientia, 29,* 1045–1058.

Daly, M., & Wilson, M. (1983). *Sex, evolution, and behavior* (2nd ed.). Boston: Willard Grant Press.

Daly, M., & Wilson, M. (1984). A sociobiological analysis of human infanticide. In G. Hausfater & S. B. Hrdy (Eds.), *Infanticide* (pp. 487–502). New York: Aldine de Gruyter.

Daly, M., & Wilson, M. (1986). A theoretical challenge to a caricature of Darwinism. *Behavioral and Brain Sciences, 9,* 189–90.

Daly, M., & Wilson, M. (1988). *Homicide.* New York: Aldine de Gruyter.

Darwin, C. (1859). *On the origin of species by means of natural selection, or the preservation of favoured races in the struggle for life.* London: Watts and Co.

Dawkins, R. (1976). *The selfish gene.* Oxford: Oxford University Press.

Dawkins, R. (1982). *The extended phenotype.* San Francisco: W. H. Freeman.

Dawkins, R. (1986). *The blind watchmaker.* New York: W.W. Norton.

Doty, R. L. (1985). The primates III: Humans. In R. E. Brown & D. W. Macdonald (Eds.), *Social odours in mammals* (Vol. 2) (pp. 804–832). Oxford: Clarendon Press.

Dunbar, R.I.M. (1988). Darwinizing man: A commentary. In L. Betzig, M. Borgerhoff Mulder, & P. Turke (Eds.), *Human reproductive behaviour: A Darwinian perspective* (pp. 161–169). New York: Cambridge University Press.

Hailman, J. P. (1980). Fitness, function, fidelity, fornication, and feminine philandering. *Behavioral and Brain Sciences, 3,* 189.

Irons, W. (1979). Natural selection, adaptation and human social behavior. In N. Chagnon & W. Irons, (Eds.), *Evolutionary biology and human social behavior: An anthropological perspective* (pp. 4–39). North Scituate, MA: Duxbury Press.

Kitcher, P. (1985). *Vaulting ambition.* Cambridge, MA: MIT Press.

Lack, D. (1954). *The natural regulation of animal numbers.* New York: Oxford University Press.

Lindblom, C. E., & Cohen D. K. (1979). *Usable knowledge.* New Haven: Yale University Press.

Lloyd, J. E. (1979). Mating behavior and natural selection. *The Florida Entomologist, 62,* 17–34.

Maynard Smith, J. (1982). Introduction. In King's College Sociobiology Study Group (Ed.), *Current problems in sociobiology* (pp. 1–3). New York: Cambridge University Press.

Mayr, E. (1983). How to carry out the adaptationist program? *The American Naturalist, 121,* 324–334.

Medawar, P. (1982). *Pluto's republic.* New York: Oxford University Press.

Murdock, G. P. (1972). Anthropology's mythology. *Proceedings of the Royal Anthropological Institute of Great Britain and Ireland for 1971* (pp. 17–24).

Pittendrigh, C. S. (1958). Adaptation, natural selection, and behavior. In A. Roe & G. G. Simpson (Eds.), *Behavior and Evolution* (pp. 390–416). New Haven: Yale University Press.

Rosenberg, A. (1980). *Sociobiology and the preemption of social science.* Baltimore: Johns Hopkins University Press.

Searle, J. (1984). *Minds, brains and science.* Cambridge, MA: Harvard University Press.

Symons, D. (1979). *The evolution of human sexuality.* New York: Oxford University Press.

Symons, D. (1987a). An evolutionary approach: Can Darwin's view of life shed light on human sexuality? In J. H. Geer & W. O'Donohue (Eds.), *Theories of human sexuality* (pp. 91–125). New York: Plenum Press.

Symons, D. (1987b). If we're all Darwinians, what's the fuss about? In C. B. Crawford, M. F. Smith, & D. L. Krebs (Eds.), *Sociobiology and psychology: Ideas, issues, and applications* (pp. 121–146). Hillsdale, NJ: Lawrence Erlbaum Associates.

Symons, D. (1987c). Reproductive success and adaptation. *Behavioral and Brain Sciences, 10,* 788–789.

Symons, D. (1989). A critique of Darwinian anthropology. *Ethology and Sociobiology, 10,* 131–144.

Thornhill, R. (1990). The study of adaptation. In M. Bekoff and D. Jamieson (Eds.), *Interpretation and explanation in the study of animal behavior* (Vol. 2, pp. 31–62). Boulder, CO: West View Press.

Tinbergen, N. (1967). Adaptive features of the black-headed gull. In D. W. Snow (Ed.), *Pro-*

ceedings of the XIV International Ornithological Congress (pp. 43–59). Oxford: Blackwell Scientific Publications.

Tooby, J. (1985). The emergence of evolutionary psychology. In D. Pines (Ed.), Emerging syntheses in science (pp. 1–6). Santa Fe, NM: Santa Fe Institute.

Tooby, J., & Cosmides, L. (1989a). Evolutionary psychologists need to distinguish between the evolutionary process, ancestral selection pressures, and psychological mechanisms. Behavioral and Brain Sciences, 12, 724–725.

Tooby, J., & Cosmides, L. (1989b). Evolutionary psychology and the generation of culture, part I: Theoretical considerations. Ethology and Sociobiology, 10, 29–49.

Tooby, J., & Cosmides, L. (1990). On the universality of human nature and the uniqueness of the individual: The role of genetics and adaptation. Journal of Personality, 58, 17–67.

Tooby, J., & DeVore, I. (1987). The reconstruction of hominid behavioral evolution through strategic modeling. In W. G. Kinzey (Ed.), The evolution of human behavior: Primate models (pp. 183–237). Albany, NY: State University of New York Press.

Townsend, J.M. (1987). Sex differences in sexuality among medical students: Effects of increasing socioeconomic status. Archives of Sexual Behavior, 16, 425–441.

Townsend, J. M. (1989). Mate selection criteria: A pilot study. Ethology and Sociobiology, 10, 241–253.

Turke, P. W., & Betzig, L. L. (1985). Those who can do: Wealth, status, and reproductive success on Ifaluk. Ethology and Sociobiology, 6, 79–87.

van den Berghe, P. L., & Frost, P. (1986). Skin color preference, sexual dimorphism, and sexual selection: A case of gene-culture co-evolution? Ethnic and Racial Studies, 9, 87–113.

Vining, D. R. (1986). Social versus reproductive success: The central theoretical problem of human sociobiology. Behavioral and Brain Sciences, 9, 167–187.

Williams, G. C. (1966). Adaptation and natural selection. Princeton, NJ: Princeton University Press.

Williams, G. C. (1985). A defense of reductionism in evolutionary biology. Oxford Surveys in Evolutionary Biology, 2, 1–27.

Wilson, E. O. (1975). Sociobiology: The new synthesis. Cambridge, MA: Harvard University Press.

Ziman, J. (1978). Reliable knowledge. Cambridge: Cambridge University Press.

II

COOPERATION

For many people, Darwin's view of life conjures up images of a Hobbesian war of all against all, where life is "solitary, poor, nasty, brutish and short," and of a "nature red in tooth and claw." Indeed, in popular discourse, to call something "Darwinian" is intended to convey the image of ruthless and exploitative competition, where individuals pursue their own interests regardless of what costs this inflicts or consequences it has on others. But for the community of evolutionary biologists, the persistence of these popular stereotypes into the 1990s is paradoxical. Far from being preoccupied with the study of competition and aggression, for more than a quarter of a century evolutionary biologists have been keenly interested in understanding such phenomena as altruism and cooperation. The desire to understand the conditions under which organisms can be expected to help one another, cooperate with one another, share with one another, and even sacrifice themselves for one another has generated some of the most important recent advances in evolutionary biology, such as Hamilton's (1964) inclusive fitness theory, Williams and Williams's (1957) work on "social donorism," Trivers's (1971) reciprocal altruism theory, and Axelrod and Hamilton's (1981) work on the evolution of cooperation. Prior to this work, most standard analyses of the evolutionary process led biologists to expect that it would be difficult or impossible for altruism to evolve, and that organisms were indeed designed to be "selfish." However, these theoretical advances have transformed our understanding both of how natural selection operates and of what kinds of adaptations it can be expected to produce. Instead of the traditional view that selfishness is "natural" and altruism is only imposed socially against natural inclination, evolutionary biology has discovered that altruism and cooperation can be as natural as selfishness. In fact, these analyses have shown that Hobbes was quite wrong: Cooperation *can* emerge in the absence of a Leviathan, and adaptations for the expression and regulation of cooperation and altruism are expected design features of social organisms.

The chapters in this part focus on investigating some of the adaptations that govern cooperation in humans and other species. Evolutionary biologists have analyzed the conditions under which adaptations for engaging in cooperative behavior can be expected to evolve. These analyses show that cognitive mechanisms for engaging in cooperation can be selected for only if they solve certain complex information-processing problems. To solve these problems efficiently, the cognitive mechanisms involved must have certain specific design features. Cosmides and Tooby review experiments that have been conducted to see whether these predicted design features actually exist. The results of these experiments strongly suggest that they do. The human cognitive architecture appears to contain algorithms that are well designed for reasoning about cooperation and that cannot be explained as the by-product of more general-purpose cognitive processes. In other words, we seem to have information-processing mechanisms that are adaptations for reasoning about cooperation.

Cosmides and Tooby used theories about selection pressures to generate hypotheses about the human cognitive architecture. In contrast, McGrew and Feistner used evidence about the adaptations governing cooperation in different primate species to generate hypotheses about the selection pressures that shaped those species. Game-theoretic analyses in evolutionary biology had already shown that some selection pressures are species-general: They apply to any species that has evolved the ability to engage in cooperation. But other selection pressures are necessarily species-specific. A species's ecology and its prior phenotypic structure can profoundly affect the design of any newly arising adaptation.

Although the adaptations governing cooperation in different species show some strking commonalities—commonalities that reflect species-general selection pressures—they also differ in some rather interesting ways. These differences should reflect species-specific histories of selection. McGrew and Feistner's goal was to reconstruct some of the selection pressures that shaped cooperation in the human line. To this end, they compared food sharing in chimpanzees, callitrichids, and humans to see whether cooperation in these species differs merely in degree or in kind. They then examined the relationship among food sharing, tool use, and the ecologies of these different species, in an attempt to shed light on the unique configuration of selection pressures that caused the human line to diverge so profoundly from that of our primate cousins.

REFERENCES

Axelrod, R., & Hamilton, W. D. (1981). The evolution of cooperation. *Science, 211*, 1390–1396.
Hamilton, W. D. (1964). The genetical evolution of social behaviour. *Journal of Theoretical Biology, 7*, 1–52.
Trivers, R. L. (1971). The evolution of reciprocal altruism. *Quarterly Review of Biology, 46*, 35–57.
Williams, G. C., & Williams, D. C. (1957). Natural selection of individually harmful social adaptations among sibs with special reference to social insects. *Evolution, 17*, 249–253.

3

Cognitive Adaptations for Social Exchange

LEDA COSMIDES AND JOHN TOOBY

INTRODUCTION

Is it not reasonable to anticipate that our understanding of the human mind would be aided greatly by knowing the purpose for which it was designed?

GEORGE C. WILLIAMS

Research Background

The human mind is the most complex natural phenomenon humans have yet encountered, and Darwin's gift to those who wish to understand it is a knowledge of the process that created it and gave it its distinctive organization: evolution. Because we know that the human mind is the product of the evolutionary process, we know something vitally illuminating: that, aside from those properties acquired by chance, the mind consists of a set of adaptations, designed to solve the long-standing adaptive problems humans encountered as hunter-gatherers. Such a view is uncontroversial to most behavioral scientists when applied to topics such as vision or balance. Yet adaptationist approaches to human psychology are considered radical—or even transparently false—when applied to most other areas of human thought and action, especially social behavior. Nevertheless, the logic of the adaptationist postion is completely general, and a dispassionate evaluation of its implications leads to the expectation that humans should have evolved a constellation of cognitive adaptations to social life. Our ancestors have been members of social groups and engaging in social interactions for millions and probably tens of millions of years. To behave adaptively, they not only needed to construct a spatial map of the objects disclosed to them by their retinas, but a social map of the persons, relationships, motives, interactions, emotions, and intentions that made up their social world.

Our view, then, is that humans have a faculty of social cognition, consisting of a rich collection of dedicated, functionally specialized, interrelated modules (i.e., functionally isolable subunits, mechanisms, mental organs, etc.), organized to collectively guide thought and behavior with respect to the evolutionarily recurrent adaptive problems posed by the social world. Nonetheless, if such a view has merit, it not only must be argued for on theoretical grounds—however compelling—but also must be substantiated by experimental evidence, as well as by converging lines of empirical support drawn from related fields such as neuroscience, linguistics, and anthropology. The eventual goal is to recover out of carefully designed experimental studies "high-reso-lution" maps of the intricate mechanisms involved. Such an approach is intended to exploit the signal virtue of cognitive psychology: With its emphasis on mechanisms, cognitive approaches allow causal pathways to be precisely specified through reference to explicitly described algorithms and representations.

Toward this end, we have conducted an experimental research program over the last eight years, exploring the hypothesis that the human mind contains algorithms (specialized mechanisms) designed for reasoning about social exchange. The topic of reasoning about social exchange was selected for several reasons. In the first place, as we will discuss, many aspects of the evolutionary theory of social exchange (also some-times called cooperation, reciprocal altruism, or reciprocation) are relatively well developed and unambiguous. Consequently, certain features of the functional logic of social exchange can be confidently relied on in constructing hypotheses about the structure of the information-processing procedures that this activity requires.

In the second place, complex adaptations are constructed in response to evolu-tionarily long-enduring problems, and it is likely that our ancestors have engaged in social exhange for at least several million years. Several converging lines of evidence support this view. Social exchange behavior is both universal and highly elaborated across all human cultures—including hunter-gatherer cultures (e.g., Cashdan, 1989; Lee & DeVore, 1968; Sharp, 1952; Wiessner, 1982)—as would be expected if it were an ancient and central part of human social life. If social exchange were merely a recent invention, like writing or rice cultivation, one would expect to find evidence of its hav-ing one or several points of origin, of its having spread by contact, and of its being extremely elaborated in some cultures and absent in others. Moreover, the nearest rel-atives to the hominid line, the chimpanzees, also engage in certain types of sophisti-cated reciprocation (de Waal, 1982; de Waal & Luttrell, 1988), which implies that some cognitive adaptations to social exchange were present in the hominid lineage at least as far back as the common ancestors that we share with the chimpanzees, five to ten million years ago. Finally, paleoanthropological evidence also supports the view that exchange behavior is extremely ancient (e.g., Isaac, 1978; McGrew & Feistner, this volume; Tooby & DeVore, 1987). These facts, plus the existence of reciprocation among members of primate species that are even more distantly related to us than chimpanzees—such as macaques and baboons (Packer, 1977; de Waal & Luttrell, 1988)—strongly support the view that situations involving social exchange have con-stituted a long-enduring selection pressure on hominids.

The third reason we selected reasoning about social exchange as the focus of this experimental series was that theories about reasoning and rationality have played a central role in both cognitive science and the social sciences. Research in this area can, as a result, function as a powerful test of certain traditional social science postulates. An adaptationist approach to human psychology is often viewed as radical or false not because of gaps in its logic or any comparative lack of evidence for its hypotheses, but because it violates certain privileged tenets of this century's dominant behavioral and social science paradigm—what we have called elsewhere the Standard Social Science Model (see Tooby & Cosmides, this volume). According to this view, all of the specific content of the human mind originally derives from the "outside"—from the environ-ment and the social world—and the evolved architecture of the mind consists solely or predominantly of a small number of general-purpose mechanisms that are *content-independent,* and which sail under names such as "learning," "induction," "intelligence," "imitation," "rationality," "the capacity for culture," or, simply, "culture." On this view, the same mechanisms are thought to govern how one acquires a language and how one acquires a gender identity. This is because the mechanisms that govern reasoning, learning, and memory are assumed to operate uniformly across all domains: They do not impart content, they are not imbued with content, and they have no features specialized for processing particular kinds of content. Hypotheses that are inconsistent with this content-free view of the mind are, a priori, not considered credible, and the data that support them are usually explained away by invoking as alternatives the operation of general-purpose processes of an unspecified nature. Strong results indicating the involvement of domain-specific adaptations in areas such as perception, language, and emotion have sometimes—though grudgingly—been accepted as genuine, but have been ghettoized as exceptional cases, not characteristic of the great majority of mental processes.

In this dialogue, reasoning has served as the paradigm case of the "general-purpose" psychological process: It has been viewed as preeminently characteristic of those processes that are purportedly the central engine of the human mind. Even vigorous advocates of modularity have held so-called central processes, such as reasoning, to be general-purpose and content-independent (e.g., Fodor, 1983). Consequently, we felt that reasoning about social exchange offered an excellent opportunity to cut to the quick of the controversy. If even human reasoning, the doctrinal "citadel" of the advocates of content-free, general-purpose processes, turns out to include a large number of content-dependent cognitive adaptations, then the presumption that psychological mechanisms are characteristically domain-general and originally content-free can no longer be accorded privileged status. Such results would jeopardize the assumption that whenever content-dependent psychological phenomena are found, they necessarily imply the prior action of cultural or environmental shaping. Instead, such results would add credibility to the contrary view that the mind is richly textured with content-specialized psychological adaptations.

Evolutionary biologists have developed useful criteria for establishing the existence of adaptations (e.g., Dawkins 1982, 1986; Symons, this volume; Thornhill, 1991; Tooby & Cosmides, 1990b; Williams, 1966, 1985), and these crtieria are helpful in evaluating experimental evidence that bears on these two positions. Adaptations can be recognized by "evidence of special design" (Williams, 1966)—that is, by recognizing that features of the evolved species-typical design of an organism are "components of some special problem-solving machinery" that solves an evolutionarily long-standing problem (Williams, 1985, p. 1). Standards for recognizing special design include factors such as economy, efficiency, complexity, precision, specialization, and reliability, which—like a key fitting a lock—render the design too good a solution to an adaptive problem to have arisen by chance (Williams, 1966). For example, the eye is extremely well suited for the detection and extraction of information presented by ambient light, and poorly designed as an orifice for ingesting food or as armor to protect the vulnerable brain from sharp objects. It displays many properties that are only plausibly interpreted as design features for solving the problem of vision. Moreover, the properties of an adaptation can be used to identify the class of problems, at the correct level of specificity or generality, that the adaptation was designed to solve. The eye allows humans to see hyenas, but that does not mean it is an adaptation that evolved particularly for hyena detection: There are no features that render it better designed for seeing hyenas than for seeing any of a far larger class of comparable objects. These principles governing adpatations can be developed into a series of methods for empirically arbitrating the dispute between traditional and domain-specific views of the mind. The Standard Social Science Model and evolutionary psychological approaches differ most strongly on the grounds of functional specialization, of content-specificity, and of evolutionary appropriateness (Tooby & Cosmides, this volume).

According to the evolutionary psychological approach to social cognition outlined here and elsewhere (Cosmides, 1985, 1989; Cosmides & Tooby, 1987, 1989; Tooby, 1985; Tooby & Cosmides, 1989, 1990b), the mind should contain organized systems of inference that are specialized for solving various families of problem, such as social exchange, threat, coalitional relations, and mate choice. Advocates of evolutionary views do not deny that humans learn, reason, develop, or acquire a culture; however, they do argue that these functions are accomplished at least in part through the operation of cognitive mechanisms that are content-specialized—mechanisms that are activated by particular content domains and that are designed to process information from those domains. Each cognitive specialization is expected to contain design features targeted to mesh with the recurrent structure of its characteristic problem type, as encountered under Pleistocene conditions. Consequently, one expects cognitive adaptations specialized for reasoning about social exchange to have some design features that are particular and appropriate to social exchange, but that are not activated by or applied to other content domains.

In contrast, the Standard Social Science Model predicts that the reasoning procedures applied to situations of social exchange should be the same reasoning procedures that are applied to other kinds of content. On this view, reasoning is viewed as the operation of content-independent procedures, such as formal logic, applied impartially and uniformly to every problem, regardless of the nature of the content involved. There should be nothing in the evolved structure of the mind—no content-sensitive procedures, no special representational format—that is more appropriate for reasoning about social exchange than about hat racks, rutabagas, warfare, Hinayana scripture, turbulence, or textuality. In other words, the standard view is that the faculty of reasoning consists of a small number of processes that are designed to solve the most inclusive and general class of reasoning problems possible—a class not defined in terms of its content, as the class includes all potential contents equally. On this view, any variability in reasoning due to content must be the product of experiential variables such as familiarity or explicit instruction.

For these reasons, the questions of interest for this experimental program include the following: Do patterns of performance on problems that require reasoning about social exchange reflect content-general rules of logic? Do patterns of performance on social exchange content, as compared with other contents, show systematic differences? If so, can these differences be explained through invoking general-purpose variables such as familiarity? Does the complexly articulated performance of subjects on social exchange problems have the detailed properties predicted in advance by an evolutionary analysis of the design features required for a cognitive adaptation to social exchange? By answering these and related questions, building from one experimental result to the next, the functional structure of human cognitive adaptations for reasoning about social exchange can begin to be delineated, and the adequacy of the Standard Social Science Model can be assessed.

Standard Analyses of the Evolution of Altruism

Natural selection is a feedback process that is driven by the differential reproduction of alternative designs. If a change in an organism's design allows it to outreproduce the alternative designs in the population, then that design change will become more common—it will be *selected for*. If this reproductive advantage continues, then over many generations that design change will spread through the population until all members of the species have it. Design changes that enhance reproduction are selected for; those that hinder reproduction relative to others are selected against and, therefore, tend to disappear. This ongoing process leads over time to the accumulation of designs organized for reproduction.

Consider, then, a design change that appears to *decrease* the reproduction of an individual who has it while simultaneously increasing the reproduction of other individuals. How could such a design change possibly spread through the population? At first glance, it would seem that a design feature that had this property would be selected against.

Yet many organisms do engage in behaviors that decrease their own reproduction while enhancing that of others. One chimpanzee will endanger itself to help another in a fight (de Waal, 1982). A vampire bat will feed blood that it has collected from its prey to a hungry conspecific (Wilkinson, 1988, 1990). A ground squirrel will warn others of the presence of a predator by emitting an alarm call that can draw the predator's attention to itself (Sherman, 1977). Among many species of social insects, workers forgo reproduction entirely in order to help raise their sisters (Wilson, 1971). People sometimes put themselves at great peril to help their fellow human beings, and carry out innumerable acts on a daily basis whose purpose is to help others. If a psychological mechanism generates such behavior on a regular basis, how could it possibly have been selected for?

Evolutionary biologists call this the "problem of altruism." An "altruistic" design feature is an aspect of the phenotype that is designed to produce some effect that enhances the reproduction of other individuals even though it may cause the individual who has it to reproduce less. The question is, how can designs that generate such behavior spread through a population until they become universal and species-typical?

So far, evolutionary biologists have provided two answers to the problem of altruism. The first, kin selection theory (or inclusive fitness theory), was proposed by W. D. Hamilton in 1964 (see also Maynard Smith, 1964; Williams & Williams, 1957). Imagine a design change that causes an individual to increase the reproduction of that individual's relatives, but that decreases the individual's own reproduction. There is some probability, r, that the kin member who receives the help has inherited that very same design change from a common ancestor. Therefore, the design change—through helping the relative to reproduce—may be spreading new copies of itself in the population, even though it is simultaneously decreasing the rate at which it creates new copies of itself through the individual it is in, by slowing the reproduction of that particular individual. Whenever a design change affects both direct reproduction and kin reproduction, there is a trade-off between these two different avenues by which a design change can be reproduced. The fate of the design change will be determined by how much it helps (or harms) the relative, how much it harms (or helps) the helper, and the probability the relative shares the design change by virtue of their sharing common ancestors. By using what was, in effect, mathematical game theory, Hamilton showed that a "helping design" can spread through the population if it causes an organism to help a kin member whenever the cost to the organism's own reproduction is offset by the benefit to the reproduction of its kin member, discounted by the probability, r, that the kin member has inherited the same helping design. Although helping under these circumstances decreases the helper's *personal* reproduction, through its effect on other individuals it causes a net increase in the reproduction of the helping design itself in the population.

Consequently, if C_i and B_i refer to costs and benefits to an individual i's own reproduction, then an altruistic design change can be selected for if it causes i to help j whenever $C_i < r_{ij}B_j$. Any design change that causes an individual to help more than this—or less than this—would be selected against. This constraint is completely general and falls out of the logic of natural selection theory: It should be true of any species on any planet at any time. A species may be solitary, and individuals may have no social interactions with their relatives; but if members of a species consistently interact socially with their relatives in ways that affect their reproduction, then they will be selected to evolve information-processing mechanisms that produce behavior that respects this constraint.

Because it suggested a rich set of hypotheses about phenotypic design, kin selection theory allowed animal behavior researchers to discover a flood of new phenomena. They began to find that the altruistic behavior of many species shows the design features that one would expect if their information-processing mechanisms had been shaped by kin selection. For example, ground squirrels are far more likely to give an alarm call if a close relative lives nearby (Sherman, 1977), and they have psychological mechanisms that allow them to discriminate full siblings from half siblings from unrelated individuals (Hanken & Sherman, 1981; Holmes & Sherman, 1982). Similarly, kinship is a major predictor of whether a vampire bat will share its food with a particular individual (Wilkinson, 1988, 1990). Most strikingly, kin selection theory (e.g., Hamilton, 1964; Williams & Williams, 1957) finally explained the existence of the sterile worker castes in the eusocial insects that had so troubled Darwin, providing an elegant set of hypotheses concerning how eusocial insects should allocate their reproductive effort among sisters, half-sisters, brothers and offspring, which have since been tested and confirmed (e.g., Frumhoff & Baker, 1988; Frumhoff & Schneider, 1987; Trivers & Hare, 1976).

The realization that a design feature can make copies of itself not only by affecting the reproductive success of its bearer, but also by affecting the reproductive success of its bearers' kin, led to a new definition of the concept of fitness. Previously, evolutionary biologists spoke of a design's "Darwinian fitness": its effect on the number of offspring produced by an individual who has the design. But since Hamilton, one speaks of a design's "inclusive fitness": its effect on the number of offspring produced by an individual who has the design *plus* its effects on the number of offspring produced by others who may have the same design—that individual's relatives—with each effect discounted by the appropriate measure of relatedness, often designated by r (Dawkins, 1982; Hamilton, 1964). Above, we used C_i and B_i to refer to effects on a design's Darwinian fitness; henceforth, we will use these variables to refer to effects on a design's inclusive fitness.[1]

Kin-directed helping behavior is common in the animal kingdom. But on occasion, one finds a species in which individuals help nonrelatives as well. How can a design feature that decreases one's own inclusive fitness while simultaneously increasing that of nonrelative be selected for? Although rare compared to kin-directed helping, such behavior does exist. For example, although kinship is a major predictor of food sharing in vampire bats, they share food with certain nonrelatives as well. Male baboons sometimes protect offspring not their own (Smuts, 1986). Unrelated chimpanzees will come to each other's aid when threatened (de Waal & Luttrell, 1988).

Williams (1966), Trivers (1971), Axelrod and Hamilton (1981), and Axelrod (1984) provided a second approach to the problem of altruism, reciprocal altruism theory, which in effect draws on the economist's concept of trade. Selection may act to create physiological or psychological mechanisms designed to deliver benefits even to nonrelatives, provided that the delivery of such benefits acts, with sufficient probability, to cause reciprocal benefits to be delivered in return. Such social exchange is easily understood as advantageous whenever there exist what economists call "gains in trade"—that is, whenever what each party receives is worth more than what it cost to deliver the reciprocal benefit to the other party. Ecologically realistic conditions, however, seldom provide opportunities in which two parties simultaneously have value to offer each other. For this reason, biologists have tended to focus on situations of deferred implicit exchange, where one party helps another at one point in time, in order to increase the probability that when their situations are reversed at some (usually) unspecified time in the future, the act will be reciprocated (hence the terms reciprocal altruism, reciprocation, or, as we prefer for the general class, social exchange).

If the reproductive benefit one receives in return is larger than the cost one incurred in rendering help, then individuals who engage in this kind of reciprocal helping behavior will outreproduce those who do not, causing this kind of helping design to spread. For example, if a vampire bat fails to find food for two nights in a row it will die, and there is high variance in food-gathering success. Sharing food allows the bats to cope with this variance, and the major predictor of whether a bat will share food with a nonrelative is whether the nonrelative has shared with that individual in the past (Wilkinson, 1988, 1990). Reciprocal altruism is simply cooperation between two or more individuals for mutual benefit, and it is variously known in the literature as social exchange, cooperation, or reciprocation. Design features that allow one to engage in reciprocal altruism can be selected for because they result in a net increase in one's own reproduction or that of one's relatives and, consequently, in the reproduction of the design features that produce this particular kind of cooperative behavior.

For example, according to reciprocal altruism theory, cognitive programs that generate food sharing among nonrelatives can be selected for only if they exhibit certain design features. By cataloging these design features, Wilkinson (1988, 1990) was able to look for—and discover—heretofore unknown aspects of the psychology and behavior of female vampire bats. Reciprocal altruism theory guided his research program:

> I needed to demonstrate that five criteria were being met: that females associate for long periods, so that each one has a large but unpredictable number of opportunities to engage in blood sharing; that the likelihood of an individual regurgitating to a roostmate can be predicted on the basis of their past association; that the roles of donor and recipient frequently reverse; that the short-term benefits to the recipient are greater than the costs to the donor; and that donors are able to recognize and expel cheaters from the system. (Wilkinson 1990, p. 77)

Like kin selection theory, reciprocal altruism theory suggested a host of hypotheses about phenotypic design, which allowed animal behavior researchers to discover many previously unsuspected phenomena. Recently, it has done the same for those who study social exchange in humans. Reciprocal altruism theory has allowed researchers to derive a rich set of hypotheses about the design features of the cognitive programs that generate cooperative behavior in humans. We will examine some of these hypotheses and the evidence for them.

This chapter is divided into three parts. In the first part (Selection Pressures) we explore some of the constraints reciprocal altruism theory places on the class of designs that can evolve in humans. These "evolvability constraints" (see Tooby & Cosmides, this volume) led us to develop a set of hypotheses about the design features of the cognitive programs that are responsible for reasoning about social exchange. In the second part (Cognitive Processes) we review research that we and others have conducted to test these hypotheses and show that the cognitive programs that govern reasoning about social exchange in humans have many of the design features one would expect if they were adaptations sculpted by the selection pressures discussed in the first part. In the third part (Implications for Culture) we discuss the implications of this work for understanding cross-cultural uniformities and variability in cooperative behavior.

SELECTION PRESSURES

Natural selection permits the evolution of only certain strategies for engaging social exchange. To be selected for, a design governing reasoning about social exchange must embody one of these strategies—in other words, it must meet an "evolvability criterion" (see Tooby & Cosmides, this volume). By studying the nature of these strategies, one can deduce many properties that human algorithms regulating social exchange must have, as well as much about the associated capabilities such algorithms require to function properly. Using this framework, one can then make empirical predictions about human performance in areas that are the traditional concern of cognitive psychologists: attention, communication, reasoning, the organization of memory, and learning. One can also make specific predictions about human performance on reasoning tests, such as the ones we will discuss in Cognitive Processes (following).

In this part, we explore the nature of the selection pressures on social exchange during hominid evolution—the relevant evolvability constraints—and see what these allow one to infer about the psychological basis for social exchange in humans.

Game-Theoretic Constraints on the Evolution of Social Exchange

> The critical act in formulating computational theories turns out to be the discovery of valid constraints on the way the world is structured. (Marr & Nishihara, 1978, p. 41)

In *Evolution and the Theory of Games,* John Maynard Smith (1982) pointed out that natural selection has a game-theoretic structure. Alternative designs are selected for or not because of the different effects they have on their "own" reproduction—that is, on the reproduction of all identical designs in the population. Some designs will outreproduce others until they become universal in the population; others will be selected out. Using game theory, one can mathematically model this process with some precision. This is true whether one is describing the alternative designs anatomically, physiologically, or cognitively. For example, it is irrelevant to the analysis whether one describes a design change in a particular region of the brain anatomically—as an increase in the density of serotonin receptors in that region—physiologically—as an increase in the rate of serotonin uptake (which was caused by the increased receptor density)—or cognitively—as a difference in how the individual who has the increased receptor density processes information. All that matters to the analysis is what effect the design change—however described—has on its own reproduction. Because our concern in this chapter is the evolution of the information-processing mechanisms that generate cooperative behavior, we will describe alternative designs cognitively, by specifying the different rules that they embody and the representations that those rules act upon.

To see how a game-theoretic analysis works, consider how one can use it to understand the ramifications of reciprocal altruism theory for the evolution of social exchange between unrelated individuals.

Designs reproduce themselves through the reproduction of the individuals who embody them. Given an individual, i, define a *benefit to i* (B_i) as the extent to which any act, entity, or state of affairs increases the inclusive fitness of that individual. Similarly, define a *cost to i* (C_i) as the extent to which any act, entity, or state of affairs decreases the inclusive fitness of individual i. Let 0_i refer to any act, entity, or state of affairs that has no effect on i's inclusive fitness. A cognitive program that causes a decrease in its own inclusive fitness while increasing that of an unrelated individual can evolve only if it has design features that embody the evolvability constraints of reciprocal altruism theory. A game-theoretic analysis allows one to explore what these constraints are. For ease of explication, the two interactants in a hypothetical social exchange will be designated "you" and "I," with appropriate possessive pronouns.

Reciprocal altruism, or social exchange, typically involves two acts: what "you" do for "me" (act 1), and what "I" do for "you" (act 2). For example, you might help me out by baby-sitting my child (act 1), and I might help you by taking care of your vegetable garden when you are out of town (act 2). Imagine the following situation: Baby-sitting my child inconveniences you a bit, but this inconvenience is more than compensated for by my watering your garden when you are out of town. Similarly, watering your garden inconveniences me a bit, but this is outweighed by the benefit to me of your baby-sitting my child. Formally put:

1. Your doing act 1 for me benefits me (B_{me}) at some cost to yourself (C_{you}).
2. My doing act 2 for you benefits you (B_{you}) at some cost to myself (C_{me}).
3. The benefit to you of receiving my act 2 is greater than the cost to you of doing act 1 for me (B_{you} of act $2 > C_{you}$ of act 1).
4. The benefit to me of receiving act 1 from you is greater than the cost to me of doing act 2 for you (B_{me} of act $1 > C_{me}$ of act 2).

If these four conditions are met—if acts 1 and 2 have this cost/benefit structure—then we would both get a net benefit by exchanging acts 1 and 2. Social exchange, or reciprocal altruism, is an interaction that has this mutually beneficial cost/benefit structure (see Table 3.1).

At first glance, one might think that natural selection would favor the emergence of cognitive programs with decision rules that cause organisms to participate in social exchange whenever the above conditions hold. After all, participation would result, by

Table 3.1 Sincere Social Contracts: Cost/Benefit Relations When One Party Is Sincere, and That Party Believes the Other Party Is Also Sincere[a]

	My offer: "If you do Act 1 for me then I'll do Act 2 for you."			
	Sincere offer		Sincere acceptance	
	I believe:		You believe:	
You do Act 1	B_{me}	C_{you}	B_{me}	C_{you}
You do not do Act 1	0_{me}	0_{you}	0_{me}	0_{you}
I do Act 2	C_{me}	B_{you}	C_{me}	B_{you}
I do not do Act 2	0_{me}	0_{you}	0_{me}	0_{you}
Profit margin	positive:	positive:	positive:	positive:
	$B_{me} > C_{me}$	$B_{you} > C_{you}$	$B_{me} > C_{me}$	$B_{you} > C_{you}$

Translation of the offer into the value systems of the participants:

| My terms | "If B_{me} then C_{me}" | "If B_{me} then C_{me}" |
| Your terms | "If C_{you} then B_{you}" | "If C_{you} then B_{you}" |

[a]B_x = benefit to x; C_x = cost to x; 0_x = no change in x's zero-level utility baseline. The zero-level utility baseline is the individual's level of well-being (including expectations about the future) at the time the offer is made, but independent of it. Benefits and costs are increases and decreases in one's utility, relative to one's zero-level utility baseline.

definition, in a net increase in the replication of such designs, as compared with alternative designs that cause one to not participate.

But there is a hitch: You can benefit *even more* by cheating me. If I take care of your garden, but you do not baby-sit my child—i.e., if I cooperate by doing act 2 for you, but you defect on the agreement by not doing act 1 for me—then you benefit more than if we both cooperate. This is because your payoff for cheating when I have cooperated (B_{you}) is greater than your payoff for mutual cooperation ($B_{you} - C_{you}$)—you have benefited from my taking care of your garden without having inconvenienced yourself by baby-sitting for me. Moreover, the same set of incentives applies to me. This single fact constitutes a barrier to the evolution of social exchange, a problem that is structurally identical to one of the most famous situations in game theory: the one-move Prisoner's Dilemma (e.g., Axelrod, 1984; Axelrod & Hamilton, 1981; Boyd, 1988; Trivers, 1971).[2]

Mathematicians and economists use game theory to determine which decision rules will maximize an individual's monetary profits or subjective utility. Consequently, they express payoffs in dollars or "utils." Such currencies are inappropriate to an evolutionary analysis, however, because the goal of an evolutionary analysis is different. Evolutionary biologists use game theory to explore evolvability constraints. The goal is to determine which decision rules can, in principle, be selected for—which will, over generations, promote their own inclusive fitness. For this purpose, units of inclusive fitness are the only relevant payoff currency. Other assumptions are minimal. The organism need not "know," either consciously or unconsciously, *why* the decision rule it executes is better or worse than others, or even *that* the rule it executes is better or worse than others. To be selected for, a decision rule must promote its inclusive fitness better than alternative rules—and that's all. It doesn't need to make one happy, it doesn't need to maximize subjective utility, it doesn't need to promote the survival of the species, it doesn't need to promote social welfare. To be selected for, it need only promote its own replication better than alternative designs.

you

		C	D
me	**C**	me: R = +3 you: R = +3	me: S = −2 you: T = +5
	D	me: T = +5 you: S = −2	me: P = 0 you: P = 0

C = Cooperate
D = Defect
R = Reward for mutual cooperation
T = Temptation to defect
S = Sucker's payoff
P = Punishment for mutual defection

Constraints: T > R > P > S; R > (T+S)/2*

*For an interated game, R > (T + S)/2. This is to prevent players from "cooperating" to maximize their utility by alternately defecting on one another.

Figure 3.1 Payoff Schedule for the Prisoner's Dilemma situation in game theory.

Mathematicians and economists have used the Prisoner's Dilemma to understand how cooperation can arise in the absence of a "Leviathan," that is, a powerful state or agency that enforces contracts. Evolutionary biologists have used it to understand the conditions under which design features that allow individuals to cooperate can be selected for. It is a game in which mutual cooperation would benefit both players, but it is in the interest of each player, individually, to defect, cheat, or inform on the other. It is frequently conceptualized as a situation in which two people who have collaborated in committing a crime are prevented from communicating with each other, while a district attorney offers each individual a lighter sentence if he will snitch on his partner. But the payoffs can represent anything for which both players have a similar preference ranking: money, prestige, points in a game—even inclusive fitness. A possible payoff matrix and the relationship that must exist between variables is shown in Figure 3.1.

Looking at this payoff matrix, one might ask: "What's the dilemma? I will be better off, and so will you, if we both cooperate—you will surely recognize this and cooperate with me." If there is only one move in the game, however, it is always in the interest of each party to defect (Luce & Raiffa, 1957). That is what creates the dilemma, as we will show below.

Figure 3.2 shows that the cost/benefit structure of a social exchange creates the same payoff matrix as a Prisoner's Dilemma: $(B_i) > (B_i − C_i) > 0 > C_i$ (i.e., T > R > P > S); and $(B_i − C_i) > (B_i − C_i)/2$ (i.e., R > T + S/2). In other words, if I cooperate on our agreement, you get B_{you} for defecting, which is greater than the $B_{you} − C_{you}$ you would get for cooperating (i.e., T > R). If I defect on our agreement, you get nothing for defecting (this is equivalent to our not interacting at all; thus P = 0 and R > P), which is better than the C_{you} loss you would incur by cooperating (i.e., P > S). The payoffs are in inclusive fitness units—the numbers listed are included simply to reinforce the analogy with Figure 3.1. In actuality, there is no reason why C_{me} must equal C_{you} (or $B_{me} = B_{you}$); an exchange will have the stucture of a Prisoner's Dilemma as long as mutual cooperation would produce a net benefit for both of us.

Now that we have defined the situation, consider two alternative decision rules:

Decision rule 1: *Always cooperate.*
Decision rule 2: *Always defect.*

you

	C	D
C	me: $R = B_{me} - C_{me} = +3$ you: $R = B_{you} - C_{you} = +3$	me: $S = C_{me} = -2$ you: $T = B_{you} = +5$
D	me: $T = B_{me} = +5$ you: $S = C_{you} = -2$	me: $P = 0_{me} = 0$ you: $P = 0_{you} = 0$

me

Figure 3.2 Social exchange sets up a Prisoner's Dilemma. B_i = Benefit to i, C_i = Cost to i, 0_i = i's inclusive fitness is unchanged.

An individual with cognitive programs that embody decision rule 1 would be an indiscriminate cooperator; an individual with cognitive programs that embody decision rule 2 would be an indiscriminate cheater.

Now imagine a population of organisms, most of whom have cognitive programs that embody decision rule 1, but a few of whom have cognitive programs that embody decision rule 2.[3] Then imagine a tournament that pits the reproduction of decision rule 1 against that of decision rule 2.

In this tournament, both sets of individuals face similar environments. For example, one might specify that both types of organisms are subject to the same payoff matrix, that each organism participates in three interactions per "generation," and that these three interactions must be with three different individuals, randomly chosen from the population. After every organism has completed its three interactions, each organism "reproduces" and then "dies." "Offspring" carry the same decision rule as the "parent," and the number of offspring produced by an individual is proportional to the payoffs it gained in the three interactions it participated in in that generation. This process repeats itself every generation.

Using this tournament, one can ask, After one generation, how many replicas of rule 1 versus rule 2 exist in the population? How many replicas of each rule exist after n generations? If one were to run a computer model of this tournament, one would find that after a few generations individuals who operate according to rule 2 ("Always defect") would, on average, be leaving more offspring than individuals operating according to rule 1 ("Always cooperate"); the magnitude of the difference between them is rule 2's "selective advantage" over rule 1. This magnitude will depend on what payoff and opportunity parameters were specified in the program used, as well as the population composition.

After a larger number of "generations," rule 1—"Always cooperate"—would be selected out. For every interaction with a cheater, rule 1 would lose two inclusive fitness points, and rule 2 would gain five. Consequently, indiscriminate cooperators would eventually be selected out, and indiscriminate cheaters would spread through the population; the number of generations this would take is a function of how many cheaters versus indiscriminate cooperators were in the initial population. In practice, a population of "cheaters" is a population of individuals who never participate in social exchange; if you "cheat" by not doing act 1 for me, and I "cheat" by not doing act 2 for you, then, in effect, we have exchanged nothing. And an indiscriminate cooperator in the midst of defectors is, in practice, always an "altruist" or victim, continually incurring costs in the course of helping others, but receiving no benefits in return.

So, after n generations, where n is a function of the magnitude of rule 2's selective advantage in the tournament's "environment" and other population parameters, one would find that rule 2 had "gone to fixation": Virtually all individuals would have rule 2, and, regardless of the population's absolute size, a vanishingly small proportion of the individuals in it would have rule 1.[4]

By using this kind of logic, one can show that if a new design coding for rule 2—"Always defect"—were to appear in a population that is dominated by individuals with rule 1—"Always cooperate"—it would spread through the population until it became fixed, and it would not be vulnerable to invasion by rule 1 (see, e.g., Axelrod, 1984). In a tournament pitting indiscriminate altruists against indiscriminate cheaters, the cheaters will come to dominate the population.

One might object that real life is not like a Prisoner's Dilemma, because real-life exchanges are simultaneous, face-to-face interactions. You can directly recognize whether I am about to cheat you or not (provided you are equipped with cognitive equipment that guides you into making this discrimination). If I show up without the item I promised, then you simply do not give me what I want. This is often true in a twentieth-century market economy, where money is used as a medium of exchange. But no species that engages in social exchange, including our own, evolved the information-processing mechanisms that enable this behavior in the context of a market economy with a medium of exchange.

Virtually any nonsimultaneous exchange increases the opportunity for defection, and in nature, most opportunities for exchange are not simultaneous. For example, a drowning man needs immediate assistance, but while he is being pulled from the water, he is in no position to help his benefactor. Opportunities for simultaneous mutual aid—and therefore for the withdrawal of benefits in the face of cheating—are rare in nature for several reasons:

The "items" of exchange are frequently acts that, once done, cannot be undone (e.g., protection from an attack and alerting others to the presence of a food source).

The needs and abilities of organisms are rarely exactly and simultaneously complementary. For example, a female baboon is not fertile when her infant needs protection, yet this is when the male's ability to protect is of most value to her.

On those occasions when repayment is made in the same currency, simultaneous exchange is senseless. If two hunters both make kills on the same day, they gain nothing from sharing their kills with each other: They would be swapping identical goods. In contrast, repayment in the same currency can be advantageous when exchange is not simultaneous, because of declining marginal utilities: The value of a piece of meat is larger to a hungry individual than to a sated one.

Thus, in the absence of a widely accepted medium of exchange, most exchanges are not simultaneous and therefore do provide opportunities for defection. You must decide whether to benefit me or not without any guarantee that I will return the favor in the future. This is why Trivers (1971) describes social exchange in nature as "reciprocal altruism." I behave "altruistically" (i.e., I incur a cost in order to benefit you) at one point in time, on the possibility that you will reciprocate may altruistic act in the future. If you do, in fact, reciprocate, then our "reciprocally altruistic" interaction is properly described as an instance of delayed mutual benefit: Neither of us has incurred a net cost; both of us have gained a net benefit.

A system of mutual cooperation cannot emerge in a one-move Prisoner's Dilemma because it is always in the interest of each player to defect. In fact, the argument is general to any known, fixed number of games (Luce & Raiffa, 1957). But selection pressures change radically when individuals play a *series* of Prisoner's Dilemma games. Mutual cooperation—and therefore social exchange—can emerge between two players when (a) there is a high probability that they will meet again, (b) neither knows for sure exactly how many times they will meet,[5] and (c) they do not value later payoffs by too much less than earlier payoffs (Axelrod, 1984; Axelrod & Hamilton, 1981). If the parties are making a series of moves rather than just one, then one party's behavior on a move can influence the other's behavior on future moves. If I defect when you cooperated, then you can retaliate by defecting on the next move; if I cooperate, then you can reward me by cooperating on the next move. In an iterated Prisoner's Dilemma game, a system can emerge that has incentives for cooperation and disincentives for defection.

The work of Trivers (1971), Axelrod and Hamilton (1981), and Axelrod (1984) has shown that indiscriminate cooperation (Decision rule 1) cannot be selected for when the opportunity for cheating exists. But *selective* cooperation can be selected for. Decision rules that cause one to cooperate with other cooperators and defect on cheaters can invade a population of noncooperators.

Consider, for example, Decision rule 3:

Decision rule 3: *Cooperate on the first move; on subsequent moves, do whatever your partner did on the previous move.*

This decision rule is known in the literature as TIT FOR TAT (Axelrod & Hamilton, 1981). If rule 3's partner cooperates on a move, rule 3 will cooperate on the next move with that partner. If rule 3's partner defects on a move, rule 3 will defect on the next move with that partner. It has been shown that rule 3 can invade a population dominated by indiscriminate cheaters (individuals who behave according to decision rule 2: "Always defect"). Using the payoff matrix in Figure 3.2, it is clear that rule 3 would outreproduce rule 2: Mutual cooperators (pairs of individuals who behave according to rule 3) would get strings of $+3$ inclusive fitness points, peppered with a few -2s from a first trial with a rule 2 cheater (after which the cooperator would cease to cooperate with that individual). In contrast, mutual defectors (pairs of individuals who behave according to rule 2) would get strings of 0s, peppered with a few $+5$s from an occasional first trial with a rule 3 cooperator (after which the cooperator would never cooperate with that individual again).

Game-theoretic analyses have shown that a decision rule embodying a cooperative strategy can invade a population of noncooperators if, and only if, it cooperates with other cooperators and excludes (or retaliates against) cheaters. If a decision rule regulating when one should cooperate and when one should cheat violates this constraint, then it will be selected against.

Axelrod (1984) has shown that there are many decision rules that do embody this constraint. All else equal (an important caveat), any of these could, in theory, have been selected for in humans. Which decision rule, out of this constrained set, is embodied in the cognitive programs that actually evolved in the human lineage is an empirical question. But note that to embody any of this class of decision rules, the cognitive programs involved would have to incorporate a number of specific design features:

1. They must include algorithms that are sensitive to cues that indicate when an exchange is being offered and when reciprocation is expected.
2. They must include algorithms that estimate the costs and benefits of various actions, entities, or states of affairs to oneself.[6]
3. They must include algorithms that estimate the costs and benefits of various actions, entities, or states of affairs to others (in order to know when to initiate an exchange).
4. They must include algorithms that estimate the probability that these actions, entities, or states of affairs will come about in the absence of an exchange.
5. They must include algorithms that compare these estimates to one another (in order to determine whether $B_i > C_i$).
6. They must include decision rules that cause i to reject an exchange offer when $B_i < C_i$.
7. They must include decision rules that cause i to accept (or initiate) an exchange when $B_i > C_i$ (and other conditions are met).
8. They must include algorithms with inference procedures that capture the intercontingent nature of exchange (see Cosmides & Tooby, 1989, pp. 81–84).
9. They must include algorithms that can translate the exchange into the value assignments appropriate to each participant.
10. They must include algorithms that can detect cheaters (these must define cheating as an illicitly taken benefit).
11. They must include algorithms that cause one to punish cheating under the appropriate circumstances.
12. They must include algorithms that store information about the history of one's past exchanges with other individuals (in order to know when to cooperate, when to defect, and when to punish defection).
13. They must include algorithms that can recognize different individuals (in order to do any of the above).
14. They need not include algorithms for detecting indiscriminate altruists, because there shouldn't be any.

Not all of these algorithms need to be part of the same "mental organ." For example, because algorithms that can do 2, 3, 4, 5, and 13 are necessary to engage in social interactions other than exchange—such as aggressive threat—these might be activated even when the algorithms that are specific to social exchange are not.

Design features 1 to 14 are just a partial listing, based on some very general constraints on the evolution of social exchange that fall out of an examination of the iterated Prisoner's Dilemma. These constraints are general in the sense that they apply to the evolution of reciprocal altruism in almost any species—from reciprocal egg trading in hermaphroditic fish (Fischer, 1988) to food sharing in humans. Other, species-specific constraints on the design of social exchange algorithms can be derived by considering how these general selection pressures would have manifested themselves in the ecological context of hominid evolution.

For example, the sharing rules that are applied to high-variance resources, such as hunted meat, should differ in some ways from those that are applied to low-variance resources, such as gathered plant foods (Kaplan & Hill, 1985; see also Implications for Culture, this chapter). This raises the possibility that human exchange algorithms have two alternative, context-specific modes of activation. Both modes would have to satisfy the general constraints listed above, but they might differ considerably in various details, such as whether one expects to be repaid in the same currency (e.g., meat for meat), whether one requires reciprocation before one is willing to help a second time, or whether one is quick to punish suspected cheaters. Another example of how ecological context can place species-specific constraints on design concerns the kind of representations that exchange algorithms can be expected to operate on. For example, exchange algorithms in humans should operate on more abstract representations than exchange algorithms in vampire bats. The reciprocation algorithms of vampire bats could, in principle, operate on representations of regurgitated blood, because this is the only item that they exchange. But item-specific representations of this kind would not make sense for the exchange algorithms of humans. Because our ancestors evolved the ability to make and use tools and to communicate information verbally, exchange algorithms that could accept a wide and ever-changing variety of goods, services, and information as input would enjoy a selective advantage over ones that were limited to only a few items of exchange. To accommodate an almost limitless variety of inputs—stone axes, meat, help in fights, sexual access, information about one's enemies, access to one's water hole, necklaces, blow guns, and so forth—representations of particular items of exchange would have to be translated into an abstract "lingua franca" that the various exchange algorithms could operate on. This constraint led us to hypothesize that an item-specific repesentation of an exchange would be translated into more abstract cost-benefit representations (like those in the last two lines of Table 3.1) at a relatively early stage in processing, and that many of the algorithms listed earlier would operate on these cost-benefit representations (Cosmides & Tooby, 1989). Because some of these species-specific constraints on the evolution of social exchange in humans have interesting implications for cultural variation, we will defer a discussion of them to the third part of this chapter (Implications for Culture), where we discuss social exchange and culture.

David Marr argued that "an algorithm is likely to be understood more readily by understanding the nature of the problem being solved than by examining the mechanism (and the hardware) in which it is embodied" (1982, p. 27). This is because the nature of the problem places constraints on the class of designs capable of solving it. The iterated Prisoner's Dilemma is an abstract description of the problem of altruism between nonrelatives. By studying it, one can derive a set of general constraints that the cognitive problems of virtually any species must satisfy to be selected for under these circumstances. By studying the ecological context in which this problem manifested itself for our Pleistocene ancestors, one can derive additional constraints. All these constraints on the evolution of social exchange—those that apply across species and those that apply just to humans—allow one to develop a task analysis or, to use Marr's term, a "computational theory" of the adaptive problem of social exchange. Cosmides and Tooby (1989) used some of these constraints to develop the beginnings of a computational theory of social exchange, which we call "social contract theory." So as not to repeat ourselves here, we refer the reader to that article for details. By constraining the class of possible designs, this theory allowed us and others to make some predictions about the design features of the algorithms and representations that evolved to solve the problem of social exchange in humans. Design features 1–14 listed earlier are a small subset of those predictions.

The computational theory we developed has guided our research program on human reasoning. We have been conducting experiments to see whether people have cognitive processes that are specialized for reasoning about social exchange. The experiments we will review in the following part were designed to test for design features 1, 9, 10, and 14, as well as some other predictions derived from the computational theory. We have been particularly interested in testing the hypothesis that humans have algorithms that are specialized for detecting cheaters in situations of social exchange.

COGNITIVE PROCESSES

Differential reproduction of alternative designs is the engine that drives natural selection: If having a particular mental structure, such as a rule of inference, allows a design to outreproduce other designs that exist in the species, then that mental structure will be selected for. Over many generations it will spread through the population until it becomes a universal, species-typical trait.

Traditionally, cognitive psychologists have assumed that the human mind includes only general-purpose rules of reasoning and that these rules are few in number and content-free. But a cognitive perspective that is informed by evolutionary biology casts doubt on these assumptions. This is because natural selection is also likely to have produced many mental rules that are specialized for reasoning about various evolutionarily important domains, such as cooperation, aggressive threat, parenting, disease avoidance, predator avoidance, object permanence, and object movement. Different adaptive problems frequently have different optimal solutions, and can therefore be solved more efficiently by the application of different problem-solving procedures. When two adaptive problems have different optimal solutions, a single general solution will be inferior to two specialized solutions. In such cases, a jack-of-all-trades will necessarily be a master of none, because generality can be achieved only by sacrificing efficiency. Indeed, it is usually more than efficiency that is lost by being limited to a general-purpose method—generality may often sacrifice the very possibility of successfully solving a problem, as, for example, when the solution requires supplemental information that cannot be sensorily derived (this is known as the "frame problem" in artificial intelligence research).

The same principle applies to adaptive problems that require reasoning: There are cases where the rules for reasoning adaptively about one domain will lead one into serious error if applied to a different domain. Such problems cannot, in principle, be solved by a single general-purpose reasoning procedure. They are best solved by different special-purpose reasoning procedures.

For example, the rules of inference of the propositional calculus (formal logic) are general-purpose rules of inference: They can be applied regardless of what subject matter one is reasoning about. Yet the consistent application of these rules of logical reasoning will not allow one to detect cheaters in situations of social exchange, because what counts as cheating does not map onto the definition of violation imposed by the propositional calculus. Suppose you and I agree to the following exchange: "If you give me your watch then I'll give you $20." You would have violated our agreement—you would have cheated me—if you had taken my $20 but not given me your watch. But according to the rules of inference of the propositional calculus, the only way this rule can be violated is by your giving me your watch but my not giving you $20.[7] If the only mental rules my mind contained were the rules of inference of the propositional calculus, then I would not be able to tell when you had cheated me. Similarly, rules of inference for detecting cheaters on social contracts will not allow one to detect bluffs or double crosses in situations of aggressive threat (Cosmides & Tooby, in prep., a). What counts as a violation differs for a social contract, a threat, a rule describing the state of the world, and so on. Because of this difference, the same reasoning procedure cannot be successfully applied to all of these situations. As a result, there cannot be a general-purpose reasoning procedure that works for all of them. If these problems are to be solved at all, they must be solved by different specialized reasoning procedures.

Given the selection pressures discussed earlier, we can define a social contract as a situation in which an individual is obligated to satisfy a requirement of some kind, usually at some cost to him- or herself, in order to be entitled to receive a benefit from another individual (or group). The requirement is imposed because its satisfaction creates a situation that benefits the party that imposed it. Thus, a well-formed social contract expresses an intercontingent situation of mutual benefit: To receive a benefit, an individual (or group) is required to provide a benefit. Usually (but not always) one incurs a cost by satisfying the requirement. But that cost is outweighed by the benefit one receives in return.

Cheating is a violation of a social contract. A cheater is an individual who illicitly benefits himself or herself by taking a benefit without having satisfied the requirement that the other party to the contract made the provision of that benefit contingent on. In this section, we review evidence that people have cognitive adaptations that are specialized for reasoning about social contracts. We will pay particular attention to the hypothesis that people have inference procedures specialized for cheater detection.

Adaptations are aspects of the phenotype that were designed by natural selection. To show that an aspect of the phenotype is an adaptation, one must produce evidence that it is well designed for solving an adaptive problem. Contrary to popular belief, developmental evidence is not criterial: Adaptations need not be present from birth (e.g., breasts), they need not develop in the absence of learning or experience (e.g., vision, language—see Pinker & Bloom, this volume),[8] and they need not be heritable (Tooby & Cosmides, 1990a). In fact, although the developmental processes that create adaptations are inherited, adaptations will usually exhibit low heritability. Differences between individuals will not be due to differences in their genes because adaptations are, in most cases, universal and species-typical—everyone has the genes that guide their development. The filter of natural selection does not sift designs on the basis of their developmental trajectory per se:[9] It doesn't matter how a design was built, only *that* it was built, and to the proper specifications.

To say that an organism has cognitive procedures that are adaptations for detecting cheaters, one must show that these procedures are well designed for detecting cheaters on social contracts. One must also show that their design features are not more parsimoniously explained as by-products of cognitive processes that evolved to solve some other kind of problem, or a more general class of problems. We approached this question by studying human reasoning. A large literature already existed that showed that people are not very good at detecting violations of conditional rules, even when these rules deal with familiar content drawn from everyday life. To show that people who ordinarily cannot detect violations of conditional rules can do so when that violation represents cheating on a social contract would constitute evidence that people have reasoning procedures that are specially designed for detecting cheaters in situations of social exchange.

The Wason Selection Task

One of the most intriguing and widely used experimental paradigms for exploring people's ability to detect violations of conditional rules has been the Wason selection task (Wason, 1966; see Figure 3.3, panel a). Peter Wason was interested in Karl Popper's view that the structure of science was hypothetico-deductive. He wondered if everyday learning was really hypothesis testing—i.e., the search for evidence that contradicts a hypothesis. Wason devised his selection task because he wanted to see whether people are well equipped to test hypotheses by looking for evidence that could potentially falsify them. In the Wason selection task, a subject is asked to see whether a conditional hypothesis of the form *If P then Q* has been violated by any one of four instances represented by cards.

A hypothesis of the form *If P then Q* is violated only when *P* is true but *Q* is false: The rule in Figure 3.3, panel a, for example, can be violated only by a card that has a D on one side and a number other than 3 on the other side. Thus, one would have to turn over the *P* card (to see if it has a *not-Q* on the back) and the *not-Q* card (to see if it has a *P* on the back)—D and 7, respectively, for the rule in Figure 3.3, panel a. Consequently, the logically correct response for a rule of the form *If P then Q* is always *P & not-Q*.

Wason expected that people would be good at detecting violations of conditional rules. Nevertheless, over the past 25 years, he and many other psychologists have found that few people actually give this logically correct answer (less than 25% for rules expressing unfamiliar relations). Most people choose either the *P* card alone or *P & Q*. Few people choose the *not-Q* card, even though a *P* on the other side of it would falsify the rule.

A wide variety of conditional rules that describe some aspect of the world ("descriptive rules") have been tested; some of these have expressed relatively familiar relations, such as "If a person goes to Boston, then he takes the subway" or "If a person eats hot chili peppers, then he will drink a cold beer." Others have expressed unfamiliar relations, such as "If you eat duiker meat, then you have found an ostrich eggshell" or "If there is an 'A' on one side of a card, then there is a '3' on the other side." In many experiments, performance on familiar descriptive rules is just as low as on unfamiliar ones. For example, rules relating food to drink, such as the hot chili pepper rule above, have never elicited logical performance higher than that elicited by unfamiliar rules, even though the typical sophomore in such experiments has had about 22,000 experiences in which he or she has had both food and drink, and even though recurrent relations between certain foods and certain drinks are common—cereal with orange juice at breakfast, red wine with red meat, coffee with dessert, and so on. Sometimes familiar rules do elicit a higher percentage of logically correct responses than unfamiliar ones, but even when they do, they typically elicit the logically correct response from fewer than half of the people tested. For example, in the Wason selection task literature, the transportation problem—"If a person goes to Boston, then he takes the

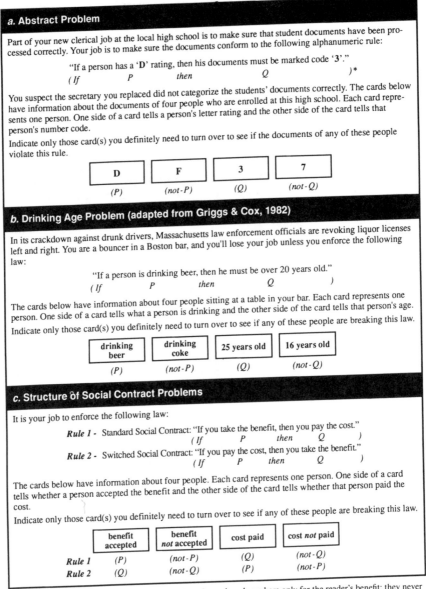

a. Abstract Problem

Part of your new clerical job at the local high school is to make sure that student documents have been processed correctly. Your job is to make sure the documents conform to the following alphanumeric rule:

"If a person has a 'D' rating, then his documents must be marked code '3'."
(*If*   *P*   *then*   *Q*   )*

You suspect the secretary you replaced did not categorize the students' documents correctly. The cards below have information about the documents of four people who are enrolled at this high school. Each card represents one person. One side of a card tells a person's letter rating and the other side of the card tells that person's number code.

Indicate only those card(s) you definitely need to turn over to see if the documents of any of these people violate this rule.

D	F	3	7
(P)	(not-P)	(Q)	(not-Q)

b. Drinking Age Problem (adapted from Griggs & Cox, 1982)

In its crackdown against drunk drivers, Massachusetts law enforcement officials are revoking liquor licenses left and right. You are a bouncer in a Boston bar, and you'll lose your job unless you enforce the following law:

"If a person is drinking beer, then he must be over 20 years old."
(*If*   *P*   *then*   *Q*   )

The cards below have information about four people sitting at a table in your bar. Each card represents one person. One side of a card tells what a person is drinking and the other side of the card tells that person's age. Indicate only those card(s) you definitely need to turn over to see if any of these people are breaking this law.

drinking beer	drinking coke	25 years old	16 years old
(P)	(not-P)	(Q)	(not-Q)

c. Structure of Social Contract Problems

It is your job to enforce the following law:

Rule 1 - Standard Social Contract: "If you take the benefit, then you pay the cost."
(*If*   *P*   *then*   *Q*   )
Rule 2 - Switched Social Contract: "If you pay the cost, then you take the benefit."
(*If*   *P*   *then*   *Q*   )

The cards below have information about four people. Each card represents one person. One side of a card tells whether a person accepted the benefit and the other side of the card tells whether that person paid the cost.

Indicate only those card(s) you definitely need to turn over to see if any of these people are breaking this law.

	benefit accepted	benefit *not accepted*	cost paid	cost *not* paid
Rule 1	(P)	(not-P)	(Q)	(not-Q)
Rule 2	(Q)	(not-Q)	(P)	(not-P)

*The logical categories (*P*s and *Q*s) marked on the rules and cards are here only for the reader's benefit; they never appear on problems given to subjects.

Figure 3.3   Content effects on the Wason selection task. The logical structures of these three Wason selection tasks are identical: They differ only in propositional content. Regardless of content, the logical solution to all three problems is the same: P & not-Q. Although <25% of college students choose both these cards for the abstract problem in panel a, about 75% do for the drinking age problem (panel b)—a familiar "standard" social contract. Panel c shows the abstract structure of a social contract problem. Cheater detection algorithms should cause subjects to choose the "benefit accepted" card and the "cost not paid" card, *regardless of which logical categories they represent*. These cards correspond to the logical categories P & not-Q for the "standard" form in rule 1, but to the logical categories *not-P & Q* for the "switched" form in rule 2.

subway"—is the familiar descriptive rule that has the best record for eliciting logically correct responses, and performance on this rule was consistently higher in Cosmides's (1989) experiments than in most others. Even so, it elicited the logically correct *P & not-Q* response from only 48% of the 96 subjects tested. Recently, Stone and Cosmides (in prep.) tested rules expressing causal relations; the pattern of results is essentially the same as for descriptive rules. Humans do not appear to be naturally equipped to seek out violations of descriptive or causal rules.

When subjects are asked to look for violations of conditional rules that express social contracts, however, their performance changes radically. Consider the "drinking age problem" in Figure 3.3, panel b. It expresses a social contract in which one is entitled to receive a benefit (beer) only if one has satisfied a requirement (being a certain age).[10] The drinking age problem elicits the logically correct response, *P & not-Q*, from about 75% of subjects (e.g., Griggs & Cox, 1982; Cosmides, 1985). On this rule, it is very easy to see why one needs to check what the 16-year-old is drinking (the *not-Q* card) and the age of the person drinking beer (*P*), and it is equally obvious why the 25-year-old (*Q*) and the coke drinker (*not-P*) need not be checked. Experiments with the drinking age problem and other familiar social contracts show that human reasoning changes dramatically depending on the subject matter one is reasoning about. Such changes are known as "content effects."

Figure 3.3, panel c, shows the abstract structure of a social contract problem. To detect cheaters on a social contract, one would always want to choose the "benefit accepted" card and the "cost not paid" card, regardless of what logical category these cards happen to correspond to. For the drinking age problem, these cards happen to correspond to the logical categories *P* and *not-Q*. Consequently, a subject who was looking for cheaters on a social contract would, by coincidence, choose the logically correct answer. But, as this figure also shows, there are situations in which the correct answer if one is looking for cheaters is not the logically correct answer. This point is important, and we will return to it later.

When we began this research in 1983, the literature on the Wason selection task was full of reports of a wide variety of content effects, and there was no satisfying theory or empirical generalization that could account for these content effects. When we categorized these content effects according to whether they conformed to social contracts, a striking pattern emerged. Robust and replicable content effects were found only for rules that related terms that are recognizable as benefits and cost/requirements in the format of a standard social contract—i.e., in the format of Rule 1 in Figure 3.3, panel c (see Cosmides, 1985, for a review). No thematic rule that was not a social contract had ever produced a content effect that was both robust and replicable. Moreover, most of the content effects reported for non-social contract rules were either weak, clouded by procedural difficulties, or had some earmarks of a social contract problem. All told, for non-social contract thematic problems, 3 experiments had produced a substantial content effect, 2 had produced a weak content effect, and 14 had produced no content effect at all. The few effects that were found did not replicate. In contrast, 16 out of 16 experiments that fit the criteria for standard social contracts—i.e., 100%—elicited substantial content effects. (Since that time, additional types of content have been tested; evolutionarily salient contents have elicited new, highly patterned content effects that are indicative of additional cognitive specializations in reasoning.)[11]

Special design is the hallmark of adaptation. As promising as these initial results were, they were not sufficient to demonstrate the existence of an adaptation for social exchange. First, although the familiar rules that were social contracts always elicited a robust effect, and the familiar rules that were not social contracts failed to elicit a robust and replicable effect, this was true *across* experiments; individual experiments usually pitted performance on a familiar social contract against performance on an unfamiliar descriptive rule. Because they confounded familiarity with whether a rule was a social contract or not, these experiments could not decide the issue.[12] Familiarity could still be causing the differences in performance in complex ways that varied across different subject populations. Second, even if it were shown that familiarity could not account for the result, these experiments could not rule out the hypothesis that social contract content simply facilitates *logical* reasoning. This is because the adaptively correct answer, if one is looking for cheaters on the social contracts tested, happened to also be the logically correct answer. To show that the results were caused by rules of inference that are specialized for reasoning about social exchange, one would have to test social contract rules in which the correct "look for cheaters" answer is *different* from the logically correct answer.

Below, we review evidence addressing these, and other, hypotheses. Our goal will be twofold: (a) to show that the reasoning procedures involved show the features of special design that one would expect if they were adaptations for social exchange, and (b) to show that the results cannot be explained as by-products of other, more general-purpose reasoning procedures.

Did the Social Contract Problems Elicit Superior Performance Because They Were Familiar?

Familiar social contracts elicited high levels of apparently logical performance. Could this result be a by-product of the familiarity of the social contract rules tested? Suppose we have general-purpose reasoning procedures whose design makes us more likely to produce logically correct answers for familiar or thematic rules. Then high levels of P & *not-Q* responses to familiar social contract rules could be a by-product of the operation of these general-purpose mechanisms, rather than the result of algorithms specialized for reasoning about social exchange.

The first family of hypotheses that we tested against was the availability theories of reasoning, which are sophisticated and detailed versions of this "by-product" hypothesis (Griggs & Cox, 1982; Johnson-Laird, 1982; Manktelow & Evans, 1979; Pollard, 1982; Wason, 1983; for a detailed review, see Cosmides, 1985). These theories come in a variety of forms with some important theoretical differences, but common to all is the notion that the subject's actual past experiences create associational links between terms mentioned in the selection task. These theories sought to explain the "now you see it, now you don't" results common for certain familiar descriptive rules, such as the transportation problem. Sometimes these rules elicited a small content effect; other times they elicited none at all. This was in spite of the fact that the *relations* tested—between, for example, destinations and means of transportation or between eating certain foods in conjunction with certain drinks—were perfectly familiar to subjects. This meant that a general familiarity with the relation itself was not sufficient to explain the results. The proposal, therefore, was that subjects who had, for example, gone to Boston more often by cab than by subway would be more likely to pick "Boston" and "cab"—i.e., P & $not\text{-}Q$—for the rule "If one goes to Boston, then one takes the subway" than those who had gone to Boston more often by subway (Q). According to the availability theories, the more exposures a subject has had to the co-occurrence of P and Q, the stronger that association will become, and the easier it will come to mind, i.e., become "available" as a response. A subject is more likely to have actually experienced the co-occurrence of P and $not\text{-}Q$ for a familiar rule, therefore familiar rules are more likely to elicit logically correct responses than unfamiliar rules. But whether a given rule elicits a content effect or not will depend on the actual, concrete experiences of the subject population tested.

Despite their differences, the various availability theories make the same prediction about unfamiliar rules. If all the terms in a task are unfamiliar, the only associational link available will be that created between P and Q by the conditional rule itself, because no previous link will exist among any of the terms. Thus P & Q will be the most common response for unfamiliar rules. P & $not\text{-}Q$ responses will be rare for all unfamiliar rules, whether they are social contracts or not. The fact that a social-contract-type *relation* might be familiar to subjects is irrelevant: Previous results had already shown that the familiarity of a relation could not, by itself, enhance performance.

We can test against this family of hypotheses because social contract theory makes very different predictions. If people do have inference procedures that are specialized for reasoning about social contracts, then these ought to function, in part, as frame or schema builders, which structure new experiences. This means they should operate in *unfamiliar* situations. No matter how unfamiliar the relation or terms of a rule, if the subject perceives the terms as representing a rationed benefit and a cost/requirement in the implicational arrangement appropriate to a social contract, then a cheater detection procedure should be activated. Social contract algorithms need to be able to operate in new contexts if one is to be able to take advantage of new exchange opportunities. Therefore, the ability to operate on social contracts even when they are unfamiliar is a design feature that algorithms specialized for reasoning about social exchange should have. Social contract theory predicts a high level of P & $not\text{-}Q$ responses on all "standard" social contract problems, whether they are familiar or not. It is silent on whether availability exerts an independent effect on non-social contract problems. A standard social contract is one that has the abstract form, *If you take the benefit, then you pay the cost* (see Rule 1, Figure 3.3, panel c).

In the first set of experiments, we pitted social contract theory against the availability family of theories by testing performance on an *unfamiliar* standard social contract—a problem for which the two hypotheses make diametrically opposite predictions (for details, see Cosmides, 1989, Experiments 1 and 2). Each subject was given four Wason selection tasks to solve: an unfamiliar social contract, an unfamiliar descriptive rule, a familiar descriptive rule (the transportation problem), and an abstract problem (as in Figure 3.3, panel a). Problem order was counterbalanced across subjects. The abstract problem was included because it is the usual standard for assessing the presence of a content effect in the Wason selection task literature; the transportation problem was included as a standard against which to judge the size of any social contract effect that might occur.

Rules such as, "If you eat duiker meat, then you have found an ostrich eggshell," or, "If a man eats cassava root, then he must have a tattoo on his face," were used for the unfamiliar problems; we felt it was safe to assume that our subjects would not have associative links between terms such as "cassava root" and "no tattoo" stored in long-term memory. An unfamiliar rule was made into a social contract or a descriptive rule by manipulating the surrounding story context. For example, a social contract version of the cassava root rule might say that in this (fictitious) culture, cassava root (P) is a much prized aphrodisiac whereas molo nuts (not-P) are considered nasty and bitter, thereby conveying that eating cassava root is considered a benefit compared to eating molo nuts, which are the alternative food. Having a tattoo on one's face (Q) means one is married; not having a tattoo (not-Q) means one is unmarried. As subjects know that marriage is a contract in which certain obligations are incurred to secure certain benefits (many of which involve sexual access), being married in this story is the cost/requirement. Finally, the story explains that because the people of this culture are concerned about sexual mores, they have created the rule, "If a man eats cassava root, then he must have a tattoo on his face." The four cards, each representing one man, would read "eats cassava root," "eats molo nuts," "tattoo," and "no tattoo." Other story contexts were invented for other rules.

The descriptive version of the unfamiliar rule would also give meaning to the terms and suggest a meaningful relation between them, but the surrounding story would not give the rule the cost/benefit structure of a social contract. For example, it might explain that cassava root and molo nuts are both staple foods eaten by the people of the fictitious culture (i.e., there is no differential benefit to eating one over the other), but that cassava root grows only at one end of the island they live on, whereas molo nuts grow only at the other end. Having a tattoo on your face or not again indicates whether a man is married, and it so happens that married men live on the side of the island where the cassava root grows, whereas unmarried men live on the side where the molo nuts grow. Note that in this version, being married has no significance as a cost/requirement; it is merely a correlate of where one lives. The story then provides a meaningful relation to link the terms of the rule, "If a man eats cassava root, then he must have a tattoo on his face," by suggesting that it simply describes the fact that men are eating the foods that are most available to them.

Subjects who were given a cassava root version of the social contract rule were given a duiker meat version of the descriptive rule, and vice versa, so that no subject encountered two versions of the exact same unfamiliar rule. The availability theories predict low levels of P & not-Q responses on both unfamiliar rules, whether they are portrayed as social contracts or not. Social contract theory predicts high levels of P & not-Q responses for the unfamiliar social contract, but not for the unfamiliar descriptive rule. The predictions of the two theories, and the results of two different sets of experiments, are shown in Figure 3.4.

The results clearly favor social contract theory. Even though they were unfamiliar and culturally alien, the social contract problems elicited a high percentage of P & not-Q responses. In fact, both we and Gigerenzer and Hug (in press) found that the performance level for unfamiliar social contracts is just as high as it usually is for familiar social contracts such as the drinking age problem—around 75% correct in our experiments. Unfamiliar social contracts elicited levels of P & not-Q responses that were even higher than those elicited by the familiar descriptive transporation problem. From our various experiments, we estimated the size of the social contract effect to be about 1.49 times larger than the size of the effect that availability has on familiar descriptive problems.

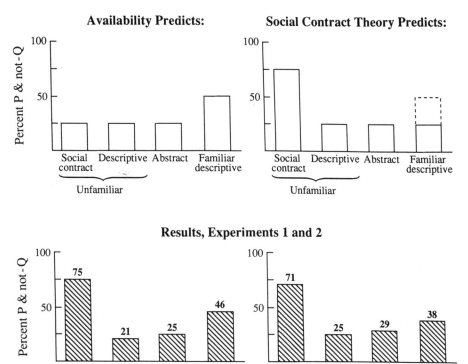

Figure 3.4   Social contract theory versus availability theory: Predictions and results for standard social contracts (from Cosmides, 1989, Experiments 1 and 2).

Familiarity, therefore, cannot account for the pattern of reasoning elicited by social contract problems. Social contract performance is not a by-product of familiarity.

Does Social Contract Content Simply Facilitate Logical Reasoning?

In the experiments just described, the adaptively correct answer if one is looking for cheaters happens to be the same as the logically correct answer—*P & not-Q*. Therefore, they cannot tell one whether performance on social contracts problems is governed by rules of inference that are specialized for reasoning about social exchange or by the rules of inference of the propositional calculus. Although we can think of no reason why this would be the case, perhaps social contract content simply facilitates logical reasoning. If so, then social contract performance could be a by-product of a logic faculty.

Two different sets of experiments show that this is not the case. The first involves "switched" social contracts (Cosmides, 1989, Experiments 3 and 4), and the second involves perspective change (Gigerenzer & Hug, in press).

Switched Social Contracts. The propositional calculus is content-independent: The combination of *P & not-Q* violates any conditional rule of the form *If P then Q*, no matter what "P" and "Q" stand for. The proposed social contract algorithms are not content-independent: Cheating is defined as accepting a benefit without paying the required cost. It does not matter what logical category these values happen to correspond to. For example, although the same social contract is expressed by both of the following statements, the proposition "you give me your watch" corresponds to the logical category P in the first rule and to Q in the second one.

Rule 1: "If you give me your watch, I'll give you \$20" (standard form).

Rule 2: "If I give you \$20, you give me your watch" (switched form).

No matter how the contract is expressed, I will have cheated you if I accept your watch but do not offer you the \$20, that is, if I accept a benefit from you without paying the required cost. If you are looking for cheaters, you should therefore choose the "benefit accepted" card (I took your watch) and the "cost not paid" card (I did not give you the \$20) no matter what logical category they correspond to. In the case of Rule 1, my taking your watch without paying you the \$20 would correspond to the logical categories P and $not\text{-}Q$, which happens to be the logically correct answer. But in the case of Rule 2, my taking your watch without giving you the \$20 corresponds to the logical categories Q and $not\text{-}P$. This is not the logically correct response. In this case, choosing the logically correct answer, P & $not\text{-}Q$, would constitute an adaptive error: If I gave you the \$20 (P) but did not take your watch ($not\text{-}Q$), I have paid the cost but not accepted the benefit. This makes me an altruist or a fool, but not a cheater.

The general principle is illustrated in Figure 3.3, panel c, which shows the cost-benefit structure of a social contract. Rule 1 ("If you take the benefit, then you pay the cost") expresses the same social contract as Rule 2 ("If you pay the cost, then you take the benefit"). A person looking for cheaters should always pick the "benefit accepted" card and the "cost not paid" card. But for Rule 1, a "standard" social contract, these cards correspond to the logical categories P and $not\text{-}Q$, whereas for Rule 2, a "switched" social contract, they correspond to the logical categories Q and $not\text{-}P$. Because the correct social contract answer is different from the correct logical answer for switched social contracts, by testing such rules we can see whether social contracts activate inference procedures of the propositional calculus, such as modus ponens and modus tollens, or inference procedures that are specialized for detecting cheaters on social contracts.

The design of the following experiments was similar to that just described. Each subject solved four Wason selection tasks, presented in counterbalanced order: an unfamiliar social contract, an unfamiliar descriptive problem, a familiar descriptive problem, and an abstract problem. The only difference was that in this case the terms of the two unfamiliar rules were "switched" within the "If-then" structure of the rule. For example, instead of reading, "If you eat duiker meat, then you have found an ostrich eggshell," the rule would read, "If you have found an ostrich eggshell, then you eat duiker meat." This was true for both the unfamiliar social contract and the unfamiliar descriptive rule. For the social contract rules, this switch in the order of the terms had the effect of putting the cost term in the "If" clause, and the benefit term in the "then" clause, giving the rule the structure of the switched social contract shown in Rule 2 of Figure 3.3, panel c.

The predictions of social contract theory and the availability theories are shown in Figure 3.5, along with the results of two experiments. *Not-P* & *Q* is an extremely rare response on the Wason selection task, but social contract theory predicts that it will be very common on switched social contracts. That is exactly what happened. In fact, as

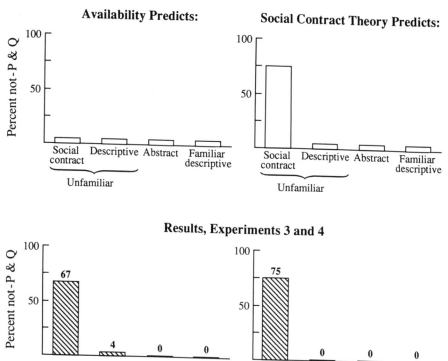

Figure 3.5 Social contract theory versus availability theory: Predictions and results for switched social contracts (from Cosmides, 1989, Experiments 3 and 4).

many people chose *not-P & Q* on the unfamiliar switched social contracts as chose *P & not-Q* on the standard social contract problems described above.

If social contract content merely facilitates logical reasoning, then subjects should have chosen *P & not-Q* on these switched social contract problems. The fact that they chose *not-P & Q*—a logically incorrect response—shows that social contract performance is not caused by the activation of a logic faculty. This is, however, the response one would expect if humans have rules of inference that are specialized for cheater detection.

Table 3.2 shows individual card choices for matching sets of experiments with standard versus switched social contracts, sorted by logical category and by social contract category. The results for non-social contract problems replicate beautifully when sorted by logical category. But not the results for the social contract problems. These results replicate when sorted by social contract category, not by logical category. This shows that the content-dependent social contract categories, not the logical categories, are psychologically real for subjects solving social contract problems. This confirms another predicted design feature of the social contract algorithms: They define cheating in terms of cost-benefit categories, not in terms of logical categories.

Manktelow and Over (1987) have pointed out that even when words such as "must" or "may" are left out of a social contract, one tends to interpret a standard social contract as meaning, "If you take the benefit, then you (must) pay the cost," whereas one tends to interpret a switched social contract as meaning, "If you pay the

Table 3.2 Selection Frequencies for Individual Cards, Sorted by Logical Category and by Social Contract Category[a]

| | Unfamiliar Descriptive | | Abstract Problem | | Familiar Descriptive | | Unfamiliar Social Contract | |
| | | | | | | | Standard | Switched |
Logical Category	Exp. 1 & 2	Exp. 3 & 4	Exp. 1 & 2	Exp. 3 & 4	Exp. 1 & 2	Exp. 3 & 4	Exp. 1 & 2	Exp. 3 & 4
P	43	40	46	46	46	45	43	3
not-P	9	11	10	11	1	2	3	36
Q	20	23	15	23	8	6	0	44
not-Q	18	20	21	25	23	32	39	3
Social Contract Category:								
Benefit accepted							43	44
Benefit not accepted							3	3
Cost paid							0	3
Cost not paid							39	36

[a]Experiments 1 and 2 tested standard versions of the two unfamiliar rules, whereas Experiments 3 and 4 tested switched versions of these rules.

cost, then you (may) take the benefit." This is, in fact, a prediction of social contract theory: A cost is something one is obligated to pay when one has accepted a benefit, whereas a benefit is something that one is entitled to take (but need not) when one has paid the required cost. Thus, the interpretive component of the social contract algorithms should cause subjects to "read in" the appropriate "musts" and "mays," even when they are not actually present in the problem (three out of four of the standard social contracts had no "must," and none of the switched social contracts had a "may"). Could it be that subjects are in fact reasoning with the propositional calculus, but applying it to these reinterpretations of the social contract rules?

No. In the propositional calculus, "may" and "must" refer to possibility and necessity, not to entitlement and obligation. On the Wason selection task, the logically correct answer for the rule, "If you pay the cost, then it is possible for you to take the benefit," is to choose no cards at all. Because this rule admits only of possibility, not of necessity, no combination of values can falsify it. The fact that most subjects chose *not-P & Q*, rather than no cards at all, shows that they were not applying the propositional calculus to a rule reinterpreted in this way.[13] To choose *not-P & Q*, one would have to be following the implicational structure of social exchange specified in Cosmides and Tooby (1989).

Perspective Change. Gigerenzer and Hug (in press) have conducted an elegant series of experiments that test another design feature of the proposed social contract algorithms, while simultaneously showing that the results cannot be explained by the propositional calculus or by permission schema theory, which is discussed later. They gave two groups of subjects Wason selection tasks in which they were to look for violations of social contract rules such as, "If an employee gets a pension, then that employee must have worked for the firm for at least 10 years." The only difference between the two groups was that one group was told "You are the employer" whereas the other group was told "You are the employee."

In social contract theory, what counts as cheating depends on one's perspective. Providing a pension is a cost that the employer incurs to benefit the employee, whereas working 10 or more years is a cost that the employee incurs to benefit the employer. Whether the event "the employee gets a pension" is considered a cost or a benefit therefore depends on whether one is taking the perspective of the employer (= cost) or the employee (= benefit). The definition of cheating as taking a benefit without paying the cost is invariant across perspectives, but the theory predicts that which events count as benefit and cost will differ across actors. From the employer's perspective, cheating is when an employee gets a pension (the employee has taken the benefit) but has not worked for the firm for at least 10 years (the employee has not paid the cost). These cards correspond to the logical categories P & $not\text{-}Q$. From the employee's perspective, cheating is when an employee has worked for at least 10 years (the employer has taken the benefit), but has not been given the pension that he or she is therefore entitled to (the employer has not paid the cost). These cards correspond to the logical categories $not\text{-}P$ & Q.

In other words, there are two different states of affairs that count as cheating in this situation, which correspond to the perspectives of the two parties to the exchange. What counts as cheating depends on what role one occupies; cheating is a well-defined concept, but its definition is preeminently content- and context-dependent.

In contrast, whether one is cued into the role of employer or employee is irrelevant to the content-independent propositional calculus. The correct answer on such a problem is P & $not\text{-}Q$ (employee got the pension but did not work 10 years), regardless of whether the subject is assigned the role of the employer or the employee.

Gigerenzer and Hug conducted four experiments using different social contract rules to test the perspective change hypothesis. The predictions of both social contract theory and the propositional calculus are shown in Figure 3.6, along with the results. The results are as social contract theory predicts: Even though it is logically incorrect, subjects answer $not\text{-}P$ & Q when these values correspond to the adaptively correct "look for cheaters" response. The hypothesis that social contract content simply facilitates logical reasoning cannot explain this result. The perspective change results and the switched social contract results show that social contract performance is not a by-product of the activation of a logic faculty.

In light of these results, it is interesting to note that although schizophrenic individuals often perform more poorly than normals on problems requiring logical reasoning, deliberation, or seriation, Maljković found that their reasoning about social contracts is unimpaired (Maljković, 1987). She argues that this result makes sense if one assumes that the brain centers that govern reasoning about social contracts are different from those that govern logical reasoning.

To show adaptation, one must both eliminate by-product hypotheses *and* show evidence of special design for accomplishing an adaptive function. Algorithms specialized for reasoning about social exchange should have certain specific design features, and the switched social contract and perspective change experiments confirm three more predictions about those design features:

1. The definition of cheating embodied in the social contract algorithms should depend on one's perspective. The perspective change experiments confirm the existence of this design feature.
2. Computing a cost-benefit representation of a social contract from one party's

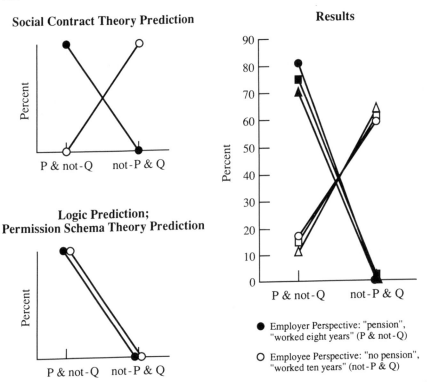

Figure 3.6 Perspective change experiments. Social contract theory versus logic facilitation hypothesis and permission schema theory: Predictions and results. Three separate experiments were conducted, testing the predictions of the theories against one another. The results of the three experiments are indicated by circles, squares, and triangles. Filled versus unfilled represents versions otherwise identical, except that the perspective (e.g., employer versus employee) is reversed (from Gigerenzer & Hug, in press).

perspective should be just as easy as computing it from the other party's perspective. There are two reasons for this prediction: First, to successfully negotiate with others, one must be able to compute the conditions under which others would feel that you had cheated them, as well as the conditions under which they had cheated you. Second, being able to understand what counts as cheating from both perspectives facilitates social learning; by watching other people's exchanges one can gather information about the values, and perhaps about the trustworthiness, of people one may interact with in the future. If people are just as good at translating a social contract into the values of one party as the other, then they should be just as good at detecting cheaters from one perspective as the other: There should be just as many *P & not-Q* responses to the "employer" version as there are *not-P & Q* responses to the "employee" version. This was, in fact, the case.

3. The implicational structure of social contract theory mandates that the statement "If an employee gets a pension, then that employee must have worked for the firm for at least 10 years" be taken to imply "If the employee has worked for 10 years, then the employer must give that employee a pension." This is

because when the employee has fulfilled his obligation to benefit the employer, the employer is obligated to benefit the employee in return. If subjects draw this implication, then they should choose *not-P & Q* in both the switched social contract experiments and in the employee version of the perspective change experiments. The fact that they did confirms this design feature of the proposed social contract algorithms.

Is There a Cheater Detection Procedure, or Are People Simply Good at Reasoning about Social Contracts?

Social contract theory posits that the mind contains inference procedures specialized for detecting cheaters and that this explains the high percentage of correct social contract answers that these problems elicit. But maybe social contract problems simply "afford" clear thinking. Perhaps they are interesting, or motivationally compelling, in a way that other problems are not. Rather than having inference procedures specialized for detecting cheaters, perhaps we form a more complete mental model of a problem space for social contract problems, and this allows us to correctly answer *any* question that we might be asked about them, whether it is about cheating or not (for descriptions of mental model theories of reasoning, see Evans, 1984; Johnson-Laird, 1983; Manktelow & Over, 1987). Although it would be difficult to reconcile the perspective change data with this hypothesis, it is still worth considering.

No one has presented any independent criteria for judging what kinds of problems ought to be "interesting," "motivationally compelling," or "easy to understand," which makes this hypothesis nebulous. Nevertheless, it can be tested by studying performance on reasoning problems in which the rule is portrayed as a social contract, but in which the subject is *not* asked to look for cheaters.

Two sets of experiments did just that. One set asked subjects to look for altruists in a situation of social exchange; the other asked subjects to look for violations of a social contract rule in a context in which looking for violations did not correspond to looking for cheaters. If people are good at detecting cheaters merely because social contract problems are easy to understand, then performance on such problems should be just as good as performance on cheater detection problems.[14] But if people are good at detecting cheaters because they have inference procedures specialized for doing so, then such problems should elicit lower levels of performance than social contract problems that require cheater detection.

Are People Good at Looking for Altruists? The game-theoretic models for the evolution of cooperation that could be reasonably applied to the range of population structures that typified hominid hunter-gatherers require the existence of some mechanism for detecting cheaters or otherwise excluding them from the benefits of cooperation. This is because the capacity to engage in social exchange could not have evolved in the first place unless the individuals involved could avoid being continually exploited by cheaters. But most models do not require the existence of a mechanism for detecting "altruists"—individuals who follow the strategy of paying the required cost (thereby benefiting the other party), but not accepting from the other party the benefit to which this act entitles them.[15] Indeed, because individuals who were consistently altruistic would incur costs but receive no compensating benefits, under most plausible scenarios they would be selected out. Because they would not be a long-enduring feature of the adaptive landscape, there would be no selection pressure for "altruist detection" mechanisms. Thus, while we did expect the existence of inference procedures specialized for detecting cheaters, we did not expect the existence of inference procedures specialized for detecting altruists.

In contrast, if people are good at detecting cheaters merely because social contract problems afford clear thinking, then performance on altruist detection problems should be just as good as performance on cheater detection problems. The mental model of the social contract would be the same in either case; one would simply search the problem space for altruists rather than for cheaters.

To see whether this was so, we tested 75 Stanford students on some of the same social contract problems that were used by Cosmides (1985, 1989), but instead of asking them to look for cheaters, they were asked to look for altruists (the procedure was the same as that described in Cosmides, 1989, for Experiments 5 through 9). Each subject was given one social contract problem, and this problem required altruist detection. There were three conditions with 25 subjects each; a different social contract problem was tested in each condition. The first two conditions tested problems that portrayed a private exchange between "Big Kiku" (a headman in a fictitious culture) and four hungry men from another band; the third condition tested a problem about a social law. The first group's problem was identical to the cheater detection problem tested by Cosmides (1989) in Experiment 2, except for the instruction to look for altruists (see Figure 3.7 for a comparison between the cheater detection version and the altruist detection version). The problem in the second condition was essentially the same, but instead of portraying Big Kiku (the potential altruist) as a ruthless character, he was portrayed as having a generous personality. The third condition tested the social law, "If you eat duiker meat, then you have found an ostrich eggshell" (see Cosmides, 1989, Experiment 1). The instructions on this problem were suitably modified to ask subjects to look for altruists rather than cheaters.[16]

Because altruists are individuals who have paid the cost but have not accepted the benefit, subjects should choose the "benefit not accepted" card (not-P) and the "cost paid" card (Q) on these problems. These values would correspond to the "no tattoo" card and the "Big Kiku gave him cassava root" card for the first two problems, and to the "does not eat any duiker meat" card and the "has found an ostrich eggshell" card for the social law problem. Table 3.3 shows that the percentage of subjects who made this response was quite low for all three altruist detection problems.

Is it possible that Stanford students simply do not know the meaning of the word "altruistic"? We thought this highly unlikely, but just to be sure, we ran another 75 Stanford students ($n = 25$ per condition) on problems that were identical to the first three, except that the word "selflessly" was substituted for the word "altruistically." The word "selfless" effectively announces its own definition—less for the self. If there are inference procedures specialized for detecting altruists—or if social contract prob-

Table 3.3 Altruist Detection: Percent Correct (not-P & Q)

| | Personal Exchange | | Law |
	Ruthless	Generous	
Altruistic	28	8	8
Selfless	40	36	12

lems merely afford clear thinking—then surely subjects should be able to perform as well on the "selfless" problems as they do on cheater detection problems.

Table 3.3 shows that this was not the case. Although performance was a bit higher on the problems that used "selfless" than on the problems that used "altruistic," performance was nowhere near the average of 74% that Cosmides (1989) found for comparable cheater detection problems.[17] In fact, performance on the selfless versions of the altruist detection problems was no better than performance on the familiar descriptive transportation problem (reported earlier). This indicates that people do not have inference procedures specialized for detecting altruists on social contracts, which is just what social contract theory predicted. More important, it casts doubt on the hypothesis that cheater detection problems elicit high levels of performance merely because social contracts afford clear thinking.

Are People Good at Looking for Violations of Social Contracts When These Do Not Indicate Cheating? Gigerenzer and Hug (in press) conducted a series of experiments designed to disentangle the concept of cheater detection from the concept of a social contract. The opportunity to illicitly benefit oneself is intrinsic to the notion of cheating. But one can construct situations in which the reason one is looking for violations of a social contract rule has nothing to do with looking for individuals who are illicitly benefiting themselves—i.e., one can construct situations in which looking for violations is *not* tantamount to looking for cheaters. Gigerenzer and Hug gave subjects Wason selection tasks in which all rules were framed as social contracts, but which varied in whether or not looking for violations constituted looking for cheaters.

Here is an example using the rule, "If one stays overnight in the cabin, then one must bring a load of firewood up from the valley." In the "cheating" version, it is explained that two Germans are hiking in the Swiss Alps and that the local Alpine Club has cabins at high altitudes that serve as overnight shelters for hikers. These cabins are heated by firewood, which must be brought up from the valley because trees do not grow at this altitude. So the Alpine Club has made the (social contract) rule, "If one stays overnight in the cabin, then one must bring along a bundle of firewood from the valley." There are rumors that the rule is not always followed. The subject is cued into the perspective of a guard whose job is to check for violations of the rule. In this version, looking for violations of the rule is the same as looking for cheaters.

In the "no cheating" version, the subject is cued into the perspective of a member of the German Alpine Association who is visiting a cabin in the Swiss Alps and wants to find out how the local Swiss Alpine Club runs the cabin. He sees people carrying loads of firewood into the cabin, and a friend suggests that the Swiss might have the same social contract rule as the Germans—"If one stays overnight in the cabin, then one must bring along a bundle of firewood from the valley." The story also mentions an alternative explanation: that members of the Swiss Alpine Club (who do not stay overnight in the cabin) bring wood, rather than the hikers. To settle the question, the subject is asked to assume that the proposed social contract rule is in effect, and then to look for violations of it. Note that the intent here is not to catch cheaters. In this situation, violations of the proposed social contract rule can occur simply because the Swiss Alpine Club never made such a rule in the first place.

In both versions, the rule in question is a social contract rule—in fact, exactly the same social contract rule. And in both versions, the subject is asked to look for violations of that rule. But in the cheating version, the subject is looking for violations

You are an anthropologist studying the Kaluame, a Polynesian people who live in small, warring bands on Maku Island in the Pacific. You are interested in how Kaluame "big men" – chieftains – wield power.

"Big Kiku" is a Kaluame big man who is known for his ruthlessness. As a sign of loyalty, he makes his own "subjects" put a tattoo on their face. Members of other Kaluame bands never have facial tattoos. Big Kiku has made so many enemies in other Kaluame bands, that being caught in another village with a facial tattoo is, quite literally, the kiss of death.

Four men from different bands stumble into Big Kiku's village, starving and desperate. They have been kicked out of their respective villages for various misdeeds, and have come to Big Kiku because they need food badly. Big Kiku offers each of them the following deal:

"If you get a tattoo on your face, then I'll give you cassava root."

Cassava root is a very sustaining food which Big Kiku's people cultivate. The four men are very hungry, so they agree to Big Kiku's deal. Big Kiku says that the tattoos must be in place tonight, but that the cassava root will not be available until the following morning.

You learn that Big Kiku hates some of these men for betraying him to his enemies. You suspect he will cheat and betray some of them. Thus, this is a perfect opportunity for you to see first hand how Big Kiku wields his power.

The cards below have information about the fates of the four men. Each card represents one man. One side of a card tells whether or not the man went through with the facial tattoo that evening and the other side of the card tells whether or not Big Kiku gave that man cassava root the next day.

Did Big Kiku get away with cheating any of these four men? Indicate only those card(s) you definitely need to turn over to see if Big Kiku has broken his word to any of these four men.

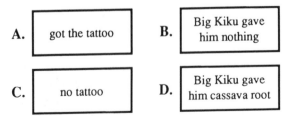

Figure 3.7 Both problems describe a social contract rule, but the problem on the left asks the subject to look for cheaters (individuals who took the benefit without paying the cost), whereas the problem on the right asks the subject to look for altruists (individuals who paid the cost but did not take the benefit to which this entitles them).

You are an anthropologist studying the Kaluame, a Polynesian people who live in small, warring bands on Maku Island in the Pacific. You are interested in how Kaluame "big men" – chieftains – wield power.

"Big Kiku" is a Kaluame big man who is known for his ruthlessness. As a sign of loyalty, he makes his own "subjects" put a tattoo on their face. Members of other Kaluame bands never have facial tattoos. Big Kiku has made so many enemies in other Kaluame bands, that being caught in another village with a facial tattoo is, quite literally, the kiss of death.

Four men from different bands stumble into Big Kiku's village, starving and desperate. They have been kicked out of their respective villages for various misdeeds, and have come to Big Kiku because they need food badly. Big Kiku offers each of them the following deal:

"If you get a tattoo on your face, then I'll give you cassava root."

Cassava root is a very sustaining food which Big Kiku's people cultivate. The four men are very hungry, so they agree to Big Kiku's deal. Big Kiku says that the tattoos must be in place tonight, but that the cassava root will not be available until the following morning.

You learn that Big Kiku hates some of these men for betraying him to his enemies. You suspect he will cheat and betray some of them. However, you have also heard that Big Kiku sometimes, quite unexpectedly, shows great generosity towards others – that he is sometimes quite altruistic. Thus, this is a perfect opportunity for you to see first hand how Big Kiku wields his power.

The cards below have information about the fates of the four men. Each card represents one man. One side of a card tells whether or not the man went through with the facial tattoo that evening, and the other side of the card tells whether or not Big Kiku gave that man cassava root the next day.

Did Big Kiku behave altruistically towards any of these four men? Indicate only those card(s) you definitely need to turn over to see if Big Kiku has behaved altruistically towards any of these four men.

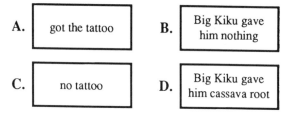

A. got the tattoo   B. Big Kiku gave him nothing

C. no tattoo   D. Big Kiku gave him cassava root

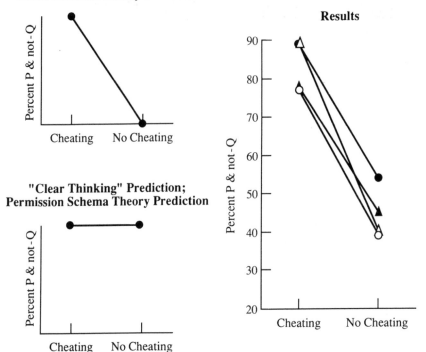

Figure 3.8 Is cheating necessary? Predictions and results. In these experiments, both conditions tested social contract rules, but in the "no cheating" condition looking for violations did not correspond to looking for cheaters. Social contract theory predicts a different pattern of results from both permission schema theory and the hypothesis that social contract content merely affords "clear thinking." Four separate experiments were conducted, testing the predictions of the theories against one another. The results of the four experiments are indicated by filled and unfilled triangles, and filled and unfilled circles (from Gigernzer & Hug, in press).

because he or she is looking for individuals who are illicitly benefiting themselves, whereas in the no cheating version the subject is looking for violations because he or she is interested in whether the proposed social contract rule is in effect. If social contract problems merely afford clear thinking, then a violation is a violation is a violation: It shouldn't matter whether the violation constitutes cheating or not.[18] In contrast, if there are inference procedures specialized for cheater detection, then performance should be much better in the cheating version, where looking for violations is looking for illicitly taken benefits. Figure 3.8 shows the predictions and the results of four such experiments. On average, 83% of subjects correctly solved the cheating version, compared with only 44% on the no cheating version. Cosmides and Tooby (in prep. b) conducted a similarly designed experiment, with similar results. In both problems, it was stipulated that a social contract rule was in effect and that the people whose job it was to enforce the rule may have violated it. But in one version violations were portrayed as due to cheating, whereas in the other version violations were portrayed as due to innocent mistakes. We found 68% of subjects correctly solved the cheating version, compared with only 27% of subjects on the "mistake" (no cheating) version. Thus, using a different "no cheating" context in which it was stipulated that the social contract rule was in effect, we were able to replicate the difference that Gigerenzer and Hug found between their cheating and no cheating versions almost exactly (39 percentage points in Gigerenzer and Hug; 41 percentage points in Cosmides and Tooby).

These data indicate that social contract problems do not merely afford clear thinking. In addition, these results provide further evidence against the availability theories and the hypothesis that social contract content merely facilitates logical reasoning.

The results of these experiments are most parsimoniously explained by the assumption that people have inference procedures specialized for detecting cheaters: individuals who have illicitly taken benefits. Because these procedures operate on the cost-benefit representation of a problem, they can detect a violation only if that violation is represented as an illicitly taken benefit. They would not be able to detect other kinds of violations, nor would they detect altruists. These results provide further confirmation of a proposed design feature of the social contract algorithms: This bundle of cognitive processes appears to include inference procedures specialized for cheater detection.

Are Cost-Benefit Representations Necessary, or Are Subjects Good at Detecting Violations of Any Rule Involving Permission?

The social contract algorithms should contain decision rules that govern when one should engage in an exchange. A proposed contract should not be accepted unless the costs one incurs by providing a benefit to the other party are outweighed by the benefits one receives by engaging in the exchange. Consequently, the first stage of processing must be the assignment of cost-benefit weightings to the "benefit term"—the event, item, or state of affairs the contract entitles one to receive—and to the "cost term"—the event, item, or state of affairs that the other party requires of one. We have been calling these terms the "benefit" term and the "cost" term for ease of explication; we do not mean to prejudge the actual values that the parties to the exchange assign to these terms. For example, satisfying the requirement may actually benefit the person who satisfies it—the offerer's belief that the benefit she is requiring the other party to provide represents a cost to that party may be erroneous. But the offerer would not propose the contract in the first place unless she believed that the state of affairs she wants to bring about would not occur in the absence of the contract—it would be silly to offer you $20 for your watch if I believed you were going to give it to me anyway. Similarly, the offerer may be mistaken in her belief that she is offering a benefit to the other party, in which case her offer will not be accepted—you might value your watch more than you would value having an extra $20. In social contract theory, costs and benefits are defined with respect to the value systems of the participants, and relative to a zero-level utility baseline that represents how each party would behave in the absence of the contract. (The cost-benefit conditions that should adhere for an offerer to offer and the acceptor to accept a social contract are spelled out in Cosmides, 1985, and Cosmides & Tooby, 1989.)

Once the contract has been translated into cost-benefit terms, the decision rules for acceptance or rejection can operate on that representation. The cheater detection procedures should also act on that representation, causing one to investigate individuals who have not fulfilled the requirement specified in the cost term and individuals who have taken the benefit offered. By operating on this relatively abstract level of representation, the cheater detection procedures can detect cheaters no matter what the actual items of exchange are. Because humans are a technological species capable of exchanging a wide array of items, we hypothesized that human cheater detection procedures would operate on this abstract cost-benefit level of representation, rather than be tied to any particular item of exchange, as discussed earlier in this chapter. In other species, the constraint that cheaters must be detected might be implemented by algorithms that operate on other levels of representation (as discussed for the vampire bat).

Thus, a predicted design feature of human social contract algorithms is that they cause one to assign cost-benefit representations to a social contract, one for each party to the interaction. The presence of this design feature was confirmed by the perspective change experiments. Another predicted design feature is that cheater detection procedures operate on cost-benefit representations. Indeed, it is difficult to see how they could do otherwise, given that cheating is defined as accepting a benefit one is not entitled to. Technically, a violation that does not illicitly benefit one is not cheating.

If cheater detection algorithms operate on cost-benefit representations, then they should not be able to operate on rules that are similar to social contracts but that have not been assigned a cost-benefit representation. To test this prediction, we investigated Wason selection task performance on permission rules that did not afford the assignment of a cost-benefit representation.

Permission rules are prescriptive rules that specify the conditions under which one is allowed to take some action. Cheng and Holyoak (1985) define them as "regulations . . . typically imposed by an authority to achieve some social purpose" (p. 398). Their "permission schema theory" proposes that people have a schema for reasoning about permissions that consists of the following four production rules:

Rule 1: If the action is to be taken, then the precondition must be satisfied.
Rule 2: If the action is not to be taken, then the precondition need not be satisfied.
Rule 3: If the precondition is satisfied, then the action may be taken.
Rule 4: If the precondition is not satisfied, then the action must not be taken.

On a Wason selection task in the linguistic format of rule 1, the first production rule would cause one to choose the "action has been taken" card (P), and the fourth production rule would cause one to choose the "precondition has not been satisfied" card ($not\text{-}Q$). Rules 2 and 3 would not cause any card to be chosen. If a Wason selection task presented a rule in the linguistic format of rule 3 (a cognate to our switched social contracts), then the same two cards would be chosen for the same reason, but they would now correspond to the logical categories Q and $not\text{-}P$.

Cheng and Holyoak (1985) and Cheng, Holyoak, Nisbett, and Oliver (1986) also propose the existence of "obligation schemas." These have the same implicational structure as permission schemas (and hence lead to the same predictions), but their representational format is "If condition C occurs, then action A must be taken." Because in any concrete case it is difficult to tell a permission from an obligation, we will refer to both kinds of rules under the rubric of permission schema theory.[19]

All social contracts are permission rules, but not all permission rules are social contracts. Social contract rules that have the form "If one takes the benefit, then one must pay the cost" are subsets of the set of all permission rules, because taking a benefit is just one kind of action that a person can take. Taking a benefit always entails taking an action, but there are many situations in which taking an action does not entail taking a benefit.

Permission schema theory has already been falsified by the experiments of Gigerenzer and Hug, described earlier. In permission schema theory, a permission rule has
been violated whenever the action has been taken but the precondition has not been
satisfied. It should not matter *why* subjects are interested in violations. Whether one
is interested in violations because one is interested in finding cheaters or because one
is interested in seeing whether the rule is in effect is irrelevant to permission schema
theory. As long as the subject recognizes that the rule is a permission rule, the permission schema should cause the "action taken" and the "precondition not met" cards to
be chosen. Consequently, permission schema theory would have to predict equally
high levels of performance for the cheating and the no cheating versions of the social
contract rules tested by Gigerenzer and Hug, as Figure 3.8 shows. Yet, even though
both problems involved looking for violations of the same (social contract) permission
rule, the cheating version elicited much higher performance than the no cheating version. Gigerenzer and Hug's perspective change experiments also falsify permission
schema theory, and for a similar reason. Only one combination of values—"action
taken" and "precondition not satisfied"—violates a permission rule in permission
schema theory. What counts as a violation does not change depending on one's perspective; indeed, permission schema theory has no theoretical vocabulary for discussing differences of perspective. Yet the subjects' definition of violation depended on
what role they were cued into, the employer's (P & *not-Q*) or the employee's (*not-P* &
Q). Permission schema theory can account for the P & *not-Q* response, but not for the
not-P & Q response.[20]

Permission schema theory and social contract theory differ in yet another way: The
permission schema operates on the more abstract and inclusive action-precondition
level of representation, whereas social contract algorithms construct and operate on
the somewhat less general cost-benefit level of representation. Consequently, permission schema theory predicts that all permission rules will elicit high levels of performance, whether they have the cost-benefit structure of a social contract or not. In contrast, social contract theory does not predict high levels of performance for permission
rules that do not afford the assignment of a cost-benefit representation. This is because
cheating is *defined* as an illicitly taken benefit; where there are no benefits, there can
be no cheating. By comparing performance on permission rules that do and do not
afford the assignment of the appropriate cost-benefit representation, we can both test
the prediction that cheater detection algorithms require a cost-benefit representation
to operate and provide yet another test between social contract theory and permission
schema theory.

In the first series of experiments (reported in Cosmides, 1989), we used the same
research strategy as before: Wason selection tasks using rules whose terms and relation
would be unfamiliar to our subjects, surrounded by a story context that either did or
did not afford the assignment of the cost-benefit structure of a social contract to the
rule. But in these experiments, the non-social-contract rule was a permission rule, not
a descriptive rule. Let us illustrate the difference with an example: the school rule.

The school rule tested in both conditions was "If a student is to be assigned to
Grover High School, then that student must live in Grover City." In the social contract
version, the story explained that Grover High School is a much better high school than
Hanover High, with a good record for getting students into college. Citizens of Grover
City pay higher taxes for education than citizens of the town of Hanover, which is why
Grover High is the better school. The story surrounding the rule thus portrays going

to Grover High as a rationed benefit that must be paid for through higher taxes. Volunteers, some of whom are mothers with high-school-age children, are processing the school assignment documents, and it is rumored that some might have cheated on the rule in assigning their own children to a school. The subject is cued into the role of someone who is supervising these volunteers and must therefore look for cheaters.

In the non-social-contract version, the same permission rule is used, but the surrounding story did not afford a cost-benefit interpretation of the terms of the rule. Grover High is not portrayed as any better than Hanover High, nor does the story mention any greater cost that is incurred by living in Grover City rather than Hanover. It is, however, explained that it is important that this rule for assigning students from various towns to the appropriate school district be followed, because the population statistics they provide allow the board of education to decide how many teachers need to be assigned to each school. If the rule is not followed, some schools could end up with too many teachers and others with too few. Thus the story context gives the rule a "social purpose." The subject is cued into the role of a person who is replacing the absent-minded secretary who was supposed to follow this rule in sorting the students' documents. Because the former secretary frequently made mistakes, the subject must check the documents to see if the rule was ever violated.

Although this rule is a permission rule—stating the conditions under which one is allowed to assign a student to Grover High School—nothing in the story affords the assignment of the cost-benefit structure of a social contract to this rule. There is no benefit to assigning someone to Grover rather than to Hanover High, and no cost associated with living in one city rather than the other. Moreover, there is no apparent way that the absent-minded secretary could have illicitly benefited from breaking the rule. Thus her mistakes would not constitute cheating. By hypothesis, cheater detection algorithms should not be able to operate on this problem, because they would have no cost-benefit representations to attach themselves to. Social contract theory, therefore, predicts a lower percent of correct responses for this version than for the social contract version. In contrast, permission schema theory predicts high levels of the correct responses for both rules, because both are permission rules: Both have the action-precondition representational format that permission schema theory requires. (For both theories, the "correct" response is *P & not-Q* for standard rules and *not-P & Q* for switched rules.)

The predictions and results of four experiments—two with standard rules, two with switched rules—are displayed in Figure 3.9. Across the four experiments, 75% of subjects chose the correct answer for the social contract permission rules, compared with only 21% for the non-social-contract permission rules.

Using unfamiliar rules with a long story context has the advantage of giving the experimenter some control over the subject's mental model of the situation. The disadvantage of this method, however, is that it is difficult to create matching stories in which only one element varies. In the matched school rules, for example, two elements that distinguish permission schema theory and social contract theory varied: (a) whether the rule was given a cost-benefit structure, and (b) whether the potential violator was portrayed as a cheater or as a person who might have broken the rule by mistake. So we tackled the question of whether the rule must have a cost-benefit structure in another way as well: We tested minimalist problems that varied only in whether the subjects' past experience would cause them to interpret the antecedent of the rule as a benefit (Cosmides & Tooby, in prep., b). These problems had virtually no context:

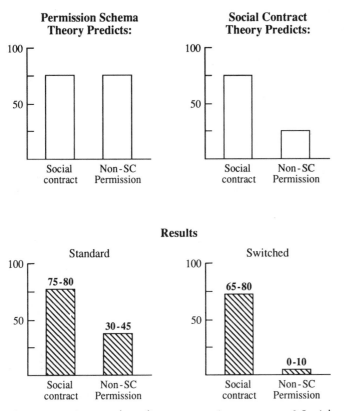

Figure 3.9 Are cost-benefit representations necessary? Social contract theory versus permission schema theory: Predictions and results (from Cosmides, 1989, Experiments 5, 6, 8, and 9).

They simply explained that among the Kalama (another fictitious culture) the elders make the laws, and one of the laws they made is, "If one is going out at night, then one must tie a small piece of red volcanic rock around one's ankle." The subject was then asked to see if any of four individuals had violated this law. The rule was based on one developed by Cheng and Holyoak (1989), but the context gave the rule no rationale or "social purpose."

We tested three versions of the rule that differed in only one respect: how much of a benefit the antecedent would seem to be. Among undergraduates, going out at night is a benefit: It represents fun, dating, adventure, and so on. Staying home at night is not as much fun, and taking out the garbage is even less so. Consequently, we compared performance on the "going out at night" rule to performance on two other rules: "If one is staying home at night, then one must tie a small piece of red volcanic rock around one's ankle" and "If one is taking out the garbage, then one must tie a small piece of red volcanic rock around one's ankle." If permission schema theory were correct, then all three of these permission rules would elicit equally high levels of P & not-Q responses.[21] But if social contract theory is correct, and the rule must have the cost-benefit structure of a social contract to elicit the effect, then performance should decline as the value of the antecedent declines. The more difficult it is to interpret the antecedent as a benefit, the more difficult it should be to see how one could illicitly benefit by breaking the rule.

The predictions and results are depicted in Figure 3.10. Performance decreases as the size of the benefit in the antecedent decreases, just as social contract theory predicts. Figure 3.10 also depicts the results of another, similar experiment with the so-called "Sears problem." As we removed the cost-benefit structure of the Sears problem, performance decreased. This experiment is also described in Cosmides and Tooby (in prep., b).

Manktelow and Over (1990; scenarios B and C) tested two obligation rules that lacked the cost-benefit structure of a social contract, with similar results. Cheng and Holyoak's theory predicts that both of these rules will elicit a high percentage of *P & not-Q* responses, yet they elicited this response from only 12% and 25% of subjects, respectively.

These experiments eliminate yet another by-product hypothesis: They show that reasoning on social contract problems is not a by-product of inference procedures for reasoning about a more general class of problems, permission problems. Permission rules elicit the effect only if the cost-benefit representation of a social contract can be assigned to them and if violating the rule would illicitly benefit the violator. These results confirm another predicted design feature: They show that cheater detection algorithms do not operate properly unless the appropriate cost-benefit representation can be assigned to the rule.

Will Any Rule That Involves the Possibility of Positive or Negative Payoffs Elicit Good Performance on the Wason Selection Task? Manktelow and Over (1990) were interested in this question when they designed the experiments just described. Their obli-

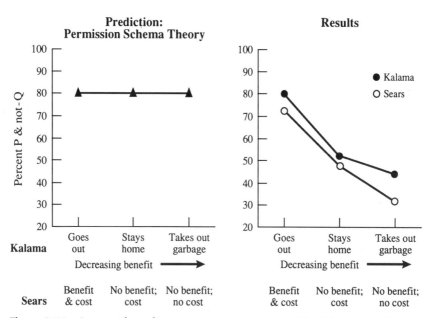

Figure 3.10 Are cost-benefit representations necessary? Social contract theory versus permission schema theory: Predictions and results (from Cosmides & Tooby, in prep., b).

gation rules lacked the cost-benefit structure of a social contract and, therefore, the property that one could illicitly benefit by cheating. But in both scenarios a person would receive high payoffs for following the rule. Violating the rule would cause a person to incur a cost; a small one in the case of one of the scenarios, a large one in the case of the other. They found that the possibility of payoffs, either positive or negative, was not sufficient to elicit high levels of *P & not-Q* responses. We found similar results in experiments testing rules concerning the possibility that a food was poisonous (Cosmides & Tooby, in prep., c), rules such as "If a person eats the red berries, then that person will vomit."

These experiments eliminate another by-product hypothesis: They show that the possibility of a payoff is not sufficient to elicit good performance on the Wason selection task and, therefore, not sufficient to explain the social contract effect. To be good at detecting violations of a social contract rule, the violation has to represent an illicitly taken benefit.

The possibility of payoffs is not sufficient to explain the social contract effect. But heuristically, it is an important dimension to consider when one is trying to discover specialized inference procedures. An evolutionary approach would lead one to investigate domains in which our foraging ancestors would have experienced positive (fitness) payoffs for "correct" reasoning and negative (fitness) payoffs for "incorrect" reasoning. We place "correct" and "incorrect" in scare quotes because these notions are defined with respect to the adaptive problem to be solved; "correct" does not necessarily mean the logically correct *P & not-Q*, just as it did not in many of the social contract experiments reported here. In addition to social contracts, there is now experimental evidence suggesting the existence of specialized inference procedures for two other domains, both of which involve large fitness payoffs: precautions (prudential obligations) and threats. Manktelow and Over (1990; scenario A) were the first to uncover the possibility of precaution schemas for rules of the general form "If one enters a hazardous situation, then one should take the appropriate precaution," and Cheng and Holyoak (1989) and we (Cosmides & Tooby, in prep., b) have provided evidence in support of the existence of precaution schemas. But the experiments to be reported in Cosmides and Tooby (in prep., b) also show that precaution rules are processed differently from social contract rules, in accordance with a different adaptive logic appropriate to that problem. This is also true of threats. Our tests of threat rules show that people are very good at detecting bluffs and double crosses in situations of threat, but again, the reasoning involved does not map on to cheater detection in social contract situations (Cosmides & Tooby, in prep., a).

Summary of Experimental Findings

Virtually all the experiments reviewed above asked subjects to detect violations of a conditional rule. Sometimes these violations corresponded to detecting cheaters on social contracts, other times they did not. The results showed that we do not have a general-purpose ability to detect violations of conditional rules. But human reasoning is well designed for detecting violations of conditional rules when these can be interpreted as cheating on a social contract. Based on our computational theory of social exchange, we predicted that reasoning about social contracts would exhibit a number of specific design features. The results of the experiments reviewed above confirm the existence of many of these design features and do not falsify any of them. They show the following:

1. The algorithms governing reasoning about social contracts include inference procedures specialized for cheater detection.
2. Their cheater detection procedures cannot detect violations that do not correspond to cheating (such as mistakes).
3. The algorithms governing reasoning about social contracts operate even in unfamiliar situations.
4. The definition of cheating that they embody depends on one's perspective.
5. They are just as good at computing the cost-benefit representation of a social contract from the perspective of one party as from the perspective of another.
6. They cannot operate so as to detect cheaters unless the rule has been assigned the cost-benefit representation of a social contract.
7. They embody implicational procedures specified by the computation theory (e.g., "If you take the benefit then you are obligated to pay the cost" implies "If you paid the cost, then you are entitled to take the benefit").
8. They do not include altruist detection procedures.

Furthermore, the highly patterned reasoning performance elicited by social contracts cannot be explained as a by-product of any of the more general-purpose reasoning procedures that have been proposed so far. The following by-product hypotheses were eliminated:

1. That familiarity can explain the social contract effect.
2. That social contract content merely activates the rules of inference of the propositional calculus.
3. That social contract content merely promotes, for whatever reason, clear thinking.
4. That permission schema theory can explain the social contract effect.
5. That any problem involving payoffs will elicit the detection of violations.
6. That a content-independent formal logic, such as the propositional calculus, quantificational logic, or deontic logic can explain the social contract effect.

These findings strongly support the hypothesis that the human mind includes cognitive procedures that are adaptations for reasoning about social exchange.

IMPLICATIONS FOR CULTURE

Cultural Forms Are Structured by Our Universal, Evolved Psychology

Wherever human beings live, their cultural forms and social life are infused with social exchange relations (e.g., Malinowski, 1922; Mauss, 1925/1967). Such relations appear in an enormous range of different guises, both simple and highly elaborated, implicit and explicit, deferred and simultaneous, practical and symbolic. The magnitude, variety, and complexity of our social exchange relations are among the most distinctive features of human social life, and differentiate us strongly from all other animal species (Tooby & DeVore, 1987).

Antiquity, universality, and cross-cultural elaboration are exactly what one expects of behaviors that are the expression of universal, evolved information-processing mechanisms. From the child who gets dessert if her plate is cleaned, to the devout Christian who views the Old and New Testaments as covenants arrived at between humans and the supernatural, to the ubiquitous exchange of women between descent groups among tribal peoples, to trading partners in the Kula Ring of the Pacific—all of these phenomena require, from the participants, the recognition and comprehension of a complex set of implicit assumptions that apply to social contract situations. Our social exchange psychology supplies a set of inference procedures that fill in all these necessary steps, mapping the elements in each exchange situation to their representational equivalents within the social contract algorithms, specifying who in the situation counts as an agent in the exchange, which items are costs and benefits and to whom, who is entitled to what, under what conditions the contract is fulfilled or broken, and so on (Cosmides & Tooby, 1989).

Without social exchange and the underlying constellation of cognitive adaptations that support it, social and mental life in every culture would be so different as to be unrecognizable as *human* life. If one removed from our evolutionary history and hence from our minds the possibility of cooperation and reciprocity—of mutually contingent benefit-benefit interactions arrived at through mutual consent—then coercion and force would loom even larger as instruments of social influence, and positive relationships would be limited primarily to self-sacrificial interactions among kin. Such conditions do, in fact, typify the social life of most other animal species. What psychological mechanisms allow humans to behave so differently?

According to the Standard Social Science Model, "culture" builds all concepts as sophisticated and as content-specific as social exchange from scratch, using only content-free general-purpose mental processes (see Tooby & Cosmides, this volume). Yet the experiments reviewed herein have shown that no general-purpose process so far proposed can produce the sophisticated pattern of reasoning needed to engage in social exchange. Moreover, these experiments cast serious doubt on the claim that content-free general-purpose mechanisms were responsible for *building* the content-specific social contract algorithms. Although we have reasoning procedures elaborately tailored for solving this ancient adaptive problem, these experiments have demonstrated that we do not have reasoning procedures that are similarly specialized for solving many familiar problems that are routinely encountered in the modern social world— such as detecting whether someone has made a mistake, detecting whether someone has broken a prescriptive rule that is not a social contract, and detecting whether a situation exists that violates a descriptive rule. Yet, if we did have a content-free psychology that could build the necessary reasoning mechanisms "as needed," then one would expect to find elaborate reasoning procedures specialized for solving such problems. Procedures specialized for solving ancient adaptive problems, such as social exchange, would be no more likely to develop than procedures specialized for solving the many evolutionarily novel problems posed by life in a modern, postindustrial culture. Evaluating a scientific hypothesis about the effects of dietary cholesterol, detecting whether someone has misfiled a document, and following the string of "if-then" directions on one's tax returns would be as effortless—and require as little explicit instruction—as detecting cheaters in a situation of social exchange among members of an unfamiliar tribal culture who value cassava root and facial tattoos.

The claim that our only evolved psychological mechanisms are general-purpose and content-free, and that "culture" must therefore supply all the specific content of our minds, is exactly the issue on which evolutionary psychological approaches diverge most sharply from more traditional ones. In our view, instead of culture manufacturing the psychology of social exchange de novo, content-specific, evolved psychologies constitute the building blocks out of which cultures themselves are manufactured (Tooby & Cosmides, this volume). These psychologies evolved to process information about ancient and important adaptive problems, such as social exchange, sexual jealousy, kin recognition, language acquisition, emotion recognition, and parenting. On this view, the social environment need not provide all of the (hypothetical) properties needed to construct a social exchange psychology from a set of content-free mental procedures (assuming this is possible at all), because the evolved architecture of the human mind already contains content-specific mechanisms that will cause a social exchange psychology to reliably develop in every normal human being. Because every human being will develop the same basic set of social contract algorithms, cultural forms that require their presence can emerge: traditions, rituals, institutions, linguistic conventions, symbols, and so forth can develop that rely on the stable features of this psychology, and that simply supply the specifics that activate and deploy it in each new situation.

Thus, there is an immediate difference between evolutionary psychological approaches and approaches that maintain that everything involving social interaction is "socially constructed." An evolutionary psychological perspective suggests that the universal evolved architecture of the human mind contains some content-specific algorithms that are shared across individuals and across cultures, and that, therefore, many things related to social exchange should be the same from place to place—as indeed they are. In contrast, a Standard Social Science Model approach would, if thoroughgoing, maintain that social exchange is culture-specific and historically contingent, existing in some places and not in others. Moreover, the Standard Model would have to predict that wherever social exchange is found to exist, it would have to be taught or communicated from the ground up. Because nothing about social exchange is initially present in the psychology of the learner, every structural feature of social exchange must be specified by the social environment as against the infinity of logically alternative branchings that could exist. It is telling that it is just this explicitness that is usually lacking in social life. Few individuals are able to articulate the assumptions that structure their own cultural forms—to do so usually requires considerable effort and is accompanied by the awareness that one is doing something unusual and artificial (Sperber, 1975). We suggest that social exchange is learned without explicit enumeration of the underlying assumptions. The surface forms are provided by specific cultures, but they are interpreted by content-specific information-processing devices containing implicit assumptions and procedures that evolved for exactly this purpose.

The issue of explicitness is particularly instructive. As Sperber and Wilson (1986) point out, if two individuals have no shared assumptions about the world, communication between them is impossible. The explicit part of the communicative process concerns those features of a situation that are known to the sender but not yet known to the receiver, but such transmissions are only interpretable because of what is already mutually understood. This point is particularly relevant to understanding human culture, because the learning of a culture itself in effect consists of successful communication between members of the culture and the individual who is learning about it. Consequently, for someone ignorant of a culture to learn it in the first place, it is necessary that she already share many assumptions with those from whom she is learning the culture. If human minds truly were initially tabula rasas, with no prior contentful structure, then no anthropologist or immigrant to a culture could ever learn about it

(Quine, 1969). More to the point, all children are immigrants, born into the world with no culturally specific knowledge at all. If the evolved architecture of the child's mind contained no meaningful content at all—that is, if it were truly content-free—then children could never learn the culture they are born into: They would have no reliable means of interpreting anything they saw and hence of placing the same construction on it that members of their culture do.

We suggest that domain-specific reasoning procedures such as social contract algorithms (Cosmides, 1989; Cosmides & Tooby, 1989) supply what is missing from traditional accounts of the acquisition of culture: that is, the necessary preexisting conceptual organization, which provides the shared assumptions humans need to interpret unfamiliar behavior and utterances and, in so doing, to acquire their culture. Such domain-specific algorithms embody both intrinsic definitions for the representations that they operate on (e.g., benefit, requirement, cheater) and cues for recognizing which elements in ontogenetically unfamiliar but evolutionarily recurrent situations correspond to those representations. This is why humans can barter or even form friendships where no common language is spoken. It is for this reason that we tested our subjects' ability to apply social contract algorithms to unfamiliar situations: Indeed, they did so as easily as they apply them to familiar contents. For all of the above reasons, we argue that such content-sensitive cognitive adaptations are a critical element in the process of cultural transmission. Culture would not be possible without them.

In any case, the way in which such a universal psychology would lead to cultural universals is straightforward, and the enumeration of cases of social exchange found cross-culturally would add little to furthering the discussion. For that reason, we would like to focus on two more interesting issues: (a) Assuming there is a universal, evolved psychology of social exchange, how might one explain and organize our understanding of cultural differences? That is, how can an underlying universal psychology express itself differently in different cultures? And (b) How can the ethnographic record inform us about additional aspects of our social exchange psychology, beyond those few algorithms and representations that have already been explored experimentally?

Intergroup Differences: Evoked Culture and Transmitted Culture

Human thought and behavior differs in an organized fashion from place to place, and these differences are typically termed "cultural" variation. In fact, it is considered a tautology by many to attribute this variation to the operation of culture—that is, to social transmission. Nevertheless, there are two complementary explanations for the existence of these within-location similarities and between-location differences in thought and behavior (Tooby & Cosmides, 1989, this volume). The traditional explanation is that these systematic differences are caused by what is usually called *transmitted culture,* that is, the process whereby the thought and behavior of some individuals (usually from the preceding generation) is passed on to other individuals, thereby causing the present pattern. Indeed, we have already touched on how domain-specific cognitive adaptations make cultural transmission possible.

There is, however, a second way in which such local similarities can be brought about. Our universal, evolved information-processing mechanisms should be sensitively context-dependent: Different informational inputs should evoke different representational and behavioral outputs. Humans share the same evolved information-processing mechanisms, but they do not all live under the same circumstances. People living in the same location are likely to experience somewhat similar circumstances, which should evoke the same kind of response from each individual; people living in different locations are likely to experience somewhat different circumstances, which should evoke different responses from each individual. To take an obvious example, people in tropical forest environments all around the world tend to be at most only lightly clothed, something that has less to do with parental example than with heat load.

The more complex the set of species-typical psychological mechanisms, the more sensitively dependent behavior may be on local circumstances. This means that cultural forms may exist that are not primarily generated by cultural transmission at all. We call similarities triggered by local circumstances *evoked culture* (Tooby & Cosmides, this volume). Of course, these two explanations are by no means mutually exclusive and usually operate together to shape the distribution of human similarities and differences. The evolutionary psychology of social exchange shapes both evoked culture and transmitted culture, and the exploration of a few of these issues serves to illustrate how an evolutionary psychological approach to culture can diverge from traditional anthropological approaches.

Open Questions in the Psychology of Social Exchange

Before proceeding to discuss how an evolved, universal psychology can generate cultural variation, it is necessary to emphasize the limited nature of what we have done so far. We have tested only a small number of hypotheses about one mode of social exchange out of many. We have not discussed, much less experimentally explored, the rest of the intricate and complex psychology of social exchange. Anyone examining his or her own human experience will immediately identify large areas of the psychology of social exchange, such as the psychology of friendship, that are not captured well by the models introduced so far. More important, the other components of our evolved faculty of social cognition—for example, the psychological mechanisms that govern sexual relations, coalitional partnerships, status, revenge, threat, and parenting—will have to be mapped out and integrated with the psychological mechanisms governing social exchange before social exchange can be fully unraveled. Each component of the faculty of social cognition can be only incompletely understood when taken in isolation.

The computational theory that we have developed specifies certain contractual relationships and implicit inferences that people can be expected to make in social contract situations (Cosmides & Tooby, 1989). It is, however, very far from a complete computational theory of social exchange and should not be mistaken for one: Instead, it is only an initial exploration of some of its major features. Many crucial questions were left unaddressed—questions that must be answered if we are to understand social exchange in all its various forms across cultures, or even within a single culture. For example, how many instances of cheating should the decision rule tolerate before it activates mechanisms causing one to sever one's relationship with an individual? Under what conditions should one cooperate with a person on a short-term basis, as opposed to a long-term basis? Should one's willingness to tolerate cheating differ in short-term versus long-term relationships? Will variance in the acquisition of the resources being exchanged affect the ways in which they are shared? What are the other bases for sharing, assistance, and friendship? What role do groups and coalitions play in shaping patterns of assistance? What is the role of aggression, retaliation, and status?

We expect that further evolutionary analyses will support the claim that different decision rules will be favored for long-term versus short-term relationships, for high-versus low-variance resources, and for other variations in conditions that we have not considered here. If this turns out to be true, then although some decision rules governing social exchange should be common across situations, others will differ in ways that are sensitively dependent on context. In other words, the social contract algorithms might contain different situation-specific modes of activation. For example, we expect the rules governing social exchange among close friends to differ somewhat from the rules governing social exchange among strangers.

Given an evolved architecture with a design of this kind, one can develop a coherent conceptual framework for understanding the ecological distribution of mental representations and expressed behaviors having to do with social exchange, both within and between groups (Sperber, 1985). Although the analysis of individual and cultural variation within the context of a universal human nature is too complex to review fully here, extended discussions can be found in Cosmides and Tooby (1987), Tooby and Cosmides (1989, 1990a), and Brown (1991). Briefly:

1. By virtue of being members of the human species, all humans are expected to have the same adaptive mechanisms, either at the level of developmental programs, which govern alternative developmental pathways, or (more likely in the case of social exchange) universally developing cognitive specializations.
2. In consequence, certain fundamental ways of thinking about social exchange will be the same everywhere, without needing to be socially transmitted.
3. As is standardly believed, social transmission may indeed shape social exchange. But it does so not by manufacturing the concept of social exchange de novo. Instead, there are probably certain specific avenues through which social transmission can act: for example, by influencing the valuation placed on items or actions, by providing information that helps one identify appropriate partners, or by providing information that allows one to identify appropriate contexts.
4. Our social exchange psychology should be highly context-dependent. Consequently, many dimensions along which social exchange varies, both within and between cultures, may be instances of evoked culture. The presence or magnitude of certain cues, conditions, or situations should cause mechanisms within our social exchange psychology to be calibrated; to be activated or inhibited; or to be switched into alternative modes of activation. When local circumstances trigger a particular mode of activation of the social contract algorithms, for example, this may cause a highly structured, multi-individual behavioral outcome. Therefore, when one sees similar patterns of social exchange in widely different parts of the world, one cannot assume that the similarity is determined primarily by social transmission; the similarity may instead by an instance of evoked culture.

One simple illustration of evoked culture involves the decision rules governing reciprocity in food sharing. As we describe below, a contextual variable—the presence or absence of "luck" in food acquisition—appears to activate different decision rules governing food sharing.

Luck and Sharing

One finding from the literature of evolutionary ecology on optimal foraging is that different kinds of sharing rules benefit the individual in different situations (see Kaplan & Hill, 1985). For example, when the variance in foraging success of the individual is greater than the variance for the band as a whole, band-wide food sharing buffers the variance. This can happen when one individual's success on a given day is unconnected to that of another.

Luck, skill, and effort all affect whether an individual finds food on a given day, but for certain food sources, luck is much more important than skill and effort. When success in finding food randomly varies a great deal from day to day, consider what would happen to a person who ate only that which he or she individually acquired. Some days that person would be starving; other days that person would have more to eat than he or she could possibly consume. It would be a feast or famine kind of life. Moreover, the temporary famines hit harder than the feasts can make up for. This is because (a) there is a zero point with food—death by starvation—and (b) the law of decreasing marginal utilities applies because we can only metabolize so much at one time—consequently, the fifth pound of meat eaten is far less valuable than the first pound. Under these circumstances, one is better off individually if one can redistribute food from periods of feast to periods of famine. There are two ways of doing this: through food storage or through pooling resources with others. Food storage is not an option for many hunter-gatherers, but pooling resources is: If two people average their returns, the variance decreases—one buys fewer periods of privation at the price of fewer periods of superabundance. By adding more individuals to the risk-pooling group, the variance may continue to decrease, making band-wide sharing an attractive system for hunter-gatherers facing certain conditions.

Thus, situations involving a random and frequent reversal of fortune can create substantial payoffs for cooperation. In effect, an individual can store food in the form of social obligations—by accepting food, others obligate themselves to reciprocate in the future. I may sacrifice by giving you some of my food today, but tomorrow I may be the one who is empty-handed and in need. For situations involving frequent, chance-driven reversals of fortune, the favored strategy involves sharing, from individuals who have food to those who do not. Luck plays an important role in hunting; consequently, hunter-gatherers frequently distribute game relatively equally to everyone in the band, no matter who found it or made a particular kill. Because it is a relatively high-variance activity, hunting may have been a particularly important driving force in the evolution of cognitive adaptations for social exchange (see Tooby & DeVore, 1987, for discussion).

By the same token, when variance in foraging success for an individual is low, the average long-term payoffs to sharing are less. If everyone reliably has access to the same goods, there is no particular benefit to sharing—one gains nothing by swapping the same goods at the same time. In this circumstance, an individual may be better off sharing just within his or her family, in accordance with kin selection, mating, and parenting principles.

Under low-variance conditions, not only might there be no benefit to sharing, there may be definite costs. When luck is eliminated as a factor, skill and effort remain. The smaller the role played by chance, the more differences between individuals in amount of food foraged will reflect differences in skill and effort. Under such circumstances, band-wide food sharing would simply redistribute food from those who expend more effort or are more skilled, to those who expend less effort or are less skilled. Sharing under these circumstances offers few—if any—intrinsic payoffs for those who have acquired more food. Without chance creating reversals of fortune, there is little reason to expect that the future will be different from the present and, therefore, little reason to expect that those with less food now will be in a better position to reciprocate in the future. Under these circumstances, then, one expects that (a) potential recipients will welcome sharing, but (b) potential donors will be more reluctant to share.

Consequently, the degree and source variance in resource acquisition were selection pressures that should have shaped the evolved architecture of our social exchange algorithms. Information about variance in foraging success should activate different modes of operation of these algorithms, with high variance due to chance triggering a psychology of sharing. To modern social scientists, factors such as variance in food acquisition may seem arcane and implausible because of their lack of connection to modern (middle-class) experience. But for our ancestors, food acquisition was a daily problem, as consequential as breathing. Daily decisions with respect to sharing had an unremitting impact on their lives and reproductive success, over hundreds of thousands of generations. In consequence, it is hard to see how our social psychology would not have been shaped by factors of this kind.

Obviously, this analysis of selection pressures is restricted to factors internal to foraging success. There are, of course, many other selection pressures that have shaped human social psychology over evolutionary time and hence many other factors that may lead to food sharing other than simple social exchange—kinship, love, parenting, sex, coercion, and status, for example. Moreover, even within the context of social exchange, the return on sharing food may be something other than food. Selection may have produced psychological mechanisms that cause highly productive foragers to share food without expecting any return of food, if, for example, by so doing others valued them highly and were therefore more disposed to render them aid when they were threatened, protect their children, grant them sexual access, and so on (Kaplan & Hill, 1985). Complicated though it may be, a more comprehensive understanding of social exchange eventually can be built up, element by element, by examining each selection pressure in turn and seeing whether our psychological mechanisms have the design features these selection pressures would lead one to expect.

In other words, the selection pressures analyzed in optimal foraging theory are one component of a task analysis, or, in David Marr's terms, a "computational theory," of the adaptive problem of foraging. It defines the nature of the problem to be solved and thereby specifies constraints that any mechanism that evolved to solve this problem can be expected to satisfy. In this case, optimal foraging theory suggests (a) that we should have content-specific information-processing mechanisms governing foraging and sharing, and (b) these mechanisms should be sensitive to information regarding variance in foraging success, causing us to prefer one set of sharing rules for high-variance items and another set for low-variance items.

The Ache: Within-Group Evidence for Evoked Culture

Kaplan and Hill's (1985) study of the Ache, a hunter-gatherer group living in eastern Paraguay, provides a particularly elegant test of the hypothesis just described, because it controls for "culture." Meat is a very high-variance food item among the Ache: On any given day, there is a 40% chance that a hunter will come back empty-handed (Kaplan, Hill, & Hurtado, 1990). Collected plant foods, in contrast, are very low-variance items. Kaplan and Hill found that the Ache engage in band-wide sharing of meat, whereas they share plant foods primarily within the nuclear family. Thus the same individuals, in the same culture, engage in different patterns of sharing for different foods, depending on the variance they experience in obtaining them.

The fact that meat is such a high-variance item also creates problems in cheater detection. If a man brings back no meat for seven days in a row, has he just had a run of bad luck, or has he been shirking? An Ache man's life and the life of his family depend on the long-term reciprocity relationships he has with the other hunters in his band. To accuse someone of cheating and ostracize him from the reciprocity network is a very serious matter. If the charge is false, then not only will the ostracized man's survival be jeopardized, but each member of the band will have lost a valuable reciprocation partner. If one is not sure, or if the suspected cheater is providing a net benefit even though it is less than he could provide if he tried harder, it might be better to continue the relationship.

The anthropologists who study the Ache know who the best hunters are, because they have recorded and weighed what each man brings back over long periods of time. Presumably, the Ache know also. But H. Kaplan (personal communication, 1991) reports that when he and his colleagues ask Ache men who the best hunters are, the question makes them very uncomfortable and they refuse to answer.[22] This is not due to a general cultural prohibition against accusing others of cheating. When the Ache are staying at a mission camp, acrimonious arguments erupt over whether various individuals are doing their fair share of work in the garden. Gardening, however, provides a low-variance source of food, making the punishment of cheaters less risky, and it occurs in a well-defined, observable location, making it easy to monitor who is, and who is not, cheating.

!Kung San Versus //Gana San: Between-Group Evidence for Evoked Culture

Resource variance can also explain differences between groups, evoking different cultures in response to different, variance-related local circumstances. For example, Cashdan (1980) found variance-related differences in sharing between groups of Kalahari San that mirror those found within Ache culture.

The Kalahari San are well known in anthropological circles for their economic and political egalitarianism. For example, the !Kung San, who experience extreme variability in the availability of food and water, have very strong social sanctions that reinforce sharing, discourage hoarding (calling someone "stingy" is a strong insult), and discourage displays of arrogance and authority. For example:

> The proper behavior of a !Kung hunter who has made a big kill is to speak of it in passing and in a deprecating manner . . . ; if an individual does not minimize or speak lightly of his own accomplishments, his friends and relatives will not hesitate to do it for him. (Cashdan, 1980, p. 116)

But it turns out that some San bands are more egalitarian than others, and their degree of egalitarianism is related to variance in their food supply. The //Gana San of the northeastern Kalahari are able to buffer themselves from variability in the food and water supply in ways that other San cannot, through a small amount of food cultivation (including a kind of melon that stores water in the desert environment) and some goat husbandry. In contrast to the !Kung, the //Gana manifest considerable economic inequality, they hoard more, they are more polygynous, and, although they have no clear-cut authority structure, wealthy, high-status //Gana men are quick to claim that they speak for others and that they are the "headman"—behavior that would be considered unconscionable among the !Kung. Again, even though the !Kung and the //Gana are culturally similar in many ways—they share the same encompassing "meme-pool," so to speak—their social rules regarding sharing and economic equality differ, and these differences track the variance in their food and water supplies.

Local Conditions and Evoked Culture

It is difficult to explain these phenomena simply as the result of cultural transmission, at least in any traditional sense. Among the Ache of Paraguay, the same individuals share food types with different variances differently. Half way around the world, in Africa, two different groups of Kalahari San manifest what appear to be the same differential sharing patterns in response to the same variable—variance. A parsimonious explanation is that these social norms and the highly patterned behaviors they give rise to are evoked by the same variable.

Because foraging and sharing are complex adaptive problems with a long evolutionary history, it is difficult to see how humans could have escaped evolving highly structured domain-specific psychological mechanisms that are well designed for solving them. These mechanisms should be sensitive to local informational input, such as information regarding variance in the food supply. This input can act as a switch, turning on and off different modes of activation of the appropriate domain-specific mechanisms. The experience of high variance in foraging success should activate rules of inference, memory retrieval cues, attentional mechanisms, and motivational mechanisms. These should not only allow band-wide sharing to occur, but should make it seem fair and appealing. The experience of low variance in foraging success should activate rules of inference, memory retrieval cues, attentional mechanisms, and motivational mechanisms that make within-family sharing possible and appealing, but that make band-wide sharing seem unattractive and unjust. These alternative modes of activation of the domain-specific mechanisms provide the core knowledge that must be mutually manifest (see Sperber & Wilson, 1986) to the various actors for band-wide or within-family sharing to occur. This core knowledge can then organize and provide points of attachment for symbolic activities that arise in these domains.

If this notion of evoked culture is correct, then one should not expect cultural variation to vary continuously along all imaginable dimensions. The free play of human creativity may assign relatively arbitrary construals to elements in some areas of life, such as the number of gods or the appropriate decoration on men's clothing. But in other areas of life one might expect there to be a limited number of recurring patterns, both within and across cultures. For certain domains of human activity, people from very different places and times may "reinvent" the same kinds of concepts, valuations, social rules, and customs (see Tooby & Cosmides, this volume). In short, such alternative modes of activation in psychological mechanisms can create alternative sets of complexly patterned social rules and activities. These will emerge independently, that is, in the absence of direct cultural transmission, in culture after culture, when the individual members are exposed to the informational cues that activate these alternative modes.

Cross-cultural studies of social exchange by Fiske provide support for this notion (Fiske, 1990, 1991a). Based on his field studies of the Moose ("Mossi") of Burkina Faso and his review of the anthropological literature, Fiske argues that the human mind contains four alternative implicit models of how sharing should be conducted, which are used to generate and evaluate social relations. These models are implicit in the sense that they are acted on unreflectively and without conscious awareness; indeed, they may never have been explicitly stated by any member of the culture. Nevertheless, "these shared but unanalyzed, tacit models for Moose social relations allow them to generate coordinated, consistent, and culturally comprehensible interactions of four contrasting types" (Fiske, 1990, pp. 180–181). For example, one of Fiske's four models is communal sharing of the kind used by the Ache in distributing hunted meat; another is "market pricing"—the kind of explicit contingent exchange that occurs when two people explicitly agree to trade, say, honey for meat or money for milk.

Varieties of Hunter-Gatherer Exchange

Whether or not Fiske's specific taxonomy of four categories is exactly the correct way to capture and characterize the limited set of modes whereby humans engage in social exchange, we very much agree with this general framework for conceptualizing cultural variation in social exchange. If human thought falls into recurrent patterns from place to place and from time to time, this is because it is the expression of, and anchored in, universal psychological mechanisms. If there is a limited set of such patterns, it is because different modes of activation of the algorithms regulating social exchange solved different adaptive problems that hunter-gatherers routinely faced. Consequently, clues as to how many modes of activation the social contract algorithms have, what the structure of each mode might be, and what kinds of circumstances can be expected to activate each mode can be found by investigating the various forms of social exchange that hunter-gatherers engage in, as well as the conditions under which each form of exchange arises.

Despite the common characterization of hunter-gatherer life as an orgy of indiscriminate, egalitarian cooperation and sharing—a kind of retro-utopia—the archaeological and ethnographic record shows that hunter-gatherers engaged in a number of different forms of social exchange (for an excellent review of hunter-gatherer economics, see Cashdan, 1989). Communal sharing does not exhaust the full range of exchange in such societies. Hunter-gatherers also engage in explicit contingent exchange—Fiske's "market pricing"—in which tools and other durable goods are traded between bands, often in networks that extend over vast areas. A common form of trade is formal gift exchanges with carefully chosen partners from other bands. For instance, aboriginal Australians traded tools such as sting ray spears and stone axes through gift exchanges with partners from neighboring bands. These partnerships were linked in a chain that extended 620 km, from the coast, where sting ray spears were produced, to the interior, where there were quarries where the stone axes could be produced. Here, environmental variation in the source of raw materials for tool making allowed gains from trade based on economic specialization, and the laws of supply and demand seemed to operate: At the coast, where sting ray spears were common, it took more of them to buy an ax than in the interior, where spears were dear and axes cheap

(Sharp, 1952). Similarly, the !Kung of the Kalahari desert engage in a system of delayed reciprocal gift giving called "hxaro" (Weissner, 1982; Cashdan, 1989), through which they trade durable goods such as blankets and necklaces.

Unpredictable variation in rainfall and game makes access to land and water resources another important "item of trade" between hunter-gatherer bands and creates situations in which a kind of implicit one-for-one reciprocity prevails (Fiske's "equality matching"). For instance, a !Kung band that is caught in a drought will "visit relatives" in a band that is camped in an area that is experiencing more rainfall (Cashday, 1989). Indeed, hxaro partners are chosen carefully, not only for their ability to confer durable goods, but also to provide alternative residences in distant places during times of local scarcity (Weissner, 1982). And before using another band's water hole or land, the !Kung are expected to ask permission; reciprocity in access to water holes is extremely important to the !Kung, who live in a desert with few permanent sources of water. Although formal permission is almost always granted, as the implicit rules of one-for-one reciprocity require, if the hosts really don't want to accommodate their guests, they make them feel unwelcome, thereby subtly encouraging them to leave (Cashdan, 1989).

Although authoritarian social relations are unusual among the few remaining modern hunter-gatherer groups, this is probably a by-product of their having been pushed into marginal environments by the peoples of agricultural and industrial cultures. Variance in the food supply is high in harsh environments like the Kalahari desert, and band-wide communal sharing is advantageous for high-variance resources. But as variance is buffered, as in the //Gana San example discussed earlier, more inequality and more authority-ranking relationships develop. This process was, for example, quite pronounced in the hunter-gatherer societies of the Pacific Northwest. The Pacific Northwest was so rich in fish and game that the hunter-gatherers living there could afford to be relatively sedentary. These people developed stable, complex societies that were so hierarchical that some of them even included a slave class formed from prisoners of war (Drucker, 1983; Donald, 1983). Of course, the distribution of goods and services that occurs between individuals of different rank is often determined by an uneasy mixture of coercion, threat, and exchange.

This is not the place to attempt a full computational theory of the various modes of activation of the social contract algorithms. But even these brief examples drawn from hunter-gatherer life provide some hints as to what might be relevant variables in such an analysis: variance in the food supply; degree of kinship; status or rank; whether a relationship is long- or short-term; whether one is in daily contact (communal sharing; implicit deferred reciprocity) or only rare contact (explicit contingent exchange); whether storage is possible; whether the group is sedentary enough for inequalities in wealth to accumulate; whether gaining a resource requires close, interdependent cooperation; whether people are trading different resources or dividing the same resource; whether an external, consensual definition of "equal portion" is feasible; whether an individual can control access to a resource, and thereby "own" it; and so on (see also McGrew & Feistner, this volume).

To understand social exchange in all its various forms, the adaptive problems that selected for different decision rules must be precisely defined in the form of computational theories. The computational theories can then be used to generate hypotheses about the design features that characterize the different modes of activation of the social contract algorithms. Psychological experiments of the kind described earlier in this chapter would allow one to test among these hypotheses and thereby develop a detailed map of the situation-specific cognitive processes that create these different modes of activation. Once we know what situational cues activate each set of decision rules, we should be able to predict a great deal of cultural variation.

Interpreting Other Cultures and Understanding Cultural Change

Significant aspects of cultural variation in social exchange can be readily reconciled with a universal human nature through applying the concept of evoked culture. The various sets of decision rules governing social exchange will be universal, but which sets are activated will differ from situation to situation within a culture, as well as between cultures. For example, in American middle-class culture different exchange rules apply to different aspects of a dinner party (Fiske, 1991b). Invitations are sometimes governed by one-for-one reciprocity—an implicit rule such as "If you had me to your home for dinner, then at some point I must invite you to dinner." But food sharing at the party is governed by the same kind of communal sharing rules that characterize Ache meat sharing. Obtaining the food that is served is governed by explicit contingent exchange at a grocery store, and seating at the dinner table is sometimes determined by rank or status (as for example, at diplomatic dinners, birthday parties, or in certain traditional families).

The point is that communal sharing, explicit contingent exchange, equality matching, and so on, are not unique to American culture: The same sets of decision rules appear in other cultures as well, but local circumstances cause them to be applied to different situations (Fiske, 1990, 1991a). Whereas all food at an American dinner party is shared communally, this is not true on Ache foraging trips: Meat is shared communally at the level of the entire band, but plant foods are not. In many cultures, men engage in explicit contingent exchange to procure wives: One man will buy another man's daughter (see Wilson & Daly, this volume). In other cultures, men do not buy wives, but instead can engage in explicit contingent exchange with a woman to gain temporary sexual access to her. In still other cultures, the use of explicit contingent exchange is illegal in both circumstances (but may still be understood and occasionally practiced).

Fiske argues that in relatively stable, traditional societies there is a tacit consensus about which decision rules to apply in which situation. To apply the wrong decision rules to a situation can be uncomfortable, insulting, or even shocking: At the end of an American dinner party, one does not pull out a wallet and offer to pay the hosts for their services. Similarly, when Americans are sitting with friends or co-workers, they might spontaneously offer to split a sandwich, but they almost never spontaneously pull out their wallets and offer money. Indeed, figuring out which decision rules a culture applies to which situations is part of what it means to understand another culture (Fiske, 1990). On this view, "interpreting another culture" is not usually a matter of absorbing wholly new systems of culturally alien semantic relations. Instead, interpreting another culture is a matter of learning how the evolved set of meanings that we have come to assign to one set of objects or elements in a situation are, in another culture, assigned to a different set.

New events of all kinds, from migrations to natural disasters to new technologies, create culturally unprecedented circumstances in which there is no within-culture consensus about which exchange rules are appropriate. In the United States, for example, there is a vigorous debate over which form of exchange should apply when a woman wants to be a surrogate mother for an infertile couple. Many women prefer explicit contingent exchange in which they are paid money for their labor (so to speak). But other Americans argue that surrogacy should occur—if at all—only among close friends and relatives who participate in informal communal sharing relationships and that women should be legally prohibited from granting access to their wombs on the basis of explicit contingent exchange.

Where do the impulses—or, more accurately—the decision rules come from that lead individuals or entire cultures to reject an existing practice or to invent or adopt something new? Transmission models can account for stable transmission of existing attitudes and cultural forms but intrinsically have no way to account for cultural change, or indeed any nonimitated individual act. The existence of a species-typical evolved psychology fills in this missing gap. It provides a basis from which one can interpret individual action, minority dissent, and the emergence of a new consensus. Dramatic new circumstances may evoke new attitudes overnight, as when the Battle of Britain changed the attitudes and sharing practices of Londoners, or when depictions of earthquakes and other natural disasters prompt people to donate food and other assistance. Even where one is dealing with the spread of new cultural forms through transmission, however, the dynamics are powerfully structured by our content-sensitive evolved psychology (for a lucid discussion of the "epidemiology" of beliefs and other representations, see Sperber, 1985, 1990).

Consider the political and moral debate concerning the homeless in the United States. Those with opposing postures concerning how much to help the homeless frame their positions in ways that exploit the structure of this evolved psychology. One persistent theme among those who wish to motivate more sharing is the idea of "there but for fortune, go you or I." That is, they emphasize the random, variance-driven dimensions of the situation. The potential recipient of aid is viewed as worthy because he or she is the unlucky victim of circumstances, such as unemployment, discrimination, or mental illness. On the other hand, those who oppose an increase in sharing with the homeless emphasize the putatively chosen or self-caused dimensions of the situation. Potential recipients are viewed as unworthy of aid because they "brought it on themselves": They are portrayed as able-bodied but lazy, or as having debilitated themselves through choosing to use alcohol and other drugs. The counterresponse from those who want to motivate more sharing is to portray drug use not as a choice, but as a sickness, and so on.

If cultural meanings were truly arbitrary, then, cross-culturally, donors would be just as likely to view people as "worthy of assistance" when they have "brought it on themselves" as when they have been "the victims of bad luck." Indeed, if all were arbitrary, somewhere one should find a culture in which potential donors are most eager to help those who are *more* fortunate than themselves, merely *because* the potential recipients are more fortunate (and not, say, because they hope for something in return).

Finally, although our cognitive mechanisms evolved to promote adaptive decisions in the Pleistocene, they do not necessarily produce adaptive decisions under evolutionarily novel modern circumstances (see Symons, this volume). For example, if individual variance in obtaining alcohol is greater than group variance for homeless alcoholics who camp out in the same alley, this circumstance might activate decision rules that promote communal sharing of alcohol, even though these people's mutual generosity would be slowly killing them.

CONCLUSIONS

Human reason has long been believed to be the paradigm case of the impartial, content-blind, general-purpose process. Further, it has been viewed as the faculty that distinguished humans from all other animals, and the very antithesis of "instinct." But if even reasoning turns out to be the product of a collection of functionally specialized, evolved mechanisms, most of which are content-dependent and content-imparting, then this has implications far beyond the study of reasoning. The presumption that psychological mechanisms are characteristically general-purpose and content-free would no longer be tenable: Such hypotheses should no longer be accorded the privileged status and the near-immunity from question that they have customarily received. Instead, domain-general and domain-specific hypotheses should be given equal footing and evaluated solely on their ability to be turned into genuine, well-specified models that actually account for observed phenomena. Guided by such tenets, we may discover that the human mind is structurally far richer than we have suspected and contains a large population of different mechanisms.

We have used as a test case the intersection of reasoning and social exchange. The results of the experiments discussed herein directly contradict the traditional view; they indicate that the algorithms and representations whereby people reason about social exchange are specialized and domain-specific. Indeed, there has been an accumulation of "evidence of special design" (Williams, 1966), indicating the presence of an adaptation. The results are most parsimoniously explained by positing the existence of "specialized problem-solving machinery" (Williams, 1966)—such as cost-benefit representations and cheater detection procedures—that are well designed for solving adaptive problems particular to social exchange. Moreover, they cannot be explained as the by-product of mechanisms designed for reasoning about classes of problems that are more general than social contracts, such as "all propositions," or even the relatively restricted class of "all permissions." In addition, the pattern of results indicate that this specialized problem-solving machinery was not *built* by an evolved architecture that is general-purpose and content-free (see Implications for Culture, this chapter, and Cosmides, 1989). In other words, the empirical record is most parsimoniously explained by the hypothesis that the evolved architecture of the human mind contains functionally specialized, content-dependent cognitive adaptations for social exchange. Such mechanisms, if they exist, would impose a distinct social contract conceptual organization on certain social situations and impart certain meanings to human psychological, social, and cultural life. We suggest these evolved algorithms constitute one functional subunit, out of many others, that are linked together to form a larger faculty of social cognition (e.g., Jackendoff, 1991).

The results of the experiments discussed herein undermine two central tenets of the Standard Social Science Model (Tooby & Cosmides, this volume). First, they undermine the proposition that the evolved architecture of the human mind contains a single "reasoning faculty" that is function-general and content-free. Instead, they support the contrary contention that human reasoning is governed by a diverse collection of evolved mechanisms, many of which are functionally specialized, domain-specific, content-imbued, and content-imparting (see Tooby & Cosmides, this volume). According to this contrary view, situations involving threat, social exchange, hazard, rigid-object mechanics, contagion, and so on each activate different sets of functionally specialized procedures that exploit the recurrent properties of the corresponding domain in a way that would have produced an efficacious solution under Pleistocene conditions. On this view, the human mind would more closely resemble an intricate network of functionally dedicated computers than a single general-purpose computer. The second tenet that these results undermine is the proposition that all contentful features of the human mind are "socially constructed" or environmentally derived. In its place, this research supports the view that the human mind imposes contentful structure on the social world, derived from specialized functional design inherent in its evolved cognitive architecture.

The conceptual integration of evolutionary biology with cognitive psychology offers something far more valuable than general arguments. The analysis of the computational requirements of specific adaptive problems provides a principled way of identifying likely new modules, mental organs, or cognitive adaptations, and thereby opens the way for extensive empirical progress. By understanding these requirements, one can make educated guesses about the design features of the information-processing mechanisms that evolved to solve them. Turning knowledge of the adaptive problems our ancestors faced over evolutionary time into well-specified computational theories can therefore be a powerful engine of discovery, allowing one to construct experiments that can capture, document, and catalog the functionally specialized information-processing mechanisms that collectively constitute much (or all) of our "central processes." In effect, knowledge of the adaptive problems humans faced, described in explicitly computational terms, can function as a kind of Rosetta Stone: It allows the bewildering array of content effects that cognitive psychologists routinely encounter—and usually disregard—to be translated into meaningful statements about the structure of the mind. The resulting maps of domain-specific information-processing mechanisms can supply the currently missing accounts of how the human mind generates and engages the rich content of human culture, behavior, and social life.

ACKNOWLEDGMENTS

We thank Jerry Barkow, David Buss, Martin Daly, Lorraine Daston, Gerd Gigerenzer, Steve Pinker, Paul Romer, Roger Shepard, Don Symons, Phil Tetlock, Dan Sperber, Valerie Stone, and Margo Wilson for productive discussions of the issues addressed in this chapter or for their useful comments on previous drafts. This chapter was prepared, in part, while the authors were Fellows at the Center for Advanced Study in the Behavioral Sciences. We are grateful for the Center's support, as well as that provided by the Gordon P. Getty Trust, the Harry Frank Guggenheim Foundation, NSF Grant BNS87-00864 to the Center, NSF Grant BNS85-11685 to Roger Shepard, and NSF Grant BNS91-57449 to John Tooby.

NOTES

1. For example, if C_i and B_i refer to decreases and increases in i's reproduction, then a decision rule that causes i to perform act Z if, and only if, C_i of doing $Z < 0$ would promote its own

Darwinian fitness, but not its inclusive fitness. In contrast, a decision rule that causes i to perform act Z if, and only if, $(C_i$ of doing $Z) < (B_j$ of i's doing $Z) \times r_{ij}$ would promote its inclusive fitness, sometimes at the expense of its Darwinian fitness. The first decision rule would be at a selective disadvantage compared with the second one, because it can make copies of itself only through its bearer, and not through its bearer's relatives. For this reason, designs that promote their own inclusive fitness tend to replace alternative designs that promote Darwinian fitness at the expense of inclusive fitness.

Although this example involves helping behavior, kin selection theory applies to the evolution of nonbehavioral design features as well, for example, to the evolution of aposematic coloration of butterfly wings. In principle, one can compute the extent to which a new wing color affects the reproduction of its bearer and its bearer's kin, just as one can compute the extent to which an *action* affects the reproduction of these individuals.

2. Other models of social exchange are possible, but they will not change the basic conclusion of this section: that reciprocation is necessary for the evolution of social exchange. For example, the Prisoner's Dilemma assumes that enforceable threats and enforceable contracts are impossibilities (Axelrod, 1984), assumptions that are frequently violated in nature. The introduction of these factors would not obviate reciprocation—in fact, they would enforce it.

3. Following Marr, 1982, we would like to distinguish between the cognitive program itself and an abstract characterization of the decision rule it embodies. Algorithms that differ somewhat in the way they process information may nevertheless embody the same decision rule. For example, the algorithms for adding Arabic numerals differ from those for adding Roman numerals, yet they both embody the same rules for addition (e.g., that $A + B = B + A$) and therefore yield the same answer (Marr, 1982).

4. These selection pressures exist *even in the absence of competition for scarce resources.* They are a consequence of the game-theoretic structure of the social interaction.

5. The game "unravels" if they do. If we both know we are playing three games, then we both know we will mutually defect on the last game. In practice, then, our second game is our last game. But we know that we will, therefore, mutually defect on that game, so, in practice, we are playing only one game. The argument is general to any known, fixed number of games (Luce & Raiffa, 1957).

6. The cost-benefit values that these algorithms assign to items of exchange should be correlated with costs and benefits to fitness in the environment in which the algorithms evolved; otherwise, the algorithms could not have been selected for. But these assigned values will not necessarily correlate with fitness in the modern world. For example, our taste mechanisms assess fat content in food and our cognitive system uses this cue to assign food value: We tend to like food "rich" (!) in fat, such as ice cream, cheese, and marbled meat. The use of this cue is correlated with fitness in a hunter-gatherer ecology, where dietary fat is hard to come by (wild game is low in fat). But in modern industrial societies, fat is cheap and plentiful, and our love of it has become a liability. The environment changed in a way that lowered the cue validity of fat for fitness. But our cognitive system, which evolved in a foraging ecology, still uses it as a cue for assigning food value.

Given the long human generation time, and the fact that agriculture represents less than 1% of the evolutionary history of the genus *Homo,* it is unlikely that we have evolved any complex adaptations to an agricultural (or industrial) way of life (Tooby & Cosmides, 1990a). Our ances · tors spent most of the last 2 million years as hunter-gatherers, and our primate ancestors before the appearance of *Homo* were foragers as well, of course. The very first appearance of agriculture was only 10,000 years ago, and it wasn't until about 5,000 years ago that a significant fraction of the human population was engaged in agriculture.

7. Interpreting the statement as a biconditional, rather than as a material conditional, will also lead to error. Consider a situation in which you gave me your watch, but you did not take my $20. This would have to be considered a violation of the rule on a biconditional interpretation, but it is not necessarily cheating. If I had not offered you the $20, then I would have cheated.

But if I had offered it and you had refused to take it, then no cheating would have occurred on either of our parts. In this situation, your behavior could be characterized as altruistic or foolish, but not as cheating. Distinctions based on notions such as "offering" or intentionality are not part of the definition of a violation in the propositional calculus.

8. Indeed, one expects learning under certain circumstances: The genome can "store" information in the environment if that information is stably present (Tooby & Cosmides, 1990a).

9. Of course, if two developmental trajectories have different reproductive consequences, one will be favored over the other.

10. The drinking-age problem is also a social contract from the point of view of those who enacted the law. Satisfying the age requirement before drinking beer provides those who enacted the law with a benefit: People feel that the roads are safer when immature people are not allowed to drink. Although satisfying the requirement in a social contract will often cause one to incur a cost, it need not do so (see Cosmides & Tooby, 1989). The requirement is imposed not because it inflicts a cost on the person who must satisfy it, but because it creates a situation that benefits the recipient, which the recipient believes would not occur if the requirement were not imposed.

Consider the following social contract: "If you are a relative of Nisa's, then you may drink from my water hole." A hunter-gatherer may make this rule because she wants to be able to call on Nisa for a favor in the future. A given person either is, or is not, Nisa's relative; it would therefore be odd to say that being Nisa's relative inflicts a cost on one. Nevertheless, it is the requirement that must be satisfied to gain access to a benefit, and it was imposed because it creates a situation that can benefit the person who imposed it. This is why Cheng and Holyoak's (1989) distinction between "true" social exchange, where the parties incur costs, and "pseudo" social exchange, where at least one party must meet a requirement that may not be costly, constitutes a misunderstanding of social contract theory and the basic evolutionary biology that underlies it. Social exchange is the reciprocal provisioning of benefits, and the fact that the delivery of a benefit may prove costly is purely a by-product.

11. So far, the evidence suggests that we also have specialized procedures for reasoning about threats and precautions, for example.

12. No criticism of the experimenters is implied; these experiments were not designed for the purpose of testing social contract theory.

13. What if people read in a "may" that refers to obligation, rather than to possibility? That is, after all, a prediction of social contract theory. Logicians have tried to create "deontic logics": rules of inference that apply to situations of obligation and entitlement. Social contract theory is, in fact, a circumscribed form of deontic logic. But could subjects be using a generalized form of deontic logic? Manktelow and Over (1987) say that the answer is not clear because deontic logicians do not yet agree: According to some, no cards should be chosen on the switched social contracts; according to others, *not-P & Q* should be chosen. Because the rules of inference in social contract theory include the concepts of entitlement and obligation, it can be thought of as a specialized, domain-specific deontic logic. But we doubt that people have a generalized deontic logic. If they did, then non-social contract problems that involve obligation should elicit equally high levels of performance. But this is not the case, as will be discussed later in the chapter.

14. Even if this hypothesis were true, one would still have to explain *why* social contract problems are easier to understand, or more interesting, than other situations. After all, there is nothing particularly complicated about the situation described in a rule such as "If a person eats red meat, then that person drinks red wine." Social contract problems could be easier to understand, or more interesting, precisely because we do have social contract algorithms that organize our experience in such situations. Consequently, showing that social contract problems afford clear thinking about a wide variety of problems would not eliminate the possibility that there are social contract algorithms; it would simply cast doubt on the more specific claim that this set of algorithms includes a procedure specialized for cheater detection.

15. We would like to point out that the relationship between psychology and evolutionary biology can be a two-way street. For example, one could imagine models for the emergence of

stable cooperation that require the evolution of a mechanism for altruism detection. If the selection pressures required by these models were present during hominid evolution, they should have left their mark on the design of our social contract algorithms. Finding that people are not good at detecting altruists casts doubt on this possibility, suggesting altruists were too rare to be worth evolving specialized mechanisms to detect, and hence gives insight into the kind of selection pressures that shaped the hominid line.

16. For example, instead of asking subjects to "indicate only those card(s) you definitely need to turn over to see if any of these boys have broken the law," the altruist version asked them to "indicate only those card(s) you definitely need to turn over to see if any of these boys have behaved altruistically with respect to this law."

17. Indeed, on the altruist detection problems in which the rule was a social law, more subjects detected *cheaters* than detected altruists! (This result was 64% in the altruist version; 44% in the selfless version.) It is almost as if, when it comes to a social law, subjects equate altruistic behavior with honorable behavior—i.e., with the absence of cheating. (This may be because for many social laws, such as the drinking age law, "society"—i.e., the individuals who enacted the law—benefits from the total configuration of events that ensues when the law is obeyed.) This was not true of the personal exchange laws, where it is easy to see how the other party benefits by your paying the cost to them but not accepting the benefit they have offered in return. (For the private exchange problems, only 16% of subjects chose the "look for cheaters" answer in the two altruist versions; 8% and 4%, in the selfless versions.)

18. Manktelow and Over (1987) point out that people do understand what conditions constitute a violation of a conditional rule, even when it is an abstract one. Hence the failure to perform well on the no cheating version cannot be attributed to subjects' not knowing what counts as a violation. (This fact may seem puzzling at first. But one can know what counts as a violation without being able to use that knowledge to generate falsifying inferences, as the failure to choose $P \& not\text{-}Q$ on abstract Wason selection tasks shows. Two separate kinds of cognitive processes appear to be involved. An analogy might be the ease with which one can recognize a name that one has been having trouble recalling.)

19. It is difficult to tell a permission from an obligation because both involve obligation and because there are no criteria for distinguishing the two representational formats ("If action is taken, then precondition must be satisfied" versus "If condition occurs, then action must be taken"). "Conditions" and "preconditions" can, after all, be "actions." The primary difference seems to be a time relation: If the obligation must be fulfilled before the action is taken, it is a permission. If the obligation can be fulfilled after a condition (which can be an "action taken") occurs, then it is an obligation. A social contract of the form, "If you take the benefit, then you must pay the cost" would be considered a permission if you were required to pay the cost before taking the benefit, but an obligation if you had first taken the benefit, thereby incurring the obligation to pay the cost.

20. To choose $not\text{-}P \& Q$, one would have to interpret "If an employee gets a pension, then that employee must have worked for the firm for at least 10 years" as also implying "If an employee has worked for the firm for at least 10 years, then that employee *must* be given a pension." Social contract theory predicts that the one statement will be interpreted as implying the other, but permission schema theory does not. In fact, its translation rules (the four production rules) bar this interpretation. The rule presented to subjects—"If an employee gets a pension, then that employee must have worked for the firm for at least 10 years"—has the linguistic format of rule 1 of the permission schema—"If the action is to be taken, then the precondition must be satisfied." Rule 1 can be taken to imply rules 2, 3, and 4, but not other rules. By rule 3, the rule stated in the problem would translate to "If an employee has worked for the firm for at least 10 years, then that employee *may* be given a pension"—not that the employee *must* be given a pension.

21. Or equally low performance. Cheng & Holyoak have provided very little theory concerning what elements in a situation can be expected to activate the permission schema.

Although they have suggested that the provision of a rationale or social purpose helps, they have never defined what counts as such, and there are (social contract) permission rules that lack rationales that nevertheless produce the effect (Cosmides, 1989). The problems that we tested here clearly stated that the rule is a law made by authorities, which ought to clarify that they are permission rules and prevent subjects from interpreting them as descriptive rules. If this is sufficient to activate a permission schema, then performance on all three problems should be equally high. But none of the problems contains or suggests a rationale. So if one were to claim that rationales are necessary, then performance on all three problems should be equally low. Either way, performance should not vary across the three problems.

22. This is the kind of situation that Nesse and Lloyd (this volume) suggest might call for benevolent self-deception. Although one memory module may be keeping an account of the other person's failure to contribute his fair share, this information might not be fed into the mechanisms that would cause an angry reaction to cheating. By preventing an angry reaction, this temporary encapsulation of the information would permit one to continue to cooperate with the suspected cheater. This situation would continue as long as one is still receiving a net benefit from the other person, or until it becomes sufficiently clear that the other person is cheating rather than experiencing a run of bad luck. At that point, the accounts kept by the one module would be fed into other modules, provoking an angry, recrimination-filled reaction.

REFERENCES

Axelrod, R., & Hamilton, W. D. (1981). The evolution of cooperation. *Science, 211,* 1390–1396.

Axelrod, R. (1984). *The evolution of cooperation.* New York: Basic Books.

Boyd, R. (1988). Is the repeated prisoner's dilemma a good model of reciprocal altruism? *Ethology and Sociobiology, 9,* 211–222.

Brown, D. (1991). *Human universals.* New York: McGraw-Hill.

Cashdan, E. (1980). Egalitarianism among hunters and gatherers. *American Anthropologist, 82,* 116–120.

Cashdan, E. (1989). Hunters and gatherers: Economic behavior in bands. In S. Plattner (Ed.), *Economic Anthropology* (pp. 21–48). Stanford: Stanford University Press.

Cheng, P., & Holyoak, K. (1985). Pragmatic reasoning schemas. *Cognitive Psychology, 17,* 391–416.

Cheng, P., & Holyoak, K. (1989). On the natural selection of reasoning theories. *Cognition, 33,* 285–313.

Cheng, P., Holyoak, K., Nisbett, R., & Oliver, L. (1986). Pragmatic versus syntactic approaches to training deductive reasoning. *Cognitive Psychology, 18,* 293–328.

Cosmides, L. (1985). *Deduction or Darwinian algorithms? An explanation of the "elusive" content effect on the Wason selection task.* Doctoral dissertation, Department of Psychology, Harvard University: University Microfilms, #86-02206.

Cosmides, L. (1989). The logic of social exchange: Has natural selection shaped how humans reason? Studies with the Wason selection task. *Cognition, 31,* 187–276.

Cosmides, L., & Tooby, J. (1987). From evolution to behavior: Evolutionary psychology as the missing link. In J. Dupré (Ed.), *The latest on the best: Essays on evolution and optimality* (p. 277–306). Cambridge, MA: MIT Press.

Cosmides, L., & Tooby, J. (1989). Evolutionary psychology and the generation of culture, part II. Case study: A computational theory of social exchange. *Ethology and Sociobiology, 10,* 51–97.

Cosmides, L., & Tooby, J. (in prep., a). The logic of threat: Evidence for another cognitive adaptation?

Cosmides, L., & Tooby, J. (in prep., b). Social contracts, precaution rules, and threats: How to tell one schema from another.

Cosmides, L., & Tooby, J. (in prep., c). Is the Wason selection task an assay for production rules in mentalese?

Dawkins, R. (1982). *The extended phenotype*. San Francisco: Freeman.

Dawkins, R. (1986). *The blind watchmaker*. New York: Norton.

Donald, L. (1983). Was Nuu-chah-nulth-aht (Nootka) society based on slave labor? Ecology and political organization in the Northwest coast of America. In E. Tooker (Ed.), *The development of political organization in native North America* (pp. 108–119). Washington, DC: American Ethnological Society.

Drucker, P. (1983). Ecology and political organization in the Northwest coast of America. In E. Tooker (Ed.), *The development of political organization in native North America* (pp. 86–96). Washington, DC: American Ethnological Society.

Evans, J.St.B.T. (1984). Heuristic and analytic processes in reasoning. *British Journal of Psychology, 75*, 457–468.

Fischer, E. A. (1988). Simultaneous hermaphroditism, tit-for-tat, and the evolutionary stability of social systems. *Ethology and Sociobiology, 9*, 119–136.

Fiske, A. P. (1990). Relativity within Moose ("Mossi") culture: Four incommensurable models for social relationships. *Ethos, 18*, 180–204.

Fiske, A. P. (1991a). *Structures of social life: The four elementary forms of human relations*. New York: Free Press.

Fiske, A. P. (1991b, June). *Innate hypotheses and cultural parameters for social relations*. Paper presented at the Society for Philosophy and Psychology, San Francisco, CA.

Fodor, J. A. (1983). *The modularity of mind*. Cambridge, MA: MIT Press.

Frumhoff, P. C., & Baker, J. (1988). A genetic component to division of labour within honey bee colonies. *Nature, 333*, 358–361.

Frumhoff, P. C., & Schneider, S. (1987). The social consequences of honey bee polyandry: The effects of kinship on worker interactions within colonies. *Animal Behaviour, 35*, 255–262.

Gigerenzer, G., & Hug, K. (in press). Reasoning about social contracts: Cheating and perspective change. Institut für Psychologie, Universität Salzburg, Austria.

Griggs, R. A., & Cox, J. R. (1982). The elusive thematic-materials effect in Wason's selection task. *British Journal of Psychology, 73*, 407–420.

Hanken, J., & Sherman, P. W. (1981). Multiple paternity in Belding's ground squirrel litters. *Science, 212*, 351–353.

Hamilton, W. D. (1964). The genetical evolution of social behaviour. Parts I, II. *Journal of Theoretical Biology, 7*, 1–52.

Holmes, W. G., & Sherman, P. W. (1982). The ontogeny of kin recognition in two species of ground squirrels. *American Zoologist, 22*, 491–517.

Isaac, G. (1978). The food-sharing behavior of protohuman hominids. *Scientific American, 238*, 90–108.

Jackendoff, R. (1991, June). Is there a faculty of social cognition? Presidential address to the Society for Philosophy and Psychology, San Francisco, CA.

Johnson-Laird, P. N. (1982). Thinking as a skill. *Quarterly Journal of Experimental Psychology, 34A*, 1–29.

Johnson-Laird, P. N. (1983). *Mental models: Towards a cognitive science of language, inference and consciousness*. Cambridge, MA: Harvard University Press.

Kaplan, H., & Hill, K. (1985). Food sharing among Ache foragers: Tests of explanatory hypotheses. *Current Anthropology, 26*, 223–239.

Kaplan, H., Hill, K., & Hurtado, A. M. (1990). Risk, foraging and food sharing among the Ache. In E. Cashdan (Ed.), *Risk and uncertainty in tribal and peasant economies*. Boulder: Westview Press.

Lee, R., & DeVore, I. (Eds.). (1968). *Man the hunter*. Chicago: Aldine.

Luce, R. D., & Raiffa, H. (1957). *Games and decisions: Introduction and critical survey.* New York: Wiley.

Malinowski, B. (1922). *Argonauts of the Western Pacific.* New York: Dutton Press.

Maljković, V. (1987). *Reasoning in evolutionarily important domains and schizophrenia: Dissociation between content-dependent and content-independent reasoning.* Unpublished undergraduate honors thesis, Department of Psychology, Harvard University.

Manktelow, K. I., & Evans, J.St.B.T. (1979). Facilitation of reasoning by realism: Effect or non-effect? *British Journal of Psychology, 70,* 477–488.

Manktelow, K. I., & Over, D. E. (1987). Reasoning and rationality. *Mind & Language, 2,* 199–219.

Manktelow, K. I., & Over, D. E. (1990). Deontic thought and the selection task. In K. J. Gilhooly, M.T.G. Keane, R. H. Logie, & G. Erdos (Eds.), *Lines of thinking* (Vol. 1). London: Wiley.

Marr, D. (1982). *Vision: A computational investigation into the human representation and processing of visual information.* San Francisco: Freeman.

Marr, D., & Nishihara, H. K. (1978, October). Visual information processing: Artificial intelligence and the sensorium of sight. *Technology Review,* 28–49.

Mauss, M. (1925/1967). *The gift: Forms and functions of exchange in archaic societies.* (I. Cunnision, Trans.). New York: Norton.

Maynard Smith, J. (1964). Group selection and kin selection. *Nature, 201,* 1145–1147.

Maynard Smith, J. (1982). *Evolution and the theory of games.* Cambridge: Cambridge University Press.

Packer, C. (1977). Reciprocal altruism in *Papio annubis. Nature, 265,* 441–443.

Pollard, P. (1982). Human reasoning: Some possible effects of availability. *Cognition, 10,* 65–96.

Quine, W.V.O. (1969). *Ontological relativity and other essays.* New York: Columbia University Press.

Sharp, L. (1952). Steel axes for stone age Australians. *Human Organization, 11,* 17–22.

Sherman, P. W. (1977). Nepotism and the evolution of alarm calls. *Science, 197,* 1246–1253.

Smuts, B. (1986). *Sex and friendship in baboons.* Hawthorne: Aldine.

Sperber, D. (1975). *Rethinking Symbolism* (A. Morton, Trans.). Cambridge: Cambridge University Press.

Sperber, D. (1985). Anthropology and psychology: Towards an epidemiology of representations. *Man (N.S.), 20,* 73–89.

Sperber, D. (1990). The epidemiology of beliefs. In C. Fraser & G. Geskell (Eds.), *Psychological studies of widespread beliefs.*

Sperber, D., & Wilson, D. (1986). *Relevance: Communication and cognition.* Oxford: Blackwell.

Stone, V., & Cosmides, L. (in prep.). Do people have causal reasoning schemas?

Thornhill, R. (1991). The study of adaptation. In M. Bekoff and D. Jamieson (Eds.), *Interpretation and explanation in the study of behavior.* Boulder: West View Press.

Tooby, J. (1985). The emergence of evolutionary psychology. In D. Pines (Ed.), *Emerging Syntheses in Science. Proceedings of the Founding Workshops of the Santa Fe Institute.* Santa Fe, NM: The Santa Fe Institute.

Tooby, J., & Cosmides, L. (1989). Evolutionary psychology and the generation of culture, part I. Theoretical considerations. *Ethology & Sociobiology, 10,* 29–49.

Tooby, J., & Cosmides, L. (1990a). On the universality of human nature and the uniqueness of the individual: The role of genetics and adaptation. *Journal of Personality, 58,* 17–67.

Tooby, J., & Cosmides, L. (1990b). The past explains the present: Emotional adaptations and the structure of ancestral environments. *Ethology and Sociobiology, 11,* 375–424.

Tooby, J., & DeVore, I. (1987). The reconstruction of hominid behavioral evolution through strategic modeling. In W. Kinzey (Ed.), *Primate Models of Hominid Behavior*. New York: SUNY Press.

Trivers, R. L. (1971). The evolution of reciprocal altruism. *Quarterly Review of Biology, 46*, 35–57.

Trivers, R. L., & Hare, H. (1976). Haplodiploidy and the evolution of social insects. *Science, 191*, 249-263.

de Waal, F.B.M. (1982). *Chimpanzee politics: Power and sex among apes*. New York: Harper & Row.

de Waal, F.B.M., & Luttrell, L. M. (1988). Mechanisms of social reciprocity in three primate species: Symmetrical relationship characteristics or cognition? *Ethology and Sociobiology, 9*, 101–118.

Wason, P. (1966). Reasoning. In B. M. Foss (Ed.), *New horizons in psychology*, Harmondsworth: Penguin.

Wason, P. (1983). Realism and rationality in the selection task. In J.St.B.T. Evans (Ed.), *Thinking and reasoning: Psychological approaches*. London: Routledge & Kegan Paul.

Weissner, P. (1982). Risk, reciprocity and social influences on !Kung San economics. In E. Leacock & R. B. Lee (Eds.), *Politics and history in band societies*. Cambridge: Cambridge University Press.

Wilkinson, G. S. (1988). Reciprocal altruism in bats and other mammals. *Ethology and Sociobiology, 9*, 85-100.

Wilkinson, G. S. (1990, February). Food sharing in vampire bats. *Scientific American*, 76–82.

Williams, G. C. (1966). *Adaptation and natural selection: A critique of some current evolutionary thought*. Princeton: Princeton University Press.

Williams, G. C. (1985). A defense of reductionism in evolutionary biology. *Oxford Surveys in Evolutionary Biology, 2*, 1–27.

Williams, G. C., & Williams, D. C. (1957). Natural selection of individually harmful social adaptations among sibs with special reference to social insects. *Evolution, 11*, 32–39.

Wilson, E. O. (1971). *The insect societies*. Cambridge, MA: Harvard University Press.

4

Two Nonhuman Primate Models for the Evolution of Human Food Sharing: Chimpanzees and Callitrichids

W. C. McGREW AND ANNA T. C. FEISTNER

INTRODUCTION: FOOD SHARING IN HOMINIZATION

Scenarios for the evolutionary transition from apes to humans, that is, for hominization, almost always include food sharing as a component (Goodall & Hamburg, 1974; Isaac, 1978; McGrew, 1979, 1981, 1992; Tanner, 1981, 1987; Tooby & DeVore, 1987; Wrangham, 1987; Zihlman, 1981). However, hypotheses on the nature, timing, and extent of food sharing in hominization vary greatly from one theory to the next.

Only the late Glynn Isaac, as far as we know, made food sharing the keystone of the process, so it is worth exploring his system in detail (see Figure 4.1). It is a three-stage sequence: His starting point is an ancestral hominoid, resembling a chimpanzee *(Pan)*, which shows small-scale hunting that leads to "tolerated scrounging" of meat. Simple tool use is present but is independent of the hunting and scrounging.

The middle stage of hominization involves food sharing as the integrator of hunting and gathering, which are organized in terms of home bases and division of labor. Supporting all of this is an infrastructure of tool making and tool using. This stage was present in the Plio-Pleistocene era of about 2 million years ago.

The third stage is what now exists in modern human societies, each with their varied versions and socioeconomic elaborations. The infrastructure includes all of technology, from milking pails to vending machines. An all-embracing superstructure of culture overlies all such food sharing, from potlatches to fast-food outlets.

Isaac's evidence for behavior was necessarily circumstantial, since only artifacts, and not actions, persist. He inferred the practice food sharing in the middle stage on the basis of concentrations of stone tools that could have been used for processing meat, found with a large mammal's fossilized bones. Stronger inference came from concentrations of tools found with the remains of several large mammals, which might indicate mass or repeated transport of prey. Why should hunters do this if not to share the proceeds? There are alternative explanations such as parallel scavenging (e.g., Shipman, 1986) or accumulated individual processing (e.g., Potts, 1987), but Isaac's explanation remains a possibility.

What is missing from Isaac's account is the catalyst for adaptation: Integrated systems such as his middle stage do not spring full-blown (Tooby & DeVore, 1987). Two

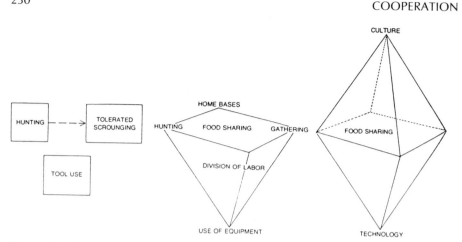

Figure 4.1 Isaac's (1978) three-stage sequence of the role of food sharing in hominid evolution.

types of explanation for the transition, one *social* and one *technological,* exemplify the two main strands of his argument.

Lovejoy (1981) placed the start of food sharing in the origins of the sexual pair-bond, 4 million years ago in the Miocene. More precisely, he outlined the likely enhancement of reproductive success that would follow from a male's provisioning his mate. By improving her physical condition, early hominid parents could shorten the birth interval between offspring and increase the likelihood of their survival. (How foraging males could enforce fidelity and ensure paternity while away hunting is not clear, however. For an alternative view that stresses the importance of food sharing at the earlier stage of courtship, see Parker, 1987).

From the more technological viewpoint, Parker and Gibson (1979) suggested that food sharing arose as a secondary adaptation from tool use. To be more exact, it was an adaptation for "extractive foraging," that is, the acquisition of embedded food items, which requires advanced sensorimotor and symbolizing intelligence. In their scheme, sharing such acquired solid food was more efficient for a mother than sharing food through the more energetically costly means of lactation. Sharing patterns thus created could generalize to other relationships. (Why chimpanzees, who are excellent extractive foragers, have not taken food sharing further in the hominid direction is not clear.)

To tackle these issues relating to the transition from nonhuman to human primate food sharing, we will proceed in stages: First, we will summarize briefly current knowledge of food sharing in other primates. Second, we will focus on the two types of primates who engage in habitual food sharing—chimpanzees and callitrichids (marmosets and tamarins). Third, we will use Isaac's (1978) 10 features of food sharing to contrast *Homo sapiens* with our nearest living relations and to contrast chimpanzees and callitrichids. These features will be reordered, modified, corrected, and updated, whenever necessary, in an effort to say whether the human-nonhuman contrasts are ones of kind or only of degree. Finally, we will return to the themes of the social and the technological, in an attempt at synthesis.

FOOD SHARING IN NONHUMAN PRIMATES

In another article, we reviewed the literature on food sharing by primates (Feistner & McGrew, 1989). Overall, we found that transfer of food items between individuals has been reported in many species of simians. However, most of these records are anecdotal or are based on a few occasional events. Except for the chimpanzees and callitrichids, there is no convincing evidence that such transfer of food is important, much less essential for survival. This contrasts markedly with several species of social carnivores, especially in the Canidae (e.g., wolves), Felidae (e.g., lions), and Hyaenidae (hyenas), in which the provisioning of mates and young is crucial to successful reproduction (e.g., Kuhme, 1965). Of the orders of mammals, the Carnivora present a more basic and consistent range of food sharing adaptations, such as reciprocal exchange, than do the Primates, humankind excepted.

When food sharing does occur commonly in nonhuman primates, it almost always takes the form of "tolerated scrounging" from parents, usually the mother, by youngsters. That is, infants or juveniles usually take leftovers, with little resistance offered by the parent. This parental tolerance may function in the weaning of offspring from milk to solid food.

There is some indication that food sharing is most developed in monogamous primates such as gibbons (*Hylobates* spp., e.g., Schlesser & Nash, 1977) and titi monkeys (*Callicebus* spp., e.g., Starin, 1978). However, no one has done an experimental study of its importance, nor have observational studies kept track of such basic variables as intake, calories, energetic or nutritional trade-offs, and so on.

For these reasons, we will focus on the chimpanzees and callitrichids, as they are the only primates in which food sharing is both important and well studied.

Chimpanzee Life

The common chimpanzee (*Pan troglodytes,* Pongidae) is among the most-studied of nonhuman primates, with ongoing field studies lasting over 30 years at sites across Africa. Studies of chimpanzees in captivity go back even further, and the settings vary from laboratories to zoos to human homes. Chimpanzees studied in captivity have ranged from those confined alone in small cages to mixed groups in naturalistic enclosures. (Neither of these generalizations applies to bonobos or pygmy chimpanzees, *Pan paniscus*, about whom little is yet known, e.g., there are as yet no empirical data on food sharing in captivity.)

There is a remarkable consensus about chimpanzee nature from these varied studies (Nishida, 1989). In the wild, the apes live in loosely structured communities or "unit-groups." Rarely, if ever, are all members together. Instead, parties of varying size and composition form and break up over hours or days. It has been said that the only lasting subgrouping is that of a mother and her dependent young (Goodall, 1968).

Parties are not random, however. Adult males (who are often related) form bonds with each other and travel together to patrol the edges of the community's home range in xenophobic vigilance. Adult females and their offspring range less widely and spend less time together. There are no lasting sexual bonds and most mating is "promiscuous," although most conceptions occur in nonrandom, temporary consortships

(Tutin, 1979). In short, males compete for access to mates, and females compete for resources for their matrilineal kin. Males stay in their natal communities, but females move out, sometimes more than once. Communities displace or avoid one another, but do not defend set territorial boundaries. There may be much overlap between the ranges of neighboring groups, especially seasonally, and some females may transfer membership between communities. In captivity, when space and facilities permit, social life shows many of the same patterns (e.g., de Waal, 1982).

Chimpanzee Food Sharing

Observations of food sharing in the wild are available only from populations of chimpanzees that are well used to humans at close range. Most come from two sites: Gombe (Goodall, 1968; McGrew, 1975; Silk, 1978) and Kasoje (Nishida, 1970), both in western Tanzania. The only nearly comparable data on bonobos are from Wamba, in Zaire (Kano, 1980; Kuroda, 1984). Observations of food sharing in captivity are fewer, despite the ubiquity of the behavioral pattern, with Silk's (1979) being the most detailed. Until recently, experimental studies were contrived or limited (Lefebvre, 1982; Nissen & Crawford, 1936), but de Waal (1989) has rectified this.

The following generalizations can be made:

1. Most food sharing of everyday plant foods is from mother to dependent offspring, and it is tolerated scrounging (see Figure 4.2).
2. Next most common is sharing by adult males with adult females, but there are not yet enough data to know if there are regular partner preferences (see Figure 4.3).
3. The nature of the food item affects the pattern of its distribution. Prized foods in short supply (e.g., meat, sugar cane) lead to high rates of interaction over food and more aggression. Common but hard-to-process foods increase mothers' sharing with infants.
4. Patterns of meat sharing are the most complicated, being influenced by age (with older individuals receiving more meat), kinship (with higher rates of sharing within matrilines), and reproductive state (with estrous females receiving more than anestrous ones).
5. The causation of transfer of food varies widely. An individual may coerce another to hand over food or, more rarely, may donate food without prompting. Most cases, however, follow on from various forms of begging by the eventual recipients, and so can be termed *responsive* rather than *spontaneous.*
6. Chimpanzees give a special call that alerts others at a distance to the presence of food. As such, this is food sharing of sorts, but it need not be interpreted as charitable. A caller faced with more than enough food will lose nothing by sharing it and may benefit later when another chimpanzee reciprocates (Wrangham, 1977).
7. The overall pattern for bonobos is apparently the same as for common chimpanzees, but age and sex differences are less clear-cut.
8. Finally, patterns of food sharing, whenever examined so far, follow evolutionarily predictable lines in terms of parent-offspring conflict, nepotism, and intersexual competition. For example, mothers share food with offspring, but come into conflict when offspring seek to continue receiving parental investment

Figure 4.2   An infant chimpanzee (left) tries to scrounge sugar cane while his mother (right) eats (at Kasoje).

beyond the point when it is in the mother's interest to redirect it to future off-spring (Trivers, 1974).

Callitrichid Life

The Callitrichidae are one of the two main families of New World monkeys (Platyr-rhini). They make up about 20 species in four genera: marmosets *(Callithrix, Cebuella)* and tamarins *(Leontopithecus, Saguinus).* The two types differ in teeth and diet, but all callitrichids have much in common, hence the collective use of the family name here.

Unlike chimpanzees, the marmosets and tamarins are as yet little known in the wild, especially in terms of their social life. Some species have not been studied at all, and of those that have, few have had identified individuals studied in the long term. Of the two species best known for food sharing, the endangered cotton-top tamarin *(Saguinus oedipus)* of northwestern Columbia has been studied in the wild only little (Neyman, 1978). The highly endangered golden lion tamarin *(Leontopithecus rosa-lia),* now reduced to a few relic populations in the Atlantic coastal forests of Brazil, is

234

Figure 4.3 An adult male chimpanzee (right) hands over a banana to an adult female (left) (at Gombe).

only now being studied in the field (Kleiman, 1981). Much more is known of both species in captivity.

What follows is a generalized description based on a synthesis of what is known across several species. As more knowledge of differences between taxa accumulates, it is harder and harder to present a "typical" callitrichid, but brevity demands it here. Callitrichids live in families usually made up of a single breeding female, one or more mates, and their offspring, adult as well as immature. The reproductive system is impressively efficient, largely because it is cooperative. The father(s) and older siblings do most of the infant rearing, leaving the mother to concentrate on gestation and lactation. The result is large neonates, twinning, and postpartum conception, which means that the mother may carry two fetuses while suckling two infants at the same time. Offspring are reproductively suppressed in the intact natal family, and incest is avoided. Family life is largely peaceful.

Callitrichids are largely arboreal and limited to forests. The families are fiercely territorial, defending small home ranges, which contain a few key resources. They resist incomers, and offspring appear to delay dispersing until "vacancies" come up in nearby territories.

Callitrichid Food Sharing

As with other aspects of social behavior, our knowledge of food sharing in callitrichid monkeys is virtually confined to a few studies in captivity. For a welcome exception, see Ferrari's (1987) study of buffy-headed marmosets, *Callithrix flaviceps,* in Brazil. The following generalizations are based mainly on golden lion tamarins (Brown & Mack, 1978; Hoage, 1982) and cotton-top tamarins (Feistner & Price, 1990) about whom the most is known:

1. Most food passes from parents or older siblings to weanling infants, within the family. Often it is *active* sharing, with possessors offering food items without any noticeable prompting.
2. Less commonly, food is shared between other family members, with the recipient being the mother, or a juvenile, or an older offspring carrying an infant.
3. Sharing rates are higher with live animal prey than with plant or artificial food items. For a few weeks in early life, infants depend upon others for food, even when they could get food directly for themselves.
4. Within the family, the network of food distribution is diffuse, with all possible combinations of age and sex categories having been seen to share. However, there are age and sex differences: Sexually mature older siblings share more than do immature older siblings (Feistner & Price, 1990), and brothers share more than do sisters (Wolters, 1978), at least in cotton-top tamarins.
5. Transferred food items are not leftovers, i.e., those no longer wanted by others who are sated. Instead, the highest rates of sharing occur soon after food is made available, and highly preferred foods are shared more than less desirable ones (Feistner & Chamove, 1986).
6. Callitrichids give a characteristic vocalization that invites sharing and accompanies offering of food. Infants quickly respond to this encouragement.
7. Food sharing is a key part of the adaptive package of cooperative breeding, along with carrying and protecting the infants. Patterns of sharing often cannot be predicted simply by degree of relatedness, since in nuclear families all but the parents are related by $r = 0.50$. Instead, patterns of sharing seem to reflect the maximizing of inclusive fitness by donors. Sexually mature older siblings waiting for breeding opportunities can recoup some of their lost reproductive value by investing in their younger siblings. Similarly, the breeding female seems to get food in accordance with the cyclical physiological demands of her reproductive state.

HUMAN VERSUS CHIMPANZEE FOOD SHARING

Many authors have used the chimpanzee as a model for early human evolution (see the list of behavior patterns mentioned earlier). Of these, Isaac (1978) dealt most comprehensively with food sharing. He presented a set of features that were meant to be *unique* to human beings, although in some cases chimpanzees show a rudimentary version of the behavioral pattern. These 10 features are given in the left-hand column of Table 4.1, where they are summarized and reordered to suit this article. The middle column gives the contrasting position for the chimpanzee, as stated or implied by Isaac, although in a few cases we have amplified the contrast to make it clear. The right-hand column gives extra information gained since Isaac's paper was published or relevant knowledge that he did not cite. What follows is a paragraph-by-paragraph reconsideration of Isaac's 10 points:

Table 4.1 Isaac's (1978) Hypothesized Major Differences Between *Homo* and *Pan* in Food Sharing and Related Activities with Updated Comments

Humans	Chimpanzees	Comments
1. Hunting and fishing, prey may weigh more than 15 kg	Less hunting, no prey weighs more than 15 kg	No real difference (Teleki, 1973)
2. Travel bipedally, carry possessions	Travel quadrupedally, possessions not carried	Chimpanzees carry hammers and nuts (Boesch & Boesch, 1983)
3. Complex, made tool kit to obtain food	Only simple tools	Tanzanian chimpanzees' tools not different from Tasmanian aborigines' (McGrew, 1987)
4. Prepare food for eating	No prepared food	Chimpanzees use hammer and anvil to smash open nuts (Boesch & Boesch, 1983)
5. Postpone eating of gathered foods	Eat food when and where found	Real difference
6. Home base as focus of ranging	No home base	Real difference
7. Long-term mating bond, economic reciprocity	No mating bonds	Sex differences in feeding (McGrew, 1979), consortship (Tutin, 1979)
8. Kinship categories regulate relations	Kinship influences relations	Seems impossible to disprove mental representation by chimpanzees of kinship relations
9. Language regulates relations	Complex communication but no language	Syntax and semanticity in natural "language"? (Menzel, 1976)
10. Active food sharing, family as crucial node in larger network	Tolerated scrounging of meat, no plant sharing	Both share natural and artificial plant foods (McGrew, 1975)

1. Human beings in nonagricultural societies devote much time and energy to seeking high-protein food (Kaplan & Hill, 1985). This takes the form of hunting or fishing, often cooperatively, for vertebrate prey. Of the primates, only *Homo* hunts big game, that is, prey weighing over 15 kg. This has direct consequences for food sharing. As no one hunter could consume the whole kill before it spoiled, surplus parts of it are available for giving away. Chimpanzees also eat birds and mammals, but individual prey are smaller, though multiple kills often mean that surplus meat is available (Boesch & Boesch, 1989). Meat is distributed, but reluctantly, and often only after coercion. Sharing is not random, but follows patterns of age, sex, and affiliation (Teleki, 1973). For this first feature, there seem to be few qualitative differences between the sharing of meat by *Homo* and *Pan*.

2. Human foragers move about bipedally, carrying with them their possessions, including food for sharing. This is made possible by the upper limbs being freed from locomotion and by the invention of containers. At the time when Isaac wrote, it could be said that chimpanzees only occasionally carried tools and that they traveled quadrupedally. Later work by Boesch and Boesch (1983) showed that wild chimpanzees may carry nuts or hammer stones and anvils for use in nut cracking, traveling tripedally.

3. The tool kits used to obtain food in even the simplest human cultures vary in type and involve complex manufacture, according to Isaac. Presumably such tools expand the menu and create surpluses for exchange (e.g., Betzig & Turke, 1986). In contrast (he wrote), chimpanzees in nature use only a few simple tools. In fact, if one uses an objective taxonomy to compare the tools used in subsistence by Tanzanian chimpanzees and by Tasmanian aborigines (who are often presented as having the simplest human technology), there is little difference (McGrew, 1987).

4. Human beings prepare food for eating by using tools to alter it, e.g., by crushing, grinding, heating, etc. Chimpanzees eat their food raw, but the Boesches (1983, 1989) have shown that some wild chimpanzees put much time and effort into hammering open hard-shelled nuts and crania and that the resulting contents are shared. Thus, making easier-to-eat foods to pass on to dependent young is found in both *Homo* and *Pan*.

5. Human beings postpone their consumption of some food items that accumulate from gathering. These collected proceeds are then taken to a meeting point for sharing or exchange with other persons. In contrast, chimpanzees almost always eat food when and where they find it. Sometimes a branch laden with fruit or a piece of a kill is carried a short distance, but this seems to be for more convenient personal consumption, not sharing. This difference between *Homo* and *Pan* seems to be a qualitative one (see also Wynn & McGrew, 1989).

6. Even the simplest gathering-and-hunting society operates from a home base of sorts, however temporary (Kaplan & Hill, 1985). This base may be moved daily or seasonally, but in any given period members of the community return from foraging to a focal point for sharing and other socializing. Chimpanzees, though somewhat territorial, have no permanent home bases. Instead, they roam widely, often alone, building a new sleeping nest each night, usually in a different place from the night before. This seems to be another qualitative difference between the two types of hominoid (but see Sept, 1992).

7. Human beings form long-term mating bonds between the sexes. Regardless of the varied forms these take in different cultures, they have common properties, among

which are reciprocal economic ties. In foraging societies, this usually means sexual division of labor by female gathering and male hunting. Chimpanzees form no lasting pair-bonds and usually mate with many partners, but the key factor evolutionarily—who impregnates females—involves mutual consent in most cases. Further, the potential for sexual division of labor is shown by sex differences in feeding on animal prey: Males do most of the hunting for mammals, and females do more gathering of insects (McGrew, 1979).

8. In human society, kin of varying degrees of relatedness are not only dealt with differently, but are explicitly so classified. These categories help to regulate social relations, including the sharing of food. Kinship, at least matrilineally, also affects relations in the chimpanzees' community, in terms of alliance, incest avoidance, and food sharing. There is no detectable categorization, but it is hard to imagine how one could rule out (or confirm) mental representations of kinship classes in wild apes.

9. Such classification and terminology may well be a prime function of language, which in human society helps to regulate social relations, including food sharing. It is not essential, however, as seen in the social relations of infants (who are preverbal) or the mentally subnormal (who may be nonverbal). Chimpanzees do have complex communication, but the extent of their linguistic ability is not clearly known. Menzel (1975) showed that signs of syntax and semanticity may exist in the communication of young chimpanzees and that this resulted in passive food sharing or its avoidance.

10. Finally, Isaac wrote that in *Homo* the acquisition of food is a corporate responsibility. Active food sharing typically occurs with the family as the crucial node in a wider social network (Betzig & Turke, 1986). Food is exchanged between adults and shared by adults with dependent others, old or young. In contrast, Isaac argued that meat sharing by chimpanzees is tolerated scrounging and not active sharing. Furthermore, he stated that vegetable foods, which make up most of the species' diet, are not shared by chimpanzees. In fact, most transfer of food by wild chimpanzees is matrilineal (Silk, 1978), and both natural and artificial plant foods change hands (McGrew, 1975). The same patterns occur in captive groups (Silk, 1979; de Waal, 1989). Distinguishing active from passive food sharing is not easy, especially if chimpanzees abandon food to forestall being pestered for it. (For a discussion of definitions, see Feistner & McGrew, 1989).

In conclusion, few of the 10 contrasting features put forward by Isaac as uniquely human are convincingly so. Some of the differences seem to be ones of degree, not of kind. Overall, this narrows the gulf between pongid and hominid in terms of reconstructing a common ancestral hominoid.

CHIMPANZEE VERSUS CALLITRICHID MODELS

Table 4.2 shows how callitrichid monkeys compare with chimpanzees on Isaac's 10 features. The contrast is soon apparent: Like chimpanzees, callitrichids also pursue high-protein animal prey, which are usually insects (Neyman, 1978). (Chimpanzees weigh about 100 times as much as typical callitrichids, that is about 45 kg versus 0.45 kg.) Such hunting is done individually. In nature and in captivity, the capture of a locust elicits intense begging and sharing.

Callitrichids use no tools, carry no objects, prepare no food items, unlike chimpanzees, but like them they do not postpone consumption.

Table 4.2 Comparison of Chimpanzee and Callitrichid on Isaac's (1978) 10 Features of Food Sharing (+ = present; − = absent)

Feature	Chimpanzee	Callitrichid
1. Animal prey	+	+
2. Carry possessions	+	−
3. Tool use	+	−
4. Prepare food	+	−
5. Postpone consumption	−	−
6. Home base	−	+
7. Mating bond	−	+
8. Kinship categories	−?	−?
9. Language	−	−
10. Active sharing	−	+

At least some tamarins have home bases: Families of golden lion tamarins use tree holes as overnight shelters for long periods (Coimbro-Filho, 1978). Cotton-top tamarins do not use tree holes, but sleep habitually in a few suitable trees (Neyman, 1978). Marmosets repeatedly use sites in certain trees, which they tap for exudates, and these gums are a major part of the diet.

In captivity, callitrichids are usually kept in nuclear families in which the parents mate for life. Only long-term studies will show how often this is true in nature, but data collected so far are largely compatible with this as an "ideal" mating system. As with sexual bonds in humans (Isaac, 1978), either monogamy or polygamy entails joint responsibility for child rearing and restrictions on sexual access, as well as some male-to-female sharing of food.

Categorization of kin is just as hard to prove or disprove in callitrichids as it is in chimpanzees. However, there is some evidence that suggests rudimentary roles within families: Eldest daughters show the most inclination to disperse, and eldest sons appear to stay on in the natal family to inherit its resources (McGrew & McLuckie, 1986).

We know of no evidence of linguistic communication in callitrichids, but their natural vocalizations may be no less complex than those of the apes. Less has been done on tamarins than on marmosets, but there is some evidence in the latter for both incipient syntax (Snowdon & Cleveland, 1984) and semanticity (Snowdon, Cleveland, & French, 1983). As noted earlier, both callitrichids and chimpanzees give specific calls in the context of sharing food.

Finally, there is a surprising difference in *active* food sharing between chimpanzees and callitrichids. The apes rarely give food without being prompted, and begging takes several forms, some of which are prolonged. Callitrichids spontaneously offer food, accompanied by a specific invitational call. Possessors give up food quickly, with little or no resistance. As with chimpanzees, however, callitrichid offspring do much tolerated scrounging, as is found in many other species of primates.

Overall, it looks as if chimpanzees and callitrichids show distinctive adaptive "packages" relating to food sharing. That of the apes is largely technologically based, in terms of the acquisition and handling of food items, and distribution follows from this. That of the callitrichids is more socially based, with food sharing being embedded in a rich, cooperative family life. The question is, Which model offers more help in understanding the evolution of food sharing during hominization? (Tooby & DeVore, 1987).

CONCLUSIONS

First, it is unlikely that either chimpanzees or callitrichids provide a tidy, single-taxon model for a stage in hominization. Nor would a simple, double-helix-style fusion of the two, however tempting, be likely to produce a believable ancestral hominid (although this remains to be explored). However, the technological theme presented by the chimpanzees and the social theme presented by the callitrichids should be of use, if only to remind us that food sharing is the bridging element between subsistence and sociality (Feistner & McGrew, 1989; Blurton Jones, 1987).

If one is to look for "Darwinian algorithms" in order to test models of cultural adaptation, subsistence and sociality seem two likely candidate areas (cf. Tooby & Cosmides, 1989). Potential food-sharers need to assess features such as resource value and divisibility, current (and future?) needs and surpluses, and relative costs and benefits of food to self and others.

Cosmides and Tooby (1989) have focused on similar problems in seeking to construct a computational theory of social exchange for human beings. Getting to the hominid state of cognitive sophistication from nonhuman origins is the challenge to evolutionary explanation. Both environmental and phylogenetic constraints need to be considered.

Chimpanzees have all the necessary technology except the container, but they are a long way on social grounds from Isaac's middle stage. Callitrichids have a true family life of rich complexity, especially in their cooperative child-rearing systems involving adult siblings, but up in the rain forest canopy, technology is a low priority.

Finally, what hypothetical selection pressures brought about the blend of social and technological factors in food sharing in hominization? How did we humans manage to get the best of both the chimpanzee and callitrichid worlds? The key seems to be in the proto-hominid's adaptation of its hunting or scavenging to include big game. Acquisition of super-surplus meat, which would otherwise be wasted, offers various adaptive opportunities: For male hunters meat provides a high-value currency for sexual (and therefore reproductive) negotiation with females. It enables possessors of meat to provision offspring or other kin and to be able to expect reciprocity accordingly (see also Blurton Jones, 1987). Risky investment in hunting or scavenging by some group members can then be offset by surplus gathering by others. With such interactions (or transactions), the potential for relationships exists, relationships that at some point are transformed into enduring reproductive bonds between the sexes. In conclusion, Isaac's focus on food sharing in the transition from ape to human was apt, but the emerging picture is richer than even he imagined.

ACKNOWLEDGMENTS

We are grateful to C. Fischler, I. de Garine, and W. Schiefenhovel, and the Werner Reimers Stiftung, for the opportunity for W. C. McGrew to take part in a symposium on food sharing at Bad Homburg where the first version of this paper was given in 1984; C. Tutin for continual collaboration in the fieldwork on chimpanzees; I. Rodgerson and E. Halloren for husbandry of the callitrichids at the University of Stirling; The Carnegie Trust for the Universities of Scotland for providing partial funding for travel to Bad Homburg; and J. Barkow, L. Cosmides, and J.

Tooby for essential critical comments and suggestions on the manuscript. We are reminded yet again in the preparation of this paper of the great loss to science from the death of Glynn Isaac.

REFERENCES

Betzig, L. L., & Turke, P. W. (1986). Food sharing on Ifaluk. *Current Anthropology, 27*, 397–400.

Blurton Jones, N. G. (1987). Tolerated theft, suggestions about the ecology and evolution of sharing, hoarding, and scrounging. *Social Science Information, 26*, 31–54.

Boesch, C., & Boesch, H. (1983). Optimisation of nut-cracking with natural hammers by wild chimpanzees. *Behaviour, 83*, 265–286.

Boesch, C., & Boesch, H. (1989). Hunting behavior of wild chimpanzees in the Tai National Park. *American Journal of Physical Anthropology, 78*, 547–573.

Brown, K., & Mack, D. S. (1978). Food sharing among captive *Leontopithecus rosalia. Folia primatologica, 29*, 268–290.

Coimbro-Filho, A. F. (1978). Natural shelters of *Leontopithecus rosalia* and some ecological implications. In D. G. Kleiman (Ed.), *The biology and conservation of the callitrichidae* (pp. 79–89). Washington, DC: Smithsonian Institution Press.

Cosmides, L., & Tooby, J. (1989). Evolutionary psychology and the generation of culture, part II: Case study: A computational theory of social exchange. *Ethology and Sociobiology, 10*, 51–97.

Feistner, A.T.C., & Chamove, A. S. (1986). High motivation towards food increases food-sharing in cotton-top tamarins. *Developmental Psychobiology, 19*, 439–452.

Feistner, A.T.C., & McGrew, W. C. (1989). Food-sharing in primates: A critical review. In P. K. Seth & S. Seth (Eds.), *Perspectives in primate biology, 3* (21–36). New Delhi: Today and Tomorrow's Printers and Publishers.

Feistner, A.T.C., & Price, E. C. (1990). Food-sharing in cotton-top tarmarins *(Saguinus oedipus). Folia primatologica, 54*, 34–45.

Ferrari, S. F. (1987). Food transfer in a wild marmoset group. *Folia primatologica, 48*, 203–206.

Goodall, J.v.L. (1968). The behaviour of free-living chimpanzees in the Gombe Stream Reserve. *Animal Behaviour Monographs, 1*, 161–311.

Goodall, J., & Hamburg, D. A. (1974). Chimpanzee behavior as a model for the behavior of early man. New evidence on possible origins of human behavior. *American Handbook of Psychiatry, 6*, 14–43.

Hoage, R. J. (1982). Social and physical maturation in captive lion tamarins, *Leontopithecus rosalia rosalia* (Primates: Callitrichidae). *Smithsonian Contributions to Zoology, 354*, 1–56.

Isaac, G. (1978). The food-sharing behavior of protohuman hominids. *Scientific American, 238*(4), 90–108.

Kano, T. (1980). Social behavior of wild pygmy chimpanzees *(Pan paniscus)* of Wamba: A preliminary report. *Journal of Human Evolution, 9*, 243–260.

Kaplan, H., & Hill, K. (1986). Food sharing among Ache foragers: Tests of explanatory hypotheses. *Current Anthropology, 26*, 233–245.

Kleiman, D. G. (1981). *Leontopithecus rosalia. Mammalian Species, 148*, 1–7.

Kuhme, W. (1965). Communal food distribution and division of labour in African hunting dogs. *Nature, 205*, 443–444.

Kuroda, S. (1984). Interaction over food among pygmy chimpanzees. In R. L. Susman (Ed.), *The pygmy chimpanzee* (pp. 301–324). New York: Plenum.

Lefebvre, L. (1982). Food exchange strategies in an infant chimpanzee. *Journal of Human Evolution, 11*, 195–204.

Lovejoy, C. O. (1981). The origin of man. *Science, 211,* 341–350.

McGrew, W. C. (1975). Patterns of plant food sharing by wild chimpanzees. In S. Kondo, M. Kawai, A. Ehara (Eds.), *Contemporary Primatology* (pp. 304–309). Basel: S. Karger.

McGrew, W. C. (1979). Evolutionary implications of sex differences in chimpanzee predation and tool use. In D. A. Hamburg & E. R. McCown (Eds.), *The great apes* (pp. 440–463). Menlo Park, CA: Benjamin/Cummings.

McGrew, W. C. (1981). The female chimpanzee as a human evolutionary prototype. In F. Dahlberg (Ed.), *Woman the gatherer* (pp. 35–73). New Haven: Yale University Press.

McGrew, W. C. (1987). Tools to get food: The subsistants of Tanzanian chimpanzees and Tasmanian aborigines compared. *Journal of Anthropological Research, 43,* 247–258.

McGrew, W. C. (1992). *Chimpanzee material culture: Implications for human evolution.* Cambridge: Cambridge University Press.

McGrew, W. C., & McLuckie, E. C. (1986). Philopatry and dispersion in the cotton-top tamarin, *Saguinus (o.) oedipus:* An attempted laboratory simulation. *International Journal of Primatology, 7,* 399–420.

Menzel, E. W. (1975). Natural language of chimpanzees. *New Scientist, 65*(932), 127–130.

Neyman, P. F. (1978). Aspects of the ecology and social organization of free-ranging cotton-top tamarins *(Saguinus oedipus)* and the conservation status of the species. In D. G. Kleiman (Ed.), *The biology and conservation of the callitrichidae* (pp. 39–71). Washington, DC: Smithsonian Institution Press.

Nishida, T. (1970). Social behavior and relationships among wild chimpanzees of the Mahale Mountains. *Primates, 11,* 47–87.

Nishida, T. (1989). Social structure and dynamics of the chimpanzee: A review. In P. K. Seth & S. Seth (Eds.), *Perspectives in primate biology, 3* (pp. 157–172). New Delhi: Today and Tomorrow's Printers and Publishers.

Nissen, H. W., & Crawford, M. P. (1936). A preliminary study of food-sharing behavior in young chimpanzees. *Journal of Comparative Psychology, 22,* 383–419.

Parker, S. T. (1987). A sexual selection model for hominid evolution. *Human Evolution, 2,* 235–253.

Parker, S. T., & Gibson, K. R. (1979). A developmental model for the evolution of language and intelligence in early hominids. *Behavioral and Brain Sciences, 2,* 367–408.

Potts, R. (1987). Reconstructions of early hominid socioecology: A critique of primate models. In W. G. Kinzey (Ed.), *The evolution of human behavior: Primate models* (pp. 28–47). Albany: S.U.N.Y. Press.

Schlesser, T., & Nash, L. T. (1977). Food sharing among captive gibbons *(Hylobates lar). Primates, 18,* 677–690.

Sept, J. M. (1992). Was there no place like home? A chimpanzee perspective on early archaeological sites. *Current Anthroplogy, 33.*

Shipman, P. (1986). Scavenging or hunting in early hominids: Theoretical framework and tests. *American Anthropologist, 88,* 27–43.

Silk, J. B. (1978). Patterns of food sharing among mother and infant chimpanzees at Gombe National Park, Tanzania. *Folia primatologica, 29,* 129–141.

Silk, J. B. (1979). Feeding, foraging and food sharing behaviour of immature chimpanzees. *Folia primatologica, 31,* 123–142.

Snowdon, C. T., & Cleveland, J. (1984). "Conversations" among pygmy marmosets. *American Journal of Primatology, 7,* 15–20.

Snowdon, C. T., Cleveland, J., & French, J. A. (1983). Responses to context- and individual-specific cues in cotton-top tamarin long calls. *Animal Behaviour, 31,* 92–101.

Starin, E. D. (1978). Food transfer by wild titi monkeys *Callicebus torquatus torquatus. Folia primatologica, 31,* 123–142.

Tanner, N. M. (1981). *On becoming human.* Cambridge: Cambridge University Press.

Tanner, N. M. (1987). The chimpanzee model revisited and the gathering hypothesis. In W. C.

Kinzey (Ed.), *The evolution of human behavior: Primate models* (pp. 3–27). Albany, S.U.N.Y. Press.

Teleki, G. (1973). The omnivorous chimpanzee. *Scientific American, 228*(1), 33–42.

Tooby, J., & Cosmides, L. (1989). Evolutionary psychology and the generation of culture, part I: Theoretical considerations. *Ethology and Sociobiology, 10,* 29–49.

Tooby, J., & DeVore, I. (1987). The reconstruction of hominid behavoral evolution through strategic modelling. In W. G. Kinzey (Ed.), *The evolution of human behavior: Primate models* (pp. 183–237). Albany: S.U.N.Y. Press.

Trivers, R. L. (1974). Parent-offspring conflict. *American Zoologist, 14,* 249–264.

Tutin, C.E.G. (1979). Mating patterns and reproductive strategies in a community of wild chimpanzees *(Pan troglodytes schweinfurthii). Behavioral Ecology and Sociobiology, 6,* 29–38.

de Waal, F.B.M. (1982). *Chimpanzee politics.* London: Jonathan Cape.

de Waal, F.B.M. (1989). Food sharing and reciprocal obligations among chimpanzees. *Journal of Human Evolution, 18,* 433–459.

Wolters, H. J. (1978). Some aspects of role taking behaviour in captive family groups of the cotton top tamarin *Saguinus oedipus oedipus.* In H. Rothe, et al. (Eds.), *Biology and behaviour of marmosets* (pp. 161–179). Gotingen: Eigenverlag H. Rothe.

Wrangham, R. W. (1977). Feeding behaviour of chimpanzees in Gombe National Park, Tanzania. In T. H. Clutton-Brock (Ed.), *Primate ecology* (pp. 503–538). London: Academic Press.

Wrangham, R. W. (1987). The significance of African apes for reconstructing human social evolution. In W. G. Kinzey (Ed.), *The evolution of human behavior: Primate models* (pp. 51–71). Albany: S.U.N.Y. Press.

Wynn, T. G., & McGrew, W. C. (1989). An ape's view of the Oldowan. *Man, 24,* 383–398.

Zihlman, A. L. (1981). Women as shapers of the human adaptation. In F. Dahlberg (Ed.), *Woman the gatherer* (pp. 75–120). New Haven: Yale University Press.

III

THE PSYCHOLOGY OF MATING AND SEX

Our scientific intuitions can be misled by our phenomenology. Consider vision. Phenomenally, seeing seems simple: We open our eyes, light hits our retina, and we see. It is effortless, automatic, reliable, fast, unconscious and requires no explicit instruction. But this apparent simplicity is deceptive. As is now known, constructing a three-dimensional representation of objects in the world from a two-dimensional retinal display poses enormously complex computational problems, and the visual system contains a vast array of dedicated, special-purpose information-processing machinery that solves these problems. Moreover, seeing is effortless, automatic, reliable, fast, and so on, precisely *because* we have this dedicated machinery.

Our phenomenal experience of an activity as "easy" or "natural" can lead us to grossly underestimate the complexity of the processes that make it possible. Doing what comes "naturally," effortlessly, or automatically is rarely simple from a computational point of view. To find someone beautiful, to fall in love, to feel jealous—all can seem as simple and automatic and effortless as opening your eyes and seeing. But this apparent simplicity is possible only because there is a vast array of complex computational machinery supporting and regulating these activities. Moreover, this machinery is and must be specialized—if a woman were to use the same taste preference mechanisms to choose a mate that she uses to choose nutritious foods, she would choose a strange mate indeed. Different adaptive problems are frequently incommensurate: They cannot, in principle, be solved by the same mechanism. To solve the adaptive problem of finding the right mate, one's choices must be guided by qualitatively different standards than when choosing the right food or the right habitat (see, e.g., Part VI, "Environmental Aesthetics"). Consequently, one's standards and preferences must be tailored to the domain and the problem. For parallel reasons, what counts as a "good mate" from an evolutionary point of view differs somewhat for women and for men. Therefore the computational processes governing mate choice in women are expected to differ in certain predictable ways from those governing mate choice in men.

The chapters by Buss and by Ellis explore the design features of the psychological mechanisms governing mate choice in humans. Ellis focuses on the mate preferences of women. Many evolutionarily informed researchers have shied away from this topic because an evolutionary analysis suggests that women's evaluative mechanisms will embody a somewhat more complicated set of standards than those of men. Ellis has reviewed the psychological literature on women's mate preferences, in an attempt to see how well certain predictions made in the evolutionary literature hold up. He also discusses how the evidence bears on certain predictions made by researchers who do not take an evolutionary perspective. Perhaps the most interesting of these is the traditional hypothesis that female and male sexual psychologies are identical, and give rise to different preferences only because female and male circumstances systematically differ. He evaluates, for example, the "structural powerlessness" hypothesis, which explains female-male differences as a response to differential power and access to resources and hence predicts that as women gain more power in society, their mating preferences will come to more closely resemble those of men.

Buss focuses on the mate preferences of both women and men. Darwin's theory of sexual selection is a rich source of hypotheses about the psychology of mate choice. In order to test some of these hypotheses, Buss has generated a truly extensive body of empirical work, both within the United States and cross-culturally. (Indeed, Buss gives the word "extensive" new meaning: One of his studies involved over 10,000 subjects.) His research shows that there is a broad cross-cultural consensus about what attributes are important in a mate. It also shows that there is a tight linkage between intersexual selection and intrasexual selection, just as Darwin predicted: In the domain of sexual attraction, the preferences of women shape the nature of competition between men, and the preferences of men shape the nature of competition between women.

For species in which both sexes cooperate to raise their joint offspring to reproductive age, there is more to the adaptive problem of mating and sex than simply finding and copulating with the right mate: There is also intense selection for mechanisms that protect the individual's investment in that cooperative relationship. To the extent that sexual access to members of the opposite sex is a reproductively limiting resource, selection will create psychologies that cause one to compete for sexual access and to defend it against rivals. For terrestrial vertebrates, it is typically the case that sexual access to females is a limiting resource for males. Consequently, selection has shaped adaptations in males of many species that facilitate success in competition for sexual access to females. These adaptations are not only anatomical, such as antlers and increased size, but psychological as well: computational systems that generate behavior whose function is to prevent females from having sexual contact with rival males.

This widespread competition for sexual access to females is further structured in species, such as humans, where males not only compete for mates, but also invest in offspring. In mammals, mothers always know who their offspring are (emergence from one's body is a very reliable cue), but paternity is far more difficult to ascertain. Paternity uncertainty poses few adaptive problems for many mammalian species, because the male does nothing to help raise their joint offspring. But paternity uncertainty does pose a serious adaptive problem for human males, who do share the burden of raising offspring: Human males must avoid investing in offspring not their own. A psychological design feature that systematically allows a male to be "cuckolded" would be intensely selected against for two reasons: (a) It would have lost an opportunity to replicate itself via that female, and (b) it would cause the male to squander scarce and valuable time and effort raising offspring that do not carry that design—time and effort that could otherwise have been spent finding new mates or helping genuine descendents. Obviously, any design feature that increases the probability that a human male will spend his time and resources raising his own children rather than those of another man would, all else equal, have a selective advantage over alternative designs. In their chapter, "The Man Who Mistook His Wife for a Chattel," Wilson and Daly propose that this constellation of selection pressures has created a male psyche that is "proprietary" about women's sexuality.

In essence, Wilson and Daly are arguing that some of the assumptions and rules of inference that evolved for reasoning about social exchange (see Cosmides & Tooby, this volume) are activated in the sexual domain as well, causing many men to view sexual access to a woman as the sort of thing a man has the right to use, monopolize, defend against trespass, modify, or dispose of—i.e., as a sort of "property." Their exploration of this dark side of the male mind reviews and organizes a broad array of cross-cultural data. Wilson and Daly are singularly well equipped to do this. The two of them have pursued what is arguably the most sustained and comprehensive empirical research program in the behavioral sciences into how selection has shaped human psychology and motivation. This has culminated in their landmark book *Homicide,* much of which deals with the conflict engendered by male motivation to gain access to or monopolize women's sexuality. Wilson, who has a degree in law as well as psychology, has been able to include evidence drawn from the comparative analysis of legal systems, as well as evidence from more standard ethnographic sources. Wilson and Daly's chapter demonstrates how a deep understanding of human psychology—and of the conflicts of interest that arise as a consequence of the design of this psychology—can illuminate the character and development of legal systems and other social institutions, even though these are the highly contingent products of a complex historical process.

5

Mate Preference Mechanisms: Consequences for Partner Choice and Intrasexual Competition

DAVID M. BUSS

Few things are more obvious than the fact that human behavior is selective and preferential. Foods rich in fat and sugar are preferred over those that are bitter or sour. Preferential avoidance of heights, snakes, darkness, and spiders—environmental hazards in human evolutionary history—are learned more readily than are fears of environmentally novel hazards such as cars or guns. Habitats offering resources and protection simultaneously, to take another example, are preferred over those lacking these attributes (Kaplan, this volume; Orians, 1980, Orians & Heerwagen, this volume). Perhaps nowhere are preferences more apparent than in human mating decisions. Nowhere do individuals prefer to mate with all members of the opposite sex equally (Buss, 1989a; Symons, 1979).

This chapter examines four fundamental premises: (a) human mate preference mechanisms are central psychological procedures that affect actual mating decisions, (b) mate preferences exert a powerful selection pressure on human intrasexual competition, (c) there exists a class of acts generated by each preference mechanism, and (d) the evolution of psychological mechanisms cannot be understood fully without identifying the class of acts generated by each psychological mechanism.

SELECTION PRESSURES CREATING MALE MATE PREFERENCES

Imagine a state in which human males had no mate preferences aside from species recognition and instead mated with females randomly. Under these conditions, males who happened to mate with females of ages falling outside the reproductive years would become no one's ancestors. Males who happened to mate with females of peak fertility, in contrast, would enjoy relatively high reproductive success. Over thousands of generations, this selection pressure would, unless constrained, fashion a psychological mechanism that inclined males to mate with females of high fertility over those of low fertility.

For males more than females, reproductive success is limited by sexual access to reproductively valuable or fertile mates (Symons, 1979; Trivers, 1972; Williams, 1975). Reproductive value refers to expected future reproduction—the extent to which persons of a given age and sex will contribute to the ancestry of future generations (Fisher, 1958). Fertility is defined as the probability of present reproduction for a given age and sex. Although cultures vary, human female fertility typically peaks in the early to middle twenties, while reproductive value peaks in the mid-teens (Thornhill & Thornhill, 1983). Both values show a sharp decrement with age.

Because male reproductive success in humans depends heavily on mating with reproductively capable females, selection over thousands of generations should favor those males who prefer to mate with reproductively capable females. Unfortunately for males seeking to solve this problem, reproductive value and fertility are not attributes that can be observed directly. This raises a crucial issue in the evolution of preference mechanisms: What *affordances* (information provided by a perceived object; Gibson, 1966) do human females yield that are correlated with reproductive capability at a sufficiently reliable level to provide a basis for male mate preferences?

Williams (1975) and Symons (1979) provide a partial answer to this question. Age of females is highly correlated with their reproductive capability—youthful females are generally more fecund than older females. But even age must be inferred, as it cannot be assessed directly (at least prior to the development of counting systems; note also that inhabitants of hunter-gatherer societies frequently do not know their age). Three classes of cues could provide, in principle, reliable guides to age and hence reproductive capability: (a) physical features (e.g., smooth, clear, and unblemished skin, lustrous hair, white teeth, absence of gray hair), (b) behavioral features (e.g., sprightly and graceful gait, high energy level, alacrity), and (c) reputation (i.e., knowledge gleaned from others regarding the age, health, condition, appearance, behavior, and prior sexual conduct of a female).

Consensual standards of attractiveness should evolve over generations to correspond to these affordances (Buss, 1987). The physical attractiveness of females, therefore, should occupy a central place in male mate preferences. This preference should be stronger for males than for females because (a) female reproductive success is not as limited by the problem of obtaining fertile mates, (b) there is less variance among males than among females in capacity to sire offspring across the lifespan (male fertility can remain high into the fifties, sixties, etc.), (c) male fertility, to the degree that it is important for female reproductive success, is less steeply age-graded than is female fertility, and (d) male fertility therefore cannot be assessed as accurately from physical appearance.

The reproductive capability of females, however, does not exhaust the preference possibilities that would afford males greater reproductive success. Another important consideration is the availability or access a particular male has to a given female's reproductive value. Observed cues or reputation suggesting that a female has diverted (or might divert) that availability to another male should be disfavored; those suggesting fidelity would be favored. One cue that appears to be afforded by reputation is chastity—the lack of prior experience in sexual intercourse. Still another consideration is nurturance—which could provide a cue to good mothering skills.

Males in our evolutionary past (prior to widespread birth control) who preferred chaste females would have enjoyed greater reproductive success through increased probability of paternity (Daly, Wilson, & Weghorst, 1982; Dickemann, 1981). Because maternity is never in doubt, there would not be analogous selective advantages for females to prefer chaste males as mates. This sexual asymmetry, however, would be compromised if prior sexual conduct of a male signaled diversion of resources away from the female and her offspring. To the degree that prior sexual experience by males provides this cue, females also should value chastity in a potential mate.

In sum, the selection pressures that have forged male mate preferences include those involving female reproductive capacity and a male's access to that capacity (see also Ellis, this volume). Because these capacities of females are not directly observable, selection has favored preferential attention to the cues that afford reliable information about them. Physical appearance, overt behavior, and reputation information are three classes of affordances that could provide the basis for the evolution of male mate preferences.

SELECTION PRESSURES CREATING FEMALE MATE PREFERENCES

Trivers's (1972) theory of parental investment and sexual selection proposes that females should seek to mate with males who show the ability and willingness to invest resources connected with parenting such as food, shelter, territory, and protection. These resources provide a selective advantage to females obtaining them because of (a) immediate material advantage to the female and her offspring, (b) increased reproductive advantage to offspring through acquired social and economic benefits, and (c) genetic reproductive advantage for the female and her offspring if variation in the qualities that lead to resource acquisition are partly heritable.

This female preference should arise only under certain ecological conditions. These include (a) where resources can be accrued, monopolized, and defended, (b) where males of the species show some nontrivial degree of parental investment or resource provisioning, (c) where males tend to control such resources, and (d) where male variance in ability or willingness to provide resources is sufficiently high. The hypothesis that females will mate preferentially with males bearing greater gifts, holding better territories, or displaying higher rank has been confirmed in many nonhuman species (Calter, 1967; Lack, 1940; Trivers, 1985).

What cues do males display that might afford a reliable basis for a female mate preference to evolve? External resources are likely to be more directly observable than internal female resources associated with reproductive value. One can directly observe territory, physical possessions, meat from the hunt, and other accoutrements of power, money, status, and prestige.

Nonetheless, humans often mate at ages before a man's potential resources are fully known. Therefore, females are often in the position of relying on cues that are only probabilistically associated with future resources. Two of the best known predictors of economic success in current human populations are sheer hard work (e.g., ambition and industriousness) and intelligence (Willerman, 1979). Assuming some continuity over human evolutionary history in these associations, females should evolve preferences for these correlates of resources more strongly than males.

The ability to acquire resources, however, does not necessarily yield information about a male's willingness to devote those resources to a particular female and her offspring. It has been speculated that expressions of love (Buss, 1987b) and kindness (Buss, 1989a) may provide reliable cues to a man's willingness to devote resources to a female and her offspring. At present, however, we lack good data on these hypotheses (but see Buss, 1987b; and Mellon, 1981 for analyses of love acts).

All of the cues associated with reproductively relevant resources for males and females, however, are amenable to manipulation. Deception and dissembling may be expected whenever reproductive advantage can be gained, on average, by doing so. This account of evolutionary selection pressures yields specific predictions about forms of male and female deception (Tooke & Camire, 1991). Specifically, because males value cues to reproductive capability and accessibility, females would be predicted to lie about their age, alter their appearance, and conceal prior sexual encounters. Because females value willingness and ability to devote resources, males would be expected to exaggerate their resource holdings, inflate perceptions of their willingness to commit, and feign love to induce a female to mate with them.

SELECTION PRESSURES EXERTED BY PREFERENCES ON TACTICS OF INTRASEXUAL COMPETITION

Mate preferences, as mechanisms affecting mate choice, may be regarded as only one component of human mating systems. Darwin's (1871) theory of sexual selection posited two processes: (a) selective choice exerted by members of one sex for members of the opposite sex possessing certain characteristics, and (b) intrasexual competition—the selection for characteristics that lead to greater success in competing with members of the same sex for access to members of the opposite sex.

Although these processes are typically examined separately, there is a powerful link between intersexual and intrasexual selection. Specifically, patterns of selective choice should influence tactics of intrasexual competition. Males should evolve over time to compete with each other most strongly to obtain those resources and display those qualities that females express in their selective choices. Females should evolve over time to compete with each other to display those cues that males express in their selective choices. In other words, mate preferences provide a potentially powerful selective force on the intrasexual component of human mating systems.

This chapter will examine two sets of predictions based on these conceptual connections. These predictions deal with: (a) tactics of human intrasexual competition to attract potential mates, and (b) tactics of intrasexual competition used to retain potential mates. Specific predictions can be generated from mate preferences about these components of human intrasexual competition.

Males should compete with each other in intrasexual mate competition by acquiring and displaying cues associated with the ability and willingness to provide resources. Females should compete with each other in intrasexual mate competition by displaying cues that males have evolved to use as indicators of reproductive capability and access to that capability.

A critical distinction in analyzing tactics of intrasexual competition is whether males and females are seeking temporary mating partners (e.g., brief affairs, one-night stands) or long-term mating or marriage partners. Displays of female fidelity, for example, are expected to be more effective in attracting long-term rather than short-term partners because of the paramount importance of paternity confidence in long-term mating bonds. Female displays of sexual openness or even promiscuity (e.g., sexy, revealing clothes) are expected to be more effective at attracting short-term mates. Although little is currently known, studies are underway to explore the importance of long-term versus short-term mating tactics (Buss & Schmitt, under review).

In the context of long-term mating strategies, mates gained must be retained to actualize the promise of reproductive effort. Effective retention tactics used by males should be those that fulfill the female's initial preference for ability and willingness to provide resources. Males who fail to deliver will risk losing an obtained mate to an intrasexual competitor. Effective female mate retention tactics should be those that correspond to a male's initial selection preferences—the ability and willingness to provide that male with access to high reproductive capacity. Females failing to deliver risk losing an obtained mate to an intrasexual competitor. It is noteworthy that female adultery and male failure to provide resources historically have been grounds for divorce in many cultures, while the reverse is far less frequent (Betzig, 1989; Daly & Wilson, 1983).

In summary, selection pressures over evolutionary time give rise to mechanisms, in this case preference mechanisms; as preference mechanisms evolve, they begin to exert selection pressure on other components of human mating. Preference mechanisms expressed by one sex should affect the patterns of intrasexual competition displayed by members of the opposite sex.

THE NATURE OF PREFERENCE MECHANISMS

Preference mechanisms, at the most general level, may be defined as psychological processes that incline or predispose organisms to selectively choose or reject stimuli in their environments. Preference mechanisms may be passive in the sense of rejecting or accepting objects that are provided, or they may be active in the sense of mobilizing behavior to seek out some objects and preemptively avoid others.

Food preferences provide an instructive example. Mechanisms associated with taste and smell incline organisms to be repulsed by certain foods, literally gagging, spitting, or vomiting out those that are inadvertently ingested. They also incline organisms to actively search for foods with properties that solve adaptive problems.

Mate preferences are presumed to operate in a similar manner. Women may reject some suitors while favoring others, or they may actively seek out those that are preferred. Mate preferences not only incline organisms to make choices; they also mobilize behavior in an active search for certain mates and an active avoidance of others (Perper & Weis, 1987).

A focus on preferences makes no presumptions about the conscious or unconscious status of the mechanisms. It is possible that men and women are aware of their preferences, yet not aware of their origins. People have no trouble expressing their views about which foods, paintings, or mates they prefer and which they find repulsive. The cliché "I don't understand modern art, but I know what I like" aptly captures this point.

EMPIRICAL FINDINGS ON MATE PREFERENCES

Major empirical findings on mate preferences, both within the United States (Buss, 1985, 1987a; Buss & Barnes, 1986) and across 33 countries (Buss, 1989a; Buss et al., 1990) have been documented elsewhere. This section will summarize the findings pertinent to the key selection pressures described earlier.

In the most extensive cross-cultural study conducted on mate preferences, Buss

(1989a, 1990) asked respondents from 33 countries located on six continents and five islands (total number equaled 10,047) to evaluate which characteristics they found most and least desirable in a potential mate. One procedure asked respondents to rate 18 characteristics on a scale ranging from 0 *(irrelevant or unimportant)* to 3 *(indispensable).* A second procedure had respondents rank 13 characteristics from 1 *(most desired in a potential mate)* to 13 *(least desired in a potential mate).* Included within the instruments were characteristics on which the central hypotheses were based: good financial prospects, good earning capacity, good looks, physical attractiveness, ambition and industriousness, and chastity (no previous experience in sexual intercourse). Respondents also were asked to state the desired age of a potential mate, relative to their own age.

These methods contain some clear limitations. First, self-reported data are subject to potential biases of self-presentation. Second, the possible range restriction inherent in the rating procedure may attenuate the magnitude of the sex differences obtained. Additional data sources (e.g., observer reports) are clearly needed to circumvent these limitations. Nonetheless, checks on the validity of these instruments suggest that we need not be pessimistic about the capacity of individuals to accurately report their mate preferences (Buss, 1989a).

As hypothesized, males placed greater value on relative youth and physical attractiveness in potential mates than did females. Although the absolute value placed on these mate characteristics varied across cultures, the sex differences remained relatively constant, transcending cultural variations. These cross-cultural findings provide powerful evidence to support the hypothesized selection pressure favoring male preference for females of relatively high fertility.

In contrast to these internationally consistent sex differences, these studies found tremendous cultural variability in (a) the absolute value placed on chastity (defined as "no prior experience in sexual intercourse"), and (b) the presence or absence of sex differences in preference for chaste mates (see Figure 5.1). In China, for example, both sexes valued chastity highly. In West Germany, chastity was not valued highly by either sex, but males nonetheless placed higher value on it than did females.

Of the 37 samples, 23 (62%) showed sex differences in the predicted direction, with males valuing chastity more than females. In the remaining 14 samples, no significant sex differences were found. In no samples did females prefer chastity in potential mates more than did males. These data, when contrasted with the relatively invariant sex differences found for youth and beauty, suggest that some preference mechanisms are highly sensitive to cultural, ecological, or mating conditions, while others transcend these variations in context.

With respect to cues correlated with resources, females in seven studies within the United States and within 36 of the 37 cross-cultural samples placed greater value on the earning capacity of a potential mate than did males (Buss, 1989a). For the characteristic of ambition-industriousness, 78% of the samples showed significant sex differences in the predicted direction, while one sample (South African Zulu) showed a significant reversal of this mate preference. Particulars of the Zulu culture involving heavy reliance on female labor for many physical tasks may account for this reversal (Buss, 1989a).

These findings speak to only the predicted sex differences in mate preferences. Are there mate preferences common to both sexes, whether predicted or not, that are highly valued? Across all 37 samples, two characteristics consistently emerged as

Chastity:

No Previous Sexual Intercourse

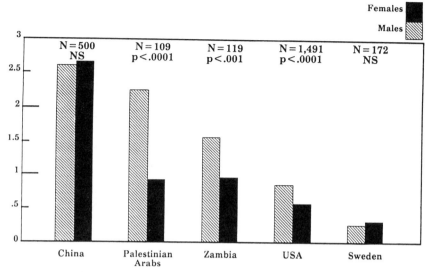

Figure 5.1 A representative sample of the value placed on chastity, defined as no previous experience in sexual intercourse. Subjects rated chastity, along with 17 other characteristics, on how important it was in a potential mate or marriage partner. A 3 represented *indispensable*, whereas a 0 represented *irrelevant or unimportant*. Figure is adapted from Buss (1989a) in the study of mate preferences in 37 cultures.

among the most highly valued mate characteristics for both sexes: (a) kind and understanding, and (b) intelligent.

As described above, kindness, particularly when directed toward a potential mate, may provide a powerful cue to the willingness of a man to devote resources to a woman and of a female to devote her reproductive resources to a particular man. The fact that both men and women worldwide place tremendous value on kindness warrants further study. Buss (1989b) speculates that kindness, as a relatively enduring personality characteristic, provides a powerful cue to a proclivity to cooperate rather than defect. Mating relationships, particularly in species that show biparental investment such as ours, are prototypical forms of reciprocal alliance formation. Two individuals who are typically unrelated to each other form a cooperative coalition. In reproductive terms, their success often hinges on the success of this coalition. The proclivity of another to cooperate, to reciprocate, and to engage in mutualism for common goals should be a central mate selection criterion for both sexes. Kindness and understanding are hypothesized to afford the most reliable cues to these proclivities.

Intelligence is a good predictor of income and so should be valued by females who seek males with good resource potential. The great value placed on intelligence by both sexes, however, warrants further study. One speculation is that "intelligence" represents a broad (and perhaps too vague) concept that subsumes a wide variety of adaptive capacities (cf. Barkow, 1989). These might include skills linked with hunting, tool making and use, and capacity to cope with changing circumstances. They may also include the ability to provide skillful parental investment and guard offspring from harm, and adeptness at handling aggressive conspecifics. More generally, intelligence could subsume a host of social skills involved in negotiating complex interpersonal arrangements such as hierarchies, intricate constellations of allies and foes, and extended kin networks (cf. Byrne & Whiten, 1988). These speculations require further study, but they suggest plausible reasons why both men and women would place great value on intelligence in a potential mate.

EFFECTS OF PREFERENCE MECHANISMS ON ACTUAL BEHAVIOR

Two sets of behavioral data are relevant to testing the potency of preference mechanisms in guiding actual behavior. The first set involves actual mate choice data—do males and females select mates who correspond to their preferences? The second set of relevant behavioral data involves patterns of intrasexual competition. Because of the causal link proposed earlier between intersexual and intrasexual selection, behavioral patterns of intrasexual competition should be predictable from mate preference mechanisms expressed by members of the opposite sex.

Before presenting the relevant empirical data corresponding to these two sets of behavioral predictions, several complexities should be noted that are likely to modify to some degree the direct effects of preferences on actual choice behavior. Preferences represent only one of the causal forces affecting actual mating decisions. Several other forces likely to be important are (a) preferences expressed by members of the opposite sex, (b) preferences expressed by parents and other family members, (c) effective sex ratio (ratio of eligible males to females), (d) particulars of the mating system (e.g., monogamous, polygynous, polyandrous, promiscuous), (e) qualities the selector brings to the potential mating (e.g., a female may reject a suitor who prefers her because he does not fulfill her preferences), and (f) inherent conflict between two or more preferences.

All these complexities are likely to attenuate the direct links between preferences and actual mating choices. Nonetheless, if the preference mechanisms proposed here have been fashioned by natural selection and carry the importance attached to them in this conceptual framework, then they must be reflected to some degree in actual mating decisions.

Two lines of evidence provide substantial support that they are. The first pertains to findings within the United States on cross-character assortment (Buss & Barnes, 1986). Because physical attractiveness is preferred more by males than by females, and earning power is preferred more by females than by males, males with earning power and females displaying beauty should be able to command the most desirable members of the opposite sex. Indeed, three studies have found that attractive females tend to marry males of high occupational status (Elder, 1969; Taylor & Glenn, 1976; Udry & Eckland, 1984). Physical attractiveness of a female turns out to be a stronger predictor of the occupational status of the man she will marry than other female characteristics such as socioeconomic status, intelligence, or education (Elder, 1969). This cross-character assortment between female attractiveness and male resources provides powerful support for the hypothesis that mate preferences are reflected in actual choice behavior.

A second line of evidence pertains to the relationship between age difference preferred between self and mate and the actual age difference between spouses at marriage. In a study of 37 samples worldwide, Buss (1989a) found that males prefer mates who are 2.66 years younger than they are on average, whereas females prefer mates who are 3.42 years older on average. Demographic data on 28 of these samples revealed that grooms were indeed older than their brides in all countries, and the mean age difference across countries was 2.99 years—approximately the average of the age difference preferred by males and females. In addition, variations across samples in preferred age differences were highly correlated with the actual age differences between brides and grooms. Again, this evidence supports the hypothesis that mate preferences affect actual mating decisions.

IMPLICATIONS FOR INTRASEXUAL COMPETITION: TACTICS FOR ATTRACTING AND RETAINING MATES

Mate preferences should have implications for the tactics deployed by members of the opposite sex in intrasexual competition. This general hypothesis has been examined in two contexts: (a) tactics used to attract mates initially, and (b) tactics used to retain acquired mates. Within each of these contexts, the implications of consensually preferred mate characteristics and sex-differentiated mate characteristics were examined.

The studies on attracting and retaining mates were all guided by the evolution-based hypotheses and predictions described earlier. Specifically, the following general and specific hypotheses were advanced about tactics of intrasexual competitions: (a) Patterns of intrasexual competition will be influenced by patterns of intersexual selection (e.g., mate preferences held by members of one sex can be used to predict the content of what members of the opposite sex compete to embody); (b) since physical appearance is a salient part of male mate preferences, providing a powerful cue to fertility or reproductive value, females will compete with one another to appear physically attractive; (c) since females value in mates their willingness and ability to devote external resources to them, providing important forms of protection and parental investment, males should compete with one another to obtain and display cues to their ability and willingness to devote resources; and (d) these predictions should obtain within the contexts of intrasexual competition to attract mates and intrasexual competition to retain mates.

The research methodologies used generally employed the following sequences of steps.

1. *Act nominations.* This procedure involves asking a large number of subjects to recall acts or behaviors that they have witnessed that were performed in order to achieve a specific goal such as attracting or retaining a mate. For example, the instructional set used to obtain act nominations for tactics of mate attraction was as follows:

> In this study, we are interested in the specific acts or tactics that people use to make themselves more attractive to members of the opposite sex. Please think of three males (females) you know well (this could include yourself). Now write down five things these males do to make themselves more attractive to members of the opposite sex. These could include: 1) actions to make themselves more attractive *relative to other males (females);* 2) actions to impress members of the opposite sex; and 3) things to increase their overall desirability or attractiveness to members of the opposite sex.

The goal of this first initial inquiry is to obtain a large and diverse set of acts that fall within a given domain such as mate attraction tactics and mate retention tactics.

By using a large number of members of the culture to obtain these nominations, it was expected that biases that might occur from purely "investigator-generated" acts would be minimized. This procedure generally produces a large number of acts, in addition to those that are directly relevant to the evolution-based hypotheses, thus generating in the "context of discovery" what the evolution-based hypotheses fail to derive deductively.

2. *Classification of acts into tactics.* This second step in the research program attempts to identify clusters of acts that are reasonably homogeneous. Toward this end, several independently working psychologists group those acts that they believe "go together" in the sense of being exemplars of essentially the same tactic. Typically, a majority consensus criterion is imposed for considering an act to belong to a tactic. These act-clusters or tactics then form the basis for further research on performance frequencies and effectiveness assessments. Sample tactics and acts are shown in Tables 5.1 and 5.2.

3. *Assessments of performance frequency.* Obtaining accurate assessments of the frequency with which each act and tactic is performed poses numerous difficult measurement issues. Many such acts are private, for example, performed in intimate contexts beyond the purview of outside observers. Indeed, some may be known only to the performer and the target of the performance. Another measurement problem is posed by time. These acts are often widely dispersed throughout the temporal stream of behavior, intercalated among numerous other acts that are irrelevant to the target tactic of the study. A third problem is posed by the inherent ambiguity in act interpretation. Was the act "He put his arm around her at the party when another man

Table 5.1   Sample Tactics and Acts for Attracting Mates

Tactics	Acts
Display resources	He flashed a lot of money to impress her.
	She drove an expensive car.
Brag about resources	He told people how important he was at work.
	She bragged about her accomplishments.
Wear sexy clothes	He wore sexy clothes.
	She wore skimpy clothes to impress guys.
Act provocative	He walked in a sexy manner.
	She acted sexy to interest him.
Flirt	He flirted verbally and visually.
	She gave encouraging glances to guys.
Wear makeup	He used makeup to accentuate his good looks.
	She wore facial makeup.
Keep clean/groomed	He washed his hair every day.
	She brushed her teeth several times a day.
Generally alter appearance	He went on a diet to improve his figure.
	She laid out in the sun to get a tan.
Wear stylish clothes	He wore stylish, fashionable clothes.
	She bought expensive clothes.
Act coy	He played hard to get.
	She tried to appear indifferent to the guy she really liked.

Table 5.2 Sample Tactics and Acts for Mate Retention

Tactics	Acts
Vigilance	He called her at unexpected times to see who she was with. He called her to make sure she was where she said she would be. He had his friends check up on her.
Concealment of mate	He did not take her to the party where other males would be present. He refused to introduce her to his same-sex friends. He took her away from the gathering where other males were present.
Monopolization of mate's time	He spent all his free time with her so that she could not meet anyone else. He insisted that she stay at home with him rather than going out. He monopolized her time at the social gathering.
Punishment for mate's threat of infidelity	He became angry when she flirted too much. He ignored his mate when she started flirting with others. He threatened to break up if she ever cheated on him.
Emotional manipulation	He cried when she said she might go out with someone else. He made her feel guilty about talking to other men. He told her he would "die" if she ever left.
Derogation of competitors	He cut down the appearance of other males. He started a bad rumor about another male. He cut down the other guy's strength.
Violence	He picked a fight with the guy who was interested in her. He got his friends to beat up the guy who was interested in her. He hit the guy who made a pass at her.

approached" performed as an act of "mate guarding," or was it simply a display of affection for his mate with no conscious or unconscious intention of mate guarding. A fourth problem is that many acts have a wide range of consequences, other than those for which the act was intended.

No single methodology can solve all of these measurement difficulties—problems that to a large extent are an intrinsic part of the phenomena under study. The following procedures, however, represent a first attempt to circumvent some of them so as to obtain reasonable gauges of performance frequency. Based on earlier steps of act nomination, act reports were formulated. These consist of retrospective reports of one's own act performance (self-reported version) or of another's act performance (observer-reported version). A specified time period was designated (e.g., one month, three months) so that a reasonable opportunity was given for the act to occur. Some studies used act reports based on self-reporting; others were based on reports by a close friend; still others were based on reports by a spouse of the performer. Although each of these data sources contains certain problems (e.g., memorial distortions; self-presentation effects), results that occur *across* data sources can be viewed with greater confidence.

4. *Evaluation of act effectiveness.* A fourth step in this program of research, applied to each domain (tactics of attraction, tactics of retention), is to obtain estimates of how effective each act and tactic is at accomplishing its goal. Ideally, this would involve first assessing the performance of each of more than 100 acts in large samples and then following up each performer over extended periods of time to see whether the goal was obtained (e.g., whether mate was retained, or mate was drawn into a mating relationship). Clearly, this ideal would be extraordinarily difficult to approach.

As an intermediate solution, judgments were obtained from panels of members of the culture with respect to how effective each act would be likely to be at accomplishing the goal. A sample instructional set in the context of mate retention is as follows:

> Below are listed acts that someone might perform to keep or retain his (her) mate, and prevent her (him) from leaving him (her) for another man (woman). In this study we are interested in your judgments about *how effective* each act would be in *keeping his (her) mate and preventing her (him) from seeing other males (females)*. Please read each act carefully, and think about its likely consequences. Then rate each act on how effective it is likely to be in keeping his (her) mate and preventing her (him) from seeing other males (females).
>
> Use this 7-point scale: a "7" means that you feel the act will be *very effective* in keeping his (her) mate and preventing her (him) from seeing other males (females). A "1" means that you feel the act will be *not very effective* in keeping his (her) mate and preventing her (him) from seeing other males (females). A "4" means that you feel the act will be *moderately effective* in keeping his (her) mate and preventing her (him) from seeing other males (females). Use intermediate numbers for intermediate judgments.

Although this procedure is clearly limited in certain ways (e.g., people may have false beliefs about what acts are effective at retaining mates, or they simply may not know), it provides a first approximation or estimation about the relative effectiveness of each act.

In a series of studies on judgments of the most effective mate attraction tactics, Buss (1988a) found that the two most effective tactics for both males and females in the United States were displaying humor (e.g., made up jokes, displayed a good sense of humor) and acting nice (e.g., was sympathetic with his/her troubles, showed good manner, offered help). These tactics showed no sex differences in either judgments of effectiveness or actual performance in two samples of subjects. The acting nice tactic corresponds closely to the most highly valued mate characteristics across cultures—kindness and understanding.

Because there exist powerful sex differences in mate preferences, it was predicted that tactics of mate attraction should show corresponding sex differences (Buss, 1988a). Thus, it was predicted that males would display tangible resources in intrasexual competition more than do females because females value this mate characteristic more than do males. Similarly, it was predicted that females would enhance their appearance more than males in intrasexual competition because physical attractiveness is a more important mate characteristic for males than for females.

These predictions were tested in two samples of American subjects. The predicted sex differences were obtained in both samples. Males used resource display (e.g., "He flashed a lot of money to impress her," "He drove an expensive car") and boasting about resources (e.g., "He told people how important he was at work," "He bragged about his accomplishments") more than did females in both samples. These tactics were also judged, in a separate study, to be more effective when used by males than when used by females. Females in both samples used a variety of appearance enhancement tactics more than males, including wearing makeup, keeping clean and groomed, altering appearance (e.g., getting a tan, going on a diet), wearing stylish clothes, and wearing jewelry. These tactics were also judged in a separate study to be more effective when used by females than when used by males.

In sum, tactics of mate attraction, both those common to males and females and those that are differentiated by sex, are predictable in at least one society from knowledge about mate preferences expressed by members of the opposite sex. There is no reason to expect that replicating this study in other cultures would provide substantively different results. Until such replication occurs, however, these conclusions must remain provisional.

In a separate series of studies, a similar set of predictions was made for tactics used to retain one's mate (Buss, 1988b). In a study of judgments of tactic effectiveness, emphasizing love and care emerged by far as the most effective tactic for retaining an acquired mate. This tactic subsumed acts such as "He told her that he loved her," "She went out of her way to be nice, kind, and caring," "He was helpful when she really needed it," and "She displayed greater affection for him." This tactic seems to correspond to the most highly valued mate characteristics, kindness and understanding, found in mate preference studies.

Do males and females differ in the predicted ways in the tactics they used to retain mates? A study of persons involved in steady heterosexual relationships found that males tended to use the resource display tactic more than females, including the acts of "Spending a lot of money on her," "buying her an expensive gift," "taking her out to a nice restaurant," and "buying her jewelry." Males also used the tactics of intrasexual threats (e.g., "threatening to hit the guy who was 'making moves on her'"), violence (e.g., "He hit the guy who made a pass at her"), and concealment of mate (e.g., "He took her away from a gathering where other males would be present") more than did females.

Only two mate retention tactics showed significantly greater female than male performance frequencies. These were enhancing appearance (e.g., making up her face to look nice, dressing nicely to maintain his interest, making herself "extra attractive" for him) and threatening infidelity (e.g., "She showed interest in other men to make him angry," "She flirted with another man to make him jealous," "She talked to another man at the party to make him jealous"). This latter tactic is particularly interesting in light of the Daly et al. (1982) proposal that sexual jealousy is a male adaptation that serves to mobilize behavior to increase probability of paternity (see also Symons, 1979). The mate retention results suggest that females are manipulating the male jealousy mechanism in the service of retaining their acquired mate.

In sum, these behavioral data largely support the hypothesis that there is a causal link between mate preferences expressed by one sex and tactics of intrasexual competition displayed by members of the opposite sex. They suggest that mate preferences are not mere whimsical fancies, but instead have potent consequences for the types of acts performed by the opposite sex.

ORIGINS OF MATE PREFERENCES: ONTOGENETIC AND SOCIOCULTURAL SPECULATIONS

Although good data are lacking on the origins of preferences mechanisms, several speculations may be offered and the alternatives outlined. One possibility is that mate preferences arise much in the same manner as food preferences—some appear to be "hard-wired" and emerge invariantly early in life, regardless of varying environmental conditions; others appear to be highly susceptible to the particulars of the sociocultural milieu (Rozin, 1976).

In the domain of food preferences, for example, infants across all cultures seem to prefer foods laden with sugar and fat, while rejecting those that are bitter or sour. No special training or socialization appears to be needed for these preferences to emerge, and the likely evolutionary origins of these invariant preferences is apparent. Other food preferences, for example those that involve spicy or bland dishes, appear to vary considerably across cultures (e.g., Mexican versus British) and require differential exposure during ontogeny to develop.

Mate preferences may also differ in the degree to which they are "hard-wired" in the sense of requiring little or no socialization to develop versus requiring extensive socialization during ontogeny. There is some fascinating evidence in studies of infants that what is considered attractive in a female face (as gauged by time spent gazing) (a) emerges in infancy, (b) shows consensus among infants, and (c) corresponds remarkably to what adults find attractive (Langois, Roggma, Casey, Ritter, Reiser-Danner, & Jenkins, 1987). Juxtaposed with the relatively invariant sex differences found across cultures in valuation of physical attractiveness, these results may suggest that this mate preference may be of the "hard-wired" variety.

In contrast, valuation of chastity shows greater cross-cultural variability than any mate characteristic yet examined. This variability occurs both with respect to the absolute value placed on chastity, and on the presence or absence of sex differences. In spite of this cross-cultural variability, the sex differences in the value placed on chastity were not infinitely plastic. Whereas in nearly two-thirds of the societies studied men valued chastity more than women, in no society was chastity valued significantly more by women than by men. Chastity differs from physical attractiveness in an important way—it is less directly observable. Even physical tests of female virginity are unreliable due to variations in the morphology of the hymen, rupture due to nonsexual causes, and deliberate alteration (Dickemann, 1981).

Sexual selection should favor preference mechanisms for cues that are reliably associated with characteristics that have fitness advantage for the mate selector. Where cues are not directly observable or cannot be reliably assessed, as in the case of chastity, it is difficult to imagine how specific preference mechanisms of the "hard-wired" sort could have been fashioned by natural selection. These considerations, of course, do not preclude selection for a somewhat more general mechanism such as sexual jealousy (Daly et al., 1982), which promotes heightened concern about females having sexual contact with other males, either prior to or after mate choice.

Could preference mechanisms be due entirely to socialization during ontogeny? On an evolutionary account, one would predict that parents would socialize their sons to prefer mates high in fertility of reproductive value and their daughters to prefer mates rich in resources. That is, parental preferences for the mate of their children should roughly coincide with those that their children would be expected to hold from an evolutionary account, although parents and their offspring may conflict and diverge with respect to some decision rules (e.g., Trivers, 1974). In this case, the evolutionary account would focus on mechanisms that have evolved in parents to socialize their children in predictable ways. Research is needed to identify the causal origins of mate preferences during ontogeny as well as the preferences that parents express for the mates of their sons and daughters.

A second ontogenetic issue concerns how intrasexual competition tactics come to be so closely calibrated to the mate preferences expressed by members of the opposite sex. There are at least two general possibilities: (a) tactics of intrasexual competition may be calibrated during ontogeny to opposite sex preferences as a result of learning through observation of what is effective, through random trial and error (general-purpose learning mechanism account), or through directed trial and error, or (b) selection has directly produced over evolutionary time mechanisms that produce tactics of intrasexual competition that have been favored because of the reproductive advantage accruing to the tactician; that is, the tactics themselves are evolved solutions to this mating problem.

The "learning during ontogeny" account has at least three variants. One is that the individual competitor, through trial and error, gets rewarded for some tactics by success (obtaining desired mate) and get punished for some tactics by failure (gets rejected by desired mate). Over ontogenetic time, individuals learn which tactics are effective and which are ineffective with members of the opposite sex. A second learning account would invoke agents of socialization (e.g., parents, siblings, uncles) who inculcate or transmit effective tactics of intrasexual mate competition. Both of these accounts invoke general-learning processes to account for the ontogeny of specific psychological mechanisms that produce mate competition tactics that are calibrated to preferences expressed by members of the opposite sex. Domain-general learning mechanisms, however, are unlikely to evolve in contexts where an important and specific adaptive problem must be solved (Cosmides & Tooby, 1987, 1989; Tooby & Cosmides, 1989).

A third learning account would involve a social learning mechanism such as "imitate those who are successful at obtaining mates." Thus, males in this culture might look to rock stars, movie stars, or other prominent males who seem to be successful and use them as models for their own behavior. Females might look to females capable of commanding high status males and emulate their apparent strategies. This social learning account is somewhat more specific than the "trial and error" account, but is still more general than the following account, which invokes direct selection for successful strategies.

Sexual selection could have directly fashioned, over the course of thousands of generations, psychological mechanisms that produce effective mate competition tactics. These mechanisms, in the present case, would be those that incline males to acquire and display resources and incline females to enhance their physical appearance and signals of fidelity because these are characteristics desired by members of the opposite sex. Our current state of knowledge does not allow us to differentiate among these alternative causal accounts of the origins of the closed calibration between mate preferences expressed by one sex and tactics of intrasexual competition displayed by members of the opposite sex.

CONCLUSION

In this chapter, I have argued that preference mechanisms are psychological processes central to human mating behavior. Hypotheses about specific selection pressures on human males and females were outlined, and the relevant empirical data were examined. Conceptually, mate preferences carry profound implications for human mating because they can affect two important classes of behavior: (a) actual mating decisions, and (b) patterns of intrasexual competition deployed by members of the opposite sex.

Empirical data from a series of studies carried out within the United States and across cultures documented support for the selection pressures proposed. Males and females show consistent sex differences in mate preferences across cultures in two major clusters: (a) females prefer mates with resources and attributes that are correlated with resource acquisition more than males do, and (b) males prefer youth and physical attractiveness, two correlates of reproductive capacity, more than females do. A somewhat less consistent finding is that males prefer chastity in potential mates more than females.

These clusters of sex differences in mate preferences apparently affect actual mating decisions in at least two contexts. First, physically attractive females were found in three separate studies to command males with high resource potential. Second, actual age differences between brides and grooms correspond closely to preferences expressed by each sex. These important findings suggest that we need not be pessimistic about the capacity of individuals to accurately report preferences that translate into actual mating decisions.

The hypothesized links between mate preferences and tactics of human intrasexual competition were examined in two series of studies within the United States. Males were found to use resource display in tactics used to attract mates as well as in tactics used to retain acquired mates. Females were found to use appearance enhancement in tactics used to attract as well as to retain mates. Consensual mate preferences for kind and understanding mates were also reflected in judgments of the most effective tactics of intrasexual competition for both sexes—displaying kindness, help, sympathy, and support.

Starting with several relatively simple hypotheses about sexually differentiated selection pressures, a host of empirical findings were discovered in different components of the human mating system. These empirical findings were robust in spite of the numerous complexities associated with other forces that could potentially attenuate the direct effects of mate preferences on actual mating behavior.

These findings point to the profound importance of two related domains of empirical work. The first is identification of psychological mechanisms in the study of evolution and human behavior (Cosmides & Tooby, 1987; Symons, 1987, 1989). Preferences are proposed here as basic psychological mechanisms central to human mating behavior. Further work is needed (a) to provide an adequate description of these basic psychological mechanisms, and (b) to identify the ontogeny of mate preferences, about which practically nothing is known.

The second is identification of classes of acts to which the psychological mechanisms correspond and in the service of which they presumably have evolved. Cognitive psychologists have tended to focus on psychological mechanisms without identifying their corresponding classes of acts. Evolutionary anthropologists have tended to focus on classes of acts without identifying the psychological mechanisms that give rise to those acts (Symons, 1989). Perhaps it is time to merge the respective strengths of these two disciplines by developing an evolutionary psychology that integrates both.

ACKNOWLEDGMENTS

Special thanks go to Jerry Barkow, Leda Cosmides, Bruce Ellis, and John Tooby for thoughtful comments on earlier versions of this chapter.

REFERENCES

Barkow, J. H. (1989). *Darwin, sex, and status: Biological approaches to mind and culture.* Toronto: University of Toronto Press.

Betzig, L. (1989). Causes of conjugal dissolution: A cross-cultural study. *Current Anthropology, 30,* 654–675.

Buss, D. M. (1985). Human mate selection. *American Scientist, 73,* 47–51.

Buss, D. M. (1987a). Sex differences in human mate selection criteria: An evolutionary perspective. In C. Crawford, M. Smith, & D. Krebs (Eds.), *Sociobiology and psychology: Ideas, issues, and applications* (335–351). Hillsdale, NJ: Erlbaum.

Buss, D. M. (1987b). Love acts: The evolutionary biology of love. In R. Sternberg & M. Barnes (Eds.), *The psychology of love.* New Haven, CT: Yale University Press.

Buss, D. M. (1988a). The evolution of human intrasexual competition: Tactics of mate attraction. *Journal of Personality and Social Psychology, 54,* 616–628.

Buss, D. M. (1988b). From vigilance to violence: Tactics of mate retention in American undergraduates. *Ethology and Sociobiology, 9,* 291–317.

Buss, D. M. (1989a). Sex differences in human mate preferences: Evolutionary hypotheses tested in 37 cultures. *Behavioral and Brain Sciences, 12,* 1–49.

Buss, D. M. (1989b, June). *A strategic theory of trait usage: Personality and the adaptive landscape.* Paper presented to the Invited Workshop on Personality Language, University of Groningen, Department of Psychology, Groningen, The Netherlands.

Buss, D. M., et al. (1990). International preferences in selecting mates: A study of 37 societies. *Journal of Cross-Cultural Psychology, 21,* 5–47.

Buss, D. M., & Barnes, M. F. (1986). Preferences in human mate selection. *Journal of Personality and Social Psychology, 50,* 559–570.

Buss, D. M. & Schmitt, D. (under review). Sexual strategies theory: A contextual evolutionary analysis of human mating.

Byrne, R. & Whiten, A. (Eds.) (1988). *Machiavellian intelligence.* Oxford: Clarendon Press.

Calter, C. (1967). Breeding behavior of the roadrunner *Geococcyx Californiaus. Aux,* 84, 577–598.

Cosmides, L., & Tooby, J. (1987). From evolution to behavior: Evolutionary psychology as the missing link. In J. Dupre (Ed.), *The latest on the best: Essays on evolution and optimality* (pp. 277–306). Cambridge: MIT Press.

Cosmides, L., & Tooby, J. (1989). Evolutionary psychology and the generation of culture, part II. Case study: A computational theory of social exchange. *Ethology and Sociobiology, 10,* 51–97.

Daly, M., & Wilson, M. (1983). *Sex, evolution, and behavior.* Boston: Willard Grant Press.

Daly, M., Wilson, M., & Weghorst, S. J. (1982). Male sexual jealousy. *Ethology and Sociobiology, 3,* 11–27.

Darwin, C. (1871). *The descent of man and selection in relation to sex.* London: Murray.

Dickemann, M. (1981). Paternal confidence and dowry competition: A biocultural analysis of purdah. In R. D. Alexander & D. W. Tinkle (Eds.), *Natural selection and social behavior* (pp. 417–438). New York: Chiron Press.

Elder, G. H., Jr. (1969). Appearance and education in marriage mobility. *American Sociological Review, 34,* 519–533.

Fisher, R. A. (1958). *The genetical theory of natural selection.* Oxford: Clarendon Press.

Gibson, S. S. (1966). *The senses considered as perceptual systems.* New York: Houghton Mifflin.

Lack, D. (1940). Pair formation in birds. *Condor. 42,* 269–286.

Langois, J., Roggma, L. A., Casey, R. J., Ritter, J. M., Rieser-Danner, L. A., & Jenkins, V. Y. (1987). Infant preferences for attractive features: Rudiments of a stereotype? *Developmental Psychology, 23,* 363–369.

Mellon, L. W. (1981). *The evolution of love.* San Francisco: W. H. Freeman.

Orians, G. (1980). Habitat selection: General theory and applications to human behavior. In J. S. Lockard (Ed.), *The evolution of human social behavior.* Chicago: Elsevier.

Perper, T., & Weis, D. L. (1987). Proceptive and rejective strategies of U.S. and Canadian college women. *The Journal of Sex Research, 23,* 455–480.

Rozin, P. (1976). Psychological and cultural determinants of food choice. In T. Silverstone (Ed.), *Appetite and food intake* (pp. 286–312). Berlin: Dahlem Konferenzen.

Symons, D. (1979). *The evolution of human sexuality.* New York: Oxford University Press.

Symons, D. (1987). If we're all Darwinians, what's the fuss about? In C. Crawford, M. Smith, & D. Krebs (Eds.), *Sociobiology and psychology: Ideas, issues, and applications.* Hillsdale, NJ: Erlbaum.

Symons, D. (1989). A critique of Darwinian anthropology. *Ethology and Sociobiology, 10,* 131–144.

Taylor, P. A., & Glenn, N. D. (1976). The utility of education and attractiveness for females' status attainment through marriage. *American Sociological Review, 41,* 484–498.

Thornhill, R., & Thornhill, N. W. (1983). Human rape: An evolutionary analysis. *Ethology and Sociobiology, 4,* 63–99.

Tooby, J., & Cosmides, L. (1989). Evolutionary psychology and the generation of culture, part I: Theoretical considerations. *Ethology and Sociobiology, 10,* 29–49.

Tooke, W. & Camire, L. (1991). Patterns of deception in intersexual and intrasexual mating strategies. *Ethology and Sociobiology, 12.*

Trivers, R. L. (1972). Parental investment and sexual selection. In B. Campbell (Ed.), *Sexual selection and the descent of man* (pp. 1871–1971). Chicago: Aldine.

Trivers, R. (1974). Parent-offspring conflict. *American Zoologist, 14,* 249–264.

Trivers, R. (1985). *Social evolution.* Menlo Park, LA: Benjamin/Cummings.

Udry, J. R., & Eckland, B. K. (1984). Benefits of being attractive: Differential payoffs for men and women. *Psychological Reports, 54,* 47–56.

Willerman, L. (1979). *The psychology of individual and group differences.* New York: Freeman.

Williams, G. C. (1975). *Sex and evolution.* Princeton, NJ: University Press.

6

The Evolution of Sexual Attraction:
Evaluative Mechanisms in Women

BRUCE J. ELLIS

> For most sexually reproducing species all conspecifics of the other sex are not equally valuable as mates: that is, they differ in "mate value." In many species selection has produced mechanisms to detect potential mates of high mate value. In other words, just as the taste of fruit varies with food value, in a natural setting, sexual attractiveness varies with mate value.
>
> DONALD SYMONS

What do women find attractive in men? Many writers who have addressed this issue have concluded that female preferences are so diverse and idiosyncratic as to defy systematic explanation. I will argue, however, that general principles guiding female mate preferences can be discerned at the appropriate level of abstraction and that the evolution-based concept of "mate value" (Symons 1987a) provides a useful heuristic in this endeavor.

Men differ in "mate value." In reproductive terms, they are not equally valuable to women as mates (Symons 1987a). Consider, for example, a woman who can choose between two husbands, A and B. Husband A is young, healthy, strong, successful, well liked, respected by his peers, and willing and able to protect and provide for her and her children; Husband B is old, weak, diseased, subordinate to other men, and unwilling and unable to protect and provide for her and her children. If she can raise more viable children with Husband A than Husband B, then his "mate value" can be said to be higher. Over evolutionary time, ancestral females who had psychological mechanisms that caused them to find males of high mate value more sexually attractive than males of low mate value, and acted on this attraction, would have outreproduced females with opposite tastes. This differential reproduction would continue until such mechanisms became universal and species-typical in women.

This logic leads one to expect that a man's sexual attractiveness to women will be a function of traits that were correlated with high mate value in our natural environment: the environment of a Pleistocene hunter-gatherer. Natural selection should have designed evaluative psychological mechanisms (information-processing rules or algorithms) in women that assess such traits and give rise to sexual and romantic attraction in response to them. In this chapter I review the psychological literature on male sexual attractiveness in order to see whether women find traits that would have signaled high mate value in our natural environment attractive in men.

SELECTION PRESSURES

The crucial question is, What traits would have been correlated with high male mate value in our natural environment? Three possible answers are as follows:

1. *The willingness and ability of a man to provide for a woman and her children.* Unlike males from most other mammalian species, who invest little in provisioning mates and offspring, human males can and do provide valuable economic and nutritional assistance to supplement what women can provide for themselves and their children. To the extent that males in the Pleistocene differed in their propensity and ability to provision their mates and children, and to the extent that this variation was signaled by observable cues, selection would have shaped female choice to favor males who displayed such cues.

2. *The willingness and ability of a man to protect a woman and her children.* Because of their smaller stature and lesser strength, women and especially children are potential victims of violence from both humans and nonhuman predators. One valuable kind of assistance males can offer their mates is protection from the negative acts of others, as well as from predation. To the extent that males in the Pleistocene differed in their propensity and ability to protect their mates and their children, and to the extent that this variation was signaled by observable cues, selection would have shaped female choice to favor males who displayed such cues.

3. *The willingness and ability of a man to engage in direct parenting activities such as teaching, nurturing, and providing social support and opportunities.* Many other acts, aside from protection and provisioning, contribute to the well-being of one's children and their eventual success in reproducing. Providing one's children with knowledge and skills, intervening on their behalf in situations of social conflict, and generally shaping conditions in ways that facilitate their health, growth, and success are dimensions of male behavior that would have had a powerful impact on a woman's reproductive success. To the extent that males in the Pleistocene differed in their propensity and ability to nurture their children, and to the extent that this variation was signaled by observable cues, selection would have shaped female choice to favor males who displayed such cues.

STATUS

Status refers to an individual's relative position in a social group; it is a measure of where one stands among one's peers and competitors. Even in hunting and gathering societies, status variations are substantial (Betzig, 1986; Lee, 1979). In general, the higher a male is in status (i.e., the higher the level of esteem and influence accorded to him by others), the greater his ability to control resources across many situations (Stone, 1989). Since control of positional resources is both a sign and a reward of status, natural selection could be expected to have favored evaluative mechanisms in women designed to detect and prefer high-status men. Forming mateships with such men could greatly enhance a woman's survival and reproductive potential through (a) elevation of her own social status, (b) immediate material and nutritional benefits, and (c) long-term access to social and economic resources. Thus, signs of current high status or future status-accruing abilities should significantly enhance female perceptions of male attractiveness.

Economic Status

The importance of male status to female perceptions of sexual attractiveness is illustrated by the lives of English tramps as described by George Orwell in *Down and Out in Paris and London*. These men lived a near sexless existence, not by choice, but by virtue of their social position: They were at the very bottom of society and had almost nothing to offer females. That American men who marry in a given year earn about 50% more money than men of the same age who do not is probably due in part to female choice for male resources (Trivers, 1985; p. 331).

Many studies of female mate preferences have focused on the relative importance women place on a man's status versus his physical attractiveness. Ford and Beach (1951), in a cross-cultural survey of sexual patterns in nearly 200 small nonurban societies drawn from the Human Relations Area Files (a collection of ethnographic materials), document dramatic variations in cultural standards of sexual attractiveness, especially along dimensions of body weight and ornamentation. Yet, "one very interesting generalization is that in most societies the physical beauty of the female receives more explicit consideration than does the handsomeness of the male. The attractiveness of a man usually depends predominantly upon his skills and prowess rather than upon his physical appearance" (p. 94). Thirty years later Gregersen (1982) extended and updated their account to include almost 300 societies, mostly from nonurban, non-Western cultures. On this subject, Gregersen's conclusion echoes his predecessors': "One generalization that can be made is that men are usually aroused more than women by physical appearance. This would seem to be true whatever sexual orientation is involved. For women the world over, male attractiveness is bound up with social status, or skills, strength, bravery, prowess, and similar qualities" (p. 186).

Western empirical studies that have investigated the relationship between status, sex, attractiveness, and physical appearance have generally confirmed the conclusions reached by both Gregersen and Ford and Beach. In a content analysis of 800 advertisements in the personals column of a national tabloid, Harrison and Saeed (1977) found that the three qualities women most often sought in men were, in descending order, sincerity (expressing concern about the potential partner's motives), age (wanting someone who was older), and financial security. Women were more than twice as likely as men to seek each one of these qualities, and women placed far more emphasis on each of these traits than on physical attractiveness. Conversely, men were more than three times as likely as women to seek "good looks." Cross-character assortment ("coupling that is based on congruent elevation of different, but similarly valued, characteristics" [Buss & Barnes, 1986; p. 560]) occurred between the stated aspirations of good-looking women and well-to-do men, a common marital pattern in real life as well: Physically attractive women are more able than less attractive women to parlay their assets into marriage with high-status men (Buss, 1987; Udry & Eckland, 1984).

The relative effects of physical attributes and socioeconomic status (SES) on female perceptions of male attractiveness have been investigated by Green, Buchanan, and Heuer (1984) and Townsend and Levy (1990a). Green et al. reviewed the dating choices of new members of a commercial dating service in Washington, DC, the Georgetown Connection. The members read profile sheets on target persons and decided—on the basis of a photograph, open-ended statements about his or her goals and interests, and demographic information (e.g., age, religion, occupation)—whether or not they wished to date the person. In other words, target persons were separated, through selection, into two categories: "winners" (those who were chosen to be a date) and "losers" (those who were not). For male targets, the strongest predictor of winning was higher status. Higher physical attractiveness was also a significant predictor. For female targets, higher physical attractiveness was the only significant predictor of winning. Joan Hendricks, president of the Georgetown Connection, commented: "Women really read over our profile forms, guys just look at the pictures" ("New Mating Game," 1986).

Various studies of dating behavior indicate that the physical attractiveness of a potential partner is very important to both sexes (Byrne, Ervin, & Lamberth, 1970; Walster, Aronson, Abrahams, & Rottman, 1966), especially in the initial phases of courtship. However, status plays an important role in early courtship as well. Townsend and Levy (1990a) investigated the relative importance of status and physical attractiveness at six levels of romantic involvement, which ranged from having a cup of coffee together to willingness to marry. Photographs of people of high, medium, and low attractiveness were paired with three levels of occupational status and income. College students viewed the different portrayals and indicated their willingness to engage in relationships of varying levels of sexual intimacy and marital potential with the targets. Townsend and Levy found that partner's SES affected women's responses more strongly than men's at all six levels of intimacy and that this sex difference increased as the sexual intimacy or marital potential of the relationship increased. Partner's physical attractiveness also affected women's willingness to enter all six relationships, but high status was able to equalize the acceptability of less physically attractive men. Corroborative results were obtained in a similar experiment by Hickling, Noel, and Yutzler (1979).

The importance of male economic status has also been studied cross-culturally as part of the International Mate Selection Project (IMSP), which investigated mate preferences in 37 cultures spread over six continents and five islands (N = 9,474). Subjects rated 18 mate characteristics on desirability. In 36 of the 37 samples, women placed significantly more value on "good financial prospect" than men did, and, overall, females valued "good financial prospect" more highly than "good looks" (Buss, 1989a; Buss et al., 1990).

In summary, cross-cultural ethnographic reports, cross-cultural empirical studies, laboratory studies on mate choice, the analysis of personal advertisements, and an examination of decisions made at a major commercial dating service coalesce on this point: Status and economic achievement are highly relevant barometers of male attractiveness, more so than physical attributes.

Ornamentation

"The study of clothes exhibits in a pure form the pursuit of status," Quentin Bell (1976, p. 17) concludes in his definitive work on the subject. "The mere fact that so purely social a consideration as the class structure of a society can to so great an extent determine our aesthetic feelings must give us pause and make us wonder how our value judgments are arrived at" (p. 185).

That people use clothes to assess class background has been convincingly demonstrated in a variety of field experiments (Molloy, 1975). If evaluative mechanisms in women were designed to detect and prefer signs of high status in men, then style of dress should provide a powerful cue to male attractiveness. Women can be expected to possess adaptive mechanisms that specify a rule such as, "Prefer ornamentation that signals high status in my culture."

Townsend and Levy (1990b) investigated the effects of male status on female willingness to engage in various romantic relationships. Male targets were prerated for physical attractiveness and divided into two categories: handsome and homely. These models, shown in 35 mm slides projected on a screen, wore one of three costumes: a designed blazer with a Rolex watch (high status), a plain white shirt (medium status), or the uniform of a Burger King employee (low status). The high-status models were described as physicians, the medium-status models were described as high school teachers, and the low-status models were described as waiters-in-training. Both undergraduates and law students viewed the slides and stated their willingness to engage in relationships with the different models at six levels of romantic involvement, ranging from casual conversation to dating, sex, and marriage. Townsend found that women were significantly more willing to engage in liaisons with the high-status/homely males than with either the medium- or low-status/handsome males at all six levels of sexual intimacy and marital potential. (In contrast, male subjects always preferred handsome females over homely females, regardless of costume, ascribed occupational status, or type of relationship proposed.)

Hill, Nocks, and Gardner (1987) manipulated physique and status displays by altering clothing tightness and skin exposure on the one hand, and styles of dress representative of different socioeconomic classes on the other. College students rated opposite-sex models in the various physique and status conditions on four different scales of attractiveness: physical, sexual, dating, and marital. In the low-physique condition, which deemphasized body form, women found high-status male dressers more attractive than low-status ones on all four scales. In the more provocative high-physique condition, which revealed more skin and accentuated body form, women found both high-status and low-status male dressers equally unattractive (each received low ratings on all four scales). Overall, high-status dress strongly inflated male attractiveness, whereas a high degree of body exposure markedly deflated it. (In appraising women, men found females in high-physique displays more attractive on the dating, sexual, and physical scales, but less attractive maritally.)

Although the basic result of this study—that high-status cues enhance male attractiveness—is clear, its details are difficult to interpret because of a possible methodological oversight: Hill, Nocks, and Gardner did not test to see whether physique and status displays were independent dimensions. Given that other studies have found that women do value physical attractiveness in men (Byrne, Ervin, & Lamberth, 1970; Walster et al., 1966), the fact the high-physique display almost completely negated the positive effect of the high-status display suggests that tight clothing and skin exposure are cues of low male status for their subject population (American university women). As stated above, style of dress is highly indicative of status; sexual advertising in male attire—even when swathed in designer labels—may connote low class, just as drab gray suits may connote high class and resource control. Unfortunately, this hypothesis cannot be tested using Hill, Nocks, and Gardner's data, because they did not obtain status ratings on high- versus low-physique displays while holding other status cues constant.

Dispositional Characteristics

If women's evaluative mechanisms were designed to detect and prefer characteristics associated with high status, then females should favor males with indicative cognitive abilities and personality traits. Relevant data were collected in a recent study on mate preferences in 6,000 American couples, including heterosexual and homosexual, married and cohabitating pairs (Howard, Blumstein, & Schwartz, 1987). Mate preferences were assessed via a factor-analyzed 14-item list of attributes that respondents rated on a 1-to-9 scale (1 = not at all important, 9 = extremely important). The factor *ambitiousness*—including the individual items "accomplished," "ambitious," "self-sufficient," and "outgoing"—emerged as a major dimension of female preference with a mean value of 6.24. Subsequent investigations of the IMSP have tended to corroborate this finding. The IMSP employed two standardized closed-form questionnaires. Samples were drawn from all parts of the world and represent a tremendous racial, ethnic, political, and religious diversity. As part of this project, Buss et al. (1990) examined mate characteristics suggestive of resource control or likely acquisition. Using a 0-to-3 scale (0 = irrelevant, 3 = indispensable), subjects rated "education and intelligence" and "ambition and industriousness" on importance and/or desirability in choosing a partner. Collapsed across the 37 samples in the international study, female subjects gave these characteristics mean scores of 2.45 and 2.15, respectively. These high ratings suggest that intelligence, the will to succeed, and the tendency to work hard are qualities strongly and universally desired by women.

Willingness to Invest

The female tendency to favor high-status males is only one part of the constellation of evaluative mechanisms expected to underlie mate choice in women. Selection should also have favored mechanisms in females designed to detect and prefer males who were willing to convert status and ability into paternal assistance. Fathers who are nurturant, as well as emotionally and economically supportive of their wives, encourage the development of achievement motivation, intellectual and social competence, psychological adjustment, and sex-typical attitudes and attributes, particularly in sons (Lamb, 1981). All else equal, therefore, women should find men who demonstrate the willingness to devote time and resources to a chosen female and her offspring more attractive than men who do not.

Recent large-scale attempts to identify major dimensions of preference in mate selection both in the United States (Howard, Blumstein, & Schwartz, 1987) and cross-nationally (Buss et al., 1990) indicate that women want someone they like and can depend on. In the study by Howard et al. (described above), the factor *expressiveness*—including the individual items "affectionate," "compassionate," "expresses feelings," and "romantic"—was by far the strongest female preference with a mean value of 7.34 on a 9-point scale. Buss et al. assessed mate preferences via an 18-item list of attributes that respondents rated on a 0-to-3 scale (0 = irrelevant, 3 = indispensable). Collapsed across the 37 samples in the IMSP, female subjects gave the following characteristics the highest ratings: mutual attraction-love (2.87), dependable character (2.69), emotional stability and maturity (2.68), and pleasing disposition (2.52). At least in these questionnaire studies, the most important qualities women sought in their mates were mutual attraction, stability, dependability, compatibility, and expressiveness. In other words, women seemed to be looking for the kind of men who would make good, willing fathers.

It is frequently observed that women are especially attracted to men they see playing nicely with young children (e.g., Remoff, 1984). This attraction appears to have nothing to do with the desire per se to have children; rather, it is autonomous. The desirability of men who show fondness for children has been documented empirically (Buss & Barnes, 1986), and further investigations are under way.

Structural Powerlessness

Many social scientists (e.g., Coombs & Kenkel, 1966; Hill, Rubin, & Peplau, 1979; Murstein, 1980) have attributed sex differences in the bases of sexual attraction to social conditioning and men's and women's differential access to power: Because women lack power, they seek in men those characteristics associated with power such as status, security, and control of resources. The primary channel for a woman to move upward in society is to marry upward in SES; hence her life chances are largely determined by the job performance of her husband. Since his occupational success depends so much on his skills and personal qualities (such as industriousness and ambition), she must choose him carefully, rigorously assessing his merits and potential. According to this theory, men do not experience the same kinds of structural constraints and, therefore, do not experience the same kinds of needs and desires. Men concern themselves with cosmetic qualities because sex is the main reward males seek in a relationship. Social conditioning, it is posited, maintains and reinforces the whole process, inculcating sex-role-appropriate values from generation to generation. Buss and Barnes (1986; p. 568) call this theory the "structural powerlessness and sex role socialization" hypothesis and note that it does not address "the question of the origins of sex role socialization practices and of the existing economic power structure." Nor does it explain the transcultural nature of sex differences.

The structural powerlessness hypothesis has led various social theorists to make this testable prediction: Women with access to power and wealth will act and fee sexually more like men (much the same as men, in fact) than women who are comparatively powerless and poor (see Dion, 1981; Rosenblatt, 1974; Murstein, 1980). Women, it is predicted, will become less sexually selective and less interested in the socioeconomic status of their mates as their own independence and socioeconomic power increases. (To my knowledge no one has suggested that men will become more selective and more concerned with personality and economic factors as their own status decreases, but this would be another implication of the theory.) In contrast, most evolutionary psychologists would predict that women will prefer high-status mates regardless of whether a female is herself low or high status.

The structural powerlessness hypothesis is directly contradicted by available data. Interview studies of both medical students and leaders in the women's movement reveal that *women's sexual tastes become more, rather than less, discriminatory as their wealth, power, and social status increases.* Fifteen feminist leaders, when asked what traits they sought in a man, recurrently used words that connote high status: "very rich" or "brilliant" or "genius." Lavish dinners, large tips, stunning suits, and so forth were regularly referred to. In short, these high-power women wanted super-powerful men (Fowler, cited in Freedman, 1979). Townsend (1987, 1989), in an investigation of second-year medical students at a northeastern university, found that female medical students often became "more selective and critical in entering and maintaining sexual relationships than they had been previously. Time spent in a relationship that had no marital potential was seen as time wasted." In contrast, male medical students were convinced that their increasing status would allow them to "seek and enjoy more transitory relationships in the future and such a course would be less damaging to their mental balance and career aspirations than would more involved relationships" (Townsend, 1987, pp. 440–441). For females, increasing SES markedly reduces the pool of acceptable romantic partners for this reason: Women, in medical school as in general, want men who are at least on a par with their own socioeconomic level, regardless of how high that level is (Blumstein & Schwartz, 1983; Goldman, Westoff, & Hammerslough, 1984; Townsend, 1987, 1989; Udry, 1981). These findings concur with those of Buss (1989b, p. 41), who found that "women who make *more* money tend to value monetary and professional status of mates *more* than those who make less money."

The evidence suggests that the female tendency to favor high-status males is not a social construct arising from context-dependent needs or women's lack of power. Rather, it suggests that the female preference for high-status males is the product of a psychological mechanism that operates whether a woman's own SES is high or low.

PHYSICAL DOMINANCE

Dominance is a measure of one individual's ability to prevail over another in competitive encounters that involve a face-to-face physical component, whether implicitly or explicitly. It is a *means* of obtaining a resource that involves imposing or threatening to impose a cost on one's competitor. The higher a male is in dominance (i.e., the greater his ability to displace others through coercion from positions or commodities they both want), the greater his access to a variety of fitness-enhancing resources.

Alexander (1987) suggests that the primary "hostile force of nature" encountered by humans is other humans. In this view of the world, conflicts of interest are pervasive, and the competitive strivings of conspecifics become the most salient feature of our adaptive landscape. A man's ability to traverse this landscape, successfully preventing others from violating his interests, depends substantially on his reputation and ability to maintain a favorable position in dominance hierarchies. Because competition is ubiquitous, and because socially dominant males (by definition) tend to fare best in face-to-face competitive encounters, natural selection can be expected to have designed evaluative mechanisms in women to detect and prefer high-dominance men. Forming mateships with such males could substantially enhance a female's survival and reproductive potential through (a) his ability to retain resources that he has or to expropriate resources from others, (b) protection from conspecifics who might otherwise harm, intimidate, or supplant her and her children, and (c) elevation of her own dominance ranking, which would increase her ability to prevail over others.

Hinde (1978) distinguishes two kinds of dominance: *dyadic* and *group*. The former refers to the pattern of imbalance within a relationship between two people; the latter, to the pattern of rank within a larger group structure. Hinde argues that these two forms of dominance are not necessarily related. This distinction is relevant here in that there is no a priori reason to predict that women will be sexually drawn to men who will dominate them in dyadic relationships. Rather, it is the male's ability to successfully negotiate the larger social hierarchy that should be attractive to females. Women pay close attention to how men interact with and are treated by other men (cf. Sadalla, Kenrick, & Vershure, 1987). And the importance of these interactions and the systems of dominance that emerge from them should not be underestimated. As Daly and Wilson (1988, p. 128) point out, "men [become] known by their fellows as 'the sort who can be pushed around' or 'the sort who won't take any shit,' as people whose word means action or people who are full of hot air, as guys whose girlfriends you can chat up with impunity or guys you don't want to mess with."

In summary, evolutionary considerations lead one to expect that women will be sexually attracted to men who display traits that are reliably correlated with social dominance. Bernstein (1980) divides these traits into three categories: (a) physical traits, such as size and physiognomy; (b) social traits such as kinship relationships and political alliances; and (c) individual behavioral traits, such as self-confidence, body language, and aggressiveness. The traits in the first category should constitute less psychologically central mate selection criteria for women than the traits in the second and third categories: While size and strength can contribute to dominance, social standing and body language signal what rank a person has actually achieved within a dominance hierarchy. For example, even a very large man may act submissive if he finds himself among giants, or among the more clever, agile, aggressive, or socially powerful. His behavioral acts (such as patterns of gesturing and deference) will reveal his real position. Nonetheless, insofar as morphological traits were associated with dominance in the Pleistocene, they should still have an impact on female perceptions of male attractiveness today.

Ideally, in an investigation of the effects of dominance on sexual attractiveness, one would look at all major traits in Bernstein's three categories and manipulate them in an experimental setting. High and low self-confidence, high and low physical strength, high and low political connectedness, etc., would each be used as independent variables, and their effect on subjects' ratings of the sexual attractiveness of target stimuli would be measured. Unfortunately, careful studies of this kind have been done on only a small set of traits, all of which fall into the first or third categories. Hence, the following review of dominance and sexual attractiveness will include only these categories.

Individual Behavioral Traits
High-Dominance Personality

The relationship between personality and "leadership," or "managerial effectiveness," has been studied extensively (e.g., Bentz, 1967; Campbell, Dunnette, Lawler, & Weick, 1970; Ghiselli, 1971; Lord, De Vader, & Alliger, 1986; Stodgill 1974), and although the relationship has proven complex and does not easily lend itself to exact interpretation, some clear trends have emerged. A number of U.S. companies have conducted cognitive and personality assessments on their employees, and these assessments have demonstrated reliable covariation between certain personality variables and rated managerial ability. In specific, those individuals who rise to the top of organizations tend to be bright, initiating, self-assured, decisive, masculine, assertive, persuasive, and ambitious (Bentz, 1967; Campbell, Dunnette, Lawler, & Weick, 1970; Ghiselli, 1971; Lord, De Vader, & Alliger, 1986; Stodgill, 1974). As Bentz (p. 118) concludes, "a cluster of psychological characteristics contributes to general executive competence that transcends the boundaries of specialized or non-specialized assignments." This cluster

closely coincides with the description of the dominant personality that emerges from
the California Psychological Inventory Dominance scale, which was designed to assess
factors of dominance, persistence, leadership ability, and social initiative and has
proven to be an effective measure of leadership potential (Gough, 1969; Megargee,
1972). Gough, McClosky, and Meehl (1951, p. 362) describe a high score on this scale:

> A careful reading of the items suggests a number of characterizations of the subjective side
> of the dominant personality. The factor which is implied by the largest number of items
> appears to be one of poise and self-assurance. The dominant personality maintains a high
> level of self-confidence, does not seem to be plagued by self-doubts or equivocations, and
> therefore appears freer to behave in an unencumbered and straightforward manner. The
> impression given is one of resoluteness and vigorous optimism. Closely related to this is
> another suggested factor of resourcefulness and efficiency. The dominant personality
> appears to move forward in a realistic, task-oriented fashion and manifests feelings of ade-
> quacy in meeting whatever obstacles may be encountered.
>
> There is also a certain element of perseverence, or even doggedness, implied. The dom-
> inant subjects admit to working on at things even when others become impatient with them,
> etc., and in general give evidence of strong completion needs.

According to the selectionist model, women should rate men who display these characteristics more favorably on measures of sexual and romantic attractiveness than
men who do not.

Preliminary work on the relationship between personality and mate preferences is
underway. Most relevant to this discussion are potential correlations between attractiveness and surgency (a major personality factor combining aspects of dominance
and extroversion). Surgency has emerged as one of the most common and replicable
dimensions in personality taxonomies (e.g., Goldberg, 1981; Norman, 1963; Wiggins,
1979). High scores on surgency are strongly correlated with a wide array of hierarchy
negotiation tactics (Kyl-Heku & Buss, n.d.), and, consistent with evolutionary predictions, high scores on this dimension are especially prized by women in potential mates
(Botwin & Buss, n.d.). Whereas the extroversion dimension of surgency is equally valued by both sexes in prospective partners, the dominance dimension comprising
power and social ascendance is far more valued by women than men (Botwin & Buss,
n.d.)

These data fall in line with the findings of Sadalla, Kenrick, and Vershure (1987),
who manipulated levels of competitive dominance-seeking behavior in an experiment
on dominance and heterosexual attraction. Targets were described as participants in
an intermediate tennis class who, despite limited training, were very coordinated players and won 60% of their matches. In the high-dominance condition, the target was
described as follows:

> His serve is very strong and his returns are extremely powerful. In addition to his physical
> abilities, he has the mental qualities that lead to success in tennis. He is extremely compet-
> itive, refusing to yield against opponents who have been playing much longer. All of his
> movements tend to communicate dominance and authority. He tends to psychologically
> dominate his opponents, forcing them off their games and into mental mistakes.

In the low-dominance condition:

> His serve and his returns are consistent and well placed. Although he plays well, he prefers
> to play for fun rather than to win. He is not particularly competitive and tends to yield to

opponents who have been playing tennis much longer. He is easily thrown off his game by opponents who play with great authority. Strong opponents are able to psychologically dominate him, sometimes forcing him off his game. He enjoys the game of tennis but avoids highly competitive situations.

Each sex read and rated descriptions of opposite sex targets. Both males and females in the high-dominance condition were rated significantly higher on the following traits: strong, hard, rugged, tough, cold, intelligent, high income, high status, and masculine. Also, dominant males were rated far higher on sexual attractiveness and dating desirability than nondominant males. (Dominance manipulations had no effects on ratings by males of female targets on these dimensions.) These findings suggest that high-dominance personality descriptions markedly enhance female perceptions of male eligibility.

High-dominance males in the Sadalla, Kenrick, and Vershure study were rated considerably more sexually attractive, but significantly less warm, likable, and tender—qualities that presumably offer cues to a male's willingness to invest in a woman and her children (see Buss, this volume). Does this mean that the very qualities that make up the ability to invest (e.g., high dominance, achievement of status) stand in opposition to qualities that indicate willingness to invest?

The solution to this apparent paradox may lie in the fact that broadly worded descriptors such as "kind" or "dominant" do not capture crucial, context-specific variation in behavior. For example, women may find "generosity" appealing in a man (Remoff, 1984), but presumably a woman wants a man who is generous with her but not with her reproductive competitors. Similarly, she may want a man who is dominant (and therefore less "warm, likable, and tender") when he is in competition with other men, but who is warm, likable, and tender toward her. This could be tested by asking subjects to rate how individuals in the high- and low-dominance conditions are likely to behave in other social contexts.

Body Language

Dominance is signaled in day-to-day life through a variety of nonverbal gestures recognizable cross-culturally. In comparison to high-dominance people, low-dominants smile more often (a gesture of appeasement), are less likely to infringe on another's personal space, and are more likely to look away from the gaze of others, eyes downcast (Maclay & Knipe, 1972; see also Mehrabian, 1969; Weisfeld & Beresford, 1982). Whereas submissives tend to exhibit a drawn-in, slouching posture, dominants tend to have an upright bearing, shoulders straight and head thrown back, and to move with a general ease and freedom of body movements (Maclay & Knipe, 1972), communicating a sense of calm and self-assurance. According to evolutionary predictions, these dominance gestures (when directed at other men) should affect female perceptions of male sexual attractiveness, with high-dominance gestures preferred.

Sadalla, Kenrick, and Vershure (1987) made silent videotapes depicting men and women engaging in either high-dominance or low-dominance behavior. Each actor played a high-dominance role in one tape and a low-dominance role in another. All videotaped interactions took place between two members of the same sex. Sadalla et al. describe the two scenes:

In the low-dominance condition, a constant male (CM) is shown seated at a desk in an office. An actor enters the room and chooses a chair near the door. . . . The actor, clutching a sheath of papers, sits in a symmetrical posture, leans slightly forward with head partially bowed, and alternately looks down at the floor and up at the CM. During an ensuing discussion, the actor engages in repetitive head nodding and lets the CM engage in longer communications.

In the high-dominance condition, the actor enters, chooses a chair closer to the CM and sits in a relaxed, asymmetrical posture. The actor's hands and legs are relaxed and his body is leaning slightly backward in the chair. During the discussion, the actor produces higher rates of gesturing and lower rates of head nodding than in the low-dominance condition.

Male subjects viewed female actors, and female subjects viewed male actors. Both men and women in the high-dominance condition were rated by subjects significantly higher on traits such as strong, hard, rugged, tough, and masculine. The participants also judged target persons on sexual attractiveness and dating desirability. High-dominance behavior significantly increased ratings of male targets on both measures. Ratings of female targets were unaffected. These data suggest that female experiences of sexual and romantic attraction are sensitive to nonverbal displays of male dominance and that this sensitivity is characteristic of female sexual psychology but not male sexual psychology.

Physical Traits
Physiognomy

Perceptions of social dominance in humans, as well as in nonhuman animals, are affected by physiognomic traits. In particular, traits associated with physical maturity (proportionately thin lips and eyes, receding hairline) and physical strength (wide face, square jaw [Guthrie, 1970]) are linked to perceptions of social dominance in people (Keating, 1985, 1987; Keating, Mazur, and Segall, 1981). This correlation emerged among a majority of observers in at least 10 of the 11 diverse cultural settings that Keating et al. studied. Consequently, if females find social dominance attractive in men, then women should rate mature male facial features more favorably than immature ones on measures of sexual attractiveness.

Keating (1985) constructed mature and immature facial composites from Identi-Kit materials (typically used by police departments to create facial composites of suspects), manipulating jaw shape and size of lips, eyes, and eyebrows. In all other ways composites were identical. The only factor distinguishing "male" and "female" faces was hairdo. Subjects rated composites on scales for dominance and attractiveness. While mature features similarly boosted the dominance ratings of both male and female composites, a divergence occurred in ratings of attractiveness. Mature male faces were rated considerably more attractive; mature female faces were rated somewhat less attractive. On a scale of 1-to-7 (1 = very unattractive, 7 = very attractive), women gave immature and mature male composites mean ratings of 2.48 and 4.40, respectively. In short, female perceptions of male attractiveness increased in response to morphological enhancements of facial dominance signals.

One would expect, however, there to be some threshold (possibly signaled by gray hair, stressed skin, baldness, etc.) after which "mature" features come to be seen as "old" and diminish in appeal. In our natural environment, male mate value must have decreased in old age when a man's ability to acquire resources and protect his family—or even to live long enough to complete the cycle of investment in a maturing child—weakened.

Height

Height is associated with power and status and in empirical studies has been shown to markedly affect one individual's ability to dominate another (Handwerker & Crosbie, 1982). In an array of situations, male height confers an economic, political, and social advantage (Gillis, 1982). Economically, taller men are more likely to be hired, tend to receive higher starting salaries, and are more likely to get promoted than shorter men. Politically, taller men tend to receive more votes in political elections (the taller candidate won 80% of U.S. presidential elections between 1904 and 1980), and voters tend to overestimate the height of their favored presidential candidate. Socially, people generally overestimate the height of individuals who are high status, whom they like, or whom they agree with. The opposite is true for individuals who are low status, disliked, or disagreed with. (These findings are discussed at length by Gillis [1982].) Most compelling, cross-cultural ethnographic reports suggest that in many parts of the world the ability to achieve positions of power is strongly rooted in relative height (Bernard, 1928; Gregor, 1979; Handwerker & Crosbie, 1982; Werner, 1982). Brown and Chia-yun (n.d.) document that the term "big man" or terms very close to it are found in aboriginal languages throughout much of the world and are used to denote persons of authority or importance. Brown and Chia-yun argue that, in fact, the term is a conflation of physical size and social rank and that "big men" are consistently *big* men, tall in stature.

This array of sociopolitical advantages presumably accrues to tall males because height constitutes a reliable cue to dominance in social interactions. Taller men are perceived as more dominant than shorter men. (For example, shorter policeman are more likely to be assaulted than taller policemen [Gregor, 1979], suggesting that the latter commands more fear and respect from adversaries.) It follows, therefore, that taller males would be preferred by other males as economic and political allies (receiving better jobs, more electoral victories, etc.) and by women as mates. If dominance is an important aspect of male mate value, then tallness should enhance female perceptions of male sexual and romantic attractiveness.

There are important constraining influences on height, however, that do not apply to most of the other traits discussed in this chapter. Whereas it is probably impossible to rank too high in status or dominance hierarchies, it is possible to be too tall. With respect to height and many other anatomical characteristics, natural selection tends to favor the population mean, weeding out the tails of the distribution (Symons, 1979). Ecological constraints probably caused natural selection to penalize very tall men (e.g., more broken bones, higher metabolic costs, poorer balance and agility, awkward gait and body carriage; see, e.g., Haldane, 1985), causing optimal male height to converge on some constrained limit. The hypothesis that males near the midpoint of the population distribution will be the most viable, and thus should be most attractive to females, has been called the "central tendency" hypothesis. In contrast, the hypothesis that females will prefer dominant males, all else equal, predicts that women will prefer men who are somewhat taller than average, although within limits set by optimal size constraints.

Some preliminary data have been collected. Graziano, Brothen, and Berscheid (1978) had short (under 5 ft., 4 in.), medium (5 ft., 4 in. to 5 ft., 6 in.), and tall (over 5 ft., 6 in.) women judge pictures of men who they believed to be short (5 ft., 6 in.), medium (5 ft., 10 in.), or tall (6 ft., 3 in.) on attractiveness and dating desirability. Women of all three height categories rated tall men more positively than short men on both measures; however, the medium-height males were clearly preferred overall. There was no interaction between the height of the evaluator and the height of the stimulus person. This preference for 5 ft., 10 in. males is consistent with the findings of Beigel (1954) and Gillis and Avis (1980), who investigated stated height preferences for ideal mates. Beigel's female sample (average height 5 ft., 3.5 in.) stated a mean preference for males who were taller than themselves by 6.7 in. Gillis and Avis's female sample (height unreported) stated a mean preference for males who were taller than themselves by 6.0 in. From this it can be calculated that both samples (assuming the latter approximates the U.S. population mean of 5 ft., 4 in. for females) preferred males who were about 5 ft., 10 in., on average. Since the population mean in the U.S. is 5 ft., 10 in. for males, these figures fit the central tendency hypothesis better than the dominance hypothesis. However, as suggested by the Graziano et al. data, deviations away from the central tendency are more tolerated by women in the tall than short direction. Cross-cultural testing is needed. In the only ethnographic study I could find on the relationship between height and sexual attractiveness, Gregor (1979) reports that tall males are strongly preferred by females as both lovers and mates among the Mehinaku Indians of central Brazil.

If the central tendency hypothesis is true, then this could have been engineered by a psychological mechanism that specifies a rule such as "Choose a man whose height is such that when you are looking him in the eyes the visual angle subtended is X." This could account for the 6 to 6.7 in. finding, as well as the well-established trend for men and women to mate assortively on height, that is, for short to marry short and tall to marry tall (Gillis & Avis, 1980; McManus & Mascie-Taylor, 1984). (Perhaps angle X is calibrated experientially to reflect the mean difference in height between men and women in the local population.)

The most salient criterion concerning height that women apply to men appears to be the "male-taller norm," which is so prevalent that it has been called the "cardinal principle of date selection" (Berscheid & Walster, 1974). Gillis and Avis (1980) examined height data collected from the bank account applications of 720 couples. Incredibly, in only one case was the woman taller than the man. Sheppard and Strathman (1989) had women view photographs of male-female dyads in which the male was pictured as approximately 5 in. taller than the female, equal in height to the female, or 5 in. shorter than the female. The same dyad was featured in each photograph. On a scale of 1-to-9 (1 = very unattractive, 9 = very attractive), the male target received mean ratings of 6.00 in the male-taller/female-shorter condition, 5.00 in the equal-height condition, and 4.10 in the female-taller/male-shorter condition. A study by Lang (1979) may shed light on some of the affective dimensions underlying these rating changes. Lang looked at women's height preferences using both the Thematic Apperception Test (TAT), where subjects are shown a picture and asked to make up a story about it, and a brief questionnaire. For the TAT, subjects were shown drawings of a man and woman of discrepant heights out on the town together. In response to the female-taller/male-shorter picture, all subjects—even those stating adamantly on the questionnaire that the size of a man made no difference to them—invented stories with negative outcomes, depicting the male as "anxious" or "weak." In the male-taller/female-shorter condition, women made up stories with bland or reasonably optimistic outcomes in which the male was typically described as confident, reassuring, and understanding. For many of the women, avowed attitudes on the questionnaire were quite different from the feelings evidenced in their stories—feelings associated with dominance perceptions of the male partner.

Assuming that (a) female height is normally distributed and (b) mean female height is less than mean male height, then a man who is shorter than his female partner will, on average, be shorter than the population average for men. Thus, the "male-taller norm" is consistent with the central tendency hypothesis. However, the TAT results suggest that dominance considerations do play a role, if only in setting a lower limit on acceptable male height. Furthermore, male height emerges as an important factor in real dating behavior. Taller men are more sought after in women's personal advertisements (Cameron, Oskamp, & Sparks, 1978), receive more responses to their own personal advertisements (Lynn & Shurgot, 1984), and tend to have prettier girlfriends (Feingold, 1982) than do shorter men. In addition, the "male-taller norm" appears to be enforced more strongly by women than by men. While Gillis and Avis (1980) and Beigel (1954) found that females preferred males who were taller by 6 in. and 6.7 in., respectively, the males in both studies sought females who were shorter by only 4.5 in. In sum, the data at this point are too sparse and potentially contradictory to decide between the central tendency and dominance hypotheses.

A MATE CHOICE PARADOX

Two of the mate characteristics examined in the IMSP (Buss et al., 1990) were "good financial prospect" and "favorable social status." Both received fairly low ratings from women (1.76 and 1.46, respectively, on a 0-to-3 scale [0 = irrelevant, 3 = indispensable], collapsed across the 37 samples). In line with this trend, the U.S. sample rated both these attributes quite low. Considering the literature reviewed in this chapter, these results seem contradictory. For instance, it is known that in the United States the men whom women actually choose to marry make 50% more money, on average, than men of the same age whom they do not choose to marry (Trivers, 1985). Along these lines, male medical students report that their increasing socioeconomic status tends to markedly enlarge their pool of available sexual and marital partners (Townsend, 1987). Moreover, in a study of selections made at a major commercial dating service, social status emerged as the single most important criterion women applied to men (Green, Buchanan, & Heuer, 1984).

A similar inconsistency emerges between women's stated preference for dominance per se and actual response to dominance signals. Buss and Barnes (1986) assessed major dimensions of mate choice in American married couples via the Marital Preferences Questionnaire, which consists of 76 adjective-items, which subjects rate on a 5-point scale of desirability. Both women and men rated "dominant" as among the *least* desirable mate characteristics. Yet, as we have seen, most women respond positively to actual dominance traits expressed in videotapes, pictures, or written descriptions, at least when dominance was expressed toward other men.

This discrepancy between avowed preferences on questionnaires, on the one hand, and actual responses in many real-life situations and experimental settings, on the other, may have numerous causes. I will discuss four possibilities.

1. *Conflation of group and dyadic dominance.* Unless a measuring instrument specifically distinguishes between group and dyadic dominance (Hinde, 1978), subjects are likely to conflate these two concepts in their responses. Women may respond so negatively to "dominance" as a mate characteristic because they perceive it as the proclivity of a male to dominate *them.* If dominance were presented in terms of a man's ability to interact confidently and successfully with other males (as it is in some of the experiments described above), then it should receive more favorable reviews.

2. *The social desirability issue.* It is possible that many women are interested in forming relationships with dominant and/or high-status men but are reluctant to admit such motives. American women who marry for money are stigmatized as "gold-diggers," and the words "status" and "dominance" have negative connotations for some people in our society. Some female writers who address the question, "What do women find attractive in men?," cloak the issues of status and dominance in euphemisms that seem designed in part to conceal the real political and socioeconomic basis of much female choice (D. Symons, personal communication, 1987). Thus Flood (1981) and Shanor (1977), for example, suggest that what matters most to women are such things as "Who a man is," "How he fits into the world," "How he handles himself," and "How he responds to other people."

3. *The preference for dominant, high-status men may be real but unconscious.* Recall the discussion of female height preferences and the incongruity between the subjects' stated attitudes and TAT outcomes. Women seemed to be responding to a dominance cue (height), while at the same time insisting that height was not important. It is possible that many women are simply not aware of their inclination to perceive and evaluate status and dominance cues favorably. Perhaps these cues evoke an emotional response in women that affects their feelings of attraction toward a male, without them being consciously aware that their response was activated by these cues. Their response might be automatic and its cause not consciously accessible.

4. *Women's questionnaire responses may be biased by a "threshold effect"* (D. Symons, personal communication, 1987). This point can be illustrated by recounting a short story about three upper-middle-class women going out to lunch together in New York City and complaining that "there are no men!" But there were men all around them: service people in the restaurant. Such working-class males were socially invisible to these upper-middle-class females because they were below the necessary threshold of social and economic status.

What could be occurring, then, is that when women fill out a questionnaire on mate preferences they are thinking only about those males who are above their status threshold and thus within their range of vision. Within that range, status differences may be relatively unimportant (as compared with other qualities such as kindness or honesty, which are consistently rated as highly desirable mate characteristics by women). However, status may still have the huge effect of setting a minimum threshold, and thereby ruling out much of the male population as potential mates for upper-middle-class women. Townsend and Levy (1990b) provide considerable empirical support for this hypothesis. This reasoning implies that in an investigation of mate preferences, researchers should have subjects not only rank the desirability of various characteristics, but also state their minimum criterion for each characteristic and rate what percentage of the population meets their minimum criterion. Kenrick, Sadalla, Groth, and Trost (1990) have begun this project.

MATE PREFERENCES IN AN UNKNOWN CULTURE

Donald Symons (1979, 1987a) has reviewed a great deal of ethnographic and psycho-logical literature on perceptions of sexual attractiveness. In summarizing the male experience of sexual attraction, Symons (1987b) asks his readers to imagine that a heretofore unknown tribal people is suddenly discovered. Drawing on touchstones of his cross-cultural review, Symons lists the characteristics that he predicts will consti-tute the ideal sexual partner of the average man in this culture. My intention is to take Symons's list full circle and to summarize my own analysis. I predict the average woman in this culture will seek the following characteristics in her ideal mate:

1. He will be dependable, emotionally stable and mature, and kind/considerate toward her.
2. He will be generous. He may communicate a spirit of caring through a will-ingness to share time and whatever commodities are valued in this culture with the woman in question.
3. He will be ambitious and perceived by the woman in question as clever or intelligent.
4. He will be genuinely interested in the woman in question, and she in him. He may express his interest through displays of concern for her well-being.
5. He will have a strong social presence and be well liked and respected by others. He will possess a strong sense of efficacy, confidence, and self-respect.
6. He will be good with, interested in, and/or show a general fondness for chil-dren.
7. He will possess whatever skills, accoutrements, physical features, and eco-nomic capabilities happen to be reliably associated with high status in this cul-ture.
8. He will possess the skills, behavioral tendencies, and physical characteristics that enable him to protect the woman in question from physical attack or intimidation and will exhibit signs that he is willing to do so.
9. He will evidence signs of health and vitality, such as firm muscle tone, clear skin, upright posture, and energetic body language. He will be taller than the woman in question, have mature physiognomic features, and display a general ease and freedom of body movements.
10. He will not be a man with whom the woman in question grew up with as a child (see Shepher, 1983).

Predictions 1, 3, 7, and 10 are "safe" in the sense that they are supported by a good deal of ethnographic and psychological data on human mate selection. Predictions 2, 4, 5, 6, 8, and 9 are as yet untested or insufficiently tested (e.g., relevant data have been collected in only one culture), though each is falsifiable. These latter six predictions are essentially derived from my interpretations of human evolutionary theory. Of course alternative evolutionary scenarios could be generated and competing hypotheses pro-posed and empirically tested against these.

The foregoing predictions are limited by their paucity and generality. The human experience is manifold. Experiences of sexual attraction in the newly discovered cul-ture will undoubtedly encompass preferences and feelings far more varied and elaborate than this limited set of predictions can capture. The Darwinian analysis scarcely illuminates the idiosyncrasies of individuals and cultures. It sheds light on basic desires—motives, drives, proclivities, aspirations—but it cannot predict the broad array of tactics deployed in our attempts to satisfy those desires. In any given culture, empirical studies are needed to identify the behavioral strategies through which desiderata are sought, as well as the proximate mechanisms that fashion those strategies.

Many of the psychological mechanisms underpinning feelings of sexual attraction in this culture will depend on ontogenetic experiences for their expression. The ability to discriminate between high- and low-status individuals and between fathers who are willing to make long-term commitments to offspring and pretenders who are not is a function of experience. Nonetheless, it is the very nature of our psychological structures that allows us to (a) extract relevant information (such as that concerning mate value) from our experiences in the first place and (b) use that information quickly and efficiently to solve adaptive problems (such as whom to feel sexual attraction toward). Our psychological mechanisms—specialized and goal-directed—allow us to learn the right things in the learning situations typically encountered in our evolutionary past. The psychological mechanisms underlying perceptions of sexual attractiveness should be sensitive to environmental cues that correlate with mate value.

In comparison to my forecasts on mate preferences among this newly discovered culture, what might a mainstream social scientist predict? Probably nothing at all. Various social scientists (e.g., Futuyama & Risch, 1984; Hoult, 1984; Simon & Gagnon, 1969) claim that humans inherit only a diffuse biological potential—a generalized sex drive that is predisposed toward nothing and, therefore, *cannot be predicted in advance* (cf. Symons, 1987b). By this reasoning, men and women could just as easily be attracted to tree trunks as other people if they were culturally conditioned to do so, and the distribution of sex differences across cultures should be random. From an adaptive viewpoint, such people could just as easily learn the wrong things as the right ones, and they could be manipulated against their selective interests by others (cf. Barkow, 1989; Cosmides & Tooby, 1987).

The central premise of this chapter—that women will respond preferentially to men displaying traits indicative of high mate value—does not imply that women consciously appraise men through the sharp eye of maternal pragmatism. When a woman experiences feelings of sexual attraction, she is not, at an unconscious level, "plotting" a reproductive strategy designed to maximize the representation of her genes in future generations. Rather, she is probably simply experiencing desire for the man in question; this desire may or may not enhance reproductive success in the milieu where it is experienced. But underlying the nature and intensity of that desire is a complex host of psychological mechanisms, and these mechanisms should have been designed by natural selection to detect and prefer male traits that in our natural environment were reliably associated with (a) the ability and willingness to provide economically, (b) the ability and willingness to protect a woman from physical attack or intimidation, and (c) the ability and willingness to engage in direct parenting activities such as teaching, nurturing, and providing social support and opportunities. Taken together, these preferences form a coherent, integrated system that throughout our evolutionary history presumably had the *effect* of causing women to choose men of high mate value.

ACKNOWLEDGMENTS

I am indebted to Jerry Barkow, David Buss, Leda Cosmides, Kelly Hardesty-Ellis, Don Symons, and John Tooby for helpful comments on this chapter.

REFERENCES

Alexander, R. D. (1987). *The biology of moral systems.* New York: Aldine de Gruyter.

Barkow, J. H. (1989). The elastic between genes and culture. *Ethology and Sociobiology, 10,* 111–129.

Beigel, H. G. (1954). Body height in mate selection. *Journal of Social Psychology, 39,* 257–268.

Bell, Q. (1976). *On human finery.* New York: Schocken Books.

Bentz, V. J. (1967). The Sears experience in the investigation, description, and prediction of executive behavior. In F. R. Wickert & D. E. McFarland (Eds.), *Measuring executive effectiveness.* New York: Appleton-Century-Crofts.

Bernard, J. (1928). Political leadership among North American Indians. *American Journal of Sociology, 34,* 296–315.

Bernstein, I. S. (1980). Dominance: A theoretical perspective for ethologists. In D. R. Omark, F. F. Strayer, & D. G. Freedman (Eds.), *Dominance relations.* New York: Garland STPM Press.

Berscheid, E., & Walster, E. (1974). Physical attractiveness. In L. Berkowitz (Ed.), *Advances in experimental social psychology,* (vol. 7). New York: Academic Press.

Betzig, L. (1986). *Depotism and differential reproduction.* New York: Aldine.

Blumstein, P., & Schwartz, P. (1983). *American couples: Money, work, and sex.* New York: William Morrow.

Botwin, M., & Buss, D. M. (n.d.). *Personality and mate preferences.*

Brown, D. E. and Chia-yun, Y. (n.d.). *"Big Man": Its distribution, meaning, and origin.*

Buss, D. M. (1987). Sex differences in human mate selection criteria: An evolutionary perspective. In C. Crawford, M. Smith, and D. Krebs (Eds.), *Sociobiology and psychology: Ideas, issues and applications.* Hillsdale, N.J.: Lawrence Erlbaum Assoc.

Buss, D. M. (1989a). Sex differences in human mate preferences: Evolutionary hypotheses tested in 37 cultures. *Behavioral and Brain Sciences, 12,* 1–14.

Buss, D. M. (1989b). Toward an evolutionary psychology of human mating. *Behavioral and Brain Sciences, 12,* 39–46.

Buss, D. M. et al. (1990). International preferences in selecting mates: A study of 37 cultures. *Journal of Cross-Cultural Psychology, 21,* 5–47.

Buss, D. M., & Barnes, M. (1986). Preference in human mate selection. *Journal of Personality and Social Psychology, 50,* 559–569.

Byrne, D., Ervin, C. R., & Lamberth, J. (1970). Continuity between the experimental study of attraction and real-life dating. *Journal of Personality and Social Psychology, 16,* 157–165.

Cambell, J. P., Dunnette, M. D., Lawler, E. E., III, & Weich, K. E., Jr. (1970). *Managerial behavior, performance, and effectiveness.* New York: McGraw-Hill.

Cameron, C., Oskamp, S., & Sparks, W. (1978). Courtship American style: Newspaper advertisements. *Family Coordinator, 26,* 27–30.

Coombs, R. H., & Kenkel, W. P. (1966). Sex differences in dating aspirations and satisfaction with computer-selected partners. *Journal of Marriage and the Family, 28,* 62–66.

Cosmides, L., & Tooby, J. (1987). From evolution to behavior: Evolutionary psychology as the missing link. In J. Dupre (Ed.), *The latest on the best: Essays on evolution and optimality.* Cambridge, MA: MIT Press.

Daly, M., & Wilson, M. (1988). *Homicide.* New York: Aldine de Gruyter.

Dion, K. (1981). Physical attractiveness, sex roles, and heterosexual attraction. In M. Cook (Ed.), *The bases of human sexual attraction.* New York: Academic Press.

Feingold, A. (1982). Do taller men have prettier girlfriends? *Psychological Reports, 50,* 810.

Flood, P. (1981, June). Body parts. *Esquire,* pp. 35–43.

Ford, C. S., & Beach, F. A. (1951). *Patterns of sexual behavior.* New York: Harper and Row.

Freedman, D. (1979). *Human sociobiology: A holistic approach.* New York: Macmillan.

Futuyama, D. J., & Risch, S. J. (1984). Sexual orientation, sociobiology, and evolution. *Journal of Homosexuality, 9,* 157–159.

Ghiselli, E. E. (1971). *Explorations in managerial talent.* Pacific Palisades, CA: Goodyear.

Gillis, J. S. (1982). *Too tall, too small.* Champaign, IL: Institute for Personality and Ability Testing.

Gillis, J. S., & Avis, W. E. (1980). The male-taller norm in mate selection. *Personality and Social Psychology Bulletin, 6,* 396–401.

Goldberg, L. R. (1981). Language and individual differences: The search for universals in personality lexicons. In L. Wheeler (Ed.), *Review of personality and social psychology* (Vol. 2). Beverly Hills: Sage.

Goldman, N., Westoff, C., & Hammerslough, C. (1984). Demography of the marriage market in the United States. *Population Index, 50,* 5–25.

Gough, H. G. (1969). *Manual for the California psychological inventory* (rev. ed.). Palto Alto, CA: Consulting Psychologists Press.

Gough, H. G., McClosky, H., & Meehl, P. E. (1951). A personality scale for dominance. *Journal of Abnormal and Social Psychology, 46,* 360–366.

Graziano, W., Brothen, T., & Berscheid, E. (1978). Height and attraction: Do men and women see eye-to-eye? *Journal of Personality, 46,* 128–145.

Green, S. K., Buchanan, D. R., & Heuer, S. K. (1984). Winners, losers, and choosers: A field investigation of dating invitation. *Personality and Social Psychology Bulletin, 10,* 502–511.

Gregersen, E. (1982). *Sexual practices: The story of human sexuality.* London: Mitchell Beazley.

Gregor, T. (1979). Short people. *Natural History, 88,* 14–23.

Guthrie, R. D. (1970). Evolution of human threat display organs. In T. Dobzhansky, M. K. Hecht, and W. C. Steere (Eds.), *Evolutionary biology 4.* New York: Appleton-Century-Crofts.

Haldane, J.B.S. (1985). *On being the right size.* New York: Oxford University Press.

Handwerker, W. P., & Crosbie, P. V. (1982). Sex and dominance. *American Anthropologist, 84,* 97–104.

Harrison, A. A., & Saaed, L. (1977). Let's make a deal: An analysis of revelations and stipulations in lonely hearts advertisements. *Journal of Personality and Social Psychology, 35,* 257–264.

Hickling, E. J., Noel, R. C., & Yutzler, P. D. (1979). Attractiveness and occupational status. *Journal of Psychology, 102,* 71–76.

Hill, C. T., Rubin, Z., & Peplau, L. A. (1979). Breakups before marriage: The end of 103 affairs. In L. Levinger and O. C. Moles (Eds.), *Divorce and separation.* New York: Basic.

Hill, E. M., Nocks, E. S., & Gardner, L. (1987). Physical attractiveness: Manipulation by physique and status displays. *Ethology and Sociobiology, 8,* 143–154.

Hinde, R. A. (1978). Dominance and role—two concepts with dual meanings. *Journal of Social and Biological Structures, 1,* 27–38.

Hoult, T. F. (1984). Human sexuality in biological perspective. *Journal of Homosexuality, 9,* 137–157.

Howard, J. A., Blumstein, P., & Schwartz, P. (1987). Social or evolutionary theories? Some observation on preferences in human mate selection. *Journal of Personality and Social Psychology, 53,* 194–200.

Keating, C. F. (1985). Gender and the physiognomy of dominance and attractiveness. *Social Psychology Quarterly, 48,* 61–70.

Keating, C. F. (1987). Human dominance signals: The primate in us. In J. F. Davidio and S. L. Ellyson (Eds.), *Power, dominance, and nonverbal communication.* New York: Springer-Verlag.

Keating, C. F., Mazur, A., & Segall, M. H. (1981). A cross-cultural exploration of physiognomic traits of dominance and happiness. *Ethology and Sociobiology, 2,* 41–48.

Kenrick, D. T., Sadalla, E. K., Groth, G., & Trost, M. R. (1990). Gender and trait requirements in a mate: An evolutionary bridge between personality and social psychology. *Journal of Personality, 58,* 97–116.

Kyl-Heku, L. M. and Buss, D. M. (n.d.). *Tactics of Hierarchy Negotiation.*

Lamb, M. E. (1981). *The role of the father in child development* (rev. ed.). New York: Wiley.

Lang, F. (1979). Mate choice in the human female: A study of height preferences. In D. Freedman, *Human sociobiology.* New York: The Free Press.

Lee, R. B. (1979). *The !Kung San: Men, women, and work in a foraging society.* New York: Cambridge University Press.

Lord, R. G., De Vader, C. L., & Alliger, G. M. (1986). A meta-analysis of the relation between personality traits and leadership perceptions. *Journal of Applied Psychology, 71,* 402–410.

Lynn, M., & Shurgot, B. A. (1984). Responses to lonely hearts advertisements: Effects of reported physical attractiveness, physique, and coloration. *Personality and Social Psychology Bulletin, 10,* 349–357.

Maclay, G., & Knipe, H. (1972). *The dominant man.* New York: Delta.

McManus, I. C., & Mascie-Taylor, C.G.N. (1984). Human assortive mating for height: Nonlinearity and heteroscendasticity. *Human Biology, 56,* 617–623.

Megargee, E. I. (1972). *The California psychological inventory handbook.* San Francisco: Jossey-Bass.

Mehrabian, A. (1969). Significance of posture and position in the communication of attitude and status relationships. *Psychological Bulletin, 71,* 359–372.

Molloy, J. T. (1975). *Dress for success.* New York: Warner Books.

Murstein, B. I. (1980). Mate selection in the 1970's. *Journal of Marriage and the Family, 42,* 777–792.

The new mating games. (1986, June 2). *Newsweek,* p. 58.

Norman, W. (1963). Toward an adequate taxonomy of personality attributes: Replicated factor structure in peer nomination personality ratings. *Journal of Abnormal and Social Psychology, 66,* 574–583.

Remoff, H. T. (1984). *Sexual choice.* New York: Dutton/Lewis.

Rosenblatt, P. C. (1974). Cross-cultural perspective on attraction. In T. L. Huston (Ed.), *Foundations of interpersonal attraction.* New York: Academic Press.

Sadalla, E. K., Kenrick, D. T., & Vershure, B. (1987). Dominance and heterosexual attraction. *Journal of Personality and Social Psychology, 52,* 730–738.

Shanor, K. (1977). *The fantasy files.* New York: Dial Press.

Shepher, J. (1983). *Incest: A biosocial view.* Orlando, FL: Academic Press.

Sheppard, J. A., & Strathman, A. J. (1989). Attractiveness and height: The role of stature in dating preference, frequency of dating, and perceptions of attractiveness. *Personality and Social Psychology Bulletin, 15,* 617–627.

Simon, W., & Gagnon, J. (1969). On psychosexual development. In D. A. Goslin (Ed.), *Handbook of socialization theory and research.* Chicago: Rand-McNally.

Stodgill, R. M. (1974). *Handbook of leadership.* New York: Free Press.

Stone, V. (1989) Perception of Status: An evolutionary analysis of nonverbal status cues. Unpublished doctoral dissertation, Dept of Psychology, Standford University.

Symons, D. (1979). *The evolution of human sexuality.* Oxford: Oxford University Press.

Symons, D. (1987a). Can Darwin's view of life shed light on human sexuality? In J. H. Geer & W. T. O'Donohue (Eds.), *Theories of human sexuality*. New York: Plenum.

Symons, D. (1987b). If we're all Darwinians, what's the fuss about? In C. Crawford, M. Smith, & D. Krebs (Eds.), *Sociobiology and psychology: Ideas, issues and applications*. Hillsdale, NJ: Lawrence Erlbaum Assoc.

Townsend, J. M. (1987). Sex differences in sexuality among medical students: Effects of increasing socioeconomic status. *Archives of Sexual Behavior, 16*, 425–441.

Townsend, J. M. (1989). Mate selection criteria: A pilot study. *Ethology and Sociobiology, 10*, 241–253.

Townsend, J. M., & Levy, G. D. (1990a). Effects of potential partners' physical attractiveness and socioeconomic status on sexuality and partner selection. *Archives of Sexual Behavior, 19*, 149–164.

Townsend, J. M., & Levy, G. D. (1990b). Effects of potential partners' costume and physical attractiveness on sexuality and partner selection. *Journal of Psychology, 124*, 371–389.

Trivers, R. L. (1985). *Social evolution*. Menlo Park, CA: Benjamin-Cummings.

Udry, J. R. (1981). Marital alternatives and marital disruption. *Journal of Marriage and the Family, 43*, 889–897.

Udry, J. R., & Eckland, B. K. (1984). Benefits of being attractive: Differential payoffs for men and women. *Psychological Reports, 54*, 47–56.

Walster, E., Aronson, V., Abrahams, D., & Rottman, L. (1966). Importance of physical attractiveness in dating behavior. *Journal of Personality and Social Psychology, 4*, 508–516.

Weisfeld, G. E., & Beresford, J. M. (1982). Erectness of posture as an indicator of dominance or success in humans. *Motivation and Emotion, 6*, 113–131.

Werner, D. (1982). Chiefs and presidents. *Ethos, 10*, 136–148.

Wiggins, J. S. (1979). A psychological taxonomy of trait-descriptive terms: The interpersonal domain. *Journal of Personality and Social Psychology, 37*, 395–412.

The Man Who Mistook His Wife
for a Chattel

MARGO WILSON AND MARTIN DALY

Men take a proprietary view of women's sexuality and reproductive capacity. In this chapter, we (a) argue that sexually proprietary male psychologies are evolved solutions to the adaptive problems of male reproductive competition and potential misdirection of paternal investments in species with mistakable paternity; (b) describe the complex interrelated design of mating and paternal decision rules in some well-studied avian examples; (c) consider the peculiarities of the human species in this context; (d) characterize some features of human male sexual proprietariness, contrasting men's versus women's perspectives and actions; and (e) review some of the diverse consequences and manifestations of this ubiquitous male mindset.

THE EVOLUTIONARY LOGIC OF MALE SEXUAL PROPRIETARINESS

By "proprietary," we mean first that men lay claim to particular women as songbirds lay claim to territories, as lions lay claim to a kill, or as people of both sexes lay claim to valuables. Having located an individually recognizable and potentially defensible resource packet, the proprietary creature proceeds to advertise and exercise the intention of defending it from rivals. Proprietariness has the further implication, possibly peculiar to the human case, of a sense of *right* or *entitlement*. Trespass by rivals provokes not only hostility but a feeling of grievance, a state of mind that apparently serves a more broadly social function. Whereas hostile feelings motivate action against one's rivals, grievance motivates appeals to other interested persons to recognize the trespass as a wrong against the property holder and hence as a justification for individual retaliation or a grounds for more collective sanctions. Proprietary entitlement thus rests upon a social contract: Property owners reciprocally acknowledge and cooperatively enforce one another's claims.

Socially recognized property rights have several components. Full proprietary entitlements include the right to sell, exchange, or otherwise dispose of one's property, to modify it without interference, and to demand redress for the theft or damage of it. People claim these entitlements with respect to inanimate objects (whether movable or not)—land, crops, livestock, and even such intangibles as investment opportunities and ideas. They have claimed the same entitlements with respect to "their" slaves, household servants, and children; and men, but not women, have regularly asserted such claims with respect to spouses.

Because claims of proprietary entitlements are responses to rivalry over limited resources, all such claims necessarily exist in an arena of actual or potential conflicts of interest. These conflicts increase in number and complexity when the "property" is a living creature with interests of its own. Efforts to exert proprietary rights over non-human animals frequently involve a conspicuous struggle of wills. Even plants exhibit thorns, toxins, and other devices that must be understood as evolved conflict tactics for thwarting those who would exploit them. But the most complexly conflictual of all proprietary claims are those in which people assert such rights over other people, for here is property that can understand its owners' purposes and weaknesses, that may have relatives and friends, that can plan escapes or attacks with as much foresight and ingenuity as its owners, and that can play rivalrous aspirant owners off against one another. And yet, with all these difficulties, men do attempt to exercise proprietary entitlements over other people (Bradley, 1984; Davis, 1966; Finley, 1980; Karras, 1988; Patterson, 1982; Sawyer, 1986), especially women.

One could imagine a psyche in which the various components of proprietariness—one's presumed rights to use, modify, and dispose of the property, and the hostility and sense of grievance aroused by trespass—were dissociated, applying independently to distinct classes of resources. But it would not be a human psyche. People merge the logically separable aspects of proprietariness and generalize their applicability, as witness the breadth of domains within which human grammars use possessive pronouns and their equivalents and within which the lexicon of ownership is applied. These considerations suggest that referring to man's view of woman as "proprietary" is more than a metaphor: Some of the same mental algorithms are apparently activated in the marital and mercantile spheres.

Why should an evolved psychology see sexual partners as a sort of property and why should this be especially characteristic of male minds? According to Trivers's (1972) influential evolutionary model of parental investment and sexual selection, whichever sex invests more efforts and material resources in offspring becomes, in effect, the limiting resource for the reproduction of the less investing sex and hence an object of competition. In mammals, including *Homo sapiens,* the female is the sex making the greater parental investment, while males devote proportionately more time and energy to mating competition. Male mammals attempt to monopolize females and their reproductive capacity just as creatures contest any limited resource. But men, unlike most male mammals, make significant parental investments, too, and the combination of this paternal investment with an asymmetrical risk of cuckoldry (misattribution of parenthood) produces a powerful selective force favoring the evolution of motives that effectively guarantee one's paternity of one's putative offspring, as opposed to merely maximizing the number of young sired. In species with internal fertilization and biparental care, males have a susceptibility to mistakes in the identification of their own offspring in a way that females lack.[1]

We propose that these selection pressures have been responsible for the evolution of psychological mechanisms whose adaptive functions are success in sexual competition and cuckoldry avoidance and that men's attitudes, emotions, and actions indicative of sexual proprietariness and the commoditization of women are products of these evolved mechanisms in the context of particular historical and cultural circumstances. There is more to human sexual proprietariness than mere mate guarding. The social complexity of our species—with its alliances based on both kinship and reciprocity, its moral systems and consequential personal reputations, and its cultural and ecological diversity—provides an arena within which male sexual proprietariness is diversely manifested (Dickemann, 1979, 1981). But we shall argue the ubiquity of a core mindset, whose operation can be discerned from numerous phenomena which are culturally diverse in their details but monotonously alike in the abstract. These phenomena include socially recognized marriage, the concept of adultery as a property violation, the valuation of female chastity, the equation of the "protection" of women with protection from sexual contact, and the special potency of infidelity as a provocation to violence.

Why Male Sexual Proprietariness?

The relationship between mates is special. Since Hamilton (1964), evolutionary theorists have considered kinship to be the bedrock of commonalities of interest: The degree of genetic concordance between two parties specifies a degree to which their genetic posterities (fitnesses) must coincide, and it is this coincidence of genetic fates that is responsible for the evolution of nepotistic solidarity. The evolutionary basis of solidarity between (unrelated) mates is analogous but distinct: The pair's genetic fates are linked not because they share genes with one another, but because both share genes with their mutual offspring, the common vehicles of their fitness. Because of this covariation between the fitnesses of sexually reproducing mates, selection shapes a relationship-specific social psychology vis à vis actual or potential mates.

Mates who are truly committed to monogamy reproduce only through shared descendants, so that insofar as descendants rather than collateral relatives are the principal beneficiaries of an organism's reproductive and/or nepotistic efforts, monogamous creatures may be expected to have evolved psyches such that individuals perceive their interests as coinciding more closely with those of their mates than with anyone else's (Alexander, 1987). But although the overlap of interests between mates is analogous to that between genetic relatives, and although this overlap has the potential of surpassing that of even the closest kin (in the case of monogamists with minimal investment in collateral kin), the relationship between mates is nevertheless more easily betrayed. Blood is thicker than water because the genetic fates of blood relatives are indissolubly linked, a consideration that favors forgiveness and reconciliation. Marital ties are thinner gruel: Because cuckolded males risk expending their lives unwittingly raising their rivals' children, the correlation between the expected fitnesses of mates can be abolished or reversed by infidelity.

The possibility of extrapair reproduction by one's mate is a major threat to fitness. Insofar as this threat has been part of a species' chronic selective milieu, we may expect to find an evolved psychology adapted to reducing its probability or impact. One strategic means for reducing the costs of cuckoldry is adjusting parental efforts according to cues indicating the likelihood of genetic parenthood; to this we shall return. But even at the prezygotic stage of mating, males confront a problem of how to budget their reproductive efforts between seeking, courting, and contesting new mates, on the one hand, and doing whatever it takes to monopolize the ones already acquired, on the other (e.g., Thornhill & Alcock, 1983; Trivers, 1972, 1985). Where to strike this balance depends in part upon the anticipated magnitude of later paternal efforts: In any species in which paternal investment is significant, one would expect males to exhibit a considerable interest in monopolizing mates, since a paternal investor is concerned not merely to maximize the numbers of his progeny, but to correctly identify them so as to avoid misdirected parental contributions (Daly & Wilson, 1987).

In the case of *Homo sapiens,* paternal investments are often substantial, including allocations of time and effort and transfers of resources over the course of decades. Thus a major threat to a man's fitness is the possibility that his mate may become pregnant by another man, especially if the cuckold should fail to detect the fact and invest in the child as his own. If there is a corresponding threat to a woman's fitness, it is not that she will be analogously cuckolded, but rather that her mate will channel resources to other women and their children. It follows that men's and women's proprietary feelings toward their mates are likely to have evolved to be qualitatively different, men being more intensely concerned with sexual infidelity per se and women more intensely concerned with the allocation of their mates' resources and attentions. We will review evidence of such a sexually differentiated psychology below, and we will argue that diverse cultural practices can be understood as manifestations of a cross-culturally ubiquitous male concern with the monopolization of female sexuality and reproductive capacity, reflecting a history of selection for a male sexual psychology effective in deterring rivals and in limiting female sexual and reproductive autonomy.

Note that male concern with sexual exclusivity need not imply a conflict of interests between mates. The fitness of both parties might benefit from female monogamy, for example, if detected infidelities were to inspire diminution of male parental efforts or if other males' mating efforts otherwise impose costs on already mated females. In such cases, a male's guarding of his mate to deter coercive copulatory attempts by rivals serves the interests of both mates (e.g., Lumpkin, 1983). But females are not always monogamously inclined, even when their mates' aggressions raise the costs of infidelity, and some guarded females expend considerable effort attempting to evade their mates. Male threats and coercion may therefore be directed at rivals, at the mate, or both. It follows that both theoretical and empirical analyses of the costs and benefits to all parties of each's alternative actions are essential if we are to understand the strategic functioning of the evolved psychological and physiological mechanisms of heterosexual transactions.

SEXUAL PROPRIETARINESS ON THE WING

Male sexual proprietariness is likely to evolve in any animal species with internal fertilization and paternal care. The best studied cases are neither human beings nor our close primate relatives, but birds. Unlike mammals, most avian species are predominantly monogamous and biparental, and they have the additional virtue of behaving sexually and parentally with relative observability. The result is that several species have been intensively studied, affording complex quantified accounts of the degree of sexual exclusivity in mated pairs, the circumstances of extrapair copulations, who the rivals are, the circumstantially contingent magnitudes of parental efforts, and (to some degree) the costs and benefits of alternative courses of action for the various actors in these dramas. Two exemplary research programs focusing on the problem of adaptive allocation of efforts by males in a biparental species are Anders Pape Møller's work on swallows and Nick Davies's work on dunnocks.

We shall discuss these avian cases in some detail. Our aim in so doing is not to provide a facile "referential model" whereby some other species is alleged to manifest a suite of characteristics identical to those that were relevant to the evolution of human traits (see Tooby & DeVore, 1987). Rather, the heuristic value of these studies for the student of human sexual proprietariness is that they illustrate the following:

1. The multiple conflicts of interest that arise in sexual rivalries (even in the absence of those nepotistic or reciprocity-based cooperative alliances among interested parties other than the mates themselves that complicate human sexual rivalry);
2. the diversity of behavioral and psychological adaptations that assess and respond to demographic and ecological cues of the expected fitness consequences of one versus another course of action ("expected" in the statistical sense of that which would be anticipated from past contingencies cumulated over evolutionary time); and, especially,
3. the ways in which behavioral ecologists have conducted a successful program of evolutionary psychological research, using apprehensions of the problems confronting the animals as guides to the search for strategic adaptations.

Swallows

The swallow *(Hirundo rustica)* nests under the roofs and eaves of barns, sometimes in isolated pairs, sometimes in colonies, which vary markedly in size according to the availability of insect food and shelter from the wind (Møller, 1987d). Female swallows, like other songbirds, are only briefly fertile. They lay their eggs one a day until a target clutch size is attained, and the intensity of incubation increases throughout laying so that the eggs hatch over a 1 to 3 day period. Female swallows incubate the eggs alone, but both parents feed the hatchlings.

Male swallows establish breeding territories in the spring, before the females arrive and choose their mates. One cue that has been experimentally demonstrated to affect female choice among potential mates is male tail length (Møller, 1988b). It is normally the older males, first to arrive in the spring, who have the longest tail feathers, and the longer the tail feathers the more attractive the male. So in effect, females use tail length as an index of age and arrival date, criteria that appear to be predictors in their turn of the male's vigor and prior breeding experience. Once established, pairs commonly remain intact through a breeding season, raising successive broods.

Male swallows behave as if cuckoldry were an imminent risk, following their mates closely while they are fertile. Once incubation begins, this mate guarding virtually ceases, and mated males reallocate their efforts toward the pursuit of whichever neighboring females are still fertile. The males of many other songbird species exhibit similar mixed strategies, dogging their own mates' every move during egg laying, and then switching to pursuing other fertile females (Birkhead, Atkin, & Møller, 1987). Mated males have even been observed to knock neighboring fertile females from the air and copulate by force. In *H. rustica,* the intensity of mate guarding is positively correlated with colony size, larger colonies entailing more persistent and intense pressure from rivalrous neighbors (Møller, 1987d).

Is alien insemination of a guarding male's mate a genuine risk? At least five lines of evidence indicate that it is. The first is that rivals indeed pursue, court, and harass his mate, and do so preferentially in her brief fertile period. Second, although female *H. rustica* have not been seen to actively solicit extrapair copulations, they accede (selectively) to some such attempts. Third, variations in the intensity of mate guarding apparently track variable risk, as noted above. Fourth, variations in the mated pair's own copulation rate can be interpreted as adaptive male responses to apprehended cuckoldry risk, the idea being that males increase insemination of their mates as an adaptive "sperm competition" tactic (Birkhead, 1988; Birkhead, Atkin, & Møller, 1987; Møller, 1988d; Smith, 1984a): The rate of copulation within mated pairs is enhanced by such cuckoldry risk cues as (a) brief experimental detentions of the male while his mate was fertile (Møller, 1987c) and (b) proximity of male neighbors (Møller, 1985). The fifth line of evidence that cuckoldry risk is genuine is that nestlings' tarsus lengths are more strongly correlated with their mothers' than with their putative fathers' dimensions; statistical analysis of these correlations suggests that as many as a quarter of nestlings are the products of extrapair copulations (Møller, 1987a). Genetic studies in progress will soon provide more precise information on the incidence of extrapair paternity in swallows.

Female infidelity is not random. Certain characteristics are preferred in adulterous sexual partners, and they seem to be the same characteristics that females seek in their permanent mates. Early-arriving males, for example, are not only chosen as mates, but also attain more extrapair copulations than the later breeding males (Møller, 1987b). One might then expect that these successful cuckolders would themselves be vulnerable to cuckoldry, as a result of leaving their mates unguarded while philandering, but early males instead hold an advantage both ways. This suggests that whatever makes an early male attractive to other females also makes his own mate less inclined to accept an extrapair copulation.

As for tail length, Møller (1988b) manipulated it experimentally, thus eliminating its association with other cues of quality. Males with artificially elongated tails were selected as mates sooner, were more likely to get two clutches completed in the season, and had significantly more fledged offspring. Moreover, as in the case of early-arriving males, those given long tails were both significantly more successful in attaining extrapair copulations and significantly less likely to be cuckolded themselves. Males who had their tails experimentally docked suffered opposite effects. These results can be interpreted to mean that a longer tail makes a male more attractive, providing him with an apparent quality marker which gains him both greater fidelity from his better satisfied mate and greater success in philandering. It is especially striking that artificial tail elongation should enhance brood size, suggesting that males with quality markers either attract the best females or inspire greater reproductive efforts from them, or both.

A further remarkable finding is that male swallows adjust their subsequent parental efforts according to cues that indicate the likelihood of having been cuckolded. Analyzing data from 38 intensively observed breeding pairs, Møller (1988c) found the male's provisioning effort during the nestling phase to be negatively correlated with the number of extrapair copulations that his mate had been observed to participate in during her fertile period two to four weeks earlier and positively correlated with the number of within-pair copulations. Møller also captured and detained males from colonies, keeping them from their mates for a few hours either in the morning (when fertile copulations normally occur) or at night; the males' proportional and absolute contributions of food to the young were reduced after detentions in which extrapair fertilizations might have occurred, but not after other detentions. Those males who were experimentally confined in the morning had good cause to act as if they had been cuckolded: Their females actually copulated with rival neighboring males whereas none of the control males' mates did so (Møller, 1987c).[2] When males with grounds for paternity doubt reduced their paternal efforts, nestling mortality increased, so it appears that copulating with a male other than her mate can cost a female valuable paternal contributions. Perhaps there are offsetting genetic benefits of such infidelity.

Additional experimental manipulations revealed that males are sensitive to another cue of cuckoldry risk in addition to separation from their mates and the proximity of rivals, namely, a partial clutch of eggs in the nest, which is ordinarily an indicator that the female is presently laying and hence fertile. Møller (1987c) showed that experimental creation of partial clutches, whether by removal or addition, inspired male mates both to guard more intensely and to initiate more copulations even though the experimental females were neither fertile nor receptive. In both of these experiments, neighboring males also became more interested in experimental females than in controls. It is unlikely that the neighboring males saw the addition or removal of eggs from the experimental nests, but they could have been cuing on the conspicuous mate guarding or copulatory behavior of the experimental males, which would ordinarily be accurate indexes of female fertility.

The many contingent responses of the male swallow described above add up to a complex anticuckoldry strategy adapted to the behavioral inclinations of both mates and rivals. Note that the male's repertoire does not include every imaginable anticuckoldry tactic. Neither male swallows nor any other songbirds seem able to discriminate own from alien conspecific offspring by comparing their phenotypes to self, for example. But the swallow's evolved domain-specific psychology of cuckoldry avoidance is quite impressive enough, combining tactics of mate guarding, of sperm competition, and of parental effort adjustment, all employed facultatively in relation to circumstantial cues indicative of cuckoldry risk.

Dunnocks

The dunnock (*Prunella modularis*) is a drab little bird found lurking in hedges of English gardens, but its mating system is anything but dull. Monogamous pairs, polyandrous trios, polygynous trios, and even polygynandrous groups are all common resolutions of battles within and between the sexes (Davies, 1989).

Dunnocks overwinter on their breeding grounds and begin to define and defend territories in late winter. As in many animals, female territory size is inversely correlated with food abundance, whereas males seem less concerned with food than with the females themselves and try to defend a superterritory overlapping that of several females (Davies & Lundberg, 1984). The final distribution of monogamous, polyandrous, and polygynous associations is the result of females' variable success in keeping female rivals at bay and males' variable success in monopolizing female territories (Davies & Houston, 1986; Houston & Davies, 1985). In a 4-year study of the dunnocks inhabiting Cambridge University's botanic garden, Davies (1989) found 62 monogamous pairs, 81 polyandrous females, 21 polygynous males, and 65 polygynandrous associations of two or more birds of each sex sharing mates. The research reports to date have primarily focused on comparing polyandrous trios and monogamous pairs.

The second ("beta") male of a polyandrous trio is subordinate to the "alpha," who by no means welcomes him. The most aggressively successful males are able to avoid polyandry, and in the polyandrous association, the alpha male apparently tolerates the beta male only after many aggressive confrontations fail to repel him. Monogamous males are slightly larger than the alphas of polyandrous trios, who are in turn significantly larger than the betas (Davies & Lundberg, 1984). Males attempting to establish themselves as betas are sometimes killed (Davies, 1989), but if they get to stay, they can get a share of paternity.

Both sexes feed nestlings and fledglings beakloads of tiny insects. Monogamous males always participate, but in the polyandrous trios, male contributions are more variable and evidently depend upon the particular male's likely genetic stake in the nest (Burke, Davies, Bruford, & Hatchwell, 1989). Beta males who were seen to enjoy some time alone with the female during her fertile period almost always helped feed the nestlings later (22 of 27 such cases in 1981 to 1988), whereas those who apparently failed to gain such access almost never helped feed (1 of 11 cases). Moreover, among those betas who helped feed, rates of food delivery increased as proportionate share of mating access increased.

Although alpha males do not welcome betas, females do. Nestlings fed by two males and a female received more feedings and weighed more than nestlings of either monogamous pairs or trios with noninvesting beta males; furthermore, the numbers fledged from such broods were greater than in any other breeding combination (Davies, 1986). From the female's point of view, then, two mates are better than one. But the alpha male's perspective is different: The brood's gain as a result of the beta male's help is insufficient to offset the loss from shared paternity, a point to which we shall return. Moreover, since a male's help at the nest is contingent upon his having mated, it behooves the female to escape the guarding alpha for trysts with the beta, and the more so since a beta male who fails to mate may become an active saboteur: Though the culprits were not observed directly, damaged or discarded eggs or chicks were peculiar to the circumstance of an unmated beta male (Davies, 1986).

A female dunnock determined to be actively polyandrous creates a problem for her mate, who has at least three distinct tactics to deal with it. The first is the preventive tactic of mate guarding. Any failure of the alpha male to stay within 5 m of the female provides an opportunity for the beta male to copulate, and so, although nest building is solely a female task, the male follows his mate closely as she gathers nesting materials and throughout her fertile period (Davies, 1985). This intensive guarding of the fertile female interferes not only with the guarding male's own feeding, but with the female's as well, yet the gain in paternity protection apparently offsets the costs from the male's perspective. However, given the female's interest in mating with both her males, the alpha's attempts to monopolize her are typically less than fully successful. Hence, the dunnock is a species characterized by sperm competition, which selects for frequent, circumstantially contingent copulation (Birkhead, 1988; Smith, 1984a), and in fact the male's second line of defense is the damage containment tactic of copulating frequently in response to cues of possible infidelity.

These two anticuckoldry tactics characterize many other songbirds, including the swallows discussed earlier. But the exceptional practice of stable polyandrous associations makes sperm competition a stronger selective force in dunnocks than in other biparental songbirds, and the male dunnock has evolved a third tactic, namely, a direct means of stimulating the female to eject a prior mate's ejaculate. In an elaborate and protracted foreplay the male pecks the female's cloaca for up to two minutes until she ejects a mass of cloacal contents, including sperm from recent matings, and then cop-

ulation occurs (Davies, 1983). Alpha males in polyandrous trios copulate more frequently than monogamous males, and they employ cloacal pecking more, too. Moreover, within polyandrous trios, both a male's copulations and his cloacal pecks become
more numerous the more time the female spends with her other mate.

Most recently, Davies and collaborators have used DNA fingerprinting to determine parenthood and assess the utility of these tactics (Burke et al., 1989). The resident
female was the mother of every one of 133 young tested, and even though neighboring
males showed interest in fertile females when their own mates were busy incubating,
only 1 of the 133 was sired by an intruder. Eleven broods of polyandrous trios were
assessed, totaling 34 chicks (including the lone chick sired by an intruder); alpha males
had a surprisingly small advantage, siring 18 chicks versus 15 by betas. From the alpha
male's perspective, the greater productivity of polyandrous trios as compared with
monogamous pairs does not nearly compensate this lost paternity, confirming the
point that the beta male's presence profits the female but not the alpha male.[3] The
other striking result was that the actual shares of paternity achieved by beta males provided no further prediction of their feeding effort beyond that provided by their mating
success; if a male copulated but happened to sire no young, he was still likely to help.
From this fact and from a lack of any evident within-brood discrimination, it seems
that males cannot identify their own offspring within a multiply sired brood, and use
proportionate sexual access as their rule of thumb for allocating parental effort.

Summary Remarks on Avian Anticuckoldry Tactics

The mating frequency, mate guarding, feeding rates, and other actions of male birds
have been shown to vary adaptively in relation to numerous predictors of cuckoldry
risk, including the timing of their mates' fertility, the mating strategies of rivals, the
degree of coloniality, one's attractiveness relative to rivals, and lapses of surveillance
of the mate. The research on swallows and dunnocks (along with work on many other
birds; e.g., Alatalo, Gottlander, & Lundberg, 1987; Beecher & Beecher, 1979; Birkhead, 1988; Birkhead, Atkin, & Møller, 1987; Bjorklund & Westman, 1983, 1986;
Carlson, Hillstrom, & Moreno, 1985; Frederick, 1987; Fujioka & Yamagishi, 1981;
Hatch, 1987; Hatchwell, 1988; Lumpkin, 1983; McKinney, Derrickson, & Mineau,
1983; Montgomerie, 1988; Morton, 1987; Ridley & Hill, 1987; Røskaft, 1983; Sherman & Morton, 1988; Smith, 1988) provides strong evidence that paternally investing
male animals have evolved sexual psychologies designed by selection to reduce both
the likelihood of cuckoldry and its costs once incurred.

We may expect no less of the evolved psyches of paternally investing *Homo
sapiens.*

THE HUMAN MALE'S PROBLEM

Unfortunately, the study of our own species has, until recently, been conducted almost
without regard for evolutionary strategic considerations. The result is that we cannot
yet answer many of the questions about human sexual proprietariness that a behavioral ecologist would consider elementary. There is a substantial social psychological

literature on "jealousy," for example, that has never addressed questions of how (or even whether) jealous feelings and actions might track the fertility of their objects. Nevertheless, sexual rivalry and proprietariness are such prominent aspects of human sociality that there is much in the literatures of the humanities and social sciences bearing upon our topic.

To elucidate the strategic organization of human male sexual psychology, we need to reconstruct the social and mating systems to which it is adapted. The evidence that human mating and parental practices have long been different from those of other hominoids is diverse, including the reduction of sexual dimorphism in hominid evolution, the sexual division of labor in foraging societies, and the cross-cultural ubiquity of marriage, of patrilineal or bilateral kinship reckoning, and of biparental care. Like songbirds, men make major parental investments in their putative offspring despite the problem of uncertain paternity (Smith, 1984b; Wilson, 1987).

But why should uncertain paternity be endemic to the human condition? Why would females in a biparental species such as ours not have evolved purely monogamous inclinations? The question is not simply sexist, since females do not gain in expected fitness by increasing their numbers of mates the way that males do (Bateman, 1948). Moreover, there are several potential costs to polyandrous mating, including wasted time in acquiring superfluous gametes, risk of disease transmission or damage during mating (Daly, 1978), possible diminution or withdrawal of paternal investments should the male mate discover that he has been cuckolded (Trivers, 1972), possible elevation of competition among one's offspring in response to cues of reduced relatedness due to mixed paternity (Holmes & Sherman, 1982), and any additional costs, such as violence, that the mate may impose in pursuing his own counterstrategy of deterring infidelity.

However, there is an equally long list of potential benefits that a paired female might gain by extrapair mating (Smith, 1984b). One possibility is superior genes; females may cuckold investing mates when other males appear to offer better genes (Benshoof & Thornhill, 1979). If this sounds farfetched, it is worth noting that the females of several "monogamous" bird species have been found to engage in surreptitious extrapair copulations mainly or exclusively with known neighbors of higher dominance rank or with other qualities that are superior to those of their mates (Frederick, 1987; Fujioka & Yamagishi, 1981; Smith, 1988; Westneat, 1987). A second reason for females to engage in extrapair mating is simply the material benefit of whatever aspirant adulterers offer (Symons, 1979). Kaplan and Hill (1985) found that the better hunters in a foraging society outreproduced poorer ones, partly because their children survived better, but also in part because hunting prowess gained men extramarital affairs with fertile women. A third potential benefit of female adultery is that the distribution of some possibility of paternity among multiple males may sometimes increase the total investments received by young from their possible fathers, as in Davies's dunnocks, or at least reduce the probability that those males will directly damage the young later (Hrdy, 1981). Other potential benefits listed by Smith (1984b) include status enhancement, fertility backup, and genetic diversification of young as a hedge against environmental change or as a means of reducing their similarity and hence their detrimental competitive effects on one another. Finally, even where extrapair mating is neither beneficial to nor sought by the female, the extrapair male's threats may make ready compliance less costly than resistance.

Strict Female Monogamy Has Not Characterized Human Sexual Selective History

A number of morphological and physiological phenomena in *Homo sapiens* appear to reflect a history of selection in which polyandrous mating and sperm competition were frequent enough to be consequential. The feature that has been most discussed in this context is "loss of estrus" (or "concealment of ovulation"): Women, unlike our nearest relatives, the chimpanzees, have no conspicuous signal of ovulation and relatively little cyclicity of sexual activity. Some of the many hypotheses proposed to account for these facts assume that the evolving human female was not strictly monogamous. Benshoof and Thornhill (1979) and Symons (1979) suggested that women have evolved to hide ovulation from their mates specifically to facilitate cuckolding them. Others have offered variants on the idea that concealing ovulation from one's mate is useful more by virtue of obliging prolonged mate guarding, which tips the male's strategic balance from extrapair mating effort to investment in spouse and offspring (Alexander & Noonan, 1979; Strassmann, 1981; Turke, 1984). Although the hypothetical female in this second scenario is more of a monogamist, her gambit can work only if she "threatens" her mate with the prospect of polyandrous mating. However, it is by no means clear that loss of estrus is adaptively linked to polyandrous inclinations in females. An alternative is that the benefit of concealed ovulation lies precisely in facilitating true monogamy, the point being to conceal her fertile state not from her monogamous mate but from other males whose high dominance status would allow them to supplant her preferred mate and who would then offer less in the way of paternal investment (Daniels, 1983; Strassmann, 1981). Burley (1979) and Symons (1979) offer additional hypotheses, which neither require nor disallow that the woman be polyandrously inclined. In view of these multiple scenarios, it is not yet clear what "loss of estrus" implies about ancestral human mating systems.[4]

More convincing vestiges of a sexual selective history in which females mated polyandrously can be found in the human male. Perhaps the clearest such vestige is testis size (Short, 1977). Men's testes are substantially larger, relative to body size, than those of gorillas, a species in which males are polygynous but females mate monogamously so that "sperm competition" within the female reproductive tract is absent. The idea is that such sperm competition selects for high sperm counts and ejaculate volumes and that men evolved in a sexual selective milieu where sperm competition was more extreme than in gorillas. Conversely, among chimpanzees, individual females frequently mate with two or more males during a single ovulatory period, engendering intense sperm competition, and the relative testis size in these animals is much greater than our own, implying that humans do not lie at this end of the sperm competition spectrum either. Broader taxonomic comparisons (Harcourt, Harvey, Larson, & Short, 1981; Kenagy & Trombulak, 1986; Møller, 1988a) verify the relationship between female polygamy and male testis size, reinforcing the conclusion that human testis size and sperm production are adapted to an ancestral mating system in which females were not so promiscuous as chimpanzees, but did not always confine themselves to a single mate either.

A distinct vestige of ancestral patterns of sexual selection is the scrotum, a structure that prolongs the shelf life of spermatozoa by placing them in a relatively cool environment, at some cost in testicular vulnerability. Bedford (1977) proposed that the scrotum is an adaptation for epididymal sperm storage in those mammals who need to produce fertile ejaculates repeatedly in short order, a demand he linked to polygyny. Smith (1984b) amended the argument, noting that the presence of scrota in chimps and people and their absence in gorillas and orangutans are more consistent with the idea that the demand derives from polyandrous matings by females and resultant sperm competition than from polygynous matings by the males.

If men have evolved under the selective pressure of sperm competition, we might furthermore expect that they, like swallows and dunnocks, will exhibit strategically variable responses to possible cues of female infidelity. Baker & Bellis (1989) had monogamous couples collect intravaginal ejaculates by condom, while maintaining diaries of their sexual and other activities. The number of spermatozoa transferred in a given copulation was not a function of the elapsed time since the last ejaculation, as might have been expected. Instead, sperm transfer increased dramatically as a function of the proportion of the time since the last copulation that the pair were out of contact. Baker and Bellis interpret this effect as a sophisticated psychophysiological adaptation to lapses in mate monitoring.[5]

MONOPOLIZING WOMEN

Men's and women's sexual psychologies have been shaped by a history of selection in an ancestral sociosexual milieu which no longer exists and which was certainly radically different from the complex agricultural and industrial societies we now occupy. But evolutionary theory, comparative data, and, above all, the common aspects of human psychology in its diverse cultural and technical settings all afford windows on the sociosexual milieu in which the human mind evolved and on the adaptive problems to which our species-typical social and sexual motives, emotions, and ways of thought constitute the solutions. Inquiring what remains stable in the face of diversity (such as sex differences in the use of violence; Daly & Wilson, 1988b, 1990) can direct us to an appropriate level of abstraction for avoiding the pitfalls of ethnocentrism in our efforts to characterize an evolved human nature. Equally revealing of evolved inclinations are phenomena that did not exist during our formative millennia and hence cannot be considered targets of selection or adaptations in their own right, yet emerge with uncanny regularity in certain circumstances. An example, discussed later, is the establishment of harems.

Our best guess about the sort of reproductive system in which the human psyche evolved, and to which it is adapted, is one in which mateships were predominantly but not exclusively monogamous, paternal investment was significant, and the variance in reproductive success was slightly greater among men than among women (*effective polygyny*). One sort of evidence supporting this surmise is the fact that such a system appears to be almost universal among relict human societies occupying nonagricultural ecological niches (although Australian aborigines, with their relatively extreme marital polygyny, are a puzzling exception). Additional evidence of mild effective polygyny during human evolution can be seen in a number of small but robust morphological and life-historical sex differences in body size, age at maturity, and rate of senescence. Sexually differentiated psychologies and sexualities are products of that same selective history.[6]

Biparental care notwithstanding, men, like other male mammals, have generally had some chance of increasing fitness by increasing their numbers of mating partners. The male psyche appears adapted to exploit any such opportunities: Men, more than women, aspire to polygamy and to extrapair sex (Symons, 1979; Daly & Wilson, 1983). One result of this chronic aspiration is that even in monogamous societies, a few relatively successful men have typically been able to be effectively polygamous, whether simultaneously or serially, by multiple marriage or concubinage. Relatively recently in human history, with the inequities engendered by agricultural surpluses and the rise of complex, role-differentiated societies, extreme polygyny and extreme sequestering of women became possible, and wherever men gained despotic power over a populace, they then used their power and resources to hoard large numbers of fertile women as wives and concubines (Dickemann, 1979; Betzig, 1986). The most despotic harem holders confined women in cells, guarded them with eunuchs, maintained records of their menstrual cycles, farmed them out to the harems of underlings when they got too old, and sometimes even killed and replaced them en masse in the event of a security failure and possible cuckoldry (Dickemann, 1979; Betzig, 1986; Busnot, 1715).

Harems are telling phenomena not because they are anything our psyches are specifically adapted to, but because they reveal evolved appetites. The widespread establishment of harems falsifies a popular hypothesis among anthropologists to the effect that the reason men aspire to polygyny is because wives are economic assets (e.g., Boserup, 1970; Grossbard-Schechtman, 1984). Proponents of this view deny, sometimes explicitly, that women are valued as goods in themselves—as the perquisites of power and status rather than the means thereto. But the occupants of harems are typically prevented from being productive, and their maintenance is enormously costly. Fantastically wealthy and powerful men neither forsake the acquisition of women nor use them to augment their wealth; they collect them. Neither can the agendas of harem holders be understood as the pursuit of mere sexual diversity: Monopolization is invariably a principal objective. Guarded harems constitute the hypertrophied manifestations of male ambitions released from the usual constraints of limited personal power, the fantastic products of a male psyche that evolved in social milieus in which extreme polygyny was impossible, but any increment in the numbers and/or the degree of monopolization of one's mating partner(s) would gain a selective advantage.

Only the richest and most powerful men could institute such elaborate arrangements to retain exclusive sexual access to many reproductively valuable women. However, millions of men have guarded and constrained "their" women by practices that seem to depart from those of despots only in degree. Veiling, chaperoning, purdah, and the literal incarceration of women are common social institutions of patrilineal societies, and it is only women of reproductive age who are confined or chaperoned. Prepubertal children and postmenopausal women enjoy considerable freedom. These practices are status graded (Dickemann, 1981): The higher the social status the more claustrated the women.[7] Chinese foot binding was another such status-graded practice, which simultaneously made an ostentation of the male owner's capacity to dispense with the woman's labor and rendered her incapable of flight. There is considerable cross-cultural variation in the severity and institutionalization of such practices, but the repeated convergent invention of claustration practices around the world and the confining and controlling behavior of men even where it is frowned upon (Daly, Wilson, & Weghorst, 1982) reflect the workings of a sexually proprietary male psychology.

Man's inventive imagination has produced countless designs for chastity belts.

Less familiar to Westerners but much more frequent are genital mutilations designed to destroy the sexual interest of young women and even their penetrability until surgically reopened. These practices range from partial through complete clitoridectomy to removal of most of the external genitalia and suturing shut of the labia majora *(infibulation)*. Such genital mutilations are still prevalent in at least 23 countries, affecting tens of millions of women (Hicks, 1986; Hosken, 1979). Claustration and chastity belts might be interpreted as responses to male-male competition whether women were polyandrously inclined or not, but practices like clitoridectomy show that the women are being "guarded" not only from predatory males but from their own inclinations.

What about mate guarding plain and simple? Are men inclined to dog their fertile mates like songbirds during egg laying? In patrilocal societies, where wives are surrounded by their husbands' relatives, a man may be content to leave his wife under the scrutiny of his mother or other kin. But in many societies—including the foraging (hunting and gathering) peoples who provide the best contemporary models of the ecological and sociopolitical contexts within which the human psyche evolved—people were mobile and group compositions variable. Men had to look out for their own interests.

We are aware of only one naturalistic study of men's mate guarding that tested evolutionary psychological hypotheses. Anthropologist Mark Flinn (1988a) recorded the identity, whereabouts, and activities of everyone he saw during standardized scan-sampling walks in a Caribbean village, where mating relationships were unstable and often nonexclusive and where men directed paternal investments selectively to their own offspring in matrifocal households of mixed paternity. Flinn's data showed the following:

1. Men whose mates were "fecund" (i.e., cycling) were more often in their company than those whose mates were "infecund" (pregnant or postreproductive).
2. A women's fecundity was associated with proportionately more agonism in her mate's interactions with her and with other men.
3. Among mated couples that had nonexclusive relationships there was more agonistic interaction than among monogamous couples.
4. Interactions between mates of a particular woman were significantly more often agonistic than other male-male interactions.

These results appear to reflect male social motives and emotions that are responsive to cues of rivalry and female fertility and that function to promote sexual exclusivity.

THE PSYCHOLOGY OF JEALOUSY

The state of being concerned that one's sexual exclusivity is or might be violated is what people usually mean when they refer to "jealousy." Social and personality psychologists have recently devoted some attention to the task of characterizing sexual jealousy (and distinguishing it from envy), but in the absence of a strategic understanding of the psyche, they have achieved little clarity.

Romantic or sexual jealousy has been conceived of as a personality characteristic, a particular emotion, a particular set of actions, or anything one feels or does in a particular sort of situation; many discussions begin with one meaning before slipping unwittingly into another. Since feelings as variable as rage and depression can be considered manifestations of "jealousy," Hupka et al. (1985) argue that someone in the situation of "threat" by a romantic "interloper" is, by definition, jealous, but they belie their own definition by elsewhere calling jealousy "a very intense and potentially destructive emotion" and by considering the intensity of jealousy to vary between people within standard situations. The problem is that the concept of jealousy cannot be captured with a definition that is purely internal (an emotion of a certain quality) *or* purely external (a situation). Sexual jealousy is a complex psychological system whose functioning is inferred from observable combinations of circumstance and response—a system that is activated by a perceived threat that a third party might usurp one's place in a sexual relationship and that generates a diversity of circumstantially contingent responses aimed at countering the threat.

Research papers on sexual jealousy often treat the subject pejoratively. After a nonevaluative introduction, for example, Buunk and Hupka (1987) consider their questionnaire respondents to have a problem if they either agree that "it would bother me if my partner frequently had satisfying sexual relations with someone else" or deny that "it is entertaining to hear the sexual fantasies my partner has about another person." With jealousy thus conceived as a character defect, its social consequences and effectiveness in promoting sexual exclusivity have hardly been explored. An exception is Mathes's (1986) prospective study, in which couples with relatively low jealousy scores were subsequently found to be relatively likely to break up, suggesting either that jealousy is actually effective in maintaining relationships or else that its intensity reflects the value that the jealous party places on the relationship. In this light, the celebrated phenomena of romantic love and the disparaged phenomena of jealousy are functionally linked aspects of individualized sexual bonding and proprietary claim, and it is hardly surprising that lovers sometimes interpret an absence of jealousy not as a sign of selfless love but as a sign of weak commitment.

Jealousy is often found to be associated with "low self-esteem" (Buunk, 1986; Mathes, Adams, & Davies, 1985; Stewart & Beatty, 1985; White, 1981) and with "emotional dependency" (Bringle & Buunk, 1985; Buunk, 1981, 1982a; White, 1981). The researchers who have demonstrated these associations have not considered whether having poor self-esteem or being emotionally dependent might reflect accurate assessments of one's own and one's mate's respective values in the heterosexual marketplace (but see Hansen, 1985) or other cost-benefit considerations relevant to the utility of maintaining the relationship, and hence might be legitimate grounds for jealous concern. (Nor do they ask the more basic evolutionary psychological question of why "self-esteem" exists and is something that people will incur costs to defend.) The impacts of factors that an evolutionist would consider crucial to the domains of mate selection and mate guarding—ages, reproductive condition, joint and separate reproductive histories, aspects of the resource circumstances of the mates and any rivals, and so forth—have yet to be addressed.

Instead, researchers have attempted to account for the domain-specific features of jealousy in terms of inadequate domain-general constructs and processes. Buunk and Bringle (1987), for example, suggest that sexual jealousy arises "in our culture" (gratuitously implying that it does not exist elsewhere) because limited opportunities to compare one's sexual talents with those of others "make the sexual realm very sensitive to insecurity and competition" (p. 130). They maintain that the seeking of privacy for sexual activity is the cause of insecurity and thus jealousy (since "a direct comparison of oneself versus others in this area is difficult"), without noticing that the seeking of privacy then requires explanation. Salovey and Rodin (1986) ask why paper-and-pencil responses to hypothetical scenarios indicate that "threats to romantic relationships have more powerful consequences than not measuring up in social-comparison situations" and suggest as an answer that "the increased affective charge in the romantic situation may be due to the involvement of two other people versus just one in the social-comparison situation" (p. 1111).

As regards sex differences, the sexual jealousy literature is even more in need of an evolutionary overview. Some researchers (e.g., Pfeiffer & Wong, 1989; Salovey & Rodin, 1986) fail to separate men's versus women's responses in their analyses. Where the sexes have been distinguished, there has been something of a controversy about which, if either, is the more jealous. The question is an empty one, arising from the practice of summing item scores to get a "scale" rather than from any theory of the nature of either jealousy or gender. Different researchers ask different things and it is clearly meaningless to simply rank the sexes, since their reactions are qualitatively different. Shettel-Neuber, Bryson, and Young (1978), for example, had students describe their own probable actions in a situation they saw portrayed on videotape: Men considered themselves likely to become angry, drunk, and threatening, women to cry and to feign indifference. Teismann and Mosher (1978) solicited the reactions of dating couples to a hypothetical jealousy-inducing situation and reported that men's concern and distress were focused on possible sexual contact between their partners and male rivals, whereas women were primarily concerned with their boyfriends' expending time, money and attention on rival females (see also Buunk 1981; Francis 1977).

Results like Teismann and Mosher's (1978) are obviously suggestive of strategic differences in sex-typical jealousy algorithms, appropriate to the somewhat different threats to fitness confronting ancestral women versus men. But one should be wary of overinterpreting the sex differences (and/or lack of same) in the social psychological literature, since the ecological validity of the studies is problematic. The bulk of the data are paper-and-pencil responses of captive undergraduates to questionnaire items which may or may not have anything to do with anything they have ever experienced. (Buunk's Dutch research is the principal exception to this excessive reliance on undergraduates, but his use of more representative subject populations does not obviate possible problems of failures of memory, candor, and self-knowledge.) In contrast with the inconclusive results of self-report studies, there is little ambiguity about sex differences in jealousy when one looks at such real-world phenomena as homicide, wife beating, initiation of divorce, and psychiatric cases of "morbid jealousy" (Daly, Wilson, & Weghorst, 1982).

The only evolutionarily informed body of self-report research on jealousy is that of Buss (1988), who defined the domain of interest in functional terms, as "mate retention tactics," and set out to characterize them. Buss asked subjects what they and others did in order "to prevent their partner from getting involved with someone else" and constructed a complex hierarchical taxonomy of the answers. His scheme first distinguishes "manipulations" aimed at partner versus rival(s) and then subdivides each. Partner-directed tactics, for example, are classified as "direct guarding" (including various threats, derogation of competitors, etc.), or "positive inducements" (including resource transfers and various behavioral accommodations to the partner's tastes). Buss then made a start on the question of what predicts a person's resorting to particular tactics by asking whether gender and the relationship histories of subjects predicted the tactics they claimed to have used, as well as asking opinons about which tactics actually work. Ecological validity remains a problem in these paper-and-pencil studies, but Buss (1988, 1989) has at least opened the way to an understanding of the tactical richness and strategic organization of the evolved psychology of sexual jealousy.

An alternative approach to the solicitation of introspections from volunteers is the analysis of archives recording manifestations of sexual jealousy. In studies of spousal homicide motives, for example, the leading substantive issue identified by police and psychiatrists is invariably "jealousy," and more specifically jealousy on the part of the man, regardless of which partner ends up dead.[8] Although the information on many cases is too sparse to attribute them to anything more specific than a domestic quarrel, diverse threads of evidence indicate that the major source of conflict in the great majority of spouse-killings is the husband's knowledge or suspicion that his wife is either unfaithful or intending to leave him (Daly & Wilson, 1988b). Studies of sublethal violence and chronic battering pinpoint the same primary issue.

A small minority of the men who kill their wives are found "unfit to stand trial" or "not guilty by reason of insanity," and many of these are considered to be suffering from a psychiatric condition called "morbid jealousy" (Mowat, 1966), diagnosed on the basis of an obsessive concern about suspected infidelity and a tendency to invoke bizarre "evidence" in support of the suspicion. However, most men who kill in a jealous rage are not considered insane. Quite the contrary: Anglo-American common law specifically deems killing upon the discovery of a wife's adultery to be the act of a "reasonable man" (Edwards, 1954) and deserving of reduced penalty "because there could not be a greater provocation" (Blackstone, 1803). Other legal traditions—European, Oriental, Native American, African, Melanesian—all concur (Daly & Wilson, 1988b; Daly, Wilson, & Weghorst, 1982). Not only is jealousy "normal," but so it seems is violent jealousy, at least if perpetrated by a man and in the heat of passion.

THE AFTERMATH OF RAPE

Reactions to rape provide a particularly revealing window on the psychology of male sexual proprietariness (Thornhill & Thornhill, 1989). Men often reject raped women as "damaged goods," sometimes accusing the victims of having provoked or enjoyed the rape (e.g., Brownmiller, 1975; Burgess & Holmstrom, 1974; Karkaria, 1972; McCahill, Meyer, & Fischman, 1979; Miller, Williams, & Bernstein, 1982; Weis & Borges, 1973). Even where there is no issue of the illicit copulation having been other than coerced, men still seem to perceive the woman as diminished in value: "She was all mine and now she's been damaged," says one participant in a therapy group for American men whose partners have been raped; "Something has been taken from me. I feel cheated. She was all mine before and now she's not," says another (Rodkin, Hunt, & Cowan, 1982).

Blaming the victim is a phenomenon that occurs more widely than just in rape cases (Ryan, 1971), so these examples may simply reflect the operation of domain-general victim-blaming-and-denigration processes (Kanekar, Kolsawalla, & D'Souza, 1981; Smith, Keating, Hester, & Mitchell, 1976). However, rape victims appear to have some special difficulties. According to Rodkin, Hunt, and Cowan (1982), "while the husband, lover, or father would seem to be a most appropriate source of comfort and understanding to whom the victim could (or should) turn, he may, in fact, be the least understanding." Any tendency for men to denigrate rape victims more than do women might be partly due to men's inability to identify with the female victims (Sorrentino & Boutilier, 1974; Weis & Borges, 1973), but this consideration cannot explain the extreme negativity of the reactions of the very men closest to the victim.

Thornhill and Thornhill (1983, 1989) have examined the aftermath of rape from an evolutionary psychological perspective. They argue that a single coerced act of extramarital sex, though representing a threat to the husband's fitness, is much less cause for his concern than a more ambiguous act, which might have involved female complicity and might therefore be predictive of further infidelities. Accordingly, they hypothesize that men's jealous concerns in the aftermath of rape will actually be alleviated by evidence of coercion, including even injury to the woman. Postrape emotional trauma and sexual dysfunction are usually interpreted as reflecting the victim's fear or repugnance of men and sex as a result of the rape itself, but the Thornhills' analysis leads to a counterintuitive prediction: Insofar as postrape emotional trauma and difficulties in relations with husbands and boyfriends arise out of men's reactions to rape victims, such problems may be less severe in the more brutal and hence less equivocal incidents. And that proves to be the case. Physical injury during the rape is associated with lesser rather than greater postrape difficulties with male partners. Note that this result is contrary not only to the commonsensical notion that greater injury would produce greater psychological trauma, but also to the idea that the aftermath of rape is an instance of a general tendency to blame victims more the greater their victimization.

Other predictors of postrape difficulties also seem to track correlates of the potential costs of infidelity to the man and hence of expected jealousy (Thornhill & Thornhill, 1989). The aftermath of rape is more problematic for women in marital or common-law unions than for single women, which is again contrary to what one might have expected from a commonsense hypothesis that rape engenders fear that is exacerbated by vulnerability, since single women seem likelier to feel especially vulnerable. Furthermore, the aftermath of rape is more problematic for women of reproductive age than for prepubertal or postmenopausal women, a result predicted by the Thornhills from evolutionary reasoning.

PATERNITY CONFIDENCE AND PATERNAL EFFORT

Parental care is costly, and selection must always have favored those psyches that allocate it discriminatively toward own offspring. But whereas female mammals normally incur no risk of misidentifying their offspring, paternity is mistakable, which partly explains why mammalian paternal care is relatively rare. As we've seen, male songbirds adjust paternal effort in relation to probability-of-paternity cues. It seems likely that the human male psyche is no less subtle, but the subject has not been studied systematically.

A particularly interesting question is whether paternal affection is influenced by the child's resemblance to self in a way that maternal affection is not (uncertain maternity having been no issue in our evolutionary history). People are profoundly sensitive to paternal resemblances of children and relatively indifferent to maternal ones (Daly & Wilson, 1982); moreover, mothers are especially keen to note such resemblances, as would be expected if such claims function as paternity assurance tactics. But no one has assessed whether paternal investments in putative offspring actually vary in relation to these probability-of-paternity cues.[9] In fact, no male animal has yet been shown to use phenotypic resemblance to self as a probabilistic paternity cue. Songbirds are apparently incapable of this feat, but men are not.

Although the impact of variations in paternity confidence upon the quality of human paternal solicitude remains unexplored, there is ample evidence of men's reluctance to assume child support obligations to the offspring of other men (Wilson, 1987). Having a child toward whom a new husband will have to assume stepparental duties diminishes rather than enhances a woman's marriageability (Becker, Landes, & Michael, 1977; Borgerhoff Mulder, 1988; Knodel & Lynch, 1985; Voland, 1988; Wilson & Daly, 1987). Moreover, steprelationships, once established, are much less satisfying to all parties than the corresponding genetic relationships (reviewed by Wilson & Daly, 1987), and much more likely to erupt in violence (Daly & Wilson, 1988a, 1988b; Wilson & Daly, 1987). Stepfathers are discriminative in their maltreatment, sparing own offspring in the same household (Lightcap, Kurland, & Burgess, 1982; Daly & Wilson, 1985). The tensions characteristic of steprelationships are cross-culturally ubiquitous (Wilson & Daly, 1987).

Some have suggested that the problems characteristic of steprelationships are incidental consequences of the creation of a "parent-offspring" relationship too late. There is no evidence in favor of this idea and at least one study that speaks against it. Flinn (1988b) found that men who coresided with stepchildren from their births were if anything even more hostile toward them than were those whose steprelationships were established later, and much more so than genetic fathers. One possible implication is that human paternal affection is "cognitively penetrable" by something like conscious knowledge of paternity or nonpaternity. (This need not imply that paternal affection is insensitive to cues like phenotypic resemblance as well, without the man's necessary awareness.)

Children can, of course, be assets regardless of genetic paternity. Their labor can create wealth for their parents, and expanding one's close kin network brings political and social clout. Some social scientists have been so impressed with these proximal utilities of children as to assert that men are not concerned about the genetic paternity of children at all, except insofar as it confers the jural rights of the "pater role." In an extreme version of such arguments, Paige and Paige (1981) proposed that the reason public acknowledgments of paternity exist is to establish paternal entitlements against rival claimants. But children impose economic costs more surely than they provide benefits, and one can find (both in our own society and in the ethnographic record) myriad disavowals of paternity and its attendant obligations for every case in which two or more men maintain that they are the sires of a disputed child. Men are profoundly concerned that the children in whose welfare they invest are their own and are often enraged to discover otherwise (Daly & Wilson, 1987, 1988b).

It does not follow that evolved paternal psyches will be such as to invest selflessly even when paternity is certain. In paternal investment, as in other spheres, creatures allocate resources in the pursuit of their perceived interests. Wherever parental care is essential but biparental care is not, desertion of the joint parental enterprise tempts whichever parent is in a position to abscond first (Maynard Smith, 1977). In mammals (though not in all animals; e.g., see Beissinger & Snyder, 1987), that means the father. Nevertheless, it is a universal apprehension of human beings that although genetic fathers may withhold investment, a man will be more inclined to pay his share when convinced that the child is his own than when he doubts it (Wilson, 1987). Moreover, people perceive it as just that a man should support his children and as an injustice when paternity is misattributed and support obligations follow.

Two legal "theories" have been proposed to justify the imposition of a child support obligation on reluctant putative fathers: the theory of delict, whereby the father's liability arises from his mere (illicit) sexual access to the mother, and the theory of descent, whereby the father's liability is based on his genetic relationship with the child (Sass, 1977). Both theories are based on the idea that beneficiaries should pay, but they differ in focusing upon benefits of greater or lesser remove from fitness, namely, sexual contact versus parenthood. Where delictual constructions have been tried, they have violated people's sense of justice and have failed (Wilson, 1987). The implication is that people perceive children to constitute benefits to their genetic parents over and above any pleasure had in conceiving them, and over and above any material and social benefits they provide to whomever occupies a parental role.

This is not to deny that patrilineal headship confers power, prestige, and resources (e.g., Paige & Paige, 1981). Most human societies are predominantly patrilineal, despite the surer links of the maternal line. This is not so paradoxical as it first appears. The point is not that people define their primary kinship links without regard to genetic relatedness, as some have claimed (e.g., Sahlins, 1976), but rather that patrilineal affiliation follows from responses to the uncertainty of paternity. In extremely patrilineal societies, patriarchal power both provides and is predicated upon confidence of paternity. A man's genetic relatives share and protect his interest in the fidelity of his wife, and patrilineage honor is defined in terms of female chastity (Dickemann, 1979, 1981). But matrilineal societies also exist, in which a man's family of primary identity consists of maternal rather than paternal relatives, and the husband/ father has less legal authority over his wife and children. In such societies, there are few cultural institutions that inhibit a woman's sexual autonomy, and men have little economic support obligation to their wives' children. The mother's brother typically plays a prominent "parental" role toward his sister's children, especially her sons, who are their maternal uncles' principal heirs *(avuncular inheritance)*. These practices tend to be associated with high rates of divorce and remarriage, maternal custody of children, and a high probability that a woman's children have different sires.

Why such social practices should exist has been called the "matrilineal puzzle." Evolutionarily sophisticated discussions of it have focused upon the lesser paternity confidence in matrilineal as compared to patrilineal societies (Alexander, 1974; Flinn, 1981; Flinn & Low, 1986; Gaulin & Schlegel, 1980; Hartung, 1985; Kurland, 1979). The risk of misattributed paternity is believed to be relatively high in matrilineal societies, and whereas a man's putative offspring may be no relative at all, his sister is at least a half-sister and her children are his kin. A problem with early evolutionary models of the adaptive rationale for avuncular inheritance is that the society-wide incidence of extramarital conception that would be required to make sisters' children closer relatives on average than wives' children, and hence men's preferred heirs, exceeds any plausible estimate for matrilineal societies. This problem is resolved, however, when one considers that rules reflect imposed resolutions of conflicts (Alexander, 1987; Flinn, 1981; Hartung, 1985; Thornhill & Thornhill, 1987). A man might prefer his putative son to his sister's son, for example, and yet be under pressure from both of his parents to prefer the latter, who is, from their perspective, the more certain grandson. The man's threshold for compliance versus resistance to such pressure should be affected by cues affecting "paternity confidence," but should not necessarily correspond to the point at which the party with greater expected relatedness to himself switches from nephew to son (Hartung, 1985).

Inheritance rules are not simple expressions of individual men's apprehensions of descent. Instead, they represent the fruits of historical successions of conflict resolutions. Typological summary characterizations of societies (as "matrilineal," "patrilineal," etc.) mask the dynamic processes of conflict resolution out of which rules and ideologies arise. The ethnographies provide countless anecdotes of individuals defying inheritance rules and incurring approbation for bestowing their resources in their own perceived interests. It may well be the case, then, that men ubiquitously adjust paternal investment in relation to paternity confidence cues. Empirical studies are needed to determine to what degree resources actually flow along the lines prescribed by the inheritance "rules" of traditional societies and to determine the correlates and causes of individual testamentary decisions.

MARITAL ALLIANCE

Marriage is a cross-culturally ubiquitous feature of human societies, notwithstanding variations in social and cultural details of the marital relationship (Flinn & Low, 1986; Murdock, 1967; van den Berghe, 1979). What this means is that men and women everywhere enter into individualized reproductive alliances in which there is some sort of mutual obligation and biparental investment in their joint progeny and that the alliance is recognized by people other than the marital partners. Most important, such alliances are characteristic of all known foraging societies. In most human societies, a minority of successful men marry polygynously, but most individual human mateships are, and probably always were, at least serially monogamous.[10] Although many writers have stressed that human marriages are economic unions or even political alliances between lineages, the marital institution has first and most basically to be understood as reproductive.

Marital alliance institutionalizes the sexual and reproductive entitlements of a man vis à vis a woman. Claude Lévi-Strauss (1969) has argued that marriage is a contract between men, whereby one or more men bestow rights in the bride upon the groom. One may quarrel that senior women as well as men often participate in negotiating marital exchanges, but that does not belie the treatment of women and their reproductive capabilities as valued and exchangeable goods. Marriage is predominantly patrilocal, the bride being incorporated into her husband's kinship group (Murdock, 1967), and even more ubiquitous is a proprietary construction of the marriage's significance. In industrial mass society, where the power of kinship has withered, a man vestigially "gives" his daughter in marriage, but in most of the human societies that have been described, getting a bride required a substantial transfer of resources to her kinsmen, usually as the explicit price of her reproductive future.[11]

Strathern (1984) has argued that it is ethnocentric to interpret bridewealth systems as the commoditization of women, because the actors neither equate women with "things" and deny them subjectivity as our conception of "property" implies nor attain individual ownership, the transactions instead occurring between lineages. But people are well capable of feeling and acting proprietarily toward other people while remaining aware of their subjectivity; it may or may not be accurate to interpret bride-wealth as the commoditization of human reproductive capacity, but it is no more eth-nocentric than interpreting slavery as the commoditization of human labor. As for Strathern's point about collective rather than individual entitlements, a man's kins-men may help him buy a bride, but she becomes only one man's wife (and she becomes part of his inheritance under the practice of levirate).

In the contemporary West, we are inclined to see the mating game as a great mar-ketplace of autonomous actors, but in kin-based societies and where power permits, people take a strong manipulative interest in the marital transactions of other people. The extensive role of kin in the arrangement and conduct of marriages makes the human case unique among animal mateships, and multiplies the potential conflicts among the parties involved. Marriages may be arranged without the principals' coun-sel, indeed before they are mature, and occasionally even before their births. A partic-ularly striking example of manipulative arranged marriages was the practice of shim-pua in Taiwan, in which parents acquired an infant girl as a bride for an immature son, and raised her to the role (Wolf & Huang, 1980). These future brides were often poorly treated as children, and their risk of dying before puberty was severalfold greater than the risk to their "adopting" in-laws' own daughters. Moreover and iron-ically, shim-pua marriages were often barren, apparently because rearing together from infancy had killed sexual interest.

WOMEN AS MEN'S LEGAL PROPERTY

Having acquired a wife, to what is a man entitled? At the least, sexual access and the chance to sire her children, and usually an exclusive right to both. Nothing more clearly reveals these proprietary entitlements than laws concerning adultery: the offense of sexual intercourse between a married woman and a man other than her hus-band.

Laws were codified by developing civilizations around the Mediterranean and in the Far East, in the Andes and Mexico, in northern Europe and throughout Africa, and although these traditions developed in ignorance of one another, they converged remarkably on this topic (Daly, Wilson, & Weghorst, 1982). All concurred in defining adultery in terms of the marital status of the woman. Whether the adulterous man was himself married was irrelevant.

Adultery is often treated explicitly as a property violation. The victim is the hus-band, who may be entitled to damages, to violent revenge, or to divorce with refund of brideprice. Where crimes (offenses against the state) are distinguished from offenses against individual plaintiffs, adultery may be criminalized as well; it was a capital crime in seventeenth-century England, for example (Quaife, 1979). Another legal sta-tus of adultery is its provision of a grounds for divorce, which is nearly universal in the case of an adulterous wife but much rarer in the opposite direction.

Still another legal status of adultery, and one especially revealing of the folk psy-chology of the subject, is its standing as a "provocation" mitigating responsibility for otherwise criminal behavior. Throughout the English-speaking world, the common law recognizes three kinds of acts as sufficiently provoking to reduce murder to man-slaughter, and they constitute a virtually exhaustive list of fundamental threats to fit-ness: assaults upon oneself, assaults upon close relatives, and sexual contact with one's wife. Several American states had statutes or rulings that made killing upon the discovery of wifely adultery no crime at all; although these were finally abolished in the 1970s, jury acquittals and discretionary refusals to prosecute persist. The violent rages of cuckolds constitute an acknowledged risk in all societies, and some sort of diminution of their criminal responsibility is apparently universal (Daly, Wilson, & Weghorst, 1982), regardless of whether the cuckold's violence is deemed a reprehensible loss of control or a praiseworthy redemption of honor (e.g., Safilios-Rothschild, 1969).

Man's ubiquitous resentment of adultery seems clearly to be functionally linked to paternity concern, but how direct are the psychological links? Men can sometimes be violently jealous of postmenopausal wives (and of homosexual partners), suggesting that jealousy is generalized. But men are far from oblivious to reproduction and paternity, and they often articulate their motives in the proprietary transactions of marriage in these terms. The greater success of descent-based than delict-based laws of paternal liability, as discussed earlier, suggests that paternity is an issue with some emotional force. In any case, whether paternity concern has direct psychological links with adultery concern or not, men have often called upon the paternity issue in order to justify adultery law and the double standard therein. A particularly striking example comes from French Revolutionary law, whose authors were much concerned with the abolition of prejudicial discriminations, including those based on sex, and yet retained sexual discrimination in this one sphere, arguing that "it is not adultery per se that the law punishes, but only the possible introduction of alien children into the family and even the uncertainty that adultery creates in this regard. Adultery by the husband has no such consequences" (our translation of Fenet, 1827, as quoted by Hadjiyannakis, 1969, p. 502).

Adultery compensations are institutionalized features of tribal bridewealth societies and were prominent in Anglo-American legal history, too. Since the earliest written codes of the Anglo-Saxons (Attenborough, 1922/1963), we have always had a variety of torts associated with unauthorized sexual relations. (A *tort* is a lawfully recognized wrong other than a crime or violation of a commercial contract, in which a plaintiff seeks redress from a defendant, usually in the form of monetary compensation.) Torts concerning wrongful sexual contacts include "adultery," "loss of consortium," "enticement," "criminal conversation," "alienation of affection," "seduction," and "abduction" (Arnold, 1985; Backhouse, 1986; Beckerman, 1981; Brett, 1955; Law Reform Commission of Ireland 1978, 1979a, 1979b; Lippman, 1930; Sinclair, 1987).

All these torts concerning sexual transgressions were created with sexually asymmetrical content to deter unauthorized sexual contact. The plaintiffs were typically husbands or fathers or men to whom women were betrothed. The requisites for a successful action make clear that the crucial damage was loss of sexual exclusivity, not lost labor: The woman's prior chastity was crucial, on the logic that a man who steals an already unchaste woman has stolen nothing. Also clearly indicative of men's proprietary construction of the women involved in these cases was the irrelevance of their consent, which did not at all mitigate the wrong against husband or father.

Sexual relations with another's wife can still cost an American man substantial court-ordered compensation. In *Chappell v. Redding* (1984), for example, a North Carolina ophthalmologist was obliged to pay his wife's ex-husband $200,000 for having alienated her affections. The interesting implication of an award like this is that a contemporary American woman is still not fully free to leave one husband for another.

When Dr. Redding went a-courting, Mrs. Chappell was not fair game. She was already taken.[12]

Besides these entitlements to compensation for infringements of their proprietary rights in their wives, Anglo-American husbands have enjoyed other owners' privileges as well. Until recently, husbands were legally entitled to confine wives against their will and to use force to enjoy their conjugal rights (Dobash & Dobash, 1979, 1981; Edwards, 1985). The origin of the term "rule of thumb" was an eighteenth-century judicial ruling that a husband was entitled to use a stick no thicker than his thumb to control an overly independent wife. Persons who gave sanctuary to a fleeing wife, including even her relatives, were legally obliged to give her up or be liable for the tort of "harboring," and Englishmen remained entitled to restrain wives intent on leaving them until a 1973 ruling made such acts kidnappings (Atkins & Hoggett, 1984). The criminalization of rape within marriage and hence the wife's legal entitlement to refuse sex has been established only very recently (Edwards, 1981; Russell, 1982).

The proprietary construction of marriage seems especially bald-faced in the English practice of wife sales at market, which ended only about a century ago. The husband would pay the market fee just as if he had a cow to sell, then lead his wife up to the auction block by a halter and call for bids (Menefee, 1981). Contemporary accounts of the practice often make it sound like the most blatant trafficking in women as commodities, but wife sales really constituted customary divorce proceedings; the woman was often—perhaps typically—sold not simply to the highest bidder but to a man who was already her lover. For the estranged husband, the transaction served as a public renunciation of his obligations to the woman and her children. But even if "wife sales" don't quite live up to the marketplace implications of their name, their trappings still provide a dramatic illustration that men conceptualized marriage in proprietary terms.

English matrimonial law evolved out of a bridewealth system. In medieval times, children could be "espoused" as early as 7 years of age, with the Christian church sanctifying the commitment (Helmholtz, 1974; Ingram, 1987; Swinburne, 1686/1985). Bidirectional property transfers were contracted at the espousal stage, and if one family opted out of the planned marriage, the other family had a grievance; although the church did not deem the marriage complete until sexual consummation after puberty, an aggrieved family could launch ecclesiastical court proceedings to recover damages. Disputes between two men who both claimed marital entitlements to the same woman constituted another frequent cause for legal actions, especially as legal powers over children, serfs, and other persons were eroded and the authority to arrange marriages became ambiguous.[13] Clearly, any recalcitrant bride who eloped with the man of her own choice before her espoused marriage was consummated could cause severe repercussions for her father, who was likely to launch proceedings against his daughter's "abductor." In 1285, the abduction of an heiress for the purpose of marriage was furthermore criminalized, and although the abductor could legalize his marriage in the eyes of the church by paying the marriage price, he might still be imprisoned for two years for his criminal violation of the father's right to bestow his daughter where he pleased; an abductor who could not or would not pay the marriage price was penalizable with life in prison. Subsequent legislation stripped eloping daughters of all claims against their familes' property. During the reign of Elizabeth I (1558–1603), abducting an heiress was a capital crime.

These legal entitlements of fathers and husbands over women within our own legal traditions are similar to provisions in stratified patrilineal societies around the world. That men should endeavor to control the sexuality of female relatives, as well as that of wives, follows from the treatment of female reproductive capacity as a valued commodity that men can own and exchange. Chaste sisters and daughters make marketable wives.

CONCLUDING REMARKS

This chapter describes various manifestations of male sexual proprietariness in different societies, but we do not pretend to have reviewed the subject exhaustively or systematically, still less to have accounted for cross-cultural diversity. In some societies, nothing is more shameful than to be cuckolded, and a violent reaction is laudable; in others, jealousy is shameful, and its violent expression is criminal. One would like to better describe and understand such diversity, but the ethnographic record is almost devoid of standardized cross-cultural data—indeed of any sort of quantification—with the result that analyses of the interrelationships among adultery rates, paternal investment, spousal violence, claustration practices, and other relevant variables are not presently possible.

We have stressed the cross-culturally general both because that is the level at which species-typical psychological mechanisms must be discerned and characterized and because the social scientific literature is riddled with exaggerated claims of cultural diversity. Many anthropologists have asserted that there are societies in which jealousy is nonexistent and sexual activity is constrained only by incest prohibitions (e.g., Ford & Beach, 1951; Mead, 1931; Stephens, 1963; Whyte, 1978), but the ethnographies cited in support of these claims explicitly contradict them (Daly, Wilson, & Weghorst, 1982). On the one hand, claustration, inheritance, and other practices relevant to the arguments in this chapter are diversely regulated and institutionalized, and the historical and ecological reasons for such variation are worthy subjects for research; on the other hand, double standards with respect to adultery are apparently universal, the many attempts to document a counterexample notwithstanding (Daly, Wilson, & Weghorst, 1982). On the one hand, rates of violence inspired by sexual rivalry and infidelity vary by orders of magnitude between times and places; on the other hand, male sexual proprietariness is the predominant motivational factor in spousal homicides wherever a sample of cases has been collected (Daly & Wilson, 1988b). Cultural diversity exists, to be sure, but its rationales will not be understood until the cross-culturally general human nature that enables it is elucidated.

ACKNOWLEDGMENTS

Our research on homicide has been supported by the Harry Frank Guggenheim Foundation, the North Atlantic Treaty Organization, the Natural Sciences & Engineering Research Council of Canada, and the Social Sciences & Humanities Research Council of Canada. This chapter was completed while the authors were Fellows of the Center for Advanced Study in the Behavioral Sciences with financial support from the John D. & Catherine T. MacArthur Foundation, the National Science Foundation #BNS87-008, the Harry Frank Guggenheim Foundation, and the Gordon P. Getty Trust, and while M. Daly was a Fellow of the John Simon Guggenheim

Foundation. We wish to thank Jerome Barkow, Leda Cosmides, John Tooby, and Anders Pape Møller for their critical comments.

NOTES

1. There are animals in which postzygotic care is biparental and both sexes are susceptible to being mistaken about the identity of their own offspring, such as the many species of birds subject to intraspecific brood parasitism (e.g., Yom-Tov, 1980). In such cases, parasitic eggs impose a cost on both partners, and both sexes have evolved tactics for reducing the risk of misdirected parental investment (e.g., Brown, 1984). But though both parents incur "cuckoldry" risk, there is still likely to be the usual sexual asymmetry: Alien insemination of the female imposes a parental burden on the male whereas the fruits of his infidelities are less likely to end up in her nest (but see Gowaty & Karlin, 1984).

2. It is interesting to note that whereas neighbors whose own mates were incubating took advantage of the detentions, those neighboring males who themselves had a fertile mate exhibited intensified mate guarding rather than courtship of the temporary "widows," apparently in response to the social disturbance and intrusions by other males inspired by the detained males' absence.

3. It seems likely that the alpha male's paternity advantage will prove larger than this 18–15 (55%) split once more data are in, since alphas enjoyed a larger 36–14 (72%) advantage over betas in polygynandrous groups, despite the seemingly greater difficulty that an alpha would incur in excluding the beta male from access to two females simultaneously. However, even a 72% paternity advantage would barely pay a monogamous male's cost of taking on a beta as "helper."

4. Rather than ask why some animals, including people, "conceal" ovulation, we might better ask why some advertise it at a distance. The principal effect, and hence the probable function, of dramatic estrous swellings in animals such as chimpanzees and baboons is apparently the incitation of male-male competition. Conspicuous advertisements of fertilizability seem to be more characteristic of primates lacking paternal investment than of pair-forming species (as is also true of birds; Montgomerie & Thornhill, 1989).

5. If sperm counts fluctuate as an adaptive response to statistical predictors of cuckoldry risk, we would expect men's sexual interest in their mates to do likewise. Observing one's mate interacting with other males, especially sexually, may be even more arousing. Drakes who witness forced copulations of their fertile mates commonly follow suit, thereby getting their own ejaculates into the competition (McKinney, Derrickson, & Mineau, 1983). Finkelhor and Yllo (1985) describe many cases of marital rape, a remarkable proportion of which followed closely upon the wife's having interacted flirtatiously with a man other than her husband (see also Russell, 1982). Whether the sexual arousal and/or insistence of human males is affected by cues of possible cuckoldry warrants investigation.

6. Discussions of ancestral human mating systems and their relevance to sex differences have typically stressed effective polygyny and its consequences (e.g., Trivers, 1972; Daly & Wilson, 1983; Alexander, 1979). But two animal species can have identical degrees of effective polygyny even though the females of one species are strictly monogamous in their behavior and inclinations while the females of the other species mate polyandrously. As noted above, the sexual psychologies of men and women apparently reflect an ancestral mating system in which neither sex was strictly monogamous, but males were the more polygamous sex.

7. This association between status and claustration may strike readers as paradoxical, both because it implies that women in some ways lose rights and privileges as they rise in status and because it is the women with the poorest mates who would seem to have the greatest incentive to stray. One interpretation is that men simply guard and constrain women as much as they can afford; moreover, there is at least some truth to the claim that guards are there to "protect" women in that highly stratified societies contain many disenfranchised, and therefore dangerous,

men (Dickemann, 1979). Furthermore, the wealthiest men have the most to lose, in misdirected inheritance, from an undetected cuckoldry (Dickemann, 1981).

8. We have used homicide archives as a sort of assay of interpersonal conflict in various analyses of the determinants of the variable intensity of such conflict (Daly & Wilson, 1988a, 1988b). Homicides are extreme, of course, but for that very reason they are surely valid manifestations of conflict and moreover are relatively free of reporting biases. We do not assume that killing is adaptive in its own right. Rather, homicide risk varies as a result of variations in the strength of human passions whose more typical and less extreme consequences are often clearly utilitarian, and this latter variation has an adaptive logic.

9. Ideally, such a study should correlate paternal solicitude with resemblance within a sample of father-child pairs confirmed by DNA fingerprinting. Otherwise, a correlation might be obtained because weak resemblance actually reflects nonpaternity and the fathers might be responding to other correlated cues thereof (e.g., detected infidelity) rather than to non-resemblance per se.

10. About 4% of human societies have institutionalized polyandrous marriage, an arrangement that seems to work only—and even then with some tension—when the cohusbands are brothers (Hiatt, 1980; Levine, 1987; Tambiah, 1966).

11. It may strike the reader that such exchanges can go either way, with dowry the mirror image of brideprice. But it is not. Dowries (which are much rarer than brideprices) usually remain with the bride for the benefit of her children (Goody & Tambiah, 1973; van den Berghe, 1979), whereas brideprice is typically a direct payment to the bride's parents without obligation that the monies will be deployed for the newlyweds or their progeny.

12. Women who are already "taken" are often marked accordingly, so an adulterer can hardly plead ignorance. In a cross-cultural study, Low (1979) showed that women are much more often obliged to display marital status markers such as wedding rings than are men; modes of address (e.g., "Mrs." versus "Miss") and the assumption of husbands' lineage names also mark women as to ownership.

13. Though bigamy had long been illegal, English marriages were not registered by a central authority until 1753 (Trumbach, 1984).

REFERENCES

Alatalo, R. V., Gottlander, K., & Lundberg, A. (1987). Extra-pair copulations and mate-guarding in the polyterritorial pied flycatcher, *Ficedula hypoleuca. Behaviour, 101,* 139–155.

Alexander, R. D. (1974). The evolution of social behavior. *Annual Review of Ecology & Systematics, 5,* 325–383.

Alexander, R. D. (1979). *Darwinism and human affairs.* Seattle, WA: University of Washington Press.

Alexander, R. D. (1987). *The biology of moral systems.* Hawthorne, NY: Aldine de Gruyter.

Alexander, R. D., & Noonan, K. M. (1979). Concealment of ovulation, parental care, and human social evolution. In N. Chagnon & W. Irons (Eds.), *Evolutionary biology and human social behavior* (p. 436–453). North Scituate, MA: Duxbury.

Arnold, M. S. (Trans. & Ed.) (1985). *Select cases of trespass from the King's Courts, 1307–1399.* London: Selden Society.

Atkins, S., & Hoggett, B. (1984). *Women and the law.* Oxford: Blackwell.

Attenborough, F. L. (1922/1963). *The laws of the earliest English kings.* New York: Russell & Russell.

Backhouse, C. (1986). The tort of seduction: Fathers and daughters in nineteenth century Canada. *Dalhousie Law Journal, 10,* 45–80.

Baker, R. R., & Bellis, M. A. (1989). Number of sperm in human ejaculates varies in accordance with sperm competition theory. *Animal Behaviour, 37,* 867–869.

Bateman, A. J. (1948). Intra-sexual selection in *Drosophila. Heredity, 2,* 349–368.

Becker, G. S., Landes, E. M., & Michael, R. T. (1977). An economic analysis of marital instability. *Journal of Political Economy, 85,* 1141–1187.

Beckerman, J. S. (1981). Adding insult to *Iniuria:* Affronts to honor and the origins of trespass. In M. S. Arnold, T. A. Green, S. A. Scully, & S. D. White (Eds.), *On the laws and customs of England.* Chapel Hill, NC: University of North Carolina Press.

Bedford, J. M. (1977). Evolution of the scrotum: The epididymis as the prime mover? In J. H. Calaby & C. H. Tyndale-Biscoe (Eds.), *Reproduction and evolution* (pp. 171–182). Canberra: Australian Academy of Science.

Beecher, M. D., & Beecher, I. M. (1979). Sociobiology of bank swallows: Reproductive strategy of the male. *Science, 205,* 1282–1285.

Beissinger, S. R., & Snyder, N.F.R. (1987) Mate desertion in the snail kite. *Animal Behaviour, 35,* 477–487.

Benshoof, L., & Thornhill, R. (1979). The evolution of monogamy and concealed ovulation in humans. *Journal of Social and Biological Structures, 2,* 95–106.

Betzig, L. L. (1986). *Depotism and differential reproduction: A Darwinian view of history.* Hawthorne, NY: Aldine de Gruyter.

Birkhead, T. R. (1988). Behavioral aspects of sperm competition in birds. *Advances in the Study of Behavior, 18,* 35–72.

Birkhead, T. R., Atkin, L., & Møller, A. P. (1987). Copulation behaviour of birds. *Behaviour, 101,* 101–138.

Bjorklund, M., & Westman, B. (1983). Extra-pair copulations in the pied flycatcher, *Ficedula hypoleuca:* A removal experiment. *Behavioral Ecology & Sociobiology, 13,* 271–275.

Bjorklund, M., & Westman, B. (1986). Mate-guarding in the great tit: Tactics of a territorial forest-living species. *Ornis Scandinavica, 17,* 99–105.

Blackstone, W. (1803). *Commentaries on the laws of England* (St. G. Tucker, Ed.). Philadelphia, PA: William Young Birch & Abraham Small.

Borgerhoff Mulder, M. (1988). Kipsigis bridewealth payments. In L. Betzig, M. Borgerhoff Mulder, & P. Turke (Eds.), *Human reproductive behavior.* Cambridge: Cambridge University Press.

Boserup, E. (1970). *Woman's role in economic development.* New York: St. Martin's Press.

Bradley, K. R. (1984). *Slaves and masters in the Roman Empire. A study in social control.* Collection Latomus. (Vol. 185). Bruxelles: Latomus Revue d'Études Latines.

Brett, P. (1955). Consortium and servitium: A history and some proposals. *Australian Law Journal, 29,* 321–328, 389–397, 428–434.

Bringle, R. G., & Buunk, B. (1985). Jealousy and social behavior: A review of person, relationship, and situational determinants. *Review of Personality and Social Psychology, 6,* 241–264.

Bringle, R. G., & Buunk, B. (1986). Examining the causes and consequences of jealousy: Some recent findings and issues. In R. Gilmour & S. Duck (Eds.), *The emerging field of personal relationships* (p. 225–240) Hillsdale, NJ: Erlbaum.

Brown, C. R. (1984). Laying eggs in a neighbor's nest: Benefit and cost of colonial nesting in swallows. *Science, 224,* 518–519.

Brownmiller, S. (1975). *Against our will.* New York: Simon & Schuster.

Burgess, A. W., & Holmstrom, L. L. (1974). *Rape, crisis and recovery.* Bowie, MD: Brady.

Burke, T., Davies, N. B., Bruford, M. W., & Hatchwell, B. J. (1989). Parental care and mating behaviour of polyandrous dunnocks *Prunella modularis* related to paternity by DNA fingerprinting. *Nature, 338,* 249–251.

Burley, N. (1979). The evolution of the concealment of ovulation. *American Naturalist, 114,* 835–858.

Busnot, F. D. (1715). *The history of the reign of Muley Ismael, the present king of Morocco, Fez, Tafilet, Sous, &c.* (Trans.) London: A. Bell. (Original work published 1714.)

Buss, D. M. (1988). From vigilance to violence: Tactics of mate retention in American under-
    graduates. *Ethology & Sociobiology, 9,* 291–317.

Buss, D. M. (1989). Conflict between the sexes: Strategic interference and the evocation of anger
    and upset. *Journal of Personality & Social Psychology, 56,* 735–747.

Buunk, B. (1981). Jealousy in sexually open marriages. *Alternative Lifestyles, 4,* 357–372.

Buunk, B. (1982a). Anticipated sexual jealousy: Its relationship to self-esteem, dependency and
    reciprocity. *Personality & Social Psychology Bulletin, 8,* 310–316.

Buunk, B. (1982b). Strategies of jealousy: Styles of coping with extramarital involvement of the
    spouse. *Family Relations, 31,* 13–18.

Buunk, B. (1986). Husbands' jealousy. In R. A. Lewis & R. E. Salt (Eds.), *Men in families* (p.
    97–114). Beverly Hills, CA: Sage.

Buunk, B., & Bringle, R. G. (1987). Jealousy in love relationships. In D. Perlman & S. Duck
    (Eds.), *Intimate relationships* (p. 123–147). Beverly Hills, CA: Sage.

Buunk, B., & Hupka, R. B. (1987). Cross-cultural differences in the elicitation of sexual jealousy.
    Journal of Sex Research, 23, 12–22.

Carlson, A., Hillstrom, L., & Moreno, J. (1985). Mate guarding in the wheatear *Oenanthe oen-
    anthe. Ornis Scandinavica, 16,* 113–120.

Chappell v. Redding, 313 S.E.2d 239 (N.C. App. 1984).

Daly, M. (1978). The cost of mating. *American Naturalist, 112,* 771–774.

Daly, M. & Wilson, M. (1982). Whom are newborn babies said to resemble? *Ethology & Socio-
    biology, 3,* 69–78.

Daly, M., & Wilson, M. (1983). *Sex, evolution and behavior.* Boston, MA: Wadsworth.

Daly, M., & Wilson, M. (1985). Child abuse and other risks of not living with both parents. *Ethol-
    ogy & Sociobiology, 6,* 197–210.

Daly, M., & Wilson, M. (1987). The Darwinian psychology of discriminative parental solicitude.
    Nebraska Symposium on Motivation, 35, 91–144.

Daly, M., & Wilson, M. (1988a). Evolutionary social psychology and family homicide. *Science,
    242,* 519–524.

Daly, M., & Wilson, M. (1988b). *Homicide.* Hawthorne, NY: Aldine de Gruyter.

Daly, M., & Wilson, M. (1990). Killing the competition. *Human Nature, 1,* 83–109.

Daly, M., Wilson, M., & Weghorst, S. J. (1982). Male sexual jealousy. *Ethology & Sociobiology,
    3,* 11–27.

Daniels, D. (1983). The evolution of concealed ovulation and self-deception. *Ethology & Socio-
    biology, 4,* 69–87.

Davies, N. B. (1983). Polyandry, cloaca-pecking and sperm competition in dunnocks. *Nature,
    302,* 334–336.

Davies, N. B. (1985). Cooperation and conflict among dunnocks, *Prunella modularis,* in a vari-
    able mating system. *Animal Behaviour, 33,* 628–648.

Davies, N. B. (1986). Reproductive success of dunnocks, *Prunella modularis,* in a variable mat-
    ing system. I. Factors influencing provisioning rate, nestling weight and fledging success.
    Journal of Animal Ecology, 55, 123–138.

Davies, N. B. (1989). The dunnock: Cooperation and conflict among males and females in a
    variable mating system. In P. Stacey & W. Koenig (Eds.), *Cooperative breeding in birds*
    (pp. 457–485). Cambridge: Cambridge University Press.

Davies, N. B., & Houston, A. I. (1986). Reproductive success of dunnocks, *Prunella modularis,*
    in a variable mating system. II. Conflicts of interest among breeding adults. *Journal of
    Animal Ecology, 55,* 139–154.

Davies, N. B., & Lundberg, A. (1984). Food distribution and a variable mating system in the
    dunnock, *Prunella modularis. Journal of Animal Ecology, 53,* 895–912.

Davis, D. B. (1966). *The problem of slavery in Western culture.* Ithaca, NY: Cornell University
    Press.

Dickemann, M. (1979). The ecology of mating systems in hypergynous dowry societies. *Social Science Information, 18,* 163–195.

Dickemann, M. (1981). Paternal confidence and dowry competition: A biocultural analysis of purdah. In R. D. Alexander & D. W. Tinkle (Eds.), *Natural selection and social behavior* (pp. 417–438). New York: Chiron Press.

Dobash, R. E., & Dobash, R. P. (1979). *Violence against wives: A case against the patriarchy.* New York: Free Press.

Dobash, R. P., & Dobash, R. E. (1981). Community response to violence against wives: Charivari, abstract justice and patriarchy. *Social Problems, 28,* 563–581.

Edwards, J.Ll.L. (1954). Provocation and the reasonable man: Another view. *Criminal Law Review, 1954,* 898–906.

Edwards, S. (1985). Male violence against women: Excusatory and explanatory ideologies in law and society. In S. Edwards (Ed.), *Gender, sex and the law* (pp. 183–213). London: Croom Helm.

Edwards, S.S.M. (1981). *Female sexuality and the law.* Oxford: Martin Robertson.

Finkelhor, D., & Yllo, K. (1985). *License to rape: Sexual abuse of wives.* New York: Holt, Rinehart & Winston.

Finley, M. I. (1980). *Ancient slavery and modern ideology.* London: Chatto & Windus.

Flinn, M. V. (1981). Uterine vs. agnatic kinship variability and associated cousin marriage preferences. In R. D. Alexander & D. W. Tinkle (Eds.), *Natural selection and social behavior* (pp. 439–475). New York: Chiron Press.

Flinn, M. V. (1988a). Mate guarding in a Caribbean village. *Ethology & Sociobiology, 9,* 1–28.

Flinn, M. V. (1988b). Stepparent and genetic parent offspring relationships in a Caribbean village. *Ethology & Sociobiology, 9,* 335–369.

Flinn, M. V., & Low, B. S. (1986). Resource distribution, social competition, and mating patterns in human societies. In D. I. Rubenstein & R. W Wrangham (Eds.), *Ecological aspects of social evolution: Birds and mammals* (pp. 217–243). Princeton, NJ: Princeton University Press.

Ford, C. S., & Beach, F. A. (1951). *Patterns of sexual behavior.* New York: Harper & Row.

Francis, J. L. (1977). Toward the management of heterosexual jealousy. *Journal of Marriage & Family Counseling, 3,* 61–69.

Frederick, P. C. (1987). Extrapair copulations in the mating system of white ibis *(Eudocimus albus). Behaviour, 100,* 170–201.

Fujioka, M., & Yamagishi, S. (1981). Extramarital and pair copulations in the cattle egret. *Auk, 98,* 134–144.

Gaulin, S.J.C., & Schlegel, A. (1980). Paternal confidence and paternal investment: A cross-cultural test of a sociobiological hypothesis. *Ethology & Sociobiology, 1,* 301–309.

Goody, J., & Tambiah, S. J. (Eds.). (1973). *Bridewealth and dowry.* Cambridge: Cambridge University Press.

Gowaty, P. A., & Karlin, A. A (1984). Multiple maternity and paternity in single broods of apparently monogamous eastern bluebirds *(Sialia sialis). Behavioral Ecology & Sociobiology, 15,* 91–95.

Grossbard-Schechtman, A. (1984). A theory of allocation of time in markets for labour and marriage. *The Economic Journal, 94,* 863–882.

Hadjiyannakis, C. (1969). *Les tendances contemporaines concernant la répression du délit d'adultère.* Thessalonika: Association Internationale du Droit Pénal.

Hamilton, W. D. (1964). The genetical evolution of social behaviour. I and II. *Journal of Theorectical Biology, 7,* 1–52.

Hansen, G. L. (1985). Perceived threats and marital jealousy. *Social Psychology Quarterly, 48,* 262–268.

Harcourt, A. H., Harvey, P. H., Larson, S. G., & Short, R. V. (1981). Testis weight, body weight and breeding system in primates. *Nature, 293,* 55–57.

Hartung, J. (1985). Matrilineal inheritance: New theory and analysis. *Behavioral and Brain Sciences, 8,* 661–688.

Hatch, S. A. (1987). Copulation and mate guarding in the northern fulmar. *Auk, 104,* 450–461.

Hatchwell, B. J. (1988). Intraspecific variation in extra-pair copulation and mate defence in common guillemots *Uria aalge. Behaviour, 107,* 157–185.

Helmholtz, R. H. (1974). *Marriage litigation in medieval England.* London: Cambridge University Press.

Hiatt, L. R. (1980). Polyandry in Sri Lanka: A test case for parental investment theory. *Man, 15,* 583–602.

Hicks, E. K. (1986). *Infibulation: Status through mutilation.* Alblasserdam: Offsetdrukkerij Kanters B. V.

Holmes, W. G., & Sherman, P. W. (1982). The ontogeny of kin recognition in two species of ground squirrels. *American Zoologist, 22,* 491–517.

Hosken, F. P. (1979). *The Hosken report. Genital and sexual mutilation of females* (2d rev. ed.). Lexington, MA: Women's International Network News.

Houston, A. I., & Davies, N. B. (1985). The evolution of cooperation and life history in dunnocks, *Prunella modularis.* In R. Sibly & R. H. Smith (Eds.), *Behavioural ecology: The ecological consequences of adaptive behavior* (pp. 471–487). Oxford: Blackwell.

Hrdy, S. B. (1981). *The woman that never evolved.* Cambridge, MA: Harvard University Press.

Hupka, R. B., Buunk, B., Falus, G., Fulgosi, A., Ortega, E., Swain, R., & Tarabrina, N. V. (1985). Romantic jealousy and romantic envy. *Journal of Cross-Cultural Psychology, 16,* 423–446.

Ingram, M. (1987). *Church courts, sex and marriage in England, 1570–1640.* Cambridge: Cambridge University Press.

Kanekar, S., Kolsawalla, M. B., & D'Souza, A. (1981). Attribution of responsibility to a victim of rape. *British Journal of Social Psychology, 20,* 165–170.

Kaplan, H., & Hill, K. (1985). Hunting ability and reproductive success among male Ache foragers: Preliminary results. *Current Anthropology, 26,* 131–133.

Karkaria, B. J. (1972, June 18). Raped women of Bangladesh, *Illustrated Weekly of India,* pp. 14–17.

Karras, R. M. (1988). *Slavery and society in medieval Scandinavia.* New Haven: Yale University Press.

Kenagy, G. J., & Trombulak, S. C. (1986). Size and function of mammalian testes in relation to body size. *Journal of Mammalogy, 67,* 1–22.

Knodel, J., & Lynch, K. A. (1985). The decline of remarriage: Evidence from German village populations in the eighteenth and nineteenth centuries. *Journal of Family History, 10,* 34–60.

Kurland, J. A (1979). Paternity, mother's brother and human sociality. In N. A. Chagnon & W. Irons (Eds.), *Evolutionary biology and human social behavior* (pp. 145–180). North Scituate, MA: Duxbury Press.

Law Reform Commission of Ireland (1978). *The law relating to criminal conversation and the enticement and harbouring of a spouse. Working paper no. 5.* Dublin, Ireland: Law Reform Commission of Ireland.

Law Reform Commission of Ireland (1979a). *The law relating to seduction and the enticement and harbouring of a child. Working Paper No. 6.* Dublin, Ireland: Law Reform Commission of Ireland.

Law Reform Commission of Ireland (1979b). *The law relating to loss of consortium and loss of services of a child. Working paper No. 7.* Dublin, Ireland: Law Reform Commission of Ireland.

Levine, N. E. (1987). Fathers and sons: Kinship value and validation in Tibetan polyandry. *Man, 22,* 267–286.

Lévi-Strauss, C. (1969). *The elementary structures of kinship.* Boston, MA: Beacon Press.

Lightcap, J. L., Kurland, J. A., & Burgess, R. L. (1982). Child abuse: A test of some predictions from evolutionary theory. *Ethology & Sociobiology, 3,* 61–67.

Lippman, J. (1930). The breakdown of consortium. *Columbia Law Review, 30,* 651–673.

Low, B. S. (1979). Sexual selection and human ornamentation. In N. A. Chagnon & W. Irons (Eds.), *Evolutionary biology and human social behavior* (pp. 463–487). North Scituate, MA: Duxbury Press.

Lumpkin, S. (1983). Female manipulation of male avoidance of cuckoldry behavior in the ring dove. In S. K. Wasser (Ed.), *Social behavior of female vertebrates* (pp. 91–112). New York: Academic Press.

Mathes, E. W. (1986). Jealousy and romantic love: A longitudinal study. *Psychological Reports, 58,* 885–886.

Mathes, E. W., Adams, H. E., & Davies, R. M. (1985). Jealousy: Loss of relationship rewards, loss of self-esteem, depression, anxiety, and anger. *Journal of Personality & Social Psychology, 48,* 1552–1561.

Maynard Smith, J. (1977). Parental investment: A prospective analysis. *Animal Behaviour, 25,* 1–9.

McCahill, T. W., Meyer, L. C., & Fischman, A. M. (1979). *The aftermath of rape.* Lexington, MA: Lexington Books.

McKinney, F., Derrickson, S. R., & Mineau, P. (1983). Forced copulation in waterfowl. *Behaviour, 86,* 250–294.

Mead, M. (1931). Jealousy: Primitive and civilised. In S. D. Schmalhausen & V. F. Calverton (Eds.), *Woman's coming of age: A symposium* (pp. 35–48). New York: Liveright.

Menefee, S. P. (1981). *Wives for sale.* Oxford: Basil Blackwell.

Miller, W. R., Williams, A. M., & Bernstein, M. H. (1982). The effects of rape on marital and sexual adjustment. *American Journal of Family Therapy, 10,* 51–58.

Møller, A. P. (1985). Mixed reproductive strategy and mate guarding in a semi-colonial passerine, the swallow *Hirundo rustica. Behavioral Ecology & Sociobiology, 17,* 401–408.

Møller, A. P. (1987a). Behavioral aspects of sperm competition in swallows *(Hirundo rustica). Behaviour, 100,* 92–104.

Møller, A. P. (1987b). Extent and duration of mate guarding in swallow *Hirundo rustica. Ornis Scandinavica, 18,* 95–100.

Møller, A. P. (1987c). Mate guarding in the swallow *Hirundo rustica. Behavioral Ecology & Sociobiology, 21,* 119–123.

Møller, A. P. (1987d). Advantages and disadvantages of coloniality in the swallow, *Hirundo rustica. Animal Behaviour, 35,* 819–832.

Møller, A. P. (1988a). Ejaculate quality, testes size and sperm competition in primates. *Journal of Human Evolution, 17,* 479–488.

Møller, A. P. (1988b). Female choice selects for male sexual tail ornaments in the monogamous swallow. *Nature, 332,* 640–642.

Møller, A. P. (1988c). Paternity and paternal care in the swallow, *Hirundo rustica. Animal Behaviour, 36,* 996–1005.

Møller, A. P. (1988d). Testes size, ejaculate quality, and sperm competition in birds. *Biological Journal of the Linnaean Society, 33,* 273–283.

Montgomerie, R. D. (Ed.). (1988). Symposium 3: Mate guarding. *Acta XIX Congressus Internationalis Ornithologici* (Vol. 1) (pp. 409–453) Ottawa: University of Ottawa Press.

Montgomerie, R., & Thornhill, R. (1989). Fertility advertisement in birds: A means of inciting male-male competition? *Ethology, 81,* 209–220.

Morton, E. S. (1987). Variation in mate guarding intensity by male purple martins. *Behaviour, 101,* 211–224.

Mowat, R. R (1966). *Morbid jealousy and murder.* London: Tavistock.

Murdock, G. P. (1967). *Ethnographic atlas.* Pittsburgh: University of Pittsburgh Press.

Paige, K. E., & Paige, J. M. (1981). *The politics of reproductive ritual.* Berkeley: University of California Press.

Patterson, O. (1982). *Slavery and social death.* Cambridge, MA: Harvard University Press.

Pfeiffer, S. M., & Wong, P.T.P. (1989). Multidimensional jealousy. *Journal of Social and Personal Relationships, 6,* 181–196.

Quaife, G. R. (1979). *Wanton wenches and wayward wives.* London: Croom Helm.

Ridley, M. W., & Hill, D. A. (1987). Social organization in the pheasant *(Phasianus colchicus):* Harem formation, mate selection and the role of mate guarding. *Journal of Zoology, 211,* 619–630.

Rodkin, L. I., Hunt E. J., & Cowan, S. D. (1982). A men's support group for significant others of rape victims. *Journal of Marital & Family Therapy, 8,* 91–97.

Røskaft, E. (1983). Male promiscuity and female adultery by the rook *Corvus frugilegus. Ornis Scandinavica, 14,* 175–179.

Russell, D.E.H. (1982). *Rape in marriage.* New York: Macmillan.

Ryan, W. (1971). *Blaming the victim.* New York: Pantheon.

Safilios-Rothschild, C. (1969). 'Honor' crimes in contemporary Greece. *British Journal of Sociology, 20,* 205–218.

Sahlins, M. D. (1976). *The use and abuse of biology.* Ann Arbor: University of Michigan Press.

Salovey, P., & Rodin, J. (1986). The differentiation of social-comparison jealousy and romantic jealousy. *Journal of Personality and Social Psychology, 50,* 1100–1112.

Sass, S. L. (1977). The defense of multiple access *(exceptio plurium concubentium)* in paternity suits: A comparative analysis. *Tulane Law Review, 51,* 468–509.

Sawyer, R. (1986). *Slavery in the twentieth century.* London: Routledge & Kegan Paul.

Sherman, P. W., & Morton, M. L. (1988). Extra-pair fertilizations in mountain white-crowned sparrows. *Behavioral Ecology & Sociobiology, 22,* 413–420.

Shettel-Neuber, J., Bryson, J. B., & Young, L. E. (1978). Physical attractiveness of the "other person" and jealousy. *Personality and Social Psychology Bulletin, 4,* 612–615.

Short, R. V. (1977). Sexual selection and the descent of man. In J. H. Calaby & C. H. Tyndale-Biscoe (Eds.), *Reproduction and evolution* (pp. 3–19). Canberra: Australian Academy of Sciences.

Sinclair, M.B.W. (1987). Seduction and the myth of the ideal woman. *Law & Inequality, 3,* 33–102.

Smith, R. E., Keating, J. P., Hester, R. K., & Mitchell, H. E. (1976). Role and justice considerations in the attribution of responsibility to a rape victim. *Journal of Research in Personality, 10,* 346–357.

Smith, R. L., (Ed). (1984a). *Sperm competition and the evolution of animal mating systems.* Orlando, FL: Academic Press.

Smith R. L. (1984b). Human sperm competition. In R. L. Smith (Ed.), *Sperm competition and the evolution of animal mating systems* (pp. 601–659). Orlando, FL: Academic Press.

Smith, S. M. (1988). Extra-pair copulations in black-capped chickadees: The role of the female. *Behaviour, 107,* 15–23.

Sorrentino, R. M., & Boutilier, R. G. (1974). Evaluation of a victim as a function of fate similarity/dissimilarity. *Journal of Experimental Social Psychology, 10,* 84–93.

Stephens, W. N. (1963). *The family in cross-cultural perspective.* New York: Holt, Rinehart & Winston.

Stewart, R. A., & Beatty, M. J. (1985). Jealousy and self-esteem. *Perceptual & Motor Skills, 60,* 153–154.

Strassmann, B. I. (1981). Sexual selection, parental care, and concealed ovulation in humans. *Ethology & Sociobiology, 2,* 31–40.

Strathern, M. (1984). Subject or object? Women and the circulation of valuables in Highlands

New Guinea. In R. Hirschon (Ed.), *Women and property—women as property* (pp. 158–175). London: Croom Helm.

Swinburne, H. (1686/1985). *A treatise of spousals, or matrimonial contracts*. London: Garland.

Symons, D. (1979). *The evolution of human sexuality*. Oxford: Oxford University Press.

Tambiah, S. J. (1966). Polyandry in Ceylon—with special reference to the Laggala region. In C. von Fürer-Haimendorf (Ed.), *Caste and kin in Nepal, India and Ceylon: Anthropological studies in Hindu-Buddhist contact zones* (pp. 264–358). New York: Asia Publishing House.

Teismann, M. W., & Mosher, D. L. (1978). Jealous conflict in dating couples. *Psychological Reports, 42,* 1211–1216.

Thornhill, R., & Alcock, J. (1983). *The evolution of insect mating systems*. Cambridge, MA: Harvard University Press.

Thornhill, R., & Thornhill, N. W. (1983). Human rape: An evolutionary analysis. *Ethology & Sociobiology, 4,* 137–183.

Thornhill, R., & Thornhill, N. W. (1987). Human rape: The strengths of the evolutionary perspective. In C. Crawford, M. Smith, & C. Krebs (Eds.), *Psychology and sociobiology* (pp. 269–291). Hillsdale, NJ: Erlbaum.

Thornhill, R., & Thornhill, N. W. (1989). The evolution of psychological pain. In R. J. Bell & N.J. Bell (Eds.), *Sociobiology and the Social Sciences* (pp. 73–103). Lubbock, TX: Texas Tech University Press.

Tooby, J., & DeVore, I. (1987). The reconstruction of hominid behavioral evolution through strategic modeling. In W. G. Kinzey, (Ed.), *The evolution of human behavior: Primate models* (pp. 183–237). Albany, NY: State University of New York Press.

Trivers, R. L. (1972). Parental investment and sexual selection. In B. Campbell (Ed.), *Sexual selection and the descent of man 1871–1971* (pp. 136–179). Chicago: Aldine.

Trivers, R. (1985). *Social evolution*. Menlo Park, CA: Benjamin/Cummings.

Trumbach, R. (1984). *Marriage, sex, and the family in England, 1660–1800*. New York: Garland.

Turke, P. W. (1984). Effects of ovulatory concealment and synchrony on protohominid mating systems and parental roles. *Ethology & Sociobiology, 5,* 33–44.

Van den Berghe, P. L. (1979). *Human family systems*. New York: Elsevier.

Voland, E. (1988). Differential infant and child mortality in evolutionary perspective: Data from late 17th to 19th century Ostfriesland (Germany). In L. Betzig, M. Borgerhoff Mulder, & P. Turke (Eds.), *Human reproductive behavior: A Darwinian perspective* (pp. 253–261). Cambridge: Cambridge University Press.

Weis, K., & Borges, S. S. (1973). Victimology and rape: The case of the legitimate victim. *Issues in Criminology, 8,* 71–115.

Westneat, D. F. (1987). Extra-pair copulations in a predominantly monogamous bird: Observations of behaviour. *Animal Behaviour, 35,* 865–876.

White, G. L. (1981). A model of romantic jealousy. *Motivation & Emotion, 5,* 295–310.

Whyte, M. K. (1978). *The status of woman in preindustrial societies*. Princeton, NJ: Princeton University Press.

Wilson, M. (1987). Impacts of the uncertainty of paternity on family law. *University of Toronto Faculty of Law Review, 45,* 216–242.

Wilson, M., & Daly, M. (1987). Risk of maltreatment of children living with stepparents. In R. J. Gelles & J. B. Lancaster (Eds.), *Child abuse and neglect: Biosocial dimensions* (pp. 215–232). Hawthorne, NY: Aldine de Gruyter.

Wolf, A. P., & Huang, C. S. (1980). *Marriage and adoption in China, 1845–1945*. Stanford, CA: Stanford University Press.

Yom-Tov, Y. (1980). Intraspecific nest parasitism in birds. *Biological Reviews, 55,* 93–108.

IV
PARENTAL CARE AND CHILDREN

Parental care is so much a part of the fabric of human life that it is sometimes difficult to remember that not all animals care for their young and that parental care would not occur at all in the absence of a number of rather specific psychological adaptations. Ironically, this form of "child neglect" spills over into the study of psychology as well: One can open almost any textbook on motivation and search in vain for a discussion of the mechanisms that motivate the complexly organized behaviors of parenting.

To put human parental care into perspective, consider the amount of effort that oysters put into raising their young. Oysters release millions of eggs and sperm into the ocean. Most of these gametes become food for some other ocean creature, but a few of them are fertilized, grow to maturity, and reproduce themselves. Oysters effectively define minimal parental care. Reptiles invest far more of their reproductive effort in their young, but compared to the mammalian norm, most reptilian species spend little time or energy caring for offspring after they have hatched. In sea turtles, for example, fertilization occurs internally, after which the female swims to shore, digs a sandy pit, and lays her eggs in it. She then covers the eggs, camouflages the spot, and goes back to sea, never to see her progeny again.

In contrast, one of the defining features of the mammals is the amount and kind of reproductive effort that females invest in their young. Not only does fertilization occur internally, but gestation does also: The embryo and then fetus develop within the mother's body cavity and are fed by the mother through the connection between her bloodstream and the placenta. Gestation can last many months, as it does in humans. After birth, the mammalian mother continues to feed her offspring from her own body through lactation. Because we live in a society in which women have access to cow's milk and commerical baby foods that infants are capable of digesting, it is easy to underestimate the importance of lactation for the human infant under more natural circumstances. But in hunter-gatherer societies, a woman must sometimes nurse her child for several years. Typically, the mother both sleeps with her baby and carries it with her on her hip when she goes gathering, so that the baby can nurse on demand; during waking hours, the baby nurses about every 15 minutes. Lactation at this level suppresses ovulation and so acts as a natural contraceptive during the period in which the baby is dependent on its mother's milk, an adaptive mechanism that promotes efficient birth spacing. Without the mother's several year commitment to gestation, lactation, and other forms of maternal care, infants in hunter-gatherer societies do not survive. In understanding parental care in humans, it is important to keep in mind that these are the circumstances in which the relevant adaptations evolved.

Although mammal species manifest high levels of maternal care, paternal care is far less common, and males in many species invest no more reproductive effort in the care of their young than oysters do. But humans are not one of these species—biparental care is common in humans, as it is in a few other primate taxa, such as the callitrichids (see, e.g., McGrew & Feistner, this volume). Gestation and lactation impose a severe metabolic drain on a woman's body, so men in hunter-gatherer societies usually provision their mates during this period, especially with meat, and other relatives may contribute as well. For many years after weaning, both parents typically continue to provision, protect, and teach their children. However, even among hunter-gatherers, male involvement in parental care is far more variable than female involvement. In short, parental care in mammals is a lengthy and complicated business, and nowhere more so than in humans. The chapters in this section examine some questions concerning parental care and child development.

To raise a child, parents must solve a whole series of different adaptive problems, which start in utero and can last for many years. Indeed, parental care that is well engineered for the needs of an embryo will not even solve the problems faced by a later-stage fetus, as Profet explains in her chapter on pregnancy sickness as an adaptation.

Plants manufacture toxic secondary compounds to discourage predators—such as ourselves—from eating them. Every time you bite into turnips or cabbage, you are ingesting a sublethal dose of naturally occurring plant toxins. We, in turn, have a number of adaptations—such as the food aversion system, the liver, and possibly certain allergy-regulating mechanisms (Profet, 1991)—that cause us to avoid ingesting these toxins, expel them from our body, or break them down into metabolites that are harmless to an adult. But toxins and toxic metabolites that are relatively harmless to an adult—or even to a late-stage fetus—can cause birth defects in an embryo during the early stages of organogenesis. The problem the mother faces is how to nourish herself yet simultaneously protect her developing embryo from exposure to the toxins that she will necessarily be ingesting during the period when the embryo is most vulnerable to their harmful effects. Profet has proposed that pregnancy sickness evolved to solve this adaptive problem. Pregnancy sickness appears to be a lowering of the threshold on the mother's normal food aversion system, which causes her to avoid or expel previously acceptable foods that can be tolerated by her body but that are too high in toxins for the developing embryo. Profet shows that many of the most characteristic features of pregnancy sickness, from the timing of its onset, to its effects on the mother's olfaction, to the timing of its offset, are elegantly explained by this hypothesis. Moreover, these design features are exceedingly difficult to explain simply as arbitrary by-products of hormonal changes during pregnancy that are unrelated to this adaptive problem. Profet has carefully addressed the key issue in evaluating whether something is an adaptation, that is, demonstrating evidence of special design for solving the adaptive problem (Williams, 1966).

In a foraging society, if an infant is born sick, or malformed, or before the mother can safely wean her previous child, the mother encounters the cruel realities of scarce resources and limiting personal circumstances: Without modern medical care and food resources, continuing to care for one child can preclude caring for or conceiving others. However any specific mother might feel about such a situation, natural selection acts over evolutionary time on the psychological mechanisms that regulate maternal attitudes and decisions about such agonizing situations. Of course, the criterion by which some design features are incorporated into our species-typical design over alternatives is fitness, which usually means (in this context) maximizing the net number of offspring raised to reproductively viable maturity. Consequently, these mechanisms will be selected to evaluate whether she should continue to care for the newborn infant or channel her remaining reproducitve effort into either producing a new child or caring for older, healthier children. The choice is often anguishing, as when a !Kung San mother gives birth to twins and must abandon one. In the harsh conditions of the Kalahari desert, a woman does not have the metabolic resources to carry and nurse two children simultaneously. Rather than let both twins slowly sicken and die, !Kung mothers sometimes kill one twin immediately after its birth and concentrate their efforts on raising its sibling.

Selection pressures this intense should have sculpted psychological mechanisms that are very sensitive to cues that were reliable in the Pleistocene, signaling the conditions under which child care should be continued or curtailed. Mann analyzes these selection pressures in her chapter, defines classes of information that should have been relevant to making this decision, and reviews evidence from the psychological and anthropological literatures concerning whether such information affects a mother's decision to nurture or neglect her child. She also reports the results of an ingenious empirical test of the hypothesis that information about an infant's health status affects such decisions. By studying the behavior of mothers toward their high-risk preterm twins, she was able to show that the mothers' level of nurturance was governed by two independent factors: (a) which twin was the healthiest, and (b) which twin cried the most. Her study provided no support for several widely believed by-product hypotheses.

Fernald has studied a brighter side of parental care: infant-directed speech, sometimes known as "baby talk" or "motherese." When speaking to infants, parents around the world slow down, pitch their voices higher, and exaggerate the prosody of their speech. Moreover, the acoustic patterns used to praise, prohibit, comfort, and elicit attention from an infant appear to be cross-culturally universal, differing only in culturally given "display rules." After reviewing the extensive body of research on infant-directed speech that she and others have generated, Fernald addresses the question, Does infant-directed speech have the design features one would expect if it had evolved as an early system for mother-infant communication?

Fernald's starting point is the current debate on language as an adaptation (see Pinker & Bloom, this volume). In discussing the conditions under which it is appropriate to invoke natural selection to explain the properties of communicative systems, she points out that if infant-directed speech is an adaptation, the infant itself is the "environment" to which it should be adapted. This "environment" includes the infant's perceptual system and its intrinsic capacity to be aroused in different ways by different stimuli. If infant-directed speech is an adaptation, it should exhibit design features that interface particularly well with the characteristics of young infants.

There is an extensive ethological literature on the evolution of animal communication systems. Evolutionary task analyses and subsequent empirical studies indicate that when animals must transmit signals in noisy environments, selection favors acoustic features that enhance signal detectability, such as conspicuousness, redundancy, small repertoires, and alerting components. Fernald argues that the immature auditory cortex of the preverbal infant makes it a "noisy receiver," and that certain acoustic patterns function as unconditioned stimuli for the newborn. She then shows that infant-directed speech has exactly those acoustic features one would expect if it were a coadapted communicative system designed to enhance signal detectability and to take advantage of the infant's intrinsic arousal system. Infant-directed speech, she concludes, is "well-engineered for effective communication with preverbal infants, and not just for communication in general."

Parents spend so much time teaching their children, it is sometimes easy to forget that children have many ways of teaching themselves. Play, for example, is often thought of as a form of practice in which children acquire skills that will be useful to them later in life. But what skills, exactly, are they practicing when they play? This is the question that Boulton and Smith raise in their chapter on rough-and-tumble play. The play fighting and chasing known as rough-and-tumble appears to be a human universal and is found in many other mammalian species as well. But what adaptive problem are children improving their ability to solve when they engage in this behavior? Some have proposed that rough-and-tumble play evolved to provide practice in hunting, others that it evolved to provide practice in predator avoidance, and still others that it evolved to provide practice in intraspecific fighting. Few researchers, however, have considered what would count as evidence for or against these various hypotheses, and this makes Boulton and Smith's work distinctive (see also Symons, 1978). Each of these hypotheses leads to different predictions about the design features of rough-and-tumble play and its prevalence in the two sexes. Boulton and Smith lay out these predictions, evaluate them against existing evidence, and specify what kinds of data would be necessary to further test among them.

However much our attention may be captured by its sham violence, rough-and-tumble play could not have evolved unless it included design features promoting cooperation and compromise, such as self-handicapping and role reversal. The constraints that natural selection theory places on the evolution of rough-and-tumble play led Boulton and Smith to predict that design features of this kind will be present whether it evolved to provide practice in hunting or predator avoidance or fighting. Whatever skills children are learning in rough-and-tumble play, they argue, it is this cooperative component that allows them to learn from one another and indeed that has allowed this behavior to evolve.

REFERENCES

Profet, M. (1991). The function of allergy: Immunological defense against toxins. *The Quarterly Review of Biology, 66*, 23–62.

Symons, D. (1978). *Play and aggression: A study of rhesus monkeys.* New York: Columbia University Press.

Williams, G. C. (1966). *Adaptation and natural selection: A critique of some current evolutionary thought.* Princeton, NJ: Princeton University Press.

8

Pregnancy Sickness as Adaptation: A Deterrent to Maternal Ingestion of Teratogens

MARGIE PROFET

The adaptationist approach to the phenomena of life, which entails recognizing certain features of organisms as specialized problem-solving mechanisms (Williams, 1985), is the basis for physiological research (Mayr, 1983). Even physiological phenomena that are commonly assumed to be anomalies or pathologies, such as human pregnancy sickness, can sometimes be illuminated by asking the adaptationist questions: Has the phenomenon been designed by natural selection to serve some function? If so, what is that function? Whether a physiological process like pregnancy sickness is an adaptation or merely a by-product of other physiological processes can be assessed by determining whether it achieves an adaptive end with sufficient precision, economy, efficiency, and complexity to indicate functional design (Williams, 1966).

Pregnancy sickness is a collection of symptoms—food aversions, nausea, and vomiting—one or all of which occur in women during the first trimester of pregnancy. Pregnancy sickness is a physiological/psychological mechanism that influences eating behavior. The human perceptual system for evaluating foods is so designed that certain tastes and smells are more aversive than others, and this system becomes much more sensitive during the first trimester of pregnancy, deterring pregnant women from eating certain foods that they would otherwise eat. Many tastes and smells that women normally find palatable become intolerable to them during the first trimester of pregnancy; and these aversive tastes and smells frequently trigger nausea and vomiting as well.

Many misconceptions surround pregnancy sickness. For example, although pregnancy sickness is often called "morning sickness," it can occur at any time throughout the day. "Sickness" implies dysfunction, yet vomiting during early pregnancy is usually a sign of health: Women who vomit or experience severe nausea during early pregnancy have lower risks of spontaneous abortion than women who experience only mild pregnancy sickness (Klebanoff, Koslowe, Kaslow, & Rhoads, 1985). Pregnant women without nausea often feel fortunate in having been spared the discomfort of pregnancy sickness, but they are actually at greater risk of harming their embryos. Many pregnant women who do not feel "sick" during early pregnancy—i.e., who neither vomit nor experience prolonged bouts of nausea—assume that they do not have pregnancy sickness. However, food aversions—whether conscious or unconscious—are integral aspects of pregnancy sickness, and many pregnant women who claim not to have pregnancy sickness are, nevertheless, repulsed by certain foods and odors that they previously found attractive.

The fitness costs incurred by women who experience pregnancy sickness can include inadequate nutrition, because of a decrease in consumption of nutritious sources of food, and lower productivity, because of feeling ill. Possible benefits that might offset these costs have rarely been explored; rather, pregnancy sickness generally has been assumed to be a side effect of pregnancy hormones and to confer no benefits per se. Although pregnancy sickness may well be triggered by hormones that signal the onset of pregnancy, the central adaptationist question is whether this triggering would have been selectively advantageous in the human environment of evolutionary adaptedness—the environment of a Plio-Pleistocene forager.

Pregnancy sickness can be illuminated by considering the functions of food aversions, nausea, and vomiting in other contexts. Bitter tastes or pungent odors in plant foods, indicating high levels of natural plant toxins, and foul odors in animal foods, indicating parasitization by toxin-producing bacteria, commonly deter ingestion of these potentially dangerous foods. If dangerous foods are ingested, they typically induce nausea, vomiting, and subsequent aversions: Vomiting expels the toxin from the stomach; nausea and subsequent aversions deter future ingestion of the toxin (Davis, Harding, Leslie, & Andrews, 1986). Food aversions are psychological mechanisms by which one learns to distinguish safe from toxic foods. Perceptual cues associated with toxins—such as certain smells—enable one to detect probable toxicity without having to risk eating the food. In this chapter I will expand and refine an earlier argument (Profet, 1988): The food aversions, nausea, and vomiting of pregnancy sickness evolved during the course of human evolution to protect the embryo against maternal ingestion of the wide array of teratogens (toxins that cause birth defects) and abortifacients (toxins that induce abortion) abundant in natural foods. (Abortifacients will be regarded here as teratogens, since most teratogens are lethal to embryos at some dose.) In particular, I will argue that pregnancy sickness represents a lowering of the usual human threshold of tolerance to toxins in order to compensate for the extreme vulnerability of the embryo to toxins during organogenesis (the embryonic period of organ differentiation and maximum vulnerability to teratogens).

Evidence for this view will include the following: (a) Many foods that are safely ingested by adults contain compounds that can induce adaptively costly or fatal malformations in embryos, (b) women with pregnancy sickness selectively avoid foods that emit cues associated with toxicity, (c) pregnancy sickness begins when the embryo becomes vulnerable to these toxins, (d) it ends when the embryo's need for calories for growth outstrips its vulnerability to these toxins, (e) the olfactory perceptual system changes during pregnancy sickness in ways that promote the selective avoidance of toxins, and (f) women who have moderate or severe pregnancy sickness, rather than mild or no pregnancy sickness, enjoy greater pregnancy success rates.

SELECTION PRESSURES FOR PREGNANCY SICKNESS

All Plants Manufacture Toxins

Plants have numerous predators (including parasites), such as herbivorous mammals, insects, and fungi. In order to defend themselves against such a wide range of predators, plants have evolved a wide range of chemical weapons—toxins—which they synthesize in their tissues to poison their predators. Broadly defined, a *toxin* is a nonnutritional substance that exerts a biodynamic effect on the body (Schultes & Hofmann, 1987). Each plant species produces a unique set of toxins—commonly a few dozen—to ward off a variety of predators, and thousands of such toxins have been discovered so far (Ames, Profet, & Gold, 1990a). Under the stress of predator attack, many plants increase their levels of toxins manyfold (Beier, 1990; Harvell, 1990). Physiological disturbances to mammals caused by plant toxins can include neurological, renal, endocrine, metabolic, hepatic, reproductive, or other impairment, which may be temporary, permanent, or lethal, depending on the doses ingested and on the metabolic defenses of the predator (Freeland & Janzen, 1974).

Toxins are ubiquitous among plants and exist even in the edible parts of the common, seemingly innocuous plants used for human foods, such as apples, bananas, cabbage, celery, cherries, nutmeg, oranges, potatoes, and soybean. Natural mutagens and carcinogens are widespread among the plants that humans eat (Ames et al., 1990a, 1990b). For example, the mutagen/carcinogen 8-methoxypsoralen is in celery (Beier, Ivie, Oertli, & Holt, 1983) and parsnip (Ivie, Holt, E. Ivey, 1981); the carcinogen allyl isothiocyanate is in cabbage, cauliflower, and brussels sprouts (Buttery, Guadagni, Ling, Seifert, & Lipton, 1976; MacLeod & Pikk, 1978); and the mutagen/carcinogen safrole in is black pepper, cocoa, and nutmeg (Richard & Jennings, 1971; Van der Wal, Kettenes, Stoffelsma, Sipma, & Semper, 1971; Archer, 1988). Many plant chemicals bind covalently (very strongly) to the DNA of mammalian cells and so are potentially mutagenic, carcinogenic, and teratogenic (Randerath, Randerath, Agrawal, Gupta, Schurdak, & Reddy, 1985). Evidence that many plant toxins are teratogenic in various mammalian species (Keeler, 1983) is crucial to the study of pregnancy sickness, as I will discuss in this chapter. For reviews of natural plant toxins see Beier, 1990; Concon, 1988; Keeler & Tu, 1983; and Rosenthal & Janzen, 1979.

Because a plant's nutrients are coupled to toxins, the predator that eats a plant consumes both; natural selection has therefore shaped plant predators to solve the problem of how to ingest a plant's nutrients without being poisoned by its toxins. To counter the chemical arsenals with which plants protect themselves and thereby to exploit plants for their nutrients, mammals have evolved elaborate defenses against toxins, particularly extensive arrays of detoxification enzymes manufactured by the liver and the surface tissues (epithelia) of various other organs, such as the skin and the lungs. The presence of toxins frequently induces cells of these organs to increase the production of enzymes that can degrade those particular toxins (Bickers & Kappas, 1980; Gram, 1980). Plant species differ in the sets of toxins they have evolved, and, hence, animals that prey on plants have evolved specialized defenses against the particular types of toxins to which they have been regularly exposed. Many plants avoid intoxicating themselves with their own defensive toxins by exploiting their predators' detoxification mechanisms: They manufacture inert compounds that become activated to their toxic forms by the predators' enzymes during the first phase of the detoxification process (Fowden & Lea, 1979). The coevolutionary struggle between plant and predator (Van Valen, 1973) has resulted in a huge spectrum of plant toxins designed to derail different parts of predator physiological systems and in complex networks of animal physiological and psychological mechanisms for detecting toxic plant compounds, for determining danger thresholds of toxin, and for detoxifying and eliminating ingested toxins.

Various mammals, for example, process plant foods to circumvent the most toxic parts (e.g., by peeling fruit); humans, in addition, cook many plant foods to destroy toxins through heat or soak them in water to leach out their toxins (Kingsbury, 1983; Stahl, 1984; Tooby & DeVore, 1987). Mammals typically diversify their diets in order to prevent overloading their systems with any one toxin. Even specialized herbivores seek dietary diversity: Koalas, for example, eat 16 different species of eucalyptus, and in captivity they will refuse to eat if offered only one species (Freeland & Janzen, 1974). Mammals initially sample a novel plant in small doses so that its toxicity can be determined and so that enzymes capable of detoxifying its particular toxins will be induced in the liver, thereby permitting a subsequent larger intake of those toxins (Freeland & Janzen, 1974). Ingested foods containing concentrations of toxin that exceed the thresholds of tolerance for that toxin may provoke nausea, vomiting, and subsequent aversions to the odors and tastes of those foods (Davis et al., 1986, Rozin & Kalat, 1971).

Both plant and predator benefit from the predator being able to detect toxins in plants—a mutual interest that has led to the coevolution of plant signals of toxicity and predator mechanisms for perceiving this toxicity. Plant toxins therefore tend to be highly correlated with odors and tastes that their predators find aversive. Many plant toxins taste or smell "bitter" or "pungent" to humans and, presumably, to other mammals. For example, most plant alkaloids and cyanogens—two large classes of plant toxins—taste bitter (Kingsbury, 1983). Chemoreceptors for detecting such toxins have been strongly favored by selection throughout the evolutionary history of plant predators and may have evolved as long ago as 500 million years (Garcia & Hankins, 1975). Nondomesticated plant predators almost universally detect plant toxins and find them unpalatable when present in poisonous concentrations (Kingsbury, 1983; Rozin & Vollmecke, 1986). Many plants even advertise their unpalatability (aposematism) by emitting volatile chemicals whose odors signal noxious tastes (Eisner & Grant, 1980). Aroma can be an especially important warning sign of plant toxicity because aromatic toxins are sometimes produced when a plant is threatened, such as when it is cut or eaten: For example, when onion tissues are ruptured, they produce their distinguishing pungent odors through enzymatic degradation of nonvolatile chemicals to volatile toxins (Cole, 1980; Schwimmer, 1968; Virtanen, 1965). Humans and other experimental plant predators, however, often learn which pungent or bitter tastes are associated with the toxins that their physiological systems tolerate well and even acquire "tastes" for some of those that signal important sources of nutrients.

Many Plants Contain Teratogens

The threshold of tolerance for toxicity sufficient to protect adults against dangerous levels of toxins may be grossly insufficient to protect developing embryos. Teratogens are threshold-limited: Below a certain dose, no malformation takes place; above this dose, the toxicity of a substance can range from teratogenic to embryolethal to maternally lethal (Williams, 1982). For many toxins, doses that have negligible effects on the adult are teratogenic or even lethal to embryos (Beck, 1973; Williams, 1982). Over one-third of the 2,800 chemicals that have been tested in various mammalian species demonstrate some measure of teratogenicity (Schardein, Schwetz, & Kenel, 1985). Although the chemical doses used in these tests are often high enough to cause some toxicity to the mother, the effects on the embryo are far more severe. Teratogens are abundant in the plant kingdom. For example, certain toxins in potatoes (the glycoal-kaloids solanine and chaconine) can cause neural tube defects in the embryos of some mammalian species, even at doses nontoxic to the mother (Renwick, Claringbold, Earthy, Few, & McLean, 1984). Potatoes are also suspected of causing neural tube defects—particularly anencephaly and spina bifida (ASB)—in humans. Ireland, a country known for heavy potato consumption, has the highest rate of ASB in the world. Potato blight significantly increases the concentration of toxins in potatos, and the severity of potato blight in Ireland is strongly correlated with the rate of ASB (Renwick et al., 1984). A toxin in eggplant (the alkaloid solasodine) likewise causes neural tube defects in hamster embryos (Beier, 1990). Toxins produced by some of the molds that parasitize nuts and grains (aflatoxins) are teratogenic in various species (Hayes, 1981; Hood & Szczech, 1983). And many pregnant domestic animals that ingest grasses, seeds, and fruits containing known teratogens give birth to malformed off-spring, yet are themselves unharmed (Keeler, 1983). Certain forage plants, such as lupine, are so teratogenic that drinking the milk from an animal that ingested the plant is a teratogenic hazard (Ames et al., 1990b). The toxin anagyrine in the plant lupine, for example, can cause severe bone abnormalities ("crooked calf syndrome"). In one rural California family, a litter of goat kids, a baby boy, and a litter of puppies all were born with this syndrome after the family's pregnant goat foraged on lupine and her milk was drunk by both the pregnant woman and the pregnant dog (Crosby, 1983; Kilgore, Crosby, Craigmill, & Poppen, 1981; Warren, 1983).

Teratogenesis can occur when low molecular weight toxins (which most plant tox-ins are) enter the maternal bloodstream and diffuse through the placenta. Toxins can initiate teratogenesis by mutating the genes, interfering with cell division, changing the characteristics of the cell membrane, or causing other disturbances that lead to disrup-tion of chemical signals, to cell death, or to lack of functional cell maturation (Hodg-son & Levi, 1987). Toxins also can harm the embryo indirectly by interfering with the placental or maternal chemicals that regulate the nourishment of the embryo (Waddell & Marlowe, 1981). Embryos thus require additional maternal defenses against dietary toxins, and maternal pregnancy sickness may be such an additional defense. When a woman becomes pregnant, lower doses of toxin become fitness threats, hence her threshold of tolerance to toxins—i.e., what she finds palatable—needs to be recali-brated.

THE NATURE OF PREGNANCY SICKNESS

Pregnancy Sickness Coincides with Organogenesis

Pregnancy sickness coincides with organogenesis, the embryonic period of maximum vulnerability to teratogens. Human organogenesis takes place from approximately day 20 to day 56 after conception and entails formation of the limbs and all the major organ systems, including the central nervous system, heart, eyes, ears, and external genitalia (Eskes & Nijdam, 1984; Hodgson & Levi, 1987). Organogenesis is therefore the period during which major morphological malformations can occur, although for some organ systems the sensitive developmental periods extend through the 14th week (Eskes & Nijdam, 1984; Hodgson & Levi, 1987). Pregnancy sickness usually begins within 2 to 4 weeks after conception (the start of organogenesis), peaks between 6 to 8 weeks after conception, falls off after 8 weeks (the end of organogenesis), and disappears completely by 14 weeks (the end of the sensitive periods of organ development) (Willson & Carrington, 1976). Pregnancy sickness—maternal aversions to teratogens—thus coincides with embryonic susceptibility to teratogens.

Neither organogenesis nor observable pregnancy sickness occurs during the first 2 weeks after conception. During these 2 weeks embryonic cells proliferate rapidly but do not differentiate much morphologically, so that chemical agents that disrupt embryonic development at this stage are more likely to cause cell death, which leads either to cell replenishment or to embryonic death, rather than malformation (Eskes & Nijdam, 1984). Furthermore, toxins that are absorbed into the maternal bloodstream are less likely to reach the embryo if the placenta, which connects the embryo to the maternal bloodstream, has not yet formed. Although the embryo implants in the uterus 6 to 7 days after conception, it does not form a placenta capable of absorbing maternal blood until 15 days after conception (Tuchmann-Duplessis, David, & Haegel, 1971). During the first week after conception, while traveling down the uterine tube, the embryo is thought to obtain nourishment by absorbing uterine gland secretions and, during implantation, by digesting uterine endometrial cells (Hamilton & Mossman, 1972). Although toxins that are absorbed into the maternal bloodstream can potentially reach these initial sources of nourishment, these sources are much less direct conduits for toxins than is the maternal blood, which nourishes the embryo from the third week until birth. Therefore, before the third week of gestation the embryo would not be expected to be highly susceptible to teratogenic harm, nor the mother to pregnancy sickness. Indeed, *in vivo* experiments have shown that very few teratogenic effects occur during the first 2 weeks after conception and that teratogenic substances ingested by the mother prior to organogenesis usually do not result in teratogenesis (Hodgson & Levi, 1987; Persaud, 1985; Spielmann & Vogel, 1987).

After organogenesis, the embryo becomes a fetus, which in general is much less susceptible to teratogens than is the embryo (Persaud, 1985). Although maternal ingestion of chronic high doses of certain toxins during the second and third trimesters can cause functional and behavioral defects in the fetus, pregnancy sickness generally ends by the second trimester; if it did not, pregnancy sickness would probably inflict nutritional costs on the fetus that outweighed the benefits of maternal aversions to food toxins. Pregnancy sickness can reduce nutrient intake because by avoiding toxins the mother avoids the nutrients coupled to those toxins. Women who experience pregnancy sickness severely enough to vomit often lose weight during the first trimester (Tierson, Olsen, & Hook, 1986). Such nutritional costs, however, are usually of small consequence to the minute embryo, who weighs only a few grams by the end of organogenesis and whose nutritional demands are therefore slight. Ovulation generally does not occur unless the woman has accumulated a certain threshold of fat reserves (Frisch, 1987), which, as discussed later, could be an adaptation to ensure a buffer against the nutritional deprivation caused by pregnancy sickness.

Pregnancy sickness might appear to be maladaptive, however, because it induces aversions to vegetables, which are the sources of folate and other vitamins and micronutrients that are essential for normal embryonic development. Certain first-trimester vitamin deficiencies have been implicated in teratogenesis. There is strong evidence, for example, that maternal folate deficiency just prior to conception and during early embryonic development increases the risk of neural tube defects (Milunsky et al., 1989; Wald, 1991). The nutritionally adverse effects of pregnancy sickness on embryonic health, however, are probably exacerbated in modern industrial societies. Pleistocene hunter-gatherer women with sufficient fat reserves to conceive are unlikely to have been vitamin deficient at conception because much of their caloric intake prior to conception would have come from vegetables and fruits. By contrast, much of the caloric intake of women in modern industrial societies comes from nonplant foods or processed foods low in vitamins. Since the normal folate stores in the liver last over 4 months (Herbert, 1990), folate reserves from a normal Pleistocene diet would have covered the temporary deficiencies incurred by the pregnant-sick Pleistocene woman.

Perhaps to compensate for decreased nutritive intake in the first trimester, maternal basal metabolic rate often decreases in the first trimester, whereas it increases considerably in the third trimester (Prentice & Whitehead, 1987). According to calculations by Prentice and Whitehead (1987), the daily energy costs of pregnancy— including the deposition of fetal, uterine, placental, and mammary tissue, the storage of maternal fat, and the maintenance of fetal and added maternal tissue—are considerably lower in the first trimester (by almost a factor of 4) than in the second and third trimesters. Even severe nutritional deprivation in the first trimester often can be compensated for by adequate nutrition in the second and third trimesters. Rush (1989) analyzed studies of the effects of famine on the fertility and pregnancy outcome of previously well-nourished populations (the World War II populations of the Netherlands, Leningrad, and Wuppertal). Although famine decreased fertility, among women who did conceive there was a significant decrease in the average birthweight of infants born during the famine, but not of infants conceived during the famine to mothers who then received adequate nutrition during the third trimester. Rush (1989) suggests that adequate nutritional intake during the third trimester can mitigate the effects of nutritional deprivation during the first trimester. (Infants conceived during famine, however, had higher rates of central nervous system abnormalities.)

Fetal nutritional requirements increase significantly during the second and third trimesters, and nutritional deficiencies incurred by the mother during the latter trimesters can lead to growth retardation of the fetus (Institute of Medicine, 1990). The mother's behavior toward food during this period should reflect the emphasis in fetal development on growth (rather than the emphasis in embryonic development on organ differentiation). Thus, the maternal threshold for tolerating toxicity, after decreasing during the first trimester, should and does increase again during the second and third trimesters. By spanning only the period during which the embryo is most vulnerable to maternal ingestion of toxins and least vulnerable to decreased maternal ingestion of nutrients, pregnancy sickness protects the embryo from toxic assault; when fetal nutritional demands require a substantial intake of maternal food, pregnancy sickness ends.

Olfactory Cues of Toxicity Elicit Pregnancy Sickness

Pregnancy sickness heightens sensitivity to noxious substances, inducing aversions, nausea, and vomiting with minimal provocation (Walters, 1987). In other words, pregnancy sickness lowers a woman's threshold for toxin detection. Women with pregnancy sickness are particularly sensitive to smells and tastes associated with toxicity, to the extent of being repulsed by previously tolerated foods. For example, obstetricians note that patients with pregnancy sickness generally become intolerant of foods with pungent or bitter odors and tastes, such as highly spiced foods and all but the most bland vegetables (Willson & Carrington, 1976). Because the various pungent aromas and bitter tastes of plant foods are generally caused by the toxins they contain (for example, the toxin allyl isothiocyanate in mustard and broccoli cause their pungent aromas [Buttery et al., 1976]), foods that stimulate the olfactory or gustatory chemo-receptors for bitter are more likely to be toxic than are bland foods (Oakley, 1986). Maternal aversions to such aromas and tastes deter maternal ingestion of these foods in favor of bland foods containing lower, less detectable concentrations of toxins and so prevent embryonic exposure to high levels of toxin. Heightened sensitivity to cues of toxins may also discourage pregnant women from experimenting with novel sources of foods, which pose a risk of potential toxicity.

Nonplant foods that emit olfactory cues of toxicity would also be expected to elicit aversions in women with pregnancy sickness. For example, pungency in meats and dairy products indicates possible parasitization by bacteria—many strains of which produce potent toxins (Alcock, 1983)—because such foods commonly acquire a pungent odor when bacteria are decomposing them. Bacterial endotoxins can cause teratogenesis as well as uterine resorption of the fetus (Gower, Baldock, O'Sullivan, Dore, Coid, & Green, 1990; Haesaert & Ornoy, 1986; Hilbelink, Chen, & Bryant, 1986; O'Sullivan, Dore, & Coid, 1988). Aversions to pungent animal-derived foods discourage the pregnant woman from inflicting bacterial toxins on her developing embryo.

The smell of cooking fumes often triggers or exacerbates the nausea, vomiting, and aversions of pregnancy sickness (Hale, 1984). Cooking fumes can signal various toxic dangers. Cooking volatilizes many plant toxins, which can then be inhaled (Buttery et al., 1976; MacLeod & MacLeod, 1970). Cooking false morel mushrooms, for example, produces highly toxic steam (Beier, 1990), which would be especially dangerous for pregnant women because pulmonary blood flow increases during pregnancy, thereby increasing the absorption rate of inhaled toxins (Hytten, 1984). Toxic cooking fumes also discourage the pregnant woman from preparing toxic foods, which is important because pregnancy increases blood flow to the skin, allowing greater absorption of toxins through the hands (Hytten, 1984). Although cooking can kill toxin-producing bacteria in meat and inactivate many toxic compounds in plants, cooking also transforms many innocuous food compounds into mutagens and carcinogens (Overvik, Nilsson, Fredholm, Levin, Nord, & Gustafsson, 1987; Sugimura, 1982). Frying or burning foods is particularly hazardous because it converts some of the protein material to mutagens; in addition, foods fried in fats absorb mutagens that are created by the thermal oxidation of the fats (Hageman, Hermans, Ten Hoor, & Kleinjans, 1990; Hageman, Kikken, Ten Hoor, & Kleinjans, 1988). In a study by Lindeskog, Overvik, Nilsson, Nord, and Gustaffson (1988), the urine of rats fed a diet of fried meat was mutagenic, whereas the urine of rats fed boiled meat was not. Since these ingested mutagens circulated to the kidneys, presumably they could have circulated as well to the placenta.

In more primitive methods of cooking, toxins may be released by the heating of cooking utensils, such as sticks or leaves, that contain resins or other toxins that can be converted into toxic vapors and inhaled. The nausea induced by inhaling noxious cooking fumes discourages pregnant women from continuing to inhale these fumes, from ingesting foods emitting the fumes, and from wanting to cook and eat these foods in the near future. By lowering the threshold for toxin detection, pregnancy sickness thus deters the pregnant woman from ingesting foods and inhaling fumes that, although not harmful to her, are potentially teratogenic to her much more vulnerable embryo.

Perceptual changes in the palatability of foods can occur during pregnancy due to changes in olfactory chemoreceptive sensitivity to substances suggestive of toxins. In general, women have lower olfactory thresholds than men (Doty, 1986) and so are able to detect toxic odors more acutely than men are. (Pleistocene selection pressures for detection of food toxins may have been greater in women than in men because women were able to transfer ingested toxins to offspring via pregnancy and lactation and because women were more likely to gather and prepare plant foods.) Preliminary studies on changes in olfactory threshold during pregnancy indicate that in the first trimester, during embryonic organogenesis, women become *hyper*osmatic (develop increased olfactory acuity) and thus are likely to develop aversions to previously tolerated foods, but by the third trimester women become *hypo*osmatic (Doty, 1976) and thus are unlikely to maintain aversions to these foods. This hypoosmaticism may be necessary to counter the food aversions that developed during the first trimester hyperosmaticism, so that the mother is not deterred from consuming important sources of nutrients during stages of rapid fetal growth. Pregnancy sickness is a temporary short-term appropriation of the mechanism for inducing food aversions that was designed for the long term (as foods containing toxins on one occasion are likely to contain them on future occasions). Consequently, overcoming first-trimester food aversions through second- and third-trimester hypoosmaticism may entail the risk of ingesting toxins during the second and third trimesters at levels that normally would cause aversions. The costs of exposure to plant toxins thus are shifted from the embryo to the fetus.

Increased olfactory sensitivity in early pregnancy might occur through several different possible mechanisms. Olfactory neurons, unlike neurons of most neuronal networks, undergo death and replenishment well into adulthood (Breipohl, Mackay-Lim, Grandt, Rehn, & Darrelmann, 1986; Wilson & Raisman, 1980). These olfactory neurons serve as the receptors for toxic and other inhaled compounds. Hormonal changes influence olfaction (Breipohl et al., 1986), and in rats (and probably other mammals), the olfactory surface tissues contain receptors for the hormone estradiol, which is secreted in large quantities during pregnancy (Vannelli & Balboni, 1982). It is therefore possible that the hormonal changes of early pregnancy induce neuron growth (or hasten neuron replenishment), thereby increasing olfactory sensitivity. Indeed, in mice, pregnancy stimulates the proliferation of olfactory neurons (Kaba, Rosser, & Keverne, 1988). (In humans, the hormonal changes of the second and third trimesters of pregnancy may reverse the effects of the first trimester, inhibiting neuron growth, and thereby decreasing olfactory sensitivity during the second and third trimesters.) In addition, because olfactory tissue contains a wide array of enzymes for detoxifying inhaled foreign compounds (Dahl, 1988), the hormonal changes of early pregnancy may affect the composition of these detoxification enzymes, just as pregnancy hormones affect the detoxification enzymes in the liver (Feuer, 1979), thereby altering the perception of toxicity of inhaled compounds. Furthermore, pregnancy raises the vascularity of the nasal mucus membrane (Hytten, 1984), as it does of other mucus membranes, which probably results in an increase in nasal uptake of toxic molecules and thus in enhanced olfactory detection of inhaled toxins. (Whether nasal vascularity then decreases after organogenesis should be investigated in order to determine whether the first-trimester increase in nasal vascularity is an adaptation for alerting the pregnant woman to toxins or merely an effect of the general vascular changes of pregnancy.) Through these and possibly other mechanisms, heightened first-trimester olfactory sensitivity to toxins would discourage maternal ingestion of teratogens.

If olfactory neuron death results from inhaling toxins that kill the neruons, then this neuron death has some interesting implications for pregnancy sickness. B. Ames (personal communication, 1989) has suggested that, as a way of detecting a broad range of toxins, mammals may have evolved mechanisms to detect neuron death, to associate neuron death with the particular toxin that caused it, and to induce aversions to that toxin henceforth. If mammals have such mechanisms, then the susceptibility of olfactory neurons to death may itself be an adaptation. Olfactory neurons are the only neurons in the body that are known to regularly die and be replenished after early childhood; yet if they are dying in response to inhaled toxins, it is to such minute quantities of toxin compared to the quantities of toxin that mammals regularly ingest that these neurons would seem to be suspiciously susceptible to death. If neuron death takes place immediately after contact with the toxins, then it might immediately deter their ingestion; if neuron death is delayed, it might be effective in deterring only their repeated ingestion. A mechanism to detect toxicity-induced death of olfactory neurons might constitute a defense primarily against toxins that affect the central nervous system—i.e., toxins that harm neurons that are not regenerated. If so, then the first-trimester increase in olfactory sensitivity—caused possibly by enhanced olfactory neuron growth—would particularly heighten maternal sensitivity to toxins that could disrupt the embryo's developing central nervous system and would selectively screen them out.

A Network of Mechanisms Minimizes Embryonic Exposure to Toxins

The need to reduce embryonic exposure to toxins is a major selection pressure. Natural selection has therefore designed various mechanisms for solving this problem. (Data on most such mechanisms are as yet sparse, because most of the studies on pregnancy-induced changes in maternal physiology compare pregnant with nonpregnant women rather than comparing the changes that occur during the course of pregnancy. Since the mother's physiological changes often reflect the needs of her embryo or fetus, and because the needs of the embryo may vastly differ from or even be opposite to those of the fetus, various aspects of the mother's physiology can change dramatically during her pregnancy.) The existence of a network of physiological mechanisms that reduce embryonic exposure to toxins supports the hypothesis that pregnancy sickness is an adaptation to protect the embryo from toxins, by supporting the hypothesis that protection from toxins during pregnancy is a major selection pressure.

Various physiological changes that occur during pregnancy, for example, concern maternal absorption and excretion of toxins. In early pregnancy, the movement of food from the stomach to the intestines (gastric motility) decreases (Calabrese, 1985). This slows the rate of absorption of foods—and toxic constituents of foods—into the bloodstream and confines foods to the stomach for a longer period of time, thus increasing the opportunity to expel foods through vomiting (Davis et al., 1986). Indeed, inhibition of gastric motility may be an adaptation to slow absorption of toxins because this inhibition also occurs in nonpregnant mammals after ingestion of substances that induce nausea (and that the body therefore recognizes as toxic) (Borison, Borison, & McCarthy, 1984). The small intestine likewise displays reduced motility during pregnancy, substantially prolonging transit times for foods (Hytten, 1984). Because the gastrointestinal absorption of toxins is spread out over a longer period of time during pregnancy, the peak toxic load on the liver is decreased so that at any given time the liver is more likely to have sufficient metabolic resources to thoroughly handle incoming toxins. (However, gastrointestinal motility is decreased throughout pregnancy, not just during the first trimester [Wald et al., 1983]. This could mean that some reduction in toxicity is important for the fetus as well as the embryo or that the decrease in gastrointestinal mobility has a function in addition to reducing toxicity, such as preventing vitamin depletion.)

Toxins entering the circulation are likely to be filtered out more quickly than usual during early pregnancy. In the normal filtration process, toxins that have been rendered water-soluble by detoxification enzymes of the liver or other organs pass through the kidneys and are excreted in the urine. In early pregnancy the maternal rate of blood flow to the kidneys increases significantly, almost doubling by the end of the second trimester and declining thereafter (Hytten, 1984). The rate at which molecules are absorbed from the bloodstream to the kidneys (the glomerular filtration rate) increases by up to 70% (Hytten, 1984), thereby increasing the rate of elimination of toxins. During late pregnancy the glomerular filtration rate generally declines (Davison & Hytten, 1974), which may prevent excess excretion of water-soluble vitamins and nutrients during the time of rapid fetal growth.

If selection pressures to reduce embryonic exposure to toxins were indeed significant during the course of human evolution, one might expect that mechanisms would have evolved to simply accelerate the rate at which the mother detoxifies toxins during the first trimester of pregnancy by, for example, increasing the number of detoxification enzymes. Mechanisms that increased the efficiency with which the liver detoxifies plant compounds might even have eliminated the need for pregnancy sickness. However, accelerating the detoxification process is problematical because a necessary step in enzymatic degradation of certain types of inert compounds to water-soluble excretable compounds entails activating them to toxic intermediate metabolites. Thus, accelerating the maternal rate of enzymatic degradation of compounds would also accelerate the rate at which these toxic metabolites are formed, which could pose various dangers for the embryo. Many of these toxic metabolites bind covalently to the DNA of cells with which they come in contact and so are potentially mutagenic and carcinogenic (Juchau, 1981). Such toxic metabolites could be especially harmful to an embryo or fetus, whose cells are proliferating at high rates. Furthermore, chemicals that bind covalently to DNA are potential teratogens (Randerath et al., 1985). Thus, merely accelerating detoxification rates might be a counterproductive mechanism for reducing toxicity to the embryo.

Pregnancy does affect the composition of the detoxification enzymes of the maternal liver (Nau, Loock, Schmidt-Gollwitzer, & Kuhnz, 1984), although it is not known whether these changes are adaptations to reduce the overall toxicity to the embryo. The effects of pregnancy on the clearance rate of toxins from the bloodstream, for example, are not consistent for all toxins or for all pregnant women. The steroids estradiol and progesterone can either inhibit or stimulate the detoxification activity of liver enzymes sufficiently to depress or enhance the rates of metabolism of many toxic compounds during pregnancy (Feuer, 1979; Juchau, 1981; Loock, Nau, Schmidt-Gollwitzer, & Dvorchik, 1988). The placenta and embryonic organs of mammals also have detoxification enzymes (Pelkonen, 1982), which are induced in response to toxins in the maternal bloodstream, such as to benzo[a]pyrene from cigarette smoke (Dey, Westphal, & Nebert, 1989). Embryonic mice that have responsive enzyme systems and whose mothers have less responsive enzyme systems suffer a greater than normal proportion of stillbirths, resorptions, and congenital defects (Shum, Jensen, & Nebert 1979). Thus, there is a trade-off between detoxifying toxins and preventing the activation of potential carcinogens, mutagens, and teratogens. These physiological constraints on the mother's ability to accelerate the rate at which she detoxifies compounds necessitate her having psychological mechanisms to alter her behavior toward toxins during early pregnancy. Consequently, pregnancy sickness—the avoidance of potential teratogens—is her best defense against toxins that could harm her embryo.

Pregnancy Sickness Probably Involves the Chemoreceptor Trigger Zone (CTZ)

Understanding the adaptive function of pregnancy sickness may help to illuminate its physiological causes. The same physiological and psychological mechanisms may underlie both pregnancy sickness and the food aversions, nausea, and vomiting that occur in response to high concentrations of food toxins in nonpregnant humans. The vomiting response in humans and other mammals often involves a part of the brain stem called the *chemoreceptor trigger zone* (CTZ), which lies in the region of the brain known as the *area postrema*. This region is involved in inducing conditioned taste aversions to a wide variety of ingested and injected substances (Bernstein, Courtney, & Braget, 1986; Borison et al., 1984). Although most parts of the brain are shielded from direct contact with blood by the blood-brain barrier, the CTZ is extensively bathed with blood and cerebrospinal fluid, which it samples for toxic constituents (Borison, 1986). Chemoreceptors of the CTZ are specific for many different types of toxic molecules and induce nausea and vomiting when levels of toxins exceed certain thresholds (Borison, 1986).

One possible mechanism for pregnancy sickness is that pregnancy hormones cause an increase in blood flow to the CTZ and, consequently, an increase in toxins that reach the CTZ from the bloodstream. This action might recalibrate the CTZ response to toxins in the bloodstream, resulting in heightened sensitivity—and the triggering of nausea and vomiting—to low concentrations of food toxins. It has been hypothesized that control of blood flow to the CTZ is regulated by receptors on the surfaces of blood vessels to the CTZ that expand or contract these vessels in response to particular substances circulating in the bloodstream (Davis et al., 1986). Thus, a role for the CTZ in inducing pregnancy sickness might be demonstrated by finding receptors on blood vessels to the CTZ that bind pregnancy hormones (discussed below). Other researchers have likewise speculated that the CTZ may play a part in pregnancy sickness (Jarnfelt-Samsioe, 1987; Walters, 1987). If pregnancy sickness does represent a recalibration of the thresholds of toxin tolerated by the CTZ, then it could have evolved fairly simply and parsimoniously: Natural selection would have had only to make a minor modification in existing physiological and psychological mechanisms for avoiding dangerous levels of toxin.

Pregnancy Sickness Is Probably Triggered Hormonally

A complex interplay of the hormones secreted during early pregnancy may trigger, sustain, and terminate pregnancy sickness. At implantation, the embryonic placenta begins to synthesize the hormone human chorionic gonadotropin (HCG). After release into the maternal bloodstream, HCG induces the maternal ovary to produce the steroid hormones estradiol and progesterone, which are necessary for stimulating uterine growth and suppressing ovulation. The placenta also synthesizes estradiol and progesterone directly and, by the 6th week after conception, synthesizes them at sufficient levels to maintain the pregnancy without hormonal support from the maternal ovary (Patillo, Hussa, Yorde, & Cole, 1983; Sadow, 1980). HCG levels in the maternal bloodstream peak by about the 8th week after conception, fall precipitously after the 10th week, and remain low from the 14th week on (Harrison, 1980). Estradiol levels rise steadily throughout pregnancy until birth (Pang, Tang, Tang, Yam, & Ng, 1987). Progesterone levels rise immediately after conception, plateau from about weeks 2 through 8, then rise steadily until birth (Aspillaga, Whittaker, Taylor, & Lind, 1983; Tulchinsky & Hobel, 1973).

The hormones traditionally viewed as the most likely candidates for triggering pregnancy sickness are HCG, because peak levels coincide with pregnancy sickness, and estradiol, because high levels cause nausea in nonpregnant women, such as in some women taking birth control pills (Jarnfelt-Samsioe, 1987). Studies correlating hormonal load with degree of pregnancy sickness have led to conflicting results (Depue, Bernstein, Ross, Judd, & Henderson, 1987; Fairweather, 1986; Jarnfelt-Samsioe, Bremme, & Eneroth, 1985; Masson, Anthony, & Chau, 1985; Soules, Hughes, Garcia, Livengood, Prystowsky, & Alexander, 1980). However, it is possible that pregnancy hormones interact synergistically or antagonistically to induce pregnancy sickness. From those just cited and other studies of plasma hormone levels during different stages of pregnancy (Guillaume, Benjamin, Sicuranza, Wand, Garcia, & Friberg, 1987; Halmesmaki, Autti, Granstrom, Stenman, & Ylikorkala, 1987; Kletzky, Rossman, Bertolli, Platt, & Mishell, 1985; Loock et al., 1988; Pang et al., 1987), the hormonal factors that stand out as most likely to influence pregnancy sickness are increases in estradiol, in the estradiol/progesterone ratio, and perhaps in HCG.

Estradiol, an estrogen, appears to profoundly affect the CTZ. Estradiol infusions in rats cause severe food aversions, whereas estradiol infusions have little effect on eating behavior when lesions of the CTZ occur (Bernstein et al., 1986). Furthermore, estrogen receptors have been found in high concentrations in the area postrema of the brain (Bernstein et al., 1986; Simerly, Chang, Muramatsu, & Swanson, 1990)—where the CTZ is located. Estradiol thus appears to increase sensitivity of the CTZ to circulating toxins, perhaps by binding to receptors that line the blood vessels of the CTZ and dilating these vessels to enable greater blood flow to the CTZ. However, estradiol levels continue to rise after pregnancy sickness ends, so if estradiol causes CTZ sensitivity during early pregnancy, the puzzling question is what shuts off pregnancy sickness after the first trimester.

The estradiol/progesterone ratio may significantly affect pregnancy sickness, because progesterone exerts some effects that are opposite to those of estradiol. Progesterone has a quiescent effect (antagonistic to estradiol's effect) on smooth muscle tissue, such as the lining of the uterus and blood vessels (Sadow, 1980); therefore, progesterone may dampen the effects that estradiol may have on blood vessels that lead to the CTZ. During the peak period of pregnancy sickness, from weeks 2 to 8 after conception, the estradiol/progesterone ratio increases between 5- and 10-fold (see Aspillaga et al., 1983; Tulchinsky & Hobel, 1973). When the rate of change of this ratio slows considerably, 8 weeks after conception, pregnancy sickness wanes. Although the estradiol/progesterone ratio slowly continues to increase after the first trimester— roughly doubling over the second and third trimesters—the rate of increase in this ratio, rather than the ratio itself, may be the more important factor in eliciting pregnancy sickness. If estradiol synthesis increases relative to progesterone synthesis and at a rate that surpasses the rate of induction of the enzymes that metabolize estradiol, then this increase in estradiol could exert a significant temporary effect on physiological systems that have target receptors for estradiol, perhaps including the CTZ.

The rise and fall of HCG levels are neatly timed with pregnancy sickness. If HCG helps to regulate pregnancy sickness, then it might do so by altering the effectiveness of estradiol and progesterone, for example, by stimulating or inhibiting the presence of receptors for these steroids on target tissues or by altering the rate of steroid metabolism in the liver. Thus, the net influence of a particular estradiol/progesterone ratio on the CTZ might be different in the presence of HCG.

If the dynamics of estradiol, progesterone, and HCG trigger pregnancy sickness, then the embryo can time pregnancy sickness to begin precisely when it needs it and shut off when it no longer needs it. At implantation, the embryo suddenly has the capacity to synthesize and release pregnancy hormones into the maternal bloodstream and to absorb toxins from the maternal bloodstream. The embryo simultaneously becomes susceptible to toxins and able to trigger pregnancy sickness. That the embryo to some extent controls pregnancy sickness points to the possibility of mother-embryo conflict of interest over the severity of pregnancy sickness—an early opportunity for parent-offspring conflict (see Trivers, 1974) that may be acted out in the control over blood flow to the CTZ. For example, the embryonic placenta and the maternal ovary might alter their production ratios of estradiol and progesterone to increase or decrease pregnancy sickness. Although the mother's genetic interests in avoiding teratogens during early pregnancy in most circumstances coincide with the embryo's interests, in some cases, such as when the mother is nursing a previous offspring on whom pregnancy sickness would inflict a nutritional cost, the mother's interests may be best served by not having severe pregnancy sickness. The potential conflict of interest between the mother, who is capable of inflicting toxins on her embryo, and the embryo, who is capable of inflicting pregnancy sickness on the mother, may be spurred by certain cues, such as the presence of lactation hormones, that stimulate mechanisms of the mother to suppress pregnancy sickness as well as mechanisms of the embryo to subvert her doing so.

"Morning Sickness"

Although pregnancy sickness can occur at any time in the course of a day, many pregnant women consistently experience pregnancy sickness upon rising in the morning, hence the term "morning sickness." One possible explanation for the presence of pregnancy sickness in the absence of food is as follows: Gastric and intestinal motility decrease during pregnancy, significantly delaying digestion and absorption of food constituents. The bacteria that extensively colonize the intestines of all mammals manufacture enzymes that interact with the contents of the intestines to produce a vast array of metabolites, many of which are toxic. During the first trimester of pregnancy the stomach becomes less acidic due to decreased secretions of hydrochloric acid, which may encourage bacterial colonization of the stomach (Calabrese, 1985). Chemoreceptors that register toxicity in the gastric region would be likely to induce nausea and vomiting in response to toxic metabolites produced from the gastrointestinal bacterial metabolism of meals eaten the night before. The CTZ would also react to the bacterially activated food toxins that are seeping into the bloodstream from the digestive tract.

To help alleviate morning sickness, physicians typically suggest eating very bland starchy food, such as soda crackers, before rising in the morning. If the gut contains toxins produced by bacterial interaction with food eaten the previous evening, then filling the stomach with foods that are extremely bland—that is, low in toxicity— dilutes the concentration of toxins in the gut and blocks the bacterially activated toxins from stimulating the chemoreceptors in the gastric region and from being absorbed into the bloodstream in high concentrations to reach the CTZ. A pregnant woman often has more success in alleviating morning sickness if she eats the first food of the day while lying supine in bed rather than after rising to an erect position, probably because the supine position minimizes the rate at which toxins circulate to the CTZ. When the human body changes from a supine to an erect posture or to locomotion, the circulatory and respiratory rates increase abruptly (Abitbol, 1988), thereby increasing the rate of blood flow to the CTZ and the rate at which residual toxins absorbed into the bloodstream are circulated to the CTZ.

The nearly constant state of nausea experienced by some women during the first trimester of pregnancy may be caused by CTZ and gastric hypersensitivity to the toxic metabolites that are manufactured constantly during digestion by the gut bacteria, for although meals tend to be eaten at discrete intervals, they may be digested continuously throughout the day and night. Also, the low levels of toxins contained in virtually all plant foods may be slowly absorbed into the circulation throughout the day, because of slower gastrointestinal absorption of food during early pregnancy, causing the hypersensitive CTZ to produce continuous nausea. Because gastrointestinal bacteria, like metabolic enzymes of the liver, inactivate some types of toxin but activate other types, the decrease in stomach acidity that enables bacteria to colonize the stomach may serve the dual function of enabling the detoxification of many compounds before they are absorbed into the circulation and of ensuring that some compounds that would have been activated to toxic metabolites in the liver are instead activated in the gut, where they can trigger chemoreceptors in the gastric region and be more easily expelled through vomiting.

Various "Anomalous" Mechanisms Protect Against Toxins

The preceding discussions have tried to shed light on the physiology of pregnancy sickness in part by considering the physiological processes that cause nausea, vomiting, and food aversions in response to toxins in nonpregnant humans. Various other seemingly anomalous mechanisms have been hypothesized to stem from defenses against toxins, and although they appear to utilize many physiological pathways different from those of pregnancy sickness, they might, nonetheless, also shed light on the mechanisms of pregnancy sickness. For example, Treisman (1977) proposed that motion sickness is a by-product of an adaptive mechanism for expelling ingested neurotoxins. Neurotoxins interfere with the normal coordination of stimuli from the eyes, ears, and other sensory organs, so a mechanism to induce nausea and vomiting when this sensory information is uncoordinated would expel recently ingested neurotoxins. However, such a mechanism would also trigger nausea and vomiting when the lack of coordination was caused by something else—such as the rocking of a boat. As another example, I argue (1991) that the capacity for allergy represents an immunological "last line of defense" against toxins. Most known allergens appear either to be themselves toxic substances—such as drugs and venoms—or to be carrier proteins that bind well to low-molecular-weight toxic substances. By causing immediate vomiting, diarrhea, coughing, sneezing, tearing, or scratching, an allergic response can expel a toxin before it circulates to target organs.

One might ask, then, whether any common denominators exist among these various mechanisms for dealing with toxicity—pregnancy sickness, motion sickness, and allergy—such that susceptibility to one is linked to susceptibility to another. Ure (1969), for example, in a study of 140 women in a gynecological ward, found a strong positive correlation between pregnancy sickness and history of allergy. One factor that seems to acutely affect all three mechanisms is psychological stress. High levels of psychological stress in pregnant women are frequently correlated with severe vomiting during pregnancy (Fitzgerald, 1984; Tylden, 1968; Uddenberg, Nilsson, & Almgren, 1971; Wolkind & Zajicek, 1978). Stress is also thought to exacerbate allergy. And stress hormones are markedly elevated during stressful motion, such as rotation, which in most persons leads to motion sickness (Kohl, 1985; Stalla, Doerr, Bildingmaier, Sippel, & von Restorff, 1985). (In the case of motion sickness, the psychological stress may be caused by the discoordination of perceptual cues.) Although numerous physiological disturbances other than pregnancy sickness, allergy, and motion sickness are correlated with psychological stress, mechanisms such as these for reducing toxic load may be designed to be especially responsive to cues of psychological stress. During periods of psychological stress, it may be necessary to avoid ingesting substances that alter one's perception of reality and that thereby make it difficult to accurately assess critical situations. Furthermore, some toxins probably cause psychological stress by directly affecting the central nervous system, in which case expulsion and avoidance of toxins would be important. It would be interesting to determine whether women with low thresholds for psychological stress are especially susceptible to pregnancy sickness, motion sickness, and allergy. Thus, the mechanisms underlying the physiological and psychological perception of toxins may evoke a wide range of protective responses, one of which—pregnancy sickness—is specialized for protecting embryos.

According to Hytten (1984, p. 12), "most of the normal physiological adaptations of pregnancy conspire together to reduce the effectiveness of many drugs," which are by definition toxic because of their biodynamic effects on the body (Brodie, Cosmides, & Rall, 1965). The ubiquitousness of naturally occurring toxins, the complexity of design of physiological processes that "conspire" against toxins, and the simplicity— from an evolutionary standpoint—of recalibrating existing mechanisms that guard against dangerous levels of toxins strongly support the hypothesis that pregnancy sickness was designed by selection to protect the embryo against teratogens.

CULTURAL IMPLICATIONS

Hunter-Gatherers Had Intense Selection Pressures for Pregnancy Sickness

Selection pressures for human pregnancy sickness would have been most intense in the human environment of evolutionary adaptedness—the environment of a Plio-Pleistocene forager—particularly during periods of scarcity. Current hunter-gatherer feeding ecology (although certainly an imperfect model of Pleistocene hunter-gatherer feeding ecology) is characterized by experimentation with a diverse array of food sources (Dunn, 1968; Woodburn, 1968). This is especially true during dry seasons or drought years when marginal environments necessitate broadening the diet to include less palatable, more bitter—and hence more toxic—plant species (Lee & Devore, 1976, p. 44). Similarly, periods of scarcity in the Pleistocene would have increased the toxic load of hunter-gatherer diets by, for example, motivating expansion into new areas with unfamiliar flora, the toxicity of which could have been determined only by experimentation (E. O'Brien, & C. Peters, personal communication, 1987). One of the dangers of venturing into unfamiliar areas to gather plant foods is the increased likelihood of mistaking poisonous species of plant for familiar, edible species. For example, although only a few of the 600 species of yam are poisonous, these are occasionally mistaken for edible species, resulting in fatal poisonings (Concon, 1988). Severe nutritional stress often prevents pregnancy, by suppressing ovulation (Frisch, 1987), or terminates it early, by causing the uterus to resorb the fetus (Prentice & Whitehead, 1987). However, under Pleistocene conditions of marginal stress that provided sufficient nutrition to permit conception and maintenance of pregnancy, pregnancy sickness would have diminished first-trimester experimentation with novel, toxic sources of food and thus traded short-term nutritional costs for long-term reproductive benefits.

Selection pressures for pregnancy sickness probably would have been stronger during the Pleistocene than since the advent of agriculture because agriculture, for the most part, has reduced the toxicity of human diets. Hook (1976) suggested that the adaptive value of pregnancy sickness may be protection against embryotoxins, but hypothesized that pregnancy sickness evolved after the advent of agriculture to protect the embryo against such cultivated drugs as caffeine, alcohol, and nicotine (which are ingested to induce the psychological effects the toxins cause rather than to provide nutrition). But Hook's own studies of maternal consumption of these recently introduced drugs did not support his hypothesis (Hook, 1976; Little & Hook, 1979), as will be discussed further on. (In two studies on food aversions during pregnancy, however, Hook mentions that these aversions might in some instances reduce maternal exposure to embryotoxins in food [Hook, 1978; Tierson, Olsen, & Hook, 1985].) Humans, like other mammals, generally seek to consume foods with minimal toxicity or foods whose toxins are readily detoxified through enzymatic degradation or processing techniques (Stahl, 1984; Tooby & DeVore, 1987). Agricultural societies have been much more effective than hunter-gatherer societies at eliminating dietary toxins. Wild foods contain much higher concentrations of toxins than do their domesticated counterparts, because agriculturalists have selectively bred more palatable, less toxic strains. For example, wild potatoes contain several times the amount of toxic glycoalkaloids (which are teratogenic in some species [Renwick et al., 1984]) than cultivated varieties of potato (Schmiediche, Hawkes, & Ochoa, 1980). Comparisons between wild and cultivated strains of cabbage, beans, strawberries, tomatoes, lima beans, cassava, lettuce, and mangos similarly indicate large reductions in toxicity through domestication (Lucas & Sotelo, 1984; Mithen, Lewis, Heaney, & Fenwick, 1987; Pyysalo, Honkanen, & Hirvi, 1979; Urbasch, 1985). By contrast, cultivated tobacco, which is bred for the toxic constituents that produce its psychotropic effects, contains at least as much nicotine as wild tobacco does (Rhoades, 1979).

Agricultural diets throughout history have been based primarily on a few domesticated staple foods rather than on the more extensive array of wild foods that characterize hunter-gatherer diets. In current rural agricultural societies, however, drought and other periods of scarcity sometimes compel agriculturalists to gather wild plants to supplement meager agricultural yields (Scudder, 1962), thus increasing dietary toxic load (E. O'Brien & C. Peters, personal communication, 1987). For example, the Gwembe Tonga agriculturalists of Zambia gather wild plants of considerable toxicity during famines. They drink a beverage made from *Tamarindus indica,* "though not without stomach aches, which are only partially avoided by adding ashes to the brew to reduce its bitterness" (Scudder, 1971, p. 29), and they eat the pods of the legume *Xeroderris stuhlmanni,* "although not without headache and other symptoms of stress" (Scudder, 1971, p. 30). Particularly unfavorable agricultural conditions may necessitate gathering novel, toxic wild plant foods. Such foraging can have severe consequences. In 1958, when 600 Gwembe Tonga were forcibly resettled by the Zambian government, the agricultural/economic hardships of relocating drove them to forage unfamiliar wild plants, and nearly 10% of the settlers died in one year from apparent poisoning by wild plants that were gathered for food (Scudder, 1971). Agriculturalists thus may greatly increase their risks of ingesting food toxins when they resort to a pre-agricultural feeding ecology.

If agriculturists generally seek to minimize toxicity by breeding less toxic strains of plant foods, one might reasonably ask why they deliberately increase their consumption of toxins by breeding certain plants for their toxic constituents—such as coffee beans for caffeine, tobacco for nicotine, and grains for making alcohol—and by flavoring foods with pungent spices—much of whose flavor results from high concentrations of toxins (see Formacek & Kubeczka, 1982; see Duke, 1988). Ingestion of toxic substances for recreational, medicinal, and other nonnutritional purposes also occurs among hunter-gatherers and other nonindustrial peoples (Duke, 1985; Schultes & Hofmann, 1987; Watt & Breyer-Brandwijk, 1962); advanced agriculturalists have merely amplified the production of plant toxins that have certain coveted effects, such as the mitigation of pain. Spicing foods is probably unique to agricultural societies; however, J. Tooby (personal communication, 1987) has pointed out that human sensory mechanisms may use toxic diversity as a cue to dietary diversity, so that adding spices to foods mimics the sensation of dietary diversity. Dietary diversity not only ensures adequate nutrition, but prevents overexposure to any one toxin. The latter is so important that mammals appear to have evolved mechanisms to partially regulate appetite in response to the diversity of food available: Humans, for example, consume more total food when offered a variety of foods than when offered only their favorite food; and rats consume more of a particular food when a variety of odorants have been added to the food (Rozin & Vollmecke, 1986). Human perceptual mechanisms are probably designed to recognize dietary diversity by diversity of flavors; since much of the flavoring of plant foods comes from their toxic constituents, toxic diversity is probably an important correlate of dietary diversity.

The invention of agriculture enabled human diets to change very rapidly, yet the psychological and physiological mechanisms that evolved to evaluate foods are adapted to Pleistocene environments and ways of life. D. Symons (personal communication, 1989) has pointed out that by breeding plant foods to be less bitter—and hence, less toxic—agriculturalists were able to eat increasingly large quantities of a small number of plant foods without overdosing on any one toxin. Although agricultural diets may lack the nutritional diversity of hunter-gatherer diets, agriculturalists can fool their sensory mechanisms into thinking that they have eaten a nutritionally diversified diet by adding spices to their bland meals. Humans in general may select foods that balance palatability (low levels of toxins) and *perceived* nutritional diversity (presence of a variety of toxins)—a balance that agriculturalists achieve by breeding low-toxicity plant foods and spicing them.

Pregnancy Sickness Is Cross-Cultural

If pregnancy sickness conferred a strong and consistent selective advantage on ancestral humans, it would be expected to be universal today, because selection (for non-frequency-dependent traits) tends to use up genetic variability. In particular, pregnancy sickness should be ubiquitous among pregnant women of hunter-gatherer societies. Although anthropological data are scant concerning first-trimester pregnancy in hunter-gatherer women, references to pregnancy sickness can be found in the literature on the !Kung of the Kalahari Desert, the Efe Pygmies of Zaire, and the Aborigines of Australia. For example, among the !Kung nausea, vomiting, and unexplained dislikes for certain foods are recognized as early signs that a woman is pregnant (Shostak, 1981, p. 178). The !Kung woman whom Shostak calls Nisa is even told by her mother-in-law, who suspects Nisa's pregnancy, "If you are throwing up like this, it means you have a little thing inside your stomach" (p. 187). An Efe Pygmy woman may realize she is pregnant when "food tastes bad" (N. Peacock, personal communication, 1987). The Australian Aboriginal woman may vomit during pregnancy (Kaberry, 1939, p. 42) and have food taboos placed on her as "a means of protecting the child developing within her womb, though she herself may sicken if she disregards them" (Kaberry, 1939, p. 241).

Evidence of the widespread occurrence of pregnancy sickness among preliterate peoples is found in the literature on geophagy (the eating of clay or dirt) and on pica (the eating of nonnutritive substances, such as clay). Geophagy has been shown to detoxify some ingested toxins: For example, when clay is eaten in conjunction with bitter potatoes, as is common among humans in South America, it binds a large percentage of the toxic potato alkaloids, thereby preventing these toxins from being absorbed into the circulation (Johns, 1986). The detoxification function of clay eating is supported by evidence that rats eat clay after being poisoned or made motion sick (the response to simulated neurotoxicity) (Morita, Takeda, Kubo, & Matsunaga, 1988). R. Wrangham (personal communication, 1989) has pointed out that the custom in many preliterate societies of geophagy during pregnancy may have the effect of protecting against teratogens. Pregnant women of cultures throughout Indonesia, Oceania, and Africa eat clay in order to prevent or counteract vomiting (Anell &

Lagercrantz, 1958). By adsorbing toxins, ingested clay reduces the toxic threat to the body and, thus, reduces the pregnant woman's need to vomit.

Thus far, the prevalence of pregnancy sickness has been systematically investigated only among women of modern industrial societies. Most such studies define pregnancy sickness as first-trimester nausea and/or vomiting and are based on interviews with pregnant women near the end of or after their first trimesters of pregnancy. The percentage of pregnant women who claim to experience first-trimester nausea in these studies ranges from about 75% (Brandes, 1967) to 89% (Tierson et al., 1986). Rates of vomiting are generally about 55% (Klebanoff et al., 1985). The pregnancy sickness rates reported in these studies are probably lower than actual rates, however, because the interviews were conducted after the peak period of pregnancy sickness and thus measure only remembered instances of nausea experienced weeks earlier (postpregnancy surveys yield still lower rates of pregnancy sickness [71%, Petitti, 1986]). But the main problem with these prevalence studies is that they define pregnancy sickness as the presence of nausea or vomiting rather than as the presence of *food aversions, nausea, or vomiting*. Were pregnancy sickness redefined to include food aversions, its prevalence would very likely approach 100%. Because the nausea of pregnancy sickness is often caused by the odor of certain foods, a pregnant woman may avoid (consciously or unconsciously) cooking and eating these foods in order not to experience nausea and vomiting. Aversions to bitter or pungent foods are the main purposes and symptoms of pregnancy sickness and therefore should be included in the definition.

If first-trimester pregnant women develop food aversions because of recalibrated thresholds for detecting and tolerating toxicity, then one would predict that the following foods and beverages would elicit aversions: (a) bitter or pungent foods—indicating high concentrations of plant toxins—such as coffee, tea, vegetables, spices and herbs, and the more pungent or bitter of the alcoholic beverages; (b) foods that emit burnt or fried odors—indicating the creation during cooking of mutagens—such as barbecued, roasted, or fried foods (as opposed to boiled foods); and (c) foods that emit smells suggestive of spoilage—indicating parasitization by toxin-producing bacteria—such as animal products that are not extremely fresh (what smells fresh to a nonpregnant woman may not necessarily smell fresh to a first-trimester pregnant woman with heightened olfactory sensitivity to cues of spoilage). On the other hand, one would predict that the best-tolerated foods would be those that have rather bland odors and tastes and that do not spoil easily, such as processed breads, cereals, and grains.

In studies that focused specifically on food aversions during pregnancy, high percentages of women were found to experience aversions to particular foods. In a series of interviews that began in the 12th week of pregnancy to determine changes in food consumption in relation to food aversions and cravings, 85% of the 400 women reported that they had experienced food aversions sometime during pregnancy (Tierson et al., 1985). The actual prevalence of food aversions among these women may have been considerably higher, of course, since pregnancy sickness would have waned by the 12th week of pregnancy, the start of the interviews. Another limitation of this study for understanding the effect of pregnancy sickness on diet is that each woman in the study was exposed to her own idiosyncratic array of foods rather than to a uniform array. Women were able to report aversions only to foods they were exposed to; consistent aversions to a highly spiced food like curry, for example, would not have been detected because most women in this country do not often eat curry. Thus, the foods eliciting aversions would to some extent reflect consumption patterns rather than toxicity. Furthermore, women in this study were assumed to be consciously aware of their food aversions; therefore, unconscious changes in choices of foods to cook or to eat would have been overlooked.

Nevertheless, the results of the study fit the above predictions fairly closely. The most pronounced aversions that the women remembered having had during the first 12 weeks of pregnancy were to coffee (129 women), meat and poultry (124 women), alcoholic beverages (79 women, who may, however, have avoided alcohol out of concern for the embryo rather than due to nausea stimulated by alcohol), and vegetables (44 women). The least pronounced aversions to commonly consumed foods were to bread (3 women) and cereals (0 women). As expected, decreases in consumption generally accompanied the food aversions, while increases in consumption generally accompanied the food cravings. However, even the few women who reported cravings for, rather than aversions to, vegetables still decreased their consumption of vegetables during the first trimester, indicating that pregnant women may have unconscious aversions to foods containing plant toxins.

Similarly, in a study of food aversions among 100 women experiencing their first pregnancies, the most frequently cited aversions were to coffee, tea, and cocoa (32 women), vegetables (18 women), and meat and eggs (16 women) (Dickens & Trethowan, 1971). Doty (1976) notes that these aversions may be related to the olfactory hyperacuity of first-trimester pregnant women. Hook (1978) also notes that food aversions during pregnancy could be mediated by changes in olfactory and taste sensitivity. Rather than interviewing women retrospectively about their food aversions during early pregnancy and relying on them to remember aversions and to accurately ascertain whether these aversions influenced their choice of foods, a more reliable method of assessing food/odor aversions during pregnancy would be to ask women in their 6th week of pregnancy to smell vials of odiferous plant or food extracts, such as garlic juice, espresso, and perfume; their reactions could be noted and compared to their reactions to the same extracts several months postpartum.

The question of whether there are peoples among whom pregnancy sickness is unknown has obvious implications for an adaptionist argument. Although references to pregnancy sickness in Western societies have existed in medical texts since ancient times (Fairweather, 1968), data on pregnancy sickness in nonindustrial cultures are exceedingly sparse. Minturn and Weiher (1984) examined data on pregnancy sickness in the Human Relations Area Files (which contain ethnographic accounts of many geographically diverse nonindustrial societies) and found specific statements by ethnographers on the presence of pregnancy sickness for 22 cultures and on the absence of pregnancy sickness for 8 cultures. However, since Minturn and Weiher made no attempt to define pregnancy sickness consistently or to establish a basis for evaluating the ethnographers' assumptions that pregnancy sickness was absent in certain cultures, the cross-cultural information that they present is questionable.

Nevertheless, their main observation of the study merits interest: Seven of the 8 cultures for which ethnographers noted an absence of pregnancy sickness, in contrast to none of the 22 cultures for which ethnographers noted the presence of pregnancy sickness, have diets based on maize. Symons (personal communication, 1989) has suggested the following hypothesis to account for these data: The nutritional deficiencies commonly found among people with maize-based diets may disrupt the physiological processes that produce pregnancy sickness. The niacin in maize is not liberated during digestion unless the maize has been treated with alkali (e.g., limestone)—a cooking procedure implemented by some maize-based New World Indian cultures—and niacin deficiency is common among peoples whose staple food is untreated maize (Katz, Hediger, & Valleroy, 1974).

Niacin deficiency causes a number of symptoms (The Merck Manual, 1987) that may affect pregnancy sickness. It causes disturbances of the central nervous system, which might directly disrupt pregnancy sickness by affecting the CTZ, a part of the brain stem. It causes gastrointestinal disorders, including nausea and vomiting, which could interfere with the absorption of foods and food toxins. Gastrointestinal disorders could also mask the symptoms of pregnancy sickness, such that a pregnant woman or ethnographic observer would not be able to distinguish pregnancy sickness from the day-to-day symptoms of niacin deficiency. Furthermore, niacin deficiency might subvert pregnancy sickness by impeding vasodilation. Because niacin causes vasodilation, which increases blood flow throughout the body—including, presumably, throughout the CTZ—it may be essential for inducing pregnancy sickness. Thus, the absence, rather than the presence, of pregnancy sickness should be viewed as pathological.

Pregnancy Sickness Reduces the Risk of Spontaneous Abortion

Although intercultural differences in degree of pregnancy sickness have not been determined, substantial intracultural variability in pregnancy sickness has been documented: The degree of pregnancy sickness can vary from woman to woman within a culture and even from pregnancy to pregnancy in the same woman. If pregnancy sickness does protect the embryo against teratogens, its protective effects should be evidenced to some extent by comparing the degree of pregnancy sickness with the outcome of pregnancy among women in similar cultures (i.e., among women with similar dietary and other backgrounds). Several extensive studies in modern industrial societies have found that women who experience severe pregnancy sickness have significantly lower rates of spontaneous abortion than do women who experience mild or no pregnancy sickness (where pregnancy sickness is defined by the presence of first-trimester nausea and/or vomiting). For example, in a study of 873 women by Weigel and Weigel (1989a), the spontaneous abortion rate during or before the 20th week of gestation for women who had experienced pregnancy sickness severe enough to cause vomiting was 1.0%, compared to 3.6% for women with nausea but no vomiting and 7.2% for women with no nausea or vomiting. In a study of 9,098 pregnant women by Klebanoff et al. (1985), the spontaneous abortion rate after the 14th week for women who had experienced pregnancy sickness severe enough to cause vomiting was 3.4%, compared to 5.3% for women who had experienced no or mild pregnancy sickness— a difference that decreased only slightly when the study was controlled for age, race, education, smoking, and number of previous pregnancies. In a study of 7,027 pregnant women by Brandes (1967), the spontaneous abortion rate before the 20th week of pregnancy for women who had experienced pregnancy sickness was 2.8%, compared to 6.6% for women who had not experienced pregnancy sickness. In a study of 3,853 pregnant women by Yerushalmy and Milkovich (1965), the abortion rate before the 20th week of pregnancy was 3.8% for women who had experienced pregnancy sickness, compared to 10.4% for women who had not experienced pregnancy sickness. This study also found that the infants of women who had experienced pregnancy sickness suffered lower rates of severe and nontrivial birth defects than did the infants of women who had not experienced pregnancy sickness. In a study of 100 pregnant women by Medalie (1957), women with severe pregnancy sickness had an abortion rate of 0%, compared to a rate of 22.9% for women with no or mild pregnancy sickness. (Weigel and Weigel [1989b] review many of the studies relating nausea and vomiting to pregnancy outcome.)

As S. Wasser (personal communication, 1987) has pointed out, however, the correlation between degree of pregnancy sickness and rate of spontaneous abortion may be spurious: If pregnancy sickness is triggered by pregnancy hormones, insufficient hormone levels may both fail to cause pregnancy sickness and cause abortion. To rule out the possibility of a spurious correlation, studies relating the degree of pregnancy sickness to the pregnancy outcome should exclude those pregnant women who have significantly lower than average levels of pregnancy hormones. A correlation between pregnancy sickness and pregnancy success among women with hormone levels sufficient for maintaining pregnancy might indeed support the hypothesis that pregnancy sickness protects the embryo against teratogens. (However, as D. Symons and L. Cosmides [personal communication, 1989] point out, this correlation could stem as well from variations in factors that have not been accounted for, such as toxicity of individual diets that would lead to different degrees of pregnancy sickness or to physiological disturbances that might mitigate or exacerbate pregnancy sickness. Furthermore, that pregnancy sickness is an adaptation does not necessarily imply that severe pregnancy sickness is adaptively superior to mild pregnancy sickness; for example, fever may be a universal adaptation, but this does not imply that the highest fevers are the most adaptive.)

Although the effect of pregnancy sickness on spontaneous abortion rates in nonindustrial cultures has not been measured, the connection between the ingestion of toxic plants and abortion is widely recognized by women of different cultures throughout the world. Many highly toxic plants have such strong abortifacient properties that they are used to deliberately induce abortions, despite their aversive tastes and odors (pregnancy sickness was not designed to deter the *deliberate* abortion of embryos). Among hunter-gatherers, for example, !Kung women try to terminate unwanted pregnancies by ingesting drinks made from toxic plants (Shostak, 1981); Efe Pygmy women claim to know of a forest plant that induces abortion (N. Peacock, personal communication, 1987); and Mbuti Pygmy women try to induce abortions by ingesting drinks made from *gorogoro* bark or *Tebvo liana* (Turnbull, 1965, p. 222) or by burning and inhaling the smoke from the highly toxic *ikanya* bark (p. 232). Among women of other nonindustrial societies, the use of plants to induce abortions is widespread. In a study of abortion in 400 preindustrial societies, Devereux (1976) gives numerous examples of cultures in which women ingest drinks from poisonous roots, leaves, or fruits in order to induce abortion: The Chamorro of Guam drink a beverage made from the bark of the pine *Ephedra vulgaris* within three months of conception, or boil and drink the grass *Cyperus kyllingia* or the root *Ceiba pentandra;* the Cahita Indians of Mexico drink a tea made by boiling the corklike pine *corcho* in water; the Miriam of the Murray Islands drink a beverage made of leaves of the *sespot, madleuer, ariari,* and *ap,* and, if ineffective, they then chew the leaves of the *tim, mikir, sorbe, bok, sem,* and *argerarger,* which causes great pain to the woman but kills the child; the Shasta Indians of California eat the root of a parasitical fern found on the tips of fir trees; and the Shortlands of the Solomon Islands swallow a certain three-lobed leaf that grows on a vine found near the seashore. In a study of women's medicine practiced in various developing countries, Newman (1985) documents the use of many plant abortifacients, such as *Borago officinalis* prepared as a tea in Costa Rica; *Turnera ulmifolia* prepared as a tea in Jamaica; *Phaseolus tuberosus* prepared in solution in Malaysia; *Pachyrhizus tuberosis* prepared as an enema in Peru; and *Mentha longifolia* prepared as a tea in Afghanistan. Schultes and Hofmann (1987, p. 103) note that the hallucinogens widely used among certain South American tribes are rarely used by women presumably because they are sufficiently toxic to cause abortion. Plant toxins that inadvertently caused or intentionally were used for abortions also have been documented in preindistrial Europe, when abortions resulted from outbreaks of ergotic poisoning from mold-infested grains (Schultes & Hofmann, 1987, p. 62). The cross-cultural perception of the connection between plant toxins and abortion thus underscores the prevalence of plants that are capable of inducing abortions and the selective significance of mechanisms, such as pregnancy sickness, that deter women from incidental ingestion of plant toxins that have abortifacient properties.

Cues of Pleistocene Toxicity Elicit Pregnancy Sickness

The environmental cues that would be expected to elicit pregnancy sickness in women today are those that were associated with toxins in the Pleistocene. Toxicity per se cannot be smelled or tasted. Defenses against plant toxins (with the exception of immunological defenses [see Profet, 1991]) tend to be general defense mechanisms, effective against a wide range of toxins. Even the most specialized nonimmunological defenses against toxins—detoxification enzymes produced by the liver and other organs—generally target classes of toxicity rather than only the specific molecular configuration of a particular toxin (see Jacoby, 1980). This nonspecificity of defense mechanisms against toxins makes adaptive sense because the array of naturally occurring toxins, produced by plants in their coevolutionary struggle with predators and by pathogens in their struggle with hosts, is so wide and continually evolving. Perceptual mechanisms are therefore designed to recognize chemical properties that are commonly associated with toxicity in the natural environment, i.e., bitterness and pungency. A mechanism, such as pregnancy sickness, that responds to these common correlates of Pleistocene toxicity is capable of recognizing a wide and ever-changing array of potential teratogens. Such a design means, however, that the symptoms of pregnancy sickness should be triggered by any substance that mimics these correlates of Pleistocene toxicity, whether or not it is toxic. For example, pungent odors indicative of plant toxins should trigger these symptoms, even if the odors are evolutionarily novel and nontoxic.

On the other hand, evolutionarily novel toxins that lack the correlates of Pleistocene toxicity should fail to trigger the symptoms of pregnancy sickness. Natural toxins that Pleistocene hominids were exposed to only in very low, nonteratogenic doses and that are neither bitter nor pungent should fail to elicit the symptoms of pregnancy sickness. Because these toxins did not pose serious threats to Pleistocene embryos, selection would not have produced mechanisms to detect and avoid them. As a result, pregnancy sickness should not be expected to protect modern women against exposure to such toxins, even when industrial processes cause them to be present at severely teratogenic levels. An example of the problems that can arise when modern pregnant women are exposed to natural teratogens at unnaturally high levels concerns methylmercury. This embryotoxin contaminates the environment in trace amounts naturally, due to bacterial conversion of metallic mercury to methylmercury; however, it can reach unnaturally high levels in industries that use inorganic mercury, which is also converted by bacteria to methylmercury (Mottet, 1981). The industrial release of methylmercury into Minamata Bay in Japan between 1954 and 1960 is believed to have been the cause of the severe neurological symptoms (fetal Minamata) of infants born to mothers who had consumed mercury-contaminated fish during pregnancy (Harada, 1986).

Other modern teratogens that have Pleistocene analogs but that may be much more dangerous in their current forms are drugs in the form of pills or injections. Most drugs are derivatives, synthetic mimics or based on prototypes of plant toxins, but are available in concentrations far more potent than plants can afford to manufacture. The toxic substances in pills and injections, in contrast to many toxins in plants, are in nonvolatile forms and so cannot be smelled by the pregnant woman; and they are usually swallowed whole in capsules or injected, rather than tasted, thereby bypassing the olfactory and gustatory chemoreceptors that trigger aversions. For example, a large percentage of first-trimester pregnant women in the late 1950s and early 1960s who took the drug thalidomide to alleviate pregnancy sickness (thalidomide was used as an antivomiting drug as well as a sedative [McBride, 1961]) later gave birth to infants with missing or severely deformed limbs. These women could not have been warned of thalidomide's teratogenicity by smelling it, even though many of them were presumably very sensitive to odors since they had pregnancy sickness severe enough to seek medicinal relief. Pregnancy sickness cannot deter women in modern industrial societies from consuming teratogens that lack Pleistocene cues of toxicity.

Because not all modern teratogens emit the cues that are necessary for triggering the aversions of pregnancy sickness and because the main Pleistocene sources of toxins—plant foods—have for the most part become less toxic through selective breeding, the selection pressures maintaining the mechanisms for detecting and avoiding Pleistocene toxins during pregnancy may have been decreasing since the advent of agriculture. (If pregnancy sickness is a simple recalibration of a complex mechanism, rather than a separate mechanism, it is probably under relatively simple genetic control and highly responsive to selective pressures [D. Symons, personal communication, 1989].) Although mechanisms for detecting and avoiding teratogens could potentially evolve to assimilate novel types of teratogens, if novel teratogens are being created much faster than mechanisms like pregnancy sickness can adapt to them, pregnancy sickness may eventually become substantially less prevalent. The apparent variability in degree of pregnancy sickness among women in industrial societies might be due either to variations in the modern environment—such as variations in dietary toxicity—or to decreasing selection pressures for detecting and avoiding substances that emit Pleistocene cues of toxicity. Comparing the variability of pregnancy sickness among women in industrial societies and among women in hunter-gatherer societies could determine whether pregnancy sickness is really more variable among the former. It is also possible that the basic human mechanisms for detecting and avoiding toxins are variable and that some of the variation in pregnancy sickness is merely a correlated by-product of the variation in these underlying mechanisms.

Agricultural Toxins Only Elicit Pregnancy Sickness if They Emit Cues of Pleistocene Toxicity

Studies showing that pregnancy sickness deters coffee drinking, yet only slightly deters alcohol consumption, and does not deter cigarette smoking (Hook, 1976; Little & Hook, 1979) support the hypothesis that the symptoms of pregnancy sickness are elicited by cues indicative of Pleistocene toxins rather than of modern cultivated toxins. (Studies of the influence of pregnancy sickness on the consumption of addictive toxic substances, which are consumed primarily for the effects of their toxins rather than for nutrition, should, however, be interpreted cautiously. It is possible that pregnancy sickness does not significantly decrease a woman's consumption of toxins to which she is addicted, even if they emit Pleistocene cues of toxicity, because the physiological/psychological unpleasantness of withdrawal from an addictive substance may outweigh the unpleasantness of the nausea induced by that substance.) Coffee, alcohol, and tobacco—which at high doses are teratogenic in some mammalian species (see Shepard, 1986)—are commonly ingested toxins in agricultural societies. In Hook's (1976) study of 295 pregnant women, 26.3% of the coffee drinkers claimed to have decreased their consumption because of nausea stimulated by coffee or loss of urge to drink it. Roasted coffee contains over 800 volatile chemicals (Ames et al., 1990a). Many coffee chemicals, such as caffeine, are bitter-tasting toxins produced by the plant for defense. Since the taste and function of these toxins conforms to the pattern of plant toxicity that was prevalent throughout the course of human evolution, coffee would be expected to elicit the nausea and aversions of pregnancy sickness.

By contrast, only 9.6% of the wine drinkers, 10.6% of the spirits drinkers, and 17.1% of the beer drinkers claimed to have decreased their consumption of alcohol because of nausea or loss of urge, and, in a study by Little and Hook (1979) involving 210 pregnant women, changes in alcohol consumption were unrelated to pregnancy sickness. Alcohol should not be expected to elicit as strong aversions as coffee does. Pure alcohol (ethanol) is a nonbitter chemical that is produced through the fermentation of plant sugars by microorganisms rather than by the plant for defense; alcohol therefore conforms far less to the pattern of Pleistocene plant toxicity than does caffeine. Ethanol may not even be recognized by the CTZ as toxic; for example, in rats, learned taste aversions to alcohol do not involve the area postrema (where the CTZ is located), as do learned taste aversions to other toxins (Hunt, Rabin, & Lee, 1987; Stewart, Perlanski, & Grupp, 1988). Furthermore, the pungency or bitterness of many alcoholic beverages stems primarily from the plant constituents of these beverages rather than from the alcohol itself. The astringent flavor of many red wines, for example, comes from the toxic tannins of the grape skins. The alcoholic beverages whose aromas frequently elicit aversions and nausea in pregnant women are likely to be those that contain especially pungent or bitter plant constituents; for example, women who normally like to drink whisky and vodka should be more averse during early pregnancy to whisky than to vodka (which is pure ethanol and water).

Alcohol is contained naturally in numerous plants, yet is an extremely weak teratogen (Ames, 1989). Although a daily dose of five alcoholic beverages during pregnancy is clearly a risk factor for mental retardation in offspring, there is no evidence that a daily dose of one alcoholic beverage is (B. Ames, personal communication, 1989). The quantities of alcohol required to cause the teratogenic condition *fetal alcohol syn-*

drome have been available to pregnant women only since the advent of agriculture; and, although contemporary hunter-gatherers may procure alcohol from neighboring agriculturalists (Tanaka, 1980), there is no evidence that Pleistocene hunter-gatherers produced alcohol (I. DeVore, personal communication, 1987). Selection pressures for aversions during pregnancy to a toxin like alcohol, which is not teratogenic in the doses consumed from the incidental ingestion of fermented plants, would have been extremely weak in the Pleistocene, unlike selection pressures for aversions to bitter plant compounds, some of which are teratogenic in minute traces. One reason that alcohol is a major cause of teratogenesis in the United States may be that pregnancy sickness is not an effective deterrent against alcohol and that alcohol, in the vast quantities consumed on a chronic basis by modern alcoholic women, can cause mental retardation in second- and third-trimester fetuses—well after the period of organogenesis and pregnancy sickness.

Studies by Little and Hook (1979) on the relation of pregnancy sickness to cigarette consumption illustrate the importance of olfactory sensitivity in the triggering of pregnancy sickness. Of 210 pregnant women, those who smoked cigarettes regularly during or prior to early pregnancy suffered much lower rates of pregnancy sickness than did nonsmokers (52% compared to 79%), leading Little and Hook to suggest that women with pregnancy sickness may have higher pregnancy success rates simply because they are less likely to be smokers. Klebanoff et al. (1985) also measured a lower incidence of first-trimester vomiting among smokers than among nonsmokers (46% compared to 58%), but found that the relationship between pregnancy sickness and pregnancy success was valid even when controlled for smoking. Although the findings regarding pregnant smokers appear to invert the expected relationship between pregnancy sickness and teratogens, they actually underscore the importance of environmental cues in eliciting pregnancy sickness: In various studies cigarette smoking has been shown to interfere with olfactory and taste chemoreception (Ahlstrom, Berglund, Berglund, Engen, & Lindvall, 1987; Arfman & Chapanis, 1962; Hubert, Fabsitz, Feinleib, & Brown, 1980) which are crucial sensory mechanisms for detecting toxins. Of particular relevance is evidence that taste thresholds for bitterness and olfactory thresholds for pungency increase for heavy smokers (Cometto-Muniz & Cain, 1982; Kaplan & Glanville, 1964). This means that pregnant smokers should be less likely than nonsmokers to register the bitter or pungent warnings of toxic plant compounds and, therefore, should be less likely to experience pregnancy sickness.

Smoking might also decrease pregnancy sickness by decreasing estrogen levels (Depue et al., 1987). Smokers excrete almost one-third less estradiol and other estrogens in their urine during the postovulatory phase of their menstrual cycles than do nonsmokers (MacMahon, 1982). The reason that smoking decreases estradiol levels might be that enzymes induced to detoxify tobacco toxins also metabolize estradiol. Benzo[a]pyrene—a major toxin in cigarette smoke—induces enzymes of the cytochrome P-450 enzyme family in maternal liver, placenta, and fetal liver, even during early pregnancy (Dey et al., 1989; Juchau, 1982; Pelkonen, 1987). Because the chemical structure of benzo[a]pyrene is similar to the chemical structure of steroids and because the cytochrome P-450 family of enzymes metabolize both benzo[a]pyrene and sex steroids (Dey et al., 1989), enzyme induction by benzo[a]pyrene might accelerate the metabolism of some steroid pregnancy hormones, like estradiol, and consequently lower the plasma levels of the hormones that are likely to trigger pregnancy sickness.

CONCLUSION

Pregnancy sickness exhibits many features of an adaptive design to deter maternal ingestion of teratogens: The timing of pregnancy sickness coincides with organogenesis, the period of maximum embryonic susceptibility to teratogens; women with pregnancy sickness have aversions to foods with high concentrations of plant toxins, such as coffee and vegetables; and olfactory sensitivity becomes hyperacute during the period of pregnancy sickness, enabling better detection of toxins in foods. Pregnancy sickness probably arose as a recalibration of existing mechanisms—involving chemoreceptors of the CTZ, gastrointestinal region, and olfactory epithelium—that induce nausea, vomiting, and aversions in response to levels of toxins that exceed set thresholds.

Comparative studies among pregnant nonhuman mammals on the detection and avoidance of toxins could provide an additional means for testing the hypothesis that human pregnancy sickness is an adaptation to prevent maternal ingestion of teratogens. Virtually all mammals as well as many other animals display innate or learned aversions to foods whose toxins they are not well equipped to handle (Freeland & Janzen, 1974; Garcia, 1990; Garcia & Hankins, 1975; Rozin & Kalat, 1971), and, as in humans, these aversions involve CTZ, gastrointestinal, and olfactory chemoreception. Although vomiting as a normal concomitant of early pregnancy has been documented only in humans (Fairweather, 1968), and nausea in nonhuman mammals may be too subjective a symptom for an observer to detect, pregnancy-induced food aversions in mammals could be detected by determining changes in dietary preferences during pregnancy. Changes in taste preference occur in pregnant rats and are thought to be hormonally triggered (Wilson, 1987); whether these changes occur during organogenesis, are accompanied by heightened olfactory sensitivity, and deter ingestion of teratogens needs to be determined. (Some domestic animals are known to alter their forage preferences during pregnancy [Heady, 1964]. However, because domestic animals are exposed to evolutionarily novel environments and artificial distortions in selection, their foraging behaviors should not necessarily be viewed as adaptations. On the contrary, they sometimes exhibit very maladaptive foraging behaviors; cows and sheep, for example, generally find locoweed unpalatable, but graze it readily during pregnancy if other green grasses are not plentiful, causing their fetuses to be malformed or aborted [James, 1983].) Furthermore, some mammals exhibit other physiological changes during pregnancy that heighten perceptual sensitivity to toxins or that reduce embryonic exposure to toxins. In mice, for example, pregnancy enhances olfactory neuron growth (Kaba et al., 1988); in guinea pigs, it decreases gastric motility (Ryan, Bhojwani, & Wang, 1987); and in gemsbuck, it increases resistance to certain drugs (Janzen, 1978).

Teratogens have been prevalent in plants throughout mammalian evolutionary history; therefore, mammals exposed to a wide and variable range of dietary toxins might have evolved at least attenuated forms of pregnancy sickness to prevent females from inflicting toxins on their embryos. Mammals that experience the most intense selection pressures for pregnancy sickness are experimental herbivores and experimental omnivores, because they ingest a wide diversity of plant toxins. Janzen (1978, pp. 77–78) notes that:

when a fetus [of an arboreal leaf-eater] is developing, the herbivorous parent is expected to become more discriminating for two reasons. First, secondary compounds that might circulate in the mother's system with little harm might be lethal to the growing fetus. Second, a large number of secondary compounds induce vomiting and other forms of contraction of the muscles in the abdomen; many naturally occurring human abortants affect the gut in exactly the same way.

Herbivores with very specialized diets experience much weaker selection pressures for pregnancy sickness, both because they are exceptionally well adapted to handle the toxins of their narrow food niches and because they would be unlikely to specialize in foods that are teratogenic for their species. Experimental insectivores may confront moderate selection pressures for pregnancy sickness because many insects produce toxins or sequester plant toxins for their own defense. Carnivores, on the other hand, would derive little benefit from pregnancy sickness (unless they scavenged rotting meat).

Humans, however, appear to face more intense selection pressures for pregnancy sickness than any other mammal because they exploit a vast array of different plants and plant parts. The invention of cooking, which occurred in the middle Pleistocene, expanded the range of plants that could be made edible and thus the range of toxins that humans could include in their diets (Stahl, 1984; Tooby & DeVore, 1987); although cooking destroys some toxins, it also creates new toxins (Buttery, 1977; Sugimura, 1982). Furthermore, the daily nutritional costs of early pregnancy sickness might be lower, and hence more affordable, in women than in other mammals. Women store a significantly higher percentage of their body weight as fat before pregnancy than most other mammals do and thus require a proportionately lower daily increase in energy intake during pregnancy (Prentice & Whitehead, 1987). Pregnancy sickness and fat storage might even be coevolved adaptations to deal with a diverse and toxic plant diet during reproduction. Women who lack a critical threshold of fat usually do not ovulate (Frisch, 1987); this threshold may be based on the nutritional stores needed to provide a buffer against the nutritional deficiencies incurred because of pregnancy sickness in the first trimester of pregnancy. Because pregnancy sickness often leads to first-trimester weight loss (and in the Pleistocene, probably commonly led to weight loss, since low-toxicity plant foods were unlikely to have been abundantly available), the presence of critical fat reserves as a condition of ovulation may ensure that pregnancy sickness does not lead to undue nutritional stress on the mother.

An understanding of the biological function of pregnancy sickness may lead to insights into its underlying physiology, which in turn may lead to the development of ways to intervene when this physiology malfunctions and produces dangerous conditions like excessive vomiting during pregnancy *(hyperemesis gravidarum).* For pregnant women, understanding the benefits conferred by pregnancy sickness may have a number of implications: Women who experience very mild pregnancy sickness may want to consciously avoid high levels of dietary toxins; women who smoke or who have recently smoked may want to select foods cautiously, since their olfactory sensitivity to toxins may be impaired; women who experience moderate to severe pregnancy sickness may want to compare the costs and benefits of pregnancy sickness before deciding to alleviate it medicinally; and finally, women who understand that pregnancy sickness is not an arbitrary affliction but rather a mechanism designed to protect their embryos against teratogens may come to a fuller acceptance of pregnancy sickness and, thus, of pregnancy.

ACKNOWLEDGMENTS

This paper was completed in July 1989 and updated in March 1991. I especially thank Don Symons for many insightful comments and editing suggestions. I also thank Bruce Ames, Jerome Barkow, Roger Bingham, Leda Cosmides, Irven DeVore, Peter Ellison, Alison Fleming, John Garcia, Kirsten Hawks, Peter Jorgensen, Gilda Morelli, Daniel Nebert, Randolph Nesse, Eileen O'Brien, Nadine Peacock, Charles Peters, Rebecca Ranninger, Eloy Rodriguez, Margaret Schoeninger, Mark Shigenaga, Marjorie Shostak, John Tooby, Samuel Wasser, George Williams, and Richard Wrangham for helpful discussions, criticisms, or references. This paper is dedicated to the memory of Harvey B. Snout.

REFERENCES

Abitbol, M. M. (1988). Effect of posture and locomotion on energy expenditure. *American Journal of Physical Anthropology, 77*, 191–199.

Ahlstrom, R., Berglund, B., Berglund, U., Engen, T., & Lindvall, T. (1987). A comparison of odor perception in smokers, nonsmokers, and passive smokers. *American Journal of Otolaryngology, 8*, 1–6.

Alcock, P. A. (1983). *Food poisoning.* London: H. K. Lewis.

Ames, B. N. (1989). Mutagenesis and carcinogenesis: Endogenous and exogenous factors. *Environmental and Molecular Mutagenesis, 14* (Suppl. 16), 66–77.

Ames, B. N., Profet, M., & Gold, L. S. (1990a). Dietary pesticides (99.99% all natural). *Proceedings of the National Academy of Sciences, 87*, 7777–7781.

Ames, B. N., Profet, M., & Gold, L. S. (1990b). Nature's chemicals and synthetic chemicals: Comparative toxicology. *Proceedings of the National Academy of Sciences, 87*, 7782–7786.

Anell, B., & Lagercrantz, S. (1958). Geophagical customs. In *Studia ethnographica upsaliensia* (Vol. 17.) (D. Burton, Trans.). Uppsala, Sweden: Almqvist & Wiksells Boktryckeri Ab.

Archer, A. W. (1988). Determination of safrole and myristicin in nutmeg and mace by high-performance liquid chromatography. *Journal of Chromatography, 438*, 117–121.

Arfman, B. L., & Chapanis, N. P. (1962). The relative sensitivities of taste and smell in smokers and non-smokers. *Journal of General Psychology, 66*, 315–320.

Aspillaga, M. O., Whittaker, P. G., Taylor, A., & Lind, T. (1983). Some new aspects of the endocrinological response to pregnancy. *British Journal of Obstetrics and Gynecology 90*, 596–603.

Beck, F. (1973). *Human embryology and genetics.* Oxford: Blackwell Scientific Publications.

Beier, R. C. (1990). Natural pesticides and bioactive components in foods. *Reviews of Environmental Contamination and Toxicology, 113*, 47–137.

Beier, R. C., Ivie, G. W., Oertli, E. H., & Holt, D. L. (1983). HPLC analysis of linear furocoumarins (psoralens) in healthy celery *(Apium graveolens). Food and Chemical Toxicology, 21*, 163–165.

Bernstein, I. L., Courtney, L., & Braget, D. J. (1986). Estrogens and the Leydig LTW(m) tumor syndrome: Anorexia and diet aversions attenuated by area postrema lesions. *Physiology and Behavior*, 159–163.

Bickers, D. R., & Kappas, A. (1980). The skin as a site of chemical metabolism. In T. E. Gram (Ed.), *Extrahepatic metabolism of drugs and other foreign compounds* (pp. 295–318) New York: S. P. Medical & Scientific Books.

Borison, H. L. (1986). Anatomy and physiology of the chemoreceptor trigger zone and area postrema. In C. J. Davis, G. V. Lake-Bakaar, & D. G. Grahame-Smith (Eds.), *Nausea and vomiting: Mechanisms and treatments* (pp. 10–17) New York: Springer-Verlag.

Borison, H. L., Borison, R., & McCarthy, L. E. (1984). Role of the area postrema in vomiting and related functions. *Federation Proceedings, 43,* 2955–2958.

Brandes, J. (1967). First trimester nausea and vomiting as related to outcome of pregnancy. *Obstetrics and Gynecology, 30,* 427–431.

Breipohl, W., Mackay-Lim, A., Grandt, D., Rehn, B., & Darrelmann, C. (1986). Neurogenesis in the vertebrate main olfactory epithelium. In W. Breipohl (Ed.), *Ontogeny of olfaction* (pp. 21–33) New York: Springer-Verlag.

Brodie, B. B., Cosmides, G. J., & Rall, D. P. (1965). Toxicology and the biomedical sciences. *Science, 148,* 1547–1554.

Buttery, R. G. (1977). The natural flavors of vegetables. *The Vortex (American Chemical Society), 38,* 7–12.

Buttery, R. G., Guadagni, D. G., Ling, L. C., Seifert, R. M., & Lipton, W. (1976). Additional volatile components of cabbage, broccoli, and cauliflower. *Journal of Agricultural and Food Chemistry, 24,* 829–832.

Calabrese, E. J. (1985). *Toxic susceptibility. Male/female differences.* New York: John Wiley & Sons.

Cole, R. A. (1980). The use of porous polymers for the collection of plant volatiles. *Journal of the Science of Food and Agriculture, 31,* 1242–1249.

Cometto-Muniz, J. E., & Cain, W. S. (1982). Perception of nasal pungency in smokers and non-smokers, *Physiology and Behavior, 29,* 727–731.

Concon, J. M. (1988). *Food toxicology: Principles and concepts.* New York: Marcel Dekker.

Crosby, D. G. (1983). Alkaloids in milk may cause birth defects. *Chemical and Engineering News, 61*(June 13), 37.

Dahl, A. R. (1988). The effect of cytochrome P-450-dependent metabolism and other enzyme activities on olfaction. In F. L. Margolis & T. V. Getchell (Eds.), *Molecular neurobiology of the olfactory system. Molecular, membranous and cytological studies* (pp. 51–70) New York: Plenum Press.

Davis, C. J., Harding, R. K., Leslie, R. A., & Andrews, P.L.R. (1986). The organisation of vomiting as a protective reflex: A commentary on the first day's discussions. In C. J. Davis, G. V. Lake-Bakaar, & D. G. Grahame-Smith (Eds.), *Nausea and vomiting: Mechanisms and treatments* (pp. 65–75) New York: Springer-Verlag.

Davison, J. M. and Hytten, F. E. (1974). Glomerular filtration during and after pregnancy. *The Journal of Obstetrics and Gynaecology of the British Commonwealth.* 81: 588–595.

Depue, R. H., Bernstein, L., Ross, R., Judd, H. L., & Henderson, B. E. (1987). Hyperemesis gravidarum in relation to estradiol levels, pregnancy outcome, and other maternal factors: A seroepidemiologic study. *American Journal of Obstetrics and Gynecology, 156,* 1137–1141.

Devereux, G. (1976). *Study of abortion in primitive societies.* New York: International Universities Press.

Dey, A., Westphal, H., & Nebert, D. W. (1989). Cell-specific induction of mouse *CYP1A1* mRNA during development. *Proceedings of the National Academy of Science, 86,* 7446–7450.

Dickens, G., & Trethowan, W. H. (1971). Cravings and aversions during pregnancy. *Journal of Psychosomatic Research, 15,* 259–268.

Doty, R. L. (1976). Reproductive endocrine influences upon human nasal chemoreception: A review. In R. L. Doty (Ed.), *Mammalian olfaction, reproductive processes, and behavior* (pp. 295–321) New York: Academic Press.

Doty, R. L. (1986). Ontogeny of human olfactory function. In W. Breipohl (Ed.), *Ontogeny of olfaction* (pp. 3–17) New York: Springer-Verlag.

Duke, J. A. (1985). *Handbook of medicinal herbs.* Boca Raton: CRC Press.

Duke, J. A. (1988). *Father Nature's Farmacy* [computer data base] Beltsville, MD: Agricultural Research Center, U.S. Department of Agriculture.

Dunn, F. L. (1968). Epidemiological factors: Health and disease in hunter-gatherers. In R. B. Lee & I. DeVore (Eds.), *Man the hunter* (pp. 221–228) Chicago: Aldine Publishing.

Eisner, T., & Grant R. (1980). Toxicity, odor aversion, and "olfactory aposematism." *Science, 213,* 476.

Eskes, T. K. A. B., & Nijdam, W. S. (1984) Epidemiology of drug intake during pregnancy. In B. Krauer, F. Krauer, F. E. Hytten & E. del Pozo (Eds.), *Drugs and pregnancy. Maternal drug handling—fetal drug exposure* (pp. 17–28) New York: Academic Press.

Fairweather, D. V. I. (1968). Nausea and vomiting during pregnancy. *American Journal of Obstetrics and Gynecology, 102,* 135–171.

Fairweather, D. V. I. (1986). Mechanisms and treatment of nausea and vomiting in pregnancy. In C. J. Davis, G. V. Lake-Bakaar, & D. G. Grahame-Smith (Eds.), *Nausea and vomiting: Mechanisms and treatments* (pp. 151–159) New York: Springer-Verlag.

Feuer, G. (1979). Action of pregnancy and various progesterones on hepatic microsomal activities. *Drug Metabolism Reviews, 9,* 147–169.

Fitzgerald, K. M. (1984). Nausea and vomiting in pregnancy. *British Journal of Medical Psychology, 57,* 159–165.

Formacek, V., & Kubeczka, K. H. (1982). *Essential oils analysis by capillary gas chromatography and carbon-13 NMR spectroscopy.* New York: John Wiley & Sons.

Fowden, L., & Lea, P. J. (1979). Mechanism of plant avoidance of autotoxicity by secondary metabolites, especially by non-protein amino acids. In G. A. Rosenthal & D. H. Janzen (Eds.), *Herbivores: Their interaction with secondary plant metabolites* (pp. 135–160) New York: Academic Press.

Freeland, W. J., & Janzen, D.H. (1974). Strategies in herbivory by mammals: The role of plant secondary compounds. *American Naturalist, 108,* 269–289.

Frisch, R. (1987). Body fat, menarche, fitness and fertility. *Human Reproduction, 2,* 521–533.

Garcia, J. (1990). Learning without memory. *Journal of Cognitive Neuroscience, 2,* 287–305.

Garcia, J., & Hankins, W. G. (1975). The evolution of bitter and the acquisition of toxiphobia. In D. A. Denton & J. P. Coghlan (Eds.), *Olfaction and taste V* (pp. 39–45) New York: Academic press.

Gower, J. D., Baldock, R. J., O'Sullivan, A. M., Dore, C. J., Coid, C. R., & Green, C. J. (1990). Protection against endotoxin-induced foetal resorption in mice by desferrioxamine and ebselen. *International Journal of Experimental Pathology, 71,* 433–440.

Gram, T. E. (1980). The metabolism of xenobiotics by mammalian lung. In T. Gram (Ed.), *Extrahepatic metabolism of drugs and other foreign compounds* (pp. 159–209) New York: S. P. Medical & Scientific Books.

Guillaume, J., Benjamin, F., Sicuranza, B., Wand, C. F., Garcia, A., & Friberg, J. (1987). Maternal serum levels of estradiol, progesterone and human chorionic gonadotropin in ectopic pregnancy and their correlation with endometrial histologic findings. *Surgery, Gynecology and Obstetrics, 165,* 9–12.

Haesaert, B., & Ornoy, A. (1986). Transplacental effects of endotoxemia on fetal mouse brain, bone, and placental tissue. *Pediatric Pathology, 5,* 167–181.

Hageman, G., Hermans, R., Ten Hoor, F., & Kleinjans, J. (1990). Mutagenicity of deep-frying fat, and evaluation of urine mutagenicity after consumption of fried potatoes. *Food and Chemical Toxicology, 28,* 75–80.

Hageman, G., Kikken, R., Ten Hoor, F., & Kleinjans, J. (1988). Assessment of mutagenic activity of repeatedly used deep-frying fats. *Mutation Research, 204,* 593–604.

Hale, R. W. (1984). Diagnosis of pregnancy and associated conditions. In R. C. Benson (Ed.), *Current obstetrics and gynecologic diagnosis and treatment* (pp. 604–613) Los Altos: Lange Medical.

Halmesmaki, E., Autti, I., Granstrom, M.-L., Stenman, U.-H., & Ylikorkala, O. (1987). Estradiol, estriol, progesterone, prolactin, and human chorionic gonadotropin in pregnant women with alcohol abuse. *Journal of Clinical Endocrinology and Metabolism, 64,* 153–156.

Hamilton, W. J., & Mossman, H. W. (1972). *Human Embryology. Prenatal Development of Form and Function,* (4th ed.). Baltimore: Williams & Wilkins.

Harada, M. (1986). Congenital minamata disease: Intrauterine methylmercury poisoning. In J. L. Sever & R. L. Brent (Eds.), *Teratogen update: Environmentally induced birth defect risks.* (pp. 123–126) New York: Alan R. Liss.

Harrison, R. F. (1980). Maternal plasma beta-HCG in early human pregnancy. *British Journal of Obstetrics and Gynecology, 87,* 705–711.

Harrison, R. F., & Kitchin, Y. (1980). Maternal plasma unconjugated oestrogens in early human pregnancy. *British Journal of Obstetrics and Gynecology, 87,* 686–694.

Harvell, C. D. (1990). The ecology and evolution of inducible defenses. *The Quarterly Review of Biology, 65,* 323–339.

Hayes, A. W. (1981). *Mycotoxin teratogenicity and mutagenicity.* Boca Raton: CRC Press.

Heady, H. F. (1964). Palatability of herbage and animal preferance. *Journal of Range Management, 17,* 76–82.

Herbert, V. (1990). Development of human folate deficiency. In M. F. Picciano, E.L.R. Stokstad, & J. F. Gregory, III (Eds.), *Health and disease* (pp. 195–210) New York: John Wiley & Sons.

Hilbelink, D. R., Chen, L. T., & Bryant, M. (1986). Endotoxin-induced hyperthermia in pregnant golden hamsters. *Teratogenesis, Carcinogenesis and Mutagenesis, 6,* 209–217.

Hodgson, E., & Levi, P. E. (1987). *A textbook of modern toxicology.* New York: Elsevier Science Publishing.

Hood, R. D., & Szczech, G. M. (1983). Teratogenicity of fungal toxins and fungal-produced antimicrobial agents. In R. F. Keeler & A. T. Tu (Eds.), *Handbook of natural toxins. Plant and fungal toxins,* (Vol. 1,) New York: Marcel Dekker.

Hook, E. B. (1976). Changes in tobacco smoking and ingestion of alcohol and caffeinated beverages during early pregnancy: Are those consequences, in part, of feto-protective mechanisms diminishing maternal exposure to embryotoxins? In S. Kelly, E. B. Hook, D. T. Janerich, & I. H. Porter (Eds.), *Birth defects: Risks and consequences* (pp. 173–183) New York: Academic Press.

Hook, E. B. (1978). Dietary cravings and aversions during pregnancy. *American Journal of Clinical Nutrition, 31,* 1355–1362.

Hubert, H. B., Fabsitz, R. R., Feinleib, M., & Brown, K. S. (1980). Olfactory sensitivity in humans: Genetic versus environmental control. *Science, 208,* 607–609.

Hunt, W. A., Rabin, B. M., & Lee, J. (1987). Ethanol-induced taste aversions: Lack of involvement of acetaldehyde and the area postrema. *Alcohol, 4,* 169–173.

Hytten, F.E. (1984). Physiological changes in the mother related to drug handling. In B. Krauer, F. Krauer, F. E. Hytten, & E. del Pozo (Eds.), *Drugs and pregnancy. Maternal drug handling—fetal drug exposure* (pp. 7–15) New York: Academic Press.

Institute of Medicine. (1990). *Nutrition during pregnancy.* Washington, DC: National Academy Press.

Ivie, G. W., Holt, D. L., & Ivey, M. C. (1981). Natural toxicants in human foods: Psoralens in raw and cooked parsnip root. *Science, 213,* 909–910.

Jacoby, W. B. (Ed.). (1980). *Enzymatic basis of detoxification.* (Vol. 1). New York: Academic Press.

James, L. F. (1983). Neurotoxins and other toxins from astragalus and related genera. In R. F. Keeler & A. T. Tu (Eds.), *Handbook of natural toxins. Plant and fungal toxins.* (Vol. 1, pp. 445–462) New York: Marcel Dekker.

Janzen, D. H. (1978). Complications in interpreting the chemical defenses of trees against tropical arboreal plant-eating vertebrates. In G. G. Montgomery (Ed.), *The ecology of arboreal folivores* (pp. 73–84) Washington, DC: Smithsonian Institution Press.

Jarnfelt-Samsioe, A. (1987). Nausea and vomiting in pregnancy: A review. *Obstetrical and Gynecological Survey, 41*, 422–427.

Jarnfelt-Samsioe, A., Bremme, K., & Eneroth, P. (1985). Steroid hormones in emetic and nonemetic pregnancy. *European Journal of Obstetrics, Gynecology and Reproductive Biology, 21*, 87–99.

Johns, T. (1986). Detoxification function of geophagy and domestication of the potato. *Journal of Chemical Ecology, 12*, 635–646.

Juchau, M. R. (1981). Enzymatic bioactivation and inactivation of chemical teratogens and transplacental carcinogens/mutagens. In M. R. Juchau (ed.), *The biochemical basis of chemical teratogenesis* (pp. 63–94) New York: Elsevier North Holland.

Juchau, M. R. (1982). The role of the placenta in developmental toxicology. In K. Snell (Ed.), *Developmental toxicology* (pp. 187–209) New York: Praeger.

Kaba, H., Rosser, A. E., & Keverne, E. B. (1988). Hormonal enhancement of neurogenesis and its relationship to the duration of olfactory memory. *Neuroscience, 24*, 93–98.

Kaberry, P. M. (1939). *Aboriginal woman.* Philadelphia: Blakiston.

Kaplan, A. R., & Glanville, E. V. (1964). Taste thresholds for bitterness and cigarette smoking. *Nature, 202*, 1366.

Katz, S. H., Hediger, M. L., & Valleroy, L. A. (1974). Traditional maize processing techniques in the New World. *Science, 184*,765–773.

Keeler, R. F. (1983). Naturally occurring teratogens from plants. In R. F. Keeler & A. T. Tu (Eds.), *Handbook of natural toxins. Plant and fungal toxins* (Vol. 1, pp. 161–191) New York: Marcel Dekker.

Keeler, R. F., & Tu, A. T. (Eds.). (1983). *Handbook of natural toxins. Plant and fungal toxins,* (Vol. 1). New York: Marcel Dekker.

Kilgore, W. W., Crosby, D. G., Craigmill, A. L., & Poppen, N. K. (1981). Toxic plants as possible human teratogens. *California Agriculture, 35*, 6.

Kingsbury, J. M. (1983). The evolutionary and ecological significance of plant toxins. In R. F. Keeler & A. T. Tu (Eds.), *Handbook of natural toxins. Plant and fungal toxins* (Vol. 1, pp. 675–706). New York: Marcel Dekker.

Klebanoff, M. A., Koslowe, P. A., Kaslow, R., & Rhoads, G. (1985). Epidemiology of vomiting in early pregnancy. *Obstetrics and Gynecology, 66*, 612–616.

Kletzky, O. A., Rossman, F., Bertolli, S. I., Platt, L. D., & Mishell, D. R. (1985). Dynamics of human chorionic gonadotropin, prolactin, and growth hormone in serum and amniotic fluid throughout normal human pregnancy. *American Journal of Obstetrics and Gynecology, 151*, 878–884.

Kohl, R. L. (1985). Endocrine correlates of susceptibility to motion sickness. *Aviation, Space, and Environmental Medicine, 56*, 1158–1165.

Lee, R., & DeVore, I. (1976). *Kalahari hunter-gatherers: Studies of the !Kung San and their neighbors.* Cambridge: Harvard University Press.

Lindeskog, P., Overvik, E., Nilsson, L., Nord, C.-E, & Gustaffson, J.-A. (1988). Influence of fried meat and fiber on cytochrome P-450 mediated activity and excretion of mutagens in rats. *Mutation Research, 204*, 553–563.

Little, R. E., & Hook, E. B. (1979). Maternal alcohol and tobacco consumption and their association with nausea and vomiting during pregnancy. *Acta Obstetrica et Gynecologica Scandinavica, 58*, 15–17.

Loock, W., Nau, H., Schmidt-Gollwitzer, M., & Dvorchik, B. H. (1988). Pregnancy-specific changes of antipyrine pharmacokinetics correlate inversely with changes of estradiol/ progesterone plasma concentration ratios. *Journal of Clinical Pharmacology, 28*, 216–221.

Lucas, B., & Sotelo, A. (1984). A simplified test for the quantitation of cyanogenic glucosides in wild and cultivated seeds. *Nutrition Reports International, 29,* 711–719.

MacLeod, A. J., & MacLeod, G. (1970). Flavor volatiles of some cooked vegetables. *Journal of Food Science, 35,* 734–738.

MacLeod, A. J., & Pikk, H. E. (1978). A comparison of the chemical flavour composition of some brussels sprouts cultivars grown at different crop spacings. *Phytochemistry, 17,* 1029–1032.

MacMahon, B. (1982). Cigarette smoking and urinary estrogens. *New England Journal of Medicine, 307,* 1062–1065.

Masson, G. M., Anthony, F., & Chau, E. (1985). Serum chorionic gonadotrophin (hCG), Schwangershaftsprotein 1 (SP1), progesterone and oestradiol levels in patients with nausea and vomiting in early pregnancy. *British Journal of Obstetrics and Gynecology, 92,* 211–215.

Mayr, E. (1983). How to carry out the adaptationist program? *American Naturalist, 121,* 324–334.

McBride, W. G. (1961). Thalidomide and congenital abnormalities. *Lancet, 2,* 1358.

Medalie, J. H. (1957). Relationship between nausea and/or vomiting in early pregnancy and abortion. *Lancet, 2,* 117–119.

The Merck manual of diagnosis and therapy (15th ed.). (1987). Rahway, NJ: Merck Sharp & Dohme Research Laboratories.

Milunsky, A., Jick, H., Jick, S. S., Bruell, C. L., MacLaughlin, D. S., Rothman, K. J., & Willett, W. (1989). Multivitamin/folic acid supplementation in early pregnancy reduces the prevalence of neural tube defects. *Journal of the American Medical Association, 262,* 2847–2852.

Minturn, L., & Weiher, A. W. (1984). The influence of diet on morning sickness: A cross-cultural study. *Medical Anthropology, 8,* 71–75.

Mithen, R. F., Lewis, B. G., Heaney, R. K., & Fenwick, G. R. (1987). Glucosinolates of wild and cultivated *Brassica* species. *Phytochemistry, 26,* 1969–1973.

Morita, M., Takeda, N., Kubo, T., & Matsunaga, T. (1988). Pica as an index of motion sickness in rats. *O.R.L. Journal for Oto-Rhino-Laryngology and Its Related Specialties, 50,* 188–192.

Mottet, N. K. (1981). Biochemical mechanisms of trace element teratogenesis. In M. R. Juchau (Ed.), *The Biochemical basis of chemical teratogenesis* (pp. 201–246) New York: Elsevier North Holland.

Nau, H., Loock, W., Schmidt-Gollwitzer, M., & Kuhnz, W. (1984). Pregnancy-specific changes in hepatic drug metabolism in man. In B. Krauer, F. Krauer, F. E. Hytten & E. del Pozo (Eds.), *Drugs and pregnancy. Maternal drug handling—fetal drug exposure* (pp. 45–62) New York: Academic Press.

Newman, L. F. (1985). Botanical index. In L. F. Newman (Ed.), *Women's medicine* (pp. 205–213) New Brunswick: Rutgers University Press.

Oakley, B. (1986). Basic taste physiology: Human perspectives. In H. L. Meiselman & R. S. Rivlin (Eds.), *Clinical measurement of taste and smell* (pp. 5–18) New York: MacMillan Publishing.

O'Sullivan, A. M., Dore, C. J., & Coid, C. R. (1988). Campylobacters and impaired fetal development in mice. *Journal of Medical Microbiology, 25,* 7–12.

Overvik, E., Nilsson, L., Fredholm, L., Levin, O., Nord, C. E., & Gustafsson, J. A. (1987). Mutagenicity of pan residues and gravy from fried meat. *Mutation Research, 187,* 47–53.

Pang, S. F., Tang, P. L., Tang, G.W.K., Yam, A.W:C., & Ng, K. W. (1987). Plasma levels of immunoreactive melatonin, estradiol, progesterone, follicle stimulating hormone, and beta-human chorionic gonadotropin during pregnancy and shortly after parturition in humans. *Journal of Pineal Research, 4,* 21–31.

Patillo, R. A., Hussa, R. O., Yorde, D. E., & Cole, L. A. (1983). Hormone synthesis by normal and neoplastic human trophoblast. In Y. W. Loke & A. Whyte (Eds.), *Biology of tropho-blast* (pp. 283–316) New York: Elsevier.

Pelkonen, O. (1982). The differentiation of drug metabolism in relation to developmental toxicology. In K. Snell (Ed.), *Developmental Toxicology* (pp. 165–186). New York: Praeger.

Pelkonen, O. (1987). Detoxification and toxification processes in the human fetoplacental unit. In B. Krauer, F. Krauer, F. E. Hytten, & E. del Pozo (Eds.), *Drugs and pregnancy. Maternal drug handling—fetal drug exposure* (pp. 63–72) New York: Academic Press.

Persaud, T.V.N. (1985). Critical phases of intrauterine development. In T. V. N. Persaud, A. E. Chudley, & R. G. Skalko (Eds.), *Basic concepts in teratology* (pp. 23–29) New York: Alan R. Liss.

Petitti, D. B. (1986). Nausea and pregnancy outcome. *Birth, 13,* 223.

Prentice, A. M., & Whitehead, R. G. (1987). The energetics of human reproduction. In A. S. I. Loudon & P. A. Racey (Eds.), *Reproductive energetics in mammals* (pp. 275–304) Oxford: Clarendon Press.

Profet, M. (1988). The evolution of pregnancy sickness as protection to the embryo against Pleistocene teratogens. *Evolutionary Theory, 8,* 177–190.

Profet, M. (1991). The function of allergy: Immunological defense against toxins. *The Quarterly Review of Biology, 66, 23–62.*

Pyysalo, T., Honkanen, E., & Hirvi, T. (1979). Volatiles of wild strawberries, *Fragaria vesca* L., compared to those of cultivated berries, *Fragaria x ananassa cv. Senga sengana. Journal of Agricultural and Food Chemistry, 27,* 19–22.

Randerath, K., Randerath, E., Agrawal, H. P., Gupta, R. C., Schurdak, M. E., & Reddy, V. M. (1985). Postlabeling methods for carcinogen-DNA adduct analysis. *Environmental Health Perspectives, 62,* 57–65.

Renwick, J. H., Claringbold, W.D.B., Earthy, M. E., Few, J. D., & McLean, A.C.S. (1984). Neural-tube defects produced in Syrian hamsters by potato glycoalkaloids. *Teratology, 30,* 371–381.

Rhoades, D. F. (1979). Evolution of plant chemical defense against herbivores. In G. A. Rosenthal & D. H. Janzen (Eds.), *Herbivores: Their interaction with secondary plant metabolites* (pp. 3–54) New York: Academic Press.

Richard, H. M., & Jennings, W. G. (1971). Volatile composition of black pepper. *Journal of Food Science, 36,* 584–589.

Rosenthal, G. A., & Janzen, D. H. (Eds.). (1979). *Herbivores: Their interaction with secondary plant metabolites.* New York: Academic Press.

Rozin, P., & Kalat, J. (1971). Specific hungers and poison avoidance as adaptive specializations of learning. *Psychological Review, 78,* 459–486.

Rozin, P., & Vollmecke, T. A. (1986). Food likes and dislikes. *Annual Review of Nutrition, 6,* 433–56.

Rush, D. (1989). Effects of changes in protein and calorie intake during pregnancy on the growth of the human fetus. In I. Chalmers, M. Enkin, & M. J. N. C. Keirse (Eds.), *Effective care in pregnancy and childbirth.* (Vol. 1, pp. 255–280) New York: Oxford University Press.

Ryan, J. P., Bhojwani, A., Wang, M. B. (1987). Effect of pregnancy on gastric motility in vivo and in vitro in the guinea pig. *Gastroenterology, 93,* 29–34.

Sadow, J.I.D. (1980). *Human reproduction: An integrated view.* Chicago: Year Book Medical Publishers.

Schardein, J. L., Schwetz, B. A., & Kenel, M. F. (1985). Species sensitivities and prediction of teratogenic potential. *Environmental Health Perspectives, 61,* 55–67.

Schmiediche, P. E., Hawkes, J. G., & Ochoa, C. M. (1980). Breeding of the cultivated potato species *Solanum x juzepczukii* Buk. and *Solanum x curtilobum Juz.* et Buk. I. A study of the natural variation of S.x *Juzepczukii,* S. x curtilobum and their wild progenitor, S. *acaule* Bitt. *Euphytica, 29,* 685–704.

Schultes, R. E., & Hofmann, A. (1987). *Plants of the gods. Origins of hallucinogenic use.* New York: Alfred van der Marck.

Schwimmer, S. (1968). Enzymatic conversion of trans-(+)-5-1-propenyl-L-cysteine S-oxide to bitter and odour bearing components of onion. *Phytochemistry, 7,* 401–404.

Scudder, T. (1962). *The ecology of the Gwembe Tonga.* Manchester, England: Manchester University Press.

Scudder, T. (1971). Gathering among African woodland savannah cultivators: A case study: The Gwembe Tonga. In *Zambian papers,* 5, 1–50.

Shepard, T. H. (1986). *Catalog of teratogenic agents* (5th ed.). Baltimore: Johns Hopkins University Press.

Shostak, M. (1981). *Nisa. The life and words of a !Kung woman.* Cambridge, MA: Harvard University Press.

Shum, S., Jensen, N. M., & Nebert, D. W. (1979). The Murine Ah Locus: In utero toxicity and teratogenesis associated with genetic differences in benzo[a]pyrene metabolism. *Teratology, 20,* 365–376.

Simerly, R. B., Chang, C., Muramatsu, M., & Swanson, L. W. (1990). Distribution of androgen and estrogen receptor mRNA-containing cells in the rat brain: an in situ hybridation study. *Journal of Comparative Neurology, 294,* 76–95.

Soules, M. R., Hughes, C. L., Garcia, J. A., Livengood, C. H., Prystowsky, M. R., & Alexander, E., III. (1980). Nausea and vomiting of pregnancy: Role of human chorionic gonadotropin and 17-hydroxyprogesterone. *Obstetrics and Gynecology, 55,* 696–700.

Spielmann, H., & Vogel, R. (1987). Transfer of drugs into the embryo before and during implantation. In H. Nau & W. J. Scott, Jr. (Eds.), *Pharmacokinetics in Teratogenesis* (Vol 1, pp. 45–53) Boca Raton: CRC Press.

Stahl, A. B. (1984). Hominid dietary selection before fire. *Current Anthropology, 25,* 151–157.

Stalla, G. K., Doerr, H. G., Bidlingmaier, F., Sippel, W. G., & von Restorff, W. (1985). Serum levels of eleven steroid hormones following motion sickness. *Aviation, Space, and Environmental Medicine, 56,* 995–999.

Stewart, R. B., Perlanski, E., & Grupp, L. A. (1988). Area postrema and alcohol: Effects of area postrema lesions on ethanol self-administration, pharmacokinetics, and ethanol-induced conditioned taste aversion. *Alcoholism, 12,* 698–704.

Sugimura, T. (1982). Mutagens in cooked food. In R. A. Fleck & A. Hollaender (Eds.), *Genetic toxicology: An agricultural perspective* (pp. 243–269). New York: Plenum Publishing.

Tanaka, T. (1980). *The San, hunter-gatherers of the Kalahari.* (D. W. Hughes, Trans.). Tokyo: University of Tokyo Press.

Tierson, F. D., Olsen, C. L., & Hook, E. B. (1985). Influence of cravings and aversions on diet in pregnancy. *Ecology of Food and Nutrition, 17,* 117–129.

Tierson, F. D., Olsen, C. L., & Hook, E. B. (1986). Nausea and vomiting of pregnancy and association with pregnancy outcome. *American Journal of Obstetrics and Gynecology, 155,* 1017–1022.

Tooby, J., & DeVore, I. (1987). The reconstruction of hominid behavioral evolution through strategic modeling. In W. G. Kinzey, Ed., *The evolution of human behavior: Primate models* (pp. 183–237). Albany: SUNY Press.

Treisman, M. (1977). Motion sickness: An evolutionary hypothesis. *Science, 197,* 493–495.

Trivers, R. L. (1974). Parent-offspring conflict. *American Zoologist, 14,* 249–264.

Tuchmann-Duplessis, H., David, G., & Haegel, P. (1971). *Illustrated human embryology* (Vol. 1). (L. Hurley, Trans.). Paris: Masson.

Tulchinsky, D., & Hobel, C. (1973). Plasma human chorionic gonadotropin, estrone, estradiol, estriol, progesterone, and 17-alpha-hydroxyprogesterone in human pregnancy. Part 3: Early normal pregnancy. *American Journal of Obstetrics and Gynecology, 117,* 884–893.

Turnbull, C. M. (1965). The Mbuti pygmies: An ethnographic survey. In *Anthropological papers of the American museum of natural history* (Vol. 50, part 3). New York: American Museum of Natural History.

Tylden, E. (1968). Hyperemesis and physiological vomiting. *Journal of Psychosomatic Research 12,* 85–93.

Uddenberg, N., Nilsson, A., & Almgren, P. (1971). Nausea in pregnancy: Psychological and psychosomatic aspects. *Journal of Psychosomatic Research, 15,* 269–276.

Urbasch, I. (1985). Produktion von 2-Tridecanon durch verschiedene Kultur- und Wildtomatenpflanzen (*Lycopersicon* spp.). *Planta Medica.* No. 6. pp. 492–494.

Ure, D.M.J. (1969). Negative association between allergy and cancer. *Scottish Medical Journal, 14,* 51–54.

Van der Wal, B., Kettenes, K. K., Stoffelsma, J., Sipma, G., & Semper, A.T.J. (1971). New volatile components of roasted cocoa. *Journal of Agricultural and Food Chemistry, 19,* 276–280.

Van Valen, L. (1973). A new evolutionary law. *Evolutionary Theory, 1,* 1–30.

Vannelli, G. B., & Balboni, G. C. (1982). On the presence of estrogen receptors in the olfactory epithelium of the rat. In W. Breiphol (Ed.), *Olfaction and endocrine regulation* (pp. 279–283) London: IRL Press.

Virtanen, A. I. (1965). Studies on organic sulphur compounds and other labile substances in plants. *Phytochemistry, 4,* 207–228.

Waddell, W. J., & Marlowe, C. (1981). Biochemical regulation of the accessibility of teratogens to the developing embryo. In M. R. Juchau (Ed.), *Biochemical basis of chemical teratogenesis* (pp. 1–62) New York: Elsevier North Holland.

Wald, A., Van Thiel, D. H., Hoechstetter, L., Gavaler, J. S., Egler, K. M., Verm, R., Scott, L. & Lester, R. (1982). Effect of pregnancy on gastrointestinal transit. *Digestive Diseases and Sciences, 27,* 1015–1018.

Walters, W.A.W. (1987). The management of nausea and vomiting during pregnancy. *Medical Journal of Australia, 147,* 290–291.

Warren, C. D. (1983). Toxic alkaloids from lupines. *Chemical and Engineering News, 61*(June 13), 3.

Watt, J. M., & Breyer-Brandwijk, M. G. (1962). *Medicinal and poisonous plants of southern and eastern Africa.* (2nd Ed.). London: E. & S. Livingstone.

Weigel, M. M., & Weigel, R. M. (1989a). Nausea and vomiting of early pregnancy and pregnancy outcome. An epidemiological study. *British Journal of Obstetrics and Gynecology, 96,* 1304–1311.

Weigel, R. M., & Weigel, M. M. (1989b). Nausea and vomiting of early pregnancy and pregnancy outcome. A meta-analytical review. *British Journal of Obstetrics and Gynecology, 96,* 1312–1318.

Williams, G. C. (1966). *Adaptation and natural selection.* Princeton: Princeton University Press.

Williams, G. C. (1985). A defense of reductionism in evolutionary biology. *Oxford Surveys in Evolutionary Biology, 2,* 1–27.

Williams, K. E. (1982). Biochemical mechanisms of teratogenesis. In K. Snell (Ed.), *Developmental toxicology* (pp. 93–121) New York: Praeger.

Willson, J. R., & Carrington, E. R. (1979). *Obstetrics and gynecology* (6th ed.). St. Louis: CV Mosby.

Wilson, C. P., & Raisman, G. (1980). Age-related changes in the neurosensory epithelium of the mouse vomeronasal organ: Extended period of post-natal growth in size and evidence for rapid cell turnover in the adult. *Brain Research, 185,* 103–113.

Wilson, J. F. (1987). Severe reduction in food intake by pregnant rats resembles a learned food aversion. *Physiology and Behavior, 41,* 291–295.

Wolkind, S., & Zajicek, E. (1978). Psycho-social correlates of nausea and vomiting in pregnancy. *Journal of Psychosomatic Research, 22,* 1–5.

Woodburn, J. (1968). An introduction to Hadza ecology. In R. B. Lee & I. DeVore (Eds.), *Man the Hunter* (pp. 49–55) Chicago: Aldine Publishing.

Yerushalmy, J., & Milkovich, L. (1965). Evaluation of the teratogenic effects of meclizine in man. *American Journal of Obstetrics and Gynecology, 93,* 553–562.

9

Nurturance or Negligence: Maternal Psychology and Behavioral Preference Among Preterm Twins

JANET MANN

A mother who had given birth to 2-pound preterm twins, one of which subsequently died, remarked to me about her surviving twin:

> I love D—so much that I can't stand to be away from him.
> You know, I hate to say this [in a whisper], but I love him
> even more than my other children.

In my study of these high-risk preterm infants, I was often struck by the extraordinary efforts some mothers made to care for, stimulate, and love their infants. In contrast, many social scientists (myself included) have also often observed parental neglect, abuse, or infanticide directed toward premature or high-risk infants. The enormous individual variation in parental care under such adverse perinatal conditions is intriguing. While some social scientists have linked infant prematurity to child mal-treatment (e.g., Burgess & Conger, 1978; Burgess & Garbarino, 1983; Friedrich & Bor-iskin, 1976), others have emphasized the increased or even excessive stimulation and nurturance that mothers of premature infants provide (e.g., Field, 1977; Goldberg & Divitto, 1983).

EVOLUTIONARY MODELS

Evolutionary biologists and psychologists have been concerned with ultimate expla-nations (adaptive function in terms of reproductive success) of parental behavior toward low-phenotypic-quality (high-risk) children. They have focused almost exclu-sively on neglect, abuse, and/or infanticide directed at high-risk infants (e.g., Daly & Wilson, 1988; Scrimshaw, 1984; also see Gelles & Lancaster, 1987). *High* parental care and investment directed toward such infants has not been addressed from this per-spective. In addition, the underlying psychological mechanisms responsible for differ-ential parental investment are largely unexplored from an evolutionary perspective. In this paper I explore some of the psychological mechanisms involved in maternal decisions to care for and invest in high-risk offspring. I also attempt to integrate psy-chological and evolutionary approaches to the study of child neglect by examining the relationship among maternal psychology, maternal behavior, and infant health characteristics in detail among a small population of extremely low-birth-weight (ELBW) preterm twins.[1]

Evolutionary biology and, in particular, parental investment theory (Trivers, 1972, 1974) provide us with valuable explanations and predictions of patterns in parental investment. From this foundation we may generate testable hypotheses concerning the dynamics of differential investment in high-risk offspring (such as preterm infants). Trivers' concept of parental investment is comprised of two essential components. First, parental investment (PI) includes all actions that contribute to the reproductive success of the offspring. On a more proximate level, this includes parental actions that contribute to the sociopsychological, physical, and cognitive well-being of the offspring. Second, investment in any particular child often compromises the ability of the parent to invest in other children (present or future). Human reproduction is limited primarily by the degree of parental effort required to successfully rear each offspring. Because natural selection favors parents who can maximize the number of offspring (or genetic representatives) surviving to reproduction, each individual child represents different costs and benefits (i.e., reproductive value) to the parent. Thus, the reproductive value of a child is considered to be the primary factor influencing parental psychology and subsequently investment. The reproductive value of a child is likely to be low when the child is unlikely to survive to reproduction or when she or he is unlikely to become reproductively competitive because of inherent physical, cognitive, or social disabilities. Children who are of low reproductive value due to adverse congenital, prenatal, or perinatal conditions are hereafter termed "high-risk" children because they are phenotypically at "high risk" for a variety of developmental delays and disabilities from birth.

The parent's adaptive problem (or dilemma) is whether to care and invest intensively in a "high-risk" child in order to improve the child's prospects or to provide minimal care and focus investment on other offspring (existing, potential, or collateral) with higher reproductive value. This problem is not new. Although current medical technology has substantially altered infant mortality and the subsequent health and developmental prospects for high-risk offspring, infant viability certainly varied extensively throughout human evolutionary history, both within and between human groups. Parents of high-risk offspring (ancestral and current) are faced with difficult choices as to how to partition their investment among existing or prospective offspring. A sick child requires more care than a healthy child. Moderate care is inadequate to rear an especially needy child to adulthood. If parents were to treat sick offspring the same as healthy offspring, both the parent and the child would frequently suffer reproductive costs. Such a child would not receive enough care to survive or become reproductively competitive, and the parent would have to absorb the investment costs that could have been directed to healthy (i.e., reproductively viable) offspring. Natural selection would operate against those parents who could not discriminate high-risk from healthy offspring and who couldn't adjust their parental behavior accordingly.

Parents have two possible routes out of their dilemma. They can either increase their investment to meet the child's greater needs or decrease their investment to minimize the costs. Intensive care is obviously very costly to the parent, but the returns on their investment (improved child quality and reproductive value) may be reproductively rewarding. Minimal care is of low cost to the parent, and other children (siblings to the sick child) may gain additional investment. (The only benefit to the high-risk offspring is the enhanced reproductive value of its siblings, potential or actual). To reiterate, the worst cost-benefit ratio investment strategy for a parent with a high-risk child may be moderate care.

Our primate cousins, along with other animals that have high PI and few offspring, share the same adaptive problem. Although primate mothers sometimes desert non-viable offspring (especially during the first few days of life), under some conditions they dramatically increase their investment in those offspring by prolonging nursing, carrying, and other forms of parental care (Altmann, 1980; Berkson, 1973; Chapman & Chapman, 1987; Fedigan & Fedigan, 1977; Furuya, 1966). It is reasonable to assume that human primate mothers shared similar evolved capacities to discriminate offspring quality and either increased or decreased care allocation, depending upon social, demographic, and environmental inputs.[2]

SELECTION PRESSURES

Throughout human evolutionary history, natural selection would have favored individuals who made "wise" reproductive decisions.[3] Such decisions could not be made in a vacuum. Individuals must have been sensitive to information relevant to inclusive fitness costs and benefits. This necessarily implies that the human psyche is designed to receive and process certain kinds of information that in turn affect our perceptions, attitudes, thoughts, and feelings about our children. Evolutionary theory informs of us of what variables should be relevant for making reproductive decisions: those decisions that affect the mother's reproductive success (i.e., her likelihood of successfully rearing current and future offspring to reproductive age).[4] These variables include the following:

1. Infant health status (probability of infant survival to adulthood and correctability of disabilities)
2. Marital status and stability (probability of parental investment)
3. Parity, age, and health of the mother (current reproductive value of the mother)
4. Abundance of and access to resources
5. Social support (relatives and/or friends that provide direct and/or indirect aid)

All of these variables cannot be discussed in detail here. Let me just provide some examples: (a) An extremely ill infant who is unlikely to survive regardless of maternal efforts is a reproductive dead-end; (b) if your mate is unable or unwilling to provide critical resources for child-rearing, then it may pay (in reproductive terms) to terminate or minimize investment in current offspring and work toward finding a more reliable and supportive mate; (c) if you are a 39-year old primiparous mother, then it would pay to invest highly in your first (and probably only) offspring (see Daly & Wilson, 1984, for further discussion). Some of these variables will be discussed in the context of how they elucidate the mechanisms involved in directing parental care strategies. Because differential parental investment (high or low) is predicted to occur with high-risk infants, then the underlying psychological mechanisms that influence parental behavior become critical to our understanding of patterns of investment and testing evolutionary hypotheses. Evolved psychological mechanisms responsible for maternal behavior toward sick offspring should take into account a variety of informational inputs. This information will affect the parent's perceptions and cognitions concerning the infant and subsequently her behavior toward the infant. First, I will elaborate on how such informational inputs can influence maternal psychology by focusing on (a) conditions that favor low investment; and (b) conditions that favor high investment. Second, I will review the acoustic and physical features of high-risk infants and their effects on parents and other adults.

Conditions Favoring Low Investment

Lowered investment may be defined as less biological and/or sociopsychological support to one child compared with the support that other same-aged children in that family would receive (Scrimshaw, 1984). Under circumstances where high-risk infants are unlikely to survive either because they are severely damaged or because the parents lack the resources to properly care for them, a low PI strategy (i.e., selective neglect or, in the extreme, infanticide) is expected to be pursued. Low PI behaviors are the outcomes of psyochological mechanisms shaped by natural selection. It is the nature of such mechanisms that I shall explore.

There is some evidence to suggest that maternal thinking can change dramatically under difficult social, economic, and demographic conditions. In a recent anthropological study of extremely poor Brazilian mothers, Scheper-Hughes (1985) found that detachment and selective neglect toward "sick" infants were common, even socially sanctioned. These mothers did not think it was wise to nurse and become attached to these infants because of their poor survival prospects. With high birth and high mortality rates, they thought it best to invest in vigorous, healthy infants (Scheper-Hughes, 1985). The psychological mechanisms of emotional detachment, rationalization, and attributions made toward the child appear to enable these mothers to exhibit selective neglect. For example, an infant born high-risk was labeled as a child who "wanted to die." Consequently the mother did not feed the child. It was therefore not really her "fault" that her child starved to death. Such attributions and rationalizations seem to shape maternal behavior. Oddly enough, these shantytown mothers were well aware of child death *a mingua* (accompanied by maternal indifference and neglect) and gave clear and correct explanations of what contributes to child mortality, but they gave different explanations for the death of their *own* child.

Probably one of the most powerful cues influencing PI psychology is infant health. In many traditional societies, when a child is born with serious developmental anomalies, it is common for all members of the society (parents, relatives, political leaders) to sanction infanticide (Daly & Wilson, 1984). In many of the societies surveyed by Daly and Wilson (from the Human Relations Area Files), infant deformity was the most common explanation for infanticide, and in many of these societies, the *only* acceptable reason for killing an infant is if it is of extremely poor quality (high-risk). Although the immediate psychological mechanisms involved in an "abandonment" decision may be feelings of "disgust" (i.e., "the child is hideous"), factors that can also influence parental beliefs include negative attributions (e.g., "the child wants to die"), personal despair (i.e., "I can't deal with this situation"), and sociocultural beliefs and rules concerning the child. In some cultures, negative attributions toward the child take the form of extreme socially sanctioned distortions. The birth of a deformed or abnormal child may be viewed as a sign that the child is evil-spirited, angry, or cursed. The Machiguenga believe that the infant born during delivery complications is angry (a devalued trait) and is therefore killed, given away, or abandoned (Korbin, 1987).

Children born with abnormalities are considered demons in some cultures (e.g., Ayoreo of Brazil; see Bugos & McCarthy, 1984) and therefore *must* be killed.

Conditions Favoring High Investment

In contrast, when conditions are relatively good, mothers may increase their investment in sick offspring (relative to care they would invest in normal offspring) either because infant health characteristics are only moderately poor or because they possess adequate resources to care for and improve their children's health status. Mothers may provide more food, care, stimulation, and protection toward "high-risk" children than to normal healthy children. Goldberg and colleagues (Goldberg & Divitto, 1983; Goldberg & Marcovitch, 1986) proposed that the increased stimulation Western mothers give to preterm infants, when contrasted with full-term infants, is an adaptive response to the greater needs of the infant. Wasserman, Allen, and Solomon (1985) found that mothers of preterm and handicapped infants worked hard at stimulating, focusing, and "revving up" their children. They also found, however, that a subset of mothers, particularly those with cosmetically handicapped children (none of the children had central nervous system damage), increase in withdrawal and rejection over a 9- to 24-month period. Extreme and obvious deformities apparently shift the parent toward a low-investment pattern. Less obvious and probably more correctable disabilities may encourage the mother to persist in working with the child, but to continually monitor the infant's progress.

Although cause and effect are not clear, there is substantial evidence demonstrating the positive effects of social support on maternal behavior, particularly if the infant is high-risk. Others have cogently argued (e.g., see Draper, 1989) that social support in childcare tasks is an important resource for the mother and can critically affect reproductive outcomes. In a study of high-risk infants, Affleck, Tennen, and Gernshman (1985) found that maternal satisfaction with emotional, tangible, and informational support was associated with more positive mother-infant interactions, positive maternal mood states, and high satisfaction with the child. In another study of high-risk infants, social support and particularly spousal support of the mother was positively related to improved mother-infant interaction, maternal satisfaction with parenting, and subsequent infant functioning, development, and attachment relationships. This study also examined mothers' cognitive adjustments following the birth of high-risk infants. Roughly 59% of the mothers described various benefits arising from the crisis of having a high-risk infant, such as closer family relationships, a "precious" child (from having survived such an ordeal), better perspective on life, and spiritual or emotional growth. Roughly 41% of the mothers saw no benefits as having arisen from the crisis. These mothers were more likely to be depressed and avoidant with their infants.

Given that a wide variety of informational inputs shape maternal thinking and behavior, what infant cues do the parents respond to? Besides obvious physical deformities, what cues might parents use to monitor infant health? How do parents perceive the physical, behavioral, and vocal features of these infants?

Perceptions of High-Risk Infant Cries

The cry is the most conspicuous and critical signal an infant has to demonstrate his or her needs to the caregiver. All infant cries are perceived as aversive, but they normally solicit parental care and intervention. Frodi and her colleagues (Frodi & Lamb, 1980; Frodi, Lamb, Leavitt, Donovan, Neff, & Sherry, 1978) have hypothesized that some preterm infant cries may pass some critical threshold of aversiveness and elicit abusive, rather than nurturant, responses. Several findings support this view. High-risk infants have distinctly more aversive cries than healthy infants (e.g., Frodi et al., 1978; Vuorenski, Lind, Wasz-Hockert, & Partenen, 1971; Zeskind & Lester, 1978). Cry sounds of fetally malnourished newborns (small for date and with a low ponderal index) show a number of acoustic and temporal deviations, such as unusual high pitch, when contrasted with cries of normal newborns (Zeskind & Lester, 1978). Abnormal infants with diverse conditions such as Down's syndrome, cri du chat, neonatal asphyxia, brain damage, hyperbilirubinemia, meningitis, and mixed or unspecified pathologies emit cries distinct from those of normal infants. The cries of abnormal infants can be distinguished from those of normal infants by either spectrograph analyses or the human ear (Vuorenski et al., 1971). Cries of newborns with birth complications were rated by parents and other adults as more urgent, grating, sick, arousing, piercing, discomforting, aversive, and distressful than the cries of low-complication newborns (Frodi et al., 1978; Zeskind & Lester, 1978). These studies suggest that cries of high-risk newborns are perceived as qualitatively different from cries of low-risk normal newborns and that the health status of the newborn is a salient factor influencing adult perceptions of the infant.

Perceptions of Physical Characteristics of High-Risk Infants

Studies using photographs, videos, and composite drawings all demonstrate that preterm and other high-risk infants have less attractive physical features than full-term infants (e.g., Frodi et al., 1978; Maier, Holmes, Slaymaker & Reich, 1984). Interestingly, it is not because the infants are perceived as immature or young; they look older in many ways. They have less "babyish" features (e.g., their foreheads are shorter relative to full-term infants). The actual dimensions of the preterm infant's face when she or he reaches what would have been term, or later, are quite different from full terms. The distances between eyes and ears, nose and mouth, and other dimensions distinguish preterm from full-term infants (Maier et al., 1984).

In addition to finding preterm infants less attractive than full-term infants, parents and nonparents alike readily attribute various negative characteristics to preterm infants regardless of whether they are aware of the infant's birth status. Based on photographs, composite drawings, or videotapes alone, the preterm infant, relative to the full-term, is characterized by raters as less fun, difficult to care for, more irritable, and less predictable. Adults say they are less willing to care for such an infant. In most of these studies, adults readily ascribed negative attributes to preterm infants whom they had never met and to whom they had minimal indirect exposure (e.g., preterm infant cries were played to subjects for 10 seconds). Negative attributions toward preterm infants consistently occurred when raters were unaware of the infant's birth status. However, if told that the infant is preterm, the effect is exacerbated (Maier et al., 1984). Stern and Hildebrandt (1986) assessed mothers' cognitive and behavioral reactions to unfamiliar full-term infants who were labeled as either preterm or full-term. These mothers described the preterm-labeled infants as less cute, less likable, and smaller, and also treated them as less mature! Clearly there are both cognitive and perceptual mechanisms that may influence parental behavior toward these infants. One wonders if the negative attributions that parents or adults make toward preterm infants facilitate neglect or abuse directed toward the infant.

This evidence leads me to suggest the following set of parental care algorithms. First, mothers have a template (or prototype) for normal, healthy infants. This template is sensitive to acoustic, physical, and some behavioral features of infants. This template is probably refined through experience, direct or indirect, with infants. Second, this template, if not matched closely, tells the mother that something is wrong with the infant or, if matched closely, tells the mother that the infant is fine. Third, subsequent decision rules concerning parental care are responsive to relevant environmental inputs. If the template is not closely matched (abnormal infant), I would expect the psychological mechanisms of care allocation to be more sensitive and reactive to social, demographic, and ecological information than if the infant was normal (matched the template). The greater the mismatch between her infant and the template, and the poorer the social and economic conditions, the greater the maternal feelings of despair,detachment, and negative infant attributions (spoiled, fussy, demonic child), and subsequently the more likely she will be to terminate or minimize investment. If the template is not closely matched, but the social and economic conditions are favorable for investment, the greater the maternal feelings of dedication and the more positive the infant attributions (special, precious miracle child), and subsequently the more likely she will be to increase maternal investment.

Given the findings and suggestions just presented, how do mothers behave with their own preterm infants? How can we explain research findings demonstrating that mothers of preterm infants give *more,* not less maternal care? How does the degree of infant illness or risk influence maternal care? Do parents tend to exhibit either high or low (bimodal) but not moderate care toward preterm infants?

WHY STUDY PRETERM TWINS?

Preterm twins provide a "natural experiment" for understanding the effects of infant health status on parental behavior and mechanisms underlying the behavior because one twin is often healthier than the other at birth. This situation presents natural controls for socioeconomic status, maternal age, demographic status, social support, and so on, because we can compare an individual mother's treatment of twins who differ primarily on health status. All other factors are roughly equivalent.

In addition, the mother has two demanding targets for investment, straining her capacity to care adequately for both. Rearing two infants simultaneously is extremely difficult. Nursing and/or carrying twins alone are impossible tasks in many traditional societies; one twin infant is often killed at birth. Usually, either the male or the healthier twin is permitted to live (Daly & Wilson, 1984; Scrimshaw, 1984). Under some circumstances, both twins are killed (as in Aranda, Oceania, and Lozi, Africa; see Daly & Wilson, 1984). Presumably, if the mother attempted to nurse both infants, both would die. In Granzberg's (1973) analysis of the Human Relations Area Files (HRAF), 18 of 70 societies did not permit one or both twins to live. He demonstrates that, in those 18 societies, the heavy maternal workloads that are necessarily endured are likely to limit the mother's ability to successfully rear both twins.

Now add to this scenerio the following facts: In 21 of 35 societies surveyed in the HRAF by Daly and Wilson (1984), infanticide or abandonment of deformed or very ill infants was common. Recall that in societies where infanticide occurs, poor infant health is the most common reason given for killing the infant. Imagine how the birth of high-risk preterm *twins* intensifies the costs of maternal care! These are the conditions under which a mother is sufficiently stressed by the costs of child care that favoritism (i.e., increased parental care) toward the healthier twin would be expected.

Approximately 20% to 40% of extremely low-birth weight (ELBW) infants (less than 1250 g at birth), such as those in the current study, are likely to have major neurodevelopmental handicaps (e.g., cerebral palsy, spastic diplegia, hypotonic quadriplegia; see Britton, Fitzhardinge, & Ashby, 1981; Buckwald, Zorn, & Egan, 1984; Hack, Caron, Rivers, & Fanaroff, 1983). The probability of rearing two normal ELBW twins is small. Western mothers do not have the option of killing, abandoning, or even obviously neglecting their infants without risking severe social sanctions (high costs). Because these mothers are required to provide at least basic care for their infants when they return from the hospital, I predict that mothers will provide equivalent basic care to both infants. However, if mothers are motivated to invest more heavily in the healthier child, then this should manifest itself in their allocation of "discretionary" investment, such as psychosocial investment (talking, affection, play, etc.). Thus, I predict that mothers will direct more psychosocial investment toward the healthier ELBW twin than toward the sicker ELBW twin. In other words, infant health characteristics are predicted to be the primary determinant of maternal preference among high-risk preterm twins.

Hypotheses

Faced with discordant ELBW twins, a mother has three alternative care strategies at a given point in time. The mother may (a) provide equal care to both twins; or (b) invest more in the sicker twin; of (c) invest more in the healthier twin.

The first option is the null hypothesis. If parents do not alter their care patterns as a function of infant health status, then either equal care would be expected or preferential treatment would vary in accordance with other factors (e.g., similarity to parent). As discussed earlier, consideration of reproductive economies where high PI is necessary leads to the prediction that unequal health status is correlated with unequal care.[5]

The second pattern, higher investment in the sicker twin, is expected if mothers simply give more care where it is needed. I would not expect selection pressures to create a maternal psyche that made such simple discriminations and decisions. Mothers who invested intensively in low-viability offspring regardless of the social, economic, and demographic conditions would leave fewer reproductive offspring than those mothers who were more selective in partitioning their investment.[6] Such a strategy would generally not pay in reproductive terms. The second hypothesis suggests that natural selection favored mothers who responded to infant need regardless of the eventual reproductive costs. Such a pattern could not be favored by natural selection unless fertility rates were extremely low (a possibility), or mates were extremely difficult to come by (unlikely).

The third hypothesis, that mothers will direct more care toward the healthy twin, appears to be most consistent with parental investment theory. For reasons already elaborated on, I predict this pattern will occur with high-risk twin populations.

If we assume for the moment that mothers favor the healthier preterm twin, what psychological mechanisms could be responsible for this pattern? In this analysis I will test four alternative, but not mutually exclusive, hypotheses concerning the nature of such mechanisms. Each of these hypotheses will be considered in turn.

1. *The Healthy Baby Hypothesis.* It is predominantly information concerning infant health characteristics that determines maternal preference in twins.
2. *The First Home Hypothesis.* Mothers are likely to favor the twin who arrives home first because that is the infant with whom they initially interact and care for most.
3. *The Fun Baby Hypothesis.* Mothers favor the healthier twin because she or he shows more positive social behaviors and fewer negative behaviors (fussing and crying) than the sicker twin.
4. *The Basic Care Hypothesis.* Both twins are likely to receive adequate maternal care necessary for physical survival (feeding and caregiving).

The Healthy Baby Hypothesis is not mutually exclusive from other psychological explanations of maternal behavior. If healthier children generally smile more, are more fun to be with, and show attachment behaviors earlier, this will certainly contribute to maternal behavior. Clearly, however, previous research on infant crying and physical characteristics has demonstrated that perceptions of high-risk infants are typically negative when contrasted with healthy infants even when those social behavioral cues (smiling, playfulness, etc.) are absent.

The First Home and Fun Baby Hypotheses have been previously proposed by psychologists, neonatologists, social workers, and the like. Previous studies (e.g., Minde, Perotta, & Corter, 1982) have demonstrated that mothers frequently prefer the healthier infant of low-birth-weight preterm twin pairs (measured by self-reports and in-hospital behavioral observations). Mothers commonly report "linkage" to one twin. The typical explanations for this phenomenon are that the healthier infant is more lively and fun to interact with and that it goes home earlier, thus having an early advantage in "bonding" with the mother. Although the healthier infant usually returns home earlier, hospital personnel frequently attempt to discharge both babies at the same time because they are concerned that the mother will "attach" more to the first one home (see Moilanen, 1987). We can test the First Home Hypothesis because twins are frequently discharged from the hospital on the same day and occasionally the healthier infant is retained in the hospital longer for minor procedures. This hypothesis is not exclusive from the first. Rather, each of these hypotheses tells us something about the nature of the mechanisms involved in parental investment decisions. However, while the Healthy and Fun Baby Hypotheses are consistent with the notion that mothers have an adaptation to invest less in high-risk infants under some conditions, the First Home Hypothesis is not. It posits that preference for the healthier baby is merely an artifact of an ordinary bonding mechanism (i.e., one that has no design features specialized for discriminating against high-risk babies).

The Fun Baby Hypothesis was motivated by the finding that premature infants take longer than full-term infants to show typical positive social behaviors such as smiling, laughing, vocal responses, and so on (Bakeman & Brown, 1980). This may be particularly true for sicker premature infants. If, however, the sicker twin is equally socially responsive when compared to the healthier twin, and the mother still prefers the healthier twin, then we must consider other mechanisms.

Alternatively, one might predict that the sicker infant cries more and receives more or less attention on that basis. Crying infants may receive more positive attention because they are crying or may receive less positive attention because they are crying (e.g., mother feeds or gives care to the infant in response to crying, rather than smiling, soothing, playing with the infant, etc.). In addition (or instead) the mother may "like" the healthier infant more because she or he cries less than the other twin, and she may favor the healthier infant for those reasons. I predicted that sicker infants would show fewer positive and more negative behaviors at 4 and 8 months of age relative to their twin siblings.

I predicted that "necessary" ministrations would not vary as a function of infant health characteristics because the mother is "required," by self and others, to perform these duties.[7] At minimum, she is expected to perform these tasks (Basic Care Hypothesis).

I made no specific predictions concerning age effects other than that I expected the results to strengthen at 8 months of age relative to 4 months. For example, I expected mothers to show greater preferences for their healthier ELBW infant relative to the sicker ELBW infant at 8 months than they did at 4 months of infant age. This is because 8-month health status is a better predictor of future outcomes than 4-month health status. I also expected the healthier twins to show more positive behaviors than their siblings at 8 months of age when contrasted with 4 months of age.

In the study I present here, I examine in detail the relationship between infant health status, infant behavior, timing of discharge, and maternal preference. All measures of maternal preference were behavioral, so I will hereafter call the phenomenon maternal behavioral preference.

To summarize, the main questions addressed in this study are the following:

1. Do mothers show behavioral preferences for one twin over the other based on health status? (Healthy Baby Hypothesis)
2. How does timing of discharge affect maternal behavioral preference? (First Home Hypothesis)
3. Do mothers provide adequate "necessary" care (feeding, changing, diapering) to both twins? (Basic Care Hypothesis)
4. How is maternal behavioral preference linked to positive infant behavior? (Fun Baby Hypothesis)
5. How is maternal behavioral preference linked to negative infant behavior? (Fun Baby Hypothesis)

Methods

Subjects

Fourteen ELBW infants (seven pairs) are the focus of this study. They are a subset of a total of 57 ELBW preterm and 62 full-term infants who were enrolled in a large ongoing longitudinal study. All families recruited for the study were living within a 50-mile radius of Ann Arbor, MI. All parents of twin pairs were married and living at home.

All ELBW infants weighed less than 1250 g at birth and received neonatal care at the Holden Intensive Care Unit at the University of Michigan Hospitals. ELBW subjects recruited for the study *did not* have a number of medical conditions and associated risk factors other than prematurity (e.g., multiple congenital anomalies or syndromes, microcephaly, congenital TORCH infections, severe intrauterine growth retardation, or maternal drug addiction). These exclusions were made in order to ensure that the sample represents ELBW infants whose primary medical conditions were associated with low-birth-weight and preterm birth, rather than with some other set of risk factors. For example, infants with congenital TORCH infections are likely to be developmentally disabled because of factors *not* due to low birth weight or prematurity.

The data presented here are but a small subset of a large longitudinal study designed to examine the cognitive, social, and physical outcomes of ELBW infants. Numerous medical, developmental, and laboratory assessments were made. ELBW infants were observed intensively during the first 14 months of life (corrected ages),[8] and periodically thereafter. For the purposes of this study, measurements of health status (medical, neurological, physical, cognitive, and developmental assessments) at birth, discharge, and at 4 and 8 months (corrected ages) are related to home observations of mother-infant interactions at 4 and 8 months of infant age.

Demographic information collected includes infant sex, maternal marital status, maternal age and parity, and socioeconomic status (Hollingshead score).[9] These data are summarized in Table 9.1. Using the medical information collected at each age, a twin was classified as relatively healthier or relatively sicker than his or her twin sibling by a Holden neonatologist who was unaware of the outcomes of the home behavioral observations. Although some infants changed in health status relative to their sibling from discharge to 4 months of age, none of the infants switched in health status from 4 to 8 months. At 8 months, one twin pair could not be discriminated easily on the basis of health status and is therefore left out of the 8-month analyses. Mothers were informed of the results of medical and developmental tests.

Procedure

Behavioral observations were conducted for one hour in the home within approximately 2 weeks of the twins' 4- and 8-month birthdays (corrected ages). Mothers were contacted by phone, the home observations procedure was described, and a time was arranged when (a) the father was not present and, (b) both twins were likely to be awake. Siblings, relatives, or friends may have been present.

It was emphasized during the phone conversation and again just prior to actual observations that we did not expect infants to "perform" or be on their best behavior.

Table 9.1 Demographic and Peripatal Characteristics of Families with ELBW Twins

	Mean	Standard Deviation
Infant		
Male	84.60%	–
Birthweight	1.09 kg	0.12
Gestational Age	28 weeks	1
Hospital Stay	79 days	45
Mother		
Primiparous	57%	–
Maternal Age	26.1 years	2.1
Hollingshead Score	2.5	1.5
Married	100%	–

We were interested in natural infant behavior. Whatever happened during the observation was fine. In addition, the mother was encouraged to go about her daily activities as if the observers were not present. She was not "expected" to interact with her babies any more than she normally would. We focused on the infants' behaviors and not the mother's. Every effort was made to relax mothers as much as possible and to discourage them from putting on a "performance" for the observers. During twin observations, two observers were present. One observer watched each twin for a maximum of 1-hour per visit. For all data used in these analyses, both twins had to be awake at the same time. At 4 months, one pair was not observed because of noncompliance on the part of the mother. At 8 months, a different mother decided to drop out of the study and was not observed. One pair was excluded at 8 months because their relative health status was ambiguous, as mentioned before. Every surviving twin pair that was enrolled in the study was observed at least once in the home. Data are available for six ELBW pairs at 4 months, and for five ELBW pairs at 8 months.

Observers

Because of the nature of the study, observers could not be totally blind to infant status. For example, socioeconomic status is difficult to conceal, given that the observers visited the subjects' homes. Observers were given only the names, addresses, and phone numbers of subjects. All observers (two female undergraduates and the author) were unaware of any specific perinatal complications, degree of infant illness, and duration of hospital stay. Observers were unaware of the relative health status of the twins, although in two cases, it was obvious which twin was doing better. The two student observers were also unaware of specific hypotheses regarding the sample, except that a great deal of variation was expected. Particular care was taken to ensure that each behavior was precisely defined, unambiguous, and easy to identify.

Reliability

Observers had more than 100 hours of training in behavioral observations and coding. This included extensive viewing of videotaped mother-infant interactions and eventually practicing with volunteer mother-infant dyads in the homes. High interobserver reliability (greater than .80 correlation coefficient) was required using live practice subjects across all behavior categories before observers were permitted to collect data on their own. Intermittent reliability checks were conducted throughout the study (over a 3-year period). Reliability was assessed with more than 10% of the infants enrolled in the study. Two behaviors (*infant explore* and *infant look at mother*) were dropped from the analyses because we did not obtain .80 reliability. Interobserver mean correlation coefficients for behaviors were .938 ($SD = .056$).

Equipment and Coding

A TRS-80 Model 100 Radio Shack portable computer with a time-event sampling program was used to collect all behavioral data. Durations and frequencies of infant and maternal behaviors were simultaneously and sequentially recorded in real time. Some behaviors were mutually exclusive (e.g., infant vocalizations and infant crying).

Behavior Categories[10]

Behaviors were categorized on four dimensions: positive maternal behaviors, "necessary" maternal behaviors, positive infant behaviors, and negative infant behaviors.

Negative maternal behaviors (taking toys away and scolding the infant) were too infrequent to include in the analyses.

Positive maternal behaviors included holding, soothing (caressing, rocking, stroking), stimulation (when the infant is not responding), mutual gazing, mutual play (stimulation when the infant is responding), affection (kissing, hugging), vocalizations ("babytalk" to infant), brief arousing events (e.g., presenting a toy to the infant), and proportion of time that mother and infant were in contact. Most, if not all developmental psychologists consider such positive behaviors critical for normal infant psychosocial development. Positive maternal behaviors included all nonnegative maternal behaviors except the minimal activities of feeding and caregiving (see discussion later). These positive behaviors were also considered critical for identifying the mothers' behavioral preferences in order to test the hypothesis that mothers consistently directed more positive behavior toward the healthier ELBW twin than to the sicker twin.

"Necessary" maternal behaviors were caregiving and feeding. These behaviors were necessary for infant physical survival, but none of the positive maternal behaviors necessarily occurred during caregiving and feeding. Caregiving included washing, diapering, brushing, changing, wiping, or adjusting the infant's clothes or position. Feeding included all instances when the infant was actually ingesting liquid or solid food.

Positive infant behaviors included infant vocalizations and responses (to mother) at 4 and 8 months; at 8 months, social initiations (toward mother), positive affect (smiles and laughter), and searching for mother (visually or motorically) were also included. Although I predicted that positive infant behavior would be more common among the healthier twins, I did not expect such behaviors in of themselves to be sufficient to elicit extra maternal care and attention (positive maternal behaviors), although they might be critical for *maintaining* a positive dyadic relationship over time.

Negative infant behaviors included fussing (brief, whiny, non-positive vocalizations) and crying.

Data Reduction and Analyses

Raw data were transferred from the TRS-80 to the Michigan Terminal System for further analyses. Behavioral states (behaviors where duration was measured) were converted into seconds per minute. Behavioral events (behaviors where frequency and time of occurrence were measured) were converted into rates (frequencies per minute). Combinations of simultaneously occurring behaviors were also calculated. Two types of combinations were used for the current analyses.

First, the rate of behavioral events (frequencies) per minute of every state was computed. For example, an infant may vocalize 100 times during a 1-hour observation period, but 30 of those may occur during the 180 seconds of holding. This would mean that the infant vocalized 10 times per minute of holding.

Second, the seconds (or rate) of each state per minute of every other state was computed. For example, the infant may cry for a total of 120 seconds during the observation. For 80 seconds of that crying, the mother may have held her infant. If the mother held her infant for a total of 4 minutes, then the infant cried for 20 seconds per minute of mother holding. Similarly we can reverse the equation. The mother held her infant for 40 seconds per minute of infant crying.

To summarize, each behavior has either a total frequency or duration and either an event-state (frequency per minute of a state) or state-state (seconds per minute of a state) rate with every other behavior.

The Wilcoxon Matched-Pairs Signed-Ranks test was used for all analyses. This test is more powerful than a Sign Test because the magnitude of differences is taken into account.

Results

Question 1: Did Mothers Show Consistent Behavioral Preferences for the Healthier ELBW Twin? (Healthy Baby Hypothesis)

Two types of behavioral preference scores were examined. First, the positive maternal behaviors were summed to achieve a *positive maternal score* for each infant for both the 4- and 8- month observations. The score of the sicker twin was subtracted from the score of the healthier twin. These were then rank-ordered across twin pairs. The Wilcoxon Matched-Pairs Signed-Ranks test was used to determine if mothers consistently directed more positive behaviors toward the healthier twin than toward the sicker twin. Because the sample size was small (five or six pairs per group at each age), all scores would have to be positive for the test to be significant. That is, every healthier twin in the ELBW group would have to have a higher positive maternal score than his or her twin sibling.

Second, maternal behavioral preference was measured across behaviors within each twin pair. This is termed *across-behavior consistency* (ABC). Difference scores are derived for each positive maternal behavior (holding, soothing, stimulation, affection, vocalizing, etc.) by subtracting the rate of the maternal behavior for the sicker twin from the rate of the same maternal behavior for the healthier twin. For example, if the mother kissed the healthier twin two times per minute, but kissed the sicker twin once every four minutes (or .25 times per min.), then the difference score would be 1.75. This was done for the nine positive maternal behaviors and then ranked for each pair. If the mother showed high across-behavior consistency, then she would have kissed, held, soothed, talked to, played with, and gazed at one baby more than the other. If the mother showed such consistency, then we may consider it a strong indicator of behavioral preference for that infant. The positive maternal score tells us something about the *group* patterns of maternal preference. The ABC score tells us something about the *individual* mother's strength of preference.

4 Months. Mothers as a group did not show clear behavioral preferences for the healthier twin at 4 months of age (corrected) as indicated by the positive maternal behavior scores. However, in three out of the six pairs, mothers showed significant (one-tailed, $p<.025$) across-behavior consistency for the positive maternal behaviors. In all three cases, the mothers showed strong behavioral preferences for the healthier twin. In no case did a mother show a behavioral preference for the sicker twin.

8 Months. Mothers as a group did show clear behavioral preferences for the healthier twin at 8 months of age, as reflected in consistently higher positive maternal scores (see Figure 9.1). All of the mothers directed more positive maternal behaviors toward the healthier twin. In two cases out of five, mothers showed significant across-behavior consistency (one-tailed, $p<.025$), and, in both cases, the mothers showed strong

Maternal Behavioral Preference Among 8-mo. Preterm Twins

⊠ Healthier Twin   ■ Sicker Twin

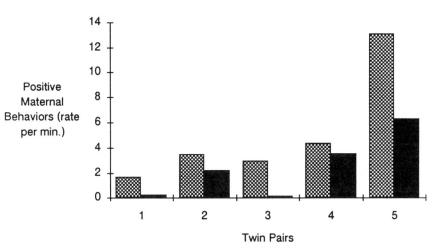

Twin Pairs

p<.05 Wilcoxon Matched Pairs Signed-Ranks Test (one-tailed)

Figure 9.1 As a group, mothers consistently demonstrated more positive maternal behaviors towards their healthier ELBW twin relative to the sicker ELBW twin (p <.05, Wilcoxon Matched Pairs Signed Ranks Test, one-tailed).

behavioral preferences for the healthier twin. One of these twin pairs showed the same ABC pattern at 4 and 8 months. In no cases did mothers show an ABC preference pattern for the sicker twin.

Question 2. How Did Timing of Discharge Affect Maternal Behavioral Preference? (First Home Hypothesis)

In one case, both twins were discharged from the hospital on the same day, yet the mother still demonstrated a behavioral preference for the healthier twin. In three cases, the sicker twin (classified at 4 and 8 months) was discharged earlier than the healthier twin, yet mothers still showed behavioral preferences for the healthier twin by 8 months of infant age. Although the sample size is quite small, these examples suggest that date of discharge was relatively unimportant and that the extra time mothers have with a particular infant does not cause her to prefer that infant.

Question 3. Do Mothers Provide Adequate, "Necessary" Care to Both Twins? (Basic Care Hypothesis)

At 4 and 8 months, mothers provided roughly equal caregiving and feeding to both twins. Mothers sometimes provided more care to the healthier twin, sometimes to the sicker one. There were no detectable differences in caregiving behaviors as a function of infant health status.

Question 4. How was Maternal Behavioral Preference Linked to Infant Behavior? (Fun Baby Hypothesis)

This question is the most complicated and deserves detailed attention. The mother-infant relationship is coactional and reciprocal. The mother does not simply "do" things to the infant; the infant initiates, responds, and helps to direct the relationship. Understanding the nature of mother-infant interactions is critical for understanding how the relationship is established, maintained, or changed over time.

Two main categories of infant behavior, *positive* infant behavior (smiles, responses, vocalizing, social initiations, attachment behaviors [following, searching for mother]) and *negative* infant behavior (fussing and crying), were examined in relation to maternal behavior.

Positive Infant Behavior at 4 Months. This behavior was not related to either infant health status or maternal behavioral preference. In two out of three cases in which there was clear across-behavior consistency (one-tailed, $p<.05$) in maternal preference for the healthier twin, the sicker one showed more positive infant behaviors. In other words, mothers spent more time interacting with the healthier twin, despite the fact that sicker infant was more vocal and responsive to his or her mother. In the three cases where there were no across-behavior consistencies in maternal preference, both twins exhibited equal amounts of positive behavior in one case; the healthier twin exhibited more positive behavior than the sicker one in another case; and the sicker twin exhibited more positive behavior than the healthier one in the third case.

Negative Infant Behavior at 4 Months. This behavior was not related to infant health status but did appear to have some relationship to maternal behavior patterns. The sicker twin was no more likely to cry than the healthier one. However, mothers did respond differently to crying infants than to noncrying infants. In all three cases where the mother showed significant across-behavior consistencies in her preference for the healthier twin, this twin cried and fussed more than the sicker twin. However, the sicker infant cried and fussed more in two of the remaining three cases, yet, even so, the mother did not show ABC preferences for the sicker twin.

In order to examine more closely the question of how negative infant behavior affects maternal behavior, I looked at the rates of positive maternal behavior *during* infant crying and fussing. By examining these behaviors contemporaneously, I can determine whether mothers are more attentive to a distressed twin when it is the healthier one. Difference scores in the rates of maternal holding, soothing, stimulation, mutual gazing, affection, and maternal vocalizations during fussing and crying were obtained for each twin pair. Five pairs were used for this analysis because one pair of twins did not fuss or cry at all. For the 4-month observations, all of the difference scores for the rates of the six positive maternal behaviors per minute of crying were ranked across five twin pairs. This would amount to 30 behaviors (5 pairs $\times$ 6 behaviors) except that in four instances, the mother did not engage in the positive behavior at all. The results indicated that mothers directed more positive maternal behaviors to their crying healthier baby than to their crying sicker baby ($N=26$ behaviors; one-tailed, $p<.025$). Additionally, in all five pairs, mothers had more eye contact with the healthier than with the sicker infant during crying (one-tailed; $p<.05$). It appears a mother is more attentive to her distressed healthier ELBW infant than to her distressed sicker ELBW infant.

Positive Infant Behavior at 8 Months. This behavior was not related to either infant health status or to maternal behavioral preference. The healthier twin showed no more positive behavior than the sicker one, and positive maternal behavior did not correlate with positive infant behavior. Again, in two out of three cases in which there was significant across-behavior consistency in maternal preference for the healthier twin, the sicker twin showed more positive infant behaviors than the healthier one. Again, in the cases where there were no significant across-behavior consistencies, positive infant behavior varied. As previously demonstrated (see Figure 9.1), mothers showed overall behavioral preferences for the healthier twin. To summarize the 8-month data, mothers directed on average 73.02% of their positive behavior toward the healthier twin ($SD = 16.4$), even though the healthier one was responsible for only 43.83% of all positive infant behavior ($SD = 21.28$). Caregiving and feeding were roughly equally distributed between the two twins (55.00% directed to the healthier twin; $SD = 27.65$).

Negative Infant Behavior at 8 Months. This behavior was rare. However, 67.04% of all fussing and crying came from the healthier twin ($SD = 38.52$). These negative infant behaviors may significantly affect maternal behavior, but there were too few twin pairs where both twins cried to make comparisons similar to those at 4 months. However, in all cases where the mothers demonstrated significant across-behavior consistency, the healthier twin cried more than the sicker twin. In all cases where mothers did not demonstrate significant across-behavior consistency, the sicker twin cried more than the healthier one.

Discussion

Although the data set is small, some intriguing results have emerged. Maternal preference patterns were clearly linked to infant health status at 8 months. At 4 months, the same preference patterns were evident only through close examination of maternal responses to infant crying. Although some mothers showed across-behavior consistency in their preferences for the healthier twin at both 4 and 8 months, by 8 months, the overall group pattern was clear. All mothers, by 8 months of infant age, directed more positive maternal behavior to the healthier twin. In no cases at either age did mothers show positive across-behavior consistency or preference for the sicker twin. The Healthy Baby Hypothesis was supported.

Date of hospital discharge did not appear to have any impact on the mother's behavioral preferences. Mothers did not show any tendencies to prefer the infant who came home first. The First Home Hypothesis was not supported.

Positive infant behavior was not related to positive maternal behavior. It did not seem to matter which infant cooed at, smiled, or followed mother more often. The hypothesis that mothers favor the healthy baby because he is the "fun" baby is not supported by these data. It could be that positive infant behavior is not a reliable health cue relative to other infant features (acoustic, physical, or other behaviors). These findings conflict with the widespread assumption that mothers favor the "fun" baby.

In fact, the opposite appeared to be true. Mothers paid more attention to the crying or fussing infant than to the smiling one. This was particularly true if the crying infant was also the healthier one. At 8 months the healthier twin received more attention regardless of his behavior, but received significantly more attention (across-behavior consistency) than his twin sibling if he cried. If the sicker infant cried more than the healthier one, mothers still paid more attention to the healthier infant, but not significantly more (no ABC preference pattern.) These data support a model in which two independent factors influence maternal behavior: (a) Mothers prefer the healthier twin, and (b) mothers respond to crying infants.

How do these data fit with laboratory studies of responses to high-risk infant cries? How do mothers perceive their ELBW twins? Recall that laboratory studies of premature infant cries found that adults and parents of preterms perceived the cries as aversive, grating, urgent, sick, etc. They are like a supernormal-stimulus. Mothers use these cues to decide how to interact and respond to their infant. I suggest that the abnormal infant's cry may elicit parental care when circumstances are relatively good (as they typically were in this study) and neglect under poor circumstances. As Frodi and Lamb (1980) proposed, the aversiveness of the cry may reach some critical threshold, as perceived by the parent, to elicit a negative rather than positive parental response. However, under the relatively favorable social and economic conditions that many Western mothers experience, mothers may actually perceive abnormal cries as more needy and urgent, thereby increasing their positive responses to the infant.

In this study, mothers responded to twin cries reliably. In preliminary analyses of singleton infant data, we have found that mothers respond on average as quickly and as appropriately to ELBW premature infant cries as to full-term infant cries, although the variance in the ELBW group is significantly greater (Mann, Plunkett, Meisels, & Bozunski, 1988). However, in this sample, mothers were more responsive to the crying healthier twin than to the crying sicker twin. As early as 4 months, it appeared as if mothers would neglect the sicker twin if the healthier twin cried, but they would *not* neglect the healthier infant if the sicker one cried. At 8 months, in all cases where mothers showed strong preferences (significant across-behavior consistency), the healthier twin cried more than the sicker one. In all cases where mothers showed a weak preference for the healthier twin (more positive behaviors directed to the healthier one overall, but no significant across-behavior consistency), the sicker infant cried much more than his twin sibling. Mothers certainly feel obligated to respond to all infant cries, but clearly they respond more to the healthier ELBW infant than to the sicker ELBW infant. It is as if the sicker twin would have to demonstrate greater need by crying more than the healthier one in order to solicit maternal attention roughly equal to that of his sibling.

The presence of observers may affect the data, but these effects are likely to reduce, not exaggerate, the results. If mothers were "performing" for observers, one would expect them to devote care and attention equally to both twins. This should lessen the observed differences between the twins. Furthermore, mothers generally interacted with both infants at relatively low rates, which is typical of home observations as opposed to laboratory observations (note Figure 9.1; see Belsky, 1980, for lab versus home comparisons). Thus if mothers were "performing," they were not doing a very good job. Although mothers may feel more obligated to respond to infant cries in our presence, this does not explain the tendency for mothers to respond more to crying by the healthier twin than to crying by the sicker twin. It is conceivable, however, that mothers would respond far less to cries of one or both twins if observers were not present.

The overall pattern of behavioral preference is clear by 8 months. However, preference tendencies are also evident at 4 months when one examines maternal responses to infant crying. Perhaps the overall pattern of maternal preference (higher positive maternal behavior scores for the healthier twin across the group) is not apparent at 4 months because mothers have not clearly formulated their preferences. One plausible explanation for this pattern is that mothers are monitoring infant health status closely during the early stages of infancy (see Daly & Wilson, 1988). Health risks are difficult to detect in early infancy. Indeed, in two cases, relative health status switched for twins from discharge to 4 months, but not from 4 to 8 months. If the mother is attempting to assess infant quality, then it might be best to wait until she has sufficient evidence to go on.

Initially a mother may stimulate her preterm infant intensively, but if her efforts are not rewarded, she may begin to disengage somewhat from the infant. If the infant continues to respond to the mother, the mother may maintain high investment for as long as necessary (until the child appears normal). Research by Goldberg and Marcovitch (1986) supports this explanation. In the three Western cultures they reviewed, mothers became more responsive and active with their premature infants as the infant's condition improved. Similar patterns are reported by them for Down's syndrome infants. Once a cycle of responses is set in motion, it will become more difficult for the mother to shift her interactive strategy (e.g., see Cohn & Tronick, 1983). By contrast, for example, if the mother ignores her infant's cries most of the time, the infant may cry *more,* not less, over time (Bell & Ainsworth, 1972), hence reinforcing the mother's negative attribution that the child is "spoiled." This conflict of interest may escalate until the mother feels she has no control over her infant, and vice versa. Mother and infant may begin to disengage from one another. This helps to explain how such patterns can begin and be maintained.

In the United States, where infant mortality is low and subsequently cultural expectations for healthy offspring are high, mothers may be psychologically prepared to invest in their infants long before they are actually born. Mothers who are anticipating good pregnancy outcomes (evident by extensive prenatal preparations, early decisions to name the infant, etc.) may be more prone to invest highly in preterm infants. In contrast, women who are economically impoverished have more to lose by a high-investment strategy and may be more prone to neglect the sicker child (e.g., Brazilian shantytown women; Scheper-Hughes, 1985).

In five of the twin pairs, mothers seemed to be providing adequate basic care (feeding and caregiving) to both infants. In two cases, the sicker ELBW twin appeared to be severely neglected. Both of these sicker twins fell at the bottom of the growth and weight charts. In both of these cases, mothers were extremely resource poor and fell into the lowest socioeconomic classes of our sample. Even when the social sanctions are severe, mothers may neglect one twin.

Differences in maternal behavior between mothers were greater than differences in maternal behavior within twin pairs. Mothers who were actively involved with the healthier twin were also active with the sicker twin, although typically less so. If mothers were ranked at 8 months of infant age according to the amount of positive behavior they directed towards their infants, the rankings would be identical for both the healthier and the sicker twin groups. In other words, the mother who interacted more with the healthier twin than any of the other mothers interacted with their healthier twins, also interacted with her sicker twin more than any of the other mothers interacted with their sicker twins. Mothers who were most involved with their twins in this study had the highest socioeconomic status (highest Hollingshead scores). What is clear, however, is that despite differing socioeconomic and demographic backgrounds, mothers demonstrated distinct behavioral preferences for their healthier twin infant. I predicted this difference because the intense stress of caring for two very sick prematures is severe enough to cause even an economically advantaged mother to favor her healthier infant. In such circumstances, mothers cannot do enough to adequately care for both without suffering tremendous costs.

Under less dire circumstances, when the mother has the means and potential to raise two normal offspring, such as when twins are born less premature and are generally healthier, I predict that mothers may favor the sicker twin instead. If one twin is doing very well and requires less care, it may pay (in reproductive terms) for the mother to turn her attentions to the sicker twin. In these conditions, economic, social, and demographic status may play a role. As infant health status improves, the cost-benefit equation shifts. When infant health status is relatively good, I would predict that mothers would have more accurate perceptions of their infants' needs and behaviors than when their infants are much sicker (as in the current study). These hypotheses can be tested with healthier samples of twins.

It is clear that infant health status influences mother-infant interactions in ways consistent with parental investment theory. I predicted that infant health would be a particularly salient feature for mothers because of the obvious consequences for reproductive fitness. Ask any pregnant woman what is her greatest wish concerning the fetus and she will probably say "that he develops normally."

Mothers of twins may unknowingly be directing more psychosocial investment toward the healthier twin. The expected outcomes of such investment are likely to include greater social benefits for the healthier twin relative to his or her twin sibling. This may increase his or her opportunities in a wide variety of social relationships. Increased maternal psychosocial investment may contribute to these outcomes, providing early relative advantages for one infant, but this does not mean the pattern is fixed, determined, or irreversible. Because the sicker twin may still receive adequate care (although less than his sibling), the long-term consequences are not yet obvious.

I proposed that mothers have an infant template they use to assess infant normality. Psychological mechanisms such as negative perceptions of sick infants' behavior imply that selection pressures have operated on our ability to discriminate and react to abnormal infants. Mothers are probably using infant physical and acoustic cues to assess infant health status (in addition to information provided by medical personnel). These cues are probably more reliable indicators of infant health status than strictly positive infant behaviors. However, the decision rules and processes by which a mother's thoughts, beliefs, and values affect her parental behavior are not apparent. It is not clear, for example, what psychological coping strategies enable mothers to replace negative perceptions of infants with positive perceptions. Although I have focused on how mothers pay less attention to the sicker twin, mothers may also invest highly in the healthier twin. In addition, mothers of singleton infants can become quite attentive and loving toward their high-risk child under some circumstances. Western mothers frequently work intensively with their premature infants and often have quite positive views of the child. Some mothers spontaneously told me that their infant was a "miracle child." Positive maternal attitudes and behavior directed at the high-risk infant would be likely to result in positive feedback from the infant (enhanced social, cognitive, and physical development), thus encouraging mothers to continue engaging and working with the infant. Initially mothers may necessarily distort their view of the high-risk infant in order to devote so much maternal care. Such devotion might be too emotionally difficult (or impossible) if mothers had more realistic appraisals of the child's prospects.

Whether differences in maternal behavior with twins reflect differences in maternal perceptions remains to be seen. As medical technologies improve and more preterm infants survive, the factors affecting parental psychology become even more critical. Parents are being pressured to care for infants that never survived in the past, and while some are thankful for such a miracle, others suffer enormous costs. In the future, the dilemma to nurture or neglect these infants is one that increasing numbers of mothers will have to face.

ACKNOWLEDGMENTS

I would like to thank Barbara Smuts, Leda Cosmides, John Tooby, Jerome Barkow, Richard Connor, and Andrew Richards for editorial and conceptual comments. Financial support was provided to the author by the Jacob Javits Congressional Award and the Evolution and Human Behavior Program at the University of Michigan. I am indebted to Lisa Chiodo and Kate Mason for their research assistance. Finally, I would like to thank those families who participated in the ELBW study and all members of the ELBW project. In particular, I would like to acknowledge the generous support of my colleagues on the ELBW project: Samuel Meisels, James Plunkett, Sarah Mangelsdorf, Mary Bozynski, Margo Dichtelmiller, and Carol Claflin.

NOTES

1. For the purposes of this study, I will focus on maternal care and attention directed toward extremely low-birth-weight, high-risk preterm infants. This discussion will be restricted to the infancy period because that is when the child is most vulnerable to both neglect or infanticide (Daly & Wilson, 1987; Johansson, 1987; Lauer, TenBroeck, & Grossman, 1975; National Center on Child Abuse and Neglect, 1981) and when the mother can most effectively ameliorate her child's condition through intensive care (e.g., see Beckwith & Parmelee, 1986; Sameroff & Chandler, 1975).

2. Maternal care will be the focus of this investigation because of its direct and profound impact on infant development. Paternal care is certainly relevant (even crucial), but cannot be adequately addressed within the limitations of this paper.

3. I do not mean to imply intention and consciousness when discussing parental reproductive interests by using terms such as decisions, pursue, choices, dilemma, value, interests, etc.

4. Variables affecting the mother's inclusive fitness (i.e., relatives and their offspring) are also important but not relevant within the limitations of this paper.

5. However, consistent with this reasoning, under some circumstances (high fertility rate, insufficient resources, etc.), the parent may also neglect both twins equally and direct more care to other children. This last hypothesis, that parents neglect both twins, is not adequately testable with the current data set because one should be able to demonstrate that the parents are giving less care to both twins than they would to healthier and/or singleton children.

6. This is necessarily the case because mothers who cut their reproductive losses sooner would (all else being equal) have another offspring sooner than mothers who continued to invest in sick offspring.

7. At minimum, I would hypothesize that a mother's internal schema for "parenting" has these features, "clean, clothe, and feed the infant" as its skeletal base. Not performing these minimal tasks would be cognitively dissonant with the mother's self-definition as a caregiver. Cultural influences must play a very strong role in the development of such a schema, but will not be addressed in detail here.

8. *Corrected age* is the age (in months) from when the infant would have been term. If an infant is born 3 months premature, then he is considered to be one month old, corrected age, four months after birth.

9. Hollingshead scores are standardized combined measures of income and education. A score of one is the highest socioeconomic class. The lowest socioeconomic score is six.

10. Details of the coding system, Mann's Observational Coding System for Parent-Infant Interactions (MOCSPI), can be obtained from the author.

REFERENCES

Affleck, G., Tennen, H., & Gernshman, K. (1985). Cognitive adaptations to high-risk infants: The search for master, meaning and protection from future harm. *American Journal of Mental Deficiency, 89*(6), 653–656.

Altmann, J. (1980). *Baboon mothers and infants.* Cambridge, MA: Harvard University Press.

Bakeman, R. & Brown, J. (1980). Early interaction: Consequences for social and mental development at 3 years. *Child Development, 51,* 437–447.

Beckwith, L. & Parmelee, A. H. (1986). EEG patterns of preterm infants, home environment and later IQ. *Child Development, 57,* 77–79.

Bell, R. Q., & Ainsworth, M. D. S. (1972). Infant crying and maternal responsiveness. *Child Development, 43,* 1171–1190.

Belsky, J. (1980). Mother-infant interaction at home and in the laboratory: A comparative study. *Journal of Genetic Psychology, 137,* 37–47.

Berkson, G. (1973). Social responses to abnormal infant monkeys. *American Journal of Physical Anthropology, 38,* 583–586.

Britton, S. B., Fitzhardinge, P. M., & Ashby, S. (1981). Is intensive care justified for infants weighing less than 801 gm at birth? *The Journal of Pediatrics, 99,* 937–943.

Buckwald, S., Zorn, W. A., & Egan, E. A. (1984). Mortality and follow-up data for neonates weighing 500 to 800 g at birth. *American Journal of Diseases of Children, 138,* 779–782.

Bugos, P. E., Jr., & McCarthy, L. M. (1984). Ayoreo infanticide: A case study. In G. Hausfater & S. B. Hrdy (Eds.), *Infanticide: Comparative and evolutionary perspectives* (pp. 503–520) New York: Aldine de Gruyter.

Burgess, R. L., & Conger, R. D. (1978). Family interaction in abusive, neglectful and normal families. *Child Development, 49,* 1163–1173.

Burgess, R. L., & Garbarino, J. (1983). Doing what comes naturally? An evolutionary perspective on child abuse. In D. Finkelhor, R. Gelles, M. Straus, & G. Hotaling (Eds.), *The dark side of the family: Current family violence research* (pp. 88–101). Beverly Hills, CA: Sage.

Chapman, C., & Chapman, L. J. (1987). Social responses to the traumatic injury of a juvenile Spider monkey *(Ateles geoffroyi). Primates, 28*(2), 271–275.

Cohn, J. F., & Tronick, E. Z. (1983). Three-month-old infants' reaction to simulated maternal depression. *Child Development, 54,* 185–193.

Daly, M. & Wilson, M. (1984). A sociobiological analysis of human infanticide. In G. Hausfater & S. B. Hrdy (Eds.), *Infanticide: Comparative and evolutionary perspectives* (pp. 487–502). New York: Aldine de Gruyter.

Daly, M., & Wilson, M. (1987). Children as homicide victims. In R. J. Gelles & J. B. Lancaster (Eds.), *Child abuse and neglect: Biosocial dimensions* (pp. 201–214). New York: Aldine.

Daly, M., & Wilson, M. (1988). *Homicide.* New York: Aldine de Gruyter.

Draper, P. (1989). African marriage systems: Perspectives from evolutionary ecology. *Ethology and Sociobiology, 10,* 145–169.

Fedigan, L. M., & Fedigan, L. (1977). The social development of a handicapped infant in a free-living troop of Japanese monkeys. In S. Chevalier-Skolnikoff & F. E. Poirier (Eds.), *Primate bio-social development* (pp. 205–222). New York: Garland Publishing.

Field, T. M. (1977). Effects of early separation, interactive deficits and experimental manipulations on infant-mother face-to-face interactions. *Child Development, 48,* 783–771.

Friedrich, W. H., & Boriskin, J. A. (1976). The role of the child in abuse: A review of the literature. *American Journal of Orthopsychiatry, 46,* 580–590.

Frodi, A. M., & Lamb, M. E. (1980). Child abusers' responses to infant smiles and cries. *Child Development, 51,* 238–241.

Frodi, A. M., Lamb, M. E., Leavitt, L. A., Donovan, C. N., Neff, C., & Sherry, D. (1978). Fathers' and mothers' responses to the faces and cries of normal and premature infants. *Developmental Psychology, 14*(5), 40–49.

Furuya, Y. (1966). On the malformation occurred in the Gagysuan troop of wild Japanese monkeys. *Primates, 7,* 488–492.

Gelles, R. J., & Lancaster, J. B. (Eds). (1987) *Child abuse and neglect: Biosocial dimensions.* New York: Aldine.

Goldberg, S., & Divitto, B. A. (1983). *Born too soon: Preterm birth and early development.* San Francisco: W. C. Freeman.

Goldberg, S., & Marcovitch, S. (1986). Nurturing under stress: The care of preterm infants and developmentally delayed preschoolers. In A. Fogel & F. Melson, (Eds.), *Origins of nurturance.* Hillsdale, NJ: Lawrence Erlbaum Associates.

Granzberg, G. (1973). Twin infanticide: A cross-cultural test of a materialistic explanation. *Ethos, 1,* 405–412.

Hack, M. H., Caron, B., Rivers, A., & Fanaroff, A. A. (1983). The very low birth weight infant: The broader spectrum of morbidity during infancy and early childhood. *Developmental and Behavioral Pediatrics, 4,* 243–248.

Johansson, S. R. (1987). Neglect, abuse and avoidable death: Parental investment and the mortality of infants and children in the European tradition. In R. J. Gelles & J. B. Lancaster (Eds.), *Child abuse and neglect: Biosocial dimensions* (pp. 57–96). New York: Aldine.

Korbin, J. E. (1987). Child maltreatment in cross-cultural perspective: Vulnerable children and circumstances. In R. J. Gelles & J. B. Lancaster (Eds.), *Child abuse and neglect: Biosocial dimensions* (pp. 31–56). New York: Aldine de Gruyter.

Lauer, B., TenBroeck, E., & Grossman, M. (1975). Battered child syndrome: Review of 130 patients with controls. *Pediatrics, 54,* 67–70.

Maier, R. A., Jr., Holmes, D. L., Slaymaker, F. L., & Reich, J. N. (1984). The perceived attractiveness of preterm infants. *Infant behavior and development, 7,* 403–414.

Mann, J., Plunkett, J. W., Meisels, S. J., & Bozynski, M. (April, 1988). *Home observations of extremely-low-birth-weight preterm infants.* Paper presented at the Sixth Biennial International Conference on Infant Studies, Washington, DC.

Minde, K. K., Perotta, M., & Corter, C. (1982). The effect of neonatal complications in same-sexed premature twins on their mothers' preference. *Journal of the American Academy of Child Psychiatry, 21,*(5), 446–452.

Moilanen, I. (1987). Dominance and submissiveness between twins: Perinatal and developmental aspects. *Acta Geneticae Medicae et Gemellologiae Twin Research, 36*(2), 249–255.

National Center on Child Abuse and Neglect. (1981). National study of the incidence and severity of child abuse and neglect. Washington, DC: Department of Health and Human Services.

Sameroff, A., & Chandler, M. T. (1975). Reproductive risk and the continuum of caretaking casualty. In F. D. Horowitz, M. Hetherington, S. Scarr-Salapetek, & G. Sieger (Eds.), *Review of child development research* (Vol. 4). Chicago: University of Chicago Press.

Scheper-Hughes, N. (1985). Culture, scarcity and maternal thinking: Maternal detachment and infant survival in a Brazilian shantytown. *Ethos, 13*(4), 291–317.

Scrimshaw, S. C. (1984). Infanticide in human populations: Societal and individual concerns. In G. Hausfater & S. B. Hrdy (Eds.), *Infanticide: Comparative and evolutionary perspectives* (pp. 439–462). New York: Aldine.

Stern, M., & Hildebrandt, K. A. (1986). Prematurity stereotyping: Effects on mother-infant interaction. *Child Development, 57,* 308–315.

Trivers, R. L. (1972). Parental investment and sexual selection. In B. Campbell (Ed.), *Sexual selection and descent of Man, 1871–1971* (pp. 136–179). Chicago: Aldine.

Trivers, R. L. (1974). Parent-offspring conflict. *American Zoologist, 14,* 249–264.

Vuorenski, B., Lind, J., Wasz-Hockert, O., & Partenen, T. (1971). Cry score: A method for evaluating the degree of abnormality in the pain cry response of the newborn young infant. *Quarterly Progress and Status Report,* (April). (Stockholm, Sweden: Speech Transmission Laboratory, Royal Institute of Technology.)

Wasserman, G. A., Allen, R., & Soloman, C. R. (1985). At-risk toddlers and their mothers: The special case of physical handicap. *Child Development, 56,* 73–83.

Zeskind, P. S., & Lester, B. M. (1978). Acoustic features and auditory perceptions of the cries of newborns with prenatal and perinatal complications. *Child Development, 49*(3), 580–589.

10

Human Maternal Vocalizations to Infants as Biologically Relevant Signals: An Evolutionary Perspective

ANNE FERNALD

When talking to infants, human mothers use vocal patterns that are unusual by the standards of normal conversation. Mothers, as well as fathers and adults who are not parents, speak consistently more slowly and with higher pitch when interacting with infants, in smooth, exaggerated intonation contours quite unlike the choppy and rapid-fire speech patterns used when addressing adults. To praise an infant, mothers typically use wide-range pitch contours with a rise-fall pattern. To elicit an infant's attention, they also use wide-range contours, but often ending with rising pitch. When soothing an infant, mothers tend to use long, smooth, falling pitch contours, in marked contrast to the short, sharp intonation patterns used in warning or disapproval. This use of exaggerated and stereotyped vocal patterns in mothers' speech to infants has been observed in numerous European, Asian, and African cultures and appears to be a universal human parental behavior.

Why do mothers make such sounds to infants? With other widespread human maternal behaviors such as nursing hungry infants and rocking distressed infants, the question "Why?" yields satisfying answers on both the proximal and ultimate levels specified by Tinbergen (1963). Nursing and rocking are understandable in terms of proximal physiological mechanisms functioning to provide nourishment and to regulate infant arousal, as well as in terms of species-typical parental strategies contributing to the survival and fitness of the offspring. Although hypotheses about the evolutionary history of particular human behaviors can always be disputed, the plausibility of ultimate explanations for such maternal actions as nursing and rocking is strengthened by compelling evidence for phylogenetic continuity. All mammals nurse their young, and almost all nonhuman primates carry their young, providing tactile and vestibular input known to have important short- and long-term regulatory effects on primate infant behavior. But ape and monkey mothers do not communicate vocally with their infants in ways that bear obvious resemblance to the melodic infant-directed speech of human mothers. Is there evidence that infant-directed speech, like nursing and rocking, is a human maternal behavior that has evolved to serve specific functions?

In this paper, I argue that the characteristic vocal melodies of human mothers' speech to infants are biologically relevant signals that have been shaped by natural selection. In the first section, I discuss the current debate about the status of human language as an evolved mechanism, as a way of examining criteria for when it is appropriate to invoke natural selection as a causal explanation for the evolution of human behavior. In particular, I examine the antiadaptationist argument that language is an "exaptation" and therefore not a product of selection, as well as the argument that adaptive features must be designed for optimal efficiency. In the next two sections, the characteristics of mothers' speech to infants are described in some detail, along with research on the communicative functions of intonation in infant-directed speech. In the final sections, I examine infant-directed speech in the context of ethological research on vocal communication and maternal behavior in nonhuman primates, in order to identify selection pressures relevant to understanding the adaptive functions of human maternal speech.

THE CURRENT DEBATE ON HUMAN LANGUAGE AND NATURAL SELECTION

The question of whether language is a product of natural selection is a subject of lively current debate among philosophers and psycholinguists armed with arguments and counterarguments from evolutionary biology. Given that we have no fossil record to document the origins and subsequent evolution of the language faculty, or any other complex behavioral capability, this question is likely never to be resolved.[1] However, the current debate has served to sharpen the focus on a number of issues fundamental to an understanding of human behavior within an evolutionary framework. This debate will be discussed in some detail here, since it is relevant to a central question in this paper. What kinds of evidence are needed to support the claim that a given human behavior reflects the influence of Darwinian natural selection? This lengthy preamble is necessary to set the stage for the subsequent discussion of the adaptive functions of human maternal speech.

Although officially banned by the Parisian Société de Linguistique in 1888, speculations on the origins of language became no less abundant or diverse as a result of this decree. In 1971, Hewes published a bibliography containing 2,600 titles of scholarly works on the topic, to which he had added another 2,500 references two years later (Hewes, 1973). Although controversy about the origins of language indeed goes back a long way, the current debate has been prompted by recent reformulations of neo-Darwinian theory within the field of biology. In two widely cited papers, Lewontin (1978) and Gould and Lewontin (1979) object to naive and inappropriate uses of the "adaptationist paradigm," which attempt to account for all characteristics of living organisms as optimal solutions to adaptive problems. Lewontin's (1979) objections, in particular, focus on the excesses of sociobiological theorizing in recent years. While sounding a valuable and timely warning about the dangers of superficial applications of evolutionary principles to the study of human behavior, these critiques of "naive adaptationism" have had an unfortunate backlash effect, contributing to widespread skepticism about the legitimacy of adaptationist reasoning at any level in relation to language and other human behavior.

Central to the antiadaptationist argument are recent discussions by Gould and his colleagues about the role of adaptation in the emergence of evolutionary novelty. By drawing a comparison with the "spandrels" of St. Marcos, the triangular spaces formed by the intersection of arches supporting the dome of the cathedral, Gould and

Lewontin (1979) make the point that apparently functional design elements can result entirely from architectural constraints. The art historian who concluded that these spandrels were the starting point, rather than a by-product, of the design would have made a serious mistake. By analogy, the conclusion that all biological forms have evolved as functional adaptations to local conditions also fails to take the nature of architectural constraints into account. In a later paper, such structural by-products are called "exaptations," defined as "characters, evolved for other usages (or for no function at all), and later 'coopted' for their current role" (Gould & Vrba, 1982, p. 6). As an example, Gould and Vrba cite Darwin's reference to unfused sutures in the skulls of young mammals, often viewed as an adaptation for facilitating passage of the neonatal skull through the birth canal. As Darwin points out, however, young birds and reptiles, which emerge from shells rather than through birth canals, also have cranial sutures, suggesting that this structure "has arisen from the laws of growth, and has been taken advantage of in the parturition of the higher animals" (Darwin, 1859). Although these unfused sutures may be vitally necessary in mammalian parturition, they are regarded as exaptations rather than adaptations, since they are not explicitly designed by natural selection to serve their present function.

The claim that language is an exaptation, a side effect of the evolution of other cognitive faculties in humans, is a central point of contention in the current debate on the evolutionary status of language. Chomsky (e.g., 1972) and Piattelli-Palmarini (1989) argue that language emerged as a qualitatively new capacity in humans and may have arisen from nonadaptationist mechanisms. Such processes, which can act as alternatives to direct selection for the establishment of particular characteristics, include genetic drift, allometry, and pleiotropic gene action (Gould & Lewontin, 1979). The fact that the design of language is in many respects arbitrary and "non-optimal" should count as further evidence for its nonadaptive origins, according to Piattelli-Palmarini. Objecting to the claim that language is an exaptation rather than an adaptation, Pinker and Bloom (1990/this volume) argue that grammar is an exquisitely complex, task-specific mechanism, not unlike other complex behavioral capabilities such as stereopsis and echolocation, which have clearly been shaped by natural selection. Two of the many interesting arguments offered by Pinker and Bloom in favor of viewing human language as a product of natural selection are particularly relevant to the concerns of this paper: first, that exaptations do not represent true alternatives to natural selection, but are in fact consistent with neo-Darwinian theory; and second, that the arbitrariness of language should not be interpreted as evidence against an adaptationist explanation.

Exaptations Are Not Inconsistent With Natural Selection

Gould and Vrba (1982) introduced the term *exaptation* in order to clarify confusion surrounding the concept of adaptation. Some evolutionary biologists designate a feature as adaptive only if it was built by natural selection to perform its current function (e.g., Williams, 1966), while others prefer a broader definition that includes any feature currently enhancing fitness, regardless of its origin (e.g., Bock, 1980). Gould and Vrba object to this broad definition of adaptiveness, on the grounds that it conflates current utility with historical genesis and neglects features that are now useful but that may have evolved originally to serve other functions. The concept of exaptation was intended to fill this gap in the terminology of evolutionary biology and to highlight the neglected role of nonadaptive aspects of form in constraining evolution. By Gould and Vrba's definition, exaptations include new roles performed by old features, as in the mammalian cranial sutures noted by Darwin, as well as features that have become specialized to perform roles very different from the ones for which they originally evolved. As an example of this latter case, feathers were initially selected for insulation and not for flight. Legs in terrestrial animals are also exaptations rather than adaptations, since they evolved from lobed fins, which were used for swimming rather than for walking.

Gould and Vrba (1982) are careful to point out that the concept of exaptation is not antiselectionist. On the contrary, they argue that their main theme is "cooptability for fitness," the process by which exaptations originate as nonadaptations, or as adaptations for another function, and then are refined by subsequent selection to perform a new role. While the limbs of salamanders, elephants, and humans indeed all arose from primordial fins, selective pressures, in the context of architectural constraints, have operated over millennia to adapt them further for efficiency in their current roles. Similarly, in the case of mammalian cranial sutures, a feature that evolved originally to accommodate skull growth in birds and reptiles became an exaptation for parturition in mammals. However, the design features of these sutures in mammalian skulls have presumably undergone complementary adaptation with features of the mammalian pelvic cavity, since otherwise parturition would be impossible. Since the evolution of all complex structures is fundamentally constrained by the characteristics of previously existing structures, as with fins and legs, the concept of exaptation does not represent a radical departure from the orthodox Darwinian account of adaptation. It could even be argued that all adaptations are in some sense exaptations, since the current functions of adaptive features inevitably differ to some extent from those of their Pliocene precursors.

Given the essential compatibility of the exaptation concept with traditional thinking about natural selection, it is surprising that it has been billed as central to a "new theory of evolution" by Piattelli-Palmarini (1989) and others. This allegedly new and better theory assumes "that full-blown evolutionary novelty can also suddenly arise, so to speak, for no reason, because novelty caused by sheer proximity between genes is not governed by function and it, therefore, eludes strict adaptationism" (Piattelli-Palmarini, 1989, p. 8). Pinker and Bloom argue that this account reflects a misunderstanding of biological principles. While novelty can indeed arise from nonadaptive features, it is utterly improbable that full-blown functional complexity will arise de novo, given that the design of complex biological structures necessarily derives from stepwise incremental modification of previously existing structures (e.g., Dawkins, 1986). The precursors to complex biological structures may emerge as exaptations, but it is only through subsequent natural selection that these structures achieve the complexity to perform their highly specialized roles.

Thus, even if the origin of linguistic competence in hominids lies in faculties initially serving other social and cognitive functions, human language has undoubtedly undergone subsequent adaptation for its specialized functions. Mammals may owe their legs to fins designed for swimming, not for walking, but the transitional evolutionary status of terrestrial fins as an exaptation obviously did not exempt mammalian limbs from the further influence of natural selection. This is an important point that has been lost in the antiadaptationist arguments about language and selection. The earliest linguistic competence may indeed have been an exaptation, but one that was eventually "coopted for fitness," in Gould and Vrba's words. As such, we can acknowledge both its possible exaptive origins and its refinement by natural selection for its current specialized function. Piattelli-Palmarini recognizes that exaptations may "later on, happen to acquire some adaptive value," such that "selection for" takes place "on top of" a previous feature that emerged through nonadaptive processes (1989, p. 11). And yet he later emphatically asserts that "adaptive constraints have no role to play in a scientific approach to language and thought" (1989, p. 20), as if the acknowledgment of exaptive origins entails rejection of any role for natural selection in the subsequent evolution of language.

It is as important to avoid "naive exaptationism" as it is to guard against superficial adaptationist stories. To identify a feature as an exaptation does not mean that feature "is not the product of natural selection"—only that it did not always serve its present function. Indeed, most human behaviors serving biological functions that enhance fitness should probably be regarded as exaptations, modified through natural selection for subsequent specialization. The manual dexterity enabling specialized human hand movements now used in food gathering, eating, caretaking, tool use, and communication derives from the ability of prosimian ancestors to grasp branches. Human emotional expressions evolved from ritualized facial and vocal displays in other ancestral species, displays derived from behaviors that originally had no signaling function (Tinbergen, 1952). The use of stereotypical vocal patterns in human mothers' speech to infants is also undoubtedly an exaptation, with evolutionary origins in ancestral nonhuman primate vocalizations used for very different purposes. The claim that such human behaviors are exaptations rather than adaptations is a claim about origins, which does not require rejection of an adaptationist explanation for the current fit between these behaviors and the biological functions they serve.

Arbitrariness and Nonoptimality Are Not Incompatible With Natural Selection

A second issue addressed in Pinker and Bloom's (1990) critique of antiadaptationist accounts of the evolution of language is whether evidence for arbitrariness in the design of natural languages should be interpreted as evidence against the formative role of natural selection. Piattelli-Palmarini argues that "adaptive constraints are typically insufficient to discriminate between real cases and an infinity of alternative, incompatible mechanisms and traits which, although abstractly compatible with the survival of a given species, are demonstrably absent" (1989, p. 19). Implicit here is the view that adaptationist arguments all rest on the Panglossian assumption of the best-of-all-possible worlds, i.e., that adaptations necessarily represent optimal solutions. As Pinker and Bloom point out, however, the idea that nature aspires to perfection was never taken seriously by modern evolutionary theorists. One reason is that trade-offs of utility among conflicting adaptive goals are inevitable and ubiquitous, because optimization for one function often compromises optimization for other concurrent functions. The gorgeous but unwieldy tail of the peacock is an obvious example. A second reason, discussed at length by Dawkins (1986), is that since evolution can exploit only existing variability and build only on existing structures, adaptations are fundamentally constrained by what is available. Thus Piattelli-Palmarini's argument that the absence of optimal design features constitutes evidence against natural selection in the evolution of language reflects a misconstrual of the adaptationist position.

The classic example of design optimization in nature is the vertebrate eye. Many evolutionary theorists have echoed Darwin in observing how elegantly the eye exemplifies adaptive engineering in the service of a well-specified function. Although several different types of eye have evolved among invertebrate species, a single basic type of image-forming eye prevails among vertebrates. The features of this basic design reflect constraints arising directly from the physical properties of light, such as the optics of lenses necessary to bend light rays in order to focus an image on the retina. Other design features reflect constraints imposed by particular environments during the course of evolution. For example, since the vertebrate eye initially evolved in water, it is sensitive only to a narrow band of wavelengths of electromagnetic radiation, the "spectrum of visible light," which is transmitted through water without significant attenuation (Fernald, 1988). Although all vertebrate eyes share this basic feature, they also differ impressively in structural details across species, reflecting the selective pressures of ecological conditions peculiar to the habitat and way of life of particular species. Nocturnal birds such as owls have retinas consisting largely of rods, sensitive to very low light levels, while birds requiring high acuity in bright sunlight, such as eagles and hawks, have retinas comprised primarily of cones. Such exquisite adaptations to specific ecological conditions, all relatively minor variations on the basic design scheme, make it clear why Darwin felt the eye to be an organ "of extreme perfection and complication," and why the eye is so often cited as a compelling example of the shaping and refinement of biological mechanisms through natural selection.

In the case of the vertebrate eye, the fact that many of the selective forces that have guided evolution are unchanging physical properties of the world accounts for the cross-species convergence on common design features that are clearly functional (see Shepard, this volume). Such "nonarbitrary" features are characteristic of all vertebrate sensory systems, as well as other biological systems strongly constrained by the physical nature of the environment. However, Piattelli-Palmarini's (1989) suggestion that nonarbitrary design is the hallmark of *all* adaptive complexity misses an important distinction. While some adaptations enable animals to cope with relatively invariant aspects of the physical world, others have evolved to enable animals to interact successfully in the rapidly shifting social world. Just as in the case of sensory systems, adaptations for social interaction occur in response to environmental pressures. The critical difference is that in the evolution of social behavior, the "environment" includes not only the immutable laws of physics but also the more malleable behavior of conspecifics.

To the extent that the evolution of social behaviors is constrained by the perceptual predispositions and behavioral response tendencies of conspecifics, which are also subject to evolutionary change, the design features of social adaptations are intrinsically more "arbitrary" than those of the visual system. In runaway sexual selection, for example, the aesthetic preferences of females act as a strong selective pressure for otherwise useless male adornments, such as extravagant tail plumage. Males also capitalize on preexisting female sensory biases, as in frog species in which female preferences for male calls can be predicted from properties of the female auditory system (Ryan, Fox, Wilczynski, & Rand, 1990). In both cases, male characteristics have been shaped in an "arbitrary" fashion, in the sense that females could conceivably have developed different visual preferences or auditory tuning curves, resulting in different selective pressures and outcomes. Of course, the evolution of such male characteristics is not entirely negotiable, since the laws of physics must still be respected if otherwise arbitrary visual and vocal signals are to be effectively transmitted and received. The important point to be made here is that evolved mechanisms that provide an interface between an organism and the physical world are fundamentally more constrained than are mechanisms that evolve to coordinate interactions among conspecifics, as in communication systems. A communication system must be *shared* by conspecifics in order to function effectively, although adaptations for communication could be designed in diverse ways while still satisfying this basic requirement.

To summarize so far, the current debate on the evolution of human language (Piatelli-Palmarini, 1989; Pinker & Bloom, 1990) has focused attention on the more general question of when it is appropriate to invoke natural selection as a causal explanation for human behavior. Skepticism about the legitimacy of adaptationist reasoning has been fueled by the misleading claim that characteristics that qualify as exaptations rather than as adaptations are exempt from the process of natural selection. The argument made here is that the origin of a human behavior as an exaptation does not preclude the influence of natural selection in the subsequent refinement of that behavior for its current specialized function. A second misguided assumption in the antiadaptationist position is that the presence of arbitrary design features provides evidence against evolution by natural slection. However, because social behaviors in general, and communicative behaviors in particular, evolve in response to relatively variable physiological and behavioral characteristics of conspecifics, they are free to be arbitrary in some respects, as long as they are shared by group members. Human maternal speech to infants will provide an example of an exaptation that emerged from vocal behaviors originally serving other functions in our primate ancestors and that has design features shaped by natural selection in the course of hominid evolution.

THE PROSODIC CHARACTERISTICS OF MOTHERS' SPEECH TO INFANTS

This section will focus on the acoustic characteristics of prosody in speech to infants in English and other languages. If the vocal modifications of infant-directed speech indeed serve important biological functions in early human development, as will be argued in subsequent sections, then the prosodic characteristics of caretakers' speech should exhibit similar features across diverse cultures. The cross-cultural universality of prosodic modifications in infant-directed vocalizations is crucial evidence for the claim that this special form of speech is a species-specific caretaking behavior.

Cross-Language Research on the Prosody of Infant-Directed Speech

Although research on parental speech to children focused initially on the simplified linguistic features typical of early language input (e.g., Snow & Ferguson, 1977), in the last decade there has been increasing interest in prosodic as well as linguistic features of child-directed speech. The use of high pitch and exaggerated intonation in speech to children had been widely observed by anthropologists and linguists in such diverse languages as Latvian (Ruke-Dravina, 1976), Japanese (Chew, 1969), Comanche (Ferguson, 1964), and Sinhala (Meegaskumbura, 1980). However, Garnica's (1977) spectrographic analysis of English speech to children provided the first systematic acoustic evidence for the reported prosodic modifications in mothers' speech. Garnica documented the higher mean fundamental frequency (F_o) and wider F_o range in speech to

2-year-old children than in speech to adults, as well as the more frequent use of rising F_0 contours. In an acoustic analysis of German mothers' speech, Fernald and Simon (1984) found that even with newborns, mothers use higher mean F_0, wider F_0 excursions, longer pauses, shorter utterances, and more highly stereotyped F_0 contours than in speech to adults, as shown in Figure 10.1. Although the use of exaggerated F_0 contours and repetitiveness is a prominent feature of mothers' speech to infants throughout the first year of life, Stern, Spieker, Barnett, and MacKain (1983) found these prosodic features to be more pronounced in speech to 4-month-old infants than in speech to younger or older infants.

The prosodic modifications most consistently observed in studies of American English (Garnica, 1977; Jacobson, Boersma, Fields, & Olson, 1983; Stern et al., 1983) and German (Fernald & Simon, 1984; Papousek, Papousek, & Haekel, 1987) include higher mean pitch, higher pitch maxima and minima, greater pitch variability, shorter vocalizations, and longer pauses. In a recent cross-language comparison of parental

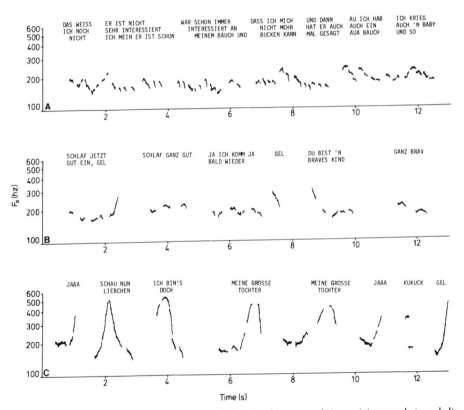

Figure 10.1   Pitch contours from one subject in three conditions: (a) speech to adults (AD speech); (b) speech addressed to absent infant (Simulated ID Speech); and (c) speech to newborn infant, held in mother's arms (ID Speech). Ordinate shows fundamental frequency (F_0) in hertz, plotted logarithmically. From "Expanded Intonation Contours in Mothers' Speech to Newborns" by A. Fernald and T. Simon, 1984, *Developmental Psychology, 20,* with permission of the American Psychological Association.

speech in French, Italian, Japanese, German, and British and American English, Fernald et al. (1989) found comparable prosodic modifications in speech to 12-month-old infants across these six language groups. Furthermore, fathers as well as mothers altered their intonation when addressing infants in these languages, as shown in Figure 10.2. Similar prosodic features have also been reported in two studies of mothers' speech to infants in Mandarin Chinese (Grieser & Kuhl, 1988; Papoušek, 1987). In the first systematic acoustic study of maternal speech in a nonindustralized cultural group, Eisen and Fernald (in preparation) have found that Xhosa-speaking mothers in South Africa also use higher mean F_0 and exaggerated F_0 contours when interacting with infants.

While acoustic analyses of parental speech in different languages have revealed global prosodic modifications that are similar across cultures, there is also evidence for cultural variability. In both Japanese (Fernald et al., 1989) and Mandarin Chinese (Papoušek, 1987), mothers seem to show less expansion of pitch range compared with American and European samples, although the differences are small and the findings inconsistent (see Grieser & Kuhl, 1988). Of the cultures studied so far, American middle-class parents show the most extreme prosodic modifications, differing significantly from other language groups in the magnitude of intonational exaggeration in infant-directed speech (Fernald et al., 1989). These findings may reflect culture-specific "display rules" governing the public expression of emotion. While in middle-class American culture, emotional expressiveness is not only tolerated but expected, in Asian cultures exaggerated facial and vocal displays are considered less acceptable (Ekman, 1972). In any event, the variations across cultures in the prosody of speech to children reported to date are relatively minor, and the few reports of cultures in which no spe-

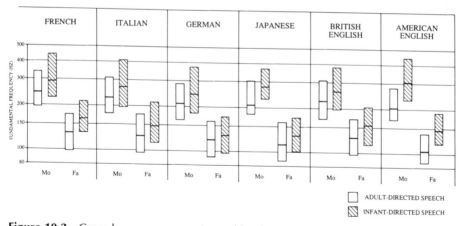

Figure 10.2 Cross-language comparison of fundamental frequency (F_0) characteristics of mothers' (Mo) and fathers' (Fa) speech to adult and infants. For each bar, the bottom line represents the mean F_0 minimum, the top line represents the mean F_0 maximum, and the intersecting line represents the mean F_0 per utterance. The extent of the bar corresponds to the F_0 range. From A. Fernald, T. Taeschner, J. Dunn, M. Papoušek, B. Boysson-Bardies, and I. Fukui, 1989, *Journal of Child Language, 16,* with permission of Cambridge University Press.

cial infant-directed speech register is used (e.g., Ratner & Pye, 1984) are difficult to interpret (see Fernald et al., 1989). The common pattern of results emerging from comparative research on parental speech is one of impressive consistency across cultures in the use of exaggerated intonation in speech to infants.

Relations Between Prosodic Form and Communicative Function in Mothers' Speech

In addition to descriptive studies that focus on the global features of parental prosody, a few studies have begun to investigate the fine structure of infant-directed intonation in relation to particular contexts of interaction. The question of interest in this research is whether specific prosodic forms in mothers' speech are regularly associated with specific communicative intentions. In particular, do mothers use context-specific intonation patterns when involved in routine caretaking activities such as soothing or comforting, eliciting attention, expressing praise or approval, or prohibiting the infant? When soothing a distressed infant, for example, mothers are more likely to use low pitch and falling pitch contours than to use high, rising contours (Fernald, Kermanschachi, & Lees, 1984; Papousek, Papousek, & Bornstein, 1985). When the mother's goal is to engage attention and elicit a response, however, rising pitch contours are more commonly used (Ferrier, 1985; Ryan, 1978). Bell-shaped pitch contours occur most frequently when the mother is attempting to maintain the infant's attention (Stern, Spieker, & MacKain, 1982). These results, all based on observations of American middle-class mothers, suggest that maternal prosody is modulated in accordance with the infant's affective state and that mothers use intonation differentially to regulate infant arousal and attention.

Are these relations between prosodic form and communicative function universal across languages? Preliminary observations suggest that certain associations of relatively stereotyped intonation patterns with particular communicative intentions share striking similarities across cultures (Fernald, 1992). For example, Approval vocalizations in English, German, French, and Italian are typically high in mean F_o and wide in F_o range, with a prominent rise-fall F_o contour, as shown in Figure 10.3. In contrast, Prohibition vocalizations in these languages are typically low in mean F_o and narrow in F_o range, as well as shorter, more intense, and more abrupt in onset. Comfort vocalizations, while similar to Prohibitions in low mean F_o and narrow F_o range, are longer, less intense, and softer in onset than Prohibitions. In musical terms, Comfort vocalizations have a smooth "legato" quality, in marked contrast to the sharp "staccato" quality of Prohibitions. It is important to note, however, that these stereotyped prosodic patterns are characterised not only by F_o characteristics such as contour shape, as shown in Figure 10.3, but also by differences in amplitude envelope and spectral composition. Thus an adequate characterization of prosodic contour types will involve developing a taxonomy based on complex functions of graded changes along these multiple acoustic dimensions, research that is currently in progress. In the meantime, these initial observations suggest that across a variety of languages, common communicative intentions are associated with particular prosodic forms in maternal speech to preverbal infants.

What would account for this correspondence between characteristic patterns of prosodic features and specific motivational states? One explanation could be that these context-specific maternal vocalizations are essentially expressions of vocal affect,

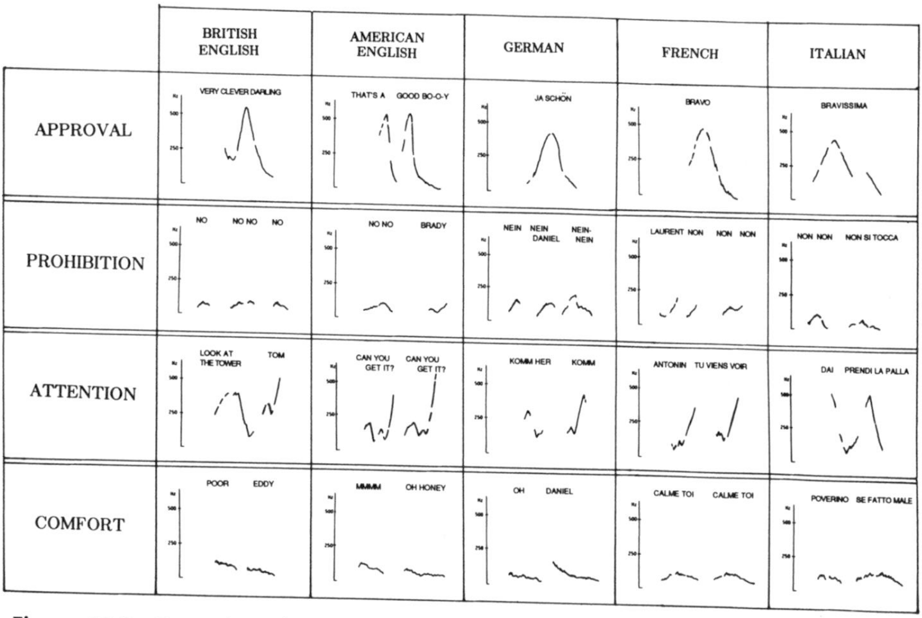

	BRITISH ENGLISH	AMERICAN ENGLISH	GERMAN	FRENCH	ITALIAN
APPROVAL	VERY CLEVER DARLING	THAT'S A GOOD BO-O-Y	JA SCHÖN	BRAVO	BRAVISSIMA
PROHIBITION	NO NO NO NO	NO NO BRADY	NEIN NEIN NEIN- DANIEL NEIN	LAURENT NON NON NON	NON NON NON SI TOCCA
ATTENTION	LOOK AT THE TOWER TOM	CAN YOU GET IT? CAN YOU GET IT?	KOMM HER KOMM	ANTONIN TU VIENS VOIR	DAI PRENDI LA PALLA
COMFORT	POOR EDDY	MMMM OH HONEY	OH DANIEL	CALME TOI CALME TOI	POVERINO SE FATTO MALE

Figure 10.3 Examples of pitch contours from Approval, Prohibition, Attention, and Comfort vocalizations in British, American, German, French, and Italian mothers' speech to 12-month-old infants.

which are similar in form across cultures just as facial expressions are universal across cultures (Ekman, 1972). In research on universal features of vocal expressions, there is general agreement that adult listeners are fairly accurate in recognizing emotions from vocal cues, although there is less consensus about which acoustic correlates differentiate among vocal expressions of different emotions (Scherer, 1986). However, several studies indicate that increases in mean F_0 and F_0 range are typical of vocal expressions of enjoyment and happiness, while decreases in mean F_0 and F_0 range, especially when accompanied by harsh voice quality, characterize expressions of irritation and anger. Scherer (1985, 1986) has proposed that these acoustic patterns are shaped by physiological and motor responses related to the underlying emotional state. Cosmides (1983) has stressed that consistent acoustic configurations in emotional expressions also yield important information about the speaker's intentions and motivations. A correspondence between particular vocal features and motivational states may even be general across different bird and mammal species. The biologist Morton (1977) found that high tonal sounds in animal repertoires are associated with fear, appeasement, or friendly approach, while low, harsh sounds are associated with threat. One interesting result consistent with such hypotheses is Tartter's (1980) finding that smiling while talking alters the shape of the human vocal tract, resulting in higher mean F_0. Whatever the underlying mechanisms influencing sound production, mothers' infant-directed approval vocalizations are high in mean F_0 and wide in F_0 range, prosodic features typical of happy vocal expressions, while prohibition vocalizations have prosodic features typical of angry expressions.

Although universal production constraints on vocal affect expressions undoubtedly do influence the sound patterns of maternal speech, the context-specific vocalizations illustrated in Figure 10.3 are not simply expressions indexing the mother's emotional or motivational state. While infant-directed praise vocalizations are positive in tone, and prohibitions are negative in tone, these vocalizations also have a strong *pragmatic* character. When the mother praises the infant, she uses her voice not only to express her own positive feelings, but also to reward and encourage the child. And whether or not the mother feels anger when producing a prohibition, she uses a sound well designed to interrupt and inhibit the child's behavior. Thus the exaggerated prosody of mothers' speech is used instrumentally to influence the child's behavior, as well as expressively, revealing information about the mother's feelings and motivational state. In this respect, the use of prosody in human maternal speech is similar to the use of vocal signals by some nonhuman primates, a proposal to be explored in later sections.

The descriptive findings on speech to human infants reviewed so far provide a foundation for the argument that the exaggerated prosodic patterns of infant-directed speech serve important developmental functions. The cross-linguistic data suggest that maternal vocalizations have common acoustic features that are potentially highly salient as auditory stimuli for the infant and that mothers tend to use specific prosodic patterns in particular contexts of interaction. However, such descriptive findings can provide only indirect support for claims about the communicative functions of prosody in early development. The next section will consider two additional kinds of relevant evidence, integrating research findings on the capabilities and predispositions of human infants with results from experimental studies directly testing hypotheses about the influence of maternal prosody on infant perception and behavior.

COMMUNICATIVE FUNCTIONS OF INTONATION IN INFANT-DIRECTED SPEECH

Why do adults engage in such unusual vocal behavior, speaking in *glissandi* sometimes exceeding two octaves in pitch range, in the presence of a young infant? Early investigations of the influence of mothers' speech on infant development focused exclusively on syntactic and semantic features of language input in relation to children's language production (see Snow, 1977). This research was motivated by the hypothesis that the primary function of the special infant-directed speech style is to teach language to a linguistic novice. However, the fact that mothers quite consistently simplify their speech and use exaggerated intonation even when interacting with newborns, long before language learning is a central developmental issue, suggests that the modifications in infant-directed speech serve prelinguistic functions as well.

A model of the multiple developmental functions of intonation in speech to infants over the first year of life is presented in Figure 10.4. According to this model, the characteristic prosodic patterns of mothers' speech serve initially to elicit infant attention, to modulate arousal and affect, and to communicate affective meaning. Only gradually, toward the end of the first year, does the prosody of mothers' speech begin to serve specifically linguistic functions, facilitating speech processing and comprehension. With its emphasis on the prelinguistic regulatory functions of intonation in mother-infant interaction, this model proposes biological predispositions rather than linguistic or cultural conventions as primary determinants of the use and effectiveness of exaggerated intonation in speech to infants. This model of the multiple developmental functions of infant-directed intonation will serve to organize support

ACOUSTIC HIGHLIGHTING OF WORDS
Prosodic marking of focused words helps the infant to identify linguistic units within the stream of speech. Words begin to emerge from the melody.

COMMUNICATION OF INTENTION & EMOTION
Vocal and facial expressions give the infant initial access to the feelings and intentions of others. Stereotyped prosodic contours occurring in specific affective contexts come to function as the first regular sound-meaning correspondences for the infant.

MODULATION OF ATTENTION, AROUSAL & AFFECT
Melodies of maternal speech become increasingly effective in directing infant attention and modulating infant arousal and emotion.

INTRINSIC PERCEPTUAL & AFFECTIVE SALIENCE
From the beginning, the infant is predisposed to respond differentially to certain prosodic characteristics of infant-directed speech. Certain maternal vocalisations function as unconditioned stimuli in alerting, soothing, pleasing, and alarming the infant.

Figure 10.4 Developmental functions of prosody in speech to infants.

for the argument that human maternal speech is an adaptive mechanism. The claim that this special vocal behavior has been selected for in evolution requires evidence on the proximal level that human maternal vocalizations have design features that are particularly advantageous in establishing communication with the young infant. This section will review the growing number of research findings demonstrating that the exaggerated melodies of mothers' speech are highly salient to young infants, influencing infant attention, arousal, emotion, and language comprehension.

The Perceptual and Affective Salience of Intonation in Infant-Directed Speech

At birth, and probably even before birth, infants are attentive to prosodic cues in speech. Very young infants are able to make discriminations based on several prosodic parameters, including frequency (Wormith, Pankhurst, & Moffitt, 1975), intensity (Steinschneider, Lipton, & Richmond, 1966), duration (Miller & Byrne, 1983), rise-time (Kearsley, 1973), and temporal pattern (Clarkson & Berg, 1983; Demany, McKenzie, & Vurpillot, 1977). Newborns show a preference for their own mother's voice over another woman's voice (DeCasper & Fifer, 1980), an effect that is presumably due to prenatal familiarization. The recent finding that newborns can discriminate between speech samples spoken in their mother's native language and in an unfamiliar language, even when the stimuli are low-pass filtered (Mehler et al., 1990), provides further evidence for the prenatal development of sensitivity to prosodic information in vocal signals.

The idea that the melodic intonation of mothers' speech is a prepotent stimulus for the human infant, effective at birth, was proposed half a century ago by Lewis (1936/1951). Lewis argued that "the voice at the outset is not a neutral stimulus; it possesses an affective character for the child—in other words, it evokes a response" (1936/1952, p. 52). This immediately compelling quality of maternal speech, because of its salient perceptual features and affective tone, underlies the most basic of the developmental functions of prosody shown in Figure 10.4. The claim that maternal intonation functions initially at this level assumes not only that prosodic features are discriminable by young infants, but also that they are qualitatively different in their effects on the infant. Several studies document such differential effects of prosodically varied auditory signals on infant behavior. For example, moderately intense sounds elicit cardiac deceleration, an orienting response, while signals higher in intensity elicit acceleration, a defensive reaction (Berg, 1975). Similarly, signals with a gradual rise-time in intensity elicit eye opening and orienting, while a more abrupt rise-time leads to eye closing and withdrawal (Kearsley, 1973).

In this sense, the characteristic infant-directed Prohibition vocalizations described earlier, high in intensity with an abrupt rise-time, function initially as *unconditioned* stimuli. That is, they elicit a defensive response from the infant that is directly attributable to acoustic rather than to linguistic features. Comfort sounds provide another example of maternal vocalizations that elicit unconditioned responses from young infants. Studies of acoustic signals most effective in soothing distressed infants have identified three types of sounds that reduce crying. First, low-frequency sounds are more effective than high-frequency sounds (Bench, 1969). Second, continuous sounds are more soothing than intermittent signals (Birns, Blank, Bridger, & Escalona, 1965). Third, white noise is particularly effective in soothing a crying infant (Watterson & Riccillo, 1983). When mothers use their voice to calm a distressed infant, they frequently incorporate all three of these acoustic features. Comfort vocalizations are typically low in pitch and continuous rather than intermittent. Mothers also intersperse these vocalizations with "shhhh" sounds, which are similar to white noise (Fernald, Kermanschachi, & Lees, 1984). Thus, in both Comfort and Prohibition vocalizations addressed to infants, mothers quite intuitively exploit infants' innate predisposition to respond differentially to particular acoustic qualities of sounds.

Maternal Intonation and Infant Attention

At the second level of the model shown in Figure 10.4, the prosody of mothers' speech functions with increasing effectiveness to engage infant attention and to modulate arousal and emotion. Over the first six months of life, the infant's visual capabilities and motor coordination improve dramatically (e.g., Aslin, 1987). By the age of 3 to 4 months, infants can recognize individual faces and orient to voices much more quickly and reliably than before. The appearance of the social smile, which is initially elicited most effectively by high-pitched voices (Wolff, 1963), is another landmark development around the age of 2 months. All of these new skills contribute to infants' growing ability to respond selectively and appropriately to the intonation of the mothers' voice over the first few months.

The argument I am making here is not just that infants are responsive in general to prosody before they are responsive to linguistic structure in speech, but that the exaggerated intonation patterns of mothers' speech are *particularly* effective in eliciting attention and affect in very young infants. Supporting evidence comes from a number of auditory preference studies in which infants could listen either to typical adult-directed (AD) speech or to infant-directed (ID) speech (e.g., Friedlander, 1968). In a listening preference study with 4-month-old infants, subjects were operantly conditioned to make a head turn to one side or the other in order to be rewarded with a short recorded segment of ID or AD speech, spoken by one of several unfamiliar women (Fernald, 1985). Infants made significantly more head turns to the side on which ID speech was available as a reinforcer. What was it about ID speech that was more appealing to infants? Since natural vocalizations were used as stimuli in this study, it was impossible to evaluate the role of any particular linguistic or acoustic feature in eliciting the infant preference. Although it could have been some prosodic characteristic of ID speech that was especially attractive to infants, it was also possible that the words themselves were more familiar in the ID speech samples.

To investigate the hypothesis that it was the prosody rather than the words of ID speech that elicited the infant listening preference, it was necessary to eliminate the lexical content of the natural speech stimuli and to isolate the three major acoustic features of intonation: fundamental frequency, amplitude, and duration. In three follow-up studies, we presented infants with synthesized signals derived from the F_o, amplitude, and temporal characteristics of the original natural speech stimuli (Fernald & Kuhl, 1987). Thus each of these experiments focused on a particular prosodic variable in the absence of phonetic variation. Our goal was to determine whether 4-month-old infants, when given the choice between listening to auditory signals derived from the prosodic characteristics of either AD or ID speech, would show a listening preference for particular prosodic features of ID speech. We found a strong infant preference for the F_o patterns of ID speech, but no preference for either the amplitude or duration patterns of ID speech. The results suggest that the F_o characteristics of mothers' speech are highly salient and rewarding to infants and may account for the listening preference for natural ID speech found in the orignal Fernald (1985) study.

One interpretation of these findings is that infants are innately predisposed to pay attention to the acoustic characteristics of ID speech. However, the finding that the pitch contours of ID speech are sufficient to elicit the infant listening preference does not preclude an interpretation based on learning. By 4 months of age, infants have had a lifetime of experience with ID speech in association with many gratifying forms of caretaking and social interaction. It could be that 4-month-old infants attend more to the pitch contours of ID speech because these melodies have become positively associated with nurturance. The recent findings of Cooper and Aslin (1990), however, suggest that early postnatal experience could not account entirely for the infant listening preference. Using a visual-fixation-based auditory preference procedure, Cooper and Aslin found that both newborns and 1-month-old infants preferred ID over AD speech. Although there was a difference in the absolute magnitude of the ID speech preference between the two ages, there was no significant difference in the relative magnitude of the effect, suggesting that the listening preference was as strong in newborns as it was at 1 month of age. These findings indicate that the preference for the exaggerated prosody of ID speech is already present at birth and does not develop gradually with postnatal experience.

Research on this early listening preference has been extended in several interesting directions by Werker and McLeod (1989). Infants at 4 and 8 months of age were shown video recordings of both male and female adults reciting identical scripts in ID and AD speech. Werker and McLeod included measures of both attentional preference and affective responsiveness in the two speech conditions. They found that infants at both ages attended longer to ID speech than to AD speech when spoken by either a male or a female. Furthermore, infants responded with significantly more positive affect to male and female ID speech than to AD speech. By demonstrating that the infant listening preference for ID speech extends to male voices, the results of this study suggest that the relatively high mean F_0 of the female voice is not the critical parameter in eliciting infant attention. Instead, it is probably the exaggerated F_0 modulation, within either the male or female F_0 range, that is the most engaging feature of ID speech.

The power of ID speech to engage infant attention has also been shown using a psychophysiological measure, in addition to the behavioral preference measures just described. In a study of infant cardiac orienting, we presented 24 4-month-old infants with vocalic sounds differing in pitch range. The stimuli consisted of repetitions of the monosyllable /a/, spoken with either the narrow F_0 range typical of AD speech or the wide F_0 range typical of ID speech, but identical in amplitude characteristics and duration. We found that infants showed reliably greater heart rate deceleration when listening to the wide-range pitch contours (Fernald & Clarkson, in preparation). These results provide convergent psychophysiological evidence for the differential attentional responsiveness of young infants to the exaggerated prosodic contours of ID speech.

Maternal Intonation and Infant Emotion

The melodies of mothers' speech not only captivate attention, but also engage the infant emotionally. In *The First Relationship: Infant and Mother,* Daniel Stern (1977) describes an interchange between a mother and her 3-month-old son, halfway through a feeding session:

> His eyes locked on to hers, and together they held motionless for an instant. The infant did not return to sucking and his mother held frozen her slight expression of anticipation. This silent and almost motionless instant continued to hang until the mother suddenly shattered

it by saying "Hey!" and simultaneously opening her eyes wider, raising her eyebrows further, and throwing her head up and toward the infant. Almost simultanteously, the baby's eyes widened. His head tilted up and, as his smile broadened, the nipple fell out of his mouth. Now she said, "Well hello!. . . heelló . . . heeellóooo!", so that her pitch rose and the "hellos" became longer and more stressed on each successive repetition. With each phrase the baby expressed more pleasure, and his body resonated almost like a balloon being pumped up, filling a little more with each breath. (p. 3)

Anyone who has interacted intensively with a young infant knows the shared pleasure of such moments, and the power of the voice to modulate and sustain that pleasure. Such observations echo those of early investigators of mother-infant interaction who reported that young infants respond differentially to positive and negative affect in the voice (Bühler & Hetzer, 1928; Lewis, 1936/1951) and smile earlier to voices than to faces (Wolff, 1963). It is therefore surprising that recent research on infants' perception of emotional signals has focused almost exclusively on the face. In fact, the ability to recognize emotional facial expressions appears to develop rather slowly over the first year. Numerous studies of infants' discrimination and categorization of facial displays indicate that it is not until 7 months of age that infants reliably recognize happy and angry facial expressions (see Nelson, 1987, for a review). Moreover, two recent studies of infants' perception of facial-plus-vocal displays (Caron, Caron, & MacLean, 1988; Werker & McLeod, 1989) suggest that in the first half year of life infants are more responsive to the voice than to the face.

In a recent series of experiments in our laboratory, we investigated infants' responsiveness to positive and negative vocalizations in unfamiliar languages as well as in English. This research was designed to address four questions: (a) Do young infants respond differentially to positive and negative affective expressions in the voice, at an age when they are not yet showing selective responsiveness to positive and negative affect in the face? (b) do infants respond appropriately to vocal expressions differing in affective tone? (c) are infants more responsive to affective vocalizations in ID speech than in AD speech? (d) are infants responsive to affective vocalizations spoken in languages with which they are completely unfamiliar?

In an auditory preference procedure, infants were presented with Approval and Prohibition vocalizations, typical of those used by mothers when praising or scolding an infant. The subjects in these studies were 5-month-old infants from monolingual English-speaking families. Each infant heard Approval and Prohibition vocalizations spoken in ID speech in German, Italian, or Japanese. English Approvals and Prohibitions were presented in both ID speech and AD speech. When listening to English, infants responded with differential and appropriate affect to these positive and negative vocal expressions in ID speech only. That is, they showed more positive affect to Approvals and more negative affect to Prohibitions in ID speech but not in AD speech. When listening to ID speech in German and Italian, infants also responded with differential and appropriate affect. In Japanese, however, infants listened with neutral affect to Approvals as well as to Prohibitions.

Why should American infants respond affectively to infant-directed Approval and Prohibition vocalizations in two unfamiliar European languages as well as in English, but not in Japanese? Although Japanese mothers elevate their pitch when speaking to infants, the pitch range used by Japanese mothers in ID speech is narrower than that used by European mothers (Fernald et al., 1989). Japanese vocalizations may be more difficult for infants to distinguish. In fact, Japanese vocal expressions of emotion are difficult for European adults to interpret (Shimoda, Argyle, & Riccibitti, 1978), perhaps for the same reason. If research on the auditory preferences of Japanese infants reveals that infants familiar with Japanese respond with appropriate affect to Approvals and Prohibitions in their language, this finding would suggest the early influence of cultural differences in mother-infant interaction. Perhaps 5-month-old American infants are already accustomed to the wider pitch modulations and dynamic range characteristic of affective vocalizations in cultures encouraging emotional expressiveness, so that German and Italian are more "familiar" to them in terms of prosodic characteristics than is Japanese. Research in progress on the prosodic characteristics of Japanese mothers' speech and on the responsiveness of Japanese infants to Approval and Prohibition vocalizations in different languages will enable us to explore further these intriguing questions.

Four major findings emerge from these studies on infants' responsiveness to affective vocal expressions. First, at the age of 5 months, when infants are not yet showing consistent selective responsiveness to positive and negative facial expressions, infants do respond differentially to positive and negative vocal expressions, suggesting that the voice is more powerful than the face as a social signal in early infancy. Second, infants respond with appropriate affect to positive and negative vocal expressions, smiling more to Approvals than to Prohibitions. Third, infants are more responsive to affective vocalizations in ID speech than in AD speech, suggesting that the exaggerated prosodic characteristics of maternal vocalizations to infants increase their salience as vocal signals. And finally, young infants are responsive to affective vocalizations spoken with infant-directed prosody even in languages that they have never heard before, providing evidence for the functional equivalence of such ID vocalizations across cultures. These findings indicate that the melodies of mothers' speech are compelling auditory stimuli, which are particularly effective in eliciting emotion in preverbal infants.

The finding that American infants differentiate maternal vocalizations in some but not all languages suggests that cultural differences in the nature and extent of emotional expressiveness may also have an early influence on infants' responsiveness to vocal signals. A process of early cultural "calibration" might account for these cross-language differences. According to this explanation, infants in all cultures are initially responsive to the same vocal cues, in that they find smooth, wide-range pitch contours of moderate loudness to be pleasing, while they find low, narrow pitch contours that are short, staccato, and loud to be more aversive. However, cultural differences in display rules governing emotional expression may determine the levels and range of emotional intensity to which the infant is routinely exposed and which the infant comes to expect in social interaction with adults.

Maternal Intonation and the Communication of Intention

At the third level of the model shown in Figure 10.4, infants gain access to the feelings and intentions of others through the intonation of infant-directed speech. That is, infants begin to learn to interpret the emotional states of others and to make predictions about the future actions of others, using information available in vocal and facial expressions. It is important to note that this third level of processing of vocal signals is quite different from the selective responsiveness of young infants to affective vocalizations described in the preceding section. When the 5-month-old infant smiles to an Approval or startles to a Prohibition, the prosody of the mother's voice influences the infant directly. The infant's differential responses in no way presuppose an ability to decode the emotions expressed by the mother. Rather, the infant listens with pleasure to pleasant sounds and with displeasure to unpleasant sounds, without necessarily understanding anything about the affective states motivating the production of these vocalizations. Such maternal vocalizations are potent signals not because they inform the infant about the mother's emotional state, but because they directly induce emotion in the infant. Of course, since pleasant-sounding praise vocalizations are often accompanied by smiles and other indexes of positive affect, while negative vocalizations co-occur with other signs of negative affect, infants have ample opportunity to learn about the contexts of occurrence of particular vocal forms, reinforcing their positive and negative associations. The important distinction to be made here is that initially these characteristic vocal patterns influence the infant by virtue of their intrinsic acoustic features, strengthened through frequent association with other forms of affective input, although they are not yet appreciated as "cues" to the emotional state of the speaker.

Evidence that infants in the second half-year of life are developing the ability to read the emotional signals of others through facial and vocal expressions comes primarily from research on "social referencing." Klinnert, Campos, Sorce, Emde, & Svejda (1983) review several studies showing that infants around the age of 8 months seek out and appropriately interpret emotional signals from adults, especially in situations of uncertainty for the infant. For example, when an 8-month-old infant is undecided as to whether to cross the visual cliff or to approach an ambiguous toy, the infant will look to the mother. If the mother responds with exaggerated facial and vocal signs of happiness, the infant will typically proceed; however, if the mother responds with exaggerated expressions of fear, the infant will withdraw. Unfortunately, the relative potency of facial and vocal cues in this situation has not been adequately investigated. Although the mother's facial expressions appear to be sufficient to regulate the infant's behavior, it is not yet known whether vocal cues are also sufficient (Klinnert et al., 1983).

Another question that has yet to be addressed is whether infants in the social referencing paradigm are responsive only to highly exaggerated emotional displays. Given the finding described earlier that younger infants respond affectively to ID speech but not to AD speech, it seems likely that the exaggerated vocal and facial expressions typically used by mothers interacting with infants would provide especially salient information about the mother's emotional state. A recent study with adult subjects provides indirect support for the hypothesis that information about emotion and communicative intent is conveyed with special clarity through the exaggerated intonation of ID speech (Fernald, 1989). Subjects listening to content-filtered ID and AD speech samples were asked to identify the communicative intent of the speaker, using only prosodic information. We found that listeners were able to use intonation to identify the speaker's intent with significantly higher accuracy in speech addressed to infants. This finding suggests that the prosodic patterns of ID speech are more distinctive and more informative than those of AD speech and may provide the preverbal infant with reliable cues to the affective state and intentions of the speaker.

Could infants potentially make use of such prosodic contours to gain access to the communicative intent of the speaker? Research on infants' perception of melodies provides indirect evidence for the salience of pitch patterns in the preverbal period. By 6 months, infants can extract the melodic contour in a tonal sequence even when the sequence is transposed into a different frequency range (Trehub, Thorpe, & Morron-giello, 1987). Infants appear to encode information about contour as opposed to absolute frequencies, since they perceive transposed melodies as similar or equivalent. This holistic mode of auditory processing, in which the relational features of a tonal sequence are retained, could enable infants to encode the prominent melodic patterns of infant-directed speech and recognize these characteristic melodies across variations in speaker, segmental content, and pitch range.

Maternal Intonation and Early Language Development

At the fourth level of the model shown in Figure 10.4, the prosodic patterns of ID speech increasingly serve linguistic functions. While the infant perceives prosodic contours holistically in the earlier preverbal period, words gradually begin to emerge from the melody toward the end of the first year. As the child develops the ability to use language to extract meaning from the mother's vocalizations, the prosody of maternal speech helps to draw the infant's attention to particular linguistic units within the continuous stream of speech. In a study of mothers' use of prosodic emphasis to mark focused words in speech to infants and adults, we found that prosodic cues in ID speech were not only more emphatic than those in AD speech, but also more consistent and more highly correlated (Fernald & Mazzie, 1991). When highlighting focused words, mothers used a distinctive prosodic strategy in speech to their infants. Focused words occurred most often on exaggerated pitch peaks in the final position in phrases in ID speech, while in AD speech the acoustic correlates of lexical stress were much more variable.

Does the use of exaggerated prosody actually help the infant to recognize focused words in continuous speech? In a recent study of the influence of intonation patterns on early lexical comprehension, we found that infants in the early stages of language acquisition are better able to recognize familiar words in ID speech than in AD speech (Fernald & McRoberts, 1991). In this study, 15-month-old infants were tested in an auditory-visual matching procedure. On each trial, infants were presented with pairs of colored slides of familiar objects, accompanied by a recorded vocalization drawing their attention to one of the two objects (e.g., "Look at the *ball!* See the *ball?*"). The vocal stimuli were presented with the target word stressed in either ID or AD intonation. The index of comprehension in this procedure was whether infants looked longer at the picture matching the target word. We found that when target words were presented in ID speech, infants recognized the target words more reliably than in AD speech. Follow-up studies suggest, however, that by 18 months infants have acquired the ability to identify familiar words equally well in AD speech and ID speech, although exaggerated intonation may still be important in acquiring new words at this age. These findings suggest that, at least in English, the acoustic highlighting of new words in mothers' speech may facilitate language comprehension in infants just beginning to learn language. Whether the dramatic pitch peaks of ID speech actually enhance the intelligibility of focused words or serve primarily as an attention-focusing device is a question still to be addressed.

To summarize so far, the exaggerated melodies of mothers' speech to infants serve many different functions in early development, prior to the acquisition of language. The distinctive prosodic patterns of maternal speech are prepotent signals for the infant at birth and are effective in the early months in eliciting and maintaining attention and in modulating arousal. Prior to the time when the mother's speech sounds can influence her child's behavior symbolically through their referential power, her intonation affects the child directly. When her intention is to arouse and delight the infant, she uses smooth, wide-range pitch contours, often with rising intonation; when her goal is to soothe, she rocks the infant and speaks with low, falling pitch. And just as the vestibular rhythms of rocking have a direct calming effect on the child (e.g., Byrne & Horowitz, 1981), the acoustic features of the mother's soothing melodies also function directly to decrease arousal and calm the infant. Thus in the first year of the infant's life, the communicative force of the mother's vocalizations derive not from their arbitrary meanings in a linguistic code, but more from their immediate musical power to arouse and alert, to calm, and to delight. Although the exaggerated pitch patterns of maternal vocalizations may eventually help the child in the second year to identify linguistic units in speech, the human voice becomes meaningful to the infant through maternal intonation much earlier in development. Through this distinctive form of vocal communication the infant begins to experience emotional communion with others, months before communion through symbols is possible.

CONSTRAINTS ON THE EVOLUTION OF VOCAL COMMUNICATION SYSTEMS

In making the argument that human maternal speech is an adaptive mechanism, it is necessary to demonstrate not only that this special form of vocal communication serves critical biological functions in early development, but also that this behavior meets other criteria for invoking a selectionist account. One question of interest is whether this human vocal behavior shares common design features with vocal behaviors in other species where natural selection has more obviously played a formative role in shaping vocal signals to serve particular functions. This section will present research findings from the ethological literature on vocal communication in nonhuman animals, focusing on studies that attempt to elucidate the functional significance of particular vocal adaptations. Three questions will be addressed: First, how do constraints on perceptual and production mechanisms, as well as characteristics of the environment, influence the evolution of vocal repertoires? Second, what selective pressures appear to favor the evolution of graded versus discrete vocal signals? And finally, how do ritualized vocal signals function to enhance signal detectability? The ethological research reviewed in the context of these three questions will suggest which forms of evidence are most relevant in evaluating the claim that the special form of infant-directed speech used by human mothers is a caretaking behavior that has also been influenced by natural selection.

In discussing the functional significance of courtship calls in certain animals, Darwin (1872) described them as "sweet to the ears of the species," referring to the special power of such species-specific signals to please and entice a potential mate. Darwin's daring speculation that communicative signals in animals are as susceptible to the influence of natural selection as are morphological features is now widely accepted. However, in the absence of a fossil record of such ephemeral phenomena as vocal sounds,[1] the evidence for adaptation in signal systems consists primarily of correlational findings relating acoustic and functional characteristics of species-specific vocalizations to other characteristics and conditions of the species that are likely to have exerted selective pressure on vocal forms. While necessarily indirect, such converging ethological evidence has become increasingly sophisticated in recent years, leading convincingly to the conclusion that the vocal repertoires of closely related species have been shaped differentially by evolution. Small variations in vocal forms between species are correlated with subtle differences in species-specific perceptual sensitivities, social organization, and habitat characteristics.

Perceptual and Production Constraints on the Evolution of Vocal Signals

The primary function of animal sounds in social species is communication. With very few exceptions (e.g., in the production of echolocation sounds by bats), the specialized sound-producing organs found across the animal kingdom have evolved to send signals to other animals (Krebs & Dawkins, 1984). The idea that the acoustic signals typical of a species are "well designed" to be detected and recognized by conspecifics is widely accepted in the ethological literature. In some cases, the responsiveness of young animals to acoustic signals emerges without social experience. Swamp sparrows reared in isolation, for example, respond with greater cardiac orienting to the swamp sparrow call than to the call of a related species, on first exposure to these sounds (Dooling & Searcy, 1980). In other avian species, prenatal exposure to the mother's call (Gottlieb, 1980) and brief postnatal experience with conspecific vocalizations (Marler, 1984) are instrumental in the development of selective responsiveness to species-specific signals. However, even when the emergence of auditory selectivity in animals requires some learning and is not, so to speak, "hardwired," genetic mechanisms are implicated (see Gould & Marler, 1987). Such selective responsiveness to particular animal sounds of high biological relevance reflects the influence of natural selection, according to Marler, Zoloth, and Dooling (1981):

> As a recurrent finding in ethological studies, developing young animals manifest responsiveness to particular environmental stimuli. These especially salient stimuli are associated with events that are fraught with special biological significance for all species members, such as predator detection or sexual communication. The selectivity of such innate responsiveness is sometimes so narrow that it is hard to imagine peripheral structures that could explain the specificity of responsiveness, especially when mediated by receptor systems known to be responsive to broader ranges of stimuli. To the extent that environmental events with special significance in the life of an organism are predictable over transgenerational time, adaptive species-specific genetic control over stimulus responsiveness becomes feasible. (p. 167)

Physiological and perceptual characteristics involved in sound production and reception are one source of evolutionary pressure shaping the form of vocal communication in a species. In some cases, the evolution of species-specific vocal signals seems to have been constrained primarily by the perceptual limitations of the conspecific recipient, as in the frog mating calls mentioned earlier. By comparing the auditory tuning curves in two closely related species of tree frog, Ryan et al. (1990) found that the auditory sensitivity of the female was the driving force in the evolution of the male mating call in one of these species. In this species, sound production in the male has been adapted to fit the preexisting perceptual bias of the female. Thus the effectiveness of species-specific vocalizations crucial to success in reproduction depends on the match between male production capabilities and female perceptual thresholds. It is important to note, however, that while the acoustic characteristics of such signals are highly constrained, given the particular physiological characteristics of the female of the species, they are also "arbitrary" in the sense that other mating signals would have evolved if, by chance, the tuning curves of the female tree frog had happened to be slightly different.

One very general constraint on the acoustic structure of vocal productions is the body size of the animal producing the signal. The frequency of phonation tends to be lower in larger animals simply because of the physics of sound production. The vocal signals of elephants and mice differ in spectral characteristics in part because resonant frequencies are directly related to the size of the animal's vocal cavities and sound producing organs. Similarly, the vocalizations of large male baboons have more energy present at lower frequencies than the vocalizations of smaller males or female baboons. Reviewing a large number of studies of avian and mammalian vocalizations, Morton (1977) observed the association mentioned earlier between acoustic structure and motivational state. Harsh, low-frequency sounds commonly signal threat or hostility, while higher-frequency, more tonal sounds signal fear, appeasement, or friendly approach. Morton explained this structural convergence in terms of a "motivation-structural rule." Because they give the impression of larger body size, harsh, lower-frequency vocalizations have been selected for across species as signals in hostile interactions, according to Morton, just as piloerection has evolved as a visual signal that increases apparent size. Conversely, the higher-frequency, tonal vocalizations used by many animals in fearful or friendly motivational states are effective in eliciting approach and support because they resemble the vocalizations of smaller, nonthreatening animals. The linguist Ohala (1984) has extended Morton's reasoning to include intonation patterns in human speech, observing that falling F_o contours are commonly associated with assertive meanings, while rising F_o contours are associated with non-assertive or conciliatory meanings. Such examples of apparent "sound symbolism" in human and nonhuman animal vocal signals reflect selection for acoustic features related to characteristics of both production mechanisms and complementary perceptual response biases.

Environmental Constraints on the Evolution of Vocal Signals

While there are numerous examples in the ethological literature in which physiological, perceptual, and cognitive characteristics of conspecifics, as well as predators, have apparently shaped the evolution of biologically significant vocal signals, the acoustic ecology of the natural habitat also plays a formative role. Marler (1965) was among the first to propose that ecological factors as well as perceptual constraints may have exerted selective pressures on the structure of acoustic communication. For example, alarm calls across a number of avian species consist of narrow-band, high-pitched calls, which appear to have two important advantages. Because of their particular acoustic structure, such alarm calls propagate widely and are very difficult for a predator to localize (Green & Marler, 1979).

Research on monkeys inhabiting East African rain forests has explored further the interaction of vocal signal characteristics and features of the environment in the evolution of communication systems. Obviously, a vocalization signaling a biologically important event can be effective only if it can be heard by the appropriate recipient. According to Brown (1989), several factors determine the audibility of vocal signals in the rain forest, including signal characteristics that facilitate transmission, signal degradation resulting from attenuation in the rain forest environment, and the intensity and spectral composition of environmental background noise, which can mask the signal. In an elegant series of field and laboratory studies, P. M. Waser, Brown, and colleagues have shown how the acoustic structure of vocal signals in various primate species has been adapted for optimal sound transmission in the particular habitat in which these signals have evolved. In a study of four species of African monkey, Waser and Waser (1977) found that the frequency spectrum of distance calls is optimal for efficient transmission through the forest canopy. Moreover, these monkeys vary their position when calling so as to maximize the audible range of their signals throughout the day, in accordance with daily fluctuations in temperature gradients.

By analyzing the attenuation properties of the rain forest habitat, Waser and Brown (1986) discovered a "sound window" ideal for the propagation of frequencies near 200 Hz. Relating these features of the habitat acoustics to the characteristics of vocalizations used by two species of monkeys indigenous to this habitat, Waser and Brown found that three of the seven species-specific calls studied made use of the optimal sound window for long-distance propagation. For the four remaining calls, the loss in audible distance resulting from the use of a dominant frequency outside the sound window appeared to be compensated for by a 10 dB increase in sound intensity at the signal source. Waser and Brown (1986) also analyzed the characteristic background noise in the rain forest, generated by birds, insects, and other species, as well as by weather, discovering a quiet zone in the ambient noise between 500–800 Hz. Here too they found an orderly relation between the habitat acoustics and the design features of indigenous monkey calls. Short-range calls appear to be constrained to some extent by the spectrum of background noise in the rain forest, while long-range calls are governed more by attenuation properties of the habitat.

In discussing these provocative findings, Brown (1989) emphasizes that the form of species-specific vocal signals is often a compromise reflecting selection for a number of different acoustic attributes. For example, the most effective frequency for signal transmission through the moist, leafy atmosphere of the rain forest may not be optimal for obscuring the location of the animal emitting the call. This selection for multiple functional attributes of the vocal signal frequently results in a form that does not reflect an optimal specification for any particular isolated attribute. Such complex interactions in the process of adaptation over evolutionary time can make it exceedingly difficult to disentangle and identify with confidence the forces at work in selection. However, by combining acoustic and functional analyses of vocal signals with both perceptual studies of auditory sensitivity and acoustic analysis of habitat characteristics, the ethological research described here provides valuable convergent evidence on the intricate fit between species-specific vocalizations and the biological functions they serve.

Graded and Discrete Vocalizations in Primate Communication

The extent of signal variability within the repertoires of different primate species is a feature of vocal communication systems that appears to be correlated not only with environmental conditions but also with type of social organization. The signal systems of lower primate species consist almost exclusively of discrete categories of vocalizations with nonoverlapping acoustic features. Among higher primates, in contrast, graded signals are more common. Graded signals exhibit considerable acoustic variability within a vocalization category and often overlap in acoustic attributes to some extent with vocalizations in other categories. While their potential ambiguity can pose problems for signal interpretation, graded signals are much more powerful than discrete signals in conveying subtle information about mood and intention.

Marler (1965, 1976) has suggested that the functional significance of discrete versus graded signal systems is related to the social structure and ecological conditions typical of different primate species. In strongly territorial groups, where a substantial portion of the vocal repertoire is dedicated to intergroup signaling, discrete repertoires are most common. The need to use loud, far-ranging calls in communication between groups may have favored the evolution of unambiguous, nonoverlapping signal categories. Discretely organized vocal repertoires are also favored when acoustic signals are not accompanied by complementary information from other sensory modalities, as is the case in long-distance calling by territorial males living in dense forests. Nonterritorial primates, in contrast, are more likely to develop graded vocalizations. Marler (1976) speculates that graded signals have been selected for in primate species that rely heavily on close-range signals within the group. Signals that function in close-range communication are less vulnerable to attenuation and distortion than are long-range calls and thus may be less constrained in the direction of discreteness and invariance. Moreover, the graded vocal signals common in close-range communication within the group can be richly supplemented by redundant visual information, including facial expressions and postural displays. Since animals interacting within the group can often see as well as hear each other, the visual signals accompanying vocalizations may help in decoding the subtleties and ambiguities of graded signals.

It is just these subtleties and potential ambiguities, however, that give graded vocal signals their enormous communicative power. Continuously varying calls can convey nuances of mood and motivation through slight shifts in pitch, intensity, spectrum, or tempo impossible with more stereotyped signals. Graded signals seem particularly appropriate to the intricate communicative acts and subtle negotiations that are so common in the complex social groups of the higher primates. For example, de Waal (1982) describes the gradual escalation in intimidation vocalizations and displays during a competition for power in a captive chimpanzee colony:

> Several times a day Nikkie would be seen sitting somewhere on his own, his hair on end, hooting. His hoot gradually swelled until it ended in a loud screech. Then he would dart across the enclosure and thump heavily on the ground or against one of the metal doors. . . . To begin with, his intimidation displays did not seem to be directed at anyone in particular, but later on they took place more and more frequently in Yeroen's vicinity. Finally he started hooting directly at Yeroen. He would sit opposite him and swing large pieces of wood in the air. (p. 121)

De Waal (1982) describes several other encounters among these chimpanzees in which the hoots of the attacking animal and the screams of the victim incite agitation and aggression, as well as supportive behavior, among other members of the troop. These episodes illustrate the potential of such continuously varying vocalizations not only to index gradual shifts in the motivational state of the vocalizing animal, but also to induce emotional changes in the listener.

Marler's (1965, 1976) speculation that human speech is much more likely to have evolved from a richly graded vocal system than from a repertoire of discrete signals is interesting in this context. Human speech is indeed a continuously varying acoustic signal, but one on which our perceptual mechanisms frequently superimpose discrete categories, as in the categorical perception of certain consonants. In its prosodic structure, however, human speech has fully retained this gradient dimension. The power of prosody in human communication to convey subtle changes in the speaker's emotions and intentions and also to engage and persuade the listener can be seen as a direct legacy from the graded vocal systems of the higher nonhuman primates.

Ritualized Vocalizations: Signal Function and Signal Detectability

A widely accepted principle of the evolution of communication systems is that social signals evolve originally from other behaviors with no signaling function. That is, signals are usually exaptations, or "derived activities" (Tinbergen, 1952), which evolve from nonsignal behaviors through the process of ritualization. Ethologists disagree, however, on how best to characterize the nature and function of communication among nonhuman animals. The idea that a central function of vocalizations is to *persuade* the listener departs from the more prevalent view of communication, which emphasizes the sending and receiving of information (e.g., Smith, 1977). In the first of two articles challenging this traditional focus on the efficiency of information reception by the receiver of a signal, Dawkins and Krebs (1978) argue that natural selection will favor signals that benefit the sender rather than the receiver. According to this more "cynical" view of the evolution of communication, animals use signals primarily in order to manipulate the behavior of other animals, rather than to convey useful information. This instrumental function of nonhuman animal signals is comparable to human advertising slogans, which are designed to persuade rather than to inform, according to Dawkins and Krebs. Thus "the evolutionary ritualisation of derived activities can be better understood in terms of selection for effective manipulation than in terms of selection for effective information transfer" (Dawkins & Krebs, 1978, p. 385).

In a later article on the adaptive functions of social signals, Krebs and Dawkins (1984) expand their previous narrow focus on manipulation in the context of exploitative communication. They argue that ritualized signals are the product of the coevolution of what they call the "manipulator" and the "mind-reader" roles. The manipulator uses signals to influence the behavior of others, usually to its own advantage, while the mind-reader uses signals to predict the future actions of others. As animals become sensitive to subtle cues allowing them to predict other animals' behavior, this increased sensitivity is favored by natural selection. Yet it is this same sensitivity that enables one animal to manipulate another. A dog baring its teeth can be the "victim" of mind reading by other animals, but can also manipulate other animals into retreating by baring its teeth with no intention of attacking. Thus mind reading and manipulation coevolve, according to Krebs and Dawkins, and communicative signals are shaped by this process of coevolution.

A second intriguing argument made by Krebs and Dawkins (1984) is that fundamentally different kinds of signals evolve depending on whether communication is cooperative or noncooperative. In exploitative situations both between and within species, the reactor benefits in general from resisting the persuasion of the manipulator, resulting in one form of arms-race coevolution. For example, skilled "salesmanship" in the use of prominent courtship displays by the males of a species may be complemented by heightened "sales-resistance" on the part of females, since females cannot afford to respond indiscriminately to sexual overtures. The features most commonly associated with signals designed for persuading unwilling victims, according to Krebs and Dawkins, are just those features most effective in advertising: bright packaging, exaggeration, rhythmic repetition, and high redundancy. In cooperative situations, however, where the reactor benefits from being influenced by the manipulator, a different kind of signal typically evolves. When signals are cooperative, as in affiliative interactions within kinship groups, evolution favors "cost-minimizing conspiratorial whispers," rather than conspicuous, repetitive signals (Krebs & Dawkins, 1984, p.391).

Regardless of whether communication is cooperative or noncooperative, effective signals must incorporate design features that ensure detectability. Krebs and Dawkins (1984) cite Wiley's (1983) intepretation of the evolution of ritualized signals in terms of signal detection theory. Wiley proposes four features that increase the reliability of signal detection in noisy environments:

1. *Redundancy,* resulting from predictable relationships among the different components of a familiar signal, facilitates accurate identification of the signal even if only part of it is heard. Repeating parts of a signal, or the entire signal, is a simple form of redundancy used to enhance detectability.
2. *Conspicuousness* by exaggeration of acoustic features enhances the signal-to-noise ratio by increasing the contrast between the signal and the irrelevant background stimulation.
3. *Small repertoires* of signals reduce the listener's uncertainty and enhance performance in signal detection tasks. With fewer and more distinctive categories in which potential signals can be classified, the opportunity for identification errors is minimized.
4. *Alerting components* at the beginning of a signal increase detectability and recognition by letting the listener know when to expect the message component of the signal.

These four adaptations for efficient communication all benefit a signaler by counteracting noise in communication. Noise can result from irrelevant background sounds, as well as from signal distortion or attenuation in the external environment, as illustrated in the research on habitat acoustics described earlier (Brown, 1989). However, noise in its technical sense can also result from characteristics of the receiver, such as the high threshold or cautiousness typical of "sales-resistant" females (Wiley, 1983). Krebs and Dawkins conclude that signal evolution will always reflect a compromise between detectability considerations and the economics of signal production. Increasing signal salience can incur additional costs, such as extra energy expenditure and enhanced risk of predation. Thus the use of such costly ritualized signals evolves only when the stakes are sufficiently high, as with signals used to attract a mate or to defend resources. In the use of cooperative signals, which are designed to influence willing reactors, selection can favor less extravagant signals, as long as they meet the criteria for detectability.

To summarize this section, ethological research suggests that many different kinds of selective pressure can influence the design of species-specific vocalizations, including physiological constraints on signal production and perception, environmental features affecting sound propagation, communicative functions of vocal signaling both within and between social groups, and engineering features ensuring signal detectability by both willing and unwilling listeners. Recent theoretical formulations of the interrelated social and ecological constraints on signal evolution (Dawkins & Krebs, 1978; Krebs & Dawkins, 1984) emphasize that animals use vocalizations for persuasion. In the complex social systems of the higher primates, where communication is focused increasingly on intricate interactions within the group, persuasive vocal signaling becomes more prevalent and more powerful through the use of graded vocalizations. Through intonation, humans also make use of graded signals, which are highly effective in interpersonal communication and which share common features with vocalizations of nonhuman primates and other species.

THE ARGUMENT FOR DESIGN IN HUMAN MATERNAL SPEECH

What kinds of evidence are needed to support the inference that the special form of speech used by human mothers with infants has evolved as a species-specific parenting behavior? It is important, although hardly sufficient, that this infant-directed speech form is widespread across human cultures and is beneficial to parents and infants in their early interactions. Two more demanding criteria need to be addressed in invoking a selectionist argument. First, it has to be shown that this maternal vocal behavior is particularly well engineered for effective communication with preverbal infants, and not just for communication in general. And second, it has to be shown how the use of infant-directed speech by human mothers could have been selected for as a parenting behavior contributing to reproductive success.

Design Features of Infant-Directed Prosody

The human infant is a noisy system, in various senses of the word, and the exaggerated prosody of infant-directed speech is exquisitely designed to boost the signal relative to the noise. The principal noise that the mother's voice must overcome is not, and never was, the attenuation characteristics of the rain forest or savannah, since mother-infant communication in higher primates takes place at very close range. Rather, the noise that interferes most with signal reception is intrinsic to the young infant, resulting from the perceptual, attentional, and cognitive limitations associated with immaturity. For example, human infants like adults are more sensitive to sounds at 500 Hz than at 200 Hz, which means that a 500-Hz signal will sound louder than a 200-Hz signal when both are presented at the same intensity. However, because young infants have higher auditory thresholds than adults (Schneider, Trehub, & Bull, 1979), sounds need to be more intense for infants in order to be detected. By elevating the fundamental frequency of the voice when addressing an infant, mothers effectively compensate for this sensory limitation by moving into a pitch range in which infants are relatively more sensitive, thus increasing the perceived loudness of the signal.

The exaggerated pitch continuity of mothers' vocalizations may also provide a processing advantage for the immature auditory system of the human infant, simplifying the initially demanding task of tracking the voice of a single speaker. Even for adults, the prosodic contours of speech enable the listener to attend selectively to one voice among many (Nooteboom, Brokx, & deRooij, 1976), although adults can use linguistic as well as acoustic structure in accomplishing this task. Infants, in contrast, have no knowledge of linguistic structure and must at first rely entirely on the prosodic coherence of the speech stream in selectively attending to a particular voice. In these and many other ways, the characteristic prosodic features of maternal speech provide the immature listener with acoustic signals that are high in perceptual salience and relatively easy to process (see Fernald, 1984). Just as the calls of birds and mammals are finely tuned to the perceptual limitations of conspecifics and ingeniously engineered to transcend the noise of the environment, the exaggerated vocalizations of human mothers' speech are well designed to accommodate the perceptual predispositions of infants and to overcome the noise in the system due to infants' initial processing limitations.

It is intriguing that the features that selection favors in enhancing signal detectability, according to Wiley (1983)—conspicuousness, redundancy, small repertoires, and alerting components—are all robustly characteristic of infant-directed vocalizations in human speech and decidedly uncharacteristic of adult-directed speech. The *conspicuousness* or perceptual prominence of ID vocalizations results from the elevation of pitch and expansion of pitch range typical of maternal speech across languages, as well as from the use of short vocalizations clearly separated by substantial pauses (Fernald et al., 1989). *Redundancy* in signaling is also strikingly prevalent in speech to infants. Stern et al. (1983) report that mothers repeat over 50% of their phrases when interacting with 2-month-old infants. Prosodic repetition is common too (Fernald & Simon, 1984), often with slight melodic variations, which keep these repetitive runs interesting as well as highly predictable for the infant (Stern, 1977). The use of *small vocal repertoires,* the third feature described by Wiley, is reflected in mothers' tendency to use relatively stereotyped prosodic contours in specific interactional contexts, as shown in Figure 10.3. Moreover, the first words learned by infants generally include "uh-oh," "bye-bye," "peek-a-boo," and other social routines marked by distinctive and highly stereotyped prosody. Finally, the use of *alerting components,* another common feature in ritualized nonhuman animal calls, is also typical of human mothers' speech. In mother-infant play with objects, American mothers frequently call the infant's name or say "Look!" or "What's that?" using elevated pitch in order to engage the infant's attention before labeling the object. Popular mother-infant games like peek-a-boo have alerting components built into the vocal routines accompanying the action (Fernald & O'Neill, in press).

These same signal-enhancing features are described by Krebs and Dawkins (1984) as "advertising" strategies especially effective in persuading unwilling reactors. Bright packaging, rhythmic repetition, exaggeration, and high redundancy also characterize the ritualized vocalizations of human mothers to their infants. In this case, however, communication is cooperative and the reactor is willing. This apparent exception to Krebs and Dawkins's claim that such conspicuous and costy vocalizations will emerge primarily in the service of noncooperative communication is actually quite consistent with their reasoning: Ritualized vocalizations are selected for when they function in biologically significant activities in noisy environments or with "cautious reactors." In mother-infant communication, the reactor is immature rather than cautious, and the noise is attributable to perceptual limitations of the infant rather than to features of the habitat. But this developmental noise is formidable, however temporary, and specially designed vocalizations are needed to ensure signal detection and recognition by an inexperienced listener.

In accord with the idea that manipulation or persuasion is a central function of animal vocalizations, the signal-enhancing features or mothers' speech also seem well designed for persuading the infant listener. As she soothes, alerts, or praises, the mother is directly influencing the infant's state and level of arousal through her voice. This instrumental function of maternal speech, described earlier in relation to the first two levels of the model shown in Figure 10.4, is indeed manipulative, as the mother intuitively and skillfully uses the music of her voice to elicit attention, modulate arousal, and induce emotion in the preverbal infant. The primitive communicative force of maternal speech in this context is close to that of the graded vocalizations used by nonhuman primates, which achieve their impact by affecting the motivational state of the listener. The analogies drawn by Krebs and Dawkins (1984) between ritualized animal signals and both advertising and musical sounds are equally apt in relation to human maternal vocalizations, which also function more immediately to modulate emotion and motivation than to transmit information.

A final argument for the special design features of infant-directed speech has to do with the neurological immaturity of the human infant's auditory system at birth. Until the auditory cortex matures, the young infant relies more on subcortical auditory structures (Whitaker, 1976), which are better suited for the holistic analysis of acoustic signals than for the high-resolution temporal analysis ultimately required for processing speech sounds. For both human and nonhuman primates, subcortical pathways are involved in the recognition of graded acoustic signals and in the production and reception of affective messages (Lamendella, 1977). Marler (1976) has observed that the retention of continuous auditory processing mechanisms for graded signals in music and in the intonation of speech is considered to be a primitive trait in humans. However, the development of limbic and hypothalamic pathways for processing continuously varying affective signals was probably a major evolutionary advance for our nonhuman primate ancestors, enabling much subtler and more differentiated forms of cooperative social communication. It is interesting to note that the earliest vocal communication between the human mother and her infant is emotional in nature and is mediated by intonation, a graded acoustic signal, which the infant can perceive using phylogenetically older and simpler auditory processing mechanisms than those that will eventually develop to process the linguistic units in speech.

Adaptive Functions of Mothers' Speech to Infants

Behaviors that are adaptations must have contributed differentially to reproductive fitness in the evolutionary past. Given their crucial biological role in enhancing reproductive success, species-typical parenting behaviors are often seen as prototypical examples of behavioral adaptations. However, parenting behaviors differ dramatically across species in the domain and time frame of their influence on the young. Some parental behaviors are adaptations for the immediate survival of the young during infancy, an obvious prerequisite for reproductive success, while others have been selected for because of their long-term contribution to the future fitness of the offspring. In arguing for the precision and efficiency of design features in human maternal speech to infants, I have focused on the adaptive functions of this parenting behavior during the infancy period. However, I would also like to argue, somewhat more speculatively, that the use of prosodically modified vocalizations in early mother-infant communication may have had adaptive advantages extending beyond the infancy period over the course of hominid evolution.

Primate parenting behaviors are among the most complex in the animal world. For most reptiles to reproduce successfully, they must survive the first year of life and mate appropriately. Mammals, in contrast, need to survive and mate, and then to nurture and protect their offspring until maturity. Across the primate order, the evolution of an increasingly long period of relative helplessness and dependency in the young has placed extensive demands on parental skills and commitment. From the point of view of basic behavioral equipment, newborn monkeys, apes, and humans are fundamentally similar. They all show reflexive rooting, sucking, and grasping responses, and they all cry when distressed and are comforted when held (Mason, 1968). But differences among primate species in the strength and persistence of these primitive infantile responses are striking. A newborn rhesus monkey placed on its back will right itself immediately, unless given something to cling to. A newborn chimpanzee in the same situation will wave its arms about irregularly, unable to right itself, until it is several weeks old. The human infant will not develop the ability to turn over until around the age of 3 months and will not be capable of locomotion until the age of 8 months.

> Viewed as a collection of responses preadapted to nursing and maintaining contact with the mother, the behavior of the chimp never displays the same reflex-like efficiency as that of the monkey. Indeed, if evolution had not brought about complementary changes in the behavior of the chimp mother that permitted her to compensate for the behavioral deficiencies of her infant, its chances of survival would be slim. (Mason, 1968)

As Mason suggests, the lengthening of the period of immaturity among higher primate species has coevolved with specialized parenting skills designed to compensate for the increased helplessness of the young. Of course, the complementary evolution of parental behaviors appropriate to the characteristics and needs of the young is evident in every species in which brood care is typical. What is remarkable about parenting behaviors in higher primates is their flexibility in accommodating the continually changing needs of the young throughout a relatively long period of dependency. An experiment by Rumbaugh (1965) demonstrated the extent to which primates will modify their behavior in unusual ways in order to accommodate the needs of the young. When the arms of an infant squirrel monkey were taped so that it was unable to cling, the mother responded with bipedal carrying and cradling of the infant, both highly atypical behaviors under normal circumstances. Berkson (1974) and Rosenblum and Youngstein (1974) report numerous other examples from field studies in which monkey mothers resourcefully compensate for infant disabilities in ways that are appropriate to the developmental status of the infant. In this respect, primates differ dramatically from reptiles, where parenting behaviors are minimal or nonexistent, as well as from most other mammals, where parenting skills are critical only during a brief period and consist of a relatively fixed and limited repertoire of behaviors.

The evolution of this primate capacity for flexible accommodation to the needs and limitations of the infant is epitomized in human parental behavior. Human mothers nurse and carry their infants, as do other primates, but with an incomparably greater diversity of means to the common biological ends, including using bottles to feed and prams to carry the infant. Human mothers' use of a special infant-directed speech style is also an accommodation to the immaturity of the infant, compensating for early perceptual, attentional, and cognitive limitations, and changing gradually as the infant develops. This species-specific parental vocal behavior is not characterized by a fixed repertoire of discrete signals, but rather by the flexible use of graded prosodic variations to influence the state and behavior of the infant and to optimize communication with an immature and inexperienced listener. Moreover, mothers of deaf infants are quite capable of translating the lively dynamic rhythms of infant-directed speech into the visual modality. What is common in mothers' interactions with both deaf and hearing infants is the use of a highly salient display that engages the infant both perceptually and affectively and that is contingent on the infant's emotional and behavioral responses. In its biological utility, its compensatory functions, and its flexibility of use, human maternal speech is similar to many other primate parenting behaviors that have evolved through natural selection.

The adaptive functions served by parental behaviors specialized for feeding, soothing, carrying, and communicating with young infants are fairly obvious in that these infant-directed behaviors enhance the likelihood of survival early in life. What is not as obvious is that such caretaking behaviors can have long-term regulatory effects on the infant's physiology and behavior extending far beyond the period of infancy. Hofer's (1981, 1987) extensive research with infant rats has revealed enduring negative effects on the infant's thermoregulation, motor behavior, and behavioral reactivity resulting from early separation from the mother. These findings suggest that under normal circumstances, the regulatory effects of maternal behavior "are exerted over long periods of time through repeated episodes of stimulation delivered by the mother to the infant during their ongoing social relationship" (Hofer, 1981, p. 93). Studies with both rats and monkeys indicate that the premature removal of these powerful regulatory influences can result in a kind of physiological withdrawal (Hofer, 1987). Hofer concludes that such biological processes hidden within the rhythmic stimulation of early human mother-infant interaction may help to explain both the formation of attachments and the intense painfulness of loss in adulthood as well as in infancy.

Hofer's (1987) ideas about the long-term regulatory functions of early mother-infant interaction suggest a psychobiological basis for the relation between maternal responsiveness and infant socioemotional development that is central to attachment theory (Ainsworth, 1973). In Bowlby's (1969) original formulation of attachment theory, he proposed that mother-infant attachment had evolved primarily as an adaptation designed to protect the infant from predators. More recent research proposes that the quality of attachment in infancy is related to social competence in childhood, as well as to the quality of emotional relationships and parenting skills in adulthood (Main, Kaplan, & Cassidy, 1985). From this broader perspective, early mother-infant attachment has adaptive consequences extending well beyond the period of infancy. It is important to remember that the prolonged period of immaturity evolved in primates not only because it takes time to grow a large and complex brain, but also because it takes time to acquire the intricate social skills prerequisite for successful reproduction in these species. In primates, reproductive success depends not only on parenting behaviors that ensure infant survival, but also on parenting behaviors that enable the infant to become a competent member of the social group and ultimately to function effectively as a parent as well. From quite different perspectives, both Hofer's (1987) psychobiological research and research on infant attachment (e.g., Main et al., 1985) suggest that human maternal behaviors in the period of evolution could have had both immediate consequences for infant survival and early development and long-term consequences related to socioemotional development later in life.

Are these conjectures reasonable, given that we have no evidence for the contribution of particular parenting behaviors to reproductive success in the era of hominid evolution? And in particular, how could the use of melodic vocalizations in mothers' interactions with infants have enhanced fitness in our hominid ancestors? Assuming, with Hofer (1987), that rhythmic tactile and vestibular stimulation has a powerful regulatory influence on all mammalian infants, rhythmic maternal vocalizations may have become a potent form of auditory stimulation for infants in a species in which speech was evolving as a primary means of communication. Mothers' use of exaggerated, highly salient vocal patterns would have been effective then, as they are now, in engaging and maintaining infants' attention and in modulating arousal and affect. And through the melodies of the mother's voice, infants could gain early access to her feelings and intentions. This experience of emotional communion through repeated episodes of rhythmic vocal stimulation may have led to what Stern (1985) calls "affect attunement" between mother and infant, giving the infant crucial early experience in mind reading and establishing the basis for effective interpersonal communication with other conspecifics. Could such early experience have provided a reproductive advantage to the offspring of mothers more skillful in establishing communication with their infants? If sensitive and responsive maternal care in infancy enhanced the ability of offspring to read accurately the social signals of conspecifics and to function more effectively in later interpersonal relationships, the offspring of more competent mothers may ultimately have been more successful in attracting desirable mates and in parenting their own infants. To the extent that early emotional experience contributed to the development of social competence, and social competence contributed to success in reproduction and parenting, maternal behaviors effective in establishing emotional communication in infancy could have had long-term as well as short-term consequences related to fitness.

Given our reliance on linguistic symbols to decode meanings in speech, it is perhaps difficult for us to appreciate just how important it is to be able to communicate through emotional signals. The prediposition to be moved by as well as to interpret the emotions of others and the ability to discern the intentions and motivations of others through expressions of the voice and face are remarkable evolutionary advances in communicative potential among the higher primates. Human mothers intuitively and skillfully use melodic vocalizations to soothe, to arouse, to warn and to delight their infants, and to share and communicate emotion. While the acquisition of language will eventually give the child an access to other minds that is immeasurably more powerful and intricate than that of other primates, human symbolic communication builds on our primate legacy, a foundation of affective communication established in the preverbal period.

NOTE

1. Although fossil evidence can help one make inferences about adaptation, it is not, of course, criterial. This is because the fossil record can preserve both adaptations and nonadaptive aspects of the phenotype. The ultimate criterion for demonstrating adaptation is good design for accomplishing a function that would have promoted reproduction in ancestral environments.

REFERENCES

Ainsworth, M.D.S. (1973). The development of infant-mother attachment. In B. M. Caldwell & H. N. Ricciuti (Eds.), *Review of child development research* (Vol. 3) (pp. 1–94). Chicago: University of Chicago Press.

Aslin, R. N. (1987). Visual and auditory development in infancy. In J. D. Osofsky (Ed.), *Handbook of infant development* (2nd ed.). New York: John Wiley & Sons.

Bench, J. (1969). Some effects of audio-frequency stimulation on the crying baby. *The Journal of Auditory Research, 9,* 122–128.

Berg, W. K. (1975). Cardiac components of defense responses in infants. *Psychophysiology, 12,* 224.

Berkson, G. (1974). Social responses of animals to infants with defects. In M. Lewis & L. A. Rosenblum (Eds.), *The effect of the infant on its caretaker.* New York: John Wiley & Sons.

Birns, B., Blank, M., Bridger, W., & Escalona, S. (1965). Behavioral inhibition in neonates produced by auditory stimuli. *Child Development, 36,* 639–645.

Bock, W. J. (1980). The definition and recognition of biological adaptation. *American Zoology, 20,* 217–227.

Bowlby, J. (1969). *Attachment.* Middlesex, England: Penguin Books.

Brown, C. W. (1989). The acoustic ecology of East African primates and the perception of vocal signals by grey-cheeked Mangabeys and blue monkeys. In R. J. Dooling & S. H. Hulse (Eds.), *The comparative psychology of audition.* Hillsdale, NJ: Erlbaum.

Bühler, C., & Hetzer, H. (1928). Das erste Verständnis für Ausdruck im ersten Lebensjahr. *Zeitschrift für Psychologie, 107,* 50–61.

Byrne, J. M., & Horowitz, F. D. (1981). Rocking as a soothing intervention: The influence of direction and type of movement. *Infant Behavior and Development, 4,* 207–218.

Caron, A. J., Caron, R. F., & MacLean, D. J. (1988). Infant discrimination of naturalistic emotional expressions: The role of face and voice. *Child Development, 59,* 604–616.

Chew, J. J. (1969). The structure of Japanese baby talk. *Association of Teachers of Japanese, 6,* 4–17.

Chomsky, N. (1972). *Language and mind.* New York: Harcourt, Brace, & World.

Clarkson, M. G., & Berg, W. K. (1983). Cardiac orienting and vowel discrimination in newborns: Crucial stimulus parameters. *Child Development, 54,* 162–171.

Cooper, R. P., & Aslin, R. N. (1990). Preference for infant-directed speech in the first month after birth. *Child Development, 61,* 1584–1595.

Cosmides, L. (1983). Invariances in the acoustic expression of emotion during speech. *Journal of Experimental Psychology, 9,* 864–881.

Darwin, C. (1859). *On the origin of species.* London: J. Murray.

Darwin, C. (1872/1956). *The expression of emotions in man and animals.* Chicago: University of Chicago Press.

Dawkins, R. (1986). *The blind watchmaker: Why the evidence of evolution reveals a universe without design.* New York: Norton.

Dawkins, R., & Krebs, J. R. (1978). Animal signals: Information or manipulation? In J. R. Krebs & N. B. Davies (Eds.), *Behavioural ecology: An evolutionary approach* (pp. 282–309). Sunderland, MA: Sinauer Associates.

DeCasper, A. J., & Fifer, W. P. (1980). *Science, 208,* 1174–1176.

Demany, L., McKenzie, B., & Vurpillot, E. (1977). Rhythm perception in early infancy. *Nature, 266,* 718–719.

de Waal, F. (1982). *Chimpanzee politics: Power and sex among apes.* Baltimore, MD: The Johns Hopkins University Press.

Dooling, R., & Searcy, M. (1980). Early perceptual selectivity in the swamp sparrow. *Developmental Psychobiology, 13,* 499–506.

Eisen, J. M., & Fernald, A. (in preparation). *Prosodic modifications in the infant-directed speech of rural Xhosa-speaking mothers.*

Ekman, P. (1972). Universals and cultural differences in facial expressions of emotion. In J. Cole (Ed.), *Nebraska symposium on motivation* (pp. 207–283). Lincoln: University of Nebraska Press.

Ferguson, C. A. (1964). Baby talk in six languages. *American Anthropologist, 66,* 103–114.

Fernald, A. (1984). The perceptual and affective salience of mothers' speech to infants. In L. Feagans, C. Garvey, & R. Golinkoff (Eds.), *The origins and growth of communication.* Norwood, N.J.: Ablex.

Fernald, A. (1985). Four-month-old infants prefer to listen to motherese. *Infant Behavior and Development, 8,* 181–195.

Fernald, A. (1989). Intonation and communicative intent: Is the melody the message? *Child Development, 60,* 1497–1510.

Fernald, A. (1992). Meaningful melodies in mothers' speech to infants. In H. Papousek, U. Jurgens, & M. Papousek (Eds.), *Nonverbal vocal communication: Comparative and developmental approaches.* Cambridge: Cambridge University Press.

Fernald, A. & Clarkson, M. (in preparation). *Infant cardiac orienting to infant-directed vocalisations.*

Fernald, A., Kermanschachi, N., & Lees, D. (1984, April). *The rhythms and sounds of soothing: Maternal vestibular, tactile, and auditory stimulation and infant state.* Paper presented at the International Conference on Infant Studies, New York.

Fernald, A., & Kuhl, P. K. (1987). Acoustic determinants of infant preference for motherese speech. *Infant Behavior and Development, 10,* 279–293.

Fernald, A., & Mazzie, C. (1991). Prosody and focus in speech to infants and adults. *Developmental Psychology 27,* 209–221.

Fernald, A., & McRoberts, G. (April 1991). *Prosody and early lexical comprehension.* Paper presented at the meeting of the Society for Research on Child Development, Seattle.

Fernald, A. & O'Neill, D. K. (in press). Peekaboo across cultures: How mothers and infants play with voices, faces, and expectations. In: K. McDonald & A. D. Pelligrini (Eds.), *Play and culture.* Buffalo, NY: SUNY Press.

Fernald, A., & Simon, T. (1984). Expanded intonation contours in mothers' speech to newborns. *Developmental Psychology, 20,* 104–113.

Fernald, A., Taeschner, T., Dunn, J., Papousek, M., Boysson-Bardies, B., & Fukui, I. (1989). A cross-language study of prosodic modifications in mothers' and fathers' speech to preverbal infants. *Journal of Child Language, 16,* 477–501.

Fernald, R. D. (1988). Aquatic adaptations in fish eyes. In J. Atema, R. R. Fay, A. N. Popper, & W. N. Tavolga (Eds.), *Sensory biology of aquatic animals* (pp. 435–466). New York: Springer-Verlag.

Ferrier, L. J. (1985). Intonation in discourse: Talk between 12-month-olds and their mothers. In K. Nelson (Ed.), *Children's language* (Vol. 5) (pp. 35–60). Hillsdale, NJ: Erlbaum.

Friedlander, B. Z. (1968). The effect of speaker identity, voice inflection, vocabulary, and message redundancy on infants' selection of vocal reinforcement. *Journal of Experimental Child Psychology, 6,* 443–459.

Garnica, O. (1977). Some prosodic and paralinguistic features of speech to young children. In C. E. Snow & C. A. Ferguson (Eds.), *Talking to children: Language input and acquisition* (pp. 63–88). Cambridge, MA: Cambridge University Press.

Gottlieb, G. (1980). Development of species identification in ducklings: VI. Specific embryonic experience required to maintain species-typical perception in Peking ducklings. *Journal of Comparative and Physiological Psychology, 94,* 499–587.

Gould, J. L., & Marler, P. (1987). Learning by instinct. *Scientific American, 255,* 74–85.

Gould, S. J., & Lewontin, R. C. (1979). The spandrels of San Marco and the Panglossian program: A critique of the adaptationist programme. *Proceedings of the Royal Society of London, 205,* 281–288.

Gould, S. J., & Vrba, E. S. (1982). Exaptation—a missing term in the science of form. *Paleobiology, 8,* 4–15.

Green, S., & Marler, P. (1979). The analysis of animal communication. In P. Marler & J. G. Vandenbergh (Eds.), *Handbook of behavioral neurobiology: III. Social behavior and communication* (pp. 73–158). New York: Plenum Press.

Grieser, D. L., & Kuhl, P. K. (1988). Maternal speech to infants in a tonal language: Support for universal prosodic features in motherese. *Developmental Psychology, 24,* 14–20.

Hewes, G. W. (1973). Primate communication and the gestural origin of language. *Current Anthropology, 14,* 5–24.

Hofer, M. A. (1981). Parental contributions to the development of their offspring. In D. J. Gubernick & P. H. Klopfer (Eds.), *Parental care in mammals* (pp. 77–115). New York: Plenum Press.

Hofer, M. A. (1987). Early social relationships: A psychobiologist's view. *Child Development, 58,* 633–647.

Jacobson, J. L., Boersma, D. C., Fields, R. B., & Olson, K. L. (1983). Paralinguistic features of adult speech to infants and small children. *Child Development, 54,* 436–442.

Kearsley, R. B. (1973). The newborn's response to auditory stimulation: A demonstration of orienting and defensive behavior. *Child Development, 44,* 582–590.

Klinnert, M., Campos, J. J., Sorce, J. F., Emae, R. N., & Svejda, M. (1983). Emotions as behavior regulators: Social referencing in infancy. In R. Plutchik & H. Kellerman (Eds.), *Emotion in early development. Vol. 2: The emotion* (pp. 57–86). New York: Academic Press.

Krebs, J. R., & Dawkins, R. (1984). Animal signals: Mind-reading and manipulation. In J. R. Krebs & N. B. Davies (Eds.), *Behavioral ecology: An evolutionary approach* (pp. 380–402). Oxford: Blackwell Scientific Publications.

Lamendella, J. T. (1977). The limbic system in human communication. In H. Whitaker & H. Whitaker (Eds.), *Studies in neurolinguistics* (Vol. 3) (pp. 157–222). New York: Academic Press.

Lewis, M. M. (1936/1951). *Infant speech: A study of the beginnings of language.* London: Routledge & Kegan Paul.

Lewontin, R. (1978). Adaptation. *Scientific American, 239,* 157–169.

Main, M., Kaplan, N., & Cassidy, J. (1985). Security in infancy, childhood, and adulthood: A move to the level of representation. In I. Bretherton & E. Waters (Eds.), *Growing points of attachment theory and research* (Monographs of the Society for Research in Child Development, Vol. 50[1–2]) (pp. 66–104). Chicago, IL: University of Chicago Press.

Marler, P. (1965). Communication in monkeys and apes. In I. DeVore (Ed.), *Primate behavior: Field studies of monkeys and apes* (pp. 544–584). New York: Holt, Rinehart & Winston.

Marler, P. (1976). Social organization, communication and graded signals: The chimpanzee and the gorilla. In P.P.G. Bateson & R. A. Hinde (Eds.), *Growing points in ethology.* Cambridge: Cambridge University Press.

Marler, P. (1984). Song learning: Innate species differences in the learning process. In P. Marler & H. S. Terrace (Eds.), *The biology of learning* (pp. 289–309). Berlin: Dahlem Konferenzen.

Marler, P., Zoloth, S., & Dooling, R. (1981). Innate programs for perceptual development: An ethological view. In G. Gollin (Ed.), *Developmental plasticity: Behavioral and biological aspects of variations in development* (pp. 135–172). New York: Academic Press.

Mason, W. A. (1968). Early social deprivation in the nonhuman primates: Implications for human behavior. In D. C. Glass (Ed.), *Environmental influence, biology and behavior series.* New York: Rockefeller University Press.

Meegaskumbura, P. B. (1980). Tondol: Sinhala baby talk. *Word, 31,* 287–309.

Mehler, J., Jusczyk, P., Lambertz, G., Halsted, N., Bertoncini, J., & Amiel-Tison, C. (1990). A precursor of language acquisition in young infants. *Cognition, 29,* 143–178.

Miller, C. L., & Byrne, J. M. (1983). Psychophysiological and behavioral response to auditory stimuli in the newborn. *Infant Behavior and Development, 6,* 369–389.

Morton, E. S. (1977). On the occurrence and significance of motivation-structural rules in some bird and mammal sounds. *American Naturalist, 111,* 855–869.

Nelson, C. A. (1987). The recognition of facial expressions in the first two years of life: Mechanisms of development. *Child Development, 58,* 889–909.

Nooteboom, S. G., Brokx, J.P.L., & DeRooij, J. J. (1976). Contributions of prosody to speech perception. *IPO Annual Progress Report, 5,* 34–54.

Ohala, J. J. (1984). An ethological perspective on common cross-language utilization of F_0 of voice. *Phonetica, 41,* 1–16.

Papousek, H. (April, 1987). *Models and messages in the melodies of maternal speech in tonal and non-tonal languages.* Paper presented at the meeting of the Society for Research in Child Development, Baltimore, MD.

Papousek, M., Papousek, H., & Bornstein, M. H. (1985). The naturalistic vocal environment of young infants: On the significance of homogeneity and variability in parental speech. In T. Field & N. Fox (Eds.), *Social perception in infants* (pp. 269–297). Norwood, NJ: Ablex.

Papousek, M., Papousek, H., & Haekel, M. (1987). Didactic adjustments in fathers' and mothers' speech to their three-month-old infants. *Journal of Psycholinguistic Research, 16,* 491–516.

Piattelli-Palmarini, M. (1989). Evolution, selection, and cognition: From "learning" to parameter setting in biology and the study of language. *Cognition, 31,* 1–44.

Pinker, S., & Bloom, P. (1990). Natural language and natural selection. *Brain and Behavioral Sciences.*

Ratner, N. B., & Pye, C. (1984). Higher pitch in BT is not universal: Acoustic evidence from Quiche Mayan. *Journal of Child Language, 2,* 515–522.

Rosenblum, L. A., & Youngstein, K. P. (1974). Developmental changes in compensatory dyadic response in mother and infant monkeys. In M. Lewis & L. A. Rosenblum (Eds.), *The effect of the infant on its caretaker* (pp. 211–226). New York: John Wiley & Sons.

Ruke-Dravina, V. (1976). Gibt es Universalien in der Ammensprache? *Salzburger Beiträge zur Linguistik, 2,* 3–16.

Rumbaugh, D. M. (1965). Maternal care in relation to infant behavior in the squirrel monkey. *Psychological Reports,* 171–176.

Ryan, M. (1978). Contour in context. In R. Campbell & P. Smith (Eds.), *Recent advances in the psychology of language* (pp. 237–251). New York: Plenum Press.

Ryan, M. J., Fox, J. H., Wilczynski, W., & Rand, A. S. (1990). Sexual selection for sensory exploitation in the frog Physalaemus pustulosus. *Nature, 343,* 66–67.

Scherer, K. R. (1985). Vocal affect signaling: A comparative approach. *Advances in the study of behavior, 15,* 189–244.

Scherer, K. R. (1986). Vocal affect expression: A review and a model for future research. *Psychological Bulletin, 99,* 143–165.

Schneider, B. A., Trehub, S. E., & Bull, D. (1979). The development of basic auditory process in infants. *Canadian Journal of Psychology, 33,* 306–319.

Shimoda, K., Argyle, M., & Riccibitti, P. (1978). The intercultural recognition of emotional expressions by three national racial groups: English, Italian and Japanese. *European Journal of Social Psychology, 8,* 169–179.

Smith, W. J. (1977). *The behavior of communicating: An ethological approach.* Cambridge, MA: Harvard University Press.

Snow, C. E. (1977). The development of conversation between mothers and babies. *Journal of Child Language, 4,* 1–22.

Snow, C. E., & Ferguson, C. E. (1977). *Talking to children: Language input and acquisition.* Cambridge, MA: Cambridge University Press.

Steinschneider, A., Lipton, E. L., & Richmond, J. B. (1966). Auditory sensitivity in the infant: Effect of intensity on cardiac and motor responsivity. *Child Development, 37,* 233–252.

Stern, D. N. (1977). *The first relationship: Infant and mother.* Cambridge: Harvard University Press.

Stern, D. N. (1985). *The interpersonal world of the infant.* New York: Basic Books.

Stern, D. N., Spieker, S., Barnett, R. K., & MacKain, K. (1983). The prosody of maternal speech: Infant age and context related changes. *Journal of Child Language, 10,* 1–15.

Stern, D. N., Spieker, S., & MacKain, K. (1982). Intonation contours as signals in maternal speech to prelinguistic infants. *Developmental Psychology, 18,* 727–735.

Tartter, V. C. (1980). Happy talk: Perceptual and acoustic effects of smiling on speech. *Perception and Psychophysics, 27,* 24–27.

Tinbergen, N. (1952). "Derived" activities: Their causation, biological significance, origin and emancipation during evolution. *Quarterly Review of Biology, 17,* 1–32.

Tinbergen, N. (1963). On aims and methods of ethology. *Zeitschrift für Tierpsychologie, 20,* 410–429.

Trehub, S. E., Thorpe, L. A., & Morrongiello, B. A. (1987). Organizational processes in infants' perception of auditory patterns. *Child Development, 58,* 741–749.

Waser, P. M., & Brown, C. H. (1986). Habitat acoustics and primate communication. *American Journal of Primatology, 10,* 135–154.

Waser, P. M., & Waser, M. S. (1977). Experimental studies of primate vocalisations: Specializations for long-distance propagation. *Zietschrift für Tierpsychologie, 43,* 239–263.

Watterson, T., & Riccillo, S. C. (1983). Vocal suppression as a neonatal response to auditory stimuli. *Journal of Auditory Research, 23,* 205–214.

Werker, J. F., & McLeod, P. J. (1989). Infant preference for both male and female infant-directed-talk: A developmental study of attentional and affective responsiveness. *Canadian Journal of Psychology, 43,* 230–246.

Whitaker, H. A. (1976). Neurobiology of language. In E. C. Carterette, & M. P. Friedman (Eds.), *Handbook of perception* (Vol. 7) (pp. 121–144). New York: Academic Press.

Wiley, R. H. (1983). The evolution of communication: Information and manipulation. In T. R. Halliday & P.J.B. Slater (Eds.), *Communication* (pp. 156–215). Oxford: Blackwell Scientific Publications.

Williams, G. C. (1966). *Adaptation and natural selection: A critique of some current evolutionary thought.* Princeton, NJ: Princeton University Press.

Wolff, P. H. (1963). Observations on the early development of smiling. In B. M. Foss (Ed.), *Determinants of infant behavior, II* (pp. 113–134). London: Methuen.

Wormith, S. J., Pankhurst, D., & Moffitt, A. R. (1975). Frequency discrimination by young infants. *Child Development, 46,* 272–275.

11

The Social Nature of Play Fighting and Play Chasing: Mechanisms and Strategies Underlying Cooperation and Compromise

MICHAEL J. BOULTON AND PETER K. SMITH

AN EVOLUTIONARY APPROACH TO THE STUDY OF ROUGH-AND-TUMBLE PLAY

Play is a widespread behavior in the young of most mammalian species. It is a noticeable feature of behavior in primates, and although there are important within- and between-species differences in the actual forms of play behaviors, it very often appears as play fighting and play chasing, collectively known as rough-and-tumble (r/t). Rough-and-tumble will be the main focus of attention in this chapter, although how it may be related to some other types of social play will also be discussed.

Defining r/t (as with other taxonomic groups of play, as well as play in general) has proved to be difficult. The problem basically lies in separating those actions that are playful from their "serious" (i.e., nonplayful) counterparts. At least at a superficial level, play fighting looks very similar in appearance to real fighting, drawing on a common repertoire of molecular action patterns such as hitting and kicking. Our recent approach (Boulton, 1988) has been to break down r/t into a taxonomy with several categories and then to apply three main identifying criteria to ensure that these behaviors can be considered playful. These criteria are as follows:

1. Characteristics of physical action per se, particularly the strength of a blow, kick, etc.
2. Presence/absence of signs of injury/distress/annoyance by recipient
3. Presence/absence of signs of regret by the perpetrators of injury/distress/annoyance

Thus, apparently, fighting behaviors are judged as playful if they do not involve powerful blows, if they do not cause injury and/or distress and/or annoyance to one or the other party, or, when they do involve these things, if the offender shows signs of regret suggesting that they were accidental. Behavior would be seen as aggressive if powerful actions that caused injury and/or distress and/or annoyance were not accompanied by signs of regret or were accompanied by insults and other negative statements.

Rough-and-tumble, as a general class of behavior, has been observed in a wide variety of cultural settings. It is well documented in the U.S.A. and U.K., and Table 11.1

Table 11.1   Studies Reporting the Occurrence of Rough-and-Tumble in non-Western
Societies

Study	Society
Blurton-Jones and Konner (1973)	Kalahari San (Zhun twa–Africa)
Eibl-Eibesfeldt (1974)	!Ko (Africa)
Fry (1987)	Zapotec (Mexico)
Mayer and Mayer (1970)	Red Xhosa (Africa)
Raum (1940)	Chaga (Tanzania)
Whiting and Whiting (1975)	Japan, the Philippines, India, Kenya, and Mexico

contains a (nonexhaustive) list of studies carried out in a number of different non-Western societies that have reported its occurrence. There are grounds for considering r/t to be a human universal, and so it would seem appropriate for researchers to consider why this should be the case. Studies of r/t might begin with the assumption that it is, or was at some period in our evolutionary past, providing some benefit to those individuals who engaged in it.

Perhaps the single most important trend in animal play research over the past two decades has been the adoption of an evolutionary approach. Such a perspective was developed by several independent investigators (Bekoff & Byers, 1981; Fagen, 1974, 1978; Konner, 1975, 1977; Symons 1974, 1978a, 1978b).

Fagen (1981) has argued that the main reason why our understanding of animal play increased only very slowly was because available theory was inadequate. He states, "recent advances in evolutionary ethology and related fields have made such theory available. Why not view play in the light of these advances? The success of this approach will ultimately be measured not only by the scientific quality of the empirical and theoretical studies it fosters, but also by the extent to which it enhances aesthetic and intellectual appreciation of the natural history of animal play" (p.37).

There is growing optimism that an evolutionary approach will yield significant gains, and more recently such an approach has been taken with the play of children (e.g., Boulton, 1988; Smith, 1982).

An unwelcomed by-product of an evolutionary approach to the study of play, especially during the 1970s, has been the emergence of a vast number of new hypotheses that are often presented in such a vague manner that testing them would be virtually impossible. In response to this plethora of hypotheses, Hinde (1975) stated that "because hard evidence is so difficult to obtain, it has become respectable to speculate about the function of behaviour in a manner that would never be permissible in studies of [proximate] causation" (p.13). It is important that hypotheses are clearly defined and predictions stemming from them made explicit.

Numerous writers have cautioned against uncritical generalizations of functional hypotheses about animal play to humans. Some (e.g., Hinde, 1974) suggest that the differences between the play of humans and other animals may be more informative sources of data than the similarities, and Smith (1983) points out that the break in the continuum is probably most apparent in terms of the presence of linguistic, fantasy, and sociodramatic play in children but their absence in all other species (with the possible exception of the chimpanzee). This is not to say that theoretical and methodological advances in the study of animal play cannot be usefully applied to humans. Martin and Caro (1985) state that "most hypotheses about the effects of human play (as opposed to biological functions, *sensu stricto*) are qualitatively different from those applied to play in other species. Notwithstanding these problems, the study of play in humans can and should be related, where possible, to our knowledge of play in other animals" (p.75).

Despite the differences between animal and human play, similarities are most marked in r/t. The forms of r/t seem to show some continuity throughout the primates, albeit varying in detail across species. We believe that there is a case for relating functional hypotheses about r/t in other species (especially primates) to our studies of the evolution of r/t behavior in humans.

SELECTION PRESSURES LEADING TO THE EMERGENCE OF HUMAN FORMS OF R/T

During hominid evolution, several different, but arguably related, environmental pressures probably led to the shaping of r/t and other forms of play into their characteristically human forms. We can gain insights into what these probable pressures might have been from a consideration of both present-day nonhuman primates and contemporary hunter-gatherer societies, whose environments are likely to have many features in common with those encountered by earlier hominids. Some of the most important candidates for these driving forces are related to skills associated with hunting, predator avoidance, and fighting. We shall consider each in turn, especially using evidence on sex differences in these forms of behavior. Sex differences in the frequency of occurrence of r/t have been observed in the play of many species. These differences may provide important insights into the function of the behavior, especially when they are related to sex differences in corresponding nonplayful behavior.

In humans, one of the most robust findings reported in the literature is that boys engage in more r/t than girls (see Humphreys & Smith, 1981; and Maccoby & Jacklin, 1974, for reviews). However, in most cases, researchers have reported only one global measure of r/t, which includes unknown amounts of play fighting and play chasing (e.g., Humphreys & Smith, 1987). These two components are not applicable to all of the hypotheses to be discussed below, and consequently, we will, where possible, consider data on sex differences in the occurrence of play fighting and play chasing separately.

Hunting

Many hypotheses about the function of r/t in nonhuman predatory species, such as cats, have suggested that this form of play provides practice for the development of adult hunting skills (Egan, 1976; Moelk, 1979). This view is not surprising given that r/t and predation both involve stalking, chasing, pouncing, pawing, and biting. Caro's (1979, 1981) investigation of the links between the social play (which often is made up of r/t) of 1- to 3-month-old kittens and their skill as hunters at 6 months provides some indirect evidence. The results were not clear-cut: While there were few significant correlations between early play and later predatory ability, "there was . . . a hint that some aspects of social play became increasingly associated with certain features of predatory behaviour in older kittens." Similar studies have been carried out with canids (e.g., Vincent & Bekoff, 1978).

A number of researchers believe that strength and skill in hunting large and difficult game were selected for in human males but not in females (e.g., Wilson, 1975; Smith, 1982). Smith stated that "hominid predation probably relied on the chasing and running down of prey, together with the increased use of weapons, both at a distance (aimed missiles) and at close quarters (sharp-edged tools, clubs)" (1982, p.151). From this analysis there are two separate components that make up hunting/predation. One is running ability directed at following a particular target, and the other is dexterity with weapons. The validity of the hypothesis that play chasing provides practice for the running aspect of hunting skill in males but not females cannot be adequately assessed. If it were true, and if the practice of hunting skills was (one of) the main function(s) of play chasing, we would predict that males should show more play chasing than females. Few researchers have provided quantitative data on this, and what there is, is inconsistent. Whereas Smith and Connolly (1972) found that nursery school girls engaged in more chasing/fleeing than boys, Boulton (1988) reported a nonsignificant difference on this dimension among 8- and 11-year-olds.

Another difficulty in assessing this prediction is that Laughlin (1968) claimed that early human hunters were more likely to quietly stalk and/or track an animal or wait in ambush for one to pass by. This alternative view would lead to a different prediction about the frequency of occurrence of chasing in the r/t of boys and girls (and one more in accord with the data so far). In general this illustrates the difficulty (but not impossibility) of getting unambiguous predictions from evolutionary theory, as alternative hypotheses can so often be suggested. In this case conclusions would be premature since more data are needed, and we would suggest that researchers also consider the characteristics of bouts of play chasing engaged in by the two sexes. If they are providing practice for males but not for females, then natural selection should have ensured that they would last longer for males as compared with females since there would be a greater need for males to build up their stamina.

The practice hunting hypothesis should also make predictions about the capability of dispatching prey once caught (in the human case, requiring physical strength and dexterity with weapons). Since these latter characteristics are also relevant to intraspecific competition, we consider it further under the later section on fighting.

Predator Avoidance

Several researchers have claimed that the play of some species, including hominids, provides practice for predator avoidance and that the actual form of play in a particular species is related to the strategies typically employed by that species to escape. Ewer (1966), for example, described "jinking play" in the African ground squirrel, in which they run fast and change direction frequently, and also a form of play in which they jump straight up in the air, turn in mid-flight, land, and run off in another direction. These play behaviors are very similar to those shown during emergency escape behavior, the latter cluster being typically shown in response to the close proximity of a snake. Ewer sees this behavior pattern as the only possible way to avoid being bitten by a snake in full strike.

Similarly, Dolhinow (1971) observed that young patas monkeys show a behavior pattern in play that adults use to escape from predators—that of running headlong into a flexible sapling or bush and catapulting themselves in another direction before running off.

One of the most detailed observational studies to address the view that some aspects of play provide practice for developing antipredator behaviors was that conducted by Symons (1978a), who took account of the design features of play chasing and serious predator-avoidance activities. He found "striking" similarities between the behaviors occurring in these two contexts and concluded that "the adaptive function of playchasing is to rehearse, or practice, and thereby perfect the specific locomotor skills used during emergencies" (p.84). Symons believes there are important reasons why such practice is necessary, particularly as he believes high-speed locomotion that occurs in the context of predator-avoidance to be "the most difficult locomotor task in the life of a rhesus monkey and probably in any animal species" (p.85). Certainly, the cost for an individual that was deficient in this skill could be high.

This view of play may also be applicable to hominids in which predator-avoidance would probably have involved fleeing from fast-moving carnivores. Aldis (1975) argued that "in most animal species, the main function of play is probably to develop strengths and skills in the young in preparation for emergency life-or-death behaviors, such as defense and flight, in adult life. This was probably also the main function of play for humans during most of the history of our species, and may remain so for the few hunting and gathering societies that linger on today" (p.2).

With the shift to bipedalism, this fleeing is unlikely to have involved retreating into the trees, a common antipredator pattern in many primate species such as langurs (Ripley, 1967), baboons (Altmann & Altmann, 1970; Saayman, 1971) and vervets (Struhsaker, 1967). Hence sustained high-speed running is likely to have been selected for. Moreover, because both males and females would have been open to predation, there would have been a need for both sexes to practice such behaviors in their play. Based on such a view, we would predict that there would be no sex difference in the frequency and form (e.g., duration, intensity) of chasing play. Note that this prediction contradicts that derived from the practice hunting hypothesis, which states that there should be a sex difference in favor of males. This sort of problem is typical of those facing researchers trying to identify the function(s) of r/t and other forms of play. Hominids would have been open to many different selection pressures, of which several might have influenced the form of r/t. Behavioral patterns typically exhibited by these species would have represented an attempt to arrive at a compromise solution to the different selection pressures.

The predator avoidance hypothesis would not predict the existence of the fighting component of r/t.

Fighting

Several researchers have reported that among primates, competition for reproductive success is higher for males than for females and that prowess in fighting is an importnat determinant of success (Symons 1978a; Wilson, 1975). Based on this proposition, a number of researchers have suggested that r/t may provide practice for real fighting skills. For example, in his detailed study of rhesus macaques, Symons (1978a) found that males engaged in significantly more r/t than females (as did van Lawick-Goodall [1971] in chimpanzees), and he suggested that rhesus play provides physical training for fighting skills and that males play more because this skill is more important to them than to females in terms of reproductive success.

Comparative data on the *form* of r/t also exist, and these, too, can be used to scrutinize functional hypotheses. Some of this evidence fails to support the practice fighting hypothesis. For example, Pellis and Pellis (1987) found that among rodents, play fighting differed from serious fighting both in terms of bodily targets (nape of the neck versus rump, respectively) and tactics, and so they concluded, "That which is 'practiced' during play-fighting is not what is most frequently used in serious fighting" (p.239). Nevertheless, in other species observations do seem to support the practice fighting hypothesis. For example, in many species, male r/t is noticeably more vigorous than that of females: Gentry (1974) reported that male sea lions pups were five times as likely to bite during play than females; Rasa (1971) found that the play fighting of male elephant seal pups exhibited similar patterns to adult male fighting whereas that of female pups contained motor patterns more characteristic of adult female aggression; Linsdale and Tomich (1953) observed that both sexes of mule deer displayed locomotor play, which they claimed provides practice for escaping from predators, but that butting with the head was shown only in the play of males. As adults, this action pattern is shown by males but not females in intraspecific fighting during the breeding season.

Several authors have also reported that the play fighting of male primates is much rougher (i.e., faster and more vigorous) than that of females (for Savanna baboons, see DeVore, 1963; Hamadryas baboons, Aldis, 1975; chimpanzees, van Lawick-Goodall, 1971; Rhesus macaques, Levy, 1979). Symons (1978a) explicitly acknowledged that the "design" or form of play must be taken into account when assessing functional hypotheses. His observations revealed that when play fighting, rhesus monkeys appear to set themselves the goal of biting their partner without themselves being bitten. Playbites were observed to be inhibited and hence noninjurious. Symons considered why young monkeys should spend so much effort in setting this goal when even if they were bitten it would be unlikely that they would be injured. He noted that adults sometimes engaged in serious and potentially lethal fights in which being bitten has serious negative consequences on fitness (both in the short and long term), but inflicting damage through biting promotes fitness. (If this were not characteristic of fights, he argued that selection would favor individuals who avoided such interactions.) Given the importance of fighting on fitness (and the apparent similarity between real fights and play fights), Symons concluded that "an interaction in which two animals simultaneously attempt to inflict and to avoid inhibited bites is the sort of interaction that is most likely to develop such skills" (p.99).

It would be useful to attempt such an analysis for humans. Wilson (1975) has proposed that human males have been selected for strength and skill for use as intraspecific fighters. This leads to the hypothesis that play fighting could function to provide practice for the development of these attributes, or at least could have during some stage of hominid evolution. Smith (1982) proposed that "the adaptive value of play-fighting as practice for adult fighting skills would have been maintained through hominid evolution" (p.151).

Data on sex differences in the frequency of occurrence of play fighting in humans generally support the practice fighting hypothesis: DiPietro (1981) reported that male preschoolers engaged in significantly more "playful physical assaults" than females, and Smith and Connolly (1972) also found that contact forms of r/t were more frequent in 2- to 4-year-old boys than girls. Boulton (1988) found a near-significant sex difference ($p < .07$) in the same direction among four classes of 8- and 11-year-old middle school children. However, any practice benefits associated with play fighting may change through ontogeny. In particular, benefits may become more important in later childhood and adolescence, as fighting itself becomes a more important skill. Evidence for this comes from Humphreys and Smith (1987), who found that, whereas at 7, 9 and 11 years, children selected well-liked classmates as r/t partners, at 11 years, but not at 7 or 9 years, initiators of r/t showed a preference for partners who were weaker, but only slightly so, than themselves. Humphreys and Smith suggest that whereas practice for fighting skills may be an important function of rough-and-tumble for the older age group, this may not be so for younger children for whom an affiliative function may be more important. Neill's (1976) observation that in 12- to 13-year-old boys some r/t was more like serious fighting, with play bouts changing into more aggressive episodes, lends some further support to this view.

SELECTION PRESSURES LEADING TO COMPROMISE AND COOPERATION IN PLAY

The relative merits and limitations of these three (and other) hypotheses about the functional nature of r/t are still being debated. This third section will consider those related selection pressures that could have led to the *social* characteristics of human (and animal) r/t, particularly those leading to compromise and cooperation. We will argue that r/t has design features for cooperation and compromise that ensure that both partners can benefit from the practice it provides, regardless of whether the function of that practice is to improve hunting, predator avoidance, or fighting ability.

In considering why selection pressures worked in favor of the evolution of compromise and cooperation among individuals, Fagen (1981) stated "a sociobiological approach to animal social play assumes that each individual behaves in its own genetic self-interest. It follows that an individual's tendencies to initiate, maintain, and terminate play are all products of natural selection. Accordingly, individuals will behave so as to play in those ways that, and with those partners who, contribute most strongly to that individual's inclusive fitness" (pp.387–388), and, "no matter what function of play is at issue, the interests of any two individuals in play will rarely coincide" (pp. 388–389). Consequently, Fagen believes that "any social play interaction necessarily involves a compromise between partners' differing optima" (p.389).

If this view is correct, individuals who adopted a strategy whereby both participants' needs were met (at least partially) would have had a distinct advantage (in terms of the hypothesized benefits they receive) over those others who sought only to maximize their own gains in the *absence* of any consideration of the benefits available to their partner. This latter strategy would not have become widespread because play partners of such individuals would quickly learn to avoid playing with them, choosing instead to play with others who, like themselves, were willing to compromise with the needs of others. The overtly "selfish" individuals would have ended up with few if any play partners and so would have received little if any benefits from social play. Thus, the cooperative strategy does not necessarily imply any measure of altruism. Individuals that adopted this approach to social play would have been acting in their own genetic self-interest because by compromising with the needs of their partner, they would have ensured that this partner would continue to play with them in the future and so would continue to provide them with play opportunities in which they could acquire some benefits.

What, then, might be the behavioral strategies that evolved to ensure compromise and cooperation in r/t? There are several candidates.

Role Reversal

Many aspects of r/t involve clearly identifiable roles; in chasing, for example, there are the complementary roles of chaser and chasee. In play fighting, this role differentiation may occur along several dimensions. Symons (1978a) distinguished between on-top and on-bottom positions and, separately, between behind and in-front positions in the play of rhesus monkeys. In a slightly different way, Levy (1979) noted that the role of attacker and the role of attacked were clearly separable in this same species.

In each case, there are probably different benefits (and costs) associated with each role. In chasing, for example, the chaser would be more likely to receive practice in hunting skills, particularly related to the pursuit of a relatively fast-moving, evasive-action-taking animal. In contrast, the chasee would be more likely to gain experiences for predator avoidance, such as skill in making sudden turns and maneuvering obstacles in between itself and its pursuer. Fry (1987) noted that among Zapotec children of Oaxaca, Mexico, fleeing children often changed their direction suddenly, an observation that supports this view. As we have seen, both skills are likely to have been useful during hominid evolution and would probably have conveyed distinct survival advantages. Role reversal may be the strategy that ensures that the different types of benefits are received by both participants, since it involves each partner compromising, as Fagen suggested.

However, with respect to r/t in humans, no one has yet provided empirical data to test these ideas. In a recent study of the behavior among two 8-year-old and two 11-year-old classes of English boys and girls in two separate schools, we set out to provide data. In all four classes together, only 3 out of a total of 358 play chases (less than 1%) involved an *immediate* swapping of the chaser/chasee roles. However, researchers have been vague concerning precise operational definitions of role reversal in r/t. For example, while most if not all researchers are likely to agree that an immediate change in roles would qualify as an instance of role reversal, no one has yet tackled the issue of longer gaps. We need to ask whether it would be appropriate to see episodes in which there is a 5 minute gap before partners change their roles as an example of role reversal per se. If so, what about gaps of 10, 15, 20 minutes, and so on? In our observations, it was clear that reversals with considerable gaps were fairly common in play chases. Perhaps the most extreme case observed, and one that we believe should qualify, was when a group of about ten 8-year-old girls chased a group of about the same number of boys for the whole of a dinnertime recess (though obviously not without pauses) with the roles being reversed on *subsequent days*. Role reversals with these lengthy delays are probably sufficient to satisfy children's apparent motivation to engage in all the different roles in their play, although more data are needed.

Self-handicapping

A second strategy leading to compromise that is related to role reversal is known as self-handicapping. This involves the deliberate attenuation of the force or intensity of an action by one or the other play partner in order to give the other a better chance of "winning" or at least to allow more evenly matched encounters. This is an important design feature of r/t, which is why we have used it as a basis of our definition of this category of behavior. The very existence of r/t, as distinct from serious activity such as fighting, is evidence for self-handicapping; if this was not the case, a play fight would quickly resemble a real fight between evenly matched opponents or quickly reach an impasse between unevenly matched opponents, where the most able participant overpowers the less able participant.

Self-handicapping may be beneficial in several respects. If introduced by the strongest partner, it may, for example, allow the reversal of the offensive and defensive roles. In our observations we saw several cases, including one in which two 11-year-old boys were wrestling together. The bout ended up with the strongest boy sitting on the weaker one and holding his partner's hands and arms to prevent him trying to escape or hitting him. This position of stalemate lasted for about 2 minutes, after which the boy on top got off and the other immediately raised himself to his feet. At the same time, the stronger boy rotated his shoulder and rolled on to the floor, enabling the weaker boy to assume the "dominant" on-top position. The encounter lasted another minute and assumed a renewed vigor. Overall, this encounter lasted for about 4 minutes, which is considerably longer than the average length of bouts of play fighting (between 10 and 15 seconds for children of this age).

Clearly, this sort of self-handicapping leads one to experience different roles that would not be the case, at least to the same degree, if play fighting was controlled by the same mechanisms and motivations as serious agonistic interactions. In the latter, for example, the stronger/dominant individual would never deliberately give up a position of advantage.

Self-handicapping and role reversal in r/t are not restricted to English children. Aldis (1975) described their occurrence in North American children, and more recently, Fry (1987) reported that these features were characteristic of the play of Zapotec children of Mexico. They also have been observed by Aldis in the play of various species, including brown bears, sloth bears, dogs, California sea lions, patas monkeys, Savanna baboons, and gibbons.

An evolutionary view of role reversal and self-handicapping must also consider the costs as well as the benefits associated with adopting a particular strategy or not doing so. With respect to role reversal, the costs include providing conspecifics with opportunities and experiences that will enhance their fitness relative to one's own. However, as we have already seen, this is a necessary cost if an individual wants to keep a pool of different playmates who may each contribute something different to his or her fitness. The costs of not showing self-handicapping are arguably even greater in this respect, as a partner would be very unwilling to continue participating if the encounters merged over into a dominance display at their expense. An associated cost could be risk of injury either to one's partner if one failed to properly self-handicap (leading to them avoiding one as a play partner in the future) or to oneself if they failed to do the same; injuries do occasionally occur in playfights (Boulton, 1988; Humphreys & Smith, 1987) but would undoubtedly be more frequent and severe if restraint was not shown, especially by the stronger child.

As was the case with role reversal, no one has yet provided quantitative data on the frequency of occurrence of *specific* types of self-handicapping. This omission in the literature is probably due to methodological difficulties; it is difficult in practice to determine whether or not a child (or animal) is showing self-handicapping. Despite the problems, we attempted to determine the proportion of play chases in which the pursuing child was judged by the observer to deliberately refrain from catching the fleeing child. This decision was based on such characteristics as the pursuing child being close enough to reach out and grasp the fleeing child but refraining from doing so, slowing down as they got close, etc. Only 10 out of 358 chases met these criteria (less than 3%), suggesting that this specific form of self-handicapping is not necessarily a common feature of play chasing for present-day children. Nevertheless, it is possible that the benefits associated with self-handicapping during chasing may not be high, at least compared with other forms of r/t (wrestling, for example), since the emergence of formal or informal rules for role reversal would have ensured that even if one individual was unable to catch another, they would still get the chance to chase and be chased.

IS R/T A FORERUNNER OF OTHER FORMS OF COOPERATIVE SOCIAL PLAY IN HUMANS?

To the casual observer, chasing occurring in the context of r/t (i.e., without explicitly articulated rules) looks very similar to chasing occurring in formally rule-bound games such as the ubiquitous "tag" (sometimes known as "tick" or "tiggy") and other team chase games.[1] Given this similarity, it is possible to argue that chasing engaged in by early hominids could have been the forerunner in the evolution of these more complex rule-governed chase games. This link, if it existed, would have been via the evolution of additional psychological mechanisms that predisposed children to engage in these rule-bound forms of chasing play. Unless these mechanisms were very culture-specific, we would expect that complex rule-governed chase games would be a common feature in the play of most societies including present day hunter-gatherer societies. The additional psychological mechanism(s) could merely be one manifestation of the capacity for social exchange that Cosmides and Tooby (1989) claim was selected for in hominids, that is, cooperation to produce mutually beneficial outcomes. These researchers propose that two features of the life history of early hominids, their longevity and their low dispersal rates, would have provided the right ecological conditions needed for the emergence of social exchange. In particular, these features would have ensured that the likelihood of two individuals having repeated meetings would be high, thus affording an individual who showed restraint and/or self-handicapping on one occasion the chance to be in a position to reap the benefits of the aid of the initial recipient. Thus it is also possible that the rule-governed aspect of chasing games is not adaptive per se, but is instead a consequence of a more general adaptive mechanism.

EVOLVED MECHANISMS

What mechanisms ensure that children enjoy play chasing and fighting and that they can take part in it in a cooperative way? Part of the answer seems to lie with hormonal mechanisms, which have been strongly implicated in the motivation to play fight in rodents (Thor & Holloway, 1984). In the human case, the evidence is tied to sex differences in r/t.

Exposure to higher levels of male sex hormone in male fetuses during pregnancy appears to have certain effects on the developing brain, one of which is to predispose the child to later enjoy r/t activities. The evidence comes mainly from the effects of anomalous amounts of prenatal sex hormones. Unusual exposure can be due to endogenous causes (such as Turner's syndrome) or exogenous causes (such as mothers being given treatment including gonadal hormones during pregnancy). Girls who were exposed prenatally to a higher than usual level of androgens are more likely to be tomboyish and to enjoy vigorous physical activity and r/t. Such a result was found by Money and Erhardt (1972) in a comparison of 25 girls who had experienced fetal androgenization with 25 normal matched controls. Erhardt and Meyer-Bahlburg (1981) suggest that intense physical energy expenditure which is a feature of r/t "seems to be an essential aspect of psychosocial development and it appears to be influenced by sex steroid variation before birth" (p.1313); they regard this as one of the "best established" influences of prenatal hormones.

Quadagno, Briscoe, and Quadagno (1977) suggested some reservations about conclusions from the Money and Erhardt (1972) study. Since the androgenized girls were born with masculinized genitalia, necessitating surgical correction, parental awareness of genital masculinity at birth might have influenced rearing practices. Specifically, the parents might encourage more tomboyish behavior in these girls. Erhardt and Meyer-Bahlburg (1981) argue that the opposite is more likely, that the parents would encourage femininity in these circumstances. However, their data are based on interviews and not on direct observations. It is therefore the perceptions of the girls and their mothers that have been shown to differ. Such perceptions could be susceptible to rearing expectations, and it would be more convincing to have behavioral data. In another review, Hines (1982) states that "studies of individuals who were exposed to unusual hormones prenatally, but who were born without abnormalities, have failed in many cases to find evidence of . . . masculinised play comparable to that reported for adrogenised girls" (p.72). However, these studies have their own weaknesses, and some investigated different hormones (Hines, 1982).

Studies of hormonal effects do not rule out environmental factors such as differential parental reinforcement or effects of gender perceptions on rearing practices. In a longitudinal study, Lamb (1981) found that fathers treated boys and girls similarly at first, but that early in the second year fathers began to direct more social behavior to sons than daughters. Thus it could be that fathers are encouraging more r/t in boys, both by modeling the behavior (and by 3 years of age children have a well-defined gender identity, Thompson, 1975), and by directing more of this behavior toward boys. In addition, parental stereotypes about the greater appropriateness of r/t for boys might suggest that both parents would reinforce r/t more in boys. The only directly relevant study gives little if any support to this last hypothesis, however. Fagot (1978), in a study of 12 boys and 12 girls aged 20 to 24 months in the home, found that parents responded to r/t in boys 91% of the time positively and 3% negatively; in girls 84% of the time positively and 2% negatively, a nonsignificant difference. She found that some other behaviors were reinforced in sex-stereotyped ways, however, and it is quite possible that parental response to r/t might become more sex-stereotyped in older children.

Children do respond very positively to father's active play (Lamb, 1981). The evidence so far is that father's r/t with sons especially increases in the third year, before differential parental reinforcement as such becomes apparent during the school years. Quite likely, the increase in father-son r/t is in part a response to the male child's greater interest in such play, which as we have seen may well have a hormonal basis.

As in many other areas, sexual stereotypes may act to magnify considerably the effects of such preexisting dispositions (see Smith, 1986, for a model of this in relation to r/t).

A further mechanism, triggered by environmental influences, may be at work to mediate the amount and extent of r/t through childhood. Fry (1987) compared patterns of play fighting that occurred within two Zapotec-speaking communities in Oaxaca, Mexico. One community was high in adult violence and the other low. Children from the former showed significantly more play aggression (and serious aggression) than those from the latter, although the nature of these interactions was similar in both locations. Overall, Fry concluded, "Whatever the functions of play fighting may be, the findings of this study support an interpretation that . . . different patterns of aggressive and prosocial behaviors are being passed from one generation to the next, as the children learn to engage in the different behavioral patterns that are modeled and accepted by adults in their respective community" (p.1017).

This evidence suggests the existence of psychological mechanisms (such as reinforcement or modeling) that function to assess how aggressive one's community is and that allow an individual to modify the level and intensity of practice based on this assessment. Given that participation in r/t carries some risk of injury (as well as time and energy costs), why else would individuals expose themselves to such an increased risk if there were no benefits available (such as improved fighting ability)? These data provide some of the best support for the practice fighting hypothesis, but are not in accord with the views that the function of r/t is to improve skill in hunting or predator avoidance. If the function of r/t was to improve skill in hunting or predator avoidance, then there is no reason why the mechanism should be sensitive to the amount of adult aggression displayed in a society.

Data on sex differences in these and other communities where endemic levels of aggression vary could provide a further test of this view. Specifically, levels of play fighting in males and females separately should be related to levels of aggression in adult males and females.

Finally, other mechanisms must be involved in the ability to accurately distinguish between play and agonistic interactions and to accurately encode playful and hostile intent, since r/t and aggressive fighting are similar at the level of the motor patterns involved. In nonhuman primates, these abilities are probably based largely on facial expressions. The existence of the play-face and the bared-teeth threat expressions has been well documented. Studies with young children suggest some degree of phylogenetic continuity. Blurton-Jones (1967, 1972) found that "laugh" and "playface" were associated with r/t, whereas "frown" and "fixate" were associated with aggression in preschool children. Smith and Lewis (1985) adopted the procedure of recording instances of both types of behavior on videotape and showing them to 4-year-old children and a small number of adults. After seeing each episode, individuals were asked whether they thought the participants were being playful or aggressive and how they could tell. Most children showed significant agreement with one another and with the adults on the judgments, and in most cases they reported basing their decisions on the physical characteristics of the behavior involved (such as the strength of a blow). Thus, by 4 years of age some children at least are skilled at distinguishing between play fighting and chasing and aggressive forms of these behaviors. We (Boulton, 1988; Boulton, in press) have carried out a similar study on a larger sample of 8- to 11-year-olds and found that almost all children were adept at making this distinction at this age. These middle school children also based their decisions largely on the physical characteristics of action, but they were also prepared to make inferences about the intentions of the participants (e.g., "he's really fighting with him 'cos he must have called him a name"). In a recent extension to this approach, with Angela Costabile and Laurence Matheson, we were able to demonstrate that this ability does not depend on the "decoders" and "actors" belonging to the same culture. English children could successfully distinguish between r/t and physical aggression shown by Italian children, and vice versa (Costabile, 1989).

There is some evidence that the ability to accurately encode affect (positive in play, negative in aggression) and to decode the affect of others may be open to practice effects. In particular, Parke and his colleagues (Parke, MacDonald, Beitel, & Bhavnagri, 1988) found positive correlations between fathers' physical play and children's abilities to correctly discriminate emotional states. A similar view is put forward by Pelligrini (1987) with respect to peer-peer r/t.

SUMMARY

In this chapter, we suggest that there are important advantages in taking an evolutionary approach to the study of r/t. Evidence suggests that this form of play may be universal to all human cultures and that, as such, it may have been shaped by natural selection acting on physiological and psycholgical mechanisms, in order to provide some specific benefits. For humans, there is some evidence, although not conclusive, that the benefits may be related to intraspecific fighting (with hunting and predator avoidance as less likely candidates). Other considerations suggest that these and other mechanisms that generate r/t have ensured that it is a cooperative activity in which participants with differing needs are willing to compromise.

NOTE

1. One common example of the latter, observed in Sheffield schools is "delavio" or "delaggio," a game with a fairly complex set of rules. Basically, it involves two teams (often boys versus girls), one of whom are the chasers, and the other the fleers. At the start of the game, the chasers count aloud for a predetermined period while the other team disperses throughout the playground. The chasers then set off, individually or in small groups, to try and catch their "opponents." When successful, a catcher escorts a victim back to a "den," an agreed-upon location in the playground, and deposits them there. However, at any time an uncaptured member of the fleeing team may run in and free those already caught by touching part of the den (often shouting "delavio/delaggio" at the same time). It is only when all the fleeing team have been caught that the roles are reversed.

REFERENCES

Aldis, O. (1975). *Play fighting.* New York: Academic Press.
Altmann, S. A., & Altmann, J. (1970). *Baboon ecology.* Chicago: University of Chicago Press.
Bekoff, M., & Byers, J. (1981). A critical reanalysis of the ontogeny and phylogeny of mammalian social and locomotor play: An ethological hornets nest. In K. Immelmann (Ed.), *Behavioral Development* (pp. 296–337). Cambridge: Cambridge University Press.
Blurton-Jones, N. (1967). An ethological study of some aspects of social behavior of children in nursery school. In D. Morris (Ed.), *Primate ethology* (pp. 347–368). London: Weidenfeld & Nicholson.

Blurton-Jones, N. (1972). Categories of child-child interaction. In N. Blurton-Jones (Ed.), *Ethological studies of child behavior* (pp. 97–127). Cambridge: Cambridge University Press.

Blurton-Jones, N., & Konner, M. J. (1973). Sex differences in behaviour of London and Bushman children. In R. P. Michael & J. H. Crook (Eds.), *Comparative ecology and behaviour of primates* (pp. 689–750). London: Academic Press.

Boulton, M. J. (1988). *A multi-methodological investigation of rough-and-tumble play, aggression, and social relationships in middle school children.* Unpublished doctoral dissertation, University of Sheffield, U.K.

Boulton, M. J. (in press). Children's abilities to distinguish between playful and aggressive fighting: A developmental perspective. *British Journal of Developmental Psychology.*

Caro, T. M. (1979). Relations between kitten behavior and adult predation. *Zeitscrift fur tierpsychol, 51,* 158–168.

Caro, T. M. (1981). Predatory behaviour and social play in kittens. *Behaviour, 76,* 1–24.

Cosmides, L., & Tooby, J. (1989). Evolutionary pscyhology and the generation of culture, part II: Case study: A computational theory of social exchange. *Ethology and Sociobiology, 10,* 51–97.

Costabile, A. (1989, August). A cross-national comparison of how children distinguish serious from playful fighting. Paper presented to 10th International Society for the Study of Behavioral Development meeting, Jyvaskyla, Finland.

DeVore, I. (1963). Mother infant relations in free-ranging baboons. In H. Rheingold (Ed.), *Maternal behaviour in mammals* (pp. 305–335). New York: Wiley.

DiPietro, J. A. (1981). Rough and tumble play: A function of gender. *Developmental Psychology, 17,* 50–58.

Dolhinow, P. J. (1971, December). At play in the fields. *Natural History,* pp. 66–71.

Egan, J. (1976). Object play in cats. In J. S. Bruner, A. Jolly, & K. Sylva (Eds.), *Play—its role in development and evolution* (pp. 161–165). New York: Basic Books.

Eibl–Eibesfeldt, I. (1974). The myth of the aggression-free hunter and gatherer society. In R. Holloway (Ed.), *Primate aggression, territoriality, and xenophobia* (pp. 435–457). New York: Academic Press.

Erhardt, A. A., & Meyer-Bahlberg, H.F.L. (1981). Effects of prenatal sex hormones on gender-related behavior. *Science, 211,* 1312–1316.

Ewer, R. F. (1966). Juvenile behaviour in the African ground squirrel, *Xerus erythropus Zeitscrift fur Tierpsychologie, 23,* 190–216.

Fagen, R. M. (1974). Selective and evolutionary aspects of animal play. *American Naturalist, 108,* 850–858.

Fagen, R. M. (1978). Evolutionary biological models of animal play behavior. In G. Burghardt & M. Bekoff (Eds.), *The development of behavior: Comparative and evolutionary aspects* (pp. 385–404). New York: Garland STPM Press.

Fagen, R. M. (1981). *Animal play behavior.* New York: Oxford University Press.

Fagot, B. I. (1978). The influence of sex of child on parental reactions to toddler children. *Child Development, 49,* 459–465.

Fry, D. P. (1987). Differences between playfighting and serious fights among Zapotec children. *Ethology and sociobiology, 8,* 285–306.

Gentry, R. L. (1974). The development of social behavior through play in the Steler sea lion. *American Zoologist, 14,* 391–403.

Hinde, R. A. (1974). *Biological basis of human social behavior.* New York: McGraw-Hill.

Hinde, R. A. (1975). The concept of function. In G. Bearends, C. Beer, & A. Manning (Eds.), *Function and evolution in behaviour* (pp. 3–15). Oxford: Clarendon Press.

Hines, M. (1982). Prenatal gonadal hormones and sex differences in human behavior. *Psychological Bulletin, 92,* 56–80.

Humphreys, A. P., & Smith, P. K. (1987). Rough-and-tumble, friendship and dominance in schoolchildren: Evidence for continuity and change with age. *Child Development, 58,* 201–212.

Klima, G. J. (1970). *The Barabaig: East African cattle herders.* New York: Holt, Rinehart & Winston.

Konner, M. S. (1975). Relations among infants and juveniles in comparative perspective. In M. Lewis & L. A. Rosenblum (Eds.), *Friendship and peer relations* (pp. 99–129). New York: Wiley.

Konner, M. S. (1977). Evolution of human behavior development. In P. H. Leiderman & S. Tulkin (Eds.), *Culture and infancy: Variations in the human experience* (pp. 69–109). New York: Academic Press.

Lamb, M. E. (1981). The development of father-infant relationships. In M. E. Lamb (ed.), *The role of the father in child development* (2nd ed.) (pp. 459–488). New York: Wiley.

Laughlin, W. S. (1968). Hunting: An integrated biobehavior system and its evolutionary importance. In R. B. Lee & I. DeVore (Eds.), *Man the hunter* (pp. 304–320). Chicago: Aldine.

Lawick-Goodall, J. van. (1971). *In the shadow of man.* London: Collins.

Levy, J. S. (1979). Play behavior and its decline during development in rhesus monkeys *(Macaca mulatta).* Unpublished doctoral dissertation, University of Chicago, Chicago.

Linsdale, J. M., & Tomich, P. Q. (1953). *A herd of mule deer.* Berkeley: University of California Press.

Maccoby, E. E., & Jacklin, C. N. (1974). *The psychology of sex differences.* Stanford: Stanford University Press.

Martin, P., & Caro, T. M. (1985). On the functions of play and its role in behavioral development. In J. S. Rosenblatt, C. Beer, M-C. Busnel, & P.J.B. Slater (Eds.), *Advances in the study of behavior* (Vol. 15) (pp. 59–103). Orlando: Academic Press.

Mayer, P., & Mayer, I. (1970). Socialization by peers. The youth organization of the red xhosa. In P. Mayer (Ed.), *Socialization: The approach from social anthropology.* London: Tavistock.

Moelk, M. (1979). The development of friendly approach behavior in the cat: A study of kitten-mother relations and the cognitive development of the kitten from birth to eight weeks. *Advances in the Study of Behavior, 10,* 163–224.

Money, J., & Erhardt, A. A. (1972). *Man and woman, boy and girl.* Baltimore: Johns Hopkins University Press.

Neill, S. R. St J. (1976). Aggressive and non-aggressive fighting in 12- to 13-year-old preadolescent boys. *Journal of Child Psychology and Psychiatry, 17,* 213–220.

Parke, R. D., MacDonald, K. B., Beitel, A., & Bhavnagri, N. (1988). The role of the family in the development of peer relationships. In R. D. Peters & R. J. McMahon (Eds.), *Social learning and systems approaches to marriage and the family,* New York: Brunner/Mazel.

Pellegrini, A. D. (1987). Rough and tumble play: Developmental and educational significance. *Educational Psychologist, 22,* 23–43.

Pellis, S. M., & Pellis, V. C. (1987). Play-fighting differs from serious fighting in both target of attack and tactics of fighting in the laboratory rat. *Rattus norvegicus. Aggressive Behavior, 13,* 227–242.

Quadagno, D. M., Briscoe, R., & Quadagno, J. S. (1977). Effects of perinatal gonadal hormones on selected nonsexual behavior patterns: A critical assessment of the nonhuman and human literature. *Psychological Bulletin, 84,* 62–80.

Rasa, O.A.E. (1971). Social interaction and object manipulation in weaned pups of the Northern elephant seal *Mirounga angustirostris. Zeitscrift fur Tierpsychologie, 29,* 82–102.

Raum, O. F. (1940). *Chaga childhood.* London: Oxford University Press.

Ripley, S. (1967). The leaping of langurs: A problem in the study of locomotor adaptation. *American Journal of Physical Anthropology, 26,* 149–170.

Saayman, G. S. (1971). Baboons' responses to predators. *African Wildlife, 25,* 46–49.

Smith, P. K. (1982). Does play matter? Functional and evolutionary aspects of animal and human play. *Behavioral and Brain Sciences, 5,* 139–184.

Smith, P. K. (1983). Differences or deficits? The significance of pretend and sociodramatic play. *Developmental Review, 3,* 6–10.

Smith, P. K. (1986). Exploration, play and social development in boys and girls. In D. J. Hargreaves & A. M. Colley (Eds.), *The psychology of sex roles* (pp. 118–141). London: Harper & Row.

Smith, P. K., & Connolly, K. J. (1972). Patterns of play and social interaction in pre-school children. In N. Blurton-Jones (Ed.), *Ethological studies of child behaviour* (pp. 65–95). Cambridge: Cambridge University Press.

Smith, P. K., & Lewis, K. (1985). Rough-and-tumble play, fighting and chasing in nursery school children. *Ethology and Sociobiology, 6,* 922–928.

Struhsaker, T. T. (1967). Behavior of vervet monkeys. *University of California Publications in Zoology, 82.*

Symons, D. A. (1974). Aggressive play and communication in rhesus monkeys *(Macaca mulatta). American Zoologist, 14,* 317–322.

Symons, D. A. (1978a). *Play and aggression: A study of rhesus monkeys.* New York: Columbia University Press.

Symons, D. A. (1978b). The question of function: Dominance and play. In E. O. Smith (Ed.), *Social play in primates* (pp. 193–230). New York: Academic Press.

Thompson, S. K. (1975). Gender labels and early sex role development. *Child Development, 46,* 339–347.

Thor, D. H., & Holloway, W. R., Jr. (1984). Social play in juvenile rats: A decade of methodological and experimental research. *Neuroscience & Biobehavioral Reviews, 8,* 455–464.

Vincent, L. E., & Bekoff, M. (1978). Quantitative analyses of the ontogeny of predatory behaviour in coyotes, *Canis Iatrans. Animal Behaviour, 26,* 225–231.

Whiting, B. B., & Whiting, J.W.M. (1975). *Children of six cultures: A psychocultural analysis.* Cambridge, MA: Harvard University Press.

Wilson, E. O. (1975). *Sociobiology: The new synthesis.* Cambridge: MA: Belknap Press.

V

PERCEPTION AND LANGUAGE AS ADAPTATIONS

Two enclaves of cognitive psychology—perception and certain areas of psycholinguistics—have rejected the assumptions of the Standard Social Science Model discussed in chapter 1. These are also the two areas of cognitive psychology that have made the most progress. We would argue that this is no coincidence. These fields focused on important adaptive problems that the human mind evolved to solve and freed themselves of the limiting Procrustean assumption that all computational problems—from vision to olfaction to phonology to syntax acquisition—are solved by the same general-purpose mechanism. By restricting their areas of inquiry in a domain-specific way, these fields were able to carve nature at the joints, discovering functionally isolable subunits that have been variously called modules, cognitive specializations, faculties, or "mental organs." The resulting models consist primarily of descriptions of modules, each of which is specialized for solving a delimited task, but that work together to produce a coordinated functional outcome such as phoneme recognition, syntax acquisition, object recognition, depth perception, or color constancy.

In the late 1950s, Noam Chomsky jolted the psychological community by demonstrating that finite state devices—the most sophisticated formal implementation of behaviorist psychological models—could not generate certain grammatical sentences in English while avoiding the generation of ungrammatical sentences. In doing this, he showed that, when it comes to the problem of generating grammatical sentences, behaviorist mechanisms could not successfully perform the task that English speakers routinely do. Therefore, these mechanisms could be ruled out as an adequate account of "verbal behavior." This demonstration was the opening shot in Chomsky's decades-long campaign against the precepts of what we have called in this volume the Standard Social Science Model.

Chomsky's detailed study of the range of grammatical structures in natural languages and how people understand them convinced him that general-purpose mechanisms could not account for our ability to use language. Just as detoxifying poisons requires different physiological machinery than pumping blood, he argued, language production requires different cognitive machinery than other kinds of information-processing tasks. Because functionally incompatible physiological problems cannot be solved by the same mechanism, the body is divided into organs, such as the liver and heart, which are functionally distinct yet interactive. By the same token, Chomsky argued, functionally distinct yet interactive "mental organs" should have evolved to solve information-processing problems that are functionally incompatible. Within psycholinguistics, a vigorous and productive research community has been following through on this proposal, which has received wide support from many converging lines of evidence, from neurobiology to formal "learnability" analyses.

However implicitly, Chomsky's argument relies on the classic evolutionary principle that form follows function for biological structures such as the inherited architecture of the human mind, a principle that can be justified only by natural selection (Dawkins, 1986; Williams, 1966). Indeed, the argument from physiological organs to mental organs fails unless one invokes natural selection: If the close fit between the structure of the liver and the functional requirements of detoxification is merely a coincidence—the result of chance or the by-product of developmental laws having nothing to do with solving functional problems such as detoxification—then there is no reason to expect this coincidence to repeat itself elsewhere in the body. The existence of functionally distinct physiological organs is legitimate grounds for positing the existence of functionally distinct mental organs only if there exists a causal process that *systematically* creates a coordination between form and function. There are only two causal processes known that can systematically bring about this kind of coordination: intelligent design (as by a human engineer) and natural selection. Because it is a feedback process that "selects" among alternative designs based on how well they solve problems that affect reproduction (such as communication with conspecifics), natural selection can systematically coordinate form and function. For the argument from physiological organs to mental organs to go forward, one must invoke either an intelligent engineer—God or an intelligent alien—or natural selection. As most scientists reject explanations that invoke either supernatural or extraterrestrial intervention, that leaves natural selection.

One would think, then, that those psycholinguists who have embraced Chomsky's notion that there is a "mental organ" that is well designed for language acquisition would view this mental organ as the product of natural selection—that is, as an adaptation. Curiously, this has not been the case. Because the behavioral and social sciences are not conceptually integrated, few psycholinguists are sufficiently well versed in neo-Darwinian theory to see that many of their arguments invoke natural selection, however implicitly, and depend crucially upon its applicability. Pinker and Bloom are notable exceptions: They are psycholinguists with an exceptionally sophisticated grasp of evolutionary biology. Their chapter closely examines the arguments that have been leveled against the notion that the language faculty is the product of natural selection and concludes that these arguments are based on misconceptions "about biology or language or both." This chapter was the subject of an intriguing public debate at MIT between Pinker and Bloom, on the one hand, and Stephen Jay Gould and Massimo Piattelli-Palmarini, on the other, and it provoked a spirited exchange in the pages of the journal *Behavioral and Brain Sciences*. We felt that no volume on the relationship between natural selection and human psychological architecture would be complete without it.

In contrast to psycholinguistics, where claims about modularity are common but still contentious, the modularity of perception has been widely accepted for a long time. Virtually no one expects the cognitive processes that support vision to be the same as those that support audition, for example. Even the most cursory task analysis is usually sufficient to convince oneself that identifying three-dimensional distal objects from a two-dimensional retinal display requires vastly different kinds of information-processing procedures than identifying sounds and their sources from air pressure waves beating against the inner ear. Moreover, the essence of the adaptive problem to be solved and the nature of the information available in the environment for solving it has been reasonably clear in the study of perception.

Even so, there has been a tendency in perception research to treat some of the cognitive processes involved as "brute facts"; relatively little thought has been given to why these processes have one structure rather than another. Shepard's chapter on color vision is a welcome exception. Ever since Newton's work on optics in 1704, we have known that the physical continuum of visible wavelengths is psychologically represented as a circle: Even though red is the longest visible wavelength of light and violet is the shortest, people view red and violet as more similar to each other than either are to green, which is of intermediate wavelength. Ever since the work of Helmholtz and Young in the 1800s, we have known that humans experience colors as varying in three-dimensions—lightness, hue, and saturation. And ever since the work of anthropologists Berlin and Kay in the 1960s, we have known that people everywhere organize the perceptual color space into categories and prototypes. But *why* are these things true? Are they arbitrary design features—mere brute facts—or are they adaptations to some enduring property of the world? Shepard argues for the latter: He makes the case that three-dimensionality and the color circle are the necessary consequence of mechanisms designed to maintain color constancy under natural variations in terrestrial illumination and that certain aspects of color categorization can be understood as a by-product of these mechanisms.

Finally, the study of spatial cognition has been closely allied to the study of visual perception. The mechanisms that allow one to imagine objects moving in space, for example, are often considered part of the perceptual system, and they are frequently viewed as critical to navigation, particularly in unknown territory. In fact, the ability to perform such quasi-geometrical mental transformations has sometimes been considered almost synonymous with spatial cognition. As a result, the fact that men tend to perform better on these tasks than women has been taken to reflect a general male advantage in spatial cognition.

Silverman and Eals dispute this claim. In their chapter on sex differences in spatial ability, they propose that human spatial cognition is not a unitary phenomenon: They propose that it is a collection of different computational processes, each designed for solving functionally different spatial tasks. In this view, the ability to solve quasi-geometrical spatial problems would be only one spatial ability among many. If Silverman and Eals are correct, then the male performance advantage on such tests would reflect a male advantage only in this one facet of spatial cognition; women might be as good, or better, at tasks that test for the presence of other spatial competences. By starting with an analysis of the kinds of spatial problems that hunter-gatherers would have had to solve, Silverman and Eals' work on spatial cognition illustrates a new approach to understanding cognitive sex differences in humans, one that has been forged by evolutionary psychology.

An adaptationist approach to sex differences is straightforward. When males and females have encountered the same adaptive problem with the same frequency over evolutionary time, their cognitive mechanisms for solving that problem should be exactly (or at least approximately) the same. Only when the two sexes have encountered adaptive problems that systematically differ over evolutionary time should their mechanisms for solving them differ. The psychology of mate selection is one obvious area where adaptive problems have differed for the two sexes over evolutionary time (see part 3, "The psychology of mating and sex"). Less well explored, however, are the implications of the sexual division of labor that persisted throughout most of human evolution, with men doing more of the hunting and women doing more of the gathering. Both of these tasks require sophisticated spatial cognition, but tracking and killing animals requires a different kind of spatial cognition than gathering plants does. Animals are mobile and often lead one on a circuitous route into relatively unfamiliar territory. When this happens, it is difficult to navigate using only known landmarks, precisely because the territory is unfamiliar. To get home, one could follow one's own trail, but this would lead one back by the same circuitous route that the animal originally traced out. Time and energy can be saved if one can compute the shortest route home through dead reckoning, a geometrical ability that is highly developed in many animals (Gallistel, 1990). In addition, hunting requires the ability to aim projectiles accurately. As it turns out, the male performance advantage is found for geometrical problems of this kind, which suggests that men have computational processes that are particularly well designed for solving spatial problems associated with hunting.

Silverman and Eals reasoned that, if natural selection had produced cognitive specializations in the male brain for solving the spatial problems associated with hunting, then it should also have produced cognitive specializations in the female brain for solving the quite different and difficult spatial problems associated with gathering. Unlike animals, plants stay in one place, and they are embedded within a complex but fairly stable array of vegetation. One needs to be able to recognize food sources within such complex arrays and to be able to relocate them, sometimes after very long periods of time. A flower among the savannah grasses in summer may indicate that a tuber will be growing underground in that spot next winter; a fig tree that is bearing no fruit when one first finds it will bear fruit in another season. To locate and relocate food sources within large, complex areas of vegetation, one must be good at remembering objects and at recalling their locations with respect to one another. The ability to collect such information automatically, without special attention, as one is going about other tasks would be an advantageous design feature as well.

Silverman and Eals predicted—and found—a substantial female performance advantage on spatial tasks designed to tap such abilities. The hypothesis that women have computational processes that are well designed for solving spatial problems associated with gathering allowed them to make this discovery. To our knowledge, this is the first time anyone has asked what kind of spatial specializations would have made human females more effective gatherers and the first time anyone has predicted a female advantage in spatial abilities. It also adds weight to the claim that "spatial cognition" is not a single ability, but actually a collection of different computational processes that are designed for solving functionally different spatial problems, some of which might be more developed in men, others of which might be more developed in women.

This research demonstrates a larger point as well: An evolutionary approach can help behavioral scientists avoid the sex bias that often emerges in theory-neutral approaches. In the absence of theory, one must rely on intuition. In the past, most psychologists were men, and the notion of what counts as spatial cognition seems to have reflected male intuitions: "Good" spatial cognition was defined as the kind of mental transformations at which males typically excel. If the tables had been turned, the field may have developed a different sex bias: On consulting their own intuitions, female psychologists might have concluded that "good" spatial cognition consists of remembering the locations of objects in complex arrays.

In contrast, the scientist who takes an evolutionary approach can go beyond his or her private intuitions. When adaptive problems differ, the cognitive mechanisms for solving them should differ, whether these mechanisms are in the same individual or in two individuals of different sex. By focusing on adaptive problems, an evolutionary approach provides principled guidelines for predicting what design features different cognitive mechanisms should have. In this case, it allowed Silverman and Eals to propose design features for solving spatial tasks related to the adaptive problem of gathering that had never been proposed before. An evolutionary approach also provides principled guidelines for predicting when cognitive sex differences should exist and when they should not. One can stop seeing males and females as "better" or "worse" versions of one another and start seeing each sex as well designed for solving the sometimes different adaptive problems that it faced during our evolutionary past.

REFERENCES

Dawkins, R. (1986). *The blind watchmaker.* New York: Norton.
Gallistel, C. R. (1990). *The organization of learning.* Cambridge, MA: MIT Press.
Williams, G. C. (1966). *Adaptation and natural selection: A critique of some current evolutionary thought.* Princeton, NJ: Princeton University Press.

12

Natural Language and Natural Selection

STEVEN PINKER AND PAUL BLOOM

INTRODUCTION

All human societies have language. As far as we know they always did; language was not invented by some groups and spread to others like agriculture or the alphabet. All languages are complex computational systems employing the same basic kinds of rules and representations, with no notable correlation with technological progress: The grammars of industrial societies are no more complex than the grammars of hunter-gatherers; Modern English is not an advance over Old English. Within societies, individual humans are proficient language users regardless of intelligence, social status, or level of education. Children are fluent speakers of complex grammatical sentences by the age of three, without benefit of formal instruction. They are capable of inventing languages that are more systematic than those they hear, showing resemblances to languages that they have never heard, and they obey subtle grammatical principles for which there is no evidence in their environments. Disease or injury can make people linguistic savants while severely retarded or linguistically impaired with normal intelligence. Some language disorders are genetically transmitted. Aspects of language skill can be linked to characteristic regions of the human brain. The human vocal tract is tailored to the demands of speech, compromising other functions such as breathing and swallowing. Human auditory perception shows complementary specializations toward the demands of decoding speech sounds into linguistic segments.

This list of facts (see Pinker, 1989a) suggests that the ability to use a natural language belongs more to the study of human biology than human culture; it is a topic like echolocation in bats or stereopsis in monkeys, not like writing or the wheel. All modern students of language agree that at least some aspects of language are due to species-specific, task-specific biological abilities, though of course there are radical disagreements about which specific aspects. A prominent position, outlined by Chomsky (1965, 1980, 1981, 1986, 1988), Fodor (1983), Lenneberg (1964, 1967), and Liberman (Liberman, Cooper, Shankweiler, & Studdert-Kennedy, 1967; Liberman & Mattingly, 1989), is that the mind is composed of autonomous computational modules—mental faculties or "organs"—and that the acquisition and representation of language are the products of several such specialized modules.

It would be natural, then, to expect everyone to agree that human language is the product of Darwinian natural selection. The only successful account of the origin of complex biological structure is the theory of natural selection, the view that the differential reproductive success associated with heritable variation is the primary organizing force in the evolution of organisms (Darwin, 1859; see Bendall, 1983, for a contemporary perspective). But surprisingly, this conclusion is contentious. Noam Chomsky, the world's best-known linguist, and Stephen Jay Gould, the world's best-known evolutionary theorist, have repeatedly suggested that language may not be the product of natural selection, but a side effect of other evolutionary forces such as an increase in overall brain size and constraints of as-yet-unknown laws of structure and growth (e.g., Chomsky, 1972, 1982a, 1982b, 1988, 1991; Gould, 1987a; Gould & Piattelli-Palmarini, 1987). Recently Massimo Piattelli-Palmarini (1989), a close correspondent with Gould and Chomsky, has done the field a service by formulating a particularly strong version of their positions and articulating it in print. Premack (1985, 1986) and Mehler (1985) have expressed similar views.

In this paper we will examine this position in detail and will come to a very different conclusion. We will argue that there is every reason to believe that language has been shaped by natural selection as it is understood within the orthodox "synthetic" or "neo-Darwinian" theory of evolution (Mayr, 1982). In one sense our goal is incredibly boring. All we argue is that language is no different from other complex abilities such as echolocation or stereopsis and that the only way to explain the origin of such abilities is through the theory of natural selection. One might expect our conclusion to be accepted without much comment by all but the most environmentalist of language scientists (as indeed it is by such researchers as Bickerton, 1981; Liberman and Mattingly, 1989; Lieberman, 1984; and, in limited respects, by Chomsky himself in some strands of his writings[1]). On the other hand, when two scholars as important as Chomsky and Gould repeatedly urge us to consider a startling contrary position, their arguments can hardly be ignored. Indeed these arguments have had a strong effect on many cognitive scientists, and the nonselectionist view has become the consensus in many circles.

Furthermore, a lot is at stake if our boring conclusion is wrong. We suspect that many biologists would be surprised at the frequent suggestion that the complexity of language cannot be explained through natural selection. For instance, Chomsky has made the following statements:

> [An innate language faculty] poses a problem for the biologist, since, if true, it is an example of true "emergence"—the appearance of a qualitatively different phenomenon at a specific stage of complexity of organization. (1972, p. 70)

> It is perfectly safe to attribute this development [of innate mental structure] to "natural selection," so long as we realize that there is no substance to this assertion, that it amounts to nothing more than a belief that there is some naturalistic explanation for these phenomena. (1972, p. 97)

> Evolutionary theory is informative about many things, but it has little to say, as of now, of questions of this nature [e.g., the evolution of language]. The answers may well lie not so much in the theory of natural selection as in molecular biology, in the study of what kinds of physical systems can develop under the conditions of life on earth and why, ultimately because of physical principles. (1988, p. 167)

> It does seem very hard to believe that the specific character of organisms can be accounted for purely in terms of random mutation and selectional controls. I would imagine that the biology of a 100 years from now is going to deal with the evolution of organisms the way it now deals with the evolution of amino acids, assuming that there is just a fairly small space of physically possible systems that can realize complicated structures. . . . Evolutionary theory appears to have very little to say about speciation, or about any kind of innovation. It

can explain how you get a different distribution of qualities that are already present, but it does not say much about how new qualities can emerge. (1982a, p. 23)

If findings coming out of the study of language forced biologists to such conclusions, it would be big news.

There is another reason to scrutinize the nonselectionist theory of language. If a current theory of language is truly incompatible with the neo-Darwinian theory of evolution, one could hardly blame someone for concluding that it is not the theory of evolution that must be questioned, but the theory of language. Indeed, this argument has been the basis of critiques of Chomsky's theories by Bates, Thal, and Marchman (1991), Greenfield (1987), and Lieberman (1984, 1989), who are nonetheless strange bedfellows with Chomsky in doubting whether an innate generative grammar could have evolved by natural selection. Since we are impressed both by the synthetic theory of evolution and by the theory of generative grammar, we hope that we will not have to choose between the two.

In this paper, we first examine arguments from evolutionary biology about when it is appropriate to invoke natural selection as an explanation for the evolution of some trait. We then apply these tests to the case of human language, and conclude that language passes. We examine the motivations for the competing nonselectionist position and suggest that they have little to recommend them. In the final section, we refute the arguments that have claimed that an innate specialization for grammar is incompatible with the tenets of a Darwinian account and thus that the two are incompatible.

THE ROLE OF NATURAL SELECTION IN EVOLUTIONARY THEORY

Gould has frequently suggested that evolutionary theory is in the throes of a scientific revolution (e.g., Eldredge & Gould, 1977; Gould, 1980). Two cornerstones of the Darwinian synthesis, adaptationism and gradualism, are, he argues, under challenge. Obviously if strict Darwinism is false in general it should not be used to explain the origin of language.

Nonselectionist Mechanisms of Evolutionary Change

In a classic paper, Gould and Lewontin (1979) warn against "naive adaptationism," the inappropriate use of adaptive theorizing to explain traits that have emerged for other reasons (see also Kitcher, 1985; Lewontin, 1978). The argument is illustrated by an analogy with the mosaics on the dome and spandrels of the San Marco basilica in Venice:

Spandrels—the tapering triangular spaces formed by the intersection of two rounded arches at right angles, . . . are necessary architectural by-products of mounting a dome on rounded arches. Each spandrel contains a design admirably fitted into its tapering space. An evangelist sits in the upper part flanked by the heavenly cities. Below, a man representing one of the four biblical rivers . . . pours water from a pitcher in the narrowing space below his feet.

The design is so elaborate, harmonious, and purposeful that we are tempted to view it as the starting point of any analysis, as the cause in some sense of the surrounding architecture. But this would invert the proper path of analysis. The system begins with an architectural constraint: the necessary four spandrels and their tapering triangular form. They provide a space in which the mosaicists worked; they set the quadripartite symmetry of the dome above.

> Such architectural constraints abound, and we find them easy to understand because we do not impose our biological biases upon them. . . . Anyone who tried to argue that the structure [spandrels] exists because of [the designs laid upon them] would be inviting the same ridicule that Voltaire heaped on Dr. Pangloss: "Things cannot be other than they are. . . . Everything is made for the best purpose. Our noses were made to carry spectacles, so we have spectacles. Legs were clearly intended for breeches, and we wear them." . . . Yet evolutionary biologists, in their tendency to focus exclusively on immediate adaptation to local conditions, do tend to ignore architectural constraints and perform just such an inversion of explanation. (pp. 147–149)

Unconvincing adaptationist explanations, which Gould and Lewontin compare to Kipling's "Just so stories," are easy to find. In the Science and Technology section of the *Boston Globe* in March 1987, an article noted that the number of teats in different mammals ought to correspond not to the average litter size but to the largest litter size that can occur for that species within some bound of probability. Since humans ordinarily bear single children but not infrequently have twins, we have an explanation for why humans have two breasts, not one. The author did not discuss the possibility that the bilateral symmetry that is so basic to the mammalian body plan makes the appearance of one-breasted humans rather unlikely.

Gould and Lewontin describe a number of nonadaptationist mechanisms that they feel are frequently not tested within evolutionary accounts: genetic drift, laws of growth and form (such as general allometric relations between brain and body size), direct induction of form by environmental forces such as water currents or gravity, the effects of accidents of history (which may trap organisms in local maxima in the adaptive landscape), and "exaptation" (Gould & Vrba, 1982), whereby new uses are made of parts that were originally adapted to some other function or of spandrels that had no function at all but were present for reasons of architecture, development, or history. They point out that Darwin himself had this pluralistic view of evolution, and that there was an "unfairly maligned" nonadaptationist approach to evolution, prominent in continental Europe, that stressed constraints on "Baupläne" (architectural plans) flowing from phyletic history and embryological development. This body of research, they suggest, is an antidote to the tendency to treat an organism as a bundle of traits or parts, each independently shaped by natural selection.

Limitations on Nonselectionist Explanations

The Gould and Lewontin argument could be interpreted as stressing that since the neo-Darwinian theory of evolution includes nonadaptationist processes it is bad scientific practice not to test them as alternatives to natural selection in any particular instance. However, they are often read as having outlined a radical new alternative to Darwin, in which natural selection is relegated to a minor role. Though Gould and Lewontin clearly eschew this view in their paper, Gould has made such suggestions subsequently (e.g., Gould, 1980), and Piattelli-Palmarini (1989, p. 1) has interpreted it as such when he talks of Darwinian natural selection being replaced by "a better evolutionary theory (one based on 'exaptation')." The reasons why we should reject this view were spelled out clearly by Williams (1966) and have been amplified recently by Dawkins (1983, 1986).

The key point that blunts the Gould and Lewontin critique of adaptationism is that *natural selection is the only scientific explanation of adaptive complexity.* "Adaptive complexity" describes any system composed of many interacting parts where the details of the parts' structure and arrangement suggest design to fulfill some function. The vertebrate eye is the classic example. The eye has a transparent refracting outer cover, a variable-focus lens, a light-sensitive layer of neural tissue lying at the focal plane of the lens, a diaphragm whose diameter changes with illumination level, muscles that move it in precise conjunction and convergence with those of the other eye, and elaborate neural circuits that respond to patterns defining edges, colors, motion, and stereoscopic disparity. It is impossible to make sense of the structure of the eye without noting that it appears as if it was designed for the purpose of seeing—if for no other reason that the man-made tool for image formation, the camera, displays an uncanny resemblance to the eye. Before Darwin, theologians, notably William Paley, pointed to its exquisite design as evidence for the existence of a divine designer. Darwin showed how such "organs of extreme perfection and complication" could arise from the purely physical process of natural selection.

The essential point is that no physical process other than natural selection can explain the evolution of an organ like the eye. The reason for this is that structures that can do what the eye does are extremely low-probability arrangements of matter. By an unimaginably large margin, most objects defined by the space of biologically possible arrangements of matter cannot bring an image into focus, modulate the amount of incoming light, respond to the presence of edges and depth boundaries, and so on. The odds that genetic drift, say, would result in the fixation within a population of just those genes that would give rise to such an object are infinitesimally small, and such an event would be virtually a miracle. This is also true of the other nonselectionist mechanisms outlined by Gould and Lewontin. It is absurdly improbable that some general law of growth and form could give rise to a functioning vertebrate eye as a by-product of some other trend such as an increase in size of some other part. Likewise, one need not consider the possibility that some organ that arose as an adaptation to some other task, or a spandrel defined by other body parts, just happened to have a transparent lens surrounded by a movable diaphragm in front of a light-sensitive layer of tissue lying at its focal plane. Natural selection—the retention across generations of whatever small, random modifications yield improvements in vision that increase chances of survival and reproduction—is the only physical process capable of creating a functioning eye, because it is the only physical process in which the criterion of being good at seeing can play a causal role. As such it is the only process that can lead organisms along the path in the astronomically vast space of possible bodies leading from a body with no eye to a body with a functioning eye.

This argument is obviously incomplete, as it relies on the somewhat intuitive notion of "function" and "design." A skeptic might accuse the proponent of circularity, asking why a lump of clay should not be considered well designed to fulfill the function of taking up exactly the region of space that it in fact takes up. But the circle can be broken in at least three ways. First, biologists need posit far fewer functions than there are biological systems; new functions are not invented for each organ of each organism. Furthermore, each legitimate function can be related via a direct plausible causal chain to other functions and—critically—to the overall function of survival and reproduction. Finally, convergent evolution and resemblance to human artifacts fulfilling the same putative function give independent criteria for design. But regardless of the precise formulation of the modern argument from design (see, e.g., Cummins, 1984), it is not controversial in practice. Gould himself readily admits that natural selection is the cause of structures such as the vertebrate eye, and he invokes the criterion of engineering design, for example, to rescue Darwinism itself from the charge of circularity (Gould, 1977a). Presumably this is why Gould and Lewontin concede that they agree with Darwin that natural selection is "the most important of evolutionary mechanisms."

What, then, is the proper relation between selectionist and nonselectionist explanations in evolution? The least interesting case involves spandrels that are not involved in any function or behavior, such as the redness of blood, the V-shaped space between a pair of fingers, the hollow at the back of a knee, the fact there are a prime number of digits on each limb, and so on. The mere presence of these *epiphenomenal spandrels,* that play no direct role in the explanation of any species-typical behavior or function, says nothing about whether the structures that they are associated with were shaped by selection. There are as many of them as there are ways of describing an organism that do not correspond to its functional parts.

Much more important are cases where spandrels are modified and put to use. However, in such cases of *modified spandrels,* selection plays a crucial role. Putting a dome on top of four arches gives you a spandrel, but it does not give you a mosaic depicting an evangelist and a man pouring water out of a pitcher. That would *really* be a miracle. To get the actual mosaic you need a designer. The designer corresponds to natural selection. Spandrels, exaptations, laws of growth, and so on can explain the basic plans, parts, and materials that natural selection works with—as Jacob (1977) put it, nature is a tinkerer, not an engineer with a clean drawing board. The best examples of structures produced entirely by nonadaptationist mechanisms are generally one-part or repetitive shapes or processes that correspond to simple physical or geometric laws, such as chins, hexagonal honeycombs, large heads on large bodies, and spiral markings. But, as Darwin stressed, when such parts and patterns are modified and combined into complex biological machines fulfilling some delicate function, these subsequent modifications and arrangements must be explained by natural selection.

The real case of evolution without selection consists of the use of *unmodified spandrels.* Gould (1987a) notes that the African heron uses its wings primarily to block reflections on the surface of water while looking for fish. The possibility that some useful structure is an unmodified spandrel is the most interesting implication of the Gould-Lewontin argument, since Darwinian natural selection would really play no role. Note, though, that unmodified spandrels have severe limitations. A wing used as a visor is a case where a structure designed for a complex engineering task that most arrangements of matter do not fulfill, such as controlled flight, is exapted to a simple engineering task that many arrangements of matter do fulfill, such as screening out reflections (we are reminded of the paperweight and aquarium depicted in *101 Uses for a Dead Computer*). When the reverse happens, such as when a solar heat exchanger is retooled as a fully functioning wing in the evolution of insects (Kingsolver & Koehl, 1985), natural selection must be the cause.

We are going over these criteria for invoking natural selection in such detail because they are so often misunderstood. We hope we have made it clear why modern evolutionary biology does *not* license Piattelli-Palmarini's conclusion that "since language and cognition probably represent the most salient and the most novel biological traits of our species, . . . it is now important to show that they may well have arisen from totally extra-adaptive mechanisms." And Piattelli-Palmarini is not alone. In many discussions with cognitive scientists we have found that adaptation and natural selection have become dirty words. Anyone invoking them is accused of being a naive adaptationist, or even of "misunderstanding evolution." Worst of all, he or she is open to easy ridicule as a Dr. Pangloss telling Just-so stories! (Premack's 1986 reply to Bickerton, 1986, is typical.) Given the uncontroversially central role of natural selection in evolution, this state of affairs is unfortunate. We suspect that many people have acquired much of their knowledge of evolutionary theory from Gould's deservedly popular essays. These essays present a view of evolution that is vastly more sophisticated than the nineteenth-century versions of Darwin commonly taught in high schools and even colleges. But Gould can easily be misread as fomenting a revolution rather than urging greater balance within current biological research, and his essays do not emphasize the standard arguments for when it is appropriate, indeed necessary, to invoke natural selection.

Also lurking beneath people's suspicions of natural selection is a set of methodological worries. Isn't adaptationism fundamentally untestable, hence unscientific, because adaptive stories are so easy to come by that when one fails, another can always be substituted? Gould and Lewontin may be correct in saying that biologists and psychologists have leapt too quickly to unmotivated and implausible adaptationist explanations, but this has nothing to do with the logic of adaptationist explanations per se. Glib, unmotivated proposals can come from all kinds of theories. To take an example close to home, the study of the evolution of language attained its poor reputation precisely because of the large number of silly nonadaptationist hypotheses that were proposed. For instance, it has been argued that language arose from mimicry of animal calls, imitations of physical sounds, or grunts of exertion (the infamous "bow-wow," "ding-dong," and "heave-ho" theories).

Specific adaptationist proposals are testable in principle and in practice (see Dennett, 1983; Kitcher, 1985; Maynard Smith, 1984; Mayr, 1982; Sober, 1984; Williams, 1966). Supplementing the criterion of complex design, one can determine whether putatively adaptive structures are correlated with the ecological conditions that make them useful, and under certain circumstances one can actually measure the reproductive success of individuals possessing them to various degrees (see, e.g., Clutton-Brock, 1983). Of course, the entire theory of natural selection may be literally unfalsifiable in the uninteresting sense that elaborations can always rescue its empirical failings, but this is true of all large-scale scientific theories. Any such theory is supported to the extent that the individual elaborations are mutually consistent, motivated by independent data, and few in number compared to the phenomena to be explained.[2]

Indeed one could argue that it is nonadaptationist accounts that are often in grave danger of vacuity. Specific adaptationist proposals may be unmotivated, but they are within the realm of biological and physical understanding, and often the problem is simply that we lack the evidence to determine which account within a set of alternative adaptive explanations is the correct one. Nonadaptationist accounts that merely suggest the possibility that there is some hitherto-unknown law of physics or constraint on form—a "law of eye-formation," to take a caricatured example—are, in contrast, empty and unfalsifiable.

Two Issues That Are Independent of Selectionism

There are two other issues that Gould includes in his depiction of a scientific revolution in evolutionary theory. It is important to see that they are largely independent of the role of selection in evolutionary change.

Gradualism

According to the theory of "punctuated equilibrium" (Eldredge & Gould, 1972; Gould & Eldredge, 1977), most evolutionary change does not occur continuously within a lineage, but is confined to bursts of change that are relatively brief on the geological time scale, generally corresponding to speciation events, followed by long periods of stasis. Gould has suggested that the theory has some very general and crude parallels with approaches to evolution that were made disreputable by the neo-Darwinian synthesis, approaches that go by the names of "saltationism," "macromutations," or "hopeful monsters" (e.g., Gould, 1981). However, he is emphatic that punctuated equilibrium is "a theory about ordinary speciation (taking tens of thousands of years) and its abrupt appearance at low scales of geological resolution, not about ecological catastrophe and sudden genetic change" (Gould 1987b, p. 234). Many other biologists see evolutionary change in an even more orthodox light. They attribute the sudden appearance of fully formed new kinds of organisms in the fossil record to the fact that speciation typically takes place in small, geographically isolated populations. Thus transitional forms, even if evolving over very long time spans, are unlikely to appear in the fossil record until they reinvade the ancestral territory; it is only the invasion that is sudden (see, e.g., Ayala, 1983; Dawkins, 1986; Mayr, 1982; Stebbins & Ayala, 1981). In any case it is clear that evolutionary change is gradual from generation to generation, in full agreement with Darwin. Thus Piattelli-Palmarini (1989, p. 8) expresses a common misunderstanding when he interprets the theory of punctuated equilibrium as showing that "many incomplete series in the fossil record are incomplete, not because the intermediate forms have been lost for *us,* but because they simply never existed."

Once again the explanation of adaptive complexity is the key reason why one should reject nongradual change as playing an important role within evolution. An important Darwinian insight, reinforced by Fisher (1930), is that the only way for complex design to evolve is through a sequence of mutations with small effects. Although it may not literally be impossible for an organ like the eye to emerge across one generation from no eye at all, the odds of this happening are unimaginably low. A random large leap in the space of possible organic forms is astronomically unlikely to land an organism in a region with a fully formed functioning eye. Only a hill-climbing process, with each small step forced in the direction of forms with better vision, can guide the lineage to such a minuscule region of the space of possible forms within the lifetime of the universe.

None of this is to deny that embryological processes can result in quite radical single-generation morphological changes. "Homeotic" mutations causing slight changes in the timing or positioning of epigenetic processes can result in radically new kinds of offspring, such as fruit flies with legs growing where their antenna should be, and it is possible that some speciation events may have begun with such large changes in structure. However, there is a clear sense in which such changes are still gradual, since they involve only a gross modification or duplication of existing structure, not the appearance of a new kind of structure (see Dawkins, 1983).

Exaptation

Exaptation is another process that is sometimes discussed as if it were incompatible both with adaptationism and with gradualism. People often wonder whether each of the "numerous, successive, slight modifications" from an ancestor lacking an organ to a modern creature enjoying the fully functioning organ leads to an improvement in the function, as it should if the necessary evolutionary sequence is to be complete. Piatelli-Palmarini cites Kingsolver and Koehl's (1985) study of qualitative shifts during the evolution of wings in insects, which are ineffective for flight below a certain size, but effective as solar heat exchange panels precisely within that range. (The homology among parts of bat wings, seal flippers, horse forelimbs, and human arms is a far older example.) Nevertheless such exaptations are still gradual and are still driven by selection; there must be an intermediate evolutionary stage at which the part can subserve both functions (Mayr, 1982), after which the process of natural selection shapes it specifically for its current function. Indeed the very concept of exaptation is essentially similar to what Darwin called "preadaptation" and played an important role in his explanation of "the incipient stages of useful structures."

Furthermore, it is crucial to understand that exaptation is merely one empirical possibility, not a universal law of evolution. Gould is often quoted as saying, "We avoid the excellent question, What good is 5 per cent of an eye? by arguing that the possessor of such an incipient structure did not use it for sight" (1977b, p. 107). (Of course no ancestor to humans literally had 5 per cent of a human eye; the expression refers to an eye that has 5 percent of the complexity of a modern eye.) In response, Dawkins (1986, p. 81) writes, "An ancient animal with 5 per cent of an eye might indeed have used it for something other than sight, but it seems to me at least as likely that it used it for 5 per cent vision. . . . Vision that is 5 percent as good as yours or mine is very much worth having in comparison with no vision at all. So is 1 per cent vision better than total blindness. And 6 per cent is better than 5, 7 per cent better than 6, and so on up the gradual, continuous series." Indeed Darwin (1859) sketched out a hypothetical sequence of intermediate forms in the evolution of the vertebrate eye, all with counterparts in living organisms, each used for vision.

In sum, the positions of Gould, Lewontin, and Eldredge should not be seen as radical revisions of the theory of evolution, but as a shift in emphasis within the orthodox neo-Darwinian framework. As such they do not invalidate gradual natural selection as the driving force behind the evolution of language on a priori grounds. Furthermore, there are clear criteria for when selectionist and nonselectionist accounts should be invoked to explain some biological structure: the presence of complex design to carry out some reproductively significant function, versus the availability of a specific physical, developmental, or random process capable of explaining the structure's existence. With these criteria in hand, we can turn to the specific problem at hand, the evolution of language.

DESIGN IN LANGUAGE

Do the cognitive mechanisms underlying language show signs of design for some function in the same way that the anatomical structures of the eye shows signs of design for the function of vision? This breaks down into three smaller questions: What is the function (if any) of language? What are the engineering demands on a system that must carry out such a function? And are the mechanisms of language tailored to meet those demands? We will suggest that language shows signs of design for the communication of propositional structures over a serial channel.

An Argument for Design in Language

Humans acquire a great deal of information during their lifetimes. Since this acquisition process occurs at a rate far exceeding that of biological evolution, it is invaluable in dealing with causal contingencies of the environment that change within a lifetime and provides a decisive advantage in competition with other species that can only defend themselves against new threats in evolutionary time (Brandon & Hornstein, 1986; Tooby & DeVore, 1987). There is an obvious advantage in being able to acquire such information about the world secondhand: By tapping into the vast reservoir of knowledge accumulated by some other individual, one can avoid having to duplicate the possibly time-consuming and dangerous trial and error process that won that knowledge. Furthermore, within a group of interdependent, cooperating individuals, the states of other individuals are among the most significant things in the world worth knowing about. Thus communication of knowledge and internal states is useful to creatures who have a lot to say and who are on speaking terms. (Later in the chapter, we discuss evidence that our ancestors were such creatures.)

Human knowledge and reasoning, it has been argued, are couched in a "language of thought" that is distinct from external languages such as English or Japanese (Fodor, 1975). The propositions in this representational medium are relational structures whose symbols pertain to people, objects, and events, the categories they belong to, their distribution in space and time, and their causal relations to one another (Jackendoff, 1983; Keil, 1979). The causal relations governing the behavior of other people are understood as involving their beliefs and desires, which can be considered as relations between an individual and the proposition that represents the content of that belief or desire (Fodor, 1975, 1987).

This makes the following kinds of contents worthy of communication among humans. We would want to be able to refer to individuals and classes; to distinguish among basic ontological categories (things, events, places, times, manners, and so on); to talk about events and states, distinguishing the participants in the event or state according to role (agents, patients, goals); and to talk about the intentional states of ourselves and others. Also, we would want the ability to express distinctions of truth value and modality (necessity, possibility, probability, factivity) and to comment on the time of an event or state including both its distribution over time (continuous, iterative, punctate) and its overall time of occurrence. One might also demand the ability to encode an unlimited number of predicates, arguments, and propositions. Further, it would be useful to be able to use the same propositional content within different speech acts, for instance, as a question, a statement, or a command. Superimposed on all of this we might ask for an ability to focus or to put into the background different parts of a proposition, so as to tie the speech act into its context of previously conveyed information and patterns of knowledge of the listener.

The vocal-auditory channel has some desirable features as a medium of communication: It has a high bandwidth, its intensity can be modulated to conceal the speaker or to cover large distances, and it does not require light, proximity, a face-to-face orientation, or tying up the hands. However, it is essentially a serial interface, lacking the full two-dimensionality needed to convey graph or tree structures and typographical devices such as fonts, subscripts, and brackets. The basic tools of a coding scheme employing it are an inventory of distinguishable symbols and their concatenation.

Thus grammars for spoken languages must map propositional structures onto a serial channel, minimizing ambiguity in context, under the further constraints that the encoding and decoding be done rapidly, by creatures with limited short-term memories, and according to a code that is shared by an entire community of potential communicants.

The fact that language is a complex system of many parts, each tailored to mapping a characteristic kind of semantic or pragmatic function onto a characteristic kind of symbol sequence, is so obvious in linguistic practice that it is usually not seen as worth mentioning. Let us list some uncontroversial facts about *substantive universals,* the building blocks of grammars that all theories of universal grammar posit either as an explicit inventory or as a consequence of somewhat more abstract mechanisms.

Grammars are built around symbols for *major lexical categories* (noun, verb, adjective, preposition) that can enter into rules specifying telltale surface distributions (e.g., verbs but not nouns generally take unmarked direct objects), inflections, and lists of lexical items. Together with *minor categories* that characteristically co-occur with the major ones (e.g., articles with nouns), the different categories are thus provided with the means of being distinguished in the speech string. These distinctions are exploited to distinguish basic ontological categories such as things, events or states, and qualities. (See, e.g., Jackendoff, 1983, 1990.)

Major phrasal categories (noun phrase, verb phrase, etc.) start off with a major lexical item, the *head,* and allow it to be combined with specific kinds of affixes and phrases. The resulting conglomerate is then used to refer to entities in our mental models of the world. Thus a noun like *dog* does not itself describe anything but it can combine with articles and other parts of speech to make noun phrases, such as *those dogs, my dog,* and *the dog that bit me,* and it is these noun phrases that are used to describe things. Similarly, a verb like *hit* is made into a verb phrase by marking it for tense and aspect and adding an object, thus enabling it to describe an event. In general, words encode abstract general categories and only by contributing to the structure of major phrasal categories can they describe particular things, events, states, locations, and properties. This mechanism enables the language user to refer to an unlimited range of specific entities while possessing only a finite number of lexical items. (See, e.g., Bloom, 1989; Jackendoff, 1977.)

Phrase structure rules (e.g., "X-bar theory" or "immediate dominance rules") force concatenation in the string to correspond to semantic connectedness in the underlying proposition and thus provide linear clues of underlying structure, distinguishing, for example, *Large trees grow dark berries* from *Dark trees grow large berries.* (See, e.g., Gazdar, Klein, Pullum, & Sag, 1985; Jackendoff, 1977.)

Rules of *linear order* (e.g., "directional parameters" for ordering heads, complements, and specifiers, or "linear precedence rules") allow the order of words within these concatenations to distinguish among the argument positions that an entity assumes with respect to a predicate, distinguishing *Man bites dog* from *Dog bites man.* (See, e.g., Gazdar, et al., 1985; Travis, 1984.)

Case affixes on nouns and adjectives can take over these functions, marking nouns according to argument role and linking noun with predicate even when the order is scrambled. This redundancy can free up the device of linear order, allowing it to be exploited to convey relations of prominence and focus, which can thus mesh with the necessarily temporal flow of attention and knowledge acquisition in the listener.

Verb affixes signal the temporal distribution of the event that the verb refers to (aspect) and the time of the event (tense); when separate aspect and tense affixes co-occur, they are in a universally preferred order (aspect closer to the verb; Bybee, 1985). Given that man-made timekeeping devices play no role in species-typical human thought, some other kind of temporal coordinates must be used, and languages employ an ingenious system that can convey the time of an event relative to the time of the speech act itself and relative to a third, arbitrary reference time (thus we can distinguish between *John has arrived, John had arrived (when Mary was speaking), John will have arrived (before Mary speaks),* and so on; Reichenbach, 1947). Verb affixes also typically agree with the subject and other arguments and thus provide another redundant mechanism that can convey predicate-argument relations by itself (e.g., in many Native American languages such as Cherokee and Navajo) or that can eliminate ambiguity left open by other mechanisms (distinguishing, e.g., *I know the boy and the girl who like chocolate* from *I know the boy and the girl who likes chocolate*).

Auxiliaries, which occur either as verb affixes (where they are distinguished from tense and aspect affixes by proximity to the verb) or in one of three sentence-peripheral positions (first, second, last), convey relations that have logical scope over the entire proposition (mirroring their peripheral position) such as truth value, modality, and illocutionary force. (See Steele, Akmajian, Demers, Jelinek, Kitagawa, Oehrle, & Wasow, 1981.)

Languages also typically contain a small inventory of phonetically reducible morphemes—*pronouns* and other *anaphoric elements*—which, by virtue of encoding a small set of semantic features such as gender and humanness and being restricted in their distribution, can convey patterns of coreference among different participants in complex relations without the necessity of repeating lengthy definite descriptions (e.g., as in *A boy showed a dog to a girl and then he/she/it touched him/her/it/himself/herself*). (See Chomsky, 1981; Wexler & Manzini, 1987.)

Mechanisms of *complementation* and *control* govern the expression of propositions that are arguments of other propositions, employing specific complementizer morphemes signaling the periphery of the embedded proposition and indicating its relation to the embedding one and licensing the omission of repeated phrases referring to participants playing certain combinations of roles. This allows the expression of a rich set of propositional attitudes within a belief-desire folk psychology, such as *John tried to come, John thinks that Bill will come, John hopes for Bill to come, John convinced Bill to come,* and so on. (See Bresnan, 1982.)

In *wh* movement (as in *wh* questions and relative clauses) there is a tightly constrained co-occurrence pattern between an empty element (a "trace" or "gap") and a sentence-peripheral quantifier (e.g., *wh* words). The quantifier word can be specific as to illocutionary force (question versus modification), ontological type (time, place, purpose), feature (animate/inanimate), and role (subject/object), and the gap can occur only in highly constrained phrase structure configurations. The semantics of such constructions allow the speaker to fix the reference of, or request information about, an entity by specifying its role within any proposition. One can refer not just to any dog but to *the dog that Mary sold __ to some students last year;* one can ask not only for the names of just any old interesting person but specifically *Who was that woman I saw you with __?* (See, e.g., Chomsky, 1981; Gazdar, Klein, Pullum, & Sag, 1985; Kaplan & Bresnan, 1982.)

And this is only a partial list, focusing on sheer expressive power. One could add to it the many syntactic constraints and devices whose structure enables them to minimize memory load and the likelihood of pursuing garden paths (incorrect guesses as to how to analyze a sentence) in speech comprehension (e.g., Berwick & Weinberg, 1984; Berwick & Wexler, 1987; Bever, 1970; Chomsky & Lasnik, 1977; Frazier, Clifton, & Randall, 1983; Hawkins, 1988; Kuno, 1973, 1974) or to ease the task of analysis for the child learning the language (e.g., Morgan, 1986; Pinker, 1984; Wexler & Culicover, 1980). On top of that there are the rules of segmental phonology that smooth out arbitrary concatenations of morphemes into a consistent sound pattern that juggles demands of ease of articulation and perceptual distinctness; the prosodic rules that disambiguate syntax and communicate pragmatic and illocutionary information; the articulatory programs that achieve rapid transmission rates through parallel encoding of adjacent consonants and vowels; and on and on. Language seems to be a fine example of "that perfection of structure and coadaptation which justly excites our admiration" (Darwin, 1859, p. 26).

As we write these words, we can hear the protests: "Pangloss! Just so stories!" Haven't we just thought up accounts about functions post hoc after examining the structure? How do we know that the neural mechanisms were not there for other reasons and that once they were there they were just put to various convenient uses by the first language users, who then conveyed their invention to subsequent generations?

Is the Argument for Language Design a Just-So Story?

First of all, there is nothing particularly ingenious, contorted, or exotic about our claims for substantive universals and their semantic functions. Any one of them could have been lifted out of the pages of linguistics textbooks. It is hardly the theory of evolution that motivates the suggestion that phrase structure rules are useful in conveying relations of modification and predicate-argument structure.

Second, it is not necessarily illegitimate to infer both special design and adaptationist origins on the basis of function itself. It all depends on the complexity of the function from an engineering point of view. If someone told you that John uses X as a sunshade or a paperweight, you would certainly be hard-pressed to guess what X is or where X came from, because all sorts of things make good sunshades or paperweights. But if someone told you that John uses X to display television broadcasts, it would be a very good bet that X is a television set or is similar in structure to one and that it was designed for that purpose. The reason is that it would be vanishingly unlikely for something that was not designed as a television to display television programs; the engineering demands are simply too complex.

This kind of reasoning is commonly applied in biology when high-tech abilities such as bat sonar are discovered. We suggest that human language is a similar case. We are not talking about noses holding up spectacles. Human language is a device capable of communicating exquisitely complex and subtle messages, from convoluted soap opera plots to theories of the origin of the universe. Even if all we knew was that humans possessed such a device, we would expect that it would have to have rather special and unusual properties suited to the task of mapping complex propositional structures onto a serial channel, and an examination of grammar confirms this expectation.

Third, the claim that language is designed for communication of propositional structures is not a logical necessity. It is easy to formulate, and reject, specific alternatives. For example, occasionally it is suggested that language evolved as a medium of internal knowledge representation for use in the computations underlying reasoning. But while there may be a languagelike representational medium—"the language of thought," or "mentalese" (Fodor, 1975)—it clearly cannot be English, Japanese, and so on. Natural languages are hopeless for this function: They are needlessly serial, rife with ambiguity (usually harmless in conversational contexts, but unsuited for long-term knowledge representation), complicated by alternations that are relevant only to discourse (e.g., topicalization), and cluttered with devices (such as phonology and much of morphology) that make no contribution to reasoning. Similarly, the facts of grammar make it difficult to argue that language shows design for "the expression of thought" in any sense that is substantially distinct from "communication." If "expression" refers to the mere externalization of thoughts, in some kind of monologue or soliloquy, it is an unexplained fact that language contains mechanisms that presuppose the existence of a listener, such as rules of phonology and phonetics (which map sentences onto sound patterns, enhance confusable phonetic distinctions, disambiguate phrase structure with intonation, and so on) and pragmatic devices that encode conversational topic, illocutionary force, discourse antecedents, and so on. Furthermore people do not express their thoughts in an arbitrary private language (which would be sufficient for pure "expression"), but have complex learning mechanisms that acquire a language highly similar in almost every detail to those of other speakers in the community.

Another example of the empirical nature of specific arguments for language design appears when we examine the specific expressive abilities that are designed into language. They turn out to constitute a well-defined set and do not simply correspond to every kind of information that humans are interested in communicating. So although we may have some a priori intuitions regarding useful expressive capacities of grammar, the matter is ultimately empirical (see, e.g., Jackendoff, 1983, 1990; Pinker, 1989b; Talmy, 1983, 1988), and such research yields results that are specific enough to show that not just any intuition is satisfied. Grammar is a notoriously poor medium for conveying subtle patterns of emotion, for example, and facial expressions and tones of voice are more informative (Ekman & Friesen, 1975; Etcoff, 1986). Although grammars provide devices for conveying rough topological information such as connectivity, contact, and containment, and coarse metric contrasts such as near/far or flat/globular, they are of very little help in conveying precise Euclidean relations: A picture is worth a thousand words. Furthermore, human grammar clearly lacks devices specifically dedicated to expressing any of the kinds of messages that characterize the vocal communication systems of cetaceans, birds, or nonhuman primates, such as announcements of individual identity, predator warnings, and claims of territory.

Finally, Williams (1966) suggests that convergent evolution, resemblance to manmade artifacts, and direct assessments of engineering efficiency are good sources of evidence for adaptation. Of course in the case of human language these test are difficult in practice: Significant convergent evolution has not occurred, no one has ever invented a system that duplicates its function (except for systems that are obviously parasitic on natural languages such as Esperanto or signed English), and most forms of experimental intervention would be unethical. Nonetheless, some tests are possible in principle, and this is enough to refute reflexive accusations of circularity.

For example, even the artificial languages that are focused on very narrow domains of content and that are not meant to be used in a natural on-line manner by people, such as computer languages or symbolic logic, show certain obvious parallels with aspects of human grammar. They needed means of distinguishing types of symbols, predicate argument relations, embedding, scope, quantification, and truth relations, and solve these problems with formal syntactic systems that specify arbitrary patterns of hierarchical concatenation, relative linear order, fixed positions within strings, and closed classes of privileged symbols. Of course there are vast dissimilarities, but the mere fact that terms like "language," "syntax," "predicate," "argument," and "state-ment" have clear meanings when applied to artificial systems, with no confusion or qualification, suggests that there are nonaccidental parallels that are reminiscent of the talk of diaphragms and lenses when applied to cameras and eyes. As for experimental investigation, in principle one could define sets of artificial grammars with and without one of the mechanisms in question, or with variations of it. The grammars would be provided or taught to pairs of communicators—formal automata, computer simula-tions, or college sophomores acting in conscious problem-solving mode—who would be required to convey specific messages under different conditions of speed, noise, or memory limitations. The proportion of information successfully communicated would be assessed and examined as a function of the presence and version of the gram-matical mechanism, and of the different conditions putatively relevant to the function in question.

Language Design and Language Diversity

A more serious challenge to the claim that grammars show evidence of good design may come from the diversity of human languages (Maratsos, 1989). Grammatical devices and expressive functions do not pair up in one-to-one fashion. For example, some languages use word order to convey who did what to whom; others use case or agreement for this purpose and reserve the use of word order to distinguish topic from comment, or do not systematically exploit word order at all. How can one say that the mental devices governing word order evolved under selection pressure for expressing grammatical relations if many languages do not use them for that purpose? Linguistic diversity would seem to imply that grammatical devices are very general-purpose tools. And a general-purpose tool would surely have a very generalized structure and thus could be a spandrel rather than an adapted machine. We begin by answering the immediate objection that the existence of diversity, for whatever reason, invalidates arguments for universal language design; at the end of the section we offer some spec-ulations as to why there should be more than one language to begin with.

First of all, the evolution of structures that serve not one but a small number of definite functions, perhaps to different extents in different environments, is common in biology (Mayr, 1982). Indeed, though grammatical devices are put to different uses in different languages, the possible pairings are very circumscribed. No language uses noun affixes to express tense or elements with the syntactic behavior of auxiliaries to express the shape of the direct object. Such universal constraints on structure and function are abundantly documented in surveys of the languages of the world (e.g., Bybee, 1985; Comrie, 1981; Greenberg, 1966; Greenberg, Ferguson, & Moravcsik, 1978; Hawkins, 1988; Keenan, 1976; and Shopen, 1985). Moreover language universals are visible in language history, where changes tend to fall into a restricted set of patterns, many involving the introduction of grammatical devices obeying characteristic constraints (Kiparsky, 1976; Wang, 1976).[3]

But accounting for the evolution of a language faculty permitting restricted variation is only important on the most pessimistic of views. Even a smidgin of grammatical analysis reveals that surface diversity is often a manifestation of minor differences in the underlying mental grammars. Consider some of the supposedly radical typological differences between English and other languages. English is a rigid word-order language; in the Australian language Warlpiri the words from different logical units can be thoroughly scrambled and case markers are used to convey grammatical relations and noun modification. Many Native American languages, such as Cherokee, use few noun phrases within clauses at all and express grammatical relations by sticking strings of agreement affixes onto the verb, each identifying an argument by a set of features such as humanness or shape. Whereas "accusative" languages like English collapse subjects of transitive and intransitive sentences, "ergative" languages collapse objects of transitives with subjects of intransitives. Whereas English sentences are built around obligatory subjects, languages like Chinese are oriented around a position reserved for the discourse topic.

However, these variations almost certainly correspond to differences in the extent to which the same specific set of mental devices is put to use, but not to differences in the kinds of devices that are put to use. English has free constituent order in strings of prepositional phrases (*The package was sent from Chicago to Boston by Mary; The package was sent by Mary to Boston from Chicago,* and so on). English has case, both in pronouns and in the genitive marker spelled *'s*. It expresses information about arguments in verb affixes in the agreement marker *-s*. Ergativity can be seen in verb alternations like *John broke the glass* and *The glass broke.* There is even a kind of topic position: *As for fish, I like salmon.* Conversely, Warlpiri is not without phrasal syntax. Auxiliaries go in second position (not unlike English, German, and many other languages). The constituents of a noun phrase must be contiguous if they are not case-marked; the constituents of a finite clause must be contiguous if the sentence contains more than one. Pinker (1984) outlines a theory of language acquisition in which the same innate learning mechanisms are put to use to different extents in children acquiring "radically" different languages.

When one looks at more abstract linguistic analyses, the underlying unity of natural languages is even more apparent. Chomsky has quipped that anything you find in one language can also be found in every other language, perhaps at a more abstract level of representation, and this claim can be justified without resorting to Procrustean measures. In many versions of his Government-Binding theory (1981), all noun phrases must be case-marked; even those that receive no overt case-marking are assigned "abstract" case by an adjacent verb, preposition, or tense element. The basic order of major phrases is determined by the value of a language-varying parameter specifying the direction in which case assignment may be executed. So in a language like Latin, the noun phrases are marked with morphological case (and can appear in any position), while in a language like English, they are not so marked and must be adjacent to a case-assigner such as a verb. Thus overt case marking in one language and word order in another are unified as manifestations of a single grammatical module. And the module has a well-specified function: In the terminology of the theory, it makes noun phrases "visible" for the assignment of thematic roles such as agent, goal, or location. Moreover, word order itself is not a unified phenomenon. Often when languages "use word order for pragmatic purposes," they are exploiting an underlying grammatical subsystem, such as stylistic rules, that has very different properties from that governing the relative order of noun phrases and their case-assigners.

Why is there more than one language at all? Here we can offer only the most tentative of speculations. For sound-meaning pairings within the lexicon, there are two considerations, First, one might suppose that speakers need a learning mechanism for labels for cultural innovations, such as *screwdriver*. Such a learning device is then sufficient for all vocabulary items. Second, it may be difficult to evolve a huge innate code. Each of tens of thousands of sound-meaning correspondences would have to be synchronized across speakers, but few words could have the nonarbitrary antecedents that would have been needed to get the standardization process started (i.e., analogous to the way bared fangs in preparation for biting evolved into the facial expression for anger). Furthermore the size of such a code would tax the time available to evolve and maintain it in the genome in the face of random perturbations from sexual recombination and other stochastic genetic processes (Williams, 1966; Tooby & Cosmides, 1990). Once a mechanism for learning sound-meaning pairs is in place, the information for acquiring any particular pair, such as *dog* for dogs, is readily available from the speech of the community. Thus the genome can store the vocabulary in the environment, as Tooby and Cosmides (1990) have put it.

For other aspects of grammar, one might get more insight by inverting the perspective. Instead of positing that there are multiple languages, leading to the evolution of a mechanism to learn the differences among them, one might posit that there is a learning mechanism, leading to the development of multiple languages. That is, some aspects of grammar might be easily learnable from environmental inputs by cognitive processes that may have been in existence prior to the evolution of grammar, for example, the relative order a of pair of sequenced elements within a bounded unit. For these aspects there was no need to evolve a fixed value, and they are free to vary across communities of speakers. Later in this chapter we discuss a simulation of evolution by Hinton and Nowlan (1987) that behaves in a way consistent with this conjecture.

Language Design and Arbitrariness

Piattelli-Palmarini (1989) presents a different kind of argument: Grammar is not completely *predictable* as an adaptation to communication, therefore it lacks design and did not evolve by selection. He writes, "Survival criteria, the need to communicate and plan concerted action, cannot account for our *specific* linguistic nature. Adaptation cannot even begin to explain any of these phenomena" (p. 25). Frequently cited examples of arbitrary phenomena in language include constraints on movement (such as subjacency), irregular morphology, and lexical differences in predicate-argument structure. For instance, it is acceptable to say *Who did John see Mary with?*, but not *Who did John see Mary and?*; *John broke the glass* but not *John breaked the glass; John filled the glass with milk,* but not *John poured the glass with milk.* The arguments that language could not be an adaptation take two forms: (a) language could be better than it is, and (b) language could be different from what it is. We show that neither form of the argument is valid, and that the facts that it invokes are perfectly consistent with language being an adaptation and offer not the slightest support to any specific alternative.

Inherent Trade-Offs

In their crudest form, arguments about the putative functionlessness of grammar run as follows: "I bet you can't tell me a function for Constraint X; therefore language is a spandrel." But even if it could be shown that one part of language had no function, that would not mean that all parts of language had no function. Recall from the section subtitled "Limitations on Nonselectionist Explanations" that many organs contain modified spandrels but this does not mean that natural selection did not assemble or shape the organ. Worse, Constraint X may not be a genuine part of the language faculty but just a description of one aspect of it, an epiphenomenal spandrel. No adaptive organ can be adaptive in every aspect, because there are as many aspects of an organ as there are ways of describing it. The recent history of linguistics provides numerous examples where a newly discovered constraint is first proposed as an explicit statement listed as part of a grammar, but is then shown to be a deductive consequence of a far more wide-ranging principle (see, e.g., Chomsky, 1981; Freidin, 1978). For example, the ungrammaticality of sentences like *John to have won is surprising,* once attributed to a filter specifically ruling out [NP-to-VP] sequences, is now seen as a consequence of the Case Filter. Although one might legitimately wonder what good "*[NP-to-VP]" is doing in a grammar, one could hardly dispense with something like the Case Filter.

Since the mere appearance of some nonoptimal feature is inconclusive, we must examine specific explanations for why the feature exists. In the case of the nonselectionist position espoused by Piattelli-Palmarini, there is none—not a hint of how any specific aspect of grammar might be explained, even in principle, as a specific consequence of some developmental process or genetic mechanism or constraint on possible brain structure. The position gains *all* its support from the supposed lack of an adaptive explanation. In fact, we will show that there is such an explanation, well motivated both within evolutionary theory and within linguistics, so the support disappears.

The idea that natural selection aspires toward perfection has long been discredited within evolutionary theory (Williams, 1966). As Maynard Smith (1984, p. 290) has put it, "If there were no constraints on what is possible, the best phenotype would live forever, would be impregnable to predators, would lay eggs at an infinite rate, and so on." Trade-offs among conflicting adaptive goals are a ubiquitous limitation on optimality in the design of organisms. It may be adaptive for a male bird to advertise his health to females with gaudy plumage or a long tail, but not to the extent that predators are attracted or flight is impossible.

Trade-offs of utility within language are also unavoidable (Bolinger, 1980; Slobin, 1977). For example, there is a conflict of interest between speaker and hearer. Speakers want to minimize articulatory effort and hence tend toward brevity and phonological reduction. Hearers want to minimize the effort of understanding and hence desire explicitness and clarity. This conflict of interest is inherent to the communication process and operates at many levels. Editors badger authors into expanding elliptical passages; parsimonious headline writers unwittingly produce *Squad Helps Dog Bite Victim* and *Stud Tires Out*. Similarly there is a conflict of interest between speaker and learner. A large vocabulary allows for concise and precise expression. But it is useful only if every potential listener has had the opportunity to learn each item. Again, this trade-off is inherent to communication; one man's jargon term is another's mot juste.

Clearly, any shared system of communication is going to have to adopt a code that is a compromise among these demands and so will appear to be arbitrary from the point of view of any one criterion. There is always a large range of solutions to the combined demands of communication that reach slightly different equilibrium points in this multidimensional space. Slobin (1977) points out that the Serbo-Croatian inflectional system is "a classic Indo-European synthetic muddle," suffixing each noun with a single affix from a paradigm full of irregularity, homophony, and zero-morphemes. As a result the system is perfected late and with considerable difficulty. In contrast the Turkish inflectional system is semantically transparent, with strings of clearly demarcated regular suffixes, and is mastered by the age of two. When it comes to production by an adult who has overlearned the system, however, Serbo-Croatian does have an advantage in minimizing the sheer number of syllables that must be articulated. Furthermore, Slobin points out that such trade-offs can be documented in studies of historical change and borrowing. For example changes that serve to enhance brevity will proceed until comprehension becomes impaired, at which point new affixes or distinctions are introduced to restore the balance (see also Samuels, 1972). A given feature of language may be arbitrary in the sense that there are alternative solutions that are better from the standpoint of some single criterion. But this does not mean that it is good for nothing at all!

Subjacency—the prohibition against dependencies between a gap and its antecedent that span certain combinations of phrasal nodes—is a classic example of an arbitrary constraint (see Freidin & Quicoli, 1989). In English you can say *What does he believe they claimed that I said?* but not the semantically parallel **What does he believe the claim that I said?*[4] One might ask why languages behave this way. Why not allow extraction anywhere, or nowhere? The constraint may exist because parsing sentences with gaps is a notoriously difficult problem and a system that has to be prepared for the possibility of inaudible elements anywhere in the sentence is in danger of bogging down by positing them everywhere. Subjacency has been held to assist parsing because it cuts down on the set of structures that the parser has to keep track of when finding gaps (Berwick & Weinberg, 1984). This bonus to listeners is often a hindrance to speakers, who struggle with resumptive pronouns in clumsy sentences such as *That's the guy that you heard the rumor about his wife leaving him.* There is nothing "necessary" about the precise English version of the constraint or about the small sample of alternatives allowed within natural language. But by settling in on a particular subset of the range of possible compromises between the demands of expressiveness and parsability, the evolutionary process may have converged on a satisfactory set of solutions to one problem within language processing.

Parity in Communications Protocols

The fact that one can conceive of a biological system being different from what it is says nothing about whether it is an adaptation (see Mayr, 1983). No one would argue that selection was not the key organizing force in the evolution of the vertebrate eye just because the compound eyes of arthropods are different. Similarly, pointing out that a hypothetical Martian language could do passivization differently is inconclusive. We must ask how well-supported specific explanations are.

In the case of features of human language structure that could have been different, again Piattelli-Palmarini presents no explanations at all and relies entirely on the putative inability of natural selection to provide any sort of motivated account. But in fact there is such an account: The nature of language makes arbitrariness of grammar itself part of the adaptive solution of effective communication *in principle.*

Any communicative system requires a coding protocol, which can be arbitrary as long as it is shared. Liberman and Mattingly (1989) call this the requirement of *parity,* and we can illustrate it with the (coincidentally named) "parity" settings in electronic communication protocols. There is nothing particularly logical about setting your printer's serial interface to the "even," as opposed to the "odd," parity setting. Nor is there any motivation to set your computer to odd as opposed to even parity. But there is every reason to set the computer and printer to the *same* parity, whatever it is, because if you don't, they cannot communicate. Indeed, standardization itself is far more important than any other adaptive feature possessed by one party. Many personal computer manufacturers in the 1980s boasted of the superior engineering and design of their product compared to the IBM PC. But when these machines were not IBM-compatible, the results are well known.

In the evolution of the language faculty many "arbitrary" constraints may have been selected simply because they defined parts of a standardized communicative code in the brains of some critical mass of speakers. Piattelli-Palmarini may be correct in claiming that there is nothing adaptive about forming yes-no questions by inverting the subject and auxiliary as opposed to reversing the order of words in the sentence. But given that language must do one or the other, it is highly adaptive for each member of a community of speakers to be forced to learn to do it the same way as all the other members. To be sure, some combination of historical accidents, epiphenomena of other cognitive processes, and neurodevelopmental constraints must have played a large role in the breaking of symmetry that was needed to get the fixation process running away in one direction or another. But it still must have been selection that resulted in the convention then becoming innately entrenched.

The requirement of parity operates at all levels of a communications protocol. Within individual languages the utility of arbitrary but shared features is most obvious in the choice of individual words: There is no reason for you to call a dog *dog* rather than *cat* except for the fact that everyone else is doing it, but that is reason enough. Saussure (1959) called this inherent feature of language "l'arbitraire du signe," and Hurford (1989), using evolutionary game theory, demonstrates the evolutionary stability of such a "Saussurean" strategy whereby each learner uses the same arbitrary signs in production that it uses in comprehension (i.e., that other speakers use in production). More generally, these considerations suggest that a preference for arbitrariness is built into the language acquisition device at two levels. The acquisition device hypothesizes rules that fall only within the (possibly arbitrary) set defined by universal grammar, and within that set, it tries to choose rules that match those used by the community, whatever they are.

The benefits of a learning mechanism designed to assess and adopt the prevailing parity settings become especially clear when we consider alternatives, such as trying to get each speaker to converge on the same standard by endogenously applying some rationale to predict form from meaning. There are many possible rationales for any form-meaning pairing, and that is exactly the problem—different rationales can impress different speakers, or the same speakers on different occasions, to different degrees. But such differences in cognitive style, personal history, or momentary interests must be set aside if people are to communicate. As mentioned, no grammatical device can simultaneously optimize the demands of talkers and hearers, but it will not do to talk in Serbo-Croatian and demand that one's listeners reply in Turkish. Furthermore, whenever cognition is flexible enough to construe a situation in more than one way, no simple correspondence between syntax and semantics can be used predictively by a community of speakers to "deduce" the most "logical" grammatical structure. For example, there is a simple and universal principle dictating that the surface direct object of a causative verb refers to an entity that is "affected" by the action. But the principle by itself is unusable. When a girl puts boxes in baskets she is literally affecting both: The boxes are changing location, and the baskets are changing state from empty to full. One would not want one perceiver interested in the boxes to say that she is *filling boxes* while another interested in the baskets to describe the same event as *filling baskets;* no one would know what went where. However, by letting different verbs idiosyncratically select different kinds of entities as "affected" (e.g., *place the box/*basket* versus *fill the basket/*box*) and forcing learners to respect the verbs' wishes, grammar can allow speakers to specify different kinds of entities as affected by putting them in the direct object position of different verbs, with minimal ambiguity. Presumably this is why different verbs have different arbitrary syntactic privileges (Pinker, 1989b), a phenomenon that Piattelli-Palmarini (1989) describes at length. Even iconicity and onomatopoeia are in the eye and ear of the beholder. The American Sign Language sign for "tree" resembles the motion of a tree waving in the wind, but in Chinese Sign Language it is the motion of sketching the trunk (Newport & Meier, 1985). In the United States, pigs go "oink"; in Japan, they go "boo-boo."

Arbitrariness and the Relation Between Language Evolution and Language Acquisition

The need for arbitrariness has profound consequences for understanding the role of communicative function in language acquisition and language evolution. Many psychologists and artificial intelligence researchers have suggested that the structure of grammar is simply the solution that every child arrives at in solving the problem of how to communicate with others. Skinner's reinforcement theory is the strongest version of this hypothesis (Skinner, 1957), but versions that avoid his behaviorism and rely instead on general cognitive problem-solving abilities have always been popular within psychology. Both Skinner and cognitive theorists such as Bates et al. (1991) explicitly draw parallels between the role of function in learning and evolution. Chomsky and many other linguists and psycholinguists have argued against functionalism in ontogeny, showing that many aspects of grammar cannot be reduced to being the optimal solution to a communicative problem; rather, human grammar has a universal idiosyncratic logic of its own. More generally, Chomsky has emphasized that people's *use* of language does not tightly serve utilitarian goals of communication but is an autonomous competence to express thought (see, e.g., Chomsky, 1975). If communicative function does not shape language in the individual, one might conclude, it probably did not shape language in the species.

We suggest that the analogy that underpins this debate is misleading. It is not just that learning and evolution need not follow identical laws, selectionist or otherwise.

(For example, as Chomsky himself has stressed, the issue never even comes up in clearer cases like vision, where nobody suggests that all infants' visual development is related to their desire to see or that visual systems develop with random variations that are selected by virtue of their ability to attain the child's goals.) In the case of language the arguments in this section (Language Design and Arbitrariness) suggest that language evolution and language acquisition not only *can* differ but that they *must* differ. Evolution has had a wide variety of equivalent communicative standards to choose from; there is no reason for it to have favored the class of languages that includes Apache and Yiddish, but not Old High Martian or Early Vulcan. But this flexibility has been used up by the time a child is born; the species and the language community have already made their choices. The child cannot learn just any useful communicative system; nor can he or she learn just any natural language. He or she is stuck with having to learn the particular kind of language the species eventually converged upon and the particular variety the community has chosen. Whatever rationales may have influenced these choices are buried in history and cannot be recapitulated in development.

Moreover, any code as complex and precise as a grammar for a natural language will not wear its protocol on its sleeve. No mortal computer user can induce an entire communications protocol or programming language from examples; that's why we have manuals. This is because any particular instance of the use of such a protocol is a unique event accompanied by a huge set of idiosyncratic circumstances, some relevant to how the code must be used, most irrelevant, and there is no way of deciding which is which. For the child, any sentence or set of sentences is compatible with a wide variety of very different grammars, only one of them correct (Chomsky, 1965, 1975, 1980, 1981; Pinker, 1979, 1984; Wexler and Culicover, 1980). For example, without prior constraints, it would be natural to generalize from input sentences like *Who did you see her with?* to **Who did you see her and?*, from *teethmarks* to **clawsmarks*, from *You better be good* to **Better you be good?* The child has no manual to consult, and presumably that is why he or she needs innate constraints.

So we see a reason why functionalist theories of the evolution of language can be true while functionalist theories of the acquisition of language can be false. From the very start of language acquisition, children obey grammatical constraints that afford them no immediate communicative advantage. To take just one example, 1- and 2-year-olds acquiring English obey a formal constraint on phrase structure configurations concerning the distinction between lexical categories and phrasal categories and as a result avoid placing determiners and adjectives before pronouns and proper names. They will use phrases like *big dog* to express the belief that a particular dog is big, but they will never use phrases like *big Fred* or *big he* to express the belief that a particular person is big (Bloom, 1990). Children respect this constraint despite the limits it puts on their expressive range.

Furthermore, despite unsupported suggestions to the contrary among developmental psychologists, many strides in language development afford the child no locally discernible increment in communicative ability (Maratsos, 1983, 1989). When children say *breaked* and *comed*, they are using a system that is far simpler and more logical than the adult combination of a regular rule and 180 irregular memorized exceptions. Such errors do not reliably elicit parental corrections or other conversational feedback (Brown and Hanlon, 1970; Morgan and Travis, 1989). There is no deficit in comprehensibility; the meaning of *comed* is perfectly clear. In fact the child's system has greater expressive power that the adult's. When children say *hitted* and *cutted,* they are distinguishing between past and nonpast forms in a manner that is unavailable to adults, who must use *hit* and *cut* across the board. Why do children eventually abandon this simple, logical, expressive system? They must be programmed so that the mere requirement of conformity to the adult code, as subtle and arbitrary as it is, wins over other desiderata.

The requirement that a communicative code have an innate arbitrary foundation ("universal grammar," in the case of humans) may have analogues elsewhere in biology. Mayr (1982, p. 612) notes the following:

> Behavior that serves as communication, for instance courtship behavior, must be stereotyped in order not to be misunderstood. The genetic program controlling such behavior must be "closed," that is, it must be reasonably resistant to any changes during the individual life cycle. Other behaviors, for instance those that control the choice of food or habitat, must have a certain amount of flexibility in order to permit the incorporation of new experiences; such behaviors must be controlled by an "open" program.

In sum, the requirement for standardization of communication protocols dictates that it is better for nature to build a language acquisition device that picks up the code of the ambient language than one that invents a code that is useful from a child's eye view. Acquiring such a code from examples is no mean feat, and so many grammatical principles and constraints must be hardwired into the device. Thus, even if the functions of grammatical devices play an important role in evolution, they may play no role in acquisition.

ARGUMENTS FOR LANGUAGE BEING A SPANDREL

Given that the criteria for being an adaptation appear to be satisfied in the case of language, we can examine the strength of the competing explanations that language is a spandrel suggested by Gould, Chomsky, and Piattelli-Palmarini.

The Mind as a Multipurpose Learning Device

The main motivation for Gould's specific suggestion that language is a spandrel is his frequently stated position that the mind is a single general-purpose computer. For example, as part of a critique of a theory of the origin of language, Gould (1979, p. 386) writes:

> I don't doubt for a moment that the brain's enlargement in human evolution had an adaptive basis mediated by selection. But I would be more than mildly surprised if many of the specific things it now can do are the product of direct selection "for" that particular behavior. Once you build a complex machine, it can perform so many unanticipated tasks. Build a computer "for" processing monthly checks at the plant, and it can also perform factor analyses on human skeletal measures, play Rogerian analyst, and whip anyone's ass (or at least tie them perpetually) in tic-tac-toe.

The analogy is somewhat misleading. It is just not true that you can take a computer that processes monthly checks and use it to play Rogerian analyst; someone has to reprogram it first. Language learning is not programming: Parents provide their children with sentences of English, not rules of English. We suggest that natural selection was the programmer.

The analogy could be modified by imagining some machine equipped with a single program that can *learn from examples* to calculate monthly checks, perform factor analyses, and play Rogerian analyst, all without explicit programming. Such a device does not now exist in artificial intelligence, and it is unlikely to exist in biological intelligence. There is no psychologically realistic multipurpose learning program that can acquire language as a special case, because the kinds of generalizations that must be made to acquire a grammar are at cross-purposes with those that are useful in acquiring other systems of knowledge from examples (Chomsky, 1975, 1982b, 1986; Pinker, 1979, 1984; Wexler & Culicover, 1980). The gross facts about the dissociability of language and other learned cultural systems, listed in the first paragraph of this paper, also belie the suggestion that language is a spandrel of any general cognitive learning ability.

Constraints on Possible Forms

The theory that the mind is an all-purpose learning device is of course anathema to Chomsky (and to Piattelli-Palmarini), making it a puzzle that they should find themselves in general agreement with Gould. Recently, Gould (1989) has described some common ground. Chomsky, he suggests, is in the Continental tradition of trying to explain evolution by structural laws constraining possible organic forms. For example, Chomsky writes:

> In studying the evolution of mind, we cannot guess to what extent there are physically possible alternatives to, say, transformational generative grammar, for an organism meeting certain other physical conditions characteristic of humans. Conceivably, there are none—or very few—in which case talk about evolution of the language capacity is beside the point. (1972, p. 97–98)

> These skills [e.g., learning a grammar] may well have arisen as a concomitant of structural properties of the brain that developed for other reasons. Suppose that there was selection for bigger brains, more cortical surface, hemispheric specialization for analytic processing, or many other structural properties that can be imagined. The brain that evolved might well have all sorts of special properties that are not individually selected; there would be no miracle in this, but only the normal workings of evolution. We have no idea, at present, how physical laws apply when 10^{10} neurons are placed in an object the size of a basketball, under the special conditions that arose during human evolution. (1982, p. 321)

> In this regard [the evolution of infinite digital systems], speculations about natural selection are no more plausible than many others; perhaps these are simply emergent physical properties of a brain that reaches a certain level of complexity under the specific conditions of human evolution. (1991, p. 50)

Although Chomsky does not literally argue for any specific evolutionary hypothesis, he repeatedly urges us to consider "physical laws" as possible alternatives to natural selection. But it is not easy to see exactly what we should be considering. It is certainly true that natural selection cannot explain all aspects of the evolution of language. But is there any reason to believe that there are as-yet undiscovered theorems of physics that can account for the intricate design of natural language? Of course human brains *obey* the law of physics, and always did, but that does not mean that their specific structure can be *explained* by such laws.

More plausibly, we might look to constraints on the possible neural basis for language and its epigenetic growth. But neural tissue is wired up by developmental processes that act in similar ways all over the cortex and to a lesser degree across the animal kingdom (Dodd & Jessell, 1988; Harrelson & Goodman, 1988). In different organisms it has evolved the ability to perform the computations necessary for pollen-source communication, celestial navigation, Doppler-shift echolocation, stereopsis, controlled flight, dam building, sound mimicry, and face recognition. The space of physically possible neural systems thus can't be all *that* small, as far as specific computational abilities are concerned. And it is most unlikely that laws acting at the level of substrate adhesion molecules and synaptic competition, when their effects are projected upward through many levels of scale and hierarchical organization, would automatically result in systems that accomplish interesting engineering tasks in a world of medium size objects.

Changes in brain quantity could lead to changes in brain quality. But mere largeness of brain is neither a necessary nor a sufficient condition for language, as Lenneberg's (1967) studies of nanencephaly and craniometric studies of individual variation have shown. Nor is there reason to think that if you simply pile more and more neurons into a circuit or more and more circuits into a brain that computationally interesting abilities would just emerge. It seems more likely that you would end up with a very big random pattern generator. Neural network modeling efforts have suggested that complex computational abilities require either extrinsically imposed design or numerous richly structured inputs during learning or both (Lachter & Bever, 1988; Pinker & Prince, 1988), any of which would be inconsistent with Chomsky's suggestions.

Finally, there may be direct evidence against the speculation that language is a necessary physical consequence of how human brains can grow. Gopnik (1990a, 1990b; Gopnik & Crago, 1991) describes a syndrome of developmental dysphasia whose sufferers lack control of morphological features such as number, gender, tense, and case. Otherwise they are intellectually normal. One 10-year-old boy earned the top grade in his mathematics class and is a respectable computer programmer. This shows that a human brain lacking components of grammar, perhaps even a brain with the capacity to manipulate other kinds of rules generating infinite digital systems, is physically and neurodevelopmentally possible.

In sum, there is no support for the hypothesis that language emerges from physical laws acting in unknown ways in a large brain. While there are no doubt *aspects* of the system that can be explained only by historical, developmental, or random processes, the most likely explanation for the complex structure of the language faculty is that it is a design imposed on neural circuitry as a response to evolutionary pressures.

THE PROCESS OF LANGUAGE EVOLUTION

For universal grammar to have evolved by Darwinian natural selection, it is not enough that it be useful in some general sense. There must have been genetic variation among individuals in their grammatical competence. There must have been a series of steps leading from no language at all to language as we now find it, each step small enough to have been produced by a random mutation or recombination, and each intermediate grammar useful to its possessor. Every detail of grammatical competence that we wish to ascribe to selection must have conferred a reproductive advantage on its speakers, and this advantage must have been large enough to have become fixed in the ancestral population. And there must be enough evolutionary time and genomic space separating our species from nonlinguistic primate ancestors.

There are no conclusive data on any of these issues. However, this has not prevented various people from claiming that each of the necessary postulates is false. We argue that what we do know from the biology of language and evolution makes each of the postulates quite plausible.

Genetic Variation

Lieberman (1984, 1989) claims that the Chomskyan universal grammar could not have evolved. He writes:

> The premises that underlie current "nativist" linguistic theory . . . are out of touch with modern biology. Ernst Mayr (1982), in his definitive work, *The Growth of Biological Thought,* discusses these basic principles that must structure any biologically meaningful nativist theory. . . . Essentialistic thinking [one of the principles] (e.g., characterizing human linguistic ability in terms of a uniform hypothetical universal grammar) is inappropriate for describing the biological endowment of living organisms. (1989, p. 203–205)

> A true nativist theory must accommodate genetic variation. A detailed genetically transmitted universal grammar that is identical for every human on the planet is outside the range of biological plausibility. (1989, p. 223)

This is part of Lieberman's argument that syntax is acquired by general-purpose learning abilities, not by a dedicated module or set of modules. But the passages quoted above contain a variety of misunderstandings and distortions. Chomskyan linguistics is the antithesis of the kind of essentialism that Mayr decries. It treats such disembodied interindividual entities as "The English Language" as unreal epiphenomena. The only scientifically genuine entities are individual grammars situated in the heads of individual speakers (see Chomsky, 1986, for extended discussion). True, grammars for particular languages, and universal grammar, are often provisionally idealized as a single kind of system. But this is commonplace in systems-level physiology and anatomy; for example the structure of the human eye is always described as if all individuals shared it and individual variation and pathology are discussed as deviations from a norm. This is because natural selection, while feeding on variation, uses it up (Ridley, 1986; Sober, 1984). In adaptively complex structures in particular, the variation we see does not consist of qualitative differences in basic design, and this surely applies to complex mental structures as well (Tooby & Cosmides, 1990).

Also, contrary to what Lieberman implies, there does exist variation in grammatical ability. Within the range that we would call "normal" we all know some individuals who habitually use tangled syntax and others who speak with elegance, some who are linguistically creative and others who lean on clichés, some who are fastidious conformists and others who bend and stretch the language in various ways. At least some of this variation is probably related to the strength or accessibility of different grammatical subsystems, and at least some, we suspect, is genetic, the kind of thing that would be shared by identical twins reared apart. More specifically, Bever, Carrithers, Cowart, and Townsend (1989) have extensive experimental data showing that right-handers with a family history of left-handedness show less reliance on syntactic analysis and more reliance on lexical association than do people without such a genetic background.

Moreover, beyond the "normal" range there are documented genetically transmitted syndromes of grammatical deficits. Lenneberg (1967) notes that specific language disability is a dominant, partially sex-linked trait with almost complete penetrance (see also Ludlow & Cooper, 1983, for a literature review). More strikingly, Gopnik (1990b, 1991; Gopnik & Crago, 1991) has found a familial selective deficit in the use of morphological features (gender, number, tense, etc.) that acts as if it is controlled by a dominant gene.

This does not mean that we should easily find cases of inherited subjacency deficiency or anaphor blindness. *Pleiotropy*—single gene changes that cause apparently unrelated phenotypic effects—is ubiquitous, so there is no reason to think that every aspect of grammar that has a genetic basis must be controlled by a single gene. Having a right hand has a genetic basis but genetic deficits do not lead to babies being born with exactly one hand missing. Moreover, even if there was a pure lack of some grammatical device among some people, it may not be easily discovered without intensive analysis of the person's perceptions of carefully constructed linguistic examples. Different grammatical subsystems can generate superficially similar constructions, and a hypothetical victim of a deficit may compensate in ways that would be difficult to detect. Indeed cases of divergent underlying analyses of a single construction are frequent causes of historical change.

Intermediate Steps

Some people have doubted that an evolutionary sequence of increasingly complex and specialized universal grammars is possible. The intermediate links, it has been suggested, would not have been viable communication systems. These arguments fall into three classes.

Nonshared Innovations

Geschwind (1980), among others, has wondered how a hypothetical "beneficial" grammatical mutation could really have benefited its possessor, given that none of the person's less evolved compatriots could have understood him or her. One possible answer is that any such mutation is likely to be shared by the individuals who are genetically related. Since much communication is among kin, a linguistic mutant will be understood by some of his or her relatives, and the resulting enhancements in information sharing will benefit each one of them relative to others who are not related.

But we think there is a more general answer. Comprehension abilities do not have to be in perfect synchrony with production abilities. Comprehension can use cognitive heuristics based on probable events to decode word sequences even in the absence of grammatical knowledge. Ungrammatical strings like *skid crash hospital* are quite understandable, and we find we can do a reasonably good job understanding Italian newspaper stories based on a few cognates and general expectancies. At the same time grammatical sophistication in such sources does not go unappreciated. We are unable to duplicate Shakespeare's complex early Modern English, but we can appreciate the subtleties of his expressions. When some individuals are making important distinctions that can be decoded with cognitive effort, it could set up a pressure for the evolution of neural mechanisms that would make this decoding process become increasingly automatic, unconscious, and undistracted by irrelevant aspects of world knowledge. These are some of the hallmarks of an innate grammatical "module"

(Fodor, 1983). The process whereby environmentally induced responses set up selection pressures for such responses to become innate, triggering conventional Darwinian evolution that superficially mimics a Lamarckian sequence, is sometimes known as the Baldwin Effect.

Not all linguistic innovations need begin with a genetic change in the linguistic abilities of speakers. Former Secretary of State Alexander Haig achieved notoriety with expressions such as *Let me caveat that* or *That statement has to be properly nuanced.* As listeners we cringe at the ungrammaticality, but we have no trouble understanding him and would be hard-pressed to come up with a concise grammatical alternative. The double standard exemplified by Haigspeak is fairly common in speech (Pinker, 1989b). Most likely this was always true, and innovations driven by cognitive processes exploiting analogy, metaphor, iconicity, conscious folk etymology, and so on, if useful enough, could set up pressures for both speakers and hearers to grammaticize those innovations. Note as well that if a single mental database is used in production and comprehension (Bresnan & Kaplan, 1982), evolutionary changes in response to pressure on one performance would automatically transfer to the other.

Categorical Rules

Many linguistic rules are categorical, all-or-none operations on symbols (see, e.g., Pinker & Prince, 1988, 1989). How could such structures evolve in a gradual sequence? Bates et al. (1991), presumably echoing Gould's "5% of an eye," (1989) write:

> What protoform can we possibly envision that could have given birth to constraints on the extraction of noun phrases from an embedded clause? What could it conceivably mean for an organism to possess half a symbol, or three quarters of a rule? (p. 31)

> Monadic symbols, absolute rules and modular systems must be acquired as a whole, on a yes-or-no basis—a process that cries out for a Creationist explanation. (p. 57)

However, two issues are being collapsed here. While one might justifiably argue that an entire system of grammar must evolve in a gradual continuous sequence, that does not mean that every aspect of every rule must evolve in a gradual continuous sequence. As mentioned, mutant fruit flies can have a full leg growing where an antenna should be, and the evolution of new taxa with different numbers of appendages from their ancestors is often attributed to such homeotic mutations. No single mutation or recombination could have led to an entire universal grammar, but it could have led a parent with an n rule grammar to have an offspring with an $n+1$ rule grammar, or a parent with an m symbol rule to have an offspring with an $m+1$ symbol rule. It could also lead to a parent with no grammatical rules at all and just rote associations to have an offspring with a single rule. Grammatical rules are symbol manipulation whose skeletal form is shared by many other mental systems. Indeed discrete symbol manipulation, free from graded application based on similarity to memorized cases, is highly useful in many domains of cognition, especially those involving socially shared information (Freyd, 1983; Pinker & Prince, 1989; Smolensky, 1988). If a genetic change caused generic copies of a nonlinguistic symbol-replacement operation to pop up within the neural system underlying communication, such protorules could be put to use as parts of encoding and decoding schemes, whereupon they could be subject to selective forces tailoring them to specific demands of language. Rozin (1976) and Shepard (1986) have argued that the evolution of intelligence was made possible by just such sequences.

Perturbations of Formal Grammars

Grammars are thought to be complex computational systems with many interacting rules and conditions. Chomsky (1981) has emphasized how grammars have a rich deductive structure in which a minor change to a single principle can have dramatic effects on the language as a whole as its effects cascade through grammatical derivations. This raises the question of how the entire system could be viable under the far more major perturbations that could be expected during evolutionary history. Does grammar degrade gracefully as we extrapolate backwards in time? Would a universal grammar with an altered or missing version of some component be good for anything, or would it result in nothing but blocked derivations, filtered constructions, and partial structures? Lieberman (1989, p. 200) claims that "the only model of human evolution that would be consistent with the current standard linguistic theory is a sudden saltation that furnished human beings with the neural bases for language." Similarly, Bates et al. (1991, p. 30) claim that "if the basic structural principles of language cannot be learned (bottom up) or derived (top down), there are only two possible explanations for their existence: either Universal Grammar was endowed to us directly by the Creator, or else our species has undergone a mutation of unprecedented magnitude, a cognitive equivalent of the Big Bang."

But such arguments are based on a confusion. While a grammar for an existing language cannot tolerate minor perturbations and still be a grammar for a language that a modern linguist would recognize, that does not mean that it cannot be a grammar at all. To put it crudely, there is no requirement that the languages of *Homo erectus* fall into the class of possible *Homo sapiens* languages. Furthermore language abilities consist of not just formal grammar but also such nonlinguistic cognitive processes as analogy, rote memory, and Haigspeak. Chomsky (1981) refers to such processes as constituting the "periphery" of grammar, but a better metaphor may put them in the "interstices," where they would function as a kind of jerry-rigging that could allow formally incomplete grammars to be used in generating and comprehending sentences.

The assertion that a natural language grammar functions either as a whole or not at all is surprisingly common. But it has no more merit than similar claims about eyes, wings, and webs that frequently pop up in the anti-Darwinian literature (see Dawkins, 1986, for examples) and that occasionally trigger hasty leaps to claims about exaptation. Pidgins, contact languages, Basic English, and the language of children, immigrants, tourists, aphasics, telegrams, and headlines provide ample proof that there is a vast continuum of viable communicative systems displaying a continuous gradation of efficiency and expressive power (see Bickerton, 1986). This is exactly what the theory of natural selection requires.

Our suggestions about interactions between learning and innate structure in evolution are supported by an interesting simulation of the Baldwin effect by Hinton and Nowlan (1987). They consider the worst imaginable scenario for evolution by small steps: a neural network with 20 connections (which can be either excitatory or inhibitory) that conveys no fitness advantage unless all 20 are correctly set. So not only is it no good to have 5% of the network; it's no good to have 95%. In a population of organisms whose connections are determined by random mutations a fitter mutant arises at a rate of only about once every million (2^{20}) genetically distinct organisms, and its advantages are immediately lost if the organism reproduces sexually. But now consider an organism where the connections are either genetically fixed to one or the other value or are settable by learning, determined by random mutation with an average of 10 connections fixed. The organism tries out random settings for the modifiable connections until it hits upon the combination that is advantageous; this is recognizable to the organism and causes it to retain those settings. Having attained that state the organism enjoys a higher rate of reproduction; the sooner it attains it, the greater the benefit. In such a population there *is* an advantage to having less than 100% of the correct network. Among the organisms with, say, 10 innate connections, the one in every thousand (2^{10}) that has the right ones will have some probability of attaining the entire network; in a thousand learning trials, this probability is fairly high. For the offspring of that organism, there are increasing advantages to having more and more of the correct connections innately determined, because with more correct connections to begin with, it takes less time to learn the rest, and the chances of going through life without having learned them get smaller.

Hinton and Nowlan confirmed these intuitions in a computer simulation, demonstrating nicely that learning can guide evolution, as the argument in this section requires, by turning a spike in fitness space into a gradient. Moreover they made an interesting discovery. Though there is always a selection pressure to make learnable connections innate, this pressure diminishes sharply as most of the connections come to be innately set, because it becomes increasingly unlikely that learning will fail for the rest. This is consistent with the speculation that the multiplicity of human languages is in part a consequence of learning mechanisms existing prior to (or at least independent of) the mechanisms specifically dedicated to language. Such learning devices may have been the sections of the ladder that evolution had no need to kick away.

Reproductive Advantages of Better Grammars

David Premack (1985, p. 281–282) writes:

> I challenge the reader to reconstruct the scenario that would confer selective fitness on recursiveness. Language evolved, it is conjectured, at a time when humans or protohumans were hunting mastodons. . . . Would it be a great advantage for one of our ancestors squatting alongside the embers, to be able to remark: "Beware of the short beast whose front hoof Bob cracked when, having forgotten his own spear back at camp, he got in a glancing blow with the dull spear he borrowed from Jack"?
>
> Human language is an embarrassment for evolutionary theory because it is vastly more powerful than one can account for in terms of selective fitness. A semantic language with simple mapping rules, of a kind one might suppose that the chimpanzee would have, appears to confer all the advantages one normally associates with discussions of mastodon hunting or the like. For discussions of that kind, syntactic classes, structure-dependent rules, recursion and the rest, are overly powerful devices, absurdly so.

Premack's rhetorical challenge captures a conviction that many people find compelling, perhaps even self-evident, and it is worth considering why. It is a good example of what Dawkins (1986) calls the Argument from Personal Incredulity. The argument draws on people's poor intuitive grasp of probabilistic processes, especially those that operate over the immensities of time available for evolution. The passage also gains intuitive force because of the widespread stereotype of prehistoric humans as grunting cave men whose main reproductive challenge was running away from tigers or hunting mastodons. The corollary would seem to be that only humans in modern industrial societies—and maybe only academics, it is sometimes implied—need to use sophisticated mental machinery. But compelling as these commonsense intuitions are, they must be resisted.

Effects of Small Selective Advantages

First one must be reminded of the fact that tiny selective advantages are sufficient for evolutionary change. According to Haldane's (1927) classic calculations, for example, a variant that produces on average 1% more offspring than its alternative allele would increase in frequency from 0.1% to 99.9% of the population in just over 4,000 generations. Even in long-lived humans this fits comfortably into the evolutionary timetable. (Needless to say fixations of different genes can go on in parallel.) Furthermore the phenotypic effects of a beneficial genetic change need not be observable in any single generation. Stebbins (1982) constructs a mathematical scenario in which a mouselike animal is subject to selection pressure for increased size. The pressure is so small that it cannot be measured by human observers, and the actual increase in size from one generation to the next is also so small that it cannot be measured against the noise of individual variation. Nonetheless this mouse would evolve to the size of an elephant in 12,000 generations, a slice of time that is geologically "instantaneous." Finally, very small advantages can also play a role in macroevolutionary successions among competing populations of similar organisms. Zubrow (1987) calculates that a 1% difference in mortality rates among geographically overlapping Neanderthal and modern populations could have led to the extinction of the former within 30 generations, or a single millennium.

Grammatical Complexity and Technology

It has often been pointed out that our species is characterized by two features—technology and social relations among nonkin—that have attained levels of complexity unprecedented in the animal kingdom. Toolmaking is the most widely advertised ability, but the knowledge underlying it is only a part of human technological competence. Modern hunter-gatherers, whose lifestyle is our best source of evidence for that of our ancestors, have a folk biology encompassing knowledge of the life cycles, ecology, and behavior of wild plants and animals "that is detailed and thorough enough to astonish and inform professional botanists and zoologists" (Konner, 1982, p. 5). This ability allows the modern !Kung San, for example, to enjoy a nutritionally complete diet with small amounts of effort in what appears to us to be a barren desert. Isaac (1983) interprets fossil remains of home bases as evidence for a life-style depending heavily on acquired knowledge of the environment as far back as 2 million years ago in *Homo habilis.* An often-noted special feature of humans is that such knowledge can accumulate across generations. Premack (1985) reviews evidence that pedagogy is a universal and species-specific human trait, and the usefulness of language in pedagogy is not something that can be reasonably doubted. As Brandon and Hornstein (1986) emphasize, presumably there is a large selective advantage conferred by being able to learn in a way that is essentially stimulus-free (Williams, 1966, made a similar point).

Children can learn from a parent that a food is poisonous or a particular animal is dangerous; they do not have to observe or experience this by themselves.

With regard to adult-to-adult pedagogy, Konner (1982, p. 171) notes that the !Kung discuss

> . . . everything from the location of food sources to the behavior of predators to the movements of migratory game. Not only stories, but great stores of knowledge are exchanged around the fire among the !Kung and the dramatizations—perhaps best of all—bear knowledge critical to survival. A way of life that is difficult enough would, without such knowledge, become simply impossible.

Devices designed for communicating precise information about time, space, predicate-argument relations, restrictive modification, and modality are not wasted in such efforts. Recursion in particular is extraordinarily useful. Premack repeats a common misconception when he uses tortuous phrases as an exemplification of recursive syntax; without recursion you can't say *the man's hat* or *I think he left*. All you need for recursion is an ability to embed a phrase containing a noun phrase within another noun phrase or a clause within another clause, which falls out of pairs of rules as simple as NP → det N PP and PP → P NP. Given such a capacity, one can now specify reference to an object to an arbitrarily fine level of precision. These abilities can make a big difference. For example, it makes a big difference whether a far-off region is reached by taking the trail that is in front of the large tree or the trail that the large tree is in front of. It makes a difference whether that region has animals that you can eat or animals that can eat you. It makes a difference whether it has fruit that is ripe or fruit that was ripe or fruit that will be ripe. It makes a difference whether you can get there if you walk for three days or whether you can get there and walk for three days.

Grammatical Complexity and Social Interactions

What is less generally appreciated is how important linguistically supported social interactions are to a hunter-gatherer way of life. Humans everywhere depend on cooperative efforts for survival. Isaac (1983) reviews evidence that a life-style depending on social interactions among nonkin was present in *Homo habilis* more than 2 million years ago. Language in particular would seem to be deeply woven into such interactions, in a manner that is not qualitatively different from that of our own "advanced" culture. Konner (1982) writes:

> War is unknown. Conflicts within the group are resolved by talking, sometimes half or all the night, for nights, weeks on end. After two years with the San, I came to think of the Pleistocene epoch of human history (the three million years during which we evolved) as one interminable marathon encounter group. When we slept in the grass hut in one of their villages, there were many nights when its flimsy walls leaked charged exchanges from the circle around the fire, frank expressions of feeling and contention beginning when the dusk fires were lit and running on until the dawn. (p. 7)

> If what lawyers and judges do is work, then when the !Kung sit up all night at a meeting discussing a hotly contested divorce, that is also work. If what psychotherapists and ministers do is work, then when a !Kung man or woman spends hours in an enervating trance trying to cure people, that is also work. (p. 371)

Reliance on such exchanges puts a premium on the ability to convey socially relevant abstract information such as time, possession, beliefs, desires, tendencies, obligations, truth, probability, hypotheticals, and counterfactuals. Once again, recursion is far from being an "overly powerful device." The capacity to embed propositions within other propositions, as in [$_S$He thinks that S] or [$_S$She said that [$_S$he thinks that S]], is essential to the expression of beliefs about the intentional states of others.

Furthermore, in a group of communicators competing for attention and sympathies there is a premium on the ability to engage, interest, and persuade listeners. This in turn encourages the development of discourse and rhetorical skills and the pragmatically relevant grammatical devices that support them. Symons's (1979) observation that tribal chiefs are often both gifted orators and highly polygynous is a splendid prod to any imagination that cannot conceive of how linguistic skills could make a Darwinian difference.

Social Use of Language and Evolutionary Acceleration

The social value of complex language probably played a profound role in human evolution that is best appreciated by examining the dynamics of cooperative interactions among individuals. As mentioned, humans, probably early on, fell into a life-style that depended on extended cooperation for food, safety, nurturance, and reproductive opportunities. This life-style presents extraordinary opportunities for evolutionary gains and losses. On the one hand, it benefits all participants by surmounting prisoners' dilemmas. On the other hand, it is vulnerable to invasion by cheaters who reap the benefits without paying the costs (Axelrod & Hamilton, 1981; Cosmides, 1989; Hamilton, 1964; Maynard Smith, 1974; Trivers, 1971). The minimum cognitive apparatus needed to sustain this life-style is memory for individuals and the ability to enforce social contracts of the form "If you take a benefit then you must pay a cost" (Cosmides, 1989). This alone puts a demand on the linguistic expression of rather subtle semantic distinctions. It makes a difference whether you understand me as saying that if you give me some of your fruit I will share meat that I will get, or that you should give me some fruit because I shared meat that I got, or that if you don't give me some fruit I will take back the meat that I got.

But this is only a beginning. Cooperation opens the door to advances in the ability of cheaters to fool people into believing that they have paid a cost or that they have not taken a benefit. This in turn puts pressure on the ability to detect subtle signs of such cheating, which puts pressure on the ability to cheat in less detectable ways, and so on. It has been noted that this sets the stage for a cognitive "arms race" (e.g., Cosmides & Tooby, 1989; Dawkins, 1976; Tooby & DeVore, 1987; Trivers, 1971). Elsewhere in evolution such competitive feedback loops, such as in the struggle between cheetahs and gazelles, have led to the rapid evolution of spectacular structures and abilities (Dawkins, 1982). The unusually rapid enlargement of the human brain, especially the frontal lobes, has been attributed to such an arms race (Alexander, 1987; Rose, 1980). After all, it doesn't take all that much brain power to master the ins and outs of a rock or to get the better of a berry. But interacting with an organism of approximately equal mental abilities whose motives are at times outright malevolent makes formidable and ever-escalating demands on cognition. This competition is not reserved for obvious adversaries. Partial conflicts of reproductive interest between male and female, sibling and sibling, and parent and offspring are inherent to the human condition (Symons, 1979; Tooby & DeVore, 1987; Trivers, 1974).

It should not take much imagination to appreciate the role of language in a cognitive arms race. In all cultures human interactions are mediated by attempts at persuasion and argument. How a choice is framed plays a huge role in determining which alternative people choose (Tversky & Kahneman, 1981). The ability to frame an offer so that it appears to present maximal benefit and minimum cost to the buyer and the ability to see through such attempts and to formulate persuasive counterproposals would have been skills of inestimable value in primitive negotiations, as they are today. So is the ability to learn of other people's desires and obligations through gossip, an apparently universal human vice (Cosmides & Tooby, 1989; Symons, 1979).

In sum, primitive humans lived in a world in which language was woven into the intrigues of politics, economics, technology, family, sex, and friendship that played key roles in individual reproductive success. They could no more live with a me-Tarzan-you-Jane level of grammar than we could.

Phyletic Continuity

Bates et al. (1991), Greenfield (1987), and Lieberman (1976, 1984) argue that if language evolved in humans by natural selection, it must have antecedents in closely related species such as chimpanzees, which share 99% of their genetic material with us and may have diverged from a common ancestor as recently as 5 to 7 million years ago (King & Wilson, 1975; Miyamoto, Slightom, & Goodman, 1987). Similarly, since no biological ability can evolve out of nothing, they claim, we should find evidence of nonlinguistic abilities in humans that are continuous with grammar. Lieberman claims that motor programs are preadaptations for syntactic rules, and Bates (1976) and Greenfield (Greenfield & Smith, 1976) suggest that communicative gestures flow into linguistic naming. As Bates et al. (1991, p. 8) put it, "we have to abandon any strong version of the discontinuity claim that has characterized generative grammar for thirty years. We have to find some way to ground symbols and syntax in the mental material that we share with other species."

The specific empirical claims have been disputed. Seidenberg and Petitto (Seidenberg, 1986; Seidenberg & Petitto, 1979, 1987) have reviewed the evidence of the signing abilities of apes and concluded that they show no significant resemblance to human language or to the process of acquiring it. In a study of the acquisition of sign language in deaf children, Petitto (1987) argues that nonlinguistic gestures and true linguistic names, even when both share the manual-visual channel, are completely dissociable. These conclusions could be fodder for the claim that natural language represents a discontinuity from other primate abilities and so could not have evolved by natural selection.

We find the Seidenberg and Petitto demonstrations convincing, but our argument is not based on whether they are true. Rather we completely disagree with the premise (not theirs) that the debate over ape signing should be treated as a referendum on whether human language evolved by natural selection. Of course human language, like other complex adaptations, could not have evolved overnight. But then there is no law of biology that says that scientists are blessed with the good fortune of being able to find evolutionary antecedents to any modern structure in some other living species. The first recognizably distinct mental system that constituted an antecedent to modern human language may have appeared in a species that diverged from the chimp-human common ancestor, such as *Australopithecus afarensis* or any of the subsequent hominid groups that led to our species. Moreover chimpanzees themselves are not generalized common ancestors but presumably have done some evolving of their own since the split. We must be prepared for the possible bad news that there just aren't any living creatures with homologues of human language, and let the chimp signing debate come down as it will.

As far as we know this would still leave plenty of time for language to have evolved: 3.5 to 5 million years, if early Australopithecines were the first talkers, or, as an absolute minimum, several hundred thousand years (Stringer & Andrews, 1988), in the unlikely event that early *Homo sapiens* was the first. (For what it's worth, Broca's area is said to be visible in cranial endocasts of 2-million-year-old fossil hominids; Falk, 1983; Tobias, 1981.) There is also no justification in trying to squeeze conclusions out of the genetic data. On the order of 40 million base pairs of DNA differ between chimpanzees and humans, and we see no reason to doubt that universal grammar would fit into these 10 megabytes with lots of room left over, especially if provisions for the elementary operations of a symbol-manipulation architecture are specified in the remaining 99% of the genome (see Seidenberg, 1986, for discussion).

In fact there is even more scope for design differences than the gross amount of nonshared genetic material suggests. The 1% difference between chimps and humans represents the fraction of base pairs that are different. But genes are long stretches of base pairs, and if even one pair is different, the entire functioning product of that gene could be different. Just as replacing one bit in every byte leads to text that is 100% different, not 12.5% different, it is possible for the differing base pairs to be apportioned so that 100% of the genes of humans and chimps are different in function. Though this extreme possibility is, of course, unlikely, it warns us not to draw any conclusions about phenotypic similarity from degree of genomic overlap.[5]

As for continuity between language and nonlinguistic neural mechanisms, we find it ironic that arguments that are touted as being "biological" do not take even the most elementary steps to distinguish between analogy and homology. Lieberman's claim that syntactic rules must be retooled motor programs, a putative case of preadaptation, is a good example. It may be right, but there is no reason to believe it. Lieberman's evidence is only that motor programs are hierarchically organized and serially ordered, and so is syntax. But hierarchical organization characterizes many neural systems, perhaps any system, living or nonliving, that we would want to call complex (Simon, 1969). And an organism that lives in real time is going to need a variety of perceptual, motor, and central mechanisms that keep track of serial order. Hierarchy and seriality are so useful that for all we know they may have evolved many times in neural systems (Bickerton, 1984, 1986, also makes this point). To distinguish true homology from mere analogy it is necessary to find some unique derived nonadaptive character shared by the relevant systems, for example, some quirk of grammar that can be seen in another system. Not only has no such shared character been shown, but the dissimilarities between syntax and motor control are rather striking. Motor control is a game of inches, so its control programs must have open continuous parameters for time and space at every level of organization. Syntax has no such analogue parameters. A far better case could be made that grammar exploited mechanisms originally used for the conceptualization of topology and antagonistic forces (Jackendoff, 1983; Pinker, 1989b; Talmy, 1983, 1988), but that is another story.

CONCLUSION

As we warned, the thrust of this paper has been entirely conventional. All we have argued is that human language, like other specialized biological systems, evolved by natural selection. Our conclusion is based on two facts that we would think would be entirely uncontroversial: Language shows signs of complex design for the communication of propositional structures, and the only explanation for the origin of organs with complex design is the process of natural selection. Although distinguished scientists from a wide variety of fields and ideologies have tried to cast doubt on an orthodox Darwinian account of the evolution of a biological specialization for grammar, upon close examination none of the arguments is compelling.

But we hope we have done more than try and set the record straight. Skepticism about the possibility of saying anything of scientific value about language evolution has a long history, beginning in the prohibition against discussing the topic by the Société de Linguistique de Paris in 1866 and culminating in the encyclopedic volume edited by Harnad, Steklis, and Lancaster (1976) that pitted a few daring speculators against an army of doubters. A suspicious attitude is not entirely unwarranted when one reads about the Age of Modifiers, Pithecanthropus alalus ("Ape-man without speech"), and the heave-ho theory. But such skepticism should not lead to equally unsupported assertions about the necessity of spandrels and saltations.

A major problem among even the more responsible attempts to speculate about the origins of language has been that they ignore the wealth of specific knowledge about the structure of grammar discovered during the past 30 years. As a result language competence has been equated with cognitive development, leading to confusions between the evolution of language and the evolution of thought, or has been expediently equated with activities that leave tangible remnants, such as tool manufacture, art, and conquest.

We think there is a wealth of respectable new scientific information relevant to the evolution of language that has never been properly synthesized. The computational theory of mind, generative grammar, articulatory and acoustic phonetics, developmental psycholinguistics, and the study of dynamics of diachronic change could profitably be combined with recent molecular, archeological, and comparative neuroanatomical discoveries and with strategic modeling of evolution using insights from evolutionary theory and anthropology (see, e.g., this volume; Bickerton, 1981; Brandon and Hornstein, 1986; Hinton & Nowlan, 1987; Hurford, 1989, 1991; Tooby & DeVore, 1987). It is certain there are many questions about the evolution of language that we will never answer. But we are optimistic that there are insights to be gained, if only the problems are properly posed.

ACKNOWLEDGMENTS

Reprinted with permission from *Behavioral and Brain Sciences, 13,* pp. 707–727.

This paper has been greatly influenced by discussions with Leda Cosmides and John Tooby. We also thank Ned Block, Susan Carey, Noam Chomsky, Nancy Etcoff, Robert Freidin, Jane Grimshaw, James Hurford, Massimo Piattelli-Palmarini, Alan Prince, Jerry Samet, Donald Symons, and several BBS reviewers for their helpful comments on earlier drafts. Needless to say all deficiencies and errors are ours. Preparation of this paper was supported by NIH Grant HD 18381; the second author was supported by a Surdna Predoctoral Fellowship.

NOTES

1. For example, he says that "language must surely confer enormous selective advantages" (Chomsky, 1980, p. 239; see also Chomsky, 1975, p. 252), and argues that "suppose that someone proposes a principle which says: The form of a language is such-and-such because having that form permits a function to be fulfilled—a proposal of this sort would be appropriate at the level of evolution (of the species, or of language), not at the level of acquisition of language by an individual (Chomsky, 1977, pp. 86–87).

2. Interestingly, Dennett (1983) argues that Gould and Lewontin's critique is remarkably similar in logic to critiques of another large-scale theory, the representational theory of mind of cognitive science, by behaviorists. Dennett sees common flaws in the critiques: Both fail to account for cases of adaptive complexity that are not direct consequences of any law of physics, and both apply the criterion of falsifiability in too literal-minded a way.

3. Note also that historical change in languages occurs very rapidly by biological standards. Wang (1976) points out, for example, that one cycle of the process whereby a language alternates between reliance on word order and reliance on affixation typically takes a thousand years. A hominid population evolving language could be exposed to the full range of linguistic diversity during a single tick of the evolutionary clock, even if no single generation was faced with all humanly possible structures.

4. In the notation standard in linguistics, an asterisk indicates a sentence sensed by the speaker of a language to be ill-formed.

5. We thank John Tooby for pointing this out to us.

REFERENCES

Alexander, R. (1987, March). Paper presented at the conference "The origin and dispersal of modern humans," Corpus Christi College, Cambridge, England. Reported in *Science, 236*, 668–669.

Axelrod, R., & Hamilton, W. D. (1981). The evolution of cooperation. *Science, 211*, 1390–1396.

Ayala, F. (1983). Microevolution and macroevolution. In D. S. Bendall (Ed.), *Evolution from molecules to man* (pp. 387–402). Cambridge: Cambridge University Press.

Bates, E. (1976). *Language and context: Studies in the acquisition of pragmatics.* New York: Academic Press.

Bates, E., Thal, D, & Marchman, V. (1991). Symbols and syntax: A Darwinian approach to language development. In N. Krasnegor, D. Rumbaugh, R. Schiefelbusch, & M. Studdert-Kennedy, (Eds.), *Biological and behavioral language development.* Hillsdale, NJ: Erlbaum.

Bendall, D. S (Ed.). (1983). *Evolution from molecules to men.* New York: Cambridge University Press.

Berwick, R. C., & Weinberg, A. S. (1984). *The grammatical basis of linguistic performance.* Cambridge, MA: MIT Press.

Berwick, R. C., & Wexler, K. (1987). Parsing efficiency, binding, c-command, and learnability. In B. Lust. Reidel (Ed.), *Studies in the acquisition of anaphora.* Dordrecht, Netherlands: Reidel, pp. 45–60.

Bever, T. G. (1970). The cognitive basis for linguistic structures. In J. R. Hayes (Ed.), *Cognition and the development of language* (pp. 279–362). New York: Wiley.

Bever, T. G., Carrithers, C., Cowart, W., & Townsend, D. J. (1989). Language processing and familial handedness. In A. M. Galaburda (Ed.), *From reading to neurons* (pp. 331–360). Cambridge, MA: MIT Press.

Bickerton, D. (1981). *The roots of language*. Ann Arbor, MI: Karoma.

Bickerton, D. (1984). The language bioprogram hypothesis. *Behavioral and Brain Sciences, 7*, 173–212.

Bickerton, D. (1986). More than nature needs? A reply to Premack. *Cognition, 23*, 73–79.

Bloom, P. (1990). Syntactic distinctions in child language. *Journal of Child Language, 17*, 353–355.

Bloom, P. (1991). *Nominals in child language*. Unpublished manuscript, Department of Psychology, University of Arizona, Tucson, AZ.

Bolinger, D. (1980). *Language: The loaded weapon*. New York: Longman.

Brandon, R. N., & Hornstein, N. (1986). From icons to symbols: Some speculations on the origin of language. *Biology and Philosophy, 1*, 169–189.

Bresnan, J. (1982). Control and complementation. In J. Bresnan (Ed.), *The mental representation of grammatical relations* (pp. 282–390). Cambridge, MA: MIT Press.

Bresnan, J., & Kaplan, R. M. (1982). Grammars as mental representations of language. In J. Bresnan (Ed.), *The mental representation of grammatical relations* (pp. xiii–lii). Cambridge, MA: MIT Press.

Brown, R., & Hanlon, C. (1970). Derivational complexity and order of acquisition in child speech. In J. R. Hayes (Ed.), *Cognition and the development of language* (pp. 11–53). New York: Wiley.

Bybee, J. (1985). *Morphology: A study of the relation between meaning and form*. Philadelphia: Benjamins.

Chomsky, N. (1965). *Aspects of the theory of syntax*. Cambridge, MA: MIT Press.

Chomsky, N. (1972). *Language and mind* (extended edition). New York: Harcourt, Brace, & Jovanovich.

Chomsky, N. (1975). *Reflections on language*. New York: Pantheon.

Chomsky, N. (1977). *Language and responsibility*. New York: Pantheon.

Chomsky, N. (1980). *Rules and representations*. New York: Columbia University Press.

Chomsky, N. (1981). *Lectures on government and binding*. Dordrecht, Netherlands: Foris.

Chomsky, N. (1982a). *Noam Chomsky on the generative enterprise: A discussion with Riny Junybregts and Henk van Riemsdijk*. Dordrecht, Netherlands: Foris.

Chomsky, N. (1982b). Discussion of Putnam's comments. In M. Piattelli-Palmarini (Ed.), *Language and learning: The debate between Jean Piaget and Noam Chomsky* (pp. 310–324). Cambridge, MA: Harvard University Press.

Chomsky, N. (1986). *Knowledge of language: Its nature, origin, and use*. New York: Praeger.

Chomsky, N. (1988a). *Language and problems of knowledge: The Managua lectures*. Cambridge, MA: MIT Press.

Chomsky, N. (1991). Linguistics and cognitive science: Problems and mysteries. In A. Kasher (Ed.), *The Chomskyan turn* (pp. 26–53). Cambridge, MA: Basil Blackwell.

Chomsky, N., & Lasnik, H. (1977). Filters and control. *Linguistic Inquiry, 8*, 425–504.

Clutton-Brock, T. H. (1983). Selection in relation to sex. In D. S. Bendall (Ed.), *Evolution from molecules to men.* (pp. 457–481). Cambridge: Cambridge University Press.

Comrie, B. (1981). *Language universals and linguistic typology: Syntax and morphology*. Cambridge, MA: Basil Blackwell.

Cosmides, L. (1989). The logic of social exchange: Has natural selection shaped how humans reason? Studies with the Wason selection task. *Cognition, 31*, 187–276.

Cosmides, L., & Tooby, J. (1989). Evolutionary psychology and the generation of culture, Part II. Case study: A computational theory of social exchange. *Ethology and Sociobiology, 10*, 51–97.

Cummins, R. (1984). Functional analysis. In E. Sober (Ed.), *Conceptual issues in evolutionary biology* (pp. 386–407). Cambridge, MA: MIT Press.

Darwin, C. (1859). *On the origin of species.* (Reprinted by Harvard University Press, Cambridge, MA, 1964.)

Dawkins, R. (1976). *The selfish gene.* New York: Oxford University Press.

Dawkins, R. (1982). *The extended phenotype: The gene as the unit of selection.* San Francisco: Freeman.

Dawkins, R. (1983). Universal Darwinism. In D. S. Bendall (Ed.), *Evolution from molecules to men.* New York: Cambridge University Press.

Dawkins, R. (1986). *The blind watchmaker: Why the evidence of evolution reveals a universe without design.* New York: Norton.

Dennett, D. C. (1983). Intentional systems in cognitive ethology: The "Panglossian Paradigm" defended. *Behavioral and Brain Sciences, 6,* 343–390.

Dodd, J., & Jessell, T. M. (1988). Axon guidance and the patterning of neuronal projections in vertebrates. *Science, 242,* 692–699.

Ekman, P., & Friesen, W. V. (1975). *Unmasking the face.* Englewood Cliffs, NJ: Prentice Hall.

Eldredge, N., & Gould, S. J. (1972). Punctuated equilibria: An alternative to phyletic gradualism. In T.J.M. Schopf (Ed.), *Models in paleobiology.* San Francisco: Freeman.

Etcoff, N. L. (1986). The neuropsychology of emotional expression. In G. Goldstein & R. E. Tarter (Eds.), *Advances in clinical neuropsychology* (Vol. 3). (pp. 127–179). New York: Plenum.

Falk, D. (1983). Cerebral cortices of East African early hominids. *Science, 221,* 1072–1074.

Fisher, R. A (1930). *The genetical theory of natural selection.* Oxford, Clarendon Press.

Fodor, J. (1975). *The language of thought.* New York: Thomas Crowell.

Fodor, J. (1983). *The modularity of mind.* Cambridge, MA: MIT Press.

Fodor, J. (1987). *Psychosemantics.* Cambridge, MA: MIT Press.

Frazier, L., Clifton, C., & Randall, J. (1983). Filling gaps: Decision principles and structure in sentence comprehension. *Cognition, 13,* 187–222.

Freidin, R. (1978). Cyclicity and the theory of grammar. *Linguistic Inquiry, 9,* 519–549.

Freidin, R., & Quicoli, C. (1989). Zero-stimulation for parameter setting. *Behavioral and Brain Sciences, 12,* 338–339.

Freyd, J. J. (1983). Shareability: The social psychology of epistemology. *Cognitive Science, 7,* 191–210.

Gazdar, G., Klein, E., Pullum, G., & Sag, I. A. (1985). *Generalized phrase structure syntax.* Cambridge, MA: Harvard University Press.

Geschwind, N. (1980). Some comments on the neurology of language. In D. Caplan (Ed.), *Biological studies of mental processes.* Cambridge, MA: MIT Press.

Gopnik, M. (1990a). Feature blindness? A case study. *Language Acquisition, 1,* 139–164.

Gopnik, M. (1990b). Dysphasia in an extended family. *Nature, 344,* 715.

Gopnik, M., & Crago, M. B. (1991). Family aggregation of a developmental language disorder. *Cognition, 39,* 1–50.

Gould, S. J. (1977a). Darwin's untimely burial. In S. J. Gould, *Ever since Darwin: Reflections on natural history.* New York: Norton.

Gould, S. J. (1977b). Problems of perfection, or how can a clam mount a fish on its rear end. In S. J. Gould, *Ever since Darwin: Reflections on natural history.* New York: Norton.

Gould, S. J. (1979). Panselectionist pitfalls in Parker & Gibson's model of the evolution of intelligence. *Behavioral and Brain Sciences, 2,* 385–386.

Gould, S. J. (1980). Is a new and general theory of evolution emerging? *Paleobiology, 6,* 119–130.

Gould, S. J. (1981). Return of the hopeful monster. In S. J. Gould, *The panda's thumb: More reflections on natural history* (pp. 186–193). New York: Norton.

Gould, S. J. (1987a, October). *The limits of adaptation: Is language a spandrel of the human brain?* Paper presented to the Cognitive Science Seminar, Center for Cognitive Science, MIT, Cambridge, MA.

Gould, S. J. (1987b). Integrity and Mr. Rifkin. In S. J. Gould, *An urchin in the storm: Essays about books and ideas* (pp. 229–239.) London: Collins Harvill.

Gould, S. J. (1989, May). *Evolutionary considerations.* Paper presented to the McDonnell Foundation conference "Selection vs. Instruction," Venice.

Gould, S. J., & Eldredge, N. (1977). Punctuated equilibria: The tempo and mode of evolution reconsidered. *Paleobiology, 3,* 115–151.

Gould, S. J., & Lewontin, R. C. (1979). The spandrels of San Marco and the Panglossian program: A critique of the adaptationist programme. *Proceedings of the Royal Society of London, 205,* 281–288.

Gould, S. J., & Piattelli-Palmarini, M. (1987). *Evolution and cognition.* Course taught at Harvard University, Cambridge, MA.

Gould, S. J., & Vrba, E. S. (1982). Exaptation—a missing term in the science of form. *Paleobiology, 8,* 4–15.

Greenberg, J. H. (Ed.). (1966). *Universals of language.* Cambridge, MA: MIT Press.

Greenberg, J. H., Ferguson, C. A., & Moravcsik, E. A. (Eds.). (1978). *Universals of human language* (Vols. 1–4). Stanford, CA: Stanford University Press.

Greenfield, P. (1987). Departmental colloquium, Department of Psychology, Harvard University, Cambridge, MA.

Greenfield, P., & Smith, J. (1976). *The structure of communication in early language development.* New York: Academic Press.

Haldane, J.B.S. (1927). A mathematical theory of natural and artificial selection. Part V. Selection and mutation. *Proceedings of the Cambridge Philosophical Society, 23,* 838–844.

Hamilton, W. D. (1964). The genetical evolution of social behavior, I & II. *Journal of Theoretical Biology, 104,* 451–471.

Harnad, S. R., Steklis, H. S., & Lancaster, J. (Eds.). (1976). Origin and evolution of language and speech. *Annals of the New York Academy of Science, 280.*

Harrelson, A. L., & Goodman, C. S. (1988). Growth cone guidance in insects: Fasciclin II is a member of the immunoglobulin superfamily. *Science, 242,* 700–708.

Hawkins, J. (Ed.). (1988). *Explaining language universals.* Basil Blackwell.

Hinton, G. E., & Nowlan, S. J. (1987). How learning can guide evolution. *Complex Systems, 1,* 495–502.

Hurford, J. R. (1989). Biological evolution of the Saussurean sign as a component of the language acquisition device. *Lingua, 77,* 187–222.

Hurford, J. R. (1991). *The evolution of the critical period for language acquisition. Cognition, 40,* 159–201.

Isaac, G. L. (1983). Aspects of human evolution. In D. S. Bendall (Ed.), *Evolution from molecules to men* (pp. 509–543). New York: Cambridge University Press.

Jackendoff, R. (1977). *X-bar syntax: A study of phrase structure.* Cambridge, MA: MIT Press.

Jackendoff, R. (1983). *Semantics and cognition.* Cambridge, MA: MIT Press.

Jackendoff, R. (1990). *Semantic structures.* Cambridge, MA: MIT Press.

Jacob, F. (1977). Evolution and tinkering, *Science, 196,* 1161–1166.

Kaplan, R. M. & Bresnan J. (1982). Lexical Functional Grammar: A formal system for grammatical representation. In J. Bresnan (Ed.), *The mental representation of grammatical relations* (pp. 173–281). Cambridge, MA: MIT Press.

Keenan, E. O. (1976). Towards a universal definition of "subject." In C. Li. (Ed.), *Subject and topic* (pp. 303–333). New York: Academic Press.

Keil, F. (1979). *Semantic and conceptual development.* Cambridge, MA: Harvard University Press.

King, M., & Wilson, A. (1975) Evolution at two levels in humans and chimpanzees. *Science, 188,* 107–116.

Kingsolver, J. G., & Koehl, M.A.R. (1985). Aerodynamics, thermoregulation, and the evolution of insect wings: Differential scaling and evolutionary change. *Evolution, 39,* 488–504.

Kiparsky, P. (1976). Historical linguistics and the origin of language. In S. R. Harnad, H.S. Steklis, & J. Lancaster (Eds.), *Annals of the New York Academy of Sciences, 280,* 97–103.

Kitcher, P. (1985). *Vaulting ambition: Sociobiology and the quest for human nature.* Cambridge, MA: MIT Press.

Konner, M. (1982). *The tangled wing: Biological constraints on the human spirit.* New York: Harper.

Kuno, S. (1973). Constraints on internal clauses and sentential subjects. *Linguistic Inquiry, 4,* 363–385.

Kuno, S. (1974). The position of relative clauses and conjunctions. *Linguistic Inquiry, 5,* 117–136.

Lachter, J., & Bever, T. G. (1988). The relation between linguistic structure and associative theories of language learning—a constructive critique of some connectionist learning models. *Cognition, 28,* 195–247.

Lenneberg, E. H. (1964). A biological perspective on language. In E. H. Lenneberg (Ed.), *New directions in the study of language* (pp. 65–88). Cambridge, MA: MIT Press.

Lenneberg, E. H. (1967). *Biological foundations of language.* New York: Wiley.

Lewontin, R. (1978). Adaptation. *Scientific American, 239,* 157–169.

Liberman, A. M., Cooper, F. S., Shankweiler, D. P., & Studdert-Kennedy, M. (1967). Perception of the speech code. *Psychological Review, 74,* 431–461.

Liberman, A. M., & Mattingly, I. G. (1989). A specialization for speech perception. *Science, 243,* 489–496.

Lieberman, P. (1976). Interactive models for evolution: Neural mechanisms, anatomy, and behavior. In S. R. Harnad, H. S. Steklis, & J. Lancaster (Eds.), Origin and evolution of language and speech. *Annals of the New York Academy of Sciences, 280,* 660–672.

Lieberman, P. (1984). *The biology and evolution of language.* Cambridge, MA: Harvard University Press.

Lieberman, P. (1989). Some biological constraints on universal grammar and learnability. In M. Rice & R. L. Schiefelbusch (Eds.), *The teachability of language* (pp. 199–225). Baltimore: Paul H. Brookes.

Ludlow, C. L., & Cooper, J. A. (1983). Genetic aspects of speech and language disorders: Current status and future directions. In C. L. Ludlow & J. A. Cooper (Eds.), *Genetic aspects of speech and language disorders* (pp. 1–19). New York: Academic Press.

Maratsos, M. (1983). Some current issues in the study of the acquisition of grammar. In P. Mussen (Ed.), *Carmichael's manual of child psychology* (4th ed.). p. 707–786). New York: Wiley.

Maratsos, M. (1988). Innateness and plasticity in language acquisition. In M. Rice & R. L. Schiefelbusch (Eds.). *The teachability of language* (pp. 105–125). Baltimore: Paul H. Brookes.

Maynard Smith, J. (1974). The theory of games and the evolution of animal conflicts. *Journal of Theoretical Biology, 47,* 209–221.

Maynard Smith, J. (1984). Optimization theory in evolution. In E. Sober (Ed.), *Conceptual issues in evolutionary biology* (pp. 289–315). Cambridge, MA: MIT Press.

Mayr, E. (1982). *The growth of biological thought.* Cambridge, MA: Harvard University Press.

Mayr, E. (1983). How to carry out the adaptationist program. *The American Naturalist, 121,* 324–334.

Mehler, J. (1985). Review of P. Lieberman's *The biology and evolution of language. Journal of the Acoustic Society of America, 80,* 1558–1560.

Miyamoto, M. M., Slightom, J. L., & Goodman, M. (1987). Phylogenetic relations of humans and African apes from DNA sequences in the $\psi\eta$-globin region. *Science, 238,* 369–373.

Morgan, J. L. (1986). *From simple input to complex grammar.* Cambridge, MA: MIT Press.

Morgan, J., & Travis, L. (1989). Limits on negative information in language input. *Journal of Child Language, 16,* 531–552.

Newport, E. L., & Meier, R. P. (1985). The acquisition of American sign language. In D. I. Slobin (Ed.), *The cross-linguistic study of language acquisition: The data* (Vol. 1). (pp. 881–938). Hillsdale, NJ: Erlbaum.

Petitto, L. (1987). On the autonomy of language and gesture: Evidence from the acquisition of personal pronouns in American Sign Language. *Cognition, 27,* 1–52.

Piattelli-Palmarini, M. (1989). Evolution, selection, and cognition: From "learning" to parameter setting in biology and the study of language. *Cognition, 31,* 1–44.

Pinker, S. (1979). Formal models of language learning. *Cognition, 7,* 217–283.

Pinker, S. (1984). *Language learnability and language development.* Cambridge, MA: Harvard University Press.

Pinker, S. (1989a). Language acquistion. In M. I. Posner (Ed.), *Foundations of cognitive science* (pp. 359–399). Cambridge, MA: MIT Press.

Pinker, S. (1989b). *Learnability and cognition: The acquisition of argument structure.* Cambridge, MA: MIT Press.

Pinker, S., & Prince, A. (1988). On language and connectionism: Analysis of a Parallel Distributed Processing model of language acquisition. *Cognition, 28,* 73–193.

Pinker, S., & Prince, A. (1989). *The nature of human concepts: Evidence from an unusual source.* Unpublished manuscript, MIT, Cambridge, MA.

Premack, D. (1985). "Gavagai!" or the future history of the animal language controversy. *Cognition, 19,* 207–296.

Premack, D. (1986). Pangloss to Cyrano de Bergerac: "Nonsense, it's perfect!" *Cognition, 23,* 81–88.

Reichenbach, H. (1947). *Elements of symbolic logic.* New York: Free Press.

Ridley, M. (1986). *The problems of evolution.* Oxford: Oxford University Press.

Rose, M. (1980). The mental arms race amplifier. *Human Ecology, 8,* 285–293.

Rozin, P. (1976). The evolution of intelligence and access to the cognitive unconscious. In L. Sprague & A. N. Epstein. (Eds.), *Progress in psychobiology and physiological psychology* (pp. 245–280). New York: Academic Press.

Samuels, M. L. (1972). *Linguistic evolution.* New York: Cambridge University Press.

Saussure, F. de, (1959). *Course in general linguistics.* New York: McGraw-Hill.

Seidenberg, M. S. (1986). Evidence from the great apes concerning the biological bases of language. In W. Demopoulos & A. Marras (Eds.), *Language learning and concept acquisition: Foundational issues* (pp. 29–53). Norwood, NJ: Ablex.

Seidenberg, M. S., & Petitto, L. A. (1979). Signing behavior in apes: A critical review. *Cognition, 7,* 177–215.

Seidenberg, M. S., & Petitto, L. A. (1987). Communication, symbolic communication, and language: Comment on Savage-Rumbaugh, McDonald, Sevcik, Hopkins, and Rupert (1986). *Journal of Experimental Psychology: General, 116,* 279–287.

Shepard, R. (1986). Evolution of a mesh between principles of the mind and regularities of the world. In J. Dupré (Ed.), *The latest on the best: Essays on optimization and evolution* (pp. 251–275). Cambridge, MA: MIT Press.

Shopen, T. (Ed.). (1985). *Language typology and syntactic description* (Vols. 1–3). New York: Cambridge University Press.

Simon, H. A. (1969). The architecture of complexity. In H. Simon (Ed.), *The sciences of the artificial* (pp. 84–118). Cambridge, MA: MIT Press.

Skinner, B. F. (1957). *Verbal behavior.* New York: Appleton.

Slobin, D. (1977). Language change in childhood and in history. In J. Macnamara (Ed.), *Language learning and thought.* New York: Academic Press.

Smolensky, P. (1988). The proper treatment of connectionism. *Behaviorial and Brain Sciences, 11*, 1–74.

Sober, E. (1984). *The nature of selection.* Cambridge, MA: MIT Press.

Stebbins, G. L. (1982). *Darwin to DNA, molecules to humanity.* San Francisco: Freeman.

Stebbins, G. L. & Ayala, F. J. (1981). Is a new evolutionary synthesis necessary? *Science, 213*, 967–971.

Steele, S., Akmajian, A., Demers, R., Jelinek, E., Kitagawa, C., Oehrle, R., & Wasow, T. (1981). *An encyclopedia of AUX: A study of cross-linguistic equivalence.* Cambridge, MA: MIT Press.

Stringer, C. B., & Andrews, P. (1988). Genetic and fossil evidence for the origin of modern humans. *Science, 239*, 1263–1268.

Symons, D. (1979). *The evolution of human sexuality.* New York: Oxford University Press.

Talmy, L. (1983). How language structures space. In H. Pick & L. Acredolo (Eds.), *Spatial orientation: Theory, research, and application* (pp. 225–282). New York: Plenum.

Talmy, L. (1988). Force dynamics in language and cognition. *Cognitive Science, 12*, 49–100.

Tobias, P. V. (1981). The emergence of man in Africa and beyond. *Philosophical Transactions of the Royal Society of London B, 292*, 43–56.

Tooby, J., & Cosmides, L. (1990). On the universality of human nature and the uniqueness of the individual: The role of genetics and adaptation. *Journal of Personality, 58*, 17–67.

Tooby, J., & DeVore, I. (1987). The reconstruction of hominid evolution through strategic modeling. In: W. G. Kinzey (Ed.), *The evolution of human behavior: Primate models* (pp. 183–237). Albany, NY: SUNY Press.

Travis, L. (1984). *Parameters and effects of word order variation.* Unpublished doctoral dissertation, MIT, Cambridge, MA.

Trivers, R. L. (1971). The evolution of reciprocal altruism. *Quarterly Review of Biology, 46*, 35–57.

Trivers, R. L. (1974). Parent-offspring conflict. *American Zoologist, 14*, 249–264.

Tversky, A., & Kahneman, D. (1981). The framing of decisions and the psychology of choice. *Science, 211*, 453–458.

Wang, W. S-Y. (1976). Language change. In S. R. Harnad, H. S. Steklis, & J. Lancaster (Eds.), Origin and evolution of language and speech. *Annals of the New York Academy of Science, 280*, 61–72.

Wexler, K., & Culicover, P. (1980). *Formal principles of language acquisition.* Cambridge, MA: MIT Press.

Wexler, K., & Manzini, R. (1984). Parameters and learnability in binding theory. In T. Roeper & E. Williams (Eds.), *Parameter setting* (pp. 41–76). Dordrecht, Netherlands: Reidel.

Williams, G. C. (1966). *Adaptation and natural selection: A critique of some current evolutionary thought.* Princeton, NJ: Princeton University Press.

Zubrow, E. (1987, March). Paper presented at the conference "The origin and dispersal of modern humans: Behavioral and biological perspectives," University of Cambridge, Cambridge, England. Reported in *Science, 237*, 1290–1291.

13

The Perceptual Organization of Colors: An Adaptation to Regularities of the Terrestrial World?

ROGER N. SHEPARD

Those taking an evolutionary approach to the behavioral, cognitive, and social sciences have been emphasizing the natural selection of mechanisms and strategies that are specific to particular species, genders, and problem domains. This emphasis on the particular is understandable as a reaction against the tendency, long dominant in these sciences, to proceed as if human and animal behavior could be explained in terms of just two things: (a) general laws of learning and cognitive constraints that hold across species, genders, and domains, and (b) the particular set of environmental (including cultural) circumstances to which each animal can adapt only by learning through its own individual experience. Evolutionary theorists' emphasis on specific adaptations may also reflect the tendency of such adaptations to catch our attention through the very diversity of their specificity. Adaptations to universal features of our world are apt to escape our notice simply because we do not observe anything with which such adaptations stand in contrast.

In this chapter I consider some characteristics of the perception and representation of colors that, although not universal in animal vision, do appear to be universal in the normal color vision of humans, prevalent in other primates, and common in a number of other quite different but also highly visual species, including the birds and the bees. The questions raised are (a) whether these characteristics of color perception and representation are merely arbitrary design features of these particular species, (b) whether these characteristics arose as specific adaptations to the particular environmental niches in which these species evolved, or (c) whether they may have emerged as advanced adaptations to some properties that prevail throughout the terrestrial environment. My discussion here will principally concern four remarkable characteristics of the colors that we so immediately and automatically experience when we look at the objects around us.

1. *The perceptual constancy of colors.* Opening our eyes seems like simply opening a window on a surrounding world of enduring objects and their inherent colors. Closer consideration shows, however, that our experience of the colors of objects depends on a process of visual analysis that, although largely unconscious, must be highly sophisticated and complex. Because most terrestrial objects do not shine by their own light, the light that strikes our eyes from those objects and by which we see them generally originates in a very different, extraterrestrial source—namely, the sun. Moreover,

depending on the circumstances in which an object is viewed, that solar illumination is subject to great variation in spectral composition. For example, light striking an object directly from an overhead sun may be strongest in middle wavelengths (yellows), light scattered to an object from clear sky may be strongest in short wavelengths (blues), and light from a setting sun may be strongest in long wavelengths (reds). Depending on such viewing conditions, the light scattered back to our eyes from any given object can accordingly be strongly biased toward the middle, shorter, or longer wavelengths and, so, toward the yellows, blues, or reds. Yet, despite these great variations in the light that a surface scatters back to our eyes under these different conditions, the color that we perceive a surface to have remains a fixed, apparently inherent property of the surface itself.

The adaptive significance of such perceptual constancy of surface color is clear. In order to respond appropriately to an object we must recognize it as the same distal object despite wide variations in the illumination and in the resulting composition of the light striking our retinas. After centuries of investigation, the question of how our visual systems achieve color constancy has been considerably clarified and, perhaps, essentially answered (see Maloney & Wandell, 1986). Although the principle of color constancy underlies virtually everything I have to say in this chapter, that principle (if not yet the specific physiological mechanism by which it is achieved) has long been accepted as fundamental in psychology and visual science. To the extent that I have any new ideas to present, these will primarily concern the following three other characteristics of human color perception and representation—characteristics that, although well established, still have no generally accepted explanations.

2. *The three-dimensional structure of colors.* In terms of its chromatic composition, the light reaching our eyes from any one point in the visual field can vary in an unlimited number of dimensions. Each such dimension would specify the amount of energy in the light that is of one particular wavelength within the range of visible wavelengths—between about 400 nm (nanometers) at the violet end of the visible spectrum and about 700 nm at the red end. Yet the surface colors experienced by humans with normal color vision (in addition to being essentially constant for each surface) vary between different surfaces in only three dimensions (Helmholtz, 1856–1866; Young, 1807). In operational terms, normal observers can match the color appearance of any given surface by adjusting three—but no fewer than three—knobs on a suitable color-mixing apparatus.

Accordingly, the surface colors experienced by normally sighted human observers are representable as points in a three-dimensional *color space,* whose three dimensions can be taken to be those called *lightness, hue,* and *saturation* (see Figure 13.1a). Why is the number of dimensions of our representation of colors exactly three—rather than, say, two or four? Furthermore, given that we experience colors as properties of the surfaces of external objects, in what sense, if any, does this three-dimensionality of our color experience reflect a three-dimensionality in the world?

3. *The circular structure of spectrally pure colors.* The colors that correspond to the hues of the rainbow—chromatically pure colors that are maximally saturated in the sense that they differ as much as possible from a (correspondingly light, medium, or dark) neutral or achromatic gray—can each be matched by light that is spectrally pure, that is, composed of a single wavelength. (The only saturated colors that cannot be matched by a single wavelength are the purples, which arise from mixtures of a long-wavelength red and a short-wavelength blue or violet.) Given that the pure colors of

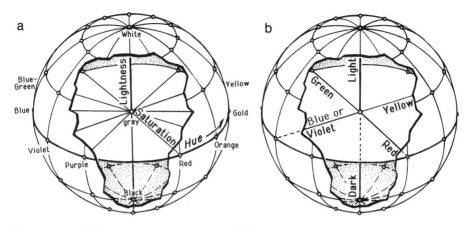

Figure 13.1 Schematic representations of three-dimensional color space in terms of three polar dimensions of lightness, hue, and saturation (a) or in terms of three dimensions of opponency between lightness versus darkness, blueness versus yellowness, and redness versus greenness (b).

the rainbow thus correspond to a single physical dimension (wavelength), it is not surprising that these colors are represented by a single dimension of three-dimensional color space—namely, the dimension we call *hue.* What does need explaining, is why this dimension of hue is not rectilinear, like the physical dimension of wavelength to which it corresponds, but circular (as schematically illustrated by the equatorial line around color space in Figure 13.1a).

The circular representation of hue, originally described by Isaac Newton (1704) and accordingly referred to as *Newton's color circle,* is entailed by an elementary perceptual fact: The hues at the two ends of the physical continuum of visible wavelengths—namely, long-wavelength red and short-wavelength violet—are experienced by individuals with normal color vision as more similar to each other than either is to intermediate-wavelength green. This psychological proximity between the physically most widely separated (red and violet) ends of the hue continuum can be achieved in color space only by bending this continuum into a circle. Is this circularity of hue a manifestation of an arbitrary characteristic of our perceptual system, then? Or might it have some deeper source, if not simply in the wavelengths of spectrally pure lights, in some more subtle regularity of the physical world?

4. *The universal organization of colors into categories and prototypes.* The three-dimensional space of surface colors (Figure 13.1)—including the one-dimensional circle of hues around the perimeter of that space—is continuous in the sense that between any two distinctly different colors, we can obtain, by color mixing, a color of intermediate appearance. At the same time, however, this continuous space seems to divide into relatively discrete regions within which the colors appear relatively more homogeneous than colors that cross the boundaries between such regions. Within such a region, moreover, the colors tend to be described by a single basic color word, such as our English words "red," "yellow," "green," or "blue." With even greater consistency, we generally agree with each other about what highly localized subregion within each such general region contains the best or most prototypical examples of the color prevailing throughout that general region.

Indeed, across human cultures, regardless of the duration and distance of their geographical separation and the number of basic color terms that their very different languages include, there seems to be a universal underlying hierarchy of regions and, especially, prototypical subregions of color space to which native speakers assign whatever basic color terms their individual languages include (Berlin & Kay, 1969). Thus, most languages have terms that native speakers apply to colors in the very same regions of color space for which we use the words "red" and "green." If, in addition, a given language also includes words corresponding (for example) to our "orange" or "purple," those word, too, are applied to essentially the same regions of color space for which we use "orange" or "purple." There is, moreover, especially great consistency in the locations of the colors that are picked out as the best or purest example of any particular color. Is this categorical organization of human color representation the manifestation of an arbitrary design feature of the human visual system? Or, again, is it a reflection of some general property of the world?

I do not claim to have final and definitive answers for these last three questions—concerning the three-dimensionality of colors, the circularity of hues, and the consistent organization of colors into categories and prototypes. Believing, however, that these questions are worth exploring from an evolutionary standpoint, I shall tentatively propose some hypotheses as to where in the world we might look for pervasive and enduring regularities to which these structural properties of the human visual system may have accommodated through natural selection. First, however, I attempt to situate my proposals within the broader context of human perceptual and cognitive constraints generally.

STRUCTURE IN HUMAN PERCEPTION AND COGNITION IN GENERAL

For over a century, psychological researchers have been probing the structures and processes of perception, memory, and thought that mediate the behaviors of humans and other animals. Typically, this probing has taken the form of behavioral experiments suggested by evidence from one or more of three sources: (a) introspections into one's own experience and inner processes, (b) information gleaned about the anatomy or physiology of the underlying physical mechanisms, and (c) results obtained from previous behavioral studies. More recently, in seeking to understand not only the nature but also the origins of psychological principles, some of us have been turning to a fourth source for guidance—namely, to the ecological properties of the world in which we have evolved and to the advantages to be realized by individuals who have genetically internalized representations of those properties.

Taken by themselves, findings based on introspective, behavioral, and physiological evidence alike, however well established and mutually consistent they may be, remain as little more than "brute facts" about the human or animal subjects studied. What such findings reveal might be merely arbitrary or ad hoc properties of the particular collection of terrestrial species investigated. Even our own perceptual and cognitive capabilities, as much as our own bodily sizes and shapes, may be the products of a history of more or less accidental circumstances peculiar to just one among uncounted evolutionary lines. Certainly, these capabilities do not appear to be wholly dictated by what is physically possible.

The following are just a few of the easily stated and well known of our perceptual/

cognitive limitations, as these have been demonstrated under highly controlled but nonnaturalistic laboratory conditions:

1. Although a physical measuring instrument can reliably identify a vast number of absolute levels of a stimulus, we reliably identify only about seven (Miller, 1956).

2. Although a physical recording instrument can register a vast number of dimensions of variation of the spectral composition of light, the colors we experience vary, as I have already noted, along only three independent dimensions (Helmholtz, 1856–1866; Young, 1807).

3. Although the red and violet spectral colors differ the most widely in physical wavelength, these colors appear more similar to each other than either does to the green of an intermediate wavelength (leading, as noted, to Newton's color circle).

4. Although a camera can record and indefinitely preserve an entire scene in a millisecond blink of a shutter, the "iconic" image that our visual system retains from a single brief exposure decays in less than a second and, during this time, we are able to encode only about four or five items for more permanent storage (Sperling, 1960).

5. Although a computer can store an essentially unlimited number of unrelated items for subsequent retrieval, following a single presentation, we can reliably recall a list of no more than about seven items (Miller, 1956).

6. Although a computer could detect correlations between events separated by any specified time interval and in either order of occurrence, in virtually all animals with nervous systems, classical conditioning generally requires that the conditioned stimulus last for a short time and either be simultaneous with the unconditioned stimulus or precede it by no more than a few seconds (Pavlov, 1927, 1928).

7. Although a computer can swiftly and errorlessly carry out indefinitely protracted sequences of abstract logical operations, we are subject to systematic errors in performing the simplest types of logical inferences (e.g., Tversky & Kahneman, 1974; Wason & Johnson-Laird, 1972; Woodworth & Sells, 1935)—at least when these inferences are not of the kind that were essential to the fitness of our hunter-gatherer ancestors during the Pleistocene era (Cosmides, 1989).

Our performance in a natural setting is, however, a very different matter. There, our perceptual and cognitive capabilities vastly exceed the capabilities of even the most advanced artificial systems. We readily parse complex and changing visual scenes and auditory streams into spatially localized external objects and sound sources. We classify those objects and sources into natural kinds despite appreciable variation in the individual instances and their contexts, positions, or conditions of illumination. We infer the likely ensuing behaviors of such natural objects—including the recognition of animals and anticipation of their approach or retreat, the recognition of faces and interpretation of their expressions, and the identification of voices and interpretation of their meanings. We recode and transfer, from one individual to another, information about arbitrary or possible states of affairs by means of a finite set of symbols (phonemes or corresponding written characters). And we plan for future courses of action and devise creative solutions to an open class of real-world problems.

To the extent that psychological science fails to identify nonarbitrary reasons or sources for these perceptual/cognitive limitations and for these perceptual/cognitive capabilities, this science will remain a merely descriptive science of this or that particular terrestrial species. This is true even if we are able to show that these limitations and capabilities are consequences of the structures of underlying neurophysiological mechanisms. Those neurophysiological structures can themselves be deemed nonarbitrary only to the extent that they can be seen to derive from some ultimately nonarbitrary source.

Where, then, should we look for such a nonarbitrary source? The answer can only be, "In the world." All niches capable of supporting the evolution and maintenance of intelligent life, though differing in numerous details, share some general—perhaps even universal—properties. It is to these properties that we must look for the ultimate, nonarbitrary sources of the regularities that we find in perception/cognition as well as in its underlying neurophysiological substrate.

Some of the properties that I have in mind here are the following (see Shepard, 1987a, 1987b, 1988, 1989): Space is three-dimensional, locally Euclidean, and endowed with a gravitationally conferred unique upward direction. Time is one-dimensional and endowed with a thermodynamically conferred unique forward direction. Periods of relative warmth and light (owing to the conservation of angular momentum of planetary rotation) regularly alternate with periods of relative coolness and darkness. And objects having an important consequence are of a particular natural kind and therefore correspond to a generally compact connected region in the space of possible objects—however much those objects may vary in their sensible properties (of size, shape, color, odor, motion, and so on).

Among the genes arising through random mutations, then, natural selection must have favored genes not only on the basis of how well they propagated under the special circumstances peculiar to the ecological niche currently occupied, but also, as I have argued previously (e.g., Shepard, 1987a), even more consistently in the long run, according to how well they propagate under the general circumstances common to *all* ecological niches. For, as an evolutionary line branches into each new niche, the selective pressures on gene propagation that are guaranteed to remain unchanged are just those pressures that are common to all niches.

Motivated by these considerations, much of my own recent work on perception and cognition in humans has sought evidence that our perceptual/cognitive systems have in fact internalized, especially deeply, the most pervasive and enduring constraints of the external world. In previous papers, I have primarily focused on evidence concerning two types of perceptual/cognitive capabilities: The first is our capability for representing rigid transformations of objects in three-dimensional Euclidean space—as revealed in experiments (a) on the perception of actually presented motions, (b) on the illusion of visual apparent motions, and (c) on the imagination of possible motions (Shepard, 1981, 1984, 1988; Shepard & Cooper, 1982; and, for a recent group-theoretic formulation, Carlton & Shepard, 1990). The second is our capability for estimating the conditional probability that a particular object is of the natural kind having some significant consequence, given that another particular object has already been found to have that consequence—as revealed in experiments on stimulus generalization and transfer of learning (Shepard, 1987b).

I turn now to the specific questions raised at the outset concerning four major characteristics of our perception and mental representation of surface colors—namely, their constancy, their three-dimensionality, their circularity with respect to hue, and their organization into categories and prototypes. In addition to reviewing the basic empirical phenomena to be explained in each case, I argue that these phenomena may arise not from merely accidental features of the human visual system but from genetic accommodations to identifiable regularities in the terrestrial world.

THE PERCEPTUAL CONSTANCY OF COLORS

Among the four major characteristics of human color perception mentioned, that of color constancy is undoubtedly of the most obvious and well-understood functional significance. This is not to say that the achievement of constancy is in any way trivial. The light that reaches our eyes from an external surface is a product of both the spectral reflectance characteristics of a surface and the spectral energy distribution of the light that happens to fall on that surface. In fact, as I shall shortly note, each of two surfaces can retain its own distinct color appearance even when the two surfaces are viewed under such different conditions of illumination that the light that each surface reflects back to our eyes is of identical composition. This could happen, for example, if a red-dish object is illuminated solely by the bluish light of the sky, while a bluish object is illuminated solely by the reddish light of the setting sun. Constancy in the appearance of surfaces under such different conditions of illumination can be achieved only to the extent that the visual system can successfully infer the separate contributions—of the surfaces and of the lighting—that have jointly given rise to the retinal stimulus.

By what mechanism does the visual system make such an inference and, hence, attain constancy in the perceived colors of objects? Two computational theories of color vision have been especially influential in the development of my own thinking about a possible evolutionary basis for the way in which humans represent colors. The first was the *retinex* theory proposed by Edwin Land (1964). Although that theory turns out to provide an incomplete solution to the problem of color constancy, it first suggested to me that the principal features of color vision may have arisen as an accom-modation to regularities in the world. The second is the general linear model for color vision more recently put forward by Maloney and Wandell (1986). In addition to pro-viding what I take to be the first satisfactory solution to the problem of color constancy, this model finally led me to an evolutionary argument as to why human vision should be trichromatic, that is, should have exactly three dimensions of color representation.

Land's Retinex Scheme for Color Constancy

The possibility that the subjective phenomena of color may reflect an objective prop-erty of the world began to intrigue me on attending Land's strikingly illustrated Wil-liam James Lectures, at Harvard University in 1967, on his *retinex theory* (see Land, 1964, 1986). Previously, textbook accounts of color vision had focused on phenomena attributable to peripheral, often retinal, sensory mechanisms. Principal examples of such phenomena are the gradual adjustment of visual sensitivity, called *adaptation,* following a change in general level of illumination, and the illusory shift in appearance of a surface (toward the lighter or darker, redder or greener, bluer or yellower), called *simultaneous* or *successive contrast,* when that surface is surrounded or preceded by a surface that is, contrastingly, darker or lighter, greener or redder, yellower or bluer, respectively. (Such phenomena of adaptation and contrast, incidentally, are characteristic of most sensory mechanisms and not specific to color or even to vision.) A major thrust of Land's approach, conditioned perhaps by his practical interest in contributing to color photography, was to look beyond the sensory mechanisms to the world outside and to consider the problem that the world presents and that the sensory mechanisms must solve. This is essentially the approach taken to vision in general by David Marr (1982) and, originally (but without the goal of also developing a computational model), by James Gibson (1950, 1966, 1979).

Land's demonstrations showed, in a particularly compelling way, that the color that we experience of a patch of surface is not determined by the composition of the light reaching our eyes from that patch. Instead, the color we experience somehow captures an invariant property of the surface itself, despite the enormous illumination-contingent variations in the composition of the light reflected back to our eyes. For example, Land illuminated a large display composed of rectangular sheets of many different colored papers (called a "Mondrian" because it resembled the abstract paintings by Mondrian) by means of three projectors—one projecting short-wavelength (violet) light, one projecting middle-wavelength (green) light, and one projecting long-wavelength (red) light. The intensity of the colored light from each projector could be independently controlled. On a screen above the Mondrian display, the projected image of the dial of a telescopic spot photometer registered the intensity of the light reflected back from any rectangular patch in the Mondrian at which the photometer was currently pointed.

With one setting of the intensities of the three illuminating projectors, Land pointed the spot photometer at a patch of the Mondrian that appeared a rust-colored reddish brown. In turn, pairs of the three projectors were turned off, and, from the projected image of the dial of the spot photometer, we could see how much light from the successively isolated short-, middle-, and long-wavelength projectors was being reflected back to our eyes from that rust-colored patch. Next, Land pointed the spot photometer at a very different, green patch. Again, all pairs of the three projectors were successively turned off (yielding, of course, very different readings of the amounts of short-, middle-, and long-wavelength light reflected from this green patch). The intensity of each projector was then adjusted so that according to the spot photometer, the very same amount of light from that projector—when it alone remained on—was reflected back to our eyes by the green patch as had previously been reflected back (from that projector) by the reddish brown patch. For example, in order to get the same amount of red light reflected back from the green patch as had previously been reflected back from the reddish brown patch, the red projector had to be turned way up; and in order to get the same amount of green light reflected back, the green projector had to be turned way down.

As a result of these adjustments, when all three projectors were then turned back on, our eyes were receiving light from the green patch that was virtually identical in physical composition to the light our eyes had previously been receiving from the reddish brown patch. Astonishingly, under this greatly altered lighting the originally reddish brown patch still appeared reddish brown, and the originally green patch still appeared green! In fact, the entire Mondrian display appeared essentially unchanged. Clearly, color is in the eye of the beholder in the sense that it does not have a one-to-one correspondence to the physical composition of the light that, striking the retina, gives rise to that color (or even, as we shall see, to the dimensionality of the variations that occur in that composition). Yet the resulting subjective experience of color, evidently, does capture an invariant property of the surface of the distal object.

Land proposed his retinex theory to account for this remarkable feat of color constancy by positing that the visual system automatically computes an independent brightness normalization within each of three color channels (which, for present purposes, we can think of as associated with the short-, middle-, and long-wavelength receptors). In effect, he assumed that the gain of each chromatic channel is increased or reduced so that the output for that channel, when that output is integrated over the entire visible scene, maintains a fixed, standard level of lightness. Thus, when the relative amount of long-wavelength light in the illumination increases—whether because the intensity of a red projector is turned up or because the sun is setting—there is a correlated increase in the amount of red light reflected from the scene as a whole. But the visual system automatically compensates for this increase in the average redness of the light reflected back from the entire scene by decreasing in the overall gain of the long-wavelength input channel. The consequence is that the scene (including a reddish brown patch and a green patch) continues to look essentially the same.

Land's approach to color vision appealed to me because it pointed toward the possibility that fundamental aspects of color vision may not be by-products of accidental design features of the visual system arising in a particular species but may represent a functional accommodation to a quite general property of the world in which we and other terrestrial animals have evolved. Land himself suggested something of this sort when he spoke of our "polar partnership" with the world around us (Land, 1978).

Still, Land's retinex theory does not in itself provide an answer to the questions of why we have specifically three chromatic input channels and hence specifically three dimensions of color vision. Moreover, the retinex theory has been shown to fall short of providing a satisfactory account of color constancy itself. Normalization of lightness within each of three color channels independently will achieve color constancy only if the whole scene viewed includes a typical representation of short-, middle-, and long-wavelength-reflecting surfaces. If the scene is strongly biased toward the reds (as in a canyon in the southwestern U.S.), toward the greens (as in a rain forest), or toward the blues (as in a sky and water scene), the proposed normalization will strive to remove the overall red, green, or blue complexion of the light reflected back from the scene even though, in each of these cases, the complexion is an inherent characteristic of the scene itself and not a bias imposed by the currently prevailing illumination. Indeed, in photography and television, color rendering is similarly attempted by separately adjusting the intensities of each of three color channels—with results whose inadequacies are all too familiar.

Maloney and Wandell's General Linear Model for Color Constancy

Building on work by Sälström (1973), Brill (1978; Brill & West, 1981; West & Brill, 1982), Buchsbaum (1980), and others, Laurence Maloney and Brian Wandell more recently put forward a general linear model for color vision that overcomes the principal limitations of Land's retinex theory (Maloney & Wandell, 1986). It was the advent of Maloney and Wandell's model at Stanford University that finally (nearly 20 years after Land's William James Lectures) suggested to me the possibility, to which I will soon turn, of a nonarbitrary basis for the three-dimensionality of human color vision.

Maloney and Wandell's theory does not require the restrictive assumption that the distribution of intrinsic colors in each scene always be the same. In their model, surface colors are estimated by applying a transformation to all three chromatic channels jointly, rather than to each channel separately. The resulting general linear transformation yields estimates of the intrinsic surface colors that approximate invariance both under natural variations of illumination and under wide variations of the distribution of surface colors (such as arise in a red canyon, a green rain forest, or a blue sky and water scene). Although the mathematical formulation of Maloney and Wandell's model is not essential here, it is worth one paragraph and one equation for those who are able to appreciate its elegance and generality.

The three central components of the model specify the dependencies on wavelength of (a) the illumination falling on the scene, (b) the light-reflecting characteristics of the surfaces making up the scene, and (c) the light-absorbing characteristics of the light-sensitive receptors in the eye viewing the scene. Specifically, for each point x of a surface in the scene and for each wavelength of light λ, the three components are: *the spectral power distribution of the illumination,* a function of wavelength $E^x(\lambda)$ specifying the amount of light (e.g., in quanta per second) of each wavelength λ falling on the surface point x; *the spectral reflectance distribution of the surface,* a function of wavelength $S^x(\lambda)$ specifying the proportion of any light quanta of wavelength λ falling on the surface at point x that will be scattered back from that surface (rather than being absorbed by the surface); and *the spectral sensitivity characteristics of the photoreceptive units,* functions of wavelength $R_k(\lambda)$ giving, for retinal receptors (cones) of each type k the sensitivity of receptors of that type to light of wavelength λ. The response of a receptor of type k to light scattered from the point x on the surface is then given by integration of the product of these three components over the entire spectrum of wavelengths:

$$\rho_k^x = \int E^x(\lambda)S^x(\lambda)R_k(\lambda)\,d\lambda \qquad k = 1, 2, \ldots, N$$

(where, for humans, who have three types of cones, $N = 3$).

Crucial to the application of this model to the problem of color constancy is the evidence, reviewed by Maloney and Wandell (1986), that the first two of these functions of wavelength, $E^x(\lambda)$ and $S^x(\lambda)$, each have only a limited number of degrees of freedom in the natural environment. That is, although the functions characterizing individual conditions of illumination and individual physical surfaces can each take on a potentially unlimited number of different shapes, in the natural terrestrial environment each of these shapes can be approximated as some linear combination of a small number of fixed underlying functions, called the *basis lighting functions* and the *basis reflectance functions,* respectively. (For those familiar with factor analysis or principal components analysis, this is merely another instance of the way in which redundancies or correlations in multivariate data permit the approximate reconstruction of those data from a relatively small number of underlying factors.)

Thus, although the complete specification of the light-reflecting characteristics of a surface must give, for each wavelength of light striking that surface, the proportion of the light of that wavelength that will be scattered back from the surface (rather than being absorbed), for most naturally occurring surfaces the spectral reflectance distributions $S^x(\lambda)$ turn out to be smooth, well-behaved functions of wavelength (Barlow, 1982; Stiles, Wyszecki, & Ohta, 1977). There is, moreover, a well-understood physical reason for this in terms of the Gaussian smoothing entailed by the very large number of quantum interactions among numerous neighboring energy states in surface atoms that mediate the absorption and re-emission of incident photons (Maloney, 1986; Nassau, 1983). In any case, the smoothness and, hence, redundancy in the empirically obtained spectral reflectance functions permits these functions to be approximated as linear combinations of a few basis functions (Cohen, 1964; Krinov, 1947; Yilmaz, 1962)—although evidently not as few as three (Buchsbaum & Gottschalk, 1984; Maloney, 1986).

For natural conditions of lighting, on the other hand, spectral energy distributions $E^x(\lambda)$ apparently can be adequately approximated as linear combinations of just three basis lighting functions (Judd, McAdam, & Wyszecki, 1964; also see Das & Sastri, 1965; Dixon, 1978; Sastri & Das, 1966, 1968; Winch, Boshoff, Kok, & du Toit, 1966). For example, Judd et al. found that 622 empirically measured spectral power distributions, measured under a great variety of conditions of weather and times of day, could be accurately approximated as weighted combinations of just the three functions plotted in Figure 13.2. (As in factor analysis, these particular basis functions are not uniquely determined by the data. Other sets of three basis functions that differ from these by a nonsingular linear transformation, such as a rotation, would account for the data just as well. The important point is that whichever of the alternative basis functions are used, they must be three in number.)

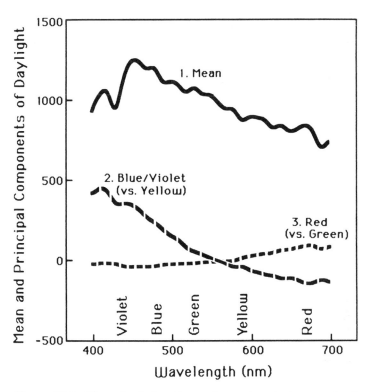

Figure 13.2 Three basis functions of wavelength that can be linearly combined to account for the major variations in the spectral compositions of natural daylight. (Based on Table 1 on page 1035 of Judd, MacAdam, & Wyszecki, 1964; copyright by the American Institute of Physics, 1964).

Maloney and Wandell (1986) showed that the unknown surface reflectance functions $S^x(\lambda)$—that is, the intrinsic colors of the surfaces—can be approximated, up to a *single* multiplicative scalar, solely from the responses of the N classes of photoreceptors, provided that two conditions are met: (a) The number of chromatically distinct classes of photoreceptors, N, must be one greater than the number of degrees of freedom of the surface reflectances; (b) the number of chromatically distinct visible surfaces in the scene viewed must be at least $N - 1$.

The single multiplicative scalar that remains indeterminate in Maloney and Wandell's model corresponds to an overall (achromatic) lightness that, according to the general linear model, cannot be unambiguously apportioned between the surfaces and the illumination falling on those surfaces. The light reflected from surfaces, alone, contains, in fact, insufficient information for any scheme of visual analysis to distinguish between darker surfaces under brighter lighting and corresponding but lighter surfaces under dimmer lighting. Maloney and Wandell nevertheless achieved a decisive advance by reducing the residual indeterminacy to a single scalar. Moreover, that single scalar can probably be adequately approximated, in most natural settings, by estimating the overall brightness of the illumination in ways not explicitly taken into account by the linear model (as originally framed solely in terms of the light scattered from surfaces). For example, the brightness of the illumination might be estimated, independently of the light scattered from surfaces, from the brightness of the sky (together with visual evidence for the presence of objects shading the surfaces viewed).

In contrast, Land's retinex model included not one but three indeterminate scalars—one for the overall lightness level within each of the three (red, green, and blue) chromatic channels. Brainard and Wandell (1986) have shown, further, that attempts to approximate surface color by correcting lightnesses within each chromatic channel separately, as in retinex schemes, rest on the unjustifiable assumption that the set of basis lighting functions $E^x(\lambda)$ can be chosen (by some linear transformation) so that variations in the weight of any one basis function affects the response of only one chromatic channel.

In fact, by striving for lightness constancy within each chromatic channel separately, retinex schemes necessarily forfeit not only constancy of colors, as such, but also constancy of the relative, purely achromatic lightnesses of different objects. In a chromatic world, even the ordering of recovered lightnesses within each channel will depend on the spectral balance of the illumination—unless the ability to see under dim illumination is seriously compromised by restricting each channel to a very narrow band of wavelengths. With broadly tuned and hence efficient channels (like those actually found in animals), a retinex system could estimate a patch of surface that reflects only shorter wavelengths to be either lighter or darker than a patch that reflects only longer wavelengths, depending on whether the illumination contains primarily shorter or longer wavelengths, respectively. This indeterminacy in the estimation of relative lightnesses of surfaces cannot be eliminated by any comparison between the outputs of separate channels, which, as in retinex schemes, each contain their own arbitrary multiplier. Such an indeterminacy can be eliminated only by transforming all three channels jointly, as advocated by Maloney and Wandell (1986).

I have gone into theories of color constancy at such length not only because color constancy is of obvious adaptive significance but also because I believe the mechanisms that have evolved for the achievement of color constancy may underlie the other major features of human color representation that I will be considering in the ensuing sections. If I am correct in this, these features, too, are not arbitrary but reflect pervasive and enduring regularities of the terrestrial environment in which we have evolved.

THE THREE-DIMENSIONALITY OF COLORS

The three-dimensionality of human color perception reveals itself, as I already noted, in the ability of a normally sighted observer to match the color appearance of any presented surface by adjusting three, but no fewer than three, unidimensional controls on a suitable color-mixing apparatus. One of these three dimensions is most naturally taken to be the (achromatic) dimension of *lightness*—as represented, in purest form, by the shades of gray ranging from black to white. The two remaining (chromatic) dimensions can be taken to be a circular dimension of *hue,* which, in cycling through the spectrally pure colors of red, orange, yellow, green, blue, and violet, forms Newton's color circle; and a radial dimension of *saturation,* which varies between each of these spectrally pure hues and a central, neutral gray of the same lightness. (Again, see Figure 13.1a.)

Alternatively, and in agreement with the opponent-process theory of color vision proposed by Ewald Hering (1878/1964) and given a modern development by Hurvich and Jameson (1957), the two chromatic dimensions can be taken to be a rectilinear red-versus-green dimension and an orthogonal yellow-versus-blue (or yellow-versus-violet) dimension—as illustrated in Figure 13.1b. Whether we thus describe the underlying three-dimensional color space using polar chromatic coordinates of hue and saturation (Figure 13.1a) or using rectangular chromatic coordinates of redness versus greenness and of yellowness versus blueness (Figure 13.1b), all colors can be obtained by an appropriate combination of values on the three chosen dimensions.

Inadequacy of Some Proposed Explanations for Trichromacy

Given that the human visual system normally provides just three dimensions for the representation of colors, a question immediately presents itself: Why exactly three dimensions, rather than, say, two or four? Three answers that might first suggest themselves all prove, on closer consideration, to be inadequate.

Is Trichromacy Explained by the Physical Properites of Surfaces?

Inherent physical properties of the surfaces in the external world do indeed determine which colors we perceive those surfaces to have—out of a three-dimensional set of alternative colors that we are capable of representing. Those physical properties of surfaces evidently do not, however, determine the dimensionality of our set of alternative color representations. What the inherent physical properties of a surface really determine is the spectral reflectance distribution of the surface; and the spectral reflectance distribution, being defined over a continuum of wavelengths, can require an unlimited number of quantities in order for us to specify the relative amount of light that would be reflected back for each wavelength between 400 and 700 nm. True, the spectral reflectance functions of wavelength for naturally occurring surfaces are generally smooth functions, and such smoothness implies the possibility of some dimensional reduction. Evidently, though, the smoothnesses are only sufficient to permit a reduction to somewhere between five and seven dimensions (Maloney, 1986). The further compression to the three dimensions of color that we experience cannot be attributed to a property of the surfaces of objects, therefore, but must be imposed by our visual system. The empirically verifiable consequence is that there are surfaces (called *meta-meric* surfaces) that have different spectral reflectance distributions according to physical measurements and yet to our eyes are perceptually indistinguishable in color.

Is Trichromacy Explained by Our Possession of Just Three Types of Cone?

As is well known, human color vision is mediated by three types of retinal cones, whose photochemical pigments have peak sensitivities tuned, respectively, to longer, middle, and shorter wavelengths in the visible range. This fact can certainly be taken to explain *how* the dimensional compression is imposed by the human visual system. But it only transforms the question of *why* there is such a dimensional compression from the question of why we experience just three dimensions of color into the still unanswered question of why our retinas possess just three classes of color sensitive units.

Is Trichromacy Explained by a Trade-off Between Chromatic and Spatial Resolution?

Possibly the majority of contemporary vision scientists regards the three-dimensionality of human color representation as a more pragmatic, somewhat arbitrary compromise between competing pressures toward the one-dimensional simplicity of achromatic (colorless) vision and the high-dimensional complexity of the truly full-color vision that would wholly capture the intrinsic spectral reflectance distribution of each and every external surface. Among our distant hunter-gatherer ancestors, those having only two types of cones and, hence, only two-dimensions of color vision might have been at a slight but (in the long run) significant disadvantage in identifying foods, mates, predators, and the like. At the same time, any ancestors that, through spontaneous mutation, might have had cones of more than three types and, hence, four or more dimensions of color vision might have lost in precision of form discrimination more than they gained in refinement of object identification through color. For, in order to accommodate an additional class of color-sensitive retinal receptors, the receptors of some class or classes would necessarily have to be less densely distributed across the retina. Indeed, there may be some validity to this third answer. Still, it is unsatisfyingly nonspecific and post hoc. It makes plausible that the number of chromatically distinct receptor classes has some optimum value, but it gives no reason why this optimum number is specifically three. Essentially the same argument could have been offered if our color vision were instead, say, two-dimensional or four- or five-dimensional.

An Alternative Basis for Trichromacy in the Degrees of Freedom of Terrestrial Illumination

In order to explain the specifically three-dimensional character of our color representation, I have been advancing a fourth, quite different possible answer (Shepard, 1990, 1991; see also, Shepard, 1987a). Given that our visual systems do not capture the full reflectance characteristics of a surface anyway, perhaps we should shift our focus away from the question of how we perceive the true color of a surface and focus, instead, on how the color that we do perceive is nevertheless the *same* color each time we view the same surface. It is after all this constancy of color that is most essential for identifying objects and, hence, for behaving appropriately with respect to them. In particular, the possibility I have been putting forward is that our visual systems analyze inputs into three color dimensions because three is the number of degrees of freedom of terrestrial illumination—degrees of freedom that must be compensated for in order to achieve constancy of perceived surface colors.

I was led to this possibility by what I took to be an implication of the general linear model for color vision proposed by Maloney and Wandell (1986). This implication concerns a hypothetical animal whose survival and reproduction depends only on its perception of the shapes and relative lightnesses of objects and not at all on their chromatic colors. It appeared to me that such an animal, although requiring only a shades-of-gray representation of its world, could nevertheless achieve constancy of its (achromatic) representation only if its visual system first analyzed the optical input into three chromatic channels. (I am assuming, again, that those channels—like actual photoreceptors—are sufficiently broad-band to make efficient use of available light.) Only in this way could the animal's estimates of the achromatic lightnesses of surrounding surfaces be stabilized against the three degrees of freedom of chromatic composition of natural illumination. (The same would be true, incidentally, for a camera, television system, or robot that was to generate a lightness-constant shades-of-gray, or "black-and-white," image.)

Maloney and Wandell (1986) themselves did not specifically address the question of the minimum number of dimensions needed to achieve color constancy. Rather, they focused on the somewhat different question of the number of dimensions needed to determine, for any surface, the full spectral reflectance function of that surface. As I noted, Maloney and Wandell concluded that, for such a determination, the number of chromatic channels N must exceed by one the number of degrees of freedom of the ensemble of reflectance functions, a number that may itself lie between five and seven (Maloney, 1986).

Suppose, however, that the principal criterion for success of the visual system's chromatic analysis is not that it represent the full complexity of the spectral reflectance function for each surface but that it ensure that the (possibly reduced) representation of each such function is constant under natural variations of illumination. We do not need to make a commitment, here, as to the particular computational method that the visual system uses to estimate the characteristics of the prevailing illumination. It could be an improved version of Land's retinex method, Maloney and Wandell's general linear method, or a method that uses (in addition) more direct information about the characteristics of the illumination (available, say, from a view of the sun, sky, shadows, or specular reflections). For any such method, the general linear model indicates that correction for the momentary characteristics of the illumination that will achieve constancy in the representation of surface colors requires that the number of chromatic channels N equal the number of degrees of freedom of natural illumination—that is, essentially three.

I am not claiming that the chromatic colors of objects themselves are of no biological significance to us, or that we experience color only as a by-product of a mechanism that has evolved to achieve invariance of lightness under changes of illumination. Perceptual constancy is important, but we undoubtedly benefit from having a perceptual representation of the chromatic colors themselves. Laboratory studies have found chromatic color to be a more effective cue for guiding visual search than either lightness or shape (see the review by Christ, 1975). And in a natural setting, a purely shades-of-gray representation of the world, even if lightness-constant, would not enable us to discriminate, say, red from green berries or even to spot ripe berries, with sufficient ease and speed, from a background of green leaves. Undoubtedly, then, a perceptual representation of chromatic surface color is itself favored by natural selection—as is suggested by the important roles it has been found to play in the behaviors of many animals (e.g., Fernald, 1984; Hardin, 1990; Jacobs, 1981; Lythgoe, 1979).

For most human purposes, however, the gain in going to the six or more chromatic channels that (according to Maloney, 1986) might be required to discriminate between all naturally occurring surface reflectance functions would probably not offset the costs of reduced spatial resolution and increased computational burden. It is, then, on the assumption that some relatively small number of color dimensions suffices for the discrimination and recognition of biologically significant objects that I propose that the precise number of dimensions of our color representation may have been determined, instead, by the number of degrees of freedom of terrestrial illumination that must be compensated for in order to obtain perceptual constancy of the colors that we do experience.

The Physical Basis of the Three Degrees of Freedom of Natural Illumination

The solar source of significant terrestrial illumination has probably been essentially invariant in the spectral composition of its output throughout recent evolutionary history. The energy distribution of the portion of that sunlight illuminating any particular terrestrial scene has nevertheless varied greatly from circumstance to circumstance (e.g., see Condit & Grum, 1964; Judd et al., 1964; McFarland & Munz, 1975a, 1975b; Sastri & Das, 1968). Yet, this variation has evidently been constrained by terrestrial conditions to essentially three independent dimensions (see Judd et al., 1964). Indeed, these three environmental degrees of freedom appear to bear a close correspondence to the three dimensions of color opponency that Hering (1878/1964) and Hurvich and Jameson (1957) proposed for the human visual system on the basis of entirely different (psychophysical rather than ecological) considerations.

The Light-Dark Variation

There is, first, the dimension of variation in overall intensity of illumination between midday and deep shade or moonlight. This variation can occur, moreover, without much shift in chromatic composition. Under natural conditions, the brightest extreme of this variation is generally provided by the full complement of light from an overhead sun on a clear day—including both the yellow-biased, middle- to long-wavelength portion that directly penetrates to a scene and the blue-biased, short-wavelength portion that is indirectly scattered to the scene by the surrounding sky. Much less bright illumination that nevertheless has a similar spectral composition can arise through the chromatically nonselective filtering that may occur when, for example, the same sort of light (from sun and sky) reaches a scene only after being scattered to it from chromatically neutral surfaces such as white clouds, grey cliffs, or the moon. Indeed, moonlight, though vastly reduced below daylight intensities, has essentially the same spectral composition, because the lunar surface reflects all solar wavelengths within the visible range about equally (Lythgoe, 1979).

The Yellow-Blue Variation

This second, chromatic dimension of variation arises primarily as a result of Raleigh scattering (see Born & Wolf, 1970; Lythgoe, 1979). The shortest (blue and violet) wavelengths of light are most scattered by the smallest particles in the atmosphere—particularly the molecules of the air itself. (As the particles in the air become larger, the distribution of scattered wavelengths becomes more uniform—so that larger water droplets, for example, give rise to the achromatic, white or gray appearance of haze, fog, or clouds.) A blue extreme in natural lighting therefore occurs when a relatively localized object (such as a tree or cloud) blocks direct sunlight from striking the scene, while permitting unobstructed illumination of that scene by the light scattered to it from an otherwise clear sky—particularly from those portions of the sky that are separated from the sun by a large celestial angle (see Lythgoe, 1979; McFarland & Munz, 1975a). The contrasting yellow extreme occurs under the complementary condition in which a portion of a scene is illuminated only by direct sunlight—as by a beam of sunlight admitted by a small opening in a cave, leafy canopy, or handmade shelter—with all skylight cut off. Such purely direct solar light does not of course consist solely of wavelengths in the yellow region of the spectrum. Rather, I call it yellow because, after removal of the shortest (blue and violet) wavelengths, the center of the remaining band of visible wavelengths falls in the yellow region of the spectrum and because these remaining wavelengths, when combined, can give rise (like the disk of the overhead sun itself) to a correspondingly yellow appearance.

The Red-Green Variation

This third dimension of variation depends both on the elevation of the sun and on the atmospheric burden of water vapor. The longest (red) wavelengths are least scattered by air molecules and other small atmospherically suspended particles (such as dust). But those wavelengths are also most absorbed by any water that they encounter. In the relative absence of water vapor, the sunlight that directly strikes a scene will have retained a relatively greater proportion of its longest-wavelength, red component as the sun approaches the horizon, and, hence, its light must penetrate a longer and (at the lower altitude) denser column of air and suspended particles before striking the scene. When the atmosphere carries a high burden of water vapor, on the other hand, the penetrating light will contain a reduced component of long-wavelength red light. (There is, incidentally, an asymmetry between the lighting conditions at sunrise and sunset because even when the sun is in the same proximity to the horizon, the preceding temperature and hence burden of water vapor will generally be different in the two cases.) For both sources of the red variation (elevation of the sun and presence of water vapor), the argument for putting green in opposition to red is essentially the same as the argument for putting yellow in opposition to blue. After removal of the longest-wavelength, red end of the visible spectrum, the center of the remaining band of visible wavelengths falls in the green portion of the spectrum.

From a purely mathematical standpoint, of course, any (nondegenerate) coordinate system is equivalent to any other in the sense that either can be obtained from the other by a suitable (nonsingular) transformation. Thus, as I have already noted, the points of three-dimensional color space can be represented by either the polar or the rectangular coordinate systems shown in Figure 13.1a and 13.1b. Similarly, as I also noted, a linear transformation could convert the three basis functions plotted in

Figure 13.2 into three differently shaped, but functionally equivalent basis functions. With respect to simplicity and efficiency of computation, however, a visual system that represents color in terms of the particular light-dark, yellow-blue, and red-green dimensions that correspond most directly to the underlying physical dimensions of variation of natural illumination in the terrestrial environment might most effectively compensate for these variations and, thus, attain color constancy.

Recapitulation and Consideration of Some Counterexamples and Methodological Issues

Recapitulation of the Argument Concerning the Source of Trichromacy

The visual system starts with a retinal input in which, at any one moment, each receptor unit (cone) receives some distribution of intensities of wavelengths between roughly 400 and 700 nm. For the purposes of identifying objects and responding adaptively to them, the visual system must analyze this complex input into two components—one corresponding to the biologically significant, invariant properties of objects (regarded as the "signal") and one corresponding to the biologically less significant momentary circumstances of viewing and lighting (regarded as the "noise"). With regard to the perception of the colors of objects, this is just the color constancy problem of analyzing the distribution of spectral energies impinging at each point of the retina into a component that captures some invariant characteristic of the surface of an object and a complementary component attributable to the momentary conditions of illumination.

I have suggested that the visual system has solved the problem of performing this analysis and hence of attaining color constancy by first analyzing the visual input into the minimum number of chromatic channels needed to carry out the required analysis, namely, three. If this suggestion is correct, the human visual system provides for three dimensions of representation of colors because there are, in fact, three corresponding degrees of freedom in the world. But these three degrees of freedom in the world are not the degrees of freedom of the intrinsic colors of surfaces, which evidently are greater than three (Maloney, 1986). Instead, the critical degrees of freedom in the world may be those of natural illumination in the terrestrial environment.

The overall (400 to 700 nm) range of spectral sensitivity of the human eye has long been regarded as an evolutionary accommodation to the range of solar wavelengths that reach us through the earth's atmosphere (and through the aqueous medium of our eyes—and, originally, of the sea from which our distant ancestors emerged) (e.g., see Jacobs, 1981; Lythgoe, 1979). Possibly, the three-dimensionality of human color vision may similarly represent an evolutionary accommodation to the essentially three degrees of freedom of terrestrial transformation of the solar wavelengths passed by the earth's atmosphere.

Based on considerations of the physical causes of these natural variations of terrestrial lighting (cf. Condit & Grum, 1964; Judd et al., 1964; McFarland & Munz, 1975a, 1975b; Sastri & Das, 1968), I have further conjectured that these three degrees of freedom may be essentially characterized as a light-dark variation (between unobstructed midday illumination versus illumination of the same composition but greatly reduced intensity that reaches the scene only by scattering from clouds, cliffs, or the moon), a yellow-blue variation (between illumination primarily reaching an object directly from an overhead sun versus indirectly by Raleigh scattering from clear sky), and a red-green variation (depending on the proximity of the sun to the horizon and the amount of water vapor in the air). If so, a selective advantage may have been conferred on genes for the development of a neural mechanism that combines and transforms the outputs of the three classes of retinal receptors into the three specific opponent-process dimensions proposed, on quite different grounds, by Hering (1878/1964) and by Hurvich and Jameson (1957).

Consideration of Some Apparent Counterexamples

Color constancy is neither perfect nor unconditional, however (e.g., see Helson, 1938; Judd, 1940; Worthy, 1985). Our own visual mechanisms have been shaped by the particular selective pressures that have operated under the natural conditions generally prevailing on planet Earth during mammalian evolution. Not surprisingly, therefore, our color constancy can break down under artificial lighting. On an evolutionary time scale, such lighting has begun to fall on our indoor or nocturnal surroundings for too short a time for selection pressures on our visual mechanisms to compensate for the new degrees of freedom introduced by such lighting. At night, we may thus fail to recognize our own car in a parking lot illuminated only by sodium or mercury vapor lamps.

To claim that the selection pressure that prevailed until very recently in the natural environment has been toward three dimensions of color representation is not, of course, to claim that this pressure has been great enough to override all other selection pressures and, hence, to ensure trichromacy in all highly visual terrestrial species. Although the once widespread belief that color vision is absent in many mammals, even including cats or dogs, is undoubtedly incorrect (Jacobs, 1981), a number of species with otherwise well-developed visual systems have been reported to lack three-dimensional color representation. Thus, while old-world monkeys evidently tend, like humans, to be trichromatic, evidence concerning new-world monkeys has pointed to dichromacy and to trichromacy in different species.

Even in humans, approximations to dichromacy occur in about 8% of the population (primarily in males, because nearly all forms of congenital color deficiencies are determined by sex-linked genes—Jacobs, 1981; Nathans, Thomas, & Hogness, 1986). Most typically, human color deficiency takes the form of some degree of collapse of color space along the red-green dimension (see Figure 13.1b). In extreme cases of this particular type of collapse *(protanopia),* human trichromacy may be reduced to pure dichromacy. Total color blindness, that is, *monochromacy,* also occurs in otherwise sighted humans—either because the color-selective retinal cones are absent (leaving only the achromatic retinal receptors, called rods) or because the neural circuitry necessary for the proper analysis of the outputs of the cones is missing or defective (see Alpern, 1974; Jacobs, 1981). Pure monochromacy of this sort is extremely rare in humans, however, as might be expected if color vision and/or lightness constancy significantly contributed to the adaptation of our ancestors.

Because a relatively large portion of the variance in the spectral distributions of natural light can be accounted for by as few as two dimensions (see, e.g., Dixon, 1978; Judd et al., 1964), a degree of color constancy that is acceptable under most natural conditions may be attainable with just two classes of cones and, hence, with just two chromatic channels of visual input. Moreover, the addition of each new class of retinal receptors entails both costs of reduced spatial resolution for each class and costs of additional neural processing structures. Nevertheless, in the long run, genes for three-dimensional color vision—genes that are already being identified and localized on human chromosomes (Nathans, Thomas, & Hogness, 1986)—may provide a sufficient contribution to color constancy so that such genes eventually tend to prevail in the most visually developed evolutionary lines. Even as remote a species as the honey bee, the insect whose color vision has been most thoroughly studied (Lythgoe, 1979), has been reported to be trichromatic (Daumer, 1956) with, in fact, color vision very much like our own (LeGrand, 1964), except that its visual sensitivity is somewhat shifted, as a whole, toward the shorter wavelengths (von Helverson, 1972).

Methodological Issues Concerning the Determination of Dimensionality

From the standpoint taken here, the dimensionality of an animal's perceptual representation of colors is perhaps the single most fundamental characteristic of that animal's color vision (cf. Jacobs, 1981, pp. 21–22). Yet, this dimensionality remains undetermined for most highly visual species. The dimensionality of color representation is in fact quite difficult to determine in nonhuman species, and it becomes more difficult in more remote species. (In a closely related primate species we might make an inference as to the dimensionality of color vision from the anatomical presence or absence of each of the three types of cones that mediate trichromacy in our own case.) As I have mentioned, the number of unidimensional variables whose adjustment is sufficient to achieve a visual match with any given color is equivalent to the dimensionality of color space for the species doing the matching. But, for a very different species, how do we know which of the potentially countless dimensions along which spectral distributions can differ are to be varied? And, although we might train an animal to make a particular response when and only when two displayed colors are indiscriminable for that animal, how do we train the animal to adjust several unidimensional controls, simultaneously, to achieve such a visual match?

In principle, the dimensionality of a space can be determined from the relations among a fixed, finite set of points (or stimuli), without requiring continuous adjustment or matching. The difficulty with this approach is that the dimensionality of the space is not a property of single points or, indeed, of pairs or of triplets of points. Rather, it is a property determined by the distances between all points (or, in the case under consideration, the dissimilarities between all colors) in sets containing a number of points (or colors) that is one greater than the number of dimensions of the space (Blumenthal, 1953). Multidimensional scaling—particularly multidimensional scaling of the so-called nonmetric variety (Kruskal, 1964; Shepard, 1962, 1980)—can permit a determination of the dimensionality of a finite set of stimuli. Such a method requires, however, obtaining estimates of the animal's perceived similarities or dissimilarities between most of the stimuli in the set. Humans readily give numerical ratings of similarity, but in the case of other animals such estimates may have to be based on the extent to which a response learned to each stimulus generalizes to each of the others or on latencies or frequencies of errors in matching-to-sample responses (e.g., see Blough, 1982; Shepard, 1965, 1986). Unfortunately, the training and data collection would require a major research effort for each species and each set of colored surfaces investigated.

Rather than attempting to determine the dimensionality of color representation in different species by means of multidimensional scaling, researchers have mostly pursued their comparative studies of the color capabilities of different species (a) anatomically or physiologically, by trying to determine the number of distinct types of receptor units (such as retinal cones having photosensitive pigments or oil-droplet filters with different wavelength characteristics), or (b) psychophysically, by measuring discriminability between neighboring spectrally pure colors at different locations along the continuum of wavelength. Neither of these methods provides, however, a reliable indication of the dimensionality of the animal's perceptual representation of colors.

From anatomical examination, one may never be sure that one has found every chromatically distinct type of receptor unit. Even if one could, there is no guarantee that the number of dimensions of color representation is as large as the number of types found. The outputs of two or more types of photoreceptors with different spectral sensitivities might be neurally combined into one chromatic channel somewhere up the line. (Indeed, humans, normally have four identified types of retinal photoreceptors—namely, cones of three types and rods of one type, each type with its peak sensitivity at a different wavelength. Yet, humans normally have only three dimensions of color representation.)

The psychophysically determined wavelength discrimination function is even less informative about the dimensionality of an animal's color representation. For example, an animal could discriminate an arbitrarily large number of wavelengths and still represent all those spectrally pure colors on a single dimension of hue, just as an individual could discriminate an arbitrarily large number of intensity levels and still represent those shades of gray on a single dimension of lightness. The question of dimensionality depends not on the number of discriminably different wavelengths but on the discriminabilities of all possible mixtures of such wavelengths. But the number of pairs of mixtures that might be tested increases radically with the number of discriminable wavelengths to be mixed.

A single example may suffice to illustrate the problem. The color capabilities of pigeons have been extensively investigated psychophysically as well as anatomically. Their wavelength discrimination function has been measured (Hamilton & Coleman, 1933; Wright, 1972), and evidence has been found for at least six chromatically distinct types of retinal cones—having different wavelength-selective combinations of photopigments and colored oil-droplet filters (Bowmaker, 1977; Bowmaker & Knowles, 1977). Nevertheless, to my knowledge, the question remains open as to whether the pigeon is trichromatic, as we are, or whether it has as many as four, five, or six dimensions of color representation. Conceivably, in the pigeon there has been sufficient selection pressure not only toward the three chromatic channels needed for the achievement of color constancy in the natural environment, but also toward the additional chromatic channels that, according to Maloney and Wandell's (1986) general linear model together with Maloney's (1986) estimate of the number of degrees of freedom of the spectral reflectance distributions of natural surfaces, might be required to capture fully the intrinsic reflectance characteristics of natural objects or foods.

THE CIRCULAR STRUCTURE OF SPECTRALLY PURE COLORS

Another one of the facts that I included in my initial, illustrative list of structural constraints on human perception and cognition is the fact that the physically rectilinear continuum of wavelength is transformed by the human visual system into the psychologically circular continuum of perceived hues—that is, into Newton's (1704) color circle, schematically portrayed as the equator of color space in Figure 13.1a. Such a circle can be recovered with considerable precision by applying methods of multidimensional scaling to human judgments of the similarities among spectral hues (Shepard, 1962; Shepard & Cooper, 1992). The circularity is implied, as I noted, by the observation that red and violet, although maximally separated in physical wavelength, appear more similar to each other than either does to green, which is of an intermediate wavelength.

I conjecture that this circularity of hue may have arisen as a consequence of the transformation by which the human visual system carries the responses of three classes of retinal photoreceptors (the cones most sensitive to long, medium, and short wavelengths) into the three opponent processes (the light-dark, yellow-blue, and red-green processes). If so, not only the three-dimensionality of color space but also the specifically circular character of the hue dimension, rather than being an arbitrary design feature of the human visual system, may be traceable to an enduring regularity in the terrestrial world in which we have evolved.

The Circularity of Hue as an Accommodation to the Degrees of Freedom of Terrestrial Illumination

In rough outline, the argument runs as follows: The extremes of the range of solar wavelengths admitted through the terrestrial "window" on the solar spectrum are, naturally, the most affected by any variations in that "window." Thus, the longest wavelengths (the red components) are both least scattered by atmospherically suspended particles and most absorbed by water vapor, and the shortest wavelengths (the blue and violet components) are most scattered by the smallest atmospheric particles—particularly by the molecules of the air itself. Now, if the variable component of longest visible wavelengths (the reds) is put in opposition to the rest of the visible wavelengths, the central tendency of those opposing wavelengths will be what we call green—not the visible wavelengths that are physically most remote from the reds (namely, the short wavelength violets). Similarly, if the independently variable component of shortest visible wavelengths (the blues and violets) are put in opposition to the rest of the visible wavelengths, the central tendency of those opposing wavelengths will be what we call yellow—not the visible wavelengths that are physically most remote from the violets and blues (namely, the long wavelength reds).

Accordingly, a transformation by the visual system from the outputs of three classes of cones to an opponent process representation has the effect of bending the physically rectilinear continuum of wavelength into a closed cycle of spectral hues. Whereas in the original physical continuum, the extreme opposites were red and violet, in the resulting cycle of hues, the opposites (schematically thought of as the diagonally opposite corners of a square) are red and green (across one diagonal) and yellow and blue/violet (across the other diagonal). The originally most remote hues of red and violet, being now separated by only one edge of the (schematic) square, become closer together than the now diagonally separated red and green or the now diagonally separated yellow and blue/violet. (The red, yellow, green, and blue "corners" are in fact quite evident in the [squarish] version of the color circle that emerged when Carroll and I applied multidimensional scaling to color naming data collected by Boynton and Gordon [1965] for 23 spectral hues [see Shepard & Carroll, 1966, Figure 6, p. 575].)

Evidence Suggestive of an Innate Structure for the Representation of Colors at Higher Levels of the Brain

Color vision requires, in any case, not only wavelength-selective receptors (the retinal cones) but also some processing machinery for estimating from the variable outputs of those receptors the invariant colors of distal objects. Presumably, moreover, the genes specifying the structure of such neuronal circuitry will generally lead to the development of that circuitry in each individual whether or not that circuitry is destined to receive normal signals from that particular individual's retinal receptors. For example, an individual who is missing the gene for one of the classes of wavelength-selective cones may, through an inability to discriminate between certain presented colors (such as red and green), give evidence of a collapsed, two-dimensional color space. But this does not preclude that individual's possession, at higher levels of the nervous system, of the circuits that have been evolutionarily shaped for representing the full, three-dimensional system of colors.

In an investigation of the representation of colors by normally sighted, color-blind, and totally blind individuals (Shepard & Cooper, 1992), Lynn Cooper and I found evidence that suggests this is the case and thus contradicts the central tenet of the British empiricist philosophers—that everything that each individual knows must have first entered through that individual's own sensory experience. We asked the individuals with the different types of normal and anomalous color vision to judge the similarities among saturated hues under two conditions: (a) when pairs of those hues were actually presented (as colored papers), and (b) when the pairs of hues were merely named (e.g., "red" compared with "orange"). We applied multidimensional scaling to the resulting similarity data for each type of individual and for each of the two conditions of presentation. Most striking were the results for the red-green color-blind individuals (*protans* and *deutans*). As expected, when the colors were actually presented to these individuals, multidimensional scaling yielded a degenerate version of Newton's color circle with the red and green sides of the circle collapsed together. Significantly, however, when only the names of the colors were presented, multidimensional scaling yielded the standard, nondegenerate color circle obtained from color-normal individuals (cf. Shepard, 1962, p. 236; 1975, p. 97).

One particularly articulate protan insisted that although he could not distinguish the (highly saturated) red and green we showed him, neither of these papers came anywhere near matching up to the vivid red and green he could imagine! In a sense, then, the internal representation of colors appears to be three-dimensional even for those who, owing to a purely sensory deficit, can only discriminate externally presented colors along two dimensions (most commonly, the dimensions of light versus dark and yellow versus blue indicated in Figure 13.1b).

Incidentally, it was in the domain of colors that the preeminent British empiricist, David Hume, acknowledged what he considered to be the one possible exception to the empiricists' central maxim. Hume, on supposing "a person to have enjoyed his sight for thirty years, and to have become perfectly well acquainted with colours of all kinds, excepting one particular shade of blue,". confessed to favoring an affirmative answer to the question of "whether 'tis possible for him, from his own imagination, to supply this deficiency, and raise up to himself the idea of that particular shade, tho' it had never been conveyed to him by his senses?" Despite this admission, Hume concluded, somewhat lamely, that "the instance is so particular and singular, that 'tis scarce worth our observing, and does not merit that for it alone we should alter our general maxim" (Hume, 1739/1896, p. 6).

THE ORGANIZATION OF COLORS INTO CATEGORIES AND PROTOTYPES

Phenomena of color vision run counter not only to the central tenet of the British empiricists but also to the views of the more recent American linguistic relativists. According to the latter, as represented particularly by Sapir (1916) and Whorf (1956), each language lexically encodes experience in a way that is unique to the culture in which that language evolved. Color offers a particularly suitable domain for testing this idea because each color, as experienced by human observers, is a relatively "unitary" or "unanalyzable" percept (Shepard, 1964, 1991; Shepard & Chang, 1963), corresponding to a single point in the color space schematically portrayed in Figure 13.1. If linguistic relativity were valid, the way continuous color space is divided into regions for the purposes of assigning discrete names to colors would be expected to differ in more or less arbitrary ways from one language to another. Empirical studies, however, have uncovered a striking degree of cross-cultural uniformity.

The Cross-Cultural Findings of Berlin and Kay

Together with their co-workers, Berlin and Kay (1969) presented native speakers of 20 diverse languages with an array of 329 Munsell color chips, including 320 maximum saturation colors differing in equally spaced steps of hue and lightness, which had originally been used in a comparison of English and Zuni color terminology by Lenneberg and Roberts (1956), and 9 additional zero-saturation shades of gray. For each basic color term, x, in an informant's native language, the informant was asked to use a black grease pencil to encircle, on an acetate sheet overlaying the array of color chips, (a) the set of colors that the informant "would under any conditions call x" and, then, (b) the subset of those colors that the informant regarded as "the best, most typical examples of x" (Berlin & Kay, 1969, p. 7).

Berlin and Kay found that the color terms in different languages did not correspond to arbitrarily overlapping regions in color space. Instead, the terms were essentially consistent with a universal underlying hierarchy of nonoverlapping regions in color space for each of the 20 languages directly investigated (and these results appeared to be consistent with the reports, by others, of the use of color terms in some 78 other languages). The languages differed primarily in the number of basic color terms they included. They differed very little in the locations of the regions corresponding to the basic color terms that they did include. In particular, the languages generally conformed with a partial order of color categories (that is, regions of color space) that—in terms of the English names that we assign to these categories—is as follows:

$$\begin{Bmatrix} \text{white} \\ \text{black} \end{Bmatrix} < \{\text{red}\} < \begin{Bmatrix} \text{green} \\ \text{yellow} \end{Bmatrix} < \{\text{blue}\} < \{\text{brown}\} < \begin{Bmatrix} \text{purple} \\ \text{pink} \\ \text{orange} \\ \text{gray} \end{Bmatrix}$$

"where, for distinct color categories (*a*, *b*), the expression $a < b$ signifies that *a* is present in every language in which *b* is present and also in some language in which *b* is not present" (Berlin & Kay, 1969, p. 4).

The regions in color space corresponding to these apparently universal color categories are (with the exception of the region for the completely unsaturated grays) conveniently displayed on the hue-by-lightness rectangle, which (as indicated in Figure 13.3) can be thought of as a Mercator projection of the maximum-saturation surface of the sphere of colors previously illustrated in Figure 13.1. (Because the achromatic color gray corresponds to the center of the sphere, it is not representable on the projection of the surface of the sphere.) The placement of the English names for the successive hues, "Red," "Orange," "Yellow," etc., is only schematic in Figure 13.3; hue and lightness turn out to be universally correlated in the locations of prototypical colors in such a way that yellow, in addition to falling between red and green in hue, is much lighter than these (and the other) basic hues.

Figure 13.4 is my adaptation of the figure in which Berlin and Kay summarized the results of their own study of color naming by native speakers of 20 different languages. The consistently found locations of the basic color categories are displayed within the two-dimensional hue-by-lightness Mercator projection of the maximum-saturation surface of color space. The encircled regions approximate the areas within which the colors chosen as "the best, most typical examples" of the corresponding basic color terms consistently fell according to informants from the various cultures. These focal color categories are labeled, for our convenience, by the basic terms that we use for these colors in English.

Surrounding each of these encircled focal regions, a larger region to which informants indicated the same color term might be extended under some condition is roughly indicated by the penumbra of corresponding initial letters, "G" for "Green," "Br" for "Brown," and so on. Again, however, some of the languages did not have basic color terms for some of the lower categories in the hierarchy, such as the categories here labeled "Pink," "Purple," or "Orange." In these cases, the penumbra surrounding the neighboring focal color regions that were represented in the language

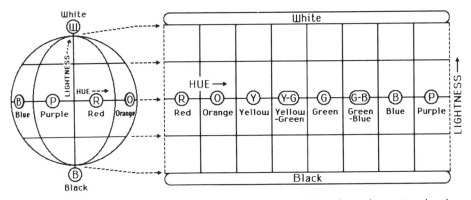

Figure 13.3 Schematic diagram of how the surface of the three-dimensional color space shown in Figure 13.1 can be unwrapped (by Mercator projection) into a rectangular map in which each point represents the most saturated color of a corresponding hue and lightness.

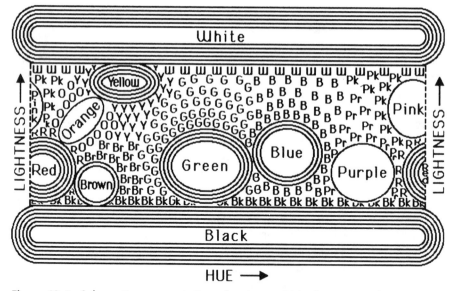

Figure 13.4 Schematic representation of regions, within the rectangular map of saturated colors (as diagramed in Figure 13.3), for which color names are commonly used in different cultures. The name printed within each region is the color word most often used by English-speaking informants. The number of lines drawn around each region indicates the cross-cultural prevalence of assigning a distinct color term specifically to that region. The penumbra of letters ("Gr" for "Green," "Br" for "Brown," etc.) indicates a larger region that the color term might be extended to under some conditions, according to informants. (Adapted from Figure 3 on page 9 of Berlin & Kay, 1969; copyright, University of California Press, 1969.)

might extend through the region lacking its own basic color term. Thus, if the language contained no basic term specifically for the colors in the category here labeled "Purple," the terms for the neighboring regions here labeled "Blue" or "Red" might be extended into the region to which we apply the word "Purple." Also, of course, the colors falling in the penumbra around a given focal color region might be distinguished by the basic color term together with a qualifier. In English, for example, "light green" or "pale green" might be used for colors falling above the focal region for green, and "blue-green" might be used for colors falling between the focal regions for blue and green. And some languages have single terms for still lower levels of the hierarchy. Thus, in English we might use the word "lavender" for light purple, "chartreuse" for yellow-green, and "navy" for dark blue (or blue-black).

I have indicated the hierarchical structure that Berlin and Kay found for the focal colors by encircling each corresponding subregion of color space by an appropriate number of lines. Thus I have drawn six lines around the universally named focal color categories corresponding to our "White" and "Black," five lines around the next most prevalently named category, corresponding to our "Red," and so on for successively lower levels in the hierarchy—"Green" and "Yellow" (four lines each), "Blue" (three lines), "Brown" (two lines), and "Pink," "Purple," and "Orange" (a single encircling line for each).

The rectangular Mercator projection of a globe is equivalent to a cylinder in that the right and left edges of the rectangle correspond to the same meridian on the sphere and, hence, to the same series of colors (here, the reds varying in lightness from darkest red to lightest pink). Thus the focal regions labeled "Red" and "Pink" in Figure 13.4 overlap both the left and right ends of the rectangle as displayed in that figure. (As I have already remarked, the surface color purple is not fully represented in the pure spectral colors of monochromatic light such as arise from a prism or in a rainbow. Surfaces that we call purple are surfaces that reflect a mixture of the longest and shortest of the visible wavelengths, that is, a mixture of red and violet.)

Additional, Corroborating Results and Implications

The locations in color space of the universal focal color regions can usefully be displayed in another way—namely, as these regions would project onto a flat plane cutting through that three-dimensional space, for example, the horizontal plane pictured as passing through the equator of the sphere in Figure 13.1. Figure 13.5 shows my adaptation of such a plot obtained by Boynton and Olson (1987) on the basis of single-word (monolexemic) color naming of 424 color samples by seven English-speaking observers. Here again, each region within which colors were given a particular color name is filled in with the initial letter of that color (e.g., with "G"s for "Green"). The point within such a region that, on average, was accepted as the best example of that

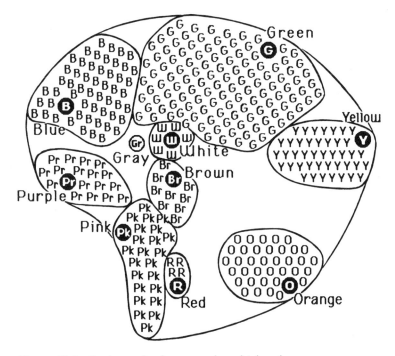

Figure 13.5 Regions of color space for which color names are commonly used, projected (by parallel projection) onto the equatorial disk cutting through the spherical model illustrated in Figure 13.1, according to data reported by Boynton and Olson (1987). (Adapted from Figure 2 on page 99 of Boynton & Olson, 1987; copyright, John Wiley & Sons, Inc., 1987.)

color, considered the focal color for that region, is indicated by the black circle (with a corresponding white letter, e.g., "G").

The dimensions of this two-dimensional projection of color space are not hue and lightness, as in the Mercator projection of Figure 13.4. Instead, the two dimensions here can be considered to be the polar-coordinate dimensions of hue (now represented as a complete circle around the perimeter) and saturation (which is now represented as radial distance out from the unsaturated white and gray near the center to the maximum saturation hues around the perimeter). The focal chromatic colors are not located at the centers of their regions but out toward the maximum saturation perimeter. As noted before, the dimensions of this same projection of color space can alternatively be taken to be the rectangular-coordinate dimensions of red-versus-green (roughly vertical in Figure 13.5) and blue-versus-yellow (roughly horizontal).

The disposition of focal color regions obtained by Boynton and Olson (1987) for this projection, though necessarily different, is quite consistent with the disposition of these regions in the Mercator type of projection presented by Berlin and Kay (1969). The representation shown in Figure 13.5, however, better accommodates the less saturated and achromatic colors (brown, gray, and white) and—in requiring no break in the pink-red region—more explicitly maintains the integrity of the hue circle (corresponding to the equator in the more schematic Figures 13.1 and 13.3). The representation in Figure 13.4, on the other hand, more clearly preserves other aspects of color representation—such as that the yellow band of the spectrum appears lighter (or brighter) than the other (red, green, or blue) bands.

The cross-cultural results such as those reported by Berlin and Kay (1969) and by Boynton and his co-workers (especially, Uchikawa & Boynton, 1987, who compared color naming by American and native Japanese observers) provide little support for the Whorfian hypothesis that language structures how we perceive the world. Instead, these results appear to be more consonant with the idea that the innate biology of our perceptual system structures the way we use language.

Physically, the spectrum of solar light seen in a rainbow or refracted by a prism (as in the original investigations by Newton, 1704) is a continuum of wavelengths. Yet we tend to perceive this continuum as divided into a few relatively discrete bands of hue, such as those we call red, orange, yellow, green, blue, and violet. (White, purple, and all unsaturated colors are obtained only by recombining these pure spectral wavelengths, while black is of course obtained by blocking off all visible wavelengths.)

Eleanor Rosch (formerly Heider) and others, in experimental studies with human observers, have shown, in addition, that the most prototypical (or focal) colors within each region of color space tend to be more rapidly and consistently perceived and remembered than nonfocal colors, which are closer to the boundaries between regions, while discrimination between colors appears to be sharper across the boundaries of these regions than within the same region (Heider, 1971, 1972; Rosch, 1973, 1975; see also Boynton & Olson, 1987, 1990; Mervis & Rosch, 1981; Nagy & Sanchez, 1988; Uchikawa & Boynton, 1987).

Moreover, the tendency toward the perceptual categorization of the hue continuum is not unique to human adults, but has been reported to take essentially the same form in human infants (Bornstein, Kessen, & Weiskopf, 1976) and in other primates (Essock, 1977; Sandell, Gross, & Bornstein, 1979). (In monkeys, incidentally, Zeki [1983] found cortical cells whose responses, like our own color-constant experience of surfaces, were determined by the intrinsic reflectance characteristics of an external surface regardless of the illumination.) There is even evidence that the spectral continuum is also perceptually partitioned by such remote species as pigeons and bees, though apparently into color categories that differ somewhat from those common to humans and monkeys (Kühn, 1927; von Frisch, 1950; Wright & Cumming, 1971). For all of these animals, including humans, visual discrimination between neighboring wavelengths has been reported to be sharpest at the boundaries between the color categories for each species (Graham, Turner, & Hurst, 1973; Jacobs & Gaylord, 1967; Nagy & Sanchez, 1988; Smith, 1971; von Helversen, 1972; Wright, 1972). And, as the evidence obtained by Berlin and Kay (1969) and others has shown in the human case, the boundaries tend to be the same across cultures that have developed on widely separated continents and that have very different languages and sets of color terms. Evidently (as concluded by Uchikawa & Boyton, 1987; and, earlier, by Ratliff, 1976), the way in which humans categorize colors is universally prewired rather than individually acquired.

But we are still left with the question of the source of the particular form of such an innate biology underlying human color categorization. Is this biology merely accidental or has it arisen as an accommodation to some identifiable regularity in the world in which we have evolved? Perhaps we should distinguish two questions: First, why is the continuous space of colors partitioned into discrete categories at all? Second, given that it is partitioned, why is it partitioned in the particular way indicated in Figures 13.4 and 13.5?

Possible Adaptive Functions of a Categorical Organization of Colors

In the case of humans, one advantage of an innate structuring of continuous color space into discrete regions, organized around prototypical or focal colors, has already been implicit in the discussion of the linguistic studies by Berlin and Kay and others. Because a discrete structuring makes possible the consistent assignment of words to colors, it facilitates communication between individuals about the colors of objects that may be of biological significance. One may readily identify a person that one is to meet for the first time on the basis of a description of a color that that person will be wearing. But it may be impossible to identify that person on the basis of a description of his or her face. The continuous space of human faces (in addition to being of higher dimensionality) does not seem to be endowed with a consistent categorical structure like that of the continuous space of colors.

Of course, this linguistic advantage of categorical representation of colors cannot be the explanation for the very similar categorical structure that has been demonstrated for the color spaces of other, nonlinguistic species. Still, for the more general purposes of biological signaling within and between species, the categorical representation of colors may have similar advantages, as has been argued by Hardin (1990). The advantages for both of these purposes—for human language and for other types of signaling—can be seen as examples of the general principle of *shareability* enunciated by Freyd (1983, 1990). When cast in evolutionary terms, that principle says that some cognitive structures or functions may arise not because they capture a biologically significant regularity in the physical world but because they facilitate or make possible the sharing of knowledge between conspecifics. Clearly, an innate predisposition toward a consistent partitioning of a biologically significant continuum into discrete categories will facilitate the sharing of knowledge about the things signified by that continuum, whether through changes in the coloration of bodily markings (as in some fish—Fernald, 1984) or through human language (Berlin & Kay, 1969; Kay & McDaniel, 1978; Lenneberg & Roberts, 1956; Miller & Johnson-Laird, 1976).

A categorical structuring can also facilitate communication within an individual—for example, from one information-processing module to another within the same individual's brain (as Freyd has suggested, personal communication, June 1990), or from one occasion to another, across time (Heider, 1972; Rosch, 1975). Certainly, to remember the color of a particular object after a single exposure may be important if that object was, say, a newly encountered predator from which one barely escaped. A clear example of the advantage of perceptual categorization has been reported in the auditory domain, where individuals differ markedly in their abilities to recognize the absolute pitch of a tone (independent of its relation to some other tone). Individuals with this ability of so-called *absolute pitch* are much more accurate in identifying whether a long-delayed second tone is or is not a repetition of an earlier tone (Bachem, 1954).

Possible Sources of the Particular Categorical Organization of Human Colors

Even if we agree that the organization of color space into categories and prototypes can serve adaptive functions, we are still left with the question of why this space is organized in the particular categorical way that it is, rather than in some other categorical way. Consideration of this question necessarily becomes still more speculative. The organization of color space that (along with its presumably genetically entrenched physiological basis) appears ubiquitous in humans and other primates may eventually be found to stem from any mixture of at least three sources: (a) The organization could be a largely accidental or arbitrary solution (out of countless other, equally satisfactory organizations) to the problem of providing a discrete basis for intraspecific signaling, language, or shareability; (b) the organization could reflect something about the natural groupings of the surface reflectance distributions of objects (into ripe versus unripe edible fruits, predators versus prey, and so on) that have long had particular biological significance for our ancestors and those of other primates (cf. Shepard, 1987b); or (c) in line with my conjecture as to the source of the three-dimensionality of color space itself, this organization could reflect something about the ways in which terrestrial lighting has most typically varied during evolutionary history.

To my knowledge, the systematic survey of the spectral reflectance distributions of a large sample of significant surfaces in the natural environment needed to support the second alternative has yet to be undertaken. In the meantime, I find some support for the last alternative in what already appears to be a nonaccidental correspondence between the most entrenched of the human color categories and the generally accepted opponent process dimensions of human color vision, which (as I have noted) seem to correspond to the natural dimensions of terrestrial lighting.

On one hand, the six focal colors that we call white, black, red, green, yellow, and blue (also termed "landmark colors" by Miller & Johnson-Laird, 1976) are highest in the hierarchy of basic color categories according to Berlin and Kay (1969), as I indicated in Figure 13.4 by encircling the corresponding English terms with three or more lines. On the other hand, these are just the six colors that are usually taken to define the ends of the three light-dark, red-green, and yellow-blue dimensions in opponent process theories. Moreover, I have suggested that just these dimensions may be best suited to compensate for the most common variations of terrestrial illumination. (Even the universally lighter appearance of focal yellow may have an ecological basis—possibly in the yellow central tendency of the direct light from an overhead sun after the shorter wavelength blues and violets have been partially filtered out by atmospheric scattering.)

The secondary focal colors, such as those that we call orange, purple, pink, brown, and gray, may be distinguished (when, in a given culture, they are distinguished) in order to fill in the main gaps in color space between the primary focal colors defined by the underlying opponent processes. Boynton has reviewed evidence that people will describe these secondary focal colors as mixtures of other more primary colors, but will not describe the primary focal colors as mixtures of any other colors (Boynton, 1988, p. 91). Thus, orange, purple, and pink will be described as containing red mixed with yellow, blue, and white, respectively, while a prototypical red itself is never described as containing any discernible contribution of a neighboring color such as orange or purple.

In any case, if categorical structuring of colors serves an adaptive function, there need be no powerful selective advantage of one structuring in order for it to prevail over others. A slight predisposition toward one organization, arising perhaps from a quite different, noncategorical source might bias the selection toward that organization over other, otherwise comparably useful categorical organizations. The source that I have been suggesting might exert such an influence is the natural variation of terrestrial illumination, for which we must compensate in order to achieve color constancy. Alternatively, of course, the largely unexplored possibility remains that the intrinsic colors of objects that have been of biological significance for our ancestors may have tended to cluster around the focal colors identified by Berlin and Kay—perhaps, red for ripe fruit (e.g., berries, apples, tomatoes, etc.), green for trees and other vegetation, and blue for water and sky. In the absence of a systematic survey of the spectral reflectance distributions of a large sample of significant surfaces in the natural environment, however, the dimensions of variation of natural illumination may provide for the most parsimonious explanation for the most salient features of the universal color categories of humans.

CONCLUDING REMARKS

The Search for a Nonarbitrary Basis for Psychological Principles

One can distinguish three ways in which we try to establish and to achieve understanding of a psychological phenomenon. Although any combination of these ways can be pursued, simultaneously or in any order, they may tend to become dominant in three successive stages of scientific investigation. The first such stage focuses on the empirical establishment of the phenomenon itself, through behavioral, psychophysical, or linguistic investigations (whether using laboratory experiments or field observations). The second stage shifts focus to the elucidation of the physiological mechanisms that may underlie that established phenomenon, through neurophysiological investigations (whether by electrophysiological recording of concomitant brain activity or by simulation of proposed connectionist models on a computer). The third stage, when indeed it is embarked upon, shifts focus, again, to the problems that the external world poses for an individual. In this stage, we try to understand both the neurophysiological mechanisms and the cognitive and behavioral functions that those mechanisms mediate not as arbitrary design features of a particular species, but as accommodations to pervasive and enduring properties of the world.

In our understanding of the human representation of colors, over three centuries of inspired empirical investigation has established many of the basic phenomena of human color vision. For present purposes, these phenomena have included the constancy, the three-dimensionality, the hue-circularity, and the categorical structure of colors. Beginning later, physiological investigations, which I have not attempted to review, have begun to pin down the neuronal mechanisms underlying a few of these phenomena (see, for example, De Valois, 1973; De Valois & De Valois, 1975; Foster, 1984; Lennie, 1984; Livingstone & Hubel, 1984; Zeki, 1983). The regularities in the world that may determine the nature of these physiological mechanisms, as well as of the perceptual-cognitive phenomena that they mediate, remain largely unexplored. Beyond the long-standing recognition of the problem of color constancy and the model recently proposed for its solution by Maloney and Wandell (1986), I have offered little more than some tentative suggestions about possibly fruitful directions for further exploration.

I began this chapter with the claim that natural selection should lead to the emergence not only of perceptual, behavioral, and cognitive mechanisms that are adapted to the specific circumstances faced by particular species, genders, and domains, but also of mechanisms that are adapted to the general circumstances faced by all so-called higher organisms. Color vision offers a less obvious case for supporting such a claim than the phenomena that I have usually used for this purpose, namely, those of spatial representation (e.g., Shepard, 1984) and of generalization (e.g., Shepard, 1987b). Color pertains to a single sensory modality (vision), whereas space and generalization are not modality specific. Moreover, the properties in the world that might determine the organization of colors are less obvious than those that might determine the representation of space. For example, whereas the three-dimensionality of perceptual space derives directly from the fundamental three-dimensionality of physical space, the three-dimensionality of color space does not so directly or obviously derive from a three-dimensionality in the world. The case for a nonarbitrary basis for the structure of colors, to the extent that it can be made, has the advantage, however, of suggesting that psychological constraints may correspond to regularities in the world even when the correspondence is not initially obvious.

Possible Universal Pressures Toward the Organization of Colors

I conclude with a brief consideration of the sense, if any, in which a tendency toward the properties of color representation I have been considering may be general—or even universal. Throughout, I have spoken of the possibility of selective pressures to which highly visual *terrestrial* animals may be subject in the *terrestrial* environment. I used the qualification "terrestrial" because conditions of illumination in other, for example, aquatic environments can be quite different. Owing to the already noted selective absorption by water of the longer wavelengths, with increasing depth in a marine environment, the available solar light, in addition to being progressively reduced in overall intensity, becomes progressively restricted in spectral range to the shorter wavelengths. This blue shift and compression in range of available wavelengths is known to be matched by a corresponding blue shift and compression in the range of spectral sensitivity of deeper dwelling marine animals (Lythgoe, 1979).

It does seem to me, however, that the wavelength dependencies of the variable height of a sun, presence of atmospheric aerosols, and direct versus indirect illumination might apply quite generally on the surfaces of planets capable of supporting the evolution of highly visual organisms. Such a planet is presumably likely (a) to circle a long-lived star that emits a stable, broad range of wavelengths, (b) to undergo regular rotation about its own axis (owing to the conservation of angular momentum), and (c) to possess an atmosphere that differentially filters the wavelengths of direct and scattered light depending on the (rotationally determined) angle at which the light enters the atmosphere and the size distribution of atmospheric particles. Moreover, the arguments for categorical representation of colors based on memorability and shareability do not depend on particular features of the terrestrial environment. In short, just as there may be universal selective pressures toward mechanisms for the representation of three-dimensional space and for an exponential law of generalization (Shepard, 1987a), there may be quite general selective pressures toward mechanisms for the representation of the surface characteristics of objects in a low-dimensional (perhaps even a three dimensional) color space, with a circular component of hue, and a categorical structure.

Possibly, behavioral and cognitive theorists should aspire to a wider scope for their science. An evolutionary theory of mind need not confine itself to the particular minds of the more or less accidental collection of species we find on planet Earth. There may be quite general or even universal principles that characterize planetary environments capable of supporting the origin and evolution of increasingly complex forms of life. If so, there may be corresponding general or even universal principles of mind that by virtue of their mesh with the principles of these environments, are favored by a process of natural selection wherever it may be taking place.

ACKNOWLEDGMENTS

Preparation of this chapter was supported by National Science Foundation Grant No. BNS 85-11685. The chapter has benefited from helpful comments on earlier drafts by a number of colleagues, including Robert Boynton, Leda Cosmides, Jennifer Freyd, Geoffrey Miller, John Tooby, and Brian Wandell.

REFERENCES

Alpern, M. (1974). What is it that confines in a world without color? *Investigative Opthamology, 13,* 648–674.

Bachem, A. (1954). Time factors in relative and absolute pitch determination. *Journal of the Acoustical Society of America, 26,* 751–753.

Barlow, H. B. (1982). What causes trichromacy? A theoretical analysis using comb-filtered spectra. *Vision Research, 22,* 635–643.

Berlin, B., & Kay, P. (1969). *Basic color terms: Their universality and evolution.* Berkeley, CA: University of California Press.

Blough, D. S. (1985). Discrimination of letters and random dot patterns by pigeons and humans. *Journal of Experimental Psychology: Animal Behavior Processes, 11,* 261–280.

Blumenthal, L. M. (1953). *Theory and applications of distance geometry.* Oxford: Clarendon Press.

Born, M., & Wolf, E. (1970). *Principles of optics.* Oxford: Pergamon Press.

Bornstein, M. H., Kessen, W., & Weiskopf, S. (1976). Color vision and hue categorization in young human infants. *Journal of Experimental Psychology: Human Perception and Performance, 2,* 115–129.

Bowmaker, J. K. (1977). The visual pigments, oil droplets and spectral sensitivities of the pigeon. *Vision Research, 17,* 1129–1138.

Bowmaker, J. K., & Knowles, A. (1977). The visual pigments and oil droplets of the chicken retina. *Vision Research, 17,* 755–764.

Boynton, R. M. (1988). Color vision. *Annual Review of Psychology, 39,* 69–100.

Boynton, R. M., & Gordon, J. (1965). Bezold-Brüke hue shift measured by color-naming technique. *Journal of the Optical Society of America, 55,* 78–86.

Boynton, R. M., & Olson, C. X. (1987). Locating basic colors in the OSA space. *COLOR research and application, 12,* 94–105.

Boynton, R. M., & Olson, C. X. (1990). Salience of chromatic basic color terms confirmed by three measures. *Vision Research, 30,* 1311–1318.

Brainard, D. H., & Wandell, B. A. (1986). Analysis of the retinex theory of color vision. *Journal of the Optical Society of America, 3,* 1651–1661.

Brill, M. H. (1978). A device for performing illuminant-invariant assessment of chromatic relations. *Journal of Theoretical Biology, 71,* 473–478.

Brill, M. H., & West, G. (1981). Contributions to the theory of invariance of color under the condition of varying illuminance, *Journal of Mathematical Biology, 11,* 337–350.

Buchsbaum, G. (1980). A spatial processor model for object color perception. *Journal of the Franklin Institution, 310,* 1–26.

Buchsbaum, G., & Gottschalk, A. (1984). Chromaticity coordinates of frequency-limited functions. *Journal of the Optical Society of America, 67,* 885–887.

Carlton, E. H., & Shepard, R. N. (1990). Psychologically simple motions as geodesic paths: I. Asymmetric objects. II. Symmetric objects. *Journal of Mathematical Psychology. 34,* 127–188, 189–228.

Christ, R. E. (1975). Review and analysis of color coding research for visual displays. *Human Factors, 17,* 542–570.

Cohen, J. (1964). Dependency of the spectral reflectance curves of the Munsell Color Chips. *Psychonomic Science, 1,* 369–370.

Condit, H. R., & Grum, F. (1964). Spectral energy distribution of daylight. *Journal of the Optical Society of America, 54,* 937–943.

Cosmides, L. (1989). The logic of social exchange: Has natural selection shaped how humans reason? Studies with the Wason selection task. *Cognition, 31,* 187–276.

Das, S. R., & Sastri, V.D.P. (1965). Spectral distribution and color of tropical daylight. *Journal of the Optical Society of America, 55,* 319.

Daumer, K. (1956). Reitzmetrische untersuchungen des Farbensehens der Bienen. *Zeitschrift für Vergleichende Physiologie, 38,* 413–478.

De Valois, R. L. (1973). Central mechanisms of color vision. In R. Jung (Ed.), *Handbook of sensory physiology* (Vol. 7/3A, pp. 209–253). Berlin: Springer Verlag.

De Valois, R. L., & De Valois, K. K. (1975). Neural coding of color. In E. C. Carterette & M. P. Friedman (Eds.), *Handbook of perception* (Vol. 5, pp. 117–166). New York: Academic Press.

Dixon, E. R. (1978). Spectral distribution of Australian daylight. *Journal of the Optical Society of America, 68,* 437–450.

Essock, S. M. (1977). Color perception and color classification. In D. M. Rumbaugh (Ed.), *Language learning by a chimpanzee: The LANA project* (pp. 207–224). New York: Academic Press.

Fernald, R. D. (1984). Vision and behavior in an African cichlid fish. *American Scientist, 72,* 58–65.

Foster, D. H. (1984). Colour vision. *Contemporary Physiology, 25,* 477–497.

Freyd, J. J. (1983). Shareability: The social psychology of epistemology. *Cognitive Science, 7,* 191–210.

Freyd, J. J. (1990). Natural selection or shareability? *Behavioral and Brain Sciences, 13,* 732–734.

Gibson, J. J. (1950). *The perception of the visual world.* Boston, MA: Houghton-Mifflin.

Gibson, J. J. (1966). *The senses considered as perceptual systems.* Boston, MA: Houghton-Mifflin.

Gibson, J. J. (1979). *The ecological approach to visual perception.* Boston, MA: Houghton-Mifflin.

Graham, E. V., Turner, M. E., & Hurst, D. C. (1973). Derivation of wavelength discrimination from color naming. *Journal of the Optical Society of America, 63,* 109–111.

Hamilton, W. F., & Coleman, T. E. (1933). Trichromatic vision in the pigeon, all illustrated by the spectral discrimination curve. *Journal of Comparative Physiology, 15,* 183–191.

Hardin, C. L. (1990). Why color? *Proceedings of the SPIE/SPSE symposium on electronic imaging: Science and technology,* 293–300.

Heider, E. R. (1971). Focal color areas and the development of color names. *Developmental Psychology, 4,* 447–455.

Heider, E. R. (1972). Universals in color naming and memory. *Journal of Experimental Psychology, 93,* 10–20.

Helmholtz, H. von (1856–1866). *Treatise on physiological optics* (Vol. 2). (J.P.C. Southall, Trans., from the third German edition). New York: Dover, 1962.

Helson, H. (1938). Fundamental problems in color vision. I. The principle governing changes in hue saturation and lightness of non-selective samples in chromatic illumination. *Journal of Experimental Psychology, 23,* 439–476.

Hering, E. (1878/1964). *Zur Lehre vom Lichtsinne.* Berlin. (Republished in English translation as *Outlines of a theory of the light sense.* Cambridge, MA: Harvard University Press.)

Hume, D. (1739/1896). A treatise of human nature. (Reprinted from the original 1739 edition, L. A. Selby-Bigge, Ed.) Oxford: Clarendon Press.

Hurvich, L. M., & Jameson, D. (1957). An opponent-process theory of color vision. *Psychological Review, 64,* 384–404.

Jacobs, G. (1981). *Comparative color vision.* New York: Academic Press.

Jacobs, G. H., & Gaylord, H. A. (1967). Effects of chromatic adaptation on color naming. *Vision Research, 7,* 645–653.

Judd, D. B. (1940). Hue, saturation and lightness of surface colors with chromatic illumination. *Journal of Experimental Psychology, 23,* 439–476.

Judd, D. B., McAdam, D. L., & Wyszecki, G. (1964). Spectral distribution of typical daylight as a function of correlated color temperature. *Journal of the Optical Society of America, 54,* 1031–1040.

Kay, P., & McDaniel, C. K. (1978). The linguistic significance of the meanings of basic color terms, *Language, 54,* 610–646.

Krinov, E. L. (1947). *Spectral reflectance properties of natural formations* (Tech. Rep. No. TT-439). Ottawa, Canada: National Research Council of Canada.

Kruskal, J. B. (1964). Multidimensional scaling by optimizing goodness of fit to a nonmetric hypothesis. *Psychometrika, 29,* 1–27.

Kühn, A. (1927). Über den Farbensinn der Bienen. *Zeitschrift für vergleichende Physiologie, 5,* 762–800.

Land, E. H. (1964). The retinex. *American Scientist, 52,* 247–264.

Land, E. H. (1978, January–February). Our "polar partnership" with the world around us. *Harvard Magazine,* 23–26.

Land, E. H. (1986). Recent advances in retinex theory and some implications for cortical computations: Color vision and the natural image. *Proceedings of the National Academy of Sciences, 80,* 5163–5169.

LeGrand, Y. (1964). Colorimétrie de l'Abeille *Apis mellifera. Vision Research, 4,* 59–62.

Lenneberg, E. H., & Roberts, J. M. (1956). The language of experience: A study in methodology. *International Journal of American Linguistics,* Memoir 13.

Lennie, P. (1984). Recent developments in the physiology of color vision. *Trends in Neuroscience, 7,* 243–248.

Livingstone, M. S., & Hubel, D. H. (1984). Anatomy and physiology of a color system in the primate visual cortex. *Journal of Neuroscience, 4,* 309–356.

Lythgoe, J. N. (1979). *The ecology of vision.* New York and London: Oxford University Press.

Maloney, L. T. (1986). Evaluation of linear models of surface spectral reflectance with small numbers of parameters. *Journal of the Optical Society of America A, 3,* 1673–1683.

Maloney, L. T., & Wandell, B. A. (1986). Color constancy: A method for recovering surface spectral reflectance. *Journal of the Optical Society of America A, 3,* 29–33.

Marr, D. (1982). *Vision.* San Francisco: W. H. Freeman.

McFarland, W. N., & Munz, F. W. (1975a). The photic environment of clear tropical seas during the day. *Vision Research, 15,* 1063–1070.

McFarland, W. N., & Munz, F. W. (1975b). The evolution of photopic visual pigments in fishes. *Vision Research, 15,* 1071–1080.

Mervis, C. B. & Rosch, E. (1981). Categorization of natural objects. *Annual Review of Psychology, 32,* 89–115.

Miller, G. A. (1956). The magical number seven, plus or minus two: Some limits on our capacity for processing information. *Psychological Review, 63,* 81–97.

Miller, G. A., & Johnson-Laird, P. N. (1976). *Language and perception.* Cambridge, MA: Harvard University Press.

Nagy, A., & Sanchez, R. (1988, November). *Color difference required for parallel visual search.* Paper presented at the Annual Meeting of the Optical Society of America, Santa Clara, CA.

Nassau, K. (1983). *The physics and chemistry of color: The fifteen causes of color.* New York: Wiley.

Nathans, J., Thomas, D., & Hogness, D. S. (1986). Molecular genetics of human color vision: The genes encoding blue, green and red pigments. *Science, 232,* 193–202.

Newton, I. (1704). *Opticks* (Book 3). London: Printed for S. Smith & B. Walford.

Pavlov, I. P. (1927). *Conditioned reflexes* (G. V. Anrep, Trans.). London: Oxford University Press.

Pavlov, I. P. (1928). Lectures on conditioned reflexes (W. H. Ganntt, Trans.). New York: International Press.

Ratliff, F. (1976). On the psychophysiological bases of universal color names. *Proceedings of the American Philosophy Society, 120,* 311–330.

Rosch, E. (1973). Natural categories. *Cognitive Psychology, 4,* 328–350.

Rosch, E. (1975). The natural mental codes for color categories. *Journal of Experimental Psychology: Human Perception and Performance, 1,* 303–322.

Sälström, P. (1973). *Colour and physics: Some remarks concerning the physical aspects of human color vision* (Report No. 73-09). Stockholm, Sweden: Institute of Physics, University of Stockholm.

Sandell, J. H., Gross, C. G., & Bornstein, M. N. (1979). Color categories in Macaques. *Journal of Comparative and Physiological Psychology, 93,* 626–635.

Sapir, E. (1916). Time perspective in aboriginal American culture. A study in method. In D. Mandelbaum (Ed.), *Selected writings of Edward Sapir in language, culture, and personality.* Berkeley, CA: University of California Press, 1949.

Sastri, U.D.P., & Das, S. R. (1966). Spectral distribution and colour of north sky at Delhi. *Journal of the Optical Society of America, 56,* 829.

Sastri, U.D.P., & Das, S. R., (1968). Typical spectral distributions and colour for tropical daylight. *Journal of the Optical Society of America, 58,* 391.

Shepard, R. N. (1962). The analysis of proximities: Multidimensional scaling with an unknown distance function (Parts 1 and 2). *Psychometrica, 27,* 125–240, 219–246.

Shepard, R. N. (1964). Attention and the metric structure of the stimulus space. *Journal of Mathematical Psychology, 1,* 54–87.

Shepard, R. N. (1965). Approximation to uniform gradients of generalization by monotone transformations of scale. In D. I. Mostofsky (Ed.), *Stimulus generalization* (pp. 94–110). Stanford, CA: Stanford University Press.

Shepard, R. N. (1975). Form, formation, and transformation of internal representations. In R. Solso (Ed.), *Information processing and cognition: The Loyola Symposium* (pp. 87–122). Hillsdale, NJ: Lawrence Erlbaum Associates.

Shepard, R. N. (1980). Multidimensional scaling, tree-fitting, and clustering. *Science, 210,* 390–398.

Shepard, R. N. (1981). Psychophysical complementarity. In M. Kubovy & J. Pomerantz (Eds.), *Perceptual organization* (pp. 279–341). Hillsdale, NJ: Lawrence Erlbaum Associates.

Shepard, R. N. (1984). Ecological constraints on internal representation: Resonant kinematics of perceiving, imagining, thinking and dreaming. *Psychological Review, 91,* 417–447.

Shepard, R. N. (1987a). Evolution of a mesh between principles of the mind and regularities of the world. In J. Dupré (Ed.), *The latest on the best: Essays on evolution and optimality* (pp. 251–275). Cambridge, MA: MIT Press/Bradford Books.

Shepard, R. N. (1987b). Toward a universal law of generalization for psychological science. *Science, 237,* 1317–1323.

Shepard, R. N. (1988). The role of transformations in spatial cognition. In J. Stiles-Davis, M. Kritchevsky, & U. Bellugi (Eds.), *Spatial cognition: Brain bases and development* (pp. 81–110). Hillsdale, NJ: Lawrence Erlbaum Associates.

Shepard, R. N. (1989). Internal representation of universal regularities: A challenge for connectionism. In L. Nadel, L. A. Cooper, P. Culicover, & R M Harnish (Eds.), *Neural connections, mental computation* (pp. 103–104). Cambridge, MA: MIT Press/Bradford Books.

Shepard, R. N. (1990). A possible evolutionary basis for trichromacy. *Proceedings of the SPIE/SPSE symposium on electronic imaging: Science and technology,* 301–309.

Shepard, R. N. (1991). Integrality versus separability of stimulus dimensions: From an early convergence of evidence to a proposed theoretical basis. In G. R. Lockhead & J. R. Pomerantz (Eds.), *Perception of structure* (pp. 53–71). Washington, DC: American Psychological Association.

Shepard, R. N. (in press). What in the world determines the structure of color space? (Commentary on Thompson, Palacios, & Varela). *Behavioral and Brain Sciences.*

Shepard, R. N. & Carroll, J. D. (1966). Parametric representation of nonlinear data structures. In P. R. Krishnaiah (Ed.), *Multivariate analysis* (pp. 561–592). New York: Academic Press.

Shepard, R. N., & Chang, J.-J. (1963). Stimulus generalization in the learning of classifications. *Journal of Experimental Psychology, 65,* 94–102.

Shepard, R. N., & Cooper, L. A. (1982). *Mental images and their transformations.* Cambridge, MA: MIT Press/Bradford Books.

Shepard, R. N., & Cooper, L. A. (1992). Representation of colors in the blind, color blind, and normally sighted. *Psychological Science, 3* (in press).

Smith, D. P., (1971). Derivation of wavelength discrimination from colour-naming data. *Vision Research, 11,* 739–742.

Sperling, G. (1960). The information available in brief visual presentations. *Psychological Monographs, 74* (11, Whole No. 498).

Stiles, W. S., Wyszecki, G., & Ohta, N. (1977). Counting metameric object-color stimuli using frequency-limited spectral reflectance functions. *Journal of the Optical Society of America, 67,* 779–784.

Tversky, A., & Kahneman, D. (1974). Judgment under uncertainty: Heuristics and biases. *Science, 185,* 1124–1131.

Uchikawa, K., & Boynton, R. M. (1987). Categorical color perception of Japanese observers: Comparison with that of Americans. *Vision Research, 27,* 1825–1833.

von Frisch, K. (1950). *Bees: Their vision, chemical senses, and language.* Ithaca, NY: Cornell University Press.

von Helverson, O. (1972). Zur spektralen Unterschiedsempfindlichkeit der Honigbiene. *Journal of Comparative Physiology, 80,* 439–472.

Wason, P. C., & Johnson-Laird, P. N. (1972). *Psychology of reasoning: Structure and content.* London: Batsford.

West, G., & Brill, M. H. (1982). Necessary and sufficient conditions for Von Kries chromatic adaptation to give color constancy. *Journal of Mathematical Biology, 15,* 249–258.

Whorf, B. L. (1956). *Language, thought, and reality.* New York: Wiley.

Winch, G. T., Boshoff, M. C., Kok, C. J., & du Toit, A. G. (1966). Spectroradiometric and colorimetric characteristics of daylight in the southern hemisphere: Pretoria, South Africa. *Journal of the Optical Society of America, 56,* 456–464.

Woodworth, R. S., & Sells, S. B. (1935). An atmosphere effect in formal syllogistic reasoning. *Journal of Experimental Psychology, 18,* 451–460.

Worthy, J. A. (1985). Limitations of color constancy. *Journal of the Optical Society of America A, 2,* 1014–1026.

Wright, A. A. (1972). Psychometric and psychophysical hue discrimination functions for the pigeon. *Vision Research, 12,* 1447–1464.

Wright, A. A., & Cumming, W. W. (1971). Color-naming functions for the pigeon. *Journal of the Experimental Analysis of Behavior, 15,* 7–17.

Yilmaz, H. (1962). Color vision and a new approach to color perception. In E. E. Bernard & M. R. Kare (Eds.) *Biological prototypes and synthetic systems* (pp. 126–141). New York, Plenum.

Young, T. (1807). On physical optics. In *A course of lectures on natural philosophy and the mechanical arts* (Vol. 1). London: Printed for Taylor and Welton, 1845.

Zeki, S. (1983). Colour pathways and hierarchies in the cerebral cortex: The responses of wavelength-selective and colour-coded cells in monkey visual cortex to changes in wavelength composition. *Neuroscience, 9,* 767–781.

14

Sex Differences in Spatial Abilities: Evolutionary Theory and Data

IRWIN SILVERMAN AND MARION EALS

Cognitive sex differences have remained a prominent topic in psychology for several decades, and one of the most consistent findings has been superior performance for males on tests of spatial abilities. Earlier attempts to account for this difference dealt mainly with socialization practices (Maccoby & Jacklin, 1974), but the generality of the phenomenon across populations and situations led to a shift in emphasis to genetic determinants. Relationships have been found for both sexes between spatial performance and hormonal variables, measured or manipulated directly or inferred from correlates such as pubertal status, physical characteristics, menstrual cycle phase, and atypical androgen levels associated with medical disorders. Additionally, similar spatial sex differences have been found in infrahuman species. (See Gaulin & Hoffman, 1988; Harris, 1978; Kimura & Hampson, 1990; Linn & Peterson, 1985; and McGee, 1979, for reviews.)

SELECTION PRESSURES FOR SPATIAL SEX DIFFERENCES

The near universality of sex differences in spatial abilities across human cultures and their occurrence in other species indicate the feasibility of an evolutionary approach, but it was not until 1986 that the first systematic attempt of this nature was reported by Gaulin and Fitzgerald. These investigators theorized that spatial abilities in males would have been selected for in polygynous species because polygynous males require navigational skills to maintain large home ranges in which to seek potential mates and/ or resources to attract mates. To test these notions they compared sex differences in range size and spatial ability between meadow voles, which are polygynous, and pine voles, which are monogamous. As predicted, male biases for both variables occurred in meadow voles, whereas pine voles showed no disparities between sexes. A follow-up study (Jacobs, Gaulin, Sherry, & Hoffman, 1990) revealed that in meadow voles, but not pine voles, males had proportionally larger hippocampi than females, which had been anticipated based on the role of the hippocampus in mediating spatial functions.

There is another measure of animal mobility, however, termed *natal dispersal* by Greenwood (1980), which is defined as the distance an animal travels from its natal site to its first breeding place. As with range size, sex differences in natal dispersal have been related to mating practices (Greenwood, 1980, 1983), but in his analysis, mating systems are dichotomized in terms of resource defense versus mate defense rather than monogamy versus polygyny.

Greenwood's theory is that in most resource defense systems, males compete for and hold territories in which they attract females; consequently, females disperse more. In mate defense, males usually locate and defend females; thus they are the greater dispersing sex. In support of these notions, Greenwood noted that birds, who tend to use resource defense, show a female bias in natal dispersion, while mammals, who mainly employ mate defense, show a male bias.

An alternative to Gaulin and Fitzgerald's model of spatial sex differences can be derived from Greenwood's concepts. It may be posited that species that employ mate defense strategies, with greater male dispersal, will show male superiority in spatial abilities; species using resource defense, with greater female dispersal, will show female superiority; and species showing neither of these patterns in typical form, and nil dispersal differences between sexes, will show no spatial sex bias.

Regarding meadow and pine voles, the former show a characteristic mate defense strategy with males dispersing more (Madison, 1980). Pine voles, on the other hand, fall into the third category above, in that they cannot be precisely designated as mate defense or resource defense. They possess a unique social structure for microtine rodents, living in highly cohesive groups comprising reproductively active members of both sexes (Fitzgerald & Madison, 1983). Thus, Gaulin and Fitzgerald's data can be explained by this extension of Greenwood's model as well as by their own theory.

On the other hand, these two theories lead to discrepant predictions regarding human spatial sex differences. By most accounts, humans are moderately polygynous (Symons, 1979), and there are cross-cultural data showing a tendency from early childhood for males to maintain larger home ranges (Gaulin & Hoffman, 1988). On the other hand, humans are resource defenders (Chagnon, 1979), with greater natal dispersal on the part of females (Koenig, 1989). Inasmuch as sex differences in spatial performance favor males, Gaulin and Fitzgerald's theory would appear to prevail in the human case.

Our own work with humans, however, was based on an alternative theory to both Gaulin and Fitzgerald's and Greenwood's. This may be a violation of parsimony inasmuch as our explanation does not extend across species as handily as the other two. On the other hand, it takes account of a particular aspect of human evolution that appears on logical grounds to be highly relevant to spatial sex differences, and it has enabled predictions that would not have emanated from the others.

We hold that the critical factor in selection for spatial dimorphism in humans was sexual division of labor between hunting and gathering during hominid evolution. Although there has, undoubtably, been overlap between sexes in these functions, archaeological and paleontological data show that across evolutionary time, males predominantly hunted and females predominantly foraged (Tooby & DeVore, 1987).

Tracking and killing animals entail different kinds of spatial problems than does foraging for edible plants; thus, adaptation would have favored diverse spatial skills between sexes throughout much of their evolutionary history. The cognitive mechanisms of contemporary *Homo sapiens* appear to reflect these differences, insofar as the various spatial measures showing male bias (e.g., mental rotations, map reading, maze learning) correspond to attributes that would enable successful hunting. Essentially, these attributes comprise the abilities to orient oneself in relation to objects or places, in view or conceptualized across distances, and to perform the mental transformations necessary to maintain accurate orientations during movement. This would enable the pursuit of prey animals across unfamiliar territory and, also, accurate placement of projectiles to kill or stun the quarry. In fact, there have been studies based on the same evolutionary notions, demonstrating direct relationships between standardized spatial test scores and throwing accuracy (Jardine & Martin, 1983; Kolakowski & Molina, 1974).

In the present paper, we have extended the premise to propose that if these attributes evolved in males in conjunction with hunting, spatial specializations associated with foraging should have, correspondingly, evolved in females. Food plants are immobile, but they are embedded within complex arrays of vegetation. Successful foraging, then, would require locating food sources within such arrays and finding them in ensuing growing seasons. These abilities entail the recognition and recall of spatial configurations of objects; that is, the capacity to rapidly learn and remember the contents of object arrays and the spatial relationships of the objects to one another. Foraging success would also be increased by peripheral perception and incidental memory for objects and their locations, inasmuch as this would allow one to assimilate such information nonpurposively, while walking about or carrying out other tasks.

In the following sections, we describe a series of studies exploring these hypothesized female spatial specializations, using student subjects from York University in Toronto.

First, however, data will be presented from other ongoing studies with the York student population in order to demonstrate that it is comparable to the population in general in regard to male biases on traditional spatial tests.

STUDIES OF MALE SPATIAL SPECIALIZATIONS

Figure 14.1 contains sample items from two widely used group-administered spatial tests that customarily show male bias: Mental Rotations (Vandenberg & Kuse, 1978) and Space Relations (Bennett, Seashore, & Wesman, 1947). The Mental Rotations test requires subjects to designate which two of the series of four drawings on the right represent the target object on the left in alternative positions. The task for Space Relations is to indicate, for each item, all of the figures on the right that could be constructed from the pattern on the left. For both tests, subjects were told they could give as many responses, per item, as they wished, but would score a point for each correct response and would have a point subtracted for each incorrect response. For each test, 20 items were used, and seven minutes were allowed.

The two tests were given to separate York samples. Space Relations was administered in individual sessions or in groups of two or three and was included in a test battery. Mental Rotations was the sole test given and was administered in larger groups. Most subjects were volunteers, recruited in classrooms or elsewhere on campus, though some Mental Rotations subjects took the test as part of a course demonstration. Findings were equivalent across all conditions.

Table 14.1 shows the results. As expected, there were significant differences favoring males for both tests. Further, sex differences on both measures appeared exceptionally large in our samples compared with extant published data (e.g., Vandenberg & Kuse, 1978), though statistical comparisons were not feasible because of differences in item composition and/or procedure.

Mental Rotations

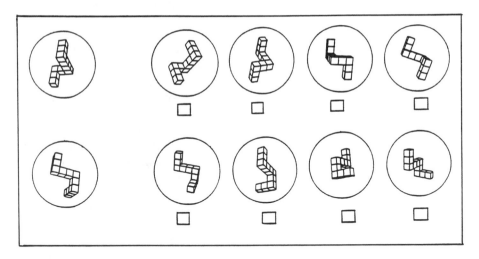

Space Relations

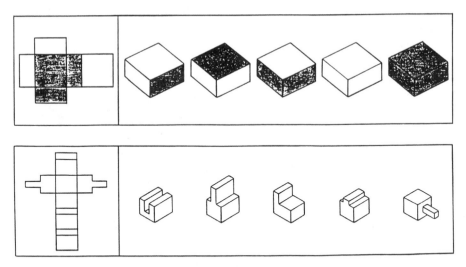

Figure 14.1 Sample Items from the Mental Rotations and Space Relations Tests.

FEMALE SPATIAL SPECIALIZATIONS

Study One: A Group Test

As in the studies just described, our initial study of foraging-related spatial abilities was administered both in individual sessions and in groups of various sizes and comprised both volunteers and students taking part in course demonstrations. Again, results were similar across all conditions.

Table 14.1 Mean Space Relations and Mental Rotations Scores, by Sex, for York University Student Samples

	Space relations				Mental rotations		
	N	Mean	*SD*		*N*	Mean	*SD*
Males	18	46.60	13.72	Males	105	15.57	9.43
Females	20	30.89	12.40	Females	98	8.66	7.48
	t = 3.69	*p* < .001			*t* = 5.80	*p* < .001	

Subjects in this study were presented first with copies of the object array (called the stimulus array) depicted in Figure 14.2 and asked to "examine the objects" for one minute.

They were then instructed to fold their copies and put them aside and were presented with copies of the array in Figure 14.3. This was identical to the stimulus array, except that a number of additional items were interspersed. Subjects were told to put a cross through all of the items that were not in the original array and that they would be allowed one minute, would be given a point for each item correctly crossed, and would have a point subtracted for each item incorrectly crossed. This served as a measure of memory for objects in an array, independent of location *(object memory)*.

Finally, subjects were shown the array in Figure 14.4. This contained the same items as the stimulus array, but some were in the same location and others were not. Subjects were asked to circle the objects that were in the same place and put a cross through those that had been moved and were scored a point for each correct response.

Figure 14.2 The stimulus array used for tests of object and location memory.

Figure 14.3 The stimulus array with added items for the object memory test.

Figure 14.4 The stimulus array with item locations changed for the object memory test.

Table 14.2 Mean Object Memory and Location Memory Scores, by Sex, for Study One

	Object Memory				Location Memory		
	N	Mean	*SD*		*N*	Mean	*SD*
Males	63	12.25	4.27	Males	83	18.45	3.58
Females	115	14.15	3.90	Females	134	20.14	4.11
	t = 2.92	*p* < .01			*t* = 3.20	*p* < .01	

This was a measure of memory for the locations of objects in an array *(location memory)*.

Table 14.2 shows the results.[1] Females scored significantly higher on both tasks; they more accurately recalled which items were in the array and where they were located.

Study Two: A Naturalistic Setting

In the following study, we attempted to replicate the findings above using an actual object array rather than a drawing, and an array presented in a naturalistic setting as opposed to an experimental context. Our criterion for naturalistic setting, following Silverman (1977), was that subjects were unaware during their exposure to the object array that it was part of a study.

We recruited volunteers for an ambiguously labeled experiment. Subjects were scheduled individually and were seen by either a male or female examiner, with the examiner's and subject's sex counterbalanced. The examiner met the subject at the

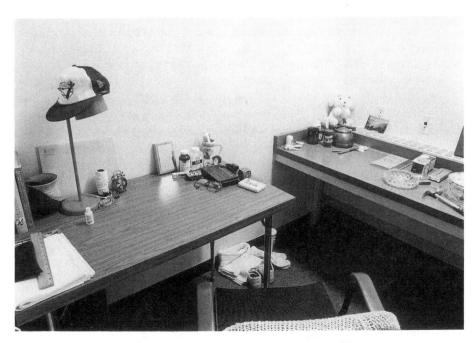

Figure 14.5 The stimulus room for the naturalistic tests of object and location memory.

Table 14.3   Mean Location Scores, by Sex, for Study Two

| | | Location memory | |
	N	Mean	*SD*
Males	21	6.80	4.34
Females	20	11.55	5.23
t = 4.04	*p* < .001		

laboratory, then led him to a cubicle-type office nearby (the stimulus room), and asked him to wait there several minutes while she completed preparations in the laboratory. Subjects were encouraged to leave books or other materials they were carrying in the laboratory, in order to prevent them from reading or studying while waiting.

Figure 14.5 is a photograph from the entrance to the stimulus room. It was outfitted as a typical graduate student office, containing a variety of work-related and personal items, and was located in an aisle of such offices. Subjects occupied the chair shown in the photograph, which was the only place to comfortably wait. The examiner returned in precisely two minutes and escorted the subject back to the laboratory.

In the laboratory subjects were told that the purpose of the study was to assess how people naturally process their environments. They were asked to name as many objects in the stimulus room as they could and, as precisely as possible, the location of each object. The examiners used prearranged probes if locations were not forthcoming or vague. In response to exit interviews, two subjects, a male and a female, indicated that they had been suspicious in the stimulus room that it was part of the study. Their scores were not atypical, however, and were kept in the data.

Subjects' full responses were tape recorded, with their permissions, and scored from written transcripts by two independent raters. Raters were unaware of which protocols were male or female. Subjects were credited if they approximated the correct location of an object; for example, "the right side of the small desk." Scoring discrepancies between raters were few and were resolved by the second rater.

Table 14.3 shows these results. Females correctly recalled significantly more items by location. Their mean score, in fact, was a robust 70% higher.

There was no measure of object memory, irrespective of location, that could be culled from these data. When subjects reported an item, they tended, with rare exceptions, to know where it was. This may reflect the manner by which people assimilate object arrays, or it may have been an artifact of the emphases on locations in the instructions and the probes.

There were, also, no sex-of-examiner effects or interactions of sex of examiner and subject.

Study Three: Incidental and Directed Learning

We then undertook a further study, using the same general procedure as Study Two, for two purposes: One was to eliminate systematic biasing of responses by examiners that may have attended the use of open-ended questions and probes in Study Two. (The examiners were not informed of the hypotheses, but were close enough to the investigator's research program to develop accurate suspicions.)

The other was to assess whether sex differences in object and location memory would be obtained in a directed learning paradigm, whereby subjects were specifically instructed to try to learn objects and their locations. In Study Two, subjects were kept unaware that the room in which they waited was part of the experiment; hence, recall was based wholly on incidental learning. In Study One, subjects were instructed to "examine the objects" in the array, and though they probably surmised that they would be tested for frequency of items remembered, it is uncertain whether they would have attempted to learn locations. Consequently, location memory in this study may have also reflected incidental learning.

In Study Three, the same laboratory and stimulus room were used as in Study Two, but the number of items in the stimulus room was doubled to bring the total to 70. The reason for this was that the dependent measures for this study were based on recognition rather than recall, which was expected to yield higher scores, and we wished to avoid a ceiling effect.

The female examiner from Study Two conducted all sessions. As in Study Two, she met subjects in the laboratory, took their books and other materials, then led them to the stimulus room. For half the subjects of each sex, determined by a prearranged, randomized order of assignment, the procedure continued exactly as previously: subjects were asked to wait in the room while the examiner completed preparations in the laboratory and were left there for two minutes. This was the incidental learning condition.

The other half of subjects were instructed when they were brought into the stimulus room to "try to memorize as many objects in the room as possible, and their approximate locations" and informed that the examiner would return to test them in two minutes. This comprised the directed learning condition.

Testing was the same in both conditions. Subjects were presented first with a list of 35 objects, 25 of which had been in the stimulus room, and were asked to indicate for each whether or not it was there. They were given a point for each correct response, including items identified as in the room and items identified as not, which comprised the measure of object memory.

Following this task, subjects were shown a schematic drawing of the room, divided into seven numbered areas, and were asked to note the area in which each item in the room was located. The number of correct responses for this task served as the measure of location memory.

Table 14.4 presents mean object and location memory scores by sex and incidental versus directed learning conditions.

The data for object memory showed a significant main effect of sex, based on higher female scores across learning conditions, and no significant interaction of sex and condition. In the analyses of simple effects, however, the sex difference for the incidental learning condition reached significance ($t = 2.25$, $p < .05$), but this difference for the directed learning condition did not ($t = 1.37$ $p = .18$). Thus, support was obtained for a female bias in incidental learning of objects, but the findings were equivocal for directed learning. The trend for the latter sex difference, however, was in the expected direction, and a larger N may bring it to a significant level.

For location memory, there was also a significant main effect of sex favoring females and no interaction of sex and condition. Analyses of simple effects revealed significant sex differences for both incidental and directed conditions ($t = 3.44$ and 4.45, respectively; $p < .001$ for both). As in Study Two, females' location memory scores based on incidental learning were more than 60% higher than males'.

Table 14.4 Mean Object Memory and Location Memory Scores, by Sex and Incidental vs. Directed Learning Conditions, for Study Three

	Object memory			
	Males		Females	
Condition	Mean	SD	Mean	SD
Incidental	19.45	(3.61)	22.20	(4.12)
Directed	25.50	(3.55)	27.05	(3.59)
F tests				
Sex	6.67	$p < .01$		
Condition	42.83	$p < .001$		
Interaction	.52	ns		
	Location memory			
	Males		Females	
Condition	Mean	SD	Mean	SD
Incidental	6.30	(3.23)	10.25	(4.00)
Directed	12.05	(2.86)	16.50	(3.44)
F tests				
Sex	30.40	$p < .001$		
Condition	62.30	$p < .001$		
Interaction	.11	ns		

($N = 20$ for each sex in each condition: Total $= 80$)

Separating Location Learning From Object Learning

In our studies, object memory and location memory were measured separately whenever possible, eschewing the possibility that these may not be independent of each other. Females may have learned more locations by virtue of their greater capacity to learn objects, or they may have learned more objects as a function of their greater capacity to learn locations. The data of Study 3 afforded an opportunity to assess sex differences in location learning with object learning controlled.

First we compared sexes on the number of objects correctly identified as *not* having been in the room in the object memory task. Means were equivalent; 7.80 ($SD = 2.14$) for males and 7.97 ($SD = 1.86$ for females). This confirmed that there were not differential tendencies between sexes to give "yes" responses, indicating that the object was in the room, when guessing. (If there had been, this would have confounded the proportional measure described next.)

Then, we took the subject's score for number of objects correctly identified as *being* in the room and divided that into the number of locations of these objects correctly identified. This proportion comprised a measure of location memory corrected for object memory.

The results are in Table 14.5. The female bias remained, as indicated by the significant main effect of sex, and simple effects tests showed that the sex difference was significant in both incidental and directed learning conditions ($t = 3.15, p < .01; t = 4.17, p < .001$, respectively).

Table 14.5 Mean Scores by Sex and Learning Condition for Locations Correctly Identified Proportional to Objects Correctly Identified (Location Memory Corrected for Object Memory)

Condition	Males			Females		
	N	Mean	SD	N	Mean	SD
Incidental	20	.53	.21	20	.71	.15
Directed	20	.71	.12	20	.87	.12
F tests						
Sex	24.33	$p < .001$				
Condition	23.99	$p < .001$				
Interaction	.09	ns				

Study Four: Hormonal Status

Are female spatial specializations related to hormonal status? One indication of the hormonal basis of the spatial abilities for which males excel is that sex differences tend to emerge most strongly after puberty (see Harris, 1978, for a review). Thus, we compared sex differences in object and location memory among school children from grades 4 through 9 (ages 8½ through 13½) to ascertain whether the female advantage would increase with grade level, as more children came into puberty.

The group test developed for Study One was used, with the same procedure described in that section. Subjects came from three junior high schools in the Toronto area and were tested in their classrooms as part of their daily routines.

Grade levels were paired in the data analysis to balance subject frequencies across conditions. This resulted in Ns of 56 versus 66, 83 versus 78, and 81 versus 86 for, respectively, males and females in grades 4–5, 6–7, and 8–9.

Results for both object and location memory scores are shown in Figure 14.6. A multiple ANOVA was performed for sex and grade level across both dependent variables.

For object memory, there were significant main effects of grade level and sex (F = 13.13 and 12.45, respectively; $p < .001$ for both) and nil interaction effect between these. In terms of simple effects, sex differences favoring females for grades 4–5 and 8–9 were significant ($t = 2.41$ and 2.08, respectively; $p < .05$ for both). Thus, a female bias in object memory was replicated with child and early adolescent subjects, but, contrary to expectations, it did not begin or increase with pubertal status.

For location memory, there was a significant main effect of grade level (F = 8.83, $p < .001$), and the main effect for sex approached significance (F = 2.66, $p = .10$). The interaction effect between sex and grade level was in the predicted direction, but did not approach significance. The sole, significant simple effect for sex difference, however, was for the higher female mean in grades 8–9 ($t = 2.24$, $p < .05$). The sex difference took a similar direction for grades 4–5 but did not approach significance, and the female mean for grades 6–7 was, in fact, slightly lower than the male.

Thus, despite the absence of a significant interaction effect, the data provide a strong suggestion that female superiority in location memory begins with puberty. Considering that some adolescents do not reach puberty by age 13½, the addition of data for grades 10 and upward would be expected to augment this trend.

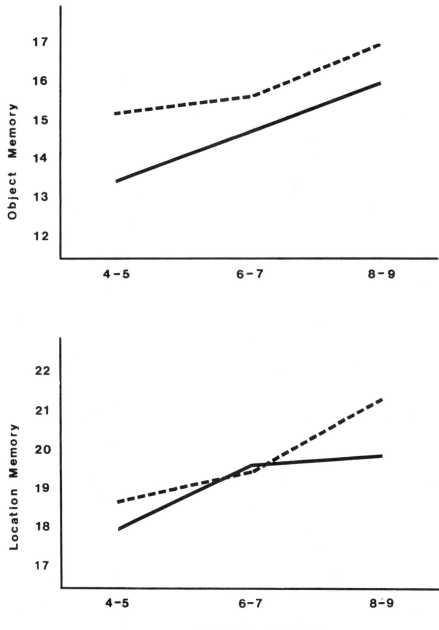

Figure 14.6   Mean object and location memory scores by sex and grade level.

Though unanticipated by our hypotheses, the finding that a female bias among prepubertal subjects occurred for object memory, but not location memory, is congruent to the data of Kail and Siegel (1977). They presented letters of the alphabet in various locations in a 4×4 matrix to males and females in third grade, sixth grade, and college, and independently measured frequencies of letters and locations recalled. For the third and sixth grades, females remembered more letters, but not more locations, than males. Similar to our first three studies, Kail and Siegel's college age females surpassed their male counterparts in locations recalled.

The discrepancy in the age at which sex differences in object and location memory become manifest is also consistent with our findings, reported in the prior section, that these are separate abilities, although there is no apparent reason for the earlier emergence of sex differences in object memory.

GENERAL CONCLUSIONS

The data of all of our studies corresponded closely to predictions from the hunter-gatherer model of spatial sex differences and consistently demonstrated a greater capacity by females to remember spatial configurations of objects. Females outperformed males in memory for both frequencies and locations of objects, in both incidental and directed learning paradigms. Sex differences for incidental learning of locations in a naturalistic setting were most striking, however; females' mean scores exceeded males' by 60 to 70%, for measures of both recognition and recall.

The findings of a female bias for both directed and incidental learning supported the specific deduction of our theory that sexes differ in perceptual style as well as learning ability. Studies with the waiting room ploy revealed that females are generally more alert than males to objects in the environment and their locations, whether or not these are perceived as relevant to a task at hand. It is often a topic of humor that the male partner is dependent on the female for locating items in the household, which is ascribed in the conventional wisdom to the greater role of the female in domestic matters. It appears, however, that this capacity is a manifestation of a global, female perceptual trait. Further, our developmental data suggested that the emergence of this trait coincides with puberty, when hormonal differentiation between sexes and male spatial specializations become pronounced.

DIRECTIONS FOR FURTHER RESEARCH

The question arises of whether the present results can be attributed to superior memory, in general, for females. The tendency, however, has been for memory tasks to show nil sex differences (Maccoby & Jacklin, 1974). Among the exceptions are several studies that reported a male bias for "spatial memory," in seeming contradiction to the present data, but the dependent variables for these were similar to traditional spatial abilities tests rather than the measures developed here.

Nevertheless, it may be informative to assess whether female superiority obtains solely for holistic learning of spatial configurations, or whether it occurs, as well, with serial presentations of objects and locations. If female spatial specializations were selected for because of their contribution to foraging, then they evolved in a holistic context. Assuming that holistic and serial learning involve different cognitive mechanisms, we may expect sex differences only for the former.

In a related question, we are exploring the suggestion in the present data that males and females employ different modes of processing the environment, which may apply both to specific learning tasks and to daily routines. The open-ended responses of Study Two were scrutinized for indications that males and females undertook the task of trying to recall the stimulus room differently. A number of strategies were detected; for example, subjects reported a distinctive object and then attempted to remember objects nearby, or focused initially on a particular location. Individuals tended to use multiple strategies, and there were no apparent sex differences, although the emphases on locations in the instructions and probes may have induced similar approaches among subjects to the task.

We are also planning follow-up studies using uncommon objects, for which subjects would not possess verbal labels. Another well-documented finding in the area of cognitive sex differences is that females excel on measures of verbal ability (Maccoby & Jacklin, 1974). The female advantage in object and location memory observed here may represent a rudimentary manifestation of superior verbal skills; specifically, a greater capacity to recall object names. If so, the female bias may not occur with uncommon objects, and if it does, we will want to investigate whether it is attributable to a greater adeptness of females at inventing verbal labels for unfamiliar stimuli.

If object and location memory are enhanced by verbal facilities in this manner, it may suggest that female verbal superiority at least its initial form, also evolved as part of division of labor. Similar to spatial differences, verbal sex differences are near universal and show hormonal correlates (Burstien, Bank, & Jarvick, 1980). Nevertheless, there has been no prior attempt to explore their ultimate causation.

Evolutionary Explanations and Proximate Mechanisms

These conjectures about spatial/verbal interactions may bear on a long-standing theory of the neuropsychology of spatial sex differences. They may also illustrate the nature of the relationship between evolutionary and proximate explanations. Evolutionary explanations are not intended to supplant proximate theories; they function, rather, to give these direction. They attempt to go beyond the question of how specific psychological, psychophysiological, or cultural mechanisms operate to elicit behavior and try to explain how these mechanisms, as opposed to all other possible mechanisms, came to exist.

The neuropsychological theory in question is based on findings that suggest that males' brains are functionally lateralized to a greater degree than females'. On this basis, it has been assumed that spatial abilities, which are primarily the province of the right hemisphere, have a larger and more homogeneous area in which to develop in the male (see McGlone, 1980, for a review).

Stated as such, without benefit of an ultimate-level causal perspective, the reasoning seems to be that male and female brains became differentially lateralized by happenstance or some circumstance unrelated to spatial processes, and spatial sex differences developed as an incidental effect of this divergence. These kinds of causal gaps necessarily pervade pure proximate theories.

Our concept of the evolution of spatial sex differences can serve to fill the gaps. The tenet of evolutionary theory is that form follows function, in regard to anatomical, physiological, behavioral, and cognitive variables. If spatial sex differences were selected for because they maximized the effectiveness of division of labor, then it would follow that sex differences in lateralization emerged as a consequence—the psychophysiological mechanism to which the selection pressures gave rise.

Our model may also bear on a problematic aspect of the lateralization theory, pointed out by Gaulin and Hoffman (1988, pp. 36–37). The theory assumes that greater specialization occurs with greater lateralization. It begs the question, however, of why the more highly lateralized brain functions of males do not render them superior to females in verbal as well as spatial abilities, inasmuch as verbal abilities are mediated mainly by the left hemisphere.

A solution to the problem may reside in the present model, in that males are not regarded as more highly spatially specialized than females, but differently specialized. Further, as contended in the prior section, the theory suggests that females' spatial specializations may interact with verbal processes and have evolved in conjunction with these, which could underlie both their enhanced verbal abilities and more heterogeneous hemispheric functions.

Relationships Among Evolutionary Models

Finally, we consider the relationships among the Gaulin and Fitzgerald model, the Greenwood model, and the present model.

To the extent that these are competing theories, there are opportunities for tests between them. Wherever polygynous species employ resource defense systems or monogamous species show mate defense, the Gaulin and Fitzgerald theory and the theory derived from Greenwood potentially lead to opposite predictions about spatial sex differences. There is also at least one case that could generate a test between our division of labor model and the two mating system theories. Lions are polygynous and possess a resource defense system in which, atypically, females hold territories and males disperse. Thus, both the Gaulin and Fitzgerald model and the Greenwood model would predict a male bias in spatial abilities. On the other hand, females do most of the hunting, which, from the concept we have presented here, would lead to the prediction of female superiority.

Such studies, however, may not render one theory prepotent. There will probably always remain cases that best fit one or another model or do not fit any. An alternative approach would eschew the concept of general spatial abilities in favor of an attempt to delineate specific spatial functions on which various species show differentiation by sex. From this standpoint, the ecological circumstances associated with these differences may be explored on a species-by-species basis. This approach may ultimately provide the most productive path to a unified theory of spatial sex differences.

ACKNOWLEDGMENTS

Gratitude is expressed to Michelle Ball, Paul Fairlie, Douglas Hardacre, and Michal Kahane for their assistance in various phases of the studies, and to Leda Cosmides, Donald Symons, and John Tooby for helpful comments on the manuscript.

Portions of this paper were presented at the meetings of the International Society for Human Ethology in Binghamton, NY, in June, 1990, the Human Behavior and Evolution Society in Los

Angeles, CA, in August, 1990, and the European Sociobiological Society in Prague in August, 1991.

NOTE

1. Table 14.2 shows smaller Ns (numbers) for object memory than location memory. In the early trials, fewer items were added to the original array for the object memory task than are depicted in Figure 14.3. Because there was an apparent ceiling effect in the scores, additional items were inserted, and the object memory data of the early trials excluded.

REFERENCES

Bennett, G. K., Seashore, H. G., & Wesman, A. G. (1947). *Differential aptitude tests.* New York: Psychological Corp.

Burstien, B., Bank, L., & Jarvick, L. F. (1980). Sex differences in cognitive functioning: Evidence, determinants, implications. *Human Development, 23,* 299–313.

Chagnon, N. A. (1979). Is reproductive success equal in egalitarian societies? In N. A. Chagnon & W. G. Irons (Eds.), *Evolutionary biology and human social behavior: An anthropological perspective.* North Scituate, MA: Duxbury Press.

Fitzgerald, R. W., & Madison, D. M. (1983). Social organization of a free-ranging population of pine voles, *Microtus pinetorum. Behavioral Ecology and Sociobiology, 13,* 183–187.

Gaulin, S.J.C., & Fitzgerald, R. W. (1986). Sex differences in spatial ability: An evolutionary hypothesis and test. *American Naturalist, 127,* 74–88.

Gaulin, S.J.C., & Hoffman, H. A. (1988). Evolution and development of sex differences in spatial ability. In L. Betzig, M. B. Mulder, & P. Turke (Eds.), *Human reproductive behavior: A Darwinian perspective* (pp. 129–152). Cambridge: Cambridge University Press.

Greenwood, P. J. (1980). Mating systems, philopatry and dispersal in birds and mammals. *Animal Behavior, 28,* 1140–1162.

Greenwood, P. J. (1983). Mating systems and the evolutionary consequences of dispersal. In I. R. Swingland & P. J. Greenwood (Eds.), *The ecology of animal movement* (pp. 116–131). Oxford: Clarendon Press.

Harris, L. J. (1978). Sex differences in spatial ability: Possible environmental, genetic and neurological factors. In M. Kinsbourn (Ed.), *Asymmetric function of the brain* (pp. 465–522). Cambridge, Cambridge University Press.

Jacobs, L. F., Gaulin, S.J.C., Sherry, D., & Hoffman, G. E. (1990). Evolution of spatial cognition: Sex-specific patterns of spatial behavior predict hippocampal size. *Proceedings of the National Academy of Science, USA, 87,* 6349–6352.

Jardine, R., & Martin, N. G. (1983). Spatial ability and throwing accuracy. *Behavior Genetics, 13,* 331–340.

Kail, R. V., Jr., & Siegel, A. W. (1977). Sex differences in retention of verbal and spatial characteristics of stimuli. *Journal of Experimental Child Psychology, 23,* 341–347.

Kimura, D., & Hampson, E. (1990). *Neural and hormonal mechanisms mediating sex differences in cognition* (Research Bulletin No. 689). London, Ontario, Canada: University of Western Ontario, Department of Psychology.

Koenig, W. D. (1989). Sex biased dispersal in the contemporary United States. *Ethology and Sociobiology, 10,* 263–278.

Kolakowski, D., & Molina, R. M. (1974). Spatial ability, throwing accuracy and man's hunting heritage. *Nature, 251,* 410–412.

Linn, M. C., & Peterson, A. C. (1985). Emergence and characterization of sex differences in spatial ability: A meta-analysis. *Child Development, 56,* 1479–1498.

Maccoby, E. E., & Jacklin, C. N. (1974). *Psychology of sex differences.* Stanford: Stanford University Press.

Madison, D. M. (1980). Space use and social structure in meadow voles, *Microtus pennsyvanicus. Behavioral Ecology and Sociobiology, 7,* 65–71.

McGee, M. G. (1979). Human spatial abilities: Psychometric studies and environmental, genetic, hormonal and neurological influences. *Psychological Bulletin, 80,* 889–918.

McGlone, J. (1980). Sex differences in human brain asymmetry: A critical survey. *Behavioral and Brain Sciences, 3,* 215–263.

Silverman, I. (1977). *The human subject in the psychological laboratory.* New York: Pergamon.

Symons, D. (1979). *The evolution of human sexuality.* Oxford: Oxford University Press.

Tooby, J., & De Vore, I. (1987). The reconstruction of hominid behavioral evolution through strategic modeling. In W. Kinzey (Ed.), *Primate models of human behavior* (pp. 183–237). New York: SUNY Press.

Vandenberg, S. G., & Kuse, A. R. (1978). Mental rotations: A group test of three-dimensional spatial visualization. *Perceptual and Motor Skills, 47,* 599–604.

VI
ENVIRONMENTAL AESTHETICS

Art and science seldom meet, but when they do, it is usually on the common ground of perception. The idea that science can shed light on how we see, and therefore on how we see art, is so compelling that it has sometimes spawned new artistic movements, such as Impressionism and Pointillism.

But can science shed light on what we *like?* Our aesthetic preferences seem so irreducibly idiosyncratic, so culturally embedded, that the idea that science could illuminate them seems absurd on the face of it.

Yet both art and science challenge us to look at familiar things with new eyes. The chapters in this section take that challenge to heart and look at environmental aesthetics from an evolutionary and ecological perspective. When viewed from this vantage point, the proposition that our aesthetic preferences in some domains are guided by universal organizing principles begins to look reasonable—in fact, inevitable.

Consider habitat selection. Organisms that cannot move of their own accord do not need many mechanisms for habitat selection—dandelion seeds and diatoms go whichever way the winds blow or the currents flow. But any organism capable of guiding its movements through the environment must have mechanisms that cause it to prefer the habitats that are best for supporting its way of life. In biology, there is a rich literature that describes the cues and mechanisms whereby birds and other animals choose the right habitat. Asking the same questions about humans (who evolved, after all, as hunter-gatherers) is simply the next step.

Supplied, as we are, with every necessity by industrial civilization, emotional responses to landscapes seem completely epiphenomenal and, indeed, seem to exemplify the kind of evanescent psychological phenomenon that must be functionless. But to understand human psychological adaptations, one must place them in the context in which they evolved. To appreciate the importance of good habitat selection for the health and safety of a hunter-gatherer, Orians and Heerwagen ask readers to "imagine you are on a camping trip that lasts a lifetime." Having to carry water from a stream and firewood from the trees, one quickly learns to appreciate the advantages of some campsites over others. Dealing with exposure on a daily basis quickly gives one an appreciation for sheltered sites, out of the wind, snow, or rain. For hunter-gatherers, there is no escape from this way of life: no opportunities to pick up food at the grocery store, no telephones, no emergency services, no artificial water supplies, no fuel deliveries, no cages, guns, or animal control officers to protect one from the predatory animals. In these circumstances, one's life depends on the operation of mechanisms that cause one to prefer habitats that provide sufficient food, water, shelter, information, and safety to support human life, and that cause one to avoid those that do not.

In their chapter on environmental aesthetics, Orians and Heerwagen first discuss evidence relevant to the "savanna hypothesis"—the hypothesis that we have evolved preferences for habitats with features characteristic of a high-quality tropical African savanna, the environment in which the human lineage is thought to have initially evolved. Then, in the spirit of David Marr, they develop a task analysis, or computational theory, specifying what kinds of decisions our ancestors would have had to make in the course of habitat selection and what kinds of environmental cues would have been reliably associated with habitat quality during the Pleistocene. They propose that habitat selection proceeds in three stages. Stage 1 is a highly affective and rapidly made decision to either avoid or explore a new habitat, Stage 2 is an information-gathering phase in which one explores the new habitat to learn more about its potential to yield needed resources, and Stage 3 is the decision to either move on or stay in the habitat long enough to carry out necessary activities. Each of these stages should be characterized by different cognitive and affective processes.

Marr argued that an essential part of any computational theory is a specification of "valid constraints on the way the world is structured—constraints that provide sufficient information to allow the processing to succeed" (Marr & Nishihara, 1978, p. 41). This is just as necessary for the problem of habitat selection as it is for the problem of visual perception. To discover the design of the cognitive processes that govern habitat selection, one must figure out which environmental cues would have provided "sufficient information to allow the processing to succeed"—that is, which would have been reliable indicators of habitat quality over time. To this end, Orians and Heerwagen catalog cues that would have been relevant to decisions that cover different time frames. For example, thunderclouds provide information about weather conditions in the short run, the first buds of spring signal a change in habitat quality that will last several months, and desert conditions indicate habitat quality over a number of years. Environmental cues that cover different time frames are relevant to different classes of decision. Those rich in information about habitat quality should captivate our attention and elicit affective reactions that impel us toward the right course of action.

Wayfinding is another example of an adaptive problem that should have shaped our environmental preferences, and the one that Kaplan's chapter focuses on. Kaplan takes the "sapiens" in *Homo sapiens sapiens* seriously, arguing that we should have evolved a thirst for certain kinds of knowledge and an attraction to situations that can provide it. For example, to develop the kind of knowledge base necessary to safely navigate through an environment, he argues that our ancestors should have been "enticed by new information, by the prospect of updating and extending their cognitive maps" while at the same time not "stray[ing] too far from the familiar, lest they be caught in a situation in which they would have been helpless due to a lack of necessary knowledge."

Using photo questionnaires, Stephen and Rachel Kaplan have been able to collect enormous amounts of data on people's aesthetic preferences. They find that people's choices on these questionnaires are strikingly patterned in a way that suggests they are drawn to scenes that promise the possibility of new information, safely obtained. The criteria that can be shown to govern these choices are abstract and deeply inferential. For example, spatial arrays that promote "legibility"—the inference that one could find one's way back if one ventured further into the scene depicted—and ones that promote "mystery"—the inference that one could acquire more information by venturing deeper into the scene and changing one's vantage point—are preferred over spatial arrays that do not promote these inferences. These deeply cognitive criteria are not open to introspection, yet they intuitively guide our attraction to new environments—cognition and affect wedded. Kaplan found that he could not understand the environmental preferences that govern wayfinding unless he could create a conceptual framework in which the usually separate studies of cognition and affect could be united. We hope his efforts inspire others to do the same.

Given that many of the criteria that govern our aesthetic preferences are complex and not open to introspection, it is perhaps not surprising that they should appear inexplicable and idiosyncratic to us. Yet a growing body of evidence suggests that this impression is illusory. Although the study of environmental preferences is still in its infancy, it already appears that our aesthetic preferences are governed by a coherent and sophisticated set of organizing principles. Just as theories of how we see have implications for how we see art, theories of what kinds of environments we prefer have implications for architectural design and urban planning, as both Kaplan and Orians and Heerwagen discuss. If these chapters are any indication, the study of environmental preferences may prove to be yet another common ground upon which science and art will meet.

REFERENCE

Marr, D., & Nishihara, H. K. (1978). Visual information processing: Artificial intelligence and the sensorium of sight. *Technology Review* (October), 28–49.

15

Evolved Responses to Landscapes

GORDON H. ORIANS AND JUDITH H. HEERWAGEN

Evolutionary approaches to aesthetics are based on the postulate that emotional responses, because they are such powerful motivators of human behavior, could not have evolved unless the behavior they evoked contributed positively, on average, to survival and reproductive success. This is why sugar is sweet and sexual activity is fun. Those of our ancestors who found consuming carbohydrates and engaging in intercourse enjoyable left more surviving descendants than those individuals who were not motivated to engage in those behaviors. Thus, an evolutionary biologist studies the actions evoked by emotional states to determine why those emotions had survival value. This is not to say that the behaviors stimulated by those emotions are always useful. Curiosity does, so we are told, sometimes kill cats. Nonetheless we believe that cats lacking curiosity fail to learn enough about their environments to function in them as well as their more curious brethren, even though being curious sometimes has unfortunate consequences.

The study of human responses to landscapes is a profitable arena in which to study the evolution of aesthetic tastes. Because selection of places in which to live is a universal animal activity, there is considerable body of theory and empirical data upon which to build hypotheses specifically oriented toward human behavior (Charnov, 1976; Cody, 1985; Levins, 1968; MacArthur & Pianka, 1966; Orians, 1980; Partridge, 1978; Rosenzweig, 1974, 1981). Also because choice of habitat exerts a powerful influence on survival and reproductive success, the behavioral mechanisms involved have been under strong selection for millennia.

In all organisms, habitat selection presumably involves emotional responses to key features of the environment. These features induce the "positive" and "negative" feelings that lead to rejection, exploration, or settlement. If the strength of these responses is a key proximate factor in decisions about where to settle, as empirical data suggest, then the ability of a habitat to evoke such emotional states should evolve to be positively correlated with expected survival and reproductive success of an organism in different habitats. Good habitats should evoke strong positive responses whereas poor habitats should evoke weaker or negative responses.

However, because a suitable habitat must provide resources for carrying out many different activities over varying time frames, evaluating habitats is a difficult process for organisms. The current state of the environment is important, but probable future states, over the entire time period the site will be occupied, may be equally or more important. For this reason, organisms evolve to use features of habitats that are good predictors of future states. Interestingly, the features used often are not the ones that directly determine success. For example, many birds use general patterns of tree density and vertical arrangement of branches as primary settling clues (Cody, 1985; Hildén, 1965; Lack, 1971) rather than attempting to assess food supplies directly. The use of features correlated with, but not actually determining, success is likely to characterize human responses to landscapes as well.

ADAPTIVE RESPONSE TO THE ENVIRONMENT: PLEISTOCENE ORIGINS

The human life cycle is characterized by long generation times and extremely long periods of offspring dependency. The habitats occupied by people during most of our evolutionary history rarely provided reliable resources for long enough times that permanent occupancy of sites was possible. Rather, during the lengthy hunter-gatherer stage of human evolution, frequent moves through the landscape were the rule (Campbell, 1985; Lee & DeVore, 1968, 1976; Lovejoy, 1981; Washburn, 1961).

To understand the importance of habitat selection to our hunting and gathering ancestors, imagine you are on a camping trip that lasts a lifetime. You wake up one morning with an empty stomach and an empty cupboard. It is time to move on. Clouds on the horizon indicate that it has rained for many days in that area, and this is where you will head to look for food. Although the rainy place is many days off, it will be lush and green with berries, vegetables, and fresh water. The animals will come to feed so hunting will be good.

The small band of adults and children gradually begins the long hike across new terrain. By midday the sun is high and hot. In the distance on a ridge crest is a cluster of big trees—they look cool and inviting, but are still several hours off. As the group continues on its hike toward the trees, one of the men spots fresh lion tracks. He stops abruptly, gestures for the group to be quiet as he climbs a rock outcropping for a better view of what's ahead. The lions are only a short distance off, almost hidden from view in the grass. The man watches the lions for hints of their future intentions. Are they hungry? Will they attack? His vast store of animal information tells him not to worry. They have just had a large meal and are resting.

By the time the group reaches the cluster of trees, the sun is low in the sky signaling an end to the unbearable heat of the day. The adults rest momentarily, knowing that soon it will be cooler. They begin setting up camp and preparing the evening meal. The rumble of thunder off in the distance is a welcomed sound. The dry season is coming to an end. Around the campfire that night, the adults break up into small groups. Several women make preparations for the next day's foraging. They discuss the route they would take, recalling where they found the best berries, fruits, and leafy greens last year. On the walk today, one woman remembered seeing Grewia flowers—there may be bushes with ripe berries nearby. Another woman talks about the large nut tree that was so productive last year. The men gather in small clusters to make arrows, all the while talking about the animal tracks they had seen. They plan tomorrow's hunt. It's an unfamiliar terrain, so they need to decide which direction looks most promising. Gradually everyone drifts off to sleep. Shortly before dawn, several of the adults awaken to a loud crashing sound in the bush. The sound recedes, and they fall back asleep. Soon all of the campers awake and begin a new day in a life-style that will last for thousands of generations.

This hypothetical scenario is intended to illustrate what daily life may have been like for our ancestors and the kind of habitat-based decisions they would have had to make. In the remaining sections of this paper, we will consider how our aesthetic reactions to landscapes may have derived, in part, from an evolved psychology that functioned to help hunter-gatherers make better decisions about when to move, where to settle, and what activities to follow in various localities. According to this argument, environmental stimuli as diverse as flowers, sunsets, clouds, thunder, snakes, and lions activate response systems of ancient origin. These systems—including perceptual, cognitive, emotional, and behavioral processes—served important functions in both day-to-day survival and the long-term fitness of early humans.

The needs of our ancestors were the same as our current needs—to find adequate food and water and to protect themselves from the physical environment, predators, and hostile conspecifics. We now seek these amenities in rather different environments than the ones in which we evolved. Nonetheless, the number of generations that we have lived in mechanized, urban environments is small in relation to the number required for substantial evolutionary changes. Therefore, it is reasonable to expect, and to search for, response patterns that evolved under conditions quite different from those we now experience but which, nonetheless, often in unconscious ways, influence our decisions today. Before we develop our argument further, it will be helpful to look at how researchers have approached environmental aesthetics from an evolutionary point of view.

EVOLUTIONARY APPROACHES TO ENVIRONMENTAL AESTHETICS

The notion that aesthetic responses to environments foster behavior that increases the ability of individuals to learn about and function effectively in the environment has been explored by investigators in a number of fields. Data with which to test these ideas are still fragmentary. Nonetheless, it is interesting that investigators from a number of disciplines have been attracted to approaches that are based upon exploring the possible adaptive significance of our patterns of response to landscapes. The research in this area falls into two broad approaches. The first approach focuses on differential response to natural biomes and, in particular, tests hypotheses related to the habitat in which people evolved. The second approach to landscape preferences is based on the notion that we prefer environments in which exploration is easy and which signal the presence of resources necessary for survival. Although these two approaches are highly compatible, the second approach is more general and does not derive its predictions from a specific habitat type.

Tests of environmental preferences have relied almost exclusively on ratings of photographs or slides of landscapes. Studies which have compared photos versus actual trips to a site show that responses do not vary significantly as a function of presentation format. This is an important finding, because photographic techniques make it possible to test a large variety of landscapes that could not be directly experienced.

The Savanna Hypothesis

The basic biological argument underlying the habitat-specific hypothesis is that natural selection should have favored individuals who were motivated to explore and settle in environments likely to afford the necessities of life but to avoid environments with poorer resources or posing higher risks (Orians, 1980, 1986). The savannas of tropical Africa, the presumed site of human origins, have high resource-providing potential for large terrestrial, omnivorous primates such as ourselves. In tropical forests, most primary productivity occurs in the canopy, and a terrestrial omnivore largely functions as a scavenger, gathering up bits of food that fall from the more productive canopy. In savannas, however, trees are scattered and much of the productivity is found within two meters of the ground where it is directly accessible to people and to grazing and browsing animals. Biomass and production of meat is also much higher in savannas than in forests (see Orians, 1986). The savannas also afford distant views and low, grassy ground cover favorable to a nomadic life-style. If we assume that the evolution of our species includes the development of psychological mechanisms that aid adaptive response to the environment, then savanna-like habitats should generate positive responses in people, much as the "right" habitat motivates exploration and settling behaviors in other species. This is because the savanna is an environment that provides what we need: nutritious food that is relatively easy to obtain; trees that offer protection from the sun and can be climbed to avoid predators; long, unimpeded views; and frequent changes in elevation that allow us to orient in space. Water is the one resource that is relatively scarce and unpredictably distributed on the African savannas. The scarcity of this critical resource plays a fundamental role in our response to environments, as will be noted later.

If we assume that habitat preferences coevolve with the intrinsic quality of habitats, then certain predictions follow. First, savanna-type environments should be favored over other biomes because of their critical role in the development of modern humans. Second, environmental features that predictably signal distinctions between high- and low-quality savanna habitats should influence preference patterns. One such feature is tree shape.

Research on landscape preferences strongly indicates that savanna-like environments are consistently better liked than other environments (see reviews in Balling & Falk, 1982; Ulrich, 1983, 1986). In the only direct test of preferences for the different biomes, Balling and Falk (1982) hypothesized that humans have an innate preference for savanna-like environments that arises from their long evolutionary history on the savannas of East Africa. They argued that an "innate predisposition" for the savanna should be more likely to be revealed in children than in adults because adults are likely to have had experience living in biomes other than savannas. Their study included six age groups (8, 11, 15, 18, 35, and 70 or over). Subjects rated how much they would like to "live in" or "visit" five natural biomes shown in slide format. The biomes included tropical forest, deciduous forest, coniferous forest, East African savanna, and desert. None of the photos used in the study included water or animals. Balling and Falk found that the 8-year-old children would rather live in as well as visit the savanna than the other habitats and that they rated the savanna higher on both factors than did all of the other age groups. From age 15 on, the savanna, deciduous forest, and coniferous forest were liked equally well, and all three biomes were preferred over the rain forest or desert. Interestingly, the desert was the least liked environment for all age groups; and two slides of the savanna during the dry season also received lower ratings than the greener savanna settings.

Because none of the respondents in the Balling and Falk study had ever been in tropical savannas, the authors postulate a developmental pattern, with innately programmed responses that later are modified by experience in particular settings (in this case, the deciduous woods of the eastern U.S.). The fact that familiar environments did not become preferred over tropical ones suggests that, although experience is important in determining aesthetic responses to environments, it does not override completely the presumably innate responses that express themselves strongly among children.

Studies we are conducting at the University of Washington also lend support to the savanna hypothesis. We have been testing people's responses to tree shapes and have found that tree shapes characteristic of high-quality savanna are preferred over those found in lower-quality savanna. To control for other features of plant architecture and general habitat, we are using photos of one species only, *Acacia tortilis*. This tree varies considerably in its shape in different areas of the savanna. In high-quality habitat, this acacia has the quintessential savanna look—a spreading, multilayered canopy and a trunk that branches close to the ground. In wetter savannas, the species has a canopy that is taller than it is broad with a high trunk, while in very dry savanna *A. tortilis* is dense and shrubby looking.

We are currently completing a cross-cultural study of aesthetic responses to trees with subjects from Seattle, Argentina, and Australia. We used a photo questionnaire similar to that used by other researchers (see Kaplan & Kaplan, 1982). Subjects were asked to rate the attractiveness of each of the trees shown in photographs. Photos used in the study were taken in Kenya by G. and E. Orians. We used black and white rather than color photography because this procedure diminishes variability in the lushness of the setting and the color of the sky, both of which are known to influence response to landscapes. A standardized photographic procedure was used to eliminate as much background variability as possible. Each photo focused on one tree, and all pictures were taken under similar daylight and weather conditions. Photos with clouds, mountains, or water were deleted from the pool of trees selected for the study. Trees selected for inclusion in the questionnaire varied in canopy density, canopy shape, trunk height, and branching pattern. Because trees tend to vary simultaneously on a number of these features, the questionnaire was designed so that each page had trees that were similar for one of the primary characteristics. For instance, one page consisted of six trees that all had high trunks. However, the canopy patterns varied among them. Some of the trees had broad, layered canopies, and the others had narrower or denser canopies. Because many of the trees appear on more than one page, we can determine if trees are rated consistently, or whether their ratings change as a result of the other trees with which they are being compared. Although our data analysis is not yet complete, we have found that the three subject groups from the U.S., Argentina, and Australia show very similar patterns of response. The trees rated as most attractive by all three groups are those in which canopies are moderately dense and trunks bifurcate near the ground. Trees with high trunks and skimpy or very dense canopies are judged as least attractive by all three groups.

The characteristic features of the tropical savanna are exploited in a variety of other ways, including landscape painting. For instance, Humphrey Repton, the nineteenth-century pioneer of British landscape gardening, regularly included scattered clumps of trees in his designs to break up straight edges dividing pastures from woods (Repton, 1907). He also regularly used animals in his famous "before and after" drawings for his potential clients, as found in his "Redbooks." In addition, he appreciated the importance of the shapes of savanna trees. For example, on page 105 of his treatise, he notes: "Those pleasing combinations of trees which we admire in forest scenery will often be found to consist of forked trees, or at least of trees placed so near each other that the branches intermix, and by a natural effort of vegetation the stems of the trees themselves are forced from that perpendicular direction which is always observable in trees planted at regular distances from each other."

Although there is evidence that savanna-like environments are positively experienced by many people, this does not mean that all cultures consider this spatial form as an ideal or preferred type. Preference is also influenced by experience. The geomorphological characteristics of many landscapes differ dramatically from those of the savanna. Personal interaction with these places over a lifetime creates a wealth of knowledge and meanings that provide the basis for emotional attachment to places (see Relph, 1976; Tuan, 1974). What we are suggesting in this paper is that people have a generalized bias toward savanna-like environments. If this bias does, indeed, exist, then people should react positively to savannas even in the absence of direct experience. Further, we predict that positive responses to other types of biomes, such as desert, steppe, and closed forest, require direct experience. In the absence of experience, these environments should be given lower aesthetic ratings than savannas.

General Evolutionary Hypotheses

Studies of adaptive responses to landscapes have looked at the spatial features and particular contents of the environment that influence preference patterns. Psychological approaches, many of which are summarized by Ulrich (1983, 1986), have found that people prefer environments that have water, large trees, a focal point, changes in elevation, semi-open space, even ground cover, distant views to the horizon, and moderate degrees of complexity. Although these features are certainly characteristic of the savanna, they are also present in other environments. Interestingly, there are no studies on the relationship between preferences and the presence of animals in the landscape. Most researchers deliberately leave animals out of photographs because they are suspected to enhance preference scores—this is true for exotic animals such as zebras or giraffes, as well as more common animals such as cows (Schauman, personal communication, 1990).

Other researchers using an evolutionary approach have looked at the features of landscapes that influence exploration and information gathering. The general argument is that safe movement through the environment requires a great deal of skill and knowledge. Landscapes that aid and encourage exploration, wayfinding, and information processing should be more favored than landscapes that impede these needs. Using slides and photo questionnaires of everyday environments, Steven and Rachel Kaplan (S. Kaplan, 1987; S. and R. Kaplan, 1982; see also S. Kaplan, this volume) have found that preferred landscapes tend to be easier to "read" than other landscapes, but not so easy that they are boring. Desirable landscapes contain moderate degrees of complexity, a sense of coherence, and a semi-open spatial configuration. These features signal ease of movement as well as the potential for gaining more information about the environment. Preferred landscapes often contain a quality the Kaplans have called "mystery"—the hint of interesting features that could be discovered if the observer were to explore the environment. Environments high in mystery contain roads or paths that bend around hills, meandering streams, or partially blocked views, all of which emotionally entice the viewer to enter the environment because there is more to be learned. The Kaplans, as well as other researchers, have consistently found that natural environments are preferred over built environments and that built environments with trees and other vegetation are more positively regarded than similar built spaces lacking vegetation (Kaplan & Kaplan, 1982; Ulrich, 1983; Wohlwill, 1983). Furthermore, Ulrich has found that people in stressful situations who view slides of nature scenes as compared to scenes of buildings show lower distress responses on a number of affective and physiological measures (Ulrich, 1979, 1986; Ulrich, Simons, Losito, Fiorito, Miles and Zelson, in press). Ulrich's work suggests that our responses to landscapes and nature can have profound effects on human well-being.

CONCEPTUAL FRAMEWORK

In this paper, we extend previous approaches to landscape aesthetics by considering several stages in people's experience of environments and novel situations. We also develop some ways of thinking about responses to environments that affect behavior on time frames ranging from very short to long-term. Our purpose is to provide a richer context than is currently available for thinking about aesthetic responses from an adaptive perspective. Most previous research has tended to focus on behavioral outcomes and, in particular, on preferences associated with different environmental settings. This paper will focus on the ways in which the context influences what we attend to, how we evaluate the situation, and how these evaluations influence variability in behavioral outcomes. In this sense, it is a "computational" framework whose purpose is to guide research into the psychological mechanisms that promote adaptive functioning in different environmental contexts (see Cosmides and Tooby, 1987, Marr, 1982). By adaptive functioning we mean what Staddon (1987) has described as programs that "enable the animal to do the right thing at the right time and place, where 'right' means 'such as to improve fitness'" (p. 103). Fitness, as Staddon notes, is difficult to measure directly; our psychological mechanisms can measure only proxies for fitness such as access to water, food, and protected places. To understand the relationship between aesthetic responses and evolution, however, we must first consider human interactions with landscapes.

Landscapes provide resources such as food, water, and safe resting and sleeping places. They are also potential sources of danger. Danger may be posed by the physical environment in the form of bad weather, physical barriers to passage, earthquakes, landslides, fires, or avalanches. Biological sources of danger include predators, parasites, toxic foods, and unfriendly conspecifics. These dangers can, of course, be reduced by not venturing out into the environment, but this is achieved at the price of forfeiting access to resources and information that improve survival and reproductive success. Organisms are expected to evolve behavioral responses that provide, on average, the best ratio of benefits to risks. Good ratios can be achieved by avoidance of environments high in potential risks relative to their resource-providing potential and concentrating exploration in those environments promising better resources that can be exploited with lower risks. The benefits and risks vary among species and within species according to current conditions and needs. Thus, there is no reason to expect an invariable ranking of risks and benefits among environments by the individuals of a particular species or for a given individual throughout its lifetime. The existence of such variability makes the task of finding general patterns more difficult but does not constitute a reason for abandoning an effort to discover them. The conceptual framework we develop in this paper is based upon habitat selection theory. A habitat selection approach that includes a consideration of the ways in which settling individuals deal with both spatial and temporal scales of assessment provides a powerful conceptual framework for thinking about a diverse array of human responses to environments.

Spatial Frame of Reference

Two complementary frames of reference are useful in approaching evolved responses to landscapes, one spatial and the other temporal. The spatial frame of reference concerns the stages of exploration of an unfamiliar landscape, or "habitat," as it is usually referred to in biological literature. Stage 1, which accompanies an initial encounter with a landscape, is the decision to either explore the landscape further or avoid it and move on to other areas. Responses at this stage are known to be highly affective, to occur almost instantly, and are believed to influence subsequent actions (Ulrich, 1983). These rapid responses can be influenced by both the innate constitution of the individual and by previous experience. The speed with which these judgments are made should not mislead us to underestimate their importance. If the initial response is negative, no further exploration of the environment is likely.

If the response in the first stage is positive, the individual may enter Stage 2, that of information gathering. In this stage the individual explores the environment to learn more about its potential to provide resources. Unlike the responses in Stage 1, cognition figures prominently in Stage 2, and the act of exploration may last many days. The individual may draw upon memories and associations between other environments and the resources they provided. Evolved responses are likely to be important at this stage to encourage exploration and to increase the likelihood that attention is given to the most relevant aspects of the habitat.

Stage 3 concerns the decision to stay in the environment to carry out a certain set of activities there. Such a decision may relate to a specific activity that is intended to last for only a short time, or it may be a permanent decision affecting all of the behavior of the individual for the rest of its life. Because organisms require many resources in their lives, often more than one habitat must be utilized to fulfill all needs. Thus, the proximity of habitats providing different components of the suite of essential resources may influence settling decisions. Research efforts must take this possibility into account.

Temporal Frame of Reference

The second frame of reference concerns the time frames of decisions. Some environmental cues pertain to conditions that are transitory. Changes in weather, perception of prey, predators, or enemies, and arrival of a prospective mate are examples of such cues. These changes in the environment demand immediate attention and evaluation and a quick response. Time for thought is minimal if the opportunity is to be seized or danger avoided.

Other environmental cues signal changes that occur more slowly and affect benefits and risks over longer time spans. Examples of these cues include seasonal changes in the vegetative and reproductive cycles of plants, and activities of animals associated with reproduction. Understanding of and response to these seasonal changes is vital for successful functioning in the environment, but rarely are immediate responses needed. Time for reflection and evaluation is readily available.

At the other extreme are cues that signal relatively permanent features of the environment. Prime among these cues are geomorphological features such as topographic relief and the presence of lakes, rivers, and streams. The general features of the vegetation of an area are also relatively permanent, although they clearly change more rapidly than cliffs erode or lakes fill with sediments.

In this paper we use these two complementary frameworks to explore a variety of environmental features known to evoke strong emotional responses in people. Throughout, our point of reference is to suggest how responses to these features enhance the ability of individuals to function in environments. We believe these frameworks can be used not only to explain known phenomena, but also to generate many new and fruitful hypotheses concerning the psychological mechanisms governing environmental aesthetics in humans.

STAGES OF HABITAT SELECTION

Responses at different stages of habitat exploration are distinguished in terms of (a) the basic kinds of decisions required; (b) the extent to which the decisions can be made automatically and rapidly or require more extensive evaluation; and (c) the features of the environment that exert the strongest influences on responses and decision making.

Stage 1

The decision made on an initial encounter with an environment is whether or not to explore the environment further or to avoid it and move on to other areas. Responses at this stage are known to be highly affective and to occur almost instantly (Ulrich, 1983; Zajonc, 1980). These rapid responses can occur with little or no conscious inference, but, nonetheless, a great deal of unconscious processing could be occurring, especially if the environments share features with others that have been previously experienced (see Kaplan, this volume). There are several compelling reasons for believing that evolutionarily molded behavioral responses should often be rapid and unconscious (Orians, 1980). Time is often important, and automatic responses leave the brain free to attend to those aspects of behavior that do require attention.

Theoretical and empirical work on preferences suggests that these rapid responses are typically made to rather general features of the environment (Baron, 1981; Zajonc, 1980). With respect to landscapes, important general features, termed "preferenda" by Zajonc (1980), include such factors as spatial configurations, gross depth cues, and certain classes of content, such as water or trees (Ulrich, 1983). Spatial configurations, such as the degree of openness of an environment (e.g., desert versus closed forest) are certainly perceived quickly. They also provide useful, and generally accurate, information regarding the ability of the space to meet human needs. An open environment, devoid of protective cover, is relatively undesirable for human occupation. So is a completely closed canopy forest within which movement and visual access are very difficult. Clearly both those types of environments are inhabited by people, but more extensive experience and learning are needed to cope with them than for dealing with environments of intermediate cover. Gross depth cues allow rapid assessment of distances, which may be of value in determining the time required to cross open spaces and the distance to potential prey or places of protection. Depth cues also facilitate judgments of the ease of movement afforded by environments. Water and trees provide useful and reasonably accurate information about the availability of basic resources and safe sites in unfamiliar environments.

Stage 2

If responses to the first stage of habitat selection are positive, the individual may enter Stage 2, that of information gathering. In this stage the individual may peruse the environment carefully and explore it to learn more about its potential to provide resources and afford safety. Features of the environment important to this stage can be divided into those that entice exploration and those that aid the ability to orient in space so that one can return safely to the point where exploration began. Exploration, according to Berlyne (1971) is whetted by such attributes as complexity, surprisingness, novelty, and incongruity. Other abstract features, such as "mystery" (Kaplan, 1987; Kaplan & Kaplan, 1982), patterns (Platt, 1961), and repeated or "rhyming" patterns (Humphrey, 1980) can entice exploration by providing inducements to gather information in an environment that is complex enough to be promising but not so complex as to be "unreadable." The horizon may also provoke exploration by stimulating imagination and the desire to know "what's over the hill" (Appleton, 1990). Way finding and long distance movements in the environment are aided by changes in elevation, or lookouts, from which the environment can be seen as a whole, allowing movements to be planned in advance. Other features include landmarks for orienting, pathways or other indicators of connections between places, and borders or edges that can be followed for some distance.

These concepts can all be accommodated in a framework that asserts that the extent of exploration of an unfamiliar environment is related to how long it takes to learn the essential features of the area (Kaplan, this volume) and whether or not the accumulated information indicates that the environment is actually rich in resources.

Although exploration is a positive emotional and cognitive experience, it brings with it potential hazards. Danger, as well as adventure, may lurk behind the bend in the road or in the cluster of trees. An exploring organism must, therefore, be constantly ready to respond to sudden stimuli and unexpected danger. This is accomplished through a reflexive orienting response to such stimuli as sudden or intense changes in sound or light levels (Bernstein, 1979; O'Gorman, 1979; Kahnemann, 1973; Sokolov, 1963). The orienting response inhibits ongoing activity, produces intense processing of stimuli, and prepares the body for future responses. Automatic fear or caution responses also occur with respect to certain classes of content, such as snakes and spiders (Öhman, 1986; Tuan, 1979; E. O. Wilson, 1984).

Some degree of risk assessment is likely in Stage 2, especially in unfamiliar environments and for movement in familiar environments under hazardous conditions (e.g., darkness, stormy weather) or when a potential new threat appears. Assessment is aided by the presence of particular environmental features such as overhangs or other formations conducive to concealment, places that afford high visual access into the space, multiple concealments that allow evaluation from different angles and distances, changes in elevation that provide expansive views of the space from above, and multiple escape routes. A newly encountered environment that contains these features should appear to us as more pleasing and desirable than a comparable environment in which they are absent. This response is not necessarily based on a conscious evaluation of these features. In fact, it is highly likely that people are not aware of or able to articulate the relationship between the presence or absence of these features in an environment and the overall impression the space provides to them. From an evolutionary perspective, an automatic risk-assessment response is highly adaptive. The focus of conscious evaluation should be directed toward the potentially dangerous event or stimulus. If we are to perform this function effectively, our mind cannot be distracted by simultaneously having to monitor and evaluate our own movements in the environment.

Blanchard and Blanchard (1988) suggest that many potentially threatening situations (including unknown places) do not involve discrete, easily discriminated sources of danger. Under these circumstances, elaborate risk-assessment and risk-avoidance behaviors are seen in many animals, such as clinging to edges or walls, refraining from unnecessary movement, leaning forward and then rapidly withdrawing ("stretched attention"), and suppression of other normal activities. If the potential dangers appear to diminish, the animal begins slowly to explore the space, and then eventually to engage in normal activities. Although we do not expect all newly encountered environments to be threatening, it is nonetheless highly advantageous for a newcomer to be receptive to environmental features that allow for investigating resources and potential dangers safely.

Stage 3

The final stage of habitat selection concerns the decision to stay in the environment to carry out a certain set of activities there. Evaluation of spatial configurations of patches in the environment may be especially important at this stage. A suitable habitat must contain a mixture of patches that provides opportunities for all of the activities required during the time interval that the habitat is to be occupied. Some of these activities, such as foraging, are extensive in nature, whereas others, such as sleeping or escape, are local. Good foraging areas may have poor escape sites and vice versa. Distances between the patches may be so great that much time and energy is lost in transit between sites. Trade-off problems of this type normally accompany habitat selection decisions by people. Once a site is selected, much effort may be devoted to improving those components most deficient (e.g., by digging a well, building a shelter).

The ways in which people handle complex trade-off decisions have been the subject of many studies (Beach, 1990; Halloway, 1979; Janis & Mann, 1977; Simon, 1957). It is beyond the scope of this paper to review and analyze these studies. For our purposes it suffices to note that the human mind clearly has evolved to be able to make such complex decisions, just as it has evolved to enable us to respond quickly to sudden changes in the environment.

TIME FRAMES OF DECISIONS

Stimuli from the environment signal features of highly variable duration. Responses to those stimuli, no matter how fast or slow they are made, also affect subsequent behavior for very different time periods. We now consider these important variations in some detail.

Environmental Cues Requiring Immediate Response

Many environmental signals emanate from events that are transitory. These transitory environmental cues—light, weather, fire, large mammals—occur regularly in most environments regardless of their overall quality from the perspective of human life. What they have in common is the requirement that adaptive responses to them must take place quickly. Hence, they must be able to override the events to which we have been paying attention. From an evolutionary perspective, therefore, the strong emotional reactions people display to these events are expected. Although appropriate responses must be exercised quickly, the response itself may also be transitory. In addition, responses to a given signal may be positive or negative, depending on the situation of the observer. The same signal may trigger quite variable responses among different individuals or by a single individual at different times or under different circumstances.

Prominent among such transitory signals are those involving weather patterns. Clouds are important signals about probable weather patterns during the next few hours. Thunder or lightening signal events impending within a few minutes. Sudden changes in the speed and direction of the wind indicate passages of frontal systems or approaching storms. It is highly likely to be adaptive for an individual to alter its behavior in response to perceiving such signals. Seeking cover, changing clothing, rounding up the kids, or protecting objects may be advisable, and to be effective these actions may need to be initiated quickly.

There is ample evidence from studies of the orienting response that people automatically attend to sudden change in signal intensity, such as sudden changes in noise levels or brightness of light (Bernstein, 1968, 1969; Sokolov, 1963). Although little attention has been paid to why we respond so strongly to these changes, from an evolutionary-ecological point of view it is clear that sudden changes in the characteristics of a stimulus are likely to be associated with potential hazards or opportunities. Bernstein (1968) suggests that sudden increases in stimulus intensity are more meaningful than changes involving a decrease in intensity. Intensity increases may represent "heightened sensation to a class of stimuli indicating 'something is approaching the organism'" (p. 128). Such sensitivity, as Bernstein notes, is highly advantageous for effective functioning in the environment.

The power of such signals to elicit attention and increase emotional arousal is well known. Even intense conversations rarely continue unabated through an outburst of thunder. Cloud patterns are among the most powerful evokers of strong emotions, both positive and negative. The development of the role of clouds in landscape painting is informative on this point. At the time of Claude Lorrain (1600–1682), when landscapes first became the subjects of attention in themselves rather than being backgrounds for the human activities that were the primary concern of the painter, skies were always bland and nonthreatening. Later, in landscapes of Constable and Turner, detailed attention is paid to clouds, and they are used to influence emotional responses to the paintings, both positive and negative. Indeed, Constable was a careful student of clouds, and his paintings reveal a detailed knowledge of the patterns of change in weather associated with different cloud patterns (Thornes, 1984).

The color of the sky and the lengthening of shadows do not necessarily reflect changes in weather patterns, but they do signal very powerfully that a change from day to night or vice versa is imminent. To diurnal animals with poor nocturnal vision like ourselves, evidence that night is approaching nearly always calls for changes in behavior that will prepare us for impending darkness. It is therefore not surprising that the color patterns in skies accompanying sunrises and sunsets should summon emotional reactions and that strong patterns of shadow should attract our attention. The flat light of midday signals that many daylight hours remain and that changes in behavior on that account are not called for. Shadows also allow better perception of depth and, hence, details of the environment. This doubtless enhances the value of attending to shadows. The dramatic effect of shadows and slanted rays of light associated with sunrise or sunset are often used by photographers to enhance the appeal of their work.

Although weather and time of day have been virtually ignored in landscape research, studies in architecture strongly suggest that building occupants are well aware of the value of windows for information on time and weather (Collins, 1975; Heerwagen, 1986; 1989; Heerwagen & Orians, 1986). Furthermore, studies of windowless intensive care units attribute poor recovery in these environments, as compared with windowed intensive care rooms, to disorientation from loss of daylight and information regarding time of day, weather, and seasonal change (Keep, 1977; Keep, James, & Inman, 1980; L. M. Wilson, 1972).

Studies of preferences for types and intensity of lighting also support the notion that shadows and variability in light levels are more emotionally pleasant and more preferred than uniformly lighted spaces (Aldworth, 1971; Flynn, Spencer, Martyniuk, & Hendrick, 1973). In fact, the more contrasting the space, the more highly pleasing it is.

Fire is another transitory environmental cue that commands our attention. Abundant archaeological evidence indicates that our ancestors used fire to provide warmth, heat, and light as well as a central focus for home base, and as a means of cooking food that would otherwise be inedible or less nutritious (Campbell, 1985). In addition, fire has been used as protection against predators, an aid in toolmaking, to influence movements of large animals, and to deflect and manage successional stages of vegetation. According to Konner (1982), the control of fire and its use as the focal point of evening social life resulted in a "quantum advance in human communication: a lengthy, nightly discussion, perhaps, of the day's events, of plans for the next day, of important occurrences in the lives of individuals and in the cultural past, and of long-term possibilities for the residence and activity of the band" (p. 50). Such conversations, as Konner notes, are more likely at night when the urgencies of the day are past and the desire is for social comfort, light, and warmth.

Although fire has many benefits, it is also a dangerous event to which immediate responses are often needed. The strong emotions evoked by fire, both positive and negative, have long been recognized. They play, for example, a key role in parts of Freudian psychology. The appeal of Smokey the Bear rests in part upon the fear that fire generates among people. Fear of fires is one of the strongest barriers to implementing the new policies of the National Park Service and the U.S. Forest Service to let fires burn naturally.

Large mammals are another source of transitory information; they are both a potential source of food and a source of danger. Although they may be present in an environment on a long-term basis, they are constantly on the move. The opportunities they provide depend upon their location at the moment, and these details, accordingly, demand immediate attention. It is not surprising, then, that we should enjoy watching large mammals and that we should find their behavior intrinsically interesting.

The power of large mammals to evoke positive emotional responses is well known. The nineteenth-century British landscape architect Humphrey Repton regularly graced the redesigned landscapes of his "Redbooks" with large ungulates as part of his strategy to acquire the business of potential clients even though the animals' presence was unrelated to any of his proposed modifications to the estates. Pastoral landscapes have long had a strong appeal to people from many cultures, and preservation of such landscapes regularly ranks high on conservation agendas. Furthermore, nature programs devote extensive time and money to portraits of large mammals, from whales and caribou to elephants and gorillas. Large mammals are also used in advertisements to add appeal to products as diverse as insurance and beer. Yet little scientific attention has been paid to the ways in which the presence of large mammals enriches environmental experience. Certainly, one of the more exciting aspects of a safari (or its surrogate in a trip to the zoo) is the sudden discovery of an animal. The visual search for "hidden" animals and other objects and the pleasure experienced in their discovery are clearly adaptive responses to our early ancestors' hunting and gathering life-style. As with sunsets, however, large hidden animals may trigger concern and fear as well as pleasure. A lion, unexpectedly encountered at a short distance, elicits fear in even the most experienced hunter. The same animal seen from a safe distance is the source of awe and delight.

Environmental Cues Associated with Seasonal Changes

Survival and reproductive success depend upon appropriate responses to seasonal changes in the environment. All natural environments, including tropical rain forests, experience important seasonal changes that affect the location and availability of resources. All traditional human cultures are rich in rituals that respond to seasonal changes and prescribe and proscribe behaviors appropriate or inappropriate to particular times of the year. The annual cycle is a powerful organizing theme for literary works. The power of Aldo Leopold's "A Sand County Almanac" results from both the ideas and concepts it espouses and the framework of changing seasons in which it is cast.

Responses to seasonal changes in the environment differ from those to the transitory events discussed earlier in four important ways. First, these responses may affect behavior for several months, as, for example, during a shift from a winter to a spring encampment. Second, the timing of the appropriate behavioral responses is highly predictable because the key environmental variables are driven by seasonal changes in day length, temperature, and rainfall whose timing varies little or not at all from year to year. Third, appropriate responses to these variables often involve anticipation of forthcoming changes and preparation for them before they are encountered. We may change to the spring encampment without having visited it that year or having any direct knowledge of conditions there. Fourth, there is usually much time in which to contemplate and plan for responses to seasonal cues.

Unfortunately, there is little research on seasonal responses to the environment. In fact, most research on environmental preferences deliberately eliminates differences in seasonal cues so that responses are made to landscapes photographed under the same conditions, usually at peak growing season.

Among the most important signals of seasonal changes are vegetative transformations of plants. The greening and browning of grass, germination and sprouting of seeds, leafing of woody plants, and autumnal coloration of deciduous trees and shrubs all provide powerful signals of changes in environmental conditions and resource availability. From an evolutionary perspective, however, we would expect the signals to be emotionally asymmetrical. That is, cues associated with productivity and harvest (greenness, budding trees, fruiting bushes) should be more positively received than cues associated with the dormant season (bare-limbed trees, brown grass). It may be difficult for many of us, with the year-round supplies of a wide array of fruits and vegetables in our supermarkets, to understand the importance of the first salad greens of the season to people throughout most of human history.

Reproductive responses of plants (e.g., flowering) also signal important changes in resource availability. For an organism that rarely eats flowers, it is perhaps surprising that we place such a high value on them and spend so much effort and money to have flowers in and around our dwellings and in city parks. The evolutionary biologist, however, sees flowers as signals of improving resources and as providing cues about good foraging sites some time in the future. In species-rich plant communities, flowers also provide the best way to determine the locations of plants that offer different resources. When not in bloom, all plants are green. (Indeed, many taxonomic distinctions can only be made by inspection of flowers.) Thus, paying attention to flowers should result in improved functioning in natural environments.

The strength of human emotional responses to flowers has been recognized and used in the developing field of therapeutic horticulture. The National Council for Therapy and Rehabilitation through Horticulture, founded in 1973, promotes and encourages the use of horticulture activities for therapy and rehabilitation. The success of these efforts is impressive (McDonald, 1976). The habit of bringing flowers to people in hospitals is not merely a friendly gesture. The presence of flowers in a hospital room may improve the mental state and rate of recovery of patients (Watson & Burlingame, 1960). Research on people's perceptions of their neighborhoods also shows the enormous appeal of flowers (Vernez-Moudon & Heerwagen, 1990). In the neighborhood study, residents accompanied the researchers on a predetermined route in three Seattle neighborhoods that varied considerably in their characteristics, especially in the type of houses. The study subjects were told to talk about what they liked and disliked about the places that were seen along the walk route. In each neighborhood, there were extensive comments on the landscapes—particularly the vegetation and flowers, which were very appealing to the subjects.

The appeal of flowers is also apparent in the enormous efforts expended on breeding plants to increase the sizes and numbers of floral parts and brilliance of coloration. Competition to excel in these efforts is widespread among both professionals and amateurs. Although the strength of our responses to flowers has long been appreciated, as far as we are aware, conceptual theories about these responses are totally lacking. An evolutionary approach to habitat selection offers a potentially powerful approach in this area of study.

Environmental Cues Influencing Long-Term Behavior

Whereas many environments are used for only short periods of time and for specific purposes, people also settle in one place for long time periods. Typically people invest heavily in construction and habitat modifications when they intend to occupy an area for a long time. This increases the costs associated with subsequent movement. For these reasons, we expect long-term settlement decisions to be influenced primarily by features of the environment that reflect its long-term safety and resource-provisioning potential. Short-term signals from the environment should be ignored unless they provide indications of longer-term consequences. For example, a current flood may indicate a high probability of future flooding.

Many features of landscapes are permanent, at least from the perspective of a human lifetime, and others change slowly enough that current conditions are reliable predictors of the relatively long-term future. Mountains, hills, valleys, rivers, and lakes do eventually erode away or fill up, but the rate at which they do so is very slow compared with the time frames relevant to our decisions about places in which to live. Vegetation changes occur more rapidly, but successional sequences from disturbance to climax may take up to several centuries. Habitat quality changes during such sequences, but rates of change are such that a good habitat today is likely to remain so for some time into the future, barring unforeseen catastrophic accidents.

Strong associations exist between vegetation patterns and availability of resources in those environments. This relationship has been exploited by Orians (1980, 1986) in his development of the "savanna theory" of environmental aesthetics, as discussed earlier. Evidence of aesthetic responses attuned to the savanna environment can be found in our manipulations of landscapes for aesthetic purposes (parks and gardens; see, e.g., Orians, 1986) and in the design of landscape paintings (Appleton, 1975).

Geological features of landscapes also exert powerful influences on human emotions. Indeed, America's national parks are centered around monumental land forms—mountains, canyons, cliffs, waterfalls, geysers, rivers, and caves. This preoccupation has been criticized by some observers who believe that parks should be established to preserve representative examples of the earth's life zones (Curry-Lindahl, 1974; Runte, 1979). Others have defended the park systems on the grounds that spectacular scenery stimulates use and that representative ecosystems, whatever their scientific value, will not serve the functions that our current types of parks do (Vale, 1988). Whatever one's perspective, it is clear that conservation and preservation values are subservient in our national parks to our preoccupation with the monumental. This is strong testimony to the powerful effect on human behavior of geological features that provide expansive views, which are important for learning about the environment and provide opportunities to view potential hazardous elements from a position of safety.

In addition to the geomorphological and vegetative features that signal the long-term quality of environments, people, as well as other organisms, also rely upon evidence of the presence of conspecifics for further information about an environment. Such information may be either positive or negative. On the positive side, signs of human occupancy suggest that other people have evaluated and selected this habitat and have survived in it. This is prima facie evidence of its suitability. Such signals as bridges and paths, suggesting safe ways of moving through the environment, also provide evidence of a suitable place. And, indeed, paths and bridges are frequently used in photography and landscape paintings to evoke positive feelings. On the negative side, the presence of other people may be associated with crowding or depleted resources. Reactions are, therefore, likely to depend on the apparent density and impacts of prior residents on the site. Simulation research by Chambers (1974) supports this prediction.

FROM SIGNAL TO SYMBOL: THE APPLICATION OF EVOLUTIONARY THEORY TO SPECIFIC ENVIRONMENTAL AESTHETIC RESPONSES

In our evolutionary past, the environmental features and events discussed in this paper warned us of hazards and threats, as well as opportunities and resources that supported life and provided comfort. If attention to these stimuli and events is a cross-cultural universal, as seems likely, many of these ecological signals should have been transformed, over time, into cultural events and artifacts that are used to manipulate aesthetic experiences. In the section that follows, we discuss specific research areas that deal with applications of evolutionary-ecological theory to environmental aesthetics.

Prospect-Refuge Theory

In its simplest terms, prospect-range theory predicts that people should prefer places and environments that allow opportunities to see without being seen (Appleton, 1975). Furthermore, under conditions of perceived hazard, the desire for refuge should be heightened. Specific predictions that result from prospect-refuge theory are that (a) people should prefer edges more than the middle of a space because edges provide the best visual access to an area; (b) spaces that provide something over the head (a roof, tree canopy, trellis, etc.) should be preferred over spaces that provide only a back or side surface; (c) spaces protected at the back or side should be preferred over those without any vertical surface (e.g., those exposed to the view of others on all sides); (d) if one wants to be seen, then sitting in the middle of a space without intervening surfaces would be preferred because this area provides the greatest exposure; (e) an environment will be judged as more pleasant if it contains a balance between prospect and refuge opportunities, with screening elements to achieve privacy and variability in desired levels of intimacy in a space (see Thiel, Harrison, & Alden, 1986); and (f) preferred spaces should contain multiple view opportunities from most locations and multiple ways of moving through the space; these features encourage environmental surveillance and escape.

Urban Spaces

Interestingly, popular small-scale urban spaces contain these features (Whyte, 1980). However, most of the research in this area has been descriptive rather than theoretical. Spaces are often chosen for analysis on the basis of their availability, rather than on variability in their spatial features or other parameters that are expected to influence preference patterns. Further, the absence of theory to guide research in the designed environment has lead to the accumulation of specific data that cannot be generalized in a way that is useful to the design of new spaces or the renovation of existing ones. The evolutionary-ecological approach we propose in this paper provides a framework for investigating these environments. However, much research needs to be done before it is possible to determine the relative contribution of different features to preference patterns. We don't know, for instance, whether a small grove of trees is a better investment than a fountain, or how to manipulate the balance between prospect and refuge opportunities to achieve different psychological experiences. These questions are amenable to study, however. Techniques such as Archea's (1984) visual-exposure/visual-access model or Benedikt's (1979) isovist process can be used to study the ways in which specific architectural or natural features influence visibility and usage patterns (Heerwagen, 1989). Analysis of the order in which people occupy particular locations will provide information on which subspaces or seating areas are most popular and how the presence of other people influences the responses of newcomers.

Architectural Application

In an intriguing psychological profile of Frank Lloyd Wright's houses, Hildebrand (1991) uses prospect-refuge theory to explain the consistent appeal of Wright's architecture. According to Hildebrand, Wright's houses have a basic motif that mixes the drama of discovery with a strong sense of hominess. Unexpected views and refuge opportunities abound, from the front gate through the backyard of a Wright house. An internal, contained fireplace with a lowered ceiling and glass doors or windows opposite gives a strong sense of refuge, balanced by the opportunity to see out and survey the surrounding environment. There are internal views of varying size and penetration throughout the houses, and terraces with deep overhangs that afford prospect opportunities from the vantage of safety. In one of Wright's most famous houses, "Falling Waters," the psychological impact of refuge is enhanced by the hazard symbolism of a gorge and waterfalls that literally surround the house.

Wright's consistent use of changes in ceiling elevation and the placement of major living spaces directly under the roof both open up the space visually and create the comfortable sensation of living under a tree canopy. The sense of refuge and protection that one feels under a spreading tree canopy is certainly consistent with an evolutionary approach to aesthetics. Trees were likely to have been used by early humans to escape from sudden hazards, such as dangerous animals, and as protection from sun and rain. In areas lacking topographic relief, a tree also provided a readily accessible viewing platform for surveying the surrounding environment. Although we may lose the tree-climbing desire as adults, trees play a significant role in children's play behaviors (Hart, 1980).

Despite the intuitive appeal of trees, very little research has dealt with human responses to trees. An evolutionary-ecological approach to aesthetics suggests that the incorporation of trees and tree forms, actual or symbolic, into the built environment should have a strong positive impact on people. In support of this prediction, Rachel Kaplan found that people whose view from an apartment window included large trees were more satisfied with both their physical and social environment than people whose views were dominated by built features or grassy expanses (R. Kaplan, 1983). Further, a study of hospital recovery indicates that patients whose windows looked out upon a small grove of deciduous trees had a more positive post-surgical recovery, including a shorter hospital stay, than a matched control group of patients whose windows looked out upon a building (Ulrich, 1984).

Further research should investigate such factors as tree shapes, the placement of trees to enhance mystery and refuge in an environment, and the symbolic use of tree shapes in the built environment. For instance, tree canopies appear symbolically in many aspects of design, including sloped ceilings, trellises, awnings, porches, and building overhangs, particularly those with pillars. We predict that the presence of these "symbolic trees' is associated with positive response to built environments.

Despite the intuitive appeal of prospect-refuge theory, research on refuge preferences has produced some mixed results (Woodcock, 1982). This is partially due to the fact that it is difficult to assess refuges in photographs (the preferred medium for landscape studies). A cluster of bushes or a thick stand of trees is certainly a refuge once one is inside, but they may be hazardous to approach, and they also block potential views in a photograph. The use of behavioral analysis to assess responses to a refuge may yield quite different results, in part, because one may have to experience a refuge to appreciate its value.

Photography and Paintings

Prospect, refuge, and other features of preferred landscapes can also be studied in other media, including paintings and photographs. Photography and landscape painting, for instance, share many common features. The views they provide are artificially bounded and two dimensional. In addition, the observer can never enter the environment even though the presentation may be designed to foster contemplation about entering it. Every person views a photograph or painting for the first time. If we assume that the artist wishes the viewer to pause on the first encounter and to want to view the scene more than once, then photos and paintings of environments should have features that evoke instantaneous positive responses. Appleton's (1975) analysis of landscape paintings shows that this is very much the case. Common photographic devices such as framing pictures (creating the impression that the observer is in a refuge), use of shadows (indicating a time of day when changes in behavior are likely to be required), and inclusion of cloud formations (signaling potential changes in weather) serve to emphasize those cues that should have evolved as immediate attention-getters. The environments portrayed need not be ones that evoke desires to engage in extensive exploration. They may be designed to evoke fear or sensations of the sublime, but above all, they must command our immediate attention.

For those photos and paintings designed to appeal to the exploratory phase of the habitat selection process (Stage 2), effective devices include evidence that exploration can take place in relative safety and evidence that abundant resources are likely to be found if exploration does ensue (e.g., presence of water, large mammals, or indicators of prior human occupation). However, to motivate the desire to view the art many times requires devices that evoke the feeling that there is always more to be learned. This component of environmental aesthetics is the heart of what the Kaplans call "mystery," creating expectations that cannot be fulfilled by repeated viewing but which may make us continually curious about what we might find if we were able to actually enter and explore the environment (see S. Kaplan, 1987 for a review).

The Role of Mystery in Environmental Aesthetics

As defined by S. Kaplan (1987), "mystery" is a characteristic of landscapes that draws the viewer into the scene, making him/her want to find out more. The winding road that goes out of sight behind a bank of trees or a small hill has proven to be an excellent predictor of landscape preferences. The desire to know what's around the hill or over the horizon is likely to have evolutionary roots. The sense of mystery would facilitate learning about the landscape, a highly adaptive behavior for our hunting and foraging ancestors who probably moved long distances on a regular basis.

Despite its positive role in environmental aesthetics, mystery has a dark side. That which is out of sight may be dangerous (see Herzog, 1988). This suggests that environments high in mystery should product a feeling of apprehension as well as of interest.

In dangerous settings, apprehension may come to the forefront, generating feelings of anxiety and stress. This is exactly what Newman (1972) found in his studies of New York housing developments. Housing residents avoided places where they did not have high visual access of the entire space. Further, crime rates and vandalism are likely to be much higher in places where visual surveillance is inadequate due to blocked views, winding paths, or changes in light availability. The little study there has been on this topic suggests that it is a fruitful area for further research on the relationship between the physical environment and crime, including vandalism (Wise, 1983) and violent crimes, particularly rapes (Stoks, 1983).

Unfortunately, the need for visual access frequently leads to extreme environments that totally lack visual appeal. Research is needed on ways of incorporating the positive aspects of mystery with the need for adequate visual surveillance of spaces. Solutions such as increasing the angle at which the path turns, using trees instead of dense shrubbery, using shrubbery with thorns to discourage hiding, and providing numerous lookout points should be studied from both a safety and aesthetic perspective. Furthermore, in environments where low levels of mystery are desirable for safety reasons, other means of increasing aesthetic responses should be encouraged, such as the use of flowers, water features, and color. At the present time, we know little about the relative value of these various aesthetic features in an environment. Questions of relative merit of design alternatives are becoming increasingly important in urban settings where spaces are limited, potential dangers are high, and funds for aesthetic features are often limited.

Age-Related Changes in Environmental Aesthetics

If our aesthetic responses are based, in part, on behavioral ecology, as we suggest in this paper, then we would expect to find fundamental age-related differences in the way humans respond to the environment. The basic perceptual and cognitive mechanisms underlying environmental response change with age as do physical abilities. An evolutionary approach begins by asking what influences survival and reproductive success most strongly at different ages. In the most general terms, both survival and success depend on how well an organism negotiates the balance between the hazards and opportunities it faces on a regular basis.

Although there is virtually no research on age-related stages of environmental aesthetics, the ecological perspective presented here predicts that age-related environmental transitions should include (a) changes in the characteristics of preferred environments; (b) changes in the kind of information that is most useful in terms of adaptive functioning; (c) differences in how environmental information is obtained and used; and (d) differences in the effects of the environment on psychological functioning. These predictions are based on the notion that humans are adapted organisms at each stage in their life: Development does not simply lead up to a final well-adapted "adult" form.

As a child's normal range of movement in the environment expands from infancy through adolescence, so, too, do the kinds of hazards and opportunities it encounters. Successful ventures into the environment depend upon paying attention to stimuli and events that are the most "useful" to a child for its age. The usefulness of information relates to changes over time, relative to the physical, cognitive, and social abilities characteristic of different life stages.

For instance, we would not expect a 3-year-old child to pay much attention to sunsets or other time cues such as lengthening shadows. This is because very young children seldom venture far from home on their own. Thus knowing what time it is, in terms of its "usefulness" in motivating needed changes in behavior, is simply not relevant to a very young child. On the other hand, it is important for a 3-year-old to be able to separate edible from inedible or toxic objects. Thus, a concern with properties and attributes of objects that separate them into good-bad, edible-inedible, ripe-not ripe would be highly adaptive. Furthermore, because young children cannot be expected to know, in advance, the consequences of eating the wrong thing, it is advantageous for them to be in close contact with an adult who does know what the expected outcomes are. The young child uses the older person's knowledge as a way of learning the appropriate response. It is not surprising then that young children are fascinated with small objects and enjoy collecting and playing with them.

Future studies need to address such basic questions as what kind of environments or environmental stimuli do people prefer at different ages? What do they reject as uninteresting? At what point in life do children begin to attend to distant stimuli such as views and sunsets? How do responses to refuge and prospect dimensions of the environment change over time? Why do adults so readily experience feelings of nostalgia and longing for the landscapes of childhood? How—and why—does environmental experience transform from the hands-on, concrete experience of childhood to the more abstract, often spiritual, quality of adult responses to the environment?

CONCLUDING REMARKS

We have attempted to illustrate the richness of concepts and hypotheses that can be generated from an evolutionary perspective on environmental aesthetics. Although many hypotheses have already been developed to varying degrees, we have just begun to scratch the surface of the possibilities inherent in this approach. Testing of these ideas has just begun, but already support for some hypotheses has been generated. Some preliminary results have demonstrated the need to modify and expand earlier predictions. This mutually stimulating interaction between theory and empirical testing can be continued profitably for a long time. Until recently, study of responses to environments has been a strongly empirical venture. The scarcity of theoretical concepts to guide research has narrowed the scope of investigations and generated a body of data that is difficult to interpret. Empirical generalizations are useful, but they lack the explanatory and predictive power of causal hypotheses.

We do not suggest that an evolutionary-adaptive approach to environmental aesthetics is the only way to proceed. However, such an approach can enrich studies from a variety of perspectives and in a wide range of topics. Indeed, we have suggested applications of this approach to the ontogeny of the aesthetic sense, to design of buildings and public spaces, and to analyses of works of art. Unless one denies that our emotional responses evolved in part to enable us to function more effectively in our environments, it is difficult to see how this approach cannot be useful. Nonetheless, the task before us is not easy. Good evolutionary theories are difficult to formulate and test. Many apparently promising ideas will be discarded on the intellectual trash heap that accompanies the progress we all seek. However, this is the pattern of all scientific research, and the discarded ideas may well have played a key role along the path to their demise.

REFERENCES

Aldworth, R. C. (1971). Design for variety in lighting. *Lighting Research and Technology, 3*(1), 8–23.

Appleton, J. (1975). *The experience of landscape.* New York: Wiley.

Appleton, J. (1990). *The Symbolism of Habitat.* Seattle, Washington: University of Washington Press.

Archea, J. (1984). *Visual access and exposure: An architectural basis for interpersonal behavior.* Unpublished doctoral dissertation, Pennsylvania State University, College Park, PA.

Balling, J. D., & Falk, J. H. (1982). Development of visual preference for natural environments. *Environment and Behavior, 14*(1), 5–28.

Baron, R. M. (1981). Social knowing from an ecological-event perspective: A consideration of the relative domains of power for cognitive and perceptual modes of knowing. In J. H. Harvey (Ed.), *Cognition, social behavior, and the environment* (pp. 61–89). Hillsdale, NJ: Erlbaum Associates.

Beach, L. R. (1990). *Image theory: Decision making in personal and organizational context.* Chichester, England: Wiley.

Benedikt, M. L. (1979). To take hold of space: Isovists and isovist fields. *Environment and Planning, 6,* 47–65.

Berlyne, D. E. (1971). *Aesthetics and psychobiology.* New York: Appleton-Century-Crofts.

Bernstein, A. S. (1968). The orienting response and direction of stimulus change. *Psychonometric Science, 12*(4), 127–128.

Bernstein, A. S. (1969). To what does the orienting response respond? *Psychophysiology, 6*(3), 338–350.

Bernstein, A. S. (1979). The orienting response as novelty *and* significance detector: Reply to O'Gorman. *Psychophysiology, 16*(3), 263–273.

Blanchard, D. C., & Blanchard, R. J. (1988). Ethoexperimental approaches to the biology of emotion. *Annual Review of Psychology, 39,* 43–68.

Campbell, B. (1985). *Human evolution* (3rd ed.). New York: Aldine.

Chambers, A. D. (1974). Assessment of environmental quality in relation to perceived density in residential settings. *Man-Environment Systems, 4*(6), 353–360.

Charnov, E. L. (1976). Optimal foraging: The marginal value theorem. *Theoretical Population Biology, 9,* 129–36.

Cody, M. L. (1985). An introduction to habitat selection in birds. In M. L. Cody (Ed.), *Habitat selection in birds.* (pp. 4–56). New York: Academic Press.

Collins, B. L. (1975). *Windows and people: A literature survey.* Building Science Series, 70. (National Bureau of Standards). Washington, DC: U.S. Government Printing Office.

Cosmides, L., & Tooby, J. (1987). From evolution to behavior: Evolutionary psychology as the missing link. In J. Dupré (Ed.), *The latest on the best: Essays on evolution and optimality.* Cambridge, MA: MIT Press.

Curry-Lindahl, K. (1974). *The global role of national parks for the world of tomorrow.* Paper given for the Fourteenth Horace M. Albright Conservation Lectureship, University of California, School of Forestry and Conservation, Berkeley, CA.

Flynn, J. E., Spencer, T. J., Martyniuk, O., & Hendrick, C. (1973, October). Interim study of the procedures for investigating the effect of light on impression and behavior. *Journal of the Illuminating Engineering Society,* pp. 87–94.

Halloway, C. A. (1979). *Decision-making under uncertainty: Models and Choices.* Englewood Cliffs, NJ: Prentice-Hall.

Hart, R. (1980). *Children's experience of place.* New York: Irvington.

Heerwagen, J. (1986). *Windowscapes: The role of nature in the view from the window.* Paper presented at the *Second International Daylighting Conference,* Long Beach, CA, November, 1986.

Heerwagen, J. (August, 1989). *The psycho-evolutionary aspects of windows and window design.* Paper presented at the American Psychological Association Conference, New Orleans, LA.

Heerwagen, J., & Orians, G. (1986). Adaptations to windowlessness: A study of the use of visual decor in windowed and windowless offices. *Environment and Behavior, 18*(5), 623–639.

Herzog, T. (1988). Danger, mystery, and environmental preference. *Environment and Behavior, 20*(3), 320–344.

Hildebrand, G. (1991). *The Wright space.* Seattle: University of Washington Press.

Hildén, O. (1965). Habitat selection in birds. *Annals, Zoologica Fennica 2,* 53–75.

Humphrey, N. K. (1980). Natural aesthetics. In B. Mikellides (Ed.), *Architecture for people* (pp. 59–73). London: Studio Vista.

Janis, I. L., & Mann, L. (1977). *Decision making: A psychological analysis of conflict, choice, and commitment.* New York: Free Press.

Kahneman, D. (1973). *Attention and effort.* Englewood Cliffs, NJ: Prentice-Hall.

Kaplan, R. (1983). The role of nature in the urban context. In I. Altman & J. F. Wohlwill (Eds.), *Behavior and the natural environment,* (pp. 127–161). New York: Plenum Press.

Kaplan, S. (1987). Aesthetics, affect, and cognition: Environmental preference from an evolutionary perspective. *Environment and Behavior, 19*(1), 3–32.

Kaplan, S., & Kaplan, R. (1982). *Cognition and environment: Functioning in an uncertain world.* New York: Praeger.

Keep, P. J. (1977). Stimulus deprivation in windowless rooms. *Anaesthesia, 32,* 598–600.

Keep, P. J., James, J., & Inman, M. (1980). Windows in the intensive therapy unit. *Anaesthesia, 35,* 257–262.

Konner, M. (1982). *The tangled wing: Biological constraints on the human spirit.* New York: Holt, Rinehart & Winston.

Lack, D. (1971). *Ecological isolation in birds.* Oxford: Blackwell.

Lee, R., & DeVore, I. (Eds.). (1968). *Man the hunter.* Chicago: Aldine.

Lee, R., & DeVore, I. (Eds.). (1976). *Kalahari hunter—gatherers.* Cambridge, MA: Harvard University Press.

Levins, R. (1968). *Evolution in changing environments.* Princeton, NJ: Princeton University Press.

Lovejoy, C. O. (1981). The origin of man. *Science, 211,* 341–350.

MacArthur, R. H., & Pianka, E. R. (1966). On the optimal use of a patchy environment. *American Naturalist, 100,* 603–609.

Marr, D. (1982). *Vision: A computational investigation into the human representation and processing of visual information.* San Francisco: Freeman.

McDonald, E. (1976). *Plants as therapy.* New York: Praeger.

Newman, O. (1972). *Defensible space.* New York: Macmillan.

O'Gorman, J. G. (1979). The orienting reflex: Novelty or significance detector? *Psychophysiology, 16*(3), 253–262.

Öhman, A. (1986). Face the beast and fear the face: Animal and social fears as prototypes for evolutionary analyses of emotion (Presidential Address, 1985). *Psychophysiology, 23*(2), 123–143.

Orians, G. (1980). Habitat selection: General theory and applications to human behavior. In J. S. Lockard (Ed.), *The evolution of human social behavior* (pp. 49–66). Chicago: Elsevier.

Orians, G. (1986). An ecological and evolutionary approach to landscape aesthetics. In E. C. Penning-Rowsell & D. Lowenthal (Eds.), *Landscape meaning and values* (pp. 3–25). London: Allen & Unwin.

Platt, J. R. (1961). Beauty: Pattern and change. In D. W. Fiske & S. R. Maddi (Eds.), *Functions of varied experience* (pp. 402–430). Homewood, IL: Dorsey Press.

Partridge, L. (1978). Habitat selection. In J. R. Krebs & N. B. Davies (Eds.), *Behavioural ecology: An evolutionary approach* (pp. 351–76) Sunderland, MA: Sinauer Associates.

Relph, E. (1976). *Place and placelessness.* London: Pion.

Repton, H. (1907). *The art of landscape gardening.* Boston: Houghton-Mifflin.

Rosenzweig, M. L. (1974). On the evolution of habitat selection. In Proceedings of the First International Congress of Ecology (pp. 401–404) Wageningen.

Rosenzweig, M. L. (1981). A theory of habitat selection. *Ecology, 62,* 327–335.

Runte, A. (1979). *National parks: The American experience.* Lincoln, NE: University of Nebraska Press.

Simon, H. A. (1957). *Models of man: Social and rational.* New York: Wiley.

Sokolov, E. N. (1963). *Perception and the conditioned reflex.* New York: MacMillan.

Staddon, J. (1987). Optimality theory and behavior. In J. Dupré (Ed.), *The latest on the best: Essays on evolution and optimality* (pp. 99–118). Cambridge, MA: MIT Press.

Stoks, F. G. (1983). Assessing urban space environments for danger of violent crime—especially rape. In D. Joiner, G. Brimilcombe, J. Daish, J. Gray and D. Kernohan (Eds.). *Proceedings of the Conference on People and the Physical Environment Research* (pp. 331–343). Ministry of Works and Development: Wellington, New Zealand.

Thiel, P., Harrison, E. D. & Alden, R. (1986). The perception of enclosure as a function of the position of architectural surfaces. *Environment and Behavior, 18*(2): 227–245.

Thornes, J. E. (1984). A reassessment of the relationship between John Constable's meterological understanding and his painting of clouds. *Landscape Research, 9,* 20–29.

Tuan, Y. (1974). *Topophilia.* Englewood Cliffs, NJ: Prentice Hall.

Tuan, Y. (1979). *Landscapes of fear.* New York: Pantheon.

Ulrich, R. (1979). Visual landscapes and psychological well-being. *Landscape Research, 4*(1): 17–23.

Ulrich, R. (1983). Aesthetic and affective response to natural environment. In I. Altman & J. F. Wohlwill (Eds.), *Behavior and the natural environment* (pp. 85–125). New York: Plenum.

Ulrich, R. (1984). View through a window may influence recovery from surgery. *Science, 224,* 420–421.

Ulrich, R. (1986). Human response to vegetation and landscapes. *Landscape and Urban Planning, 13,* 29–44.

Ulrich, R., Simons, R. F., Losito, B. D., Fiorito, E., Miles, M. A. & Zelson, M., (in press). Stress recovery during exposure to natural and urban environments. *Environmental Psychology.*

Vale, T. (1988). No romantic landscapes for our national parks? *Natural Areas Journal, 8,* 115–117.

Vernez-Moudon, A. & Heerwagen, J. (1990). *Resident's attitudes toward design diversity.* Research report. Washington, DC: National Endowment for the Arts.

Washburn, S. L. (Ed.). (1961). *Social life of early man.* Chicago: Aldine.

Watson, D., & Burlingame, A. W. (1960). *Therapy through horticulture.* New York: Macmillan.

Whyte, W. H. (1980). *The social life of small urban spaces.* Washington, DC: Conservation Foundation.

Wilson, E. O. (1984). *Biophilia.* Cambridge, MA: Harvard University Press.

Wilson, L. M. (1972). The effects of outside deprivation on a windowless intensive care unit. *Archives of Internal Medicine, 130,* 225–226.

Wise, J. A. (1983, January-February). Designing against vandalism: A threefold way. *The Recreation Reporter,* pp. 8–10.

Wohlwill, J. F. (1983). The concept of nature. In I. Altman & J. F. Wohwill (Eds.), *Behavior and the natural environment* (pp. 5–37). New York: Plenum.

Woodcock, D. (1982). *A functionalist approach to environmental preference.* Unpublished doctoral dissertation, University of Michigan, Ann Arbor, MI.

Zajonc, R. (1980). Feeling and thinking: Preferences need no inferences. *American Psychologist, 35*(2), 151–175.

16

Environmental Preference in a Knowledge-Seeking, Knowledge-Using Organism

STEPHEN KAPLAN

If an individual from outer space were to hire a consultant in order to learn about what sort of creatures these earthlings be, the particular scholarly expertise of the consultant could have a substantial effect on the nature of the report. Given the current configuration of academic disciplines, it is not hard to imagine that a consultant affiliated with one group of scholars might emphasize such themes as sexuality, warfare, and deception. Alternatively it would not be surprising to find a consultant who would place primary emphasis on humans as processors of information, as builders of knowledge structures. Such a scholar might emphasize themes such as attention, memory, and limited processing capacity.

At one level it might be considered remarkable that both consultants thought they were talking about the same organism. But the contrast between these positions is at least as instructive as it is remarkable. Of particular interest is the fact that the difference between these two hypothetical scholarly groups is multidimensional; it concerns not only content, but explanatory principles as well (Table 16.1).

Several aspects of Table 16.1 are noteworthy. The unlabeled cells represent areas of potential integration that have been largely neglected. The evolution and cognition domains represent two apparently polar positions that have only recently begun to receive attention (cf. Cosmides & Tooby, 1987; Tooby & Cosmides, this volume). The table also reflects (accurately, I believe) the relatively sharp line that is often drawn between cognition and affect. This too leads to ignoring some fascinating and poten-

Table 16.1 A Framework for Comparing Alternative Approaches to the Understanding of Human Behavior

Basis of explanation	Content emphasis	
	Affect	Cognition
Evolution	Sociobiology	
Information processing		Cognitive Science

tially integrating topics. One such topic, the study of environmental preference, is the subject of this paper. By intentionally straddling the implicit boundaries in Table 16.1, this paper attempts to demonstrate the value of a more synthetic and inclusive approach to the understanding of human behavior.

As a preparation for building these various bridges the discussion begins with an analysis of the role of information in human evolution. Wayfinding is then introduced as an important activity of early humans with interesting informational properties. The discussion of an extensive program of research on human environmental preference will serve as a window on the motivational inclinations that encourage the acquisition and utilization of wayfinding information.

HUMANS, EVOLUTION, AND INFORMATION: TOWARD A SYNTHESIS

As treated here, the topic of preference falls under the larger domain of affective biases toward patterns of information. In this framework not only information in its own right, but the concern for information is considered a basic part of the human makeup. "Concern" as it is used here is intended to cover a wide spectrum of human affective relationships. There is the motivation to seek information, to be "the first to know," and to discuss it with others. There is the distress that comes from struggling with information that is hard to understand, or inappropriate to the current situation, or not what one expected. There is the joy of recognizing and predicting despite uncertainty, or using information confidently and effectively. And there are undoubtedly numerous other human affective relationships to information that remain to be identified and conceptualized.

It is reasonable to question whether this concern for information is a human characteristic of long standing, a matter of widespread interest and importance for the species at large, or, alternatively, simply a characteristic perversion of a small group of scholars. Is there any basis for believing that knowledge and the concern for knowledge have any role in the evolutionary scheme of things?

Perhaps an appropriate place to start is with the selection pressures facing our human ancestors. A common caricature of the early human is as a dim-witted hunter, an individual more concerned with the power of muscle than the power of mind. The view presented by Laughlin (1968) provides a striking contrast to this caricature. Laughlin argues that the weapons of early humans were limited to an effective range of less than 30 feet. It thus became essential to approach relatively closely to the intended quarry, placing considerable emphasis on skill in stalking. This, in turn, required vast knowledge on the part of the hunter:

> The hunter is concerned with the freshness of the track and the direction in which he is moving. He wants all possible information on the quarry's condition: its age, sex, size, rate of travel, and a working estimate of the distance by which the animal leads him. In the final stages, when he is closing with the animal, the hunter employs his knowledge of animal behavior and situational factors relevant to that behavior in a crucial fashion. For all birds, animals, and fish the hunter must estimate flight distance, the point at which they will take flight or run away. Conversely, with animals that are aggressive, he needs to interpret any signs, raising or lowering of tail, flexing of muscles, blowing, or salivation, etc., that indicate an attack rather than a flight. The variations are innumerable. (pp. 308–309)

From this description it is apparent that concern for knowledge acquisition cannot be motivationally neutral. A comparable argument has been made by Flannery (1955) with respect to gathering. Knowing where to find potential food, knowing when to search for it so that it would be ripe but not yet taken by competing species, and finding one's way back home again are all central informational aspects of the gathering process.

Peters and Mech (1975) suggest that predation by early humans might have emphasized trapping rather than hunting. In either case, however, they point to the centrality of cooperative behavior, the use of strategy, and environmental knowledge. All three of these components are, in their analysis, essential to what they call the "intellectual hunter" niche. All three, in other words, rely heavily on information.

A strikingly similar emphasis on information has been described among the Kalahari Bushmen. Tulkin and Konner (1973) indicate not only that the Bushmen possess a remarkable amount of knowledge about animal behavior, but that their methods of evaluating knowledge are equally impressive. As an example, they describe the distinctions made among the following types of evidence:

> (1) I saw it with my own eyes; (2) I didn't see it with my eyes but I saw the tracks. Here is how I inferred it from the tracks; (3) I didn't see it with my own eyes or see the tracks but I heard it from many people who saw it (or a few people who saw it, or one person who saw it); (4) It's not certain because I didn't see it with my eyes or talk directly with people who saw it. (p. 35)

In *The Creative Explosion,* Pfeiffer (1982) puts cave paintings into a similar context. He offers a persuasive argument for the hypothesis that these were means of storing and transmitting information central to the cultures that produced them.

These various pieces of evidence, while far from overwhelming, suggest the likely significance of information to human evolution. Given the relative silence of many investigators with respect to this question, one might wish for stronger evidence. On the other hand, the name *Homo sapiens* presumably refers to a quality that arose not by chance, but because of selection pressure.

These bits and pieces of evidence supporting the significance of information in human evolution are intended to serve two functions: On the one hand, they point to the capacity of early humans to carry out such information-based processes as recognition, prediction, evaluation, and action (S. Kaplan, 1973). On the other hand, they suggest the likelihood that strong affect can be associated with these processes.

Consider, for example, the prototypic information-processing activity of attempting to recognize an object. A mismatch between an internal representation and the sensory pattern arising from the object (or, for that matter, a good match achieved under difficult conditions) has powerful motivational implications. Such signal events attract *attention,* increase *arousal,* and affect the *pleasure/pain* system, thus involving all three domains classically covered under the "motivation and emotion" heading. Speaking more broadly, informational states such as being lost, confused, or disoriented not only have affective consequences, but must have such consequences in terms of their adaptive implications. An information-oriented organism that did not find confusion disturbing might be content to spend considerable time confused. Such an organism would, in the words of a colleague, be "easy to eat."

THE LANDSCAPE AS AN EVOLUTIONARILY SIGNIFICANT
INFORMATIONAL PATTERN

The evolutionary significance of landscape arises out of the convergence of a number of factors. That early humans were knowledge-dependent is only part of the story. Being a far-ranging and yet home-based organism placed considerable priority on way-finding. Covering a territory of 100 square miles or more made that skill not only important, but challenging. Exploration of one's environment in order to know it well enough to range widely and yet not get lost was thus an important element in a larger survival pattern.

Locomotion through an unfamiliar environment brought with it survival implications in addition to the demands of wayfinding. Such an activity would necessarily be accompanied by considerable potential uncertainty. Where one ventured could be survival enhancing; it could also be survival threatening. Such issues as how well one could see and how easily one could be seen may also have entered into the human relationship to landscape.

The biases and inclinations serving these wayfinding concerns lack the dramatic impact of fight or flight, or of food and sex. They would not necessarily have *direct* and *immediate* payoff in terms of differential survival. Nonetheless, in the long run these biases could have subtle, yet pervasive implications. Wayfinding across wide territories would be enhanced. Knowledge about the location of critical resources such as water or edible plant material would be more extensive. Knowing the terrain might lead to a faster and more appropriate response in an emergency; it would also make it far more likely that an individual would be able to return to home base.

The emphasis on knowledge acquisition contrasts sharply with the so-called drive-reduction position once so dominant in theories of human behavior. According to the latter position, the organism is motivated to respond to the pressure of whatever drive is dominant at the moment. Lacking such internal pressure, the organism presumably could curl up and go to sleep. In terms of the perspective taken here, however, when drives are satisfied, when nothing else is going on would be precisely the time when the individual could be obtaining information about the environment for use at a later time. Such information need not be related to any particular outcome, or to outcomes of any kind. Humans store knowledge about their physical environment even when its use is not necessarily evident at the time. The potential use is, however, vast. The knowledge about a particular environment, gained at leisure, might be used many times over in the course of an individual's lifetime, while some knowledge so gained might never be used at all. Acquisition of knowledge is thus, in this sense, speculative.

As we have seen, learning about the surrounding environment was likely to have been a matter of considerable interest and concern to early humans. At the same time the very importance of this knowledge created a potential conflict. On the one hand, acquisition of important new information would be greatly facilitated if the individual preferred routes that would enhance the chances for such learning. In other words, it would have been adaptive if, given a choice of routes, the individual were to favor the one offering something new, rather than traveling on an already familiar route.

On the other hand, the dependence upon knowledge for functioning effectively can be particularly problematic when exploring new environments, where knowledge or comprehension of one's surroundings is lacking. There is necessarily an adaptive pay-off to avoiding environments where one would not know how to navigate appropriately. This potential conflict between seeking knowledge and avoiding what is new and hard to comprehend is what Hebb (1972) identified as "man's ambivalent nature." At the very least it requires sensitivity to settings that offer new information at a controlled rate and within a broad framework that adequately supports one's competence. And, from an affective point of view, this dual concern places as high a premium on the understanding of an environment as it does on the environment's capacity to offer new information.[1]

One further factor affecting the adaptive relationship of the early human to landscape is the limited processing capacity of the human mind. Considerable evidence points to the fact that contemporary humans have a limited capacity to comprehend and respond to information; this was presumably true of our ancestors as well. The early human, as we have seen, had many things to think about while traversing the landscape.

Because of this capacity limitation, being preoccupied with environmental choice and decision making about routes could interfere with other information processing of adaptive importance. An individual so preoccupied would be less vigilant and would have less processing capacity available for other matters. It would seem reasonable that selection favored a capacity to assess environments not only rapidly but also in a way that would not compete with conscious processing. This capacity should, in other words, be under automatic control most of the time, very much like the control of the breathing process. The assessment of environments should be immediate and intuitive rather than conscious and deliberate.

A MECHANISM FOR APPROPRIATELY RESPONDING TO ENVIRONMENTS

There is, thus, reason to believe that selection pressures in early humans favored acquiring new information about one's environment while not straying too far from the known. If it is adaptive to make such choices, one would expect them to be part of the human affective makeup. They should, in other words, be matters about which humans have strong feelings. Further, whatever mechanism is responsible for this frequently updated evaluation of the environment needs to function rapidly and unobtrusively.

One domain that considers the affective aspects of informational patterns is aesthetics. Although aesthetics in the narrow sense is sometimes viewed as an elitist concern, in the current context it refers to a broad and widely shared inclination that is concerned not with the contents of museums but with the realities of the outdoor environment. In other words, aesthetics is seen as applying not only to a broad population but to a broad class of stimulus patterns as well. Aesthetics in this perspective is a functionally based way of responding to the environment.

The original focus of the body of research described here was to understand landscape aesthetics—to study, in other words, what humans find beautiful in the outdoor environment. The evolutionary perspective that now characterizes our work in this area was not a part of the original plan; rather, it emerged gradually as the data accumulated and the theory began to crystalize. The research discussed here has been guided by Rachel Kaplan and myself and includes work by many of our students.

The Preference Model

While the survival requirements of contemporary humans differ in many ways from those of our ancestors, in many respects the story has not changed dramatically. One must still negotiate the physical environment, assess lurking threats and dangers, and concern oneself with finding one's way back. Nor have humans ceased to be information-based animals, continuously struggling to make sense of their surroundings and exploring new adventures.

Although in retrospect these functional and evolutionary issues seem to fit the findings of research on preference to a remarkable degree, they were by no means as obvious 20 years ago. Rather than evolutionary concerns, two other major questions motivated the early research. The first involved what we called "content"—did the category or type of environment matter in what people liked? The second issue focused on our doubt that Complexity, the attribute that had received substantial psychological attention (Berlyne, 1960; Day, 1967; Vitz, 1966), was the single most important predictor of preference. Although psychologists had done relatively little research focused on the environment, they had studied "experimental aesthetics." Wohlwill's (1968) study was an exciting venture because he included scenes of the physical environment instead of the artificially constructed stimulus patterns that had dominated this research area. His conclusion, however, surprised us. Based on only 14 scenes of diverse settings he reported that environmental patterns, much like experimentally contrived ones, show highest preference at middling levels of Complexity. Despite its long history in experimental aesthetics, such an "optimal complexity" finding seemed unsatisfactory both because it ignored the environmental content of the scenes and because it was hard to believe that preference could be explained that simply.[2]

Our discomfort with these conclusions led to a study (Kaplan, Kaplan & Wendt, 1972) that was procedurally similar to Wohlwill's. To examine the issue of content more closely scenes were selected to reflect a continuum ranging from natural to urban. Concern with adequate sampling of the environment led to inclusion of 56 scenes. Study participants were asked to rate each of these scenes in terms of both Complexity and their preference, using a 5-point scale. The results of this initial study confirmed our suspicions. Complexity was not a powerful predictor of environmental preference, but the content of the scene was a major contributor. Even though none of the scenes was of extraordinary places, those with nature content were consistently preferred. Even with this highly preferred domain, however, the variation in preference was considerable. While the initial study generated some hypotheses about such differences in preference, a framework for understanding these patterns—the Preference Model—was the result of substantial further research.

Before discussing this framework it may be useful to describe the 30 studies that form its empirical base. (Synopses of these can be found in Appendix B of R. Kaplan & Kaplan, 1989.) In each of these studies close attention was paid to adequate sampling of the kinds of environments that were studied. The environments were represented visually, using black and white photographs or color slides. Study participants were asked to rate each scene using a 5-point scale that represented how much they liked the pictured setting. Generally, they were also asked other questions as part of the study. Additionally, in many cases other individuals were asked to rate the same scenes in terms of other attributes.

The studies differed in terms of the kinds of environments represented, both geographically and with respect to land cover or land uses. While the vast majority of the scenes were of places in the United States, five studies were based on environments in other parts of the world (Western Australia, Egypt, Korea, British Columbia, and, in the case of Woodcock [1982], a diverse range of countries and continents). Several studies focused on waterscapes—including urban waterfronts, a storm drain in a residential setting, coastal areas, and relatively pristine rivers and marshes. A few others focused on scenic areas, including a botanic area, landscaped gardens, mountains and canyons, and locally scenic highways. In many of the studies the settings were in the context of the built environment, including residential areas, urban scenes, and natural areas juxtaposed with buildings. Forests and fields were also the contents of several of the studies, often depicting the common countryside of the north central states.

Participants in the studies represented a considerable diversity in terms of demographic dimensions. College students were the respondents in half of these studies. Teen-aged youths were included in three studies, each of which also included adults. The remaining samples consisted of residents, visitors to the respective areas, or employees. Two studies involved ethnic comparisons, and three compared samples from different cultures (Korean/Western; Australian/American). The question of the familiarity of the environments to the participants was specifically addressed in several cases, and the studies differed in the use of familiar and unfamiliar settings.

Many of these studies confirm the initial results related to content. Natural, as opposed to human-influenced environments, repeatedly emerges as preferred. Trees and water in particular tend to enhance preference. Since environments lacking these contents are less likely to support human survival, selection pressures in this direction would hardly be surprising.

The findings across these studies, however, cannot be explained sufficiently on the basis of content alone. Not all nature scenes are highly preferred; even waterscapes show considerable variation in preference. What became apparent is that in their rapid assessment of what they liked, participants were drawing inferences based on the spatial information in the scene. More specifically, there seemed to be an implicit assessment of how one could function in the space represented by the scene.

The Preference Model that has emerged from this research program focuses on these spatial patterns. It identifies two major classes of information. One of these entails the human requirements to *Understand* and to *Explore*. The other factor concerns how much processing it takes to draw inferences about the setting. For practical purposes this is divided into a relatively *immediate* category and one that requires greater *inference*. The combination of these two classes of information generates a 2 × 2 matrix (Table 16.2). The variable in each of the cells of the matrix is a potential predictor of preference. Every environmental scene can have a score for each of these

Table 16.2. The Preference Matrix

	Understanding	Exploration
Immediate	Coherence	Complexity
Inferred	Legibility	Mystery

variables, although in many studies scores were obtained for various subsets of the four.

Understanding is enhanced by many aspects of a scene. *Coherence* refers to the ease with which one can grasp the organization of the scene. Repeating elements permit a very rapid assessment of how the scene hangs together. A scene with a modest number of distinctive regions that are relatively uniform within themselves and clearly different from each other will tend to have a high level of coherence.

Japanese rock gardens provide a convenient example of coherence. A quick glance is likely to yield a few distinct rocks against a background of gravel that creates a relatively smooth, uniform texture. Both the repeated elements (the rocks) and the uniform textures that are easy to treat as regions are characteristic factors enhancing coherence. Upon closer examination of our hypothetical Japanese garden, it may turn out that some of the major "rocks" are actually a cluster of rocks, a large one with smaller ones associated with it. Grouping of this kind, which facilitates visual organization, is also characteristic of high-coherence scenes. A random distribution of rocks, by contrast, would destroy the coherence of the scene. (It would not, however, alter the complexity, since the same number of different objects would be present.)

Even though a scene is communicated in a two-dimensional form, it is useful to consider the difference between the picture plane itself and the representation of a third dimension, or depth. It is this distinction that is implied in the Preference Model's attention to the degree of processing. Thus *Legibility* calls for an inference based on this third dimension. It is the assessment of how well one could find one's way within the depicted scene. Legibility concerns the inference that one will be able to maintain one's orientation, that one will find one's way there and back, as one wanders more deeply into the scene. A scene that is open enough to offer visual access, but with distinct and varied objects to provide landmarks is high in Legibility.

Just as Coherence involves an assessment of the picture plane with respect to understanding what is there, *Complexity* entails an assessment of this two-dimensional aspect of the scene in terms of exploring. Complexity involves the richness or the number of different objects in the scene. For example, a scene composed solely of an extended plowed field and the sky is low in Complexity and unlikely to provide much to look at. A scene that is relatively high in Complexity, by contrast, can still maintain Coherence depending on how the different portions of the scene are arranged.

We have defined *Mystery* as the promise of more information if one can venture deeper into the scene. In other words, it is the inference that one could learn more about the scene if one could explore its third dimension by changing one's vantage point. Mystery is enhanced by such characteristics as screening in the foreground, or a winding path, or other features that suggest the presence of more information while at the same time partially obscuring it.

Supporting Data

The Preference Matrix evolved from examination of empirical results. By examining what scenes in different studies were most and least preferred, common themes emerged. Mystery was clearly the most consistent predictor of preference. A consistent pattern that is highly preferred involves smooth ground textures with widely distributed trees. The trees give the scene depth and thus increase the Legibility. Equally consistent have been low-preference ratings for scenes that lack Coherence or Complexity.

This pattern of findings suggests that the inferential aspect plays an important role. When the characteristics that are immediately apparent, that require minimal inference, are absent, the result is a scene low in preference. Lacking Coherence, the scene is difficult to understand; lacking Complexity, the scene offers little to look at. On the other hand, it is not clear that high degrees of either of these attributes increases preference linearly.[3] With higher Coherence or Complexity, there does not appear to be a further increase in preference.

With the more inferred attributes, by contrast, there is a stronger suggestion of linearity. Mystery seems to predict preference across its entire range. In all but one of the studies that involved ratings of the scenes in terms of this predictor, it was a positive, significant factor with partial correlations between .31 and .56. The pattern with Legibility is not as well documented. It is the most recently developed of the predictors and as such has undergone changes in its definition. Ellsworth's (1982) study, for example, included Legibility, but the ratings did not correspond to the description here. While the ratings that were used did not predict preference, Ellsworth mentions in the text that scenes that were preferred were characterized precisely by the qualities that we refer to as Legibility.

Such was the process of theory formulation. The research results have guided the theory development, and it, in turn, has led to further research. While the experimenter's desire for well-controlled, uniform conditions goes unfulfilled, insights emerge from repeated attempts, even where many aspects keep changing—the scenes, the definitions, the individuals producing the ratings. Despite all this variation the Preference Model has continued to be a useful tool.

The intention of the Model is to inform intuition. It is a framework that has led to new insights and continues to undergo changes. It suggests that the needs for Understanding and Exploration are both important; one cannot replace the other. The environment is constantly reviewed in terms of how these needs are likely to be met. The setting that is immediately before one provides essential cues to effective functioning. Furthermore, humans assess what is likely to happen; they must infer what lies ahead in terms of both the continuing ability to make sense of it and in terms of the promise of new information.

TOWARD AN EVOLUTIONARY INTERPRETATION

As is evident from this brief sketch of our early research in the area of environmental preference, the initial focus of the research program was on preference and its predictors; there was only the slightest hint of an evolutionary interpretation. As the findings began to accumulate, however, this theme became increasingly evident. Interpretation of new findings repeatedly suggested parallels between what people preferred and the environmental circumstances under which humans evolved. Ultimately, considerations about human evolution played a role not only in the interpretation of results, but also in the way the studies were designed and the content chosen as stimulus material.

An important step toward an evolutionary interpretation concerned the preference concept itself. We initially chose preference as a convenient measure, as a fairly simple and direct dependent variable. It soon became clear that preference was much more than that. Participants made preference judgments rapidly and easily.[4] They even seemed to enjoy the process. The results were not wildly idiosyncratic as folklore seemed to imply, but were remarkably stable and repeatable across groups with widely varying backgrounds.

Increasingly we have come to see preference as an expression of an intuitive guide to behavior, an inclination to make choices that would lead the individual away from inappropriate environments and toward desirable ones. As we have seen, the centrality of information in these decisions is quite appropriate. If humans are indeed organisms whose survival through the course of evolution required the construction and use of cognitive maps (S. Kaplan, 1972, 1973), then being attracted by information would seem thoroughly adaptive. In particular, people should be enticed by new information, by the prospect of updating and extending their cognitive maps (Barkow, 1983). At the same time, however, as was clear from our prior discussion of selection pressures, early humans could not stray too far from the familiar, lest they be caught in a situation in which they would have been helpless due to a lack of necessary knowledge (for a more extensive discussion of these themes, cf. S. Kaplan & Kaplan, 1982). These two vectors, which seem reasonable on theoretical grounds, match well the two categories of predictor variables, Exploration and Understanding, which have been shown to be reasonably effective in understanding preference data.

Another interesting aspect of the findings concerns the inaccessibility of the preference process to introspection. Despite the ease with which participants in preference studies are able to make their judgments, and despite the highly regular and meaningful pattern of the results, participants are generally unable to explain their choices. When questioning participants about the bases of their preferences, we have found that they are usually quite unaware of the variables that proved so effective in predicting what they would prefer.

Those inclined to emphasize the role of consciousness in human thought and action might find such a state of affairs discomforting. On the other hand, if these findings in humans constitute an appropriate parallel to habitat selection in other animals, then such unconscious processing by humans is neither unreasonable nor unprecedented. Perhaps there is an evolved bias in humans favoring preference for certain kinds of environments, just as there is an evolved bias favoring reproductive activities. In the case of sexual behavior we do not expect people to be able to explain their inclinations on adaptive grounds, although such inclinations must ultimately derive from an adaptive basis. Indeed, as Hebb (1972) has so eloquently pointed out, "The primary reason that human beings engage in sex behavior is not to produce another generation of troublemakers in this troubled world but because human beings like sex behavior" (p. 131).

Thus both the nature of the predictor variables and the nature of the preference response itself tended to support the existence of an evolved bias toward certain landscape configurations. This is not, however, an uncontroversial interpretation of the data of environmental preference. It is, in fact, in sharp contrast to the position taken by several investigators in this area. Lyons (1983), for example, expresses doubt as to the existence of what she refers to as "an innate, heritable component of landscape preference." Tuan (1971) emphasizes the lack of underlying consistency in which landscapes are preferred by different people at different times. While such concerns led us to be cautious in making an evolutionary interpretation, subsequent work has tended to support this position.

The first research finding bearing directly on the possibility of an evolved bias for landscape configurations was reported by Balling and Falk (1982). They studied preferences of individuals of different ages for various kinds of environments. While a variety of settings had been studied in previous work in environmental preference, this was the first study to systematize the range of environments represented. Five different biomes were used, namely, desert, rain forest, savanna, mixed hardwoods, and boreal forest. Since familiarity and, hence, experience have been shown to be important factors in preference, one would expect that any evolutionary influences on biome preference would be most evident at a fairly young age, when familiarity factors might be exerting only a limited influence.

This pattern was in fact obtained. There was a substantial preference on the part of young children (8- and 11-year-olds) for savanna over all other biomes. (The scenes representing this category in the Balling and Falk study were of African savanna, a relatively open landscape with smooth ground texture and widely spaced trees.) By the age of 15 the hardwood forest had risen in preference to equal the savanna level. Since the participants in this research were from the eastern United States, the increased preference for hardwoods with increasing age is a predictable outcome of increasing familiarity. Comparably, such a familiarity effect would be least strong in young children, given their limited experience with any environment. Thus the younger children's decided preference for savanna, the environment in which the human species is believed to have evolved, is consistent with an evolutionary interpretation. This is, in fact, the interpretation that Balling and Falk adopt.

A second study pointing to a possible role of adaptive specialization in human preference was reported by Orians (1985). He suggested that manipulated landscapes, such as ornamental gardens, might reflect a preference for patterns characteristic of the savanna biome. More specifically, he studied the tree forms selected out of all forms available to the Japanese gardener. He found that both selection and pruning practices favor the shapes characteristic of savanna.

A third kind of converging evidence came from a quite unexpected source. Jay Appleton, a British geographer, produced a thoughtful study of English landscape painting. *The Experience of Landscape* (1975) puts forward a theory of preference whose two main components are Prospect and Refuge. *Prospect* refers to having a grand view, an overview, as it were, of the landscape. *Refuge* refers to having a safe place to hide, a place from which one can see without being seen. Preferring settings with these properties might sound as if it would confer an adaptive advantage, and that is indeed the interpretation Appleton favors. Although this does not constitute empirical support in the usual sense of the term, it is interesting to discover a totally independent type of scholarly work arriving at a strikingly similar conclusion, namely, that human landscape preferences are concerned with information and, more particularly, with the gathering of information on the one hand and the danger of being at an informational disadvantage on the other.

An empirical examination of this tantalizing perspective was not long in coming. Woodcock (1982), in a landmark study, wanted to test both the Prospect/Refuge and the Understanding/Exploration approaches in the same context. He also was concerned to see the operation of these variables in the context of certain critical biomes. He chose three of the biomes Balling and Falk had included: savanna, mixed hardwoods, and rain forest. As their study had included only four scenes to represent each of their five biomes, there was reason to question the adequacy of the environmental sampling. By restricting the study to only three biomes, Woodcock (1982) could sample each of them far more thoroughly, using 24 examples for each.[5]

To represent the domains of Understanding and Exploration, Woodcock chose the predictor variables of Legibility and Mystery. In order to have well-defined variables representing the Prospect/Refuge theory, he decided to examine each of these concepts in terms of a "primary" and "secondary" version. In each case "primary" involved a view that showed the desired quality had been achieved, while "secondary" involved seeing a place from which it could be achieved. Thus *Primary Prospect* was defined as a good view or vista, while *Secondary Prospect* was defined in terms of a hill or other vantage point from which one might expect to have a good view. Likewise, *Primary Refuge* involved a view from cover, where one could see without being seen. *Secondary Refuge,* then, was defined in terms of a view of some area from which one could see without being seen. All six predictor variables were rated by a panel of judges; the 72 scenes were rated on a 5-point preference scale by 200 participants.

Three of the predictor variables fared well in this study. Mystery, Legibility, and Primary Prospect were all strong predictors of preference. Further analysis of these data suggests a complex relationship among these variables.[6] Mystery and Legibility interact such that a high Mystery, high Legibility scene is even more preferred than would be expected given the independent contribution of these variables. Primary Prospect and Legibility interact in a quite different fashion. A high Primary Prospect scene will tend to be preferred whether it is legible or not, but preference for a scene low in Primary Prospect is dependent upon Legibility. Perhaps the grand vista is so engaging that the possibility of getting there and back is not a consideration. By contrast, lacking such a vista the focus may shift to such practical matters as moving through the terrain without getting lost.

Although Primary Prospect was a strong predictor, the other predictors based on Appleton's theory fared less well. Secondary Prospect and Secondary Refuge predicted preference significantly only in the savanna biome. Primary Refuge behaved in a fashion strikingly counter to expectations. In Woodcock's (1982) study, scenes high in Primary Refuge were characteristically views from woods or bushy areas, looking out toward a clearing or more open area. Such scenes tended to be less preferred, although not significantly so. Primary Refuge turned out to be, if anything, a negative predictor. Densely vegetated environments are likely to hinder locomotion and/or the capacity to see what is approaching before one is upon it. Apparently, the problem with a hiding place in the woods is that one is in the woods.

In attempting to determine whether the unexpected result with respect to Primary Refuge could be explained on some other basis, Woodcock tried a number of post hoc predictor variables. (It should be noted that this is a feasible strategy in research of this kind. Since post hoc ratings can be realiably made by judges, independent of any knowledge of the preference results, the fact that the data have already been collected is not contaminating.) Of the various variables explored, only one proved to have a significant effect. This variable was named *Agoraphobia,* implying the discomfort associated with wide open areas, largely lacking in protective cover. When the Primary Refuge and Agoraphobia results are combined it becomes clear that neither being out in the open nor being in deep woods is favored. An extensive reanalysis of previous studies in this area confirms this impression (R. Kaplan, 1985). There is a pervasive bias toward tree cover in open land and toward more openness in forested environments. These themes reflect the spirit, if not the letter, of Appleton's proposal. Being in the open makes one easy to see and, hence, more vulnerable. Being in a dense, visually impenetrable environment makes it hard to see where one is going or what one is getting into.

In summary, the results of these more specifically evolutionarily oriented studies are complementary to the results obtained in the context of the Preference Model. It seems clear that the hypothesis of an evolved bias favoring certain characteristic spatial configurations has received further support. These studies also make another important contribution: they point to the compatibility of cultural and evolutionary influences in landscape preference. Balling and Falk (1982) present results suggesting the significance of the savanna pattern in preference, a pattern that is repeated in park design in both England and the United States. At the same time, Orians (1985) analyzes the Japanese garden as a savanna derivative. Yet these two versions of savanna differ in many respects. There appears to be ample room for cultural influences as well as for echoes of early human experience in the landscapes people prefer.

COGNITION AND AFFECT FROM THE PERSPECTIVE OF ENVIRONMENTAL AESTHETICS

Although much work needs to be done before a fully adequate theory of environmental preference is available, theoretical developments thus far seem reasonably encouraging. An informational approach based on the broad categories of Understanding and Exploration seems to be a useful tool. An evolutionary interpretation also seems to hold considerable promise. Fortunately, although these two viewpoints are generally not found together, they are thoroughly compatible (Cosmides & Tooby, 1987; S. Kaplan, 1972; S. Kaplan & Kaplan, 1978; Lachman & Lachman, 1979).

Another conclusion one can draw from the growing research literature on environmental preference is that there appears to be substantial information processing occurring in the course of arriving at a preference judgment. Thus environmental preference might indeed offer a useful new perspective on the cognition/affect relationship. Interest in the role of preference in the ecology of the mind has been given a substantial boost by Zajonc's (1980) argument that preference can occur directly, without cognitive intervention. This interpretation has unquestionably helped focus attention on the relationship of cognition and affect and has appropriately challenged the assumption of a conscious assessment as a precursor to preference. At the same time, Zajonc's position is not without its problems. These difficulties have been ably documented by Seamon and his colleagues (Seamon, Brody & Kauff, 1983; Seamon, Marsh & Brody, 1984), Holyoak and Gordon (1984), and Lazarus (1984). It is not the purpose of this chapter to add to these reservations. It does, however, seem appropriate to point out ways in which environmental preference research and theory can lead to a reconceptualization of Zajonc's arguments that might at the same time be both more fruitful and less controversial.

It is important to recognize that in his assertion that "preferences need no inferences," Zajonc is not stating that preference never relies on cognition but rather that there exist instances (which may be rather commonly encountered) in which preference occurs without the intervention of any cognition whatsoever. In effect this establishes a suggested dichotomy between two classes of preferences, those that are cognitively mediated and those that are direct and unmediated. From the perspective of research and theory in environmental preference, however, there appear to be not two, but a whole spectrum of different relationships between input and affect with the cognitive component varying considerably across this spectrum.

In order to discuss the range of roles that cognition might play in preference, it is necessary to be clear on the range of mediating processes appropriately called cognitive. At present there is a widespread tendency to assume that cognition is by definition a conscious calculational process (Lazarus, 1982). Such a perspective is limiting and distorting; fortunately it has not gone unchallenged (e.g., Bowers, 1981; Dreyfus, 1972; Posner & Snyder, 1974). It also flies in the face of one of psychology's most important contributions to intellectual history, namely, that many important psychological processes (not only affective ones) are not accessible to conscious observation. Thus in the discussion that follows it is assumed neither that cognition is necessarily a conscious process nor that it necessarily involves calculation.

With consciousness eliminated as a criterion for whether a process is cognitive or not, one is left with the nature and quantity of processing involved. The distinction in the Preference Model based on the degree of inference is useful to examine in this context. The two Exploration predictors, Complexity and Mystery, show this contrast dramatically. It might be helpful to explore this contrast in terms of what a computer model of the two predictors might involve. Complexity, in these terms, is relatively straightforward. Whether there are many different things in the scene, or few, can be determined based on the information provided in the stimulus array. A count of the number of different objects in the scene would yield the desired index.

Mystery, by contrast, is quite another matter. This predictor is, as we have seen, based on the assessment of whether one could learn more by proceeding deeper into the scene. A key issue here is the fact that there are a large number of different environmental patterns for which this can be the case. Scenes high in Mystery can vary widely. Some contain a bending path. Others contain a brightly lit area partially obscured by light foliage. Others are dominated by visually impenetrable foliage, but with a hint of a gap where one could pass through. Even slight undulations of the terrain can contribute substantially to Mystery.

What these scenes share in common is a complex relationship that exists between the observer and the environment. That relationship cannot be detected directly in terms of any simple counting procedure. Even an ingenious combination of features would be unlikely to yield a consistently valid index for Mystery.

How, then, might one approach the development of a computer model to measure this comparatively subtle construct? The solution appears to require breaking the problem down into a number of steps, rather than attempting to solve it directly. First, feature information from the scene could be used to construct a rough conceptual model of the three-dimensional space represented by the scene. Then, by simulating a hypothetical organism moving about within this hypothetical space, it could be determined if more information would be acquired as the organism proceeded deeper into the scene. Three aspects of such an approach are pertinent here: (a) It captures the relational nature of the construct; (b) it mirrors what the judges are asked to do when rating scenes in terms of Mystery; and (c) it is inferential, reflecting the *potential* information in the scene. Thus Mystery seems to call upon a reasonably complex, albeit unconscious, inferential process.

The point of all this is not to say that preferences do (or do not) require inferences.

Rather there appears to be a considerable diversity of cognitive processes involved for the different predictors. Preference does not depend upon conscious calculation or even on calculation of any kind in the usual sense of the term. But it often does depend on cognitive processes of varying kind and varying amount. Surely this very variation is worthy of further study.

AESTHETICS REVISITED: SOME CONCLUDING COMMENTS

From an evolutionary perspective there appear to be great advantages in making a quick, automatic prediction about the informational possibilities of a place one is approaching. In moving about through varied terrain, individuals must continually reevaluate the appropriate direction to be taken, as the very process of moving through the landscape opens up new vistas and new possibilities. Speed of processing is thus essential if one is to keep up with new information and respond accordingly. In this way preference would help keep the individual in an environment where orientation and access to new information can be maintained easily—quite apart from the particular purposes that individual was pursuing at that moment.

Thus environmental aesthetics can be seen as both efficient and economical. And the impact of this process is far from trivial; by influencing choice and guiding locomotion, it plays an important role in determining the environment in which the individual is located. From this perspective environmental aesthetics is not a special case of aesthetics but a reflection of a broad and pervasive function. In fact, some of the more traditional aesthetic domains may be derivative of this more basic function.

Research in environmental preference not only has yielded insight into the aesthetics of landscape, but also turns out to have considerable theoretical interest that extends beyond the environmental context:

1. The way preference feels to the perceiver stands in sharp contrast to the process that underlies it. Preference is experienced as direct and immediate. There is no hint in consciousness of the complex, inferential process that appears to underlie the judgment of preference. Given the range of variables that are being assessed, the underlying process must be carried out with remarkable speed and efficiency.
2. It is now quite clear that there is more to experimental aesthetics than optimal complexity. Further, since the additional components that have been discovered concern adaptive functioning in a complex environment, they are of some theoretical interest. These components also point to the high premium placed on information. Both the acquisition of new information and its comprehension turn out to be central themes underlying the preference process.
3. Aesthetic reactions reflect neither a casual nor a trivial aspect of the human makeup. Aesthetics is not the reflection of a whim that people exercise when they are not otherwise occupied. Rather, such reactions appear to constitute a guide to human behavior that has far-reaching consequences. Many everyday behaviors, such as organizing one's workspace and arranging and maintaining one's home, may reflect factors of this kind. Even in patterns of thought, avoiding certain directions and approaching others may be based as much on feelings of Mystery, Coherence, and the like, as on the specific content involved. Aes-

thetics could thus be seen as a set of inclinations, however intuitive or unconscious, that might influence the direction people choose not only in the physical environment but in other domains as well.

4. Environmental preference may constitute a useful conceptual link in analyzing the impact of evolution on behavior. Until recently, discussion of potential evolutionary influences on human behavior has generally ignored psychological mechanisms. Here preference could play a useful bridging function. It is a domain where, based on animal studies, an evolutionary role might be expected. The factors that have been demonstrated empirically to be predictors of preference are consistent with such an evolutionary interpretation.

5. Gradually a better understanding is emerging of what sort of environments are appropriate for humans. Certainly there are implications here for the design and management of landscapes. One might suspect that appropriate environments would be healthier environments, and thus there are implications for the design of hospitals as well. Studies of the impact of the view out the window on health support this suspicion (Moore, 1981; Ulrich, 1984; Verderber, 1986). The concept of environment can be taken more broadly still: since some of the basic factors involved are informational, they might well apply to such abstract environments as the human-computer interface. Initial work of this kind in fact appears to be promising (Leventhal, 1988). Thus work in environmental preference may even offer a useful framework for characterizing settings and circumstances that are compatible with and supportive of the human species (S. Kaplan, 1983).

ACKNOWLEDGMENTS

A portion of this chapter previously appeared in *Environment and Behavior* (1987), *19*, 3–32. The invaluable assistance of Rachel Kaplan in the writing of this chapter is gratefully acknowledged. I also want to thank Warren Holmes, Leda Cosmides, and John Tooby for their thoughtful comments and suggestions. Preparation of the article and much of the work reported in it were supported, in part, by the U.S. Forest Service, North Central Forest Experiment Station, Urban Forestry Project through several cooperative agreements.

NOTES

1. It is important to realize that placing a premium on understanding does not require one to remain in familiar environments. Much as familiarity facilitates understanding, the structure of some environments makes them far more understandable, or "easier to read," than others.

2. Despite the rather widespread acceptance of the optimality hypothesis, this position has serious theoretical difficulties (Martindale, 1984a) and fared badly in an ingenious series of direct experimental tests (Martindale, 1984b).

3. There are a variety of ways of assessing any particular relationship, each with advantages and disadvantages. Examining the predictor values for scenes rated at the high and low end of the preference scale provides a straightforward and intuitive means of identifying such relationships. Regression analyses with preference as the dependent variable and ratings for each predictor as independent variables can yield a different view of the results. When viewed in terms of multiple regression analyses across several studies, coherence plays a stronger role than this discussion suggests and does, in fact, approximate a linear relationship (R. Kaplan & Kaplan, 1989).

4. Participants not only made a quick response; they were capable of doing so after only a

brief glimpse of the stimulus array. R. Kaplan (1975) studied preference ratings made of scenes presented for 10, 40, and 200 ms. These brief exposure ratings correlated .97 with ratings of scenes presented for 15 s.

5. Through the kind cooperation of John Balling and John Falk their scenes constituted a subset of the scenes used for each of the three biomes under study.

6. These analyses were performed subsequent to the completion of the dissertation; David Woodcock kindly made available the data on which the additional analyses are based.

REFERENCES

Appleton, J. (1975). *The experience of landscape.* London: Wiley.

Balling, J. D., & Falk, J. H. (1982). Development of visual preference for natural environments. *Environment and Behavior, 14,* 5–28.

Barkow, J. H. (1983). Begged questions in behavior and evolution. In G.C.L. Davey (Ed.), *Animal models of human behavior* (pp. 205–222). Chichester, England: Wiley.

Berlyne, D. E. (1960). *Conflict, arousal and curiosity.* New York: McGraw-Hill.

Bowers, K. S. (1981). Knowing more than we can say leads to saying more than we can know: On being implicitly informed. In D. Magnusson (Ed.), *Towards a psychology of situations* (pp. 179–194). Hillsdale, NJ: Erlbaum.

Cosmides, L., & Tooby, J. (1987). From evolution to behavior: Evolutionary psychology as the missing link. In J. Dupré (Ed.), *The latest and the best: Essays on evolution and optimality* (pp. 277–306). Cambridge, MA: MIT Press.

Dreyfus, H. L. (1972). *What computers can't do.* New York: Harper & Row.

Day, H. (1967). Evaluation of subjective complexity, pleasingness and interestingness for a series of random polygons varying in complexity. *Perception and Psychophysics, 2,* 281–286.

Ellsworth, J. C. (1982). *Visual assessment of rivers and marshes: An examination of the relationship of visual units, perceptual variables and preference.* Unpublished master's thesis, Utah State University, Logan, UT.

Flannery, K. V. (1955). The ecology of early food production in Mesopotamia. *Science, 147,* 1247–1256.

Hebb, D. O. (1972). *A textbook of psychology.* Philadelphia: Saunders.

Holyoak, K. J., & Gordon, P. C. (1984). Information processing and social cognition. In R. S. Wyer & T. K. Srull (Eds.), *Handbook of social cognition* (Vol. 1). Hillsdale, NJ: Erlbaum.

Kaplan, R. (1975). Some methods and strategies in the prediction of preference. In E. H. Zube, R. O. Brush, & J. G. Fabos (Eds.), *Landscape assessment* (pp. 118–129). Stroudsburg, PA: Dowden, Hutchinson & Ross.

Kaplan, R. (1985). The analysis of perception via preference: A strategy for studying how the environment is experienced. *Landscape Planning, 12,* 161–176.

Kaplan, R., & Kaplan, S. (1989). *The experience of nature: A psychological perspective.* New York: Cambridge.

Kaplan, S. (1972). The challenge of environmental psychology: A proposal for a new functionalism. *American Psychologist, 27,* 140–143.

Kaplan, S. (1973). Cognitive maps in perception and thought. In R. M. Downs & D. Stea (Eds.), *Image and environment* (pp. 63–78). Chicago: Aldine.

Kaplan, S. (1983). A model of person-environment compatibility. *Environment and Behavior, 15,* 311–332.

Kaplan, S. (1987). Aesthetics, affect and cognition: Environmental preference from an evolutionary perspective. *Environment and Behavior, 19,* 3–22.

Kaplan, S., & Kaplan, R. (1978). *Humanscape: Environments for people.* Ann Arbor, MI: Ulrich's.

Kaplan, S., & Kaplan, R. (1982). *Cognition and environment: Functioning in an uncertain world.* New York: Praeger.

Kaplan, S., Kaplan, R., & Wendt, J. S. (1972). Rated preference and complexity for natural and urban visual material. *Perception and Psychophysics, 12,* 354–356.

Lachman, J. L., & Lachman, R. (1979). Theories of memory organization and human evolution. In C. R. Puff (Ed.), *Memory organization and structure* (pp. 133–190). New York: Academic.

Laughlin, W. S. (1968). Hunting: An integrating biobehavior system and its evolutionary importance. In R. B. Lee & I. DeVore (Eds.), *Man the hunter.* Chicago: Aldine.

Lazarus, R. S. (1982). Thoughts on the relations between emotion and cognition. *American Psychologist, 37,* 1019–1024.

Lazarus, R. S. (1984). On the primary of cognition. *American Psychologist, 39,* 124–129.

Leventhal, L. M. (1988). Experience of programming beauty: Some patterns of programming aesthetics. *International Journal of Man-Machine Studies, 28,* 525–550.

Lyons, E. (1983). Demographic correlates of landscape preference. *Environment and Behavior, 15,* 487–511.

Martindale, C. (1984a). The pleasure of thought: A theory of cognitive hedonics. *Journal of Mind and Behavior, 5,* 49–80.

Martindale, C. (August, 1984b). The decline and fall of Berlyne's theory of aesthetics. Paper presented at the Symposium on the Current Status of Berlyne's New Empirical Aesthetics, American Psychological Association, Toronto, Canada.

Moore, E. O. (1981). A prison environment's effect on health care service demands. *Journal of Environmental Systems, 11,* 17–34.

Orians, G. H. (1985). *An ecological and evolutionary approach to landscape esthetics.* Unpublished manuscript, University of Washington, Seattle.

Peters, R. P., & Mech, L. D. (1975). Behavioral and cultural adaptations to the hunting of large animals in selected mammalian predators. In R. Tuttle (Ed.), *Antecedents of man and after.* Chicago: University of Chicago Press.

Pfeiffer, J. E. (1982). *The creative explosion: An inquiry into the origins of art and religion.* New York: Harper & Row.

Posner, M. I., & Snyder, C.R.R. (1974). Attention and cognitive control. In R. Solso (Ed.), *Information processing and cognition: The Loyola Symposium.* Hillsdale, NJ: Erlbaum.

Seamon, J. G., Brody, N., & Kauff, D. M. (1983). Affective discrimination of stimuli that are not recognized: Effects of shadowing, masking and cerebral laterality. *Journal of Experimental Psychology: Learning, Memory and Cognition, 9,* 544–555.

Seamon, J. G., Marsh, R. L., & Brody, N. (1984). Critical importance of exposure duration for affective discrimination of stimuli that are not recognized. *Journal of Experimental Psychology: Learning, Memory and Cognition, 10,* 465–469.

Tuan, Y-F. (1971). *Man and nature* (Resource Paper No. 10). Washington, DC: Association of American Geographers.

Tulkin, S. R., & Konner, M. J. (1973). Alternative conceptions of intellectual functioning. *Human Development, 16,* 33–52.

Ulrich, R. S. (1984). View through a window may influence recovery from surgery. *Science, 224.* 420–421.

Verderber, S. (1986). Dimensions of person-window transactions in the hospital environment. *Environment and Behavior, 18,* 450–466.

Vitz, P. C. (1966). Affect as a function of stimulus variation. *Journal of Experimental Psychology, 71,* 74–79.

Wohlwill, J. F. (1968). Amount of stimulus exploration and preference as differential functions of stimulus complexity. *Perception and Psychophysics, 4,* 307–312.

Woodcock, D. M. (1982). *A functionalist approach to environmental preference.* Unpublished doctoral dissertation, University of Michigan, Ann Arbor, MI.

Zajonc, R. B. (1980). Feeling and thinking: Preferences need no inferences. *American Psychologist, 35,* 151–175.

VII
INTRAPSYCHIC PROCESSES

Modularity—the proposition that the mind contains many different, functionally isolable subunits, each specialized to process different kinds of information—has been gaining increasing empirical support in recent years from research in both cognitive psychology and cognitive neuroscience. But the possibility of a multimodular mind raises the problem of when different modules should "communicate" their output to one another. What happens when two modules that are working on different aspects of the same adaptive problem output information that implies two mutually contradictory courses of action?

Earlier in this volume, for example, Cosmides and Tooby argued that the mind contains a complex set of algorithms specialized for reasoning about social exchange. These different algorithms may form different, functionally isolable subunits: a cheater detection module; a module monitoring one's history of interaction with another person; a module that decides, based on information provided by these other modules, whether to terminate a reciprocation relationship with a particular person; and so on. But what happens when the cheater detection module and the history of interaction module "disagree"? Imagine, for example, that your cheater detection module determines that your friend has been cheating you recently, but the module that keeps track of your history of interaction with that person determines that your friend has been a good reciprocator in the past and that you have, on average, received considerable benefits by your continued association with her. What, then, should your decision module do? Ordinarily, the output from your cheater detection module would cause the decision module to activate procedures that would cause you to somehow punish your friend's cheating, perhaps by threatening to end your friendship if she doesn't make amends. But the output of your history of interaction module would ordinarily cause the decision module to activate procedures that would cause you to act in an affiliative way, thereby continuing the relationship. How can this intermodule conflict be resolved—that is, what would a system that is well designed for solving this adaptive problem do? Should the decision module feed the information about the recent cheating episode into procedures that would cause you to punish the cheating, thereby jeopardizing your friendship? Or should it cause this information to be "repressed": stored by the history of interaction module, but not fed into procedures that would cause you to jeopardize your relationship?

This situation could well be described as one involving an "intrapsychic conflict" that can be "resolved" by "repression." These are terms of art of psychodynamic theory. Yet modern models of a multimodular mind may give these terms more rigorous definitions and a new relevance to cognitive psychology.

Nesse and Lloyd argue vigorously for that position in their chapter on the evolution of psychodynamic mechanisms. In their own words:

> The sharpening focus of psychological research on the information-processing mechanisms that regulate human behavior may give new importance to psychodynamic psychology. Although psychodynamic theory has proven difficult to test, and is based, in part, on outmoded biology, some psychodynamic traits may turn out to closely match functional subunits of the mind that are currently being sought by cognitive and evolutionary psychology. The merit of this conjecture can be appraised by considering how repression and other psychodynamic capacities could have offered a fitness advantage. Such capacities seem maladaptive because they distort reality, but if there are situations in which distortion offers an advantage over accuracy, then natural selection might well have shaped mental mechanisms that systematically distort conscious experience. Interpersonal relationships may be such situations, and repression and the psychological defenses may be such mechanisms. The capacity for repression may facilitate self-deception and deception of others in ways that offered a variety of selective advantages. Attempts to integrate psychodynamics and evolutionary psychology may advance the developing information-processing models of the mind and may offer a revised biological foundation for psychoanalysis.

In cognitive science and experimental psychology, it is common to ignore psychodynamic theory, to view it as a tangle of vague concepts and inconsistently defined phenomena. Indeed, some of *The Adapted Mind*'s editors share these concerns. Rather than abandoning the enterprise, however, Nesse and Lloyd's article exemplifies one promising approach to salvaging what is of value in psychodynamic theory and for recovering the significance of the clinical phenomena that sustain it.

17

The Evolution of Psychodynamic Mechanisms

RANDOLPH M. NESSE AND ALAN T. LLOYD

If . . . deceit is fundamental to animal communication, then there must be strong selection to spot deception and this ought, in turn, to select for a degree of self-deception.

ROBERT TRIVERS

Indeed, a great part of psychoanalysis can be described as a theory of self-deception.

HEINZ HARTMANN

As cognitive psychologists ask more about the origins and functions of mental mechanisms, they turn to evolutionary theory (Boden, 1987; Buss, 1984; Cosmides & Tooby, 1987; Tooby, 1985). As evolutionists ask more about the mental mechanisms shaped by natural selection, they turn to cognitive psychology (Barkow, 1984; Crawford, Smith, & Krebs, 1987; Symons, 1987, 1989). This converging focus on human information-processing mechanisms may give new significance to psychodynamics. Cognitive and evolutionary psychologists may find in psychodynamics careful descriptions of traits that may closely match the functional subunits of the mind that they are seeking. Psychodynamic psychologists and psychiatrists may find in evolutionary psychology new possibilities for a theoretical foundation in biology.

This potential integration is viable only if some of the traits described by psychoanalysts are legitimate objects of evolutionary explanation. There are several reasons to believe that repression and other psychodynamic traits may be mental mechanisms shaped by natural selection: (a) They appear to be fairly uniform in humans (although more cross-cultural study of this issue would be welcome); (b) their developmental patterns appear related to the tasks faced at each phase of life; (c) their complexity and careful regulation imply that they serve significant functions; (d) the behaviors they mediate, such as patterns of relationships, social communication, courtship behavior, inhibitions, guilt, and cognitive representations of the external world, are important to reproductive success; and (e) when they function abnormally, fitness often decreases. Although these factors justify an evolutionary analysis (Mayr, 1988, pp. 148–160), an evolutionary explanation of these traits requires demonstration of specific functions and ways in which they enhance fitness. This is often difficult, even for a physical trait. Nonetheless, this is what must be attempted, because confidence that a trait has been shaped by natural selection usually is based on a demonstration that its details match its function (Williams, 1966).

Although several promising and interesting forays have been made (Badcock,

1986, 1988; Leak & Christopher, 1982; Rancour-Laferriere, 1985; Slavin, 1987; Wenegrat, 1984), it remains uncertain whether a genuine integration of psychoanalysis and evolutionary biology is essential or inevitable. We will, therefore, defend the more modest thesis that such an integration may be possible and valuable. We argue not that psychoanalytic concepts will turn out to exactly match the evolved functional subunits of the mind, but only that they offer the best available starting point. Our method will be to examine a variety of phenomena that are well accepted by psychoanalysts—repression, psychological defenses, intrapsychic conflict, conscience, transference, and childhood sexuality—in order to compare their characteristics to the predictions made by various hypotheses about their possible functions.

OBSTACLES

Before considering the ways in which psychoanalysis and evolutionary psychology might inform each other, several obstacles must be acknowledged. The first is that psychoanalytic theory was built on several doctrines that are now known to be mistaken (MacDonald, 1986): Freud was overtly Lamarckian (S. Freud, 1912–1914, 1915b/1920; 1987; Gruberich-Simitis, 1987), he followed Haeckel's dictum that "ontogeny recapitulates phylogeny" (S. Freud, 1912–1914, 1915b/1987; Gould, 1977), and he was a group selectionist (S. Freud, 1915a, 1920, 1923). These errors, although common in Freud's time and derived partly from Darwin himself (Ritvo, 1964), have led many modern biologists to disregard psychoanalytic theory altogether. This is understandable, but somewhat ironic, because Freud was one of the few in his time who went beyond proximate explanations and tried to understand the origins and functions of mental traits (Gay, 1988; Sulloway, 1979). In fact, the enduring vitality of Freud's work may result from his attempts to explain the adaptive significance of mental phenomena. He recognized that an explanation of the psyche depended on understanding the adaptive significance of its components. He tried to explain how these components were shaped by events in the distant past. He recognized the central importance of reproduction to mental life. And, he unflinchingly documented the selfish, aggressive, and sexual impulses he found at the root of human motivation. Few other theorists in his time tried to understand the functions of high-level mental structures in such an explicitly biological way. Until the advent of evolutionary psychology, those who sought ultimate explanations for the origins and adaptive significance of high-level mental traits turned often to psychoanalysis.

A second major problem is the scientific acceptability of psychoanalytic data. Some scientists dismiss all subjective reports and accept only "objective" behavioral observations of the sort that can be obtained with other species. This stance takes the high ground of scientific legitimacy, but severely constrains the study of subjective experience and our capacities for empathy, insight, and self-deception—all human traits worthy of study. Psychoanalytic data are also criticized because it is so difficult to disentangle empirical observations from the complex theory in which they are often embedded. This criticism is justified. Psychoanalytic methods can, nonetheless, provide a unique window on high levels of mental organization.

Another barrier to linking psychoanalysis with mainstream science arises from the repugnance of some psychoanalytic discoveries. People are reluctant to admit that they are motivated by socially offensive unconscious wishes. Interestingly, biologists have, especially in the past decade, encountered similar objections to the discovery that natural selection does not shape behavior for the benefit of the group or the species, but for the benefit of the individual and its genes (Dawkins, 1976, 1982; Williams, 1966). People prefer not to acknowledge the underlying "selfishness" in many apparently altruistic patterns of behavior. Equally discomforting is the corollary that deception is not an anomaly in a harmonious natural world, but an expected strategy in a world of individuals acting on behalf of their genes (Dawkins, 1982; Mitchell & Thompson, 1986; Trivers, 1985; Wallace, 1973).

THE BENEFITS OF SELF-DECEPTION—A MISSING LINK?

A possible evolutionary function for repression comes from a proposal made by Alexander (1975, 1979) and by Trivers (1976, 1985): The capacity for self-deception may offer a selective advantage by enhancing the ability to deceive others. "Selection has probably worked against the understanding of such selfish motivations becoming a part of human consciousness, or perhaps even being easily acceptable" (Alexander, 1975, p.96). "There must be strong selection to spot deception and this ought, in turn, to select for a degree of self-deception, rendering some facts and motives unconscious so as not to betray—by the subtle signs of self-knowledge—the deception being practised" (Trivers, 1976, p. vi.). Although neither Alexander nor Trivers explicitly connect their hypothesis with psychoanalysis, it nonetheless offers a potential explanation of how natural selection could have shaped traits that systematically distort conscious mental experience. If the ability to deceive increases fitness and if self-deception increases the ability to deceive others, then "the conventional view that natural selection favors nervous systems which produce ever more accurate images of the world must be a very naive view of mental evolution" (Trivers, 1976, p. vi).

Evolutionary psychology came to focus on self-deception through deductive reasoning. The functions of animal communication are well understood (Wilson, 1975, pp. 176–241), and the evolved capacities for deception in plants and animals are well recognized (Dawkins, 1982; Krebs & Dawkins, 1984; Mitchell & Thompson, 1986; Wallace, 1973). With the demise of group selectionism (Dawkins, 1976, 1982; Williams, 1966), the rise of kin-selection theory (Hamilton, 1964), and the subsequent recognition of the crucial role of reciprocity relationships for Darwinian fitness (Axelrod, 1984; Axelrod & Hamilton, 1981; Trivers, 1971), research on deception (Mitchell & Thompson, 1986) and self-deception (Lockard & Paulhus, 1988) has grown rapidly. Biologists have generally not recognized, however, that self-deception is the focus of decades of psychoanalytic study (Leak & Christopher, 1982).

Psychoanalysis came to focus on self-deception by inductive reasoning. It began with the observation of symptoms and uncensored reports of thoughts and emotions. Attempts to understand this data have consistently resulted in explorations of the meanings and mechanisms of self-deception. Psychoanalytic theory, unsatisfactory as it may be, constitutes the best available proximate explanation of the mechanisms of human self-deception.

REPRESSION—THE PHENOMENON

Freud asked his patients to say whatever came to mind, no matter how embarrassing or seemingly irrelevant. He found, by this "free-association method," that much in the mind is not what it seems to be. Sometimes, professed love conceals hatred, indignant morality conceals perverse wishes, and flagrant nonconformism conceals profound guilt. Much in the mind is unconscious, not just because it is not brought to consciousness, but because it cannot be brought to consciousness, no matter what the effort. *Repression* (in the general sense) is a psychological mechanism that keeps unacceptable thoughts and wishes unconscious (Fenichel, 1945, p. 17). The psychological defenses are the devices that distort cognition in ways that facilitate repression.

Confusion often results because the term "unconscious" sometimes refers generally to anything that is outside of conscious awareness and sometimes refers to the more specific "dynamic unconscious," a special repository for mental contents that would be accessible to consciousness, except that they are actively repressed. Freud was not the first to recognize the existence of the dynamic unconscious, but he was one of the first to systematically explore and describe it (Ellenberger, 1970). Similar confusion results because "repression" describes two things: (a) the general capacity for keeping things unconscious (the meaning we will use), and (b) the more specific defense mechanism of simply "forgetting" things that are unacceptable (Erdelyi, 1985, pp. 218–225; A. Freud, 1966).

The study of unconscious mental events has inherent problems. As Erdelyi (1985, p. 65) notes, "The problem of the unconscious poses special challenges for scientific psychology. Not only are unconscious processes inaccessible to public observation, but they are excluded, by definition, from private subjective experience as well. How then can the unconscious be known—if, indeed, there is such a thing as the unconscious?"

Clinical evidence for the existence of repression comes from symptoms, dreams, slips, and posthypnotic suggestion (Brenner, 1974; Fenichel, 1972; Freud, 1915a, 1917; Horowitz, 1988). Freud originally derived the concept from observations of dramatic symptoms that expressed unacceptable unconscious thoughts and wishes. A woman who wants to stab her husband develops paralysis of her right arm. A woman who wants to be taken into the arms of a certain man suddenly faints in front of him. A man with unconscious homosexual wishes develops unfounded fears that others are saying that he is homosexual. Evidence for unconscious phenomena also comes from more commonplace slips. Who has not decided to remember the birthday of a dubious friend and nonetheless forgotten? Or resolved to conceal something and nonetheless made revealing slips of the tongue?

Modern laboratory studies of unconscious information processing and their implications for psychoanalysis are critically reviewed by Erdelyi (1985). Early studies demonstrated that stimuli can induce changes in affect even when they are presented so briefly that the subject cannot recognize the image (Fisher & Greenberg, 1977). More recent studies (Erdelyi, 1985, pp. 65–105; Horowitz, 1988; Kihlstrom, 1987; Shevrin & Dickman, 1980) have further confirmed the cognitive processing of subliminal stimuli. Additional studies are based on hypnotic phenomena, and on split-brain research. In a series of ingenious social psychological experiments, Lewicki (1986) has convincingly demonstrated the nonconscious processing of social information. Such demonstrations may seem distant from clinical material and may demonstrate phenomena somewhat different from active repression, but they do offer replicable demonstrations of unconscious information processing.

PSYCHOANALYTIC AND COGNITIVE EXPLANATIONS OF REPRESSION

The psychoanalytic explanation of repression emphasizes the role it plays in regulating affects and impulses. Repression decreases anxiety by decreasing awareness of painful facts and wishes. Repression also inhibits the expression of impulses by keeping unacceptable wishes out of consciousness. This is a correct description, but there are several reasons to think it is an incomplete explanation. The main limitation is that it does not explain why other simpler mechanisms for regulating affect and impulse are not sufficient. For instance, why aren't unacceptable thoughts and wishes completely eliminated from the mind? Not only do repressed mental contents remain in the mind, they remain close to centers of motivation and they influence behavior. The usual psychoanalytic explanation also has difficulty explaining the complexity and delicacy of repression. Why doesn't the system work better? And, why are there so many distinct defense mechanisms? The disadvantages of repression—distortion of experience, expenditure of mental energy, distraction from other tasks, and the costs of developing and maintaining complex and delicate mechanisms—suggest that these costs must be outweighed by some other functions that give substantial fitness advantages.

Cognitive psychology has made substantial progress in the study of nonconscious mental processing and has confirmed that most information processing in the mind goes on outside of consciousness and that multiple processors operate in parallel (Goleman, 1985). These findings undercut the presumption that all information-processing mechanisms should be conscious. If many stimuli were represented simultaneously in the central processing unit, mental life would be chaos. Furthermore, the mind has limited processing power. To be effective, it must focus on a limited number of tasks at once. These factors can explain the existence of mechanisms that limit access to the central processing unit and thus, the phenomenon of attention and the associated suppression of certain mental events. This is not the same, however, as explaining psychodynamic repression.

Especially prone to repression is any mental content that causes anxiety, guilt, or other painful emotions. But if mental pain, like physical pain, is a useful evolved capacity, then blocking it should be maladaptive. One intriguing suggestion is that the capacity for self-deception may be an adaptation to control mental pain in the same way that the endorphin system limits physical pain (Goleman, 1985, p. 30–43). There are, however, major differences between the systems. Endorphins operate in emergency situations to nonspecifically down-regulate the pain system, while repression operates constantly to specifically admit and exclude particular cognitive items. Nonetheless, the parallel is useful. Perhaps repression does keep painful stimuli out of consciousness when the pain would serve no purpose. Desires that cannot be fulfilled and transgressions that cannot be undone might both be better kept out of consciousness to reserve processing power for useful tasks. Objective facts that would cause hopelessness and low self-esteem might better be kept from awareness.

While cognitive explanations for repression are important for understanding unconscious mental processing and attention, it is not clear that they are sufficient to explain psychodynamic repression. Repression does not just draw attention away from certain mental contents, it actively blocks attempts by the self or others to bring them to consciousness. If it were not for repression, the content of the dynamic unconscious would be accessible to consciousness. It is this active process of repression that is the focus of explanation.

REPRESSION AND SELF-DECEPTION

As noted previously, Alexander and Trivers have proposed that self-deception could increase fitness by increasing the ability to pursue selfish motives without detection. The full argument has several stages: Human reproductive success requires human social success, social success requires success in reciprocity relationships, success in reciprocity relationships comes from getting a bit more than you give, getting a bit more than you give requires the ability to deceive others, and the ability to deceive others is enhanced by the ability to deceive oneself. In short, people who incorrectly experience themselves as altruists will be better at exploiting others through deceptive means.

If repression gives an advantage by increasing the ability to deceive others and if being deceived is sometimes disadvantageous, then natural selection should increase the ability to detect deception. This will, in turn, shape ever more subtle abilities to deceive, which will shape still more sophisticated abilities to detect deception. Such evolutionary "arms races" between the ability to deceive and the ability to detect deception are well known in other species (Alexander, 1979, in press; Dawkins, 1982, pp. 55–80; Trivers, 1985, pp. 395–420). The complexity of the firefly's signal, for instance, has been shaped, in part, by the presence of cannibalistic fireflies that imitate females in an attempt to lure males close enough to capture them (J. Lloyd, 1986). Such an "arms-race" between deception and ability to detect deception might help to explain the extraordinary complexity of the human mind, and the difficulty of formulating simple principles of human psychology.

A number of publications have addressed the possible benefits of repression. The fundamental point was stated well by Barash in 1982: "There seems little doubt that the unconscious, although poorly understood, is real, and that in certain obscure ways it influences our behavior. We can, therefore predict that it is a product of our evolution, and, especially insofar as it is widespread and 'normal,' that it should be an adaptive product as well" (p. 211). Trivers (1985) reviews studies of deception and self-deception and emphasizes work by Gur and Sackheim (1979) that demonstrates motivated self-deception. In this series of experiments, people listened to recordings of themselves or someone else speaking and then guessed if the voice was their own or not. Skin conductance was measured, since it increases in response to hearing one's own voice and decreases when listening to another voice. What is remarkable is that the greatest skin conductance changes occurred in subjects who tended not to recognize their own voices. What is more, after a manipulation that decreased self-esteem, the tendency to acknowledge one's own voice decreased, but skin conductance increased—findings that are consistent with the existence of self-deception.

Slavin (1987) has emphasized the advantages of deception in the negotiation of the parent-offspring conflict. This conflict was described by Trivers (1974) in a seminal paper that outlines the inevitability of conflicts between parents with remaining reproductive capacity and their offspring. Offspring are selected to attempt to manipulate parents to provide more resources and help than is in the parent's best reproductive interests (for instance, by prolonged nursing). Conversely, offspring may be manipulated to behave in ways that are not in their best interests (for instance, by parental sanctions against sibling conflict). Slavin observes that deception (and, therefore, self-deception) is the best strategy for the otherwise powerless child. The child's wishes that are acceptable to the parent remain conscious, while those that would be punished are pursued unconsciously. Repression is a "built-in mechanism to help ensure autonomy from the family environment" (p. 425). His argument may be a special case of more general functions of repression. It has the particular merit of explaining why children obey moral principles that are not in their interests (in order to simultaneously placate and manipulate their parents). When these tendencies persist into adulthood they may, he suggests, help to explain adult neurosis. Indeed, the syndrome of neurosis includes preoccupation with following rules, trying hard to please others, being unaware of personal desires, and experiencing others as if they were parents. This view of neurosis may, we suspect, prove to be of enormous value.

Badcock (1986), in a review of the implications of evolutionary biology for psychoanalysis, emphasizes the role of deception in negotiating relationships. Lockard's (1980) careful review covers the history of research on self-deception and data-oriented studies that bear on it. A recent edited volume contains a wealth of material on self-deception (Lockard & Paulhus, 1988). In addition to a synthetic chapter (Sackheim, 1988), the volume offers several chapters that take an evolutionary approach, including one that presents data confirming the operation of self-deception in human social networks (Essock-Vitale, McGuire, & Hooper, 1988).

THE FUNCTIONS OF REPRESSION

The core of the Alexander/Trivers hypothesis is that self-deception offers advantages by concealing covertly pursued motives. This insight is valuable because it offers a way to explain the apparent anomaly of mechanisms that distort mental experience. Its scope is unduly narrowed, however, by overemphasis on the benefits of self-deception in facilitating cheating and underemphasis on the benefits of "benevolent" self-deception in facilitating long-term cooperative relationships (Nesse, 1990).

A taxonomy of the functions of repression begins by distinguishing situations in which repressed motives are pursued from those in which they are not. Motives may be repressed even as they are being pursued, for instance, when seductive content is unknowingly inserted into ordinary conversation. Or, repression may hide alternative strategies that are held in reserve for possible use at another time. For instance, a dissatisfied spouse may embark on a campaign to please the mate, while repressing the simultaneous unconscious consideration of plans to leave the relationship.

In other stuations, however, repression is valuable even though the repressed wishes are not pursued at all. Some atavistic wishes are concealed in order to appear to more closely match the social ideal. For instance, cannibalistic wishes, which are sometimes uncovered during psychoanalysis, are generally not repressed just so that they can be pursued later. In other situations, opportunities for cheating or defection are repressed, not so they can be acted on later, but so that neither the self nor the other is aware that the possibility ever was considered. People rarely think of stealing from the home of a friend, and most would be outraged at the suggestion that they might have sexual desires for their stepchildren. In such situations, repression functions, as analysts have long recognized, to inhibit conscious recognition of socially unacceptable impulses.

Repression can conceal the motives of others as well as those of the self. Repression makes it easier to overlook a friend's transgression. A personal slight might have been a misunderstanding instead of a defection. Even if it was a defection, it might best be ignored in order to maintain the relationship. This function has been suggested by A. Lloyd (1984) and Lockard (1980). In this case, it is not one's own selfish motives that are repressed, but those of someone else. This is especially valuable in hierarchical relationships, where it may be advantageous for the less powerful person to present himself as less capable than he actually is in order to avoid the attacks and resource deprivation that might result if the more powerful person's position was threatened. This deception is most effective if the individual actually believes that he lacks ability. This strategy may account for some behaviors that seem self-defeating (Hartung, 1988). Deception remains at the root of these functions of repression, but the motive is not short-term selfish gain, but the maintenance of long-term relationships. This might be called "benevolent self-deception," to reflect the self-sacrifice that it requires in the short run (Nesse, 1990).

In other situations, positive feelings are repressed. For instance, when a person threatens to leave a relationship, the threat lacks conviction unless warm feelings are repressed. This may explain the dramatic swings between passionate love and bitter hatred that occur when couples are in the process of ending a relationship. Similarly, when it is best to fight without ambivalence or when benefit comes from brazenly denying the selfish nature of an action, awareness of guilt is disadvantageous. In such circumstances, it is not selfish motives, but inhibitions and guilt that are kept unconscious.

Much more could be done to develop a taxonomy of the functions of repression. Our point here is simply that, although repression may well have evolved to facilitate self-deception and thus the deception of others, current benefits of deception are far broader than merely to facilitate cheating—it also facilitates strategies that require short-term sacrifice for the sake of maintaining a long-term relationship. We will consider the hypothesis that the evolutionary function of repression is to increase the effectiveness of the deception of others in ways that give diverse benefits, including not only those that come from cheating in reciprocity relationships, but also those that result from concealing impulses for cheating and aggression in order to preserve long-term relationships. This hypothesis offers a possible explanation of the distinctive aspects of repression—the active distortion and limitation of conscious experience. In the absence of a strong competing hypothesis, it is tempting to simply accept it, but, are its predictions consistent with what we know?

One prediction is that people regularly deceive each other; this is obvious enough. Another prediction is deception can be detected. This has been amply confirmed with modern techniques (Depaulo & Rosenthal, 1979, p. 225–227). Although visual cues communicate more information about emotions than auditory cues (DePaulo & Rosenthal, 1979, p. 208–209), deception can be detected more reliably from vocal cues and body language than facial cues (Eckman & Friesen, 1974). A third prediction is that self-deception assists in the deception of others. If self-deception conceals any of the cues used to detect deception (as it almost certainly must), then this prediction is supported. Measures of ability to deceive others are well developed (DePaulo & Rosenthal, 1979) as are measures of ability to detect lying (DePaulo, Lanier & Dans, 1983). If these instruments were combined with a measure of capacity for self-deception (perhaps by assessing susceptibility to cognitive dissonance or by a method that measures tendencies toward social conformity), it should be possible to test the hypotheses that people with a high capacity for self-deception have superior ability to deceive others and that people with psychiatric conditions that involve failures of repression have superior abilities to detect deception. The same issues could also be addressed by assessing the personality characteristics of people with especially high and low abilities to deceive others and especially high and low abilities to detect deception. The ability to detect deception is independent of the ability to decode pure or consistent cues (Rosenthal, Jall, DiMatteo, Rogers, and Archer, 1979) and is negatively correlated with scores on a scale of Machiavellianism, but is uncorrelated with scores on scales that assess self-monitoring and the subject's estimation of the complexity of human nature (DePaulo & Rosenthal, 1979, p. 229–230). It is of interest that women are better than men at decoding body language, but men are better at noting discrepancies between communication channels that may indicate deception.

A strong test would be made possible by studying variations in the ability and tendency of people to use repression. We cannot think of a ready way to experimentally vary the tendency to repress, but some groups of people vary in the extent to which they use repression. In particular, people with neuroses are described as "repressed" because so many of their feelings remain unconscious, while people with schizophrenia are said to demonstrate inadequate repression. Clinicians have long puzzled over the uncanny ability of certain schizophrenics to apprehend secret and unsavory motives in others; at times it seems as if they really can read minds. It may be, however, that schizophrenia somehow interferes with the normal ability to adaptively deceive oneself about the motives of others. It would be worthwhile to arrange situations that aroused socially unacceptable wishes and to then see if the ability to deceive self and others declines as one moves from neurotics to normals to schizophrenics.

A recent review of cognition in depression concluded that normal people consistently distort reality in ways that make it less threatening and more positive than it is, while depressed people are far more accurate in their judgements about themselves and their situations (Taylor & Brown, 1988). The strength of these findings suggests that there might well be some selective advantage to this normal tendency toward systematic distortion.

Studies of people who have been psychoanalyzed would be of special interest. If psychoanalytic treatment facilitates access to repressed material and if maintenance of repression is useful in negotiating and maintaining relationships, then psychoanalyis might be expected to decrease, not increase, the ability to have normal relationships. The apparent contradiction may be explained because many who seek analysis start with excessive repression. Also, psychoanalysis does not remove defenses; it replaces them with more mature and flexible defenses. Nonetheless, it would be interesting to consider the possibility that increased access to unconscious material (of the sort that results from a training analysis) may cause difficulties in maintaining ordinary relationships.

One final prediction is that if deception offers particular benefits in certain situations, then specialized patterns of cognition and behavior might have been shaped to make deception especially effective in such situations. If such patterns exist, their characteristics should make sense as strategies shaped specifically to facilitate deception of others. Defenses may be such patterns.

THE DEFENSES

A *defense,* in the psychoanalytic sense, is a mental process that keeps unacceptable or painful thoughts, wishes, impules, or memories from consciousness and thereby reduces painful affects. Dorpat (1985) has convincingly argued that repression is made possible by the primary defensive process of denial—the disavowal of stimuli that arouse unacceptable thoughts or feelings. He further argues that other defenses consist of degrees of denial combined with other psychological maneuvers. These more complex and specific defenses—reaction formation, projection, rationalization, and others—involve degrees of compromise between impulses and inhibitions. The most basic form of denial is simply not consciously acknowledging things one would rather not, such as unwelcome sexual innuendo. *Rationalization* is a more sophisticated form of denial in which events are acknowledged but their meaning and importance is denied. This is illustrated by a person who acknowledges the reality of criticism by the leader of the group, but attributes the criticism to the leader's irritable mood in order to deny its personal significance. In *reaction formation,* unacceptable feelings are replaced by opposite feelings, for instance, when unconscious sexual wishes are covered by conscious disgust with sexuality.

About two dozen kinds of defenses have been recognized. Why are there so many? Psychoanalysis describes how each defense regulates impulses and protects consciousness from painful thoughts and feelings. This provides a proximate explanation for the defenses, but still may not explain why there are so many. Why didn't natural selection strengthen repression so it could do the job by itself? Also, the existence of multiple defenses gives rise to situations in which defenses conflict with each other. The existence of many and elaborate ego defenses is not necessarily explained by their functions of internal regulation.

We will consider the hypothesis that the specific ego defenses are specialized strategies for deceiving others. Certain defenses may regularly be elicited by certain conflicts because they are especially effective strategies in certain situations. The various aspects of a defensive pattern may be consistently associated because they are parts of a coherent deceptive strategy. If this hypothesis is correct, then it should be possible to demonstrate how the aspects of each defense facilitate deception. We will, therefore, analyze the characteristics of a variety of defenses to see if they can be understood as facilitators of deception of others.

The Defenses as Deceptive Strategies

Regression is reversion to earlier patterns of behavior. In a time of stress, a four-year-old may begin wetting the bed again, an eight-year-old may renew temper tantrums, and an adult may act uncharacteristically dependent or manipulative. Some functions of regression are clear. When a child is sick or hurt, it is in the parent's genetic interests to provide extra resources. This may have led to regression becoming a general signal of a need for aid, perhaps one used routinely by adults as well as children. But, as noted by Trivers (1985) and Slavin (1987), children can manipulate this system to get extra resources by acting younger than they really are, thus setting in motion an arms race between deception and ability to detect deception. Slavin deemphasizes the benefits of deception in adult interactions and implies that regression is a reversion to earlier modes of interaction. It would be reassuring to find that regression and other deceptive strategies are used mainly by children and pathological adults, but, in fact, children's ability to use regression in the service of deception may be only an early and relatively crude precursor of manipulation skills that become so practiced and natural in normal adults that they are easily overlooked. Perhaps anger at adults who complain and act helpless reflects an intuition that such strategies are often exploitative.

Reaction formation, the tendency to experience and express the exact opposite of an unconscious wish or feeling, may be more effective than simple repression at promoting deception in certain circumstances. The man who is aroused by a proposition from an attractive woman may loudly proclaim that he is starting an antipornography group, thus effectively hiding his secret. If he overdoes it people will recognize that "He doth protest too much," but people are remarkably reluctant to consider impure motives in loud moralists. People who use reaction formation extensively are prone, in certain situations, to suddenly and apparently unaccountably express the repressed impulse—witness the preacher who rails about sexual dissolution but then finds himself patronizing prostitutes (unaccountably or not, depending on the degree of self-deception involved). Even if the impulse is not acted on, repression can still be adaptive, by distracting others from recognizing socially unacceptable impulses.

Projection is the defense of denying unacceptable aspects of the self and simultaneously attributing them to others. For instance, a person who feels intellectually inferior may attack others for being "stupid." Accusations distract people from looking too closely at the accuser. For instance, unconscious homosexual wishes are commonly manifested by expressions of revulsion toward homosexuals. This may lead to serious problems, but it usually effectively conceals the secret. In a conscious version of the same mechanism, Cyril Burt, the psychologist who invented data to confirm his beliefs about the heritability of intelligence, made vitriolic attacks on the integrity of other scientists. Projection offers a further advantage because a person who experiences his own unsavory motives as if they were in others will often accurately anticipate other people's plans. Empathy, the ability to experience the feelings of other people as if they were one's own, is the converse of projection. It is highly valued because it allows people to accurately anticipate the needs of others, but it can also facilitate effective manipulation.

Identification and *introjection* are the psychological mechanisms by which values and characteristics of others are taken into the self. Early in life, children identify with their parents and introject norms and beliefs, thus facilitating the transmission of culture. Later in life, a tendency to unconsciously absorb a leader's wishes as one's own offers substantial benefits if the leader distributes status and rewards to those who support his beliefs. If a leader punishes those who are seen as opponents, identification offers additional advantages, while independent perception may have disastrous consequences (Barkow, 1976, 1980). If this is correct, people should identify most with those who are wealthy, powerful, opinionated, and punitive, a prediction that seems likely to be correct, but is difficult to derive from other theories.

Identification with the aggressor occurs when people accept, as their own, the wishes of someone who is exploiting or abusing them. This defense is often thought to be pathological, but it offers benefits in many situations. If a person has been captured by another group, the capacity to psychologically join the captors might be life-saving (Barkow, 1976). Within a group, identification with a powerful leader offers substantial advantages, even if it means accepting exploitation and humiliation at times. Individuals who accurately perceive exploitation may be at a considerable disadvantage. A common clinical problem is the difficulty of getting an abused spouse to acknowledge the abuse. Often there is an apparent loyalty to the abuser and a belief that the abuse is justified. In modern societies where women have some legal protection and opportunities for alternative mates, this tendency often simply perpetuates abuse, but in many traditional societies, such identification with the aggressor may prevent an even worse fate.

Splitting is a recently described and somewhat controversial defense (Dorpat, 1985; Kernberg, 1975) in which some people are idealized and others are depreciated. Patients who use this pattern often disrupt the relationships among psychiatric hospital staff by idealizing some and depreciating others. Though immature and unsavory, splitting is a powerful strategy in triangular competitions. Idealization strengthens the bond with one person, while derogation of others disengages the idealized person from previous allies. This seemingly pathological defense may be especially useful when most alliance partners are already committed. Children may learn to rely on splitting when it works especially well, for instance, when they can get support from one parent mainly by depreciating the other. If this speculation is correct, splitting should be especially common in children who have been emotional pawns in bitter divorces.

Rationalization and intellectualization are relatively mature defenses that are used regularly by normal people. *Rationalization* consists of making up alternative explanations that distract attention from true motives. This can be used for manipulation or to maintain relationships. *Intellectualization* is similar, in that the facts of a situation are acknowledged, but here the emotional content is kept carefully separate. This makes it possible to acknowledge the facts of a situation while avoiding disadvantageous displays of pleasure or anger.

Then there are the defenses that are seen as the most mature of all—humor and sublimation. *Humor* turns problematic confrontations into play, so that neither party is required to compete altogether seriously, with the risks that would entail. It allows graceful yielding without admitting inferior status. It can also be used to subtly insult a third party so as to define those present as an in-group in contrast to an inferior outgroup (Alexander, 1986). In *sublimation,* the wish is partially satisfied in displacement and some derivative satisfaction is achieved. Sublimation allows the partial satisfaction of forbidden wishes in socially acceptable ways.

Each of these defenses seems to have characteristics that are unnecessary for internal regulatory functions. In addition to being covert cognitive manipulations, they are also overt behavior patterns aimed at influencing others. Furthermore, the aspects of each defense seem to be designed to give benefits in certain situations. To the extent that the individual defenses can be shown to facilitate adaptive deception of others, this supports the broader hypothesis about repression in general.

How can this hypothesis about the specific defenses be further tested? It predicts that the habitual use of certain defenses should be closely related to certain social situations and relationship strategies. For instance, reaction formation should be common when expression of prohibited impulses is enforced by strict social norms in a tightly knit group. Identification with the aggressor should be common in people who have no alternative but to submit to a powerful figure. Splitting should be associated with situations in which an individual has few close relationships. The literature that relates personality types and defensive styles might offer more specific predictions to test the hypothesis that certain defenses are particularly useful in conjunction with certain interpersonal strategies.

MENTAL CONFLICT

One of the more widely accepted observations of psychoanalysis is the central role of conflict in mental life. This conflict does not seem to simply reflect mere competition between various possible behaviors. Instead, intrapsychic conflicts usually seem to have two sides, with impulses on one side and inhibitions on the other. This pattern so consistently describes the observations of analysts that they call the source of impulses the *id* and the modules that inhibit the expression of impulses, because of external and internal constraints, the *ego* and *superego*, respectively (Leak & Christopher, 1982; Trivers, 1985). The superego can be thought of as the conscience, while the ego is the locus of executive functions that balance satisfaction of impulses with anticipated internal and external costs.

This model of mental conflict poses a challenge for evolutionary psychology. It is most curious that conflict should occupy such an important place in the mind. The allocation of substantial mental-processing time to conflict seems unwieldy and inefficient. Why hasn't natural selection shaped a simpler algorithm to prioritize various behavioral options without all the complexity, anxiety, and symptoms that attend intrapsychic conflicts? Perhaps phylogenetic constraints make a simpler system impossible, but before accepting this hypothesis, we must consider the possible functional significance of mental conflict.

What important categories of decisions could be reflected by the pattern of mental conflict reported by psychoanalysts? One important category of decisions is whether to invest in direct reproduction or in kin. This does not, however, match the reports of analysts. For instance, the mother's desire to be with her baby seems to arise as much from id as from ego and superego. Other important categories of decisions concern relative investments in reproductive versus somatic effort or in defense versus foraging (Townsend & Calow, 1981). But the match between the psychoanalytic model and these tasks also seems poor. The id seems to motivate behavior that will bring individual satisfaction in the short run, while the superego motivates normative behavior that has short-term costs to the individual and benefits to others. What important resource allocation decision would be reflected in this dichotomy?

Human Darwinian fitness depends profoundly on success in social relationships, and success in social relationships depends upon the ability to correctly decide when to cooperate and when not to cooperate. Those who are too selfish have no friends and lose the social competition. Those who are indiscriminately generous are exploited and also lose. Natural selection may have divided human social motivation into two streams, one to advocate for each strategy. Conflict may be at the core of the mind because it is so essential for humans to correctly decide, at every moment, whether to invest in a relationship or a group that may offer long-term benefits or whether to directly pursue individual benefits. A person who invests substantial processing time to these decisions and makes them well will have a substantial fitness advantage over a person who makes them poorly.

A similar division of motivation has been proposed by Margolis (1982) to explain economic allocation patterns that cannot be accounted for by traditional theory. He postulates the Darwinian origins of separate mental agencies that motivate self-interested and group-interested spending in order to explain the economic conundrum of voluntary contributions to the public good. He does not note the striking parallel to psychoanalytic theory.

Frank (1988) explains Adam Smith's "moral emotions" as Darwinian solutions to the commitment problem. This problem refers to the advantages and difficulties of arranging a system of rewards and punishments to ensure a certain pattern of behavior in the future. By motivating behavior that has an immediate cost, a moral emotion can allow greater benefits later. These go beyond the benefits of reciprocity relationships to include the benefits of intimidating bullies by demonstrations of spiteful behavior.

The conflict at the core of the mind may thus be viewed as conflict between strategies that have short-term and long-term payoffs, between selfish and altruistic motivation, between pleasure seeking and normative behavior, and between individual and group interests. The functions of the id match the first half of each of these pairs, while the functions of the ego/superego match the second half. The association of long-term strategies with altruism, normative behavior, and group interests into a specialized motivational stream (the superego) may result from the importance and delayed nature of benefits from social relationships.

If, as seems to be the case, the capacity for negotiating and maintaining relationships has become steadily more important to hominid reproductive success, then the capacity for inhibiting and repressing "selfish" impulses has likely been increasing during human evolution, and the clinical description of such impulses as "primitive" may be more than just a figure of speech. In modern societies, however, with their cultural diversity, large fluid groups of nonkin, and transient relationships, the benefits of benevolent self-deception may decrease. In fact, tendencies to deceive oneself about the motives of others increase vulnerability to exploitation. Perhaps the rapid growth of psychotherapy in recent decades results, in part, from its ability to weaken the evolved tendency toward benevolent self-deception in societies where it is less useful.

This brings us back to the problem of repression. Why not be conscious of both sides of the conflict? In any given situation, whether an impulse is inhibited or expressed, the alternative is usually repressed. Why is this? The social benefits of a generous act are severely compromised if it is accompanied by indications that a selfish alternative was seriously considered. Repression conceals the rejected alternative and thus gives an advantage. When an aggressive impulse is expressed tentatively, it may be useless. Repression blocks the restraints of conscience so decisive action is possible. When repression is ineffective, neither side of the conflict may gain ascendency, and pathological ambivalence may paralyze effective action, as is often the case in obsessive compulsive disorder and schizophrenia.

CONSCIENCE, GUILT, AND NEUROSIS

This model of mental conflict offers a perspective on how conscience could have evolved. *Conscience*, the mental agency that punishes behavior that deviates from internal and external norms, is difficult to explain because behavior guided by norms is less flexible in changing situations and because so many norms promote altruistic behavior and thus increase vulnerability to exploitation. Granted, people's norms can change somewhat, and cultures vary in the rigidity with which norms regulate behavior, but people do seem to have a special mechanism for absorbing and maintaining certain norms. If the contents of conscience are regulated by a special system, this suggests that they may serve special functions.

One function of norms is to transmit cultural knowledge and conventions whose benefits are not obvious. Those who follow such conventions benefit both by following concrete rules (such as not eating pork) and by participating in social conventions (Barkow, 1976; Boyd & Richerson, 1985; Lumsden & Wilson, 1981; Tooby & Cosmides, 1989). Cultural norms may also provide a "language for relationships," a set of rules that facilitate relationships (Cosmides & Tooby, 1989; Tooby & Cosmides, 1989). Some such rules are arbitrary—it makes little difference if we drive on the right or left side of the road or if we greet each other with right or left hand, but such conventions, once established, are stable. Other rules are constrained by the tendencies of a brain that has been shaped by natural selection as a result of success and failure in dealing with just such situations. People follow both kinds of subtle rules of relationships as intuitively as they follow grammatical rules. If norms provide an emotional language for conducting relationships, then a fundamental bias toward social conservatism and conformity might well offer an advantage.

There has been much debate about other functions for conscience, but no consensus is in sight (Alexander, 1987; Axelrod, 1986; Campbell, 1975). Alexander has emphasized the possibility that the capacity for conscience evolved in special circumstances (stable kin groups competing with other groups) and that its current everyday manifestations are mainly attempts to manipulate others. This usefully emphasizes the manipulative uses of guilt and the importance of intergroup conflict in shaping norms. It does not, however, fully explain the manifestations of conscience in everyday circumstances. Conscience must have some benefit other than as a way we are manipulated by others for their benefit, or conscience would be selected against. And, if it functioned mainly to enforce group cooperation in the face of competition with other groups, our concern for everyday individual behavior within the group would occasion less moral concern.

Important moral rules tend to be about social behavior that requires short-term sacrifice. But, how can such behavior increase fitness? We hypothesize that the capacity for conscience may have been shaped by natural selection to promote and preserve reciprocity relationships. This explains why many moral principles require self-sacrifice for the sake of some relationship partner. It is also consistent with certain emotional inclinations. People who subordinate their own satisfactions to those of friends are valuable partners; those who cooperate only when a quick benefit is available are much less desirable (Barkow, 1980; Trivers, 1971, 1981). We do not seek reciprocity relationships that involve the mere trading of favors. Instead, we seek relationships based on apparently irrational emotional bonds. Because friends allow debts far beyond the available collateral, they provide help when times are hard, which is, of course, when it is needed most. This may be an example of how emotions help to solve the commitment problem (Frank, 1988).

Violation of an internalized moral rule sometimes offers substantial benefits. Such situations pose difficult moral dilemmas. Adaptive behavior in such circumstances often requires strict repression of either the wish or the norm, and symptoms often result. The anticipation of experiencing guilt weakens impulses to violate moral principles. When they are violated, guilt and self-destructive behavior may arise, even if the violation remains unconscious. Conversely, pride is aroused when an opportunity for taking a short-term gain is forgone.

Are some moral principles universal? Specific moral rules differ vastly in different cultures, but the attempt to find deep commonalties in moral systems has long been a concern of moral philosophy. Moral beliefs may share a deep structure that constrains them, as is the case for language (Chomsky, 1975; Cosmides and Tooby, 1989). Could the common patterns of kin and reciprocity relationships (Alexander, in press; Axelrod, 1986; Axelrod & Hamilton, 1981; Hamilton, 1964) shape a deep moral structure of the mind? Understanding such structures may be a necessary precursor to a valid cross-cultural psychodynamics.

These conjectures about the functions of conscience and guilt provide a framework for considering neurosis. People with *neurosis* remain unaware of many of their own impulses, strictly follow internal norms, and try hard to please others. They experience considerable anxiety about possible transgressions and guilt about past transgressions, and they are reluctant to express anger when others cheat or defect. In reciprocity terms, neurotics cooperate too consistently. Some neuroses must simply be strategies to attract and hold fair reciprocity partners, but neurosis can also be an exploitative strategy (Alexander, 1989; Hartung, 1988; Slavin, 1987). If especially altruistic people exist, it might be profitable to try to convince others that one is such a person in order to get them to try to establish relationships with you (Alexander, 1987). This strategy is facilitated by the ability to systematically exclude impure motives from consciousness, an ability that requires the subtle use of many defenses. When relationship partners prove untrustworthy, many neurotics do not leave the relationship. Instead, they induce guilt, demand retributions, and inflict subtle revenge (often unconsciously). In order to keep the partner from seeking relationships elsewhere, extensive and subtle attempts may be made to lower the partner's self-esteem. In psychotherapy, it is crucial, but usually difficult, to get such patients to admit that they have impulses that are less than pure. This is not surprising, since any such admission would cripple their main strategy for relating to others. Making the transition to a strategy based more on ordinary reciprocity is difficult to do gradually; one cannot simultaneously cooperate and defect. This may help to explain the period of unstable functioning that often is observed in the process of an ultimately successful psychotherapy for neurosis. Neurosis itself is not likely to be a product of evolution, but its pattern tends to confirm the hypothesis that guilt evolved to preserve reciprocity relationships.

TRANSFERENCE

Transference is the phenomenon of transferring (displacing) feelings about one relationship onto another. The psychoanalytic conceptualization of transference emphasizes the importance of early childhood relationships in shaping feelings about people later in life. The conclusion of psychoanalytic studies is that many strong feelings in adult relationships arise, not from current circumstances, but from expectations based on transference. Thus, we once again are faced with the problem of explaining a psychological mechanism that distorts reality.

The least controversial explanation for transference is that children's first relationships serve as templates for those that follow. Repeated experiences with primary care-takers build up mental representations of other people. Some of these expectations are taken for granted in healthy people: for instance, the expectation that others have separate motives, feelings, and movements. Other expectations, however, are shaped much more by the individual child's experiences. For example, a well-loved child may find it easy to believe that a new acquaintance wants to be his friend, while an abused child may expect to be disliked. Such mental models of the social world are useful and not at all surprising. What is surprising is that they tend to be so inflexible. Psycho-analytic studies find not only that people's transference expectations persist remark-ably in the face of contrary evidence, but that people seem to act in ways that induce others to take on the role of the original transference object. How could such an inflex-ible system offer an advantage?

One explanation is that early relationship experiences result in facility with certain strategies of conducting relationships, and practice makes these strategies more and more rewarding. Because some strategies are mutually exclusive (such as influencing others using threats or affection), personality patterns are further stabilized. These strategies are still further perpetuated by choosing partners who offer complementary roles and by the self-fulfilling nature of expectations. In short, early relationship strat-egies and transferences lead to the formation of stable personality characteristics (Blos, 1962; Buss, 1984).

Another explanation for the persistent nature of transference expectations is that patterns of interaction were probably much more stable in the hunter-gatherer bands that characterized most of our evolutionary past. In a stable culture, an individual who makes global assumptions about others based on sparse clues may better predict the behavior of others than those who wait to rely on "objective" learning. In our fluid and diverse society, transference expectations are less likely to be accurate and therefore less likely to be adaptive. While some aspects of transference may be mental patterns for relationships that are part of our "prepared learning," the more individualized and complex aspects of transference are a cause of character neuroses and other pathology. In earlier times, such tendencies might well have provided a grammar for the conduct of relationships and accurate anticipation of what could be expected from another per-son. Advantages arising from these functions could provide a modern biological expla-nation for the phylogeny of transference, which Freud (1915b/1987) so long sought.

CHILDHOOD SEXUALITY

The power of the concept of transference is closely connected to childhood sexuality because early attachments usually involve sexual wishes. Such conviction about the importance and prevalence of infantile sexuality often seems preposterous to those without psychoanalytic experience. Not only is there a lack of quantitative documen-tation, but the incest taboo makes such statements disturbing and, given the disad-vantages of inbreeding, biologically implausible. Nonetheless, psychodynamic inves-tigators have, for several generations, reported that children have intense conscious and unconscious sexual wishes toward their parents. How might this phenomenon be explained?

The first possibility is that children experience sexual wishes toward parents simply because they are the first available potential sexual partners. This view, advocated by

Bowlby (1969), sees childhood sexuality as an early stage in typical primate development, a precursor to adult exogamous sexuality. A refinement is offered by Rancour-Laferriere (1985) who proposes that childhood sexual interest in the opposite sex parent can offer a model for later appropriate mate choice. Since the parent is successful, choosing someone similar may offer an advantage. A related explanation of how childhood sexual identification may affect adult sexual behavior is offered by Draper and Harpending (1982).

Badcock (1986, 1988) offers a complementary view of childhood sexuality. He suggests that children may manipulate parents by precocious sexual signaling, a capacity that may be especially valuable in the light of sibling competition and parent-offspring conflict. Just as a woman may use sexual cues to get a male friend to offer assistance, a girl may use a similar strategy with her father, and a boy with his mother. Practicing these strategies with parents may also give additional benefits later in life. Badcock also suggests that such patterns of manipulation exist between members of the same sex (such as a son behaving in a submissive pseudofeminine manner in order to increase his father's cooperation) and that this might explain some unconscious homosexual feelings that are uncovered in the course of psychoanalysis.

Badcock (1989) further suggests that children may elicit still more investment from parents if their precocious sexuality indicates special competence in adulthood. He argues that sons will benefit more from this strategy, especially in polygamous cultures, because of the Trivers-Willard effect (1973). Such preferential investment in sons would arouse envy in daughters, envy that could explain certain nuances of the Oedipus complex (Badcock, 1989). This predicts that Oedipal wishes and envy will be more prominent in girls who have brothers and in families that are of high status.

One final speculation is the possibility that early transference patterns may provide the foundations for self-deceptions such as the idealization needed to fall in love or the unrealistic devaluation that occurs when people defect from an established relationship. In such instances, the seemingly unnecessary complexity of "incestuous" infantile sexuality may be parts of proximate mechanisms for regulation of adult sexual strategies.

As far as we know, there have not been attempts to objectively measure the intensity of Oedipal phenomena and to correlate them with family variables that might allow a test of such hypotheses. Whether these speculations on the possible functions of childhood sexuality turn out to be wrong or right, we believe they suggest enough possibilities to make it worthwhile to consider the possible functions of childhood sexuality.

IMPLICATIONS FOR PSYCHOANALYIS AND PSYCHIATRY

Psychoanalysis and psychiatry may derive important benefits from an integration with evolutionary biology. The most important is the possibility of a solid theoretical foundation in the natural sciences (Bowlby, 1981; Sulloway, 1979). Efforts to link psychoanalysis to proximate neurophysiology have not brought psychoanalysis into the scientific mainstream, but evolutionary psychobiology, with its growing emphasis on the adaptive functions of subunits of the mind, may offer a natural foundation for psychoanalysis. Such a foundation provides an excellent first test for any aspect of psychoanalytic theory; that is, is it consistent with evolution? This immediately weeds out ideas based on outmoded or implausible biology. Other previously problematic ideas gain legitimacy when viewed from an evolutionary perspective. For instance, Freud's emphasis on the sexual origins of human motivation as reflected in the concept of "libido" is remarkably congruent with the evolutionary psychobiologist's recognition of the crucial importance of reproductive success to human motivation.

An evolutionary approach gives rise to new questions for those who study psychodynamics: Why is there repression? How has natural selection shaped the framework for intrapsychic conflict? Why are there so many defenses, instead of just repression? What are the evolutionary benefits of conscience? Is childhood sexuality a strategy of parental manipulation? We do not pretend to have answered these questions, but we hope that we have convinced the reader that they are worth further study.

Evolutionary theory also can contribute to our understanding of psychopathology (McGuire & Essock-Vitale, 1981; McGuire & Fairbanks, 1977; Nesse, 1984; Wenegrat, 1984). The organization of clinical information according to evolutionarily significant concepts may clarify diagnosis and explain the patterns of some syndromes (McGuire & Essock-Vitale, 1981). The evolutionary significance of the capacities for anxiety (Marks, 1987; Nesse, 1987, 1988) and mood (Gardner, 1982; Sloman & Price, 1987) are being explored. Personality disorders may be interpreted as exaggerations of adaptive strategies for negotiating interpersonal relationships (Buss, 1984). And psychotherapy, the technique of intervention that has developed from proximate psychodynamic theories, may be clarified and made more effective when studied from an evolutionary perspective (A. Lloyd, 1990; Slavin, 1988).

Evolutionary theory may also provide the long-sought common language for psychoanalysis, neuroscience, and cognitive science. The "bio-psycho-social model" was first brought to American psychiatry by Adolf Meyer in an attempt to bring Darwinism and the adaptive significance of individual life events to a profession that mainly considered genetic and physical factors (Willmuth, 1986). Psychiatry is again preoccupied with proximate mechanisms and will again benefit from a scientific approach to adaptation. The goal of integration is now widely accepted in psychiatry, but still missing is a framework for linking studies at different levels of organization. By analyzing the adaptive functions of traits on all levels, an evolutionary perspective may provide such a framework.

IMPLICATIONS FOR EVOLUTIONARY PSYCHOLOGY

Psychoanalysis offers a perspective on mental life that has, so far, not been incorporated by evolutionary psychology. Its research method, free association, offers unique opportunities for naturalistic observation of the operation of psychological mechanisms. It also offers a wealth of information about human self-deception, the importance of which is just now being recognized by evolutionary biologists. Finally, it offers a theory of high-level mental mechanisms, one that is fragmented and sometimes obscure but that is descriptively rich and derived from clinical material, independent of modern evolutionary insights.

Evolutionists who want to use the data of psychoanalysis face many difficulties. The database of public clinical material is sparse. Objective observations are hard to disentangle from theoretical doctrines. And, there is a tendency for fields that lack full recognition as sciences, such as psychoanalysis and evolutionary psychology, to avoid associations with other fields whose scientific identity is also insecure. We believe it will be worth the effort to surmount these obstacles in order to tap psychoanalytic descriptions of mental mechanisms at high levels of abstraction. They describe patterns of self-deception and a rudimentary map through territory in which the evolutionist might otherwise become hopelessly lost. While evolutionists can propose mechanisms they expect to find based on the tasks that the mind must perform, psychoanalysts can offer their observations about the mechanisms they have observed.

CONCLUSION

Psychoanalysts and evolutionary psychologists share a view of the mind as a system of domain-specific mechanisms. The concepts used by psychoanalysts—repression, defenses, intrapsychic conflict, childhood sexuality, and transference—may not turn out to be the best categories for scientific research, but they are currently the best available at this level of mental organization. The reports of psychoanalysts about mental phenomena may be similar to those of early ethologists—rich sources of observation that are mixed, often haphazardly, with concepts and theories, some idiosyncratic, others prescient, with no simple way to distinguish between the two. The systematic application of evolutionary principles has transformed ethology into a mature science; perhaps it can do the same for psychoanalysis. If so, psychoanalysts will someday be recognized as pioneering naturalists of the mind.

ACKNOWLEDGMENTS

Preparation of this chapter was supported by the University of Michigan Evolution and Human Behavior Program, and the Psychiatry and Evolutionary Psychobiology Project. For thoughtful and provocative comments that have helped to develop this manuscript, we would like to thank C. Badcock, J. Barkow, L. Betzig, L. Cosmides, M. Daly, A. Eisen, E. Hill, K. Kerber, M. McGuire, M. Root, A. Tapp, J. Tooby, P. Turke, and M. Wilson.

REFERENCES

Alexander, R. D. (1975). The search for a general theory of behavior. *Behavioral Sciences, 20,* 77–100.

Alexander, R. D. (1979). *Darwinism and human affairs.* Seattle: University of Washington Press.

Alexander, R. D. (1986). Ostracism and indirect reciprocity: The reproductive significance of humor. *Ethology and Sociobiology, 7,* 253–270.

Alexander, R. D. (1987). *The biology of moral systems.* Hawthorne, NY: Aldine de Gruyter Press.

Alexander, R. D. (1989). Evolution of the human psyche. In P. Mellars & C. Stringer. (Eds.), *Origins and dispersal of modern humans* (pp. 455–513). Princeton, NJ: Princeton University Press.

Axelrod, R. (1984). *The evolution of cooperation.* New York: Basic Books.

Axelrod, R. (1986). An evolutionary approach to norms. *American Political Science Review, 80,* 1095–1111.

Axelrod, R., & Hamilton, W. D. (1981). The evolution of cooperation. *Science, 211,* 1390–1396.

Badcock, C. R. (1986). *The problem of altruism.* Oxford: Basil Blackwell.

Badcock, C. R. (1988). *Essential Freud.* Oxford: Basil Blackwell.

Badcock, C. R. (1989). *Oedipus in evolution: Three essays on the new theory of sexuality.* Oxford: Basil Blackwell.

Barash, D. P. (1982). *The whisperings within* (2nd ed.). New York: Harper & Row.

Barkow, J. (1976). Attention structure and internal representations. In M. R. Chance & R. R. Larsen (Eds.), *The social structure of attention* (pp. 203–219). London: Wiley.

Barkow, J. (1980). Biological evolution of culturally patterned behavior. In J. S. Lockard & D. L. Paulhus (Eds.), *The evolution of human social behavior* (pp. 277–296). Engelwood, NJ: Prentice Hall.

Barkow, J. (1984). The distance between genes and culture. *Journal of Anthropological Research, 40,* 367–379.

Blos, P. (1962). *On adolescence: A psychoanalytic interpretation.* New York: The Free Press.

Boden, M. A. (1987). *Artificial intelligence and the natural mind.* New York: Basic Books.

Bowlby, J. (1969). *Attachment and loss: Attachment* (Vol. 1). New York: Basic Books.

Bowlby, J. (1981). Psychoanalysis as a natural science. *International Review of Psycho-Analysis, 8,* 243–256.

Boyd, R., & Richerson, P. J. (1986). *Culture and the evolutionary process.* Chicago: The University of Chicago Press.

Brenner, C. (1974). *An elementary textbook of psychoanalysis* (rev. ed.). New York: Doubleday Press.

Buss, D. M. (1984). Evolutionary biology and personality psychology: Towards a conception of human nature and individual differences. *American Psychologist, 39,* 1135–1147.

Campbell, D. T. (1975). On the conflicts between biological and social evolution and between psychology and the moral tradition. *American Psychologist, 30,* 1103–1126.

Chomsky, N. (1975). *Reflections on language.* New York: Random House.

Cosmides, L., & Tooby, J. (1987). From evolution to behavior: Evolutionary psychology as the missing link. In John Dupré (Ed.), *The latest on the best: Essays on evolution and optimality* (pp. 277–306). Cambridge, MA: MIT Press.

Cosmides, L., & Tooby, J. (1989a). Evolutionary psychology and the generation of culture. Part I: Theoretical considerations. *Ethology and Sociobiology, 10,* 51–98.

Cosmides, L., & Tooby, J. (1989). Evolutionary psychology and the generation of culture. Part II: Case study: A computational theory of social exchange. *Ethology and Sociobiology, 10,* 51–98.

Crawford, C., Krebs, D., & Smith, M. (Eds.). (1987). *Sociobiology and psychology.* Hillsdale, NJ: Erlbaum.

Dawkins, R. (1976). *The selfish gene.* New York: Oxford Press.

Dawkins, R. (1982). *The extended phenotype: The gene as the unit of selection.* San Francisco: W. H. Freeman and Company.

DePaulo, B. M., & Rosenthal, R. (1979). Ambivalence, discrepancy, and deception. In R. Rosenthal (Ed.), *Skill in nonverbal communication: Individual differences.* Cambridge, MA: Oellgeschlager, Gunn & Hain.

DePaulo, B. M., Lanier, K., & Dans, T. (1983). Detecting the deceit of the motivated lier. *Journal of Personality and Social Psychology, 45,* 1096–1103.

Dorpat, T. L. (1985). *Denial and defense in the therapeutic situation.* New York: Jason Aronson.

Draper, P., & Harpending, H. (1982). Father absence and reproductive strategies: An evolutionary perspective. *Journal of Anthropological Research, 38,* 255–273.

Eckman, P., & Friesen, W. V. (1974). Detecting deception from the body or face. *Journal of Personality and Social Psychology, 29,* 288–298.

Ellenberger, H. F. (1970). *The discovery of the unconscious: The history and evolution of dynamic psychiatry.* New York: Basic Books.

Essock-Vitale, S. M., McGuire, M. T., & Hooper, B. (1988). Self-deception in social-support networks. In J. S. Lockard & D. L. Paulhus (Eds.), *Self-deception: An adaptive mechanism?* (pp. 200–211). Engelwood Cliffs, NJ: Prentice Hall.

Erdelyi, M. H. (1985). *Psychoanalysis: Freud's cognitive psychology.* New York: W. H. Freeman.

Fenichel, O. (1945). *The psychoanalytic theory of neurosis.* New York: W. W. Norton.

Fisher, R. P., & Greenberg, S. (1977). *The scientific credibility of Freud's theories and therapy.* New York: Basic Books.

Frank, R. H. (1988). *Passions within reason: The strategic role of the emotions.* New York: W. W. Norton.

Freud, A. (1936/1966). *The ego and the mechanisms of defense.* New York: International Universities Press.

Freud, S. (1912–1914). *Totem and taboo* (standard ed., Vol. 14). London: Hogarth Press.

Freud, S. (1915a). Instincts and their vicissitudes (standard ed.) (Vol. 14). London: Hogarth Press.

Freud, S. (1915b/1987) *A phylogenetic fantasy: Overview of the transference neuroses* (I. Gruberich-Simitis, Trans. & Ed.). Cambridge, MA: Belknap Press.

Freud, S. (1917). Mourning and melancholia (standard ed.) (Vol. 14). London: Hogarth Press.

Freud, S. (1920). Beyond the pleasure principle (standard ed.) (Vol. 18). London: Hogarth Press.

Freud, S. (1923). The ego and the id (standard ed.) (Vol. 19). London: Hogarth Press.

Gardner, R., Jr. (1982). Mechanisms in manic-depressive disorder. *Archives of General Psychiatry, 39*:1436–1441.

Gay, P. (1988). *Freud: A life for our times.* New York: W. W. Norton.

Goleman, D. (1985). *Vital lies, simple truths.* New York: Simon & Schuster.

Gould, S. J. (1977). *Ontogeny and phylogeny.* Cambridge, MA: Harvard University Press.

Gruberich-Simitis, I. (1987). Metapsychology and metabiology. Introduction to I. Gruberich-Simitis (Ed.), *A phylogenetic fantasy: Overview of the transference neuroses.* Cambridge, MA: Belknap Press.

Gur, C. R., & Sackheim, H. A. (1979). Self-Deception: A concept in search of a phenomenon. *Journal of Personality and Social Psychology, 37,* 147–169.

Hamilton, W. (1964). The genetical evolution of social behavior. *Journal of Theoretical Biology, 7,* 1–52.

Hartmann, H. (1939). *Ego psychology and the problem of adaptation.* New York: International Universities Press.

Hartung, J. (1988). Deceiving down. In J. S. Lockard & D. L. Paulhus (Eds.), *Self-deception: An adaptive mechanism?* (pp. 170–185). Engelwood Cliffs, NJ: Prentice Hall.

Horowitz, M. J. (Ed.). (1988). *Psychodynamics and cognition.* Chicago: University of Chicago Press.

Kernberg, O. F. (1975). *Borderline conditions and pathological narcissism.* New York: Jason Aronson.

Kihlstrom, J. F. (1987). The cognitive unconscious. *Science, 237,* 1445–1452.

Krebs, J. R., & Dawkins, R. (1984). Animal signals: Mind-reading and manipulation. In J. R. Krebs & N. B. Davies (Eds.), *Behavioral ecology* (pp. 380–402). Sunderland, MA: Sinauer Associates.

Leak, G. K., & Christopher, S. B. (1982). Freudian psychoanalysis and sociobiology: A synthesis. *American Psychologist, 37,* 313–322.

Lewicki, P. (1986). *Nonconscious social information processing.* Orlando, FL: Academic Press.

Lloyd, A. T. (1984). *On the Evolution of instincts: Implications for psycho-analysis.* Unpublished manuscript.

Lloyd, A. T. (1990). Implications of an evolutionary metapsychology for clinical psychoanalysis. *Journal of American Academic Psychoanalysis, 8*:286–306.

Lloyd, J. E. (1986). Firefly communication and deception: Oh, what a tangled web. In R. W.

Mitchell & N. S. Thompson (Eds.), *Deception: Perspectives on human and nonhuman deceit.* Albany, NY: SUNY Press.

Lockard, J. (1980). Speculations on the adaptive significance of self-deception. In J. Lockard (Ed.), *The evolution of human behavior.* New York: Elsevier.

Lockard, J. S., & Paulhus, D. L. (Eds.). (1988). *Self-deception: An adaptive mechanism?* Engelwood Cliffs, NJ: Prentice Hall.

Lumsden, C. J., & Wilson, E. O. (1981). *Genes, mind, and culture.* Cambridge, MA: Harvard University Press.

MacDonald, K. (1986). Civilization and its discontents revisited: Freud as an evolutionary biologist. *Journal of Social and Biological Structures, 9*, 307–318.

Margolis, H. (1982). *Selfishness, altruism, and rationality: A theory of social choice.* Chicago: University of Chicago Press.

Marks, I. M. (1987). *Fears, phobias, and rituals.* New York: Oxford University Press.

Mayr, E. (1988). *Towards a new philosophy of biology.* Cambridge, MA: Belknap Press.

McGuire, M. T., & Essock-Vitale, S. (1981). Psychiatric disorders in the context of evolutionary biology: A functional classification of behavior. *Journal of Nervous and Mental Disease 169*, 672–686.

McGuire, M. T., & Fairbanks, L. A. (Eds.). (1977). Ethological psychiatry: *Psychopathology in the context of evolutionary biology.* New York: Grune & Stratton.

Mitchell, R. W., & Thompson, N. S. (Eds.). (1986). *Deception: Perspectives on human and non-human deceit.* Albany, NY: SUNY Press.

Nesse, R. M. (1984). An evolutionary perspective on psychiatry. *Comprehensive Psychiatry, 25*, 575–580.

Nesse, R. M. (1987). An evolutionary perspective on panic disorder and agoraphobia. *Ethology and Sociobiology 8*, 73s–85s.

Nesse, R. M. (1988). Panic disorder: An evolutionary view. *Psychiatric Annals, 18*, 478–483.

Nesse, R. M. (1990). The evolutionary functions of repression and the ego defenses. *Journal of American Academy of Psychoanalysis. 18*(2), 260–285.

Rancour-Laferriere, D. (1985). *Signs of the flesh.* New York: Aldine de Gruyter.

Ritvo, L. B. (1964). Darwin as the source of Freud's neo-Lamarkianism. *Journal of the American Psychoanalitical Association, 46*, 499–517.

Rosenthal, R., Jall, J. A., DiMatteo, M. R., Rogers, P. L., & Archer, D. (1979). *Sensitivity to nonverbal communication: The PONS test.* Baltimore, MD: Johns Hopkins University Press.

Sackheim, H. A. (1988). Self-deception: A synthesis. In J. S. Lockard & D. L. Paulhus (Eds.), *Self-deception: An adaptive mechanism?* (pp. 146–165). Englewood Cliffs, NJ: Prentice Hall.

Shevrin, H., & Dickman, S. (1980). The psychological unconscious: A necessary assumption for all psychological theory? *American Psychologist, 35*, 421–434.

Slavin, M. O. (1987). The origins of psychic conflict and the adaptive functions of repression: An evolutionary biological view. *Psychoanalysis Contemporary Thought, 8*, 407–440.

Slavin, M. O. (1988, May 7). *Parent/offspring conflict and the evolution of repression.* Presentation to The American Academy of Psychoanalysis Annual Meeting, Montreal, Canada.

Sloman, L., & Price, J. S. (1987). Losing behavior (yielding subroutine) and human depression: Proximate and selective mechanisms. *Ethology and Sociobiology, 8*, 99s–109s.

Sulloway, F. J. (1979). *Freud, biologist of the mind.* New York: Basic Books.

Symons, D. (1987). If we're all Darwinians, what's all the fuss about? In C. Crawford, D. Krebs, & M. Smith (Eds.), *Sociobiology and psychology.* Hillsdale, NJ: Erlbaum.

Symons, D. (1989). A critique of Darwinian anthropology. *Ethology and Sociobiology, 10:* 131–144.

Taylor, S. E., & Brown, J. (1988). Illusion and well being: A social psychological perspective on mental health. *Psychological Bulletin, 103*, 193–210.

Tooby, J. (1985). The emergence of evolutionary psychology. In D. Pines (Ed.), *Emerging syntheses in science.* Santa Fe: Santa Fe Institute.

Tooby, J. & Cosmides, L. (1989). Evolutionary psychology and the generation of culture. Part I: Theoretical considerations. *Ethology and Sociobiology, 10:* 51–98.

Townsend, C. R., & Calow, P. (Eds.). (1981). *Physiological ecology: An evolutionary approach to resource use.* Sunderland, MA: Sinauer Associates.

Trivers, R. L. (1971). The evolution of reciprocal altruism. *Quarterly Review of Biology, 46,* 35–57.

Trivers, R. L. (1974). Parent-offspring conflict. *American Zoology, 14,* 249–264.

Trivers, R. L. (1976). Foreword in R. Dawkins (Ed.), *The selfish gene.* New York: Oxford Press.

Trivers, R. L. (1981). Sociobiology and politics. In E. White (Ed.), *Sociobiology and politics* (pp. 1–43). Lexington, MA: Lexington Books, D. C. Heath.

Trivers, R. L. (1985). *Social evolution.* California: Benjamin/Cummings.

Trivers, R. L., & Willard, D. E. (1973). Natural selection of parental ability to vary sex ratio of offspring. *Science, 179,* 90–92.

Wallace, B. (1973). Misinformation, fitness and selection. *American Naturalist, 107,* 1–7.

Wenegrat, B. (1984). *Sociobiology and mental disorder: A new view.* Menlo Park, CA: Addison-Wesley.

Williams, G. C. (1966). *Adaptation and natural selection. A critique of some current evolutionary thought.* Princeton, NJ: Princeton University Press.

Willmuth, L. R. (1986). A retrospective evaluation—Darwin comes to American psychiatry: Evolutionary biology and Adolf Meyer. *Journal of Social and Biological Structures, 9,* 279–287.

Wilson, E. O. (1975). *Sociobiology.* Cambridge, MA: Harvard University Press.

VIII

NEW THEORETICAL APPROACHES TO CULTURAL PHENOMENA

What is hardest to grasp about the functional patterns forged by natural selection is that their analysis is relentlessly past-oriented: An evolved adaptation exists now only because it served a function in the past, regardless of its consequences in the present (see Barkow, 1989a; Symons, this volume). In contrast, most branches of functional analysis, from everyday teleology to social science functionalism, are present or future oriented. When functionalists, from Merton to Malinowski to Harris, ask what the function of something is, they mean: How is it explained by the utility of its consequences?

However appealing, functionalism in the social sciences has fallen into disfavor, largely under the weight of two lines of criticism. Simply put, the first is that causes precede consequences, so that (ordinarily) the consequences of a phenomenon can be neither the cause of the phenomenon nor its explanation. Causal explanations must necessarily focus on antecedent conditions. The second criticism is the observation that human action doesn't look all that functional anyway, regardless of what your standard of functionality is. People commit suicide, live for their music, engage in convoluted rituals, read obsessively about famous people, bleed themselves when they get sick, and otherwise participate in an endless parade of what, from a utilitarian point of view, appears to be madness and folly. Human behavior appears to reflect an odd mixture of complexly functional and elaborately nonfunctional conduct. Attempts to explain away the large classes of apparently nonfunctional behavior as subtly functional usually seem strained and unpersuasive.

As Barkow's article demonstrates, although traditional functionalism is susceptible to these two lines of criticism, a conceptually integrated approach is not. In the first place, Darwin's theory of natural selection provides an explanation of how functional design can emerge from a nonforesightful causal process (Dawkins, 1986). The recurrent consequences of a design feature on reproduction cause it to be selected out or, alternatively, to spread throughout the population until it has become incorporated into the standard design of the species. In other words, a design feature's functional consequences *in earlier generations* explain its presence in this one. The evolution of adaptations by natural selection is one of the few cases in which the consequences of something can be properly used to explain its existence (Dawkins, 1986).

Conveniently, the fact that the evolutionary standard of functionality refers to past conditions and not present conditions supplies an answer to the second line of criticism as well. Why does human behavior appear to be functionally patterned in some respects, but obviously nonfunctional, or even maladaptive, in others? The answer lies in the fact that mechanisms designed to perform well under one set of conditions will often perform poorly under changed conditions. Our adaptations were designed (i.e., selected) to operate in the Pleistocene context in which they evolved, regardless of what that design would lead to in changed circumstances. Modern conditions differ in many important respects from the Pleistocene world of hunter-gatherers, and so many of our psychological mechanisms are operating outside of the envelope of conditions in which they can be expected to perform functionally. In consequence, our psychological mechanisms will often produce behavior that is maladaptive (Barkow, 1989a; 1989b).

As Barkow makes clear, evolutionary functionalism, properly understood, is fundamentally different from traditional social science functionalist approaches, and free of their defects. There is no Procrustean burden to show how the present consequences of behavior are functional in the modern context. Instead, an evolutionary functionalist approach is completely open to the possibility that any specific behavior, from watching soap operas on television, to reading about the famous, to engaging in symbolically intricate and costly rituals, is presently nonfunctional. But for the same reason, no modern behavior is exempt from or beyond the scope of evolutionary functionalist analysis either, not even behavior that is currently nonfunctional or evolutionarily novel. All behavior is the product of our evolved psychology (together with other factors), and so even the most novel of modern behaviors needs to be related to the design of our psychological architecture—a design that makes functional sense when located in the context of Pleistocene hunter-gatherers. In this final section, Barkow argues that even the most novel cultural forms, such as soap operas and social stratification, are expressions of our Pleistocene psychology taking "biologically unanticipated" forms. Freed from the Standard Social Science Model's stricture to locate all causes outside of psychology, modern approaches to culture can explore the ancient psychological mechanisms that underlie and explain recently emerging cultural phenomena. As Barkow succinctly puts it, "beneath new culture is old psychology."

REFERENCES

Barkow, J. H. (1989a). *Darwin, sex, and status: Biological approaches to mind and culture.* Toronto: University of Toronto Press.

Barkow, J. H. (1989b). The elastic between genes and culture. *Ethology and Sociobiology, 10,* 111–129.

Dawkins, R. (1986). *The blind watchmaker.* New York: Norton.

18

Beneath New Culture Is Old Psychology: Gossip and Social Stratification

JEROME H. BARKOW

How do we apply an evolutionary perspective to human culture? After all, much of post-Pleistocene society is *evolutionarily unanticipated,* that is, many of its most prominent features could not have existed during the Pleistocene. Even contemporary agrarian communities differ drastically from the foraging economies of our ancestors, and modern industrial societies are utterly exotic in Pleistocene perspective, with their vast scale, literacy and libraries, mass communication and media, ease of travel, general-purpose money, control of indoor climate, ubiquity of strangers, reliance on external institutions to accomplish heretofore familial responsibilities, availability of contraception, juxtaposition of great wealth with great poverty, and relatively low infant and child mortality rates. But does all this novelty mean that evolution is irrelevant to an understanding of our current way of life? Key aspects of most contemporary societies could have emerged, after all, only long after human psychology had largely reached its present form and then either ceased to evolve entirely or to evolve at a very slow rate.[1]

Fortunately, the concepts of vertical integration and evolutionary psychology (discussed in the Introduction and in Barkow, 1989, 1991; Tooby & Cosmides, 1989) suggest that there is little human that is really "evolutionarily unanticipated." Our evolved psychology underlies even the most novel and complex of sociocultural forms.[2] Beneath new culture is old psychology. Psychological traits originally selected for because (presumably) they led to adaptive behavior in earlier environments today underlie complex and novel sociocultural forms. As Symons points out in this volume, seeking to measure the current adaptive value of such forms is of much less theoretical interest than is understanding their relationship to our evolved psychology.

The two "biologically unanticipated" social phenomena this chapter deals with are (a) gossip, soap operas, and "celebrities"; and (b) social stratification. I have deliberately chosen disparate examples of evolutionary novelty in order to illustrate how the evolving field of evolutionary psychology permits powerful, vertically integrated explanations of major sociocultural phenomena.

SELECTION FOR ATTENDING TO ONE'S OWN: GOSSIP, SOCIAL CONTROL, REPUTATION, AND *PEOPLE MAGAZINE*

It was the late Max Gluckman (1963) who taught us that gossip defines the social group: If no one tells you the gossip, you are an outsider. Gossip from an anthropologist's perspective is a means of social control, a sanction that forces one to adhere more closely to social norms than one would otherwise be inclined. Reputation is determined by gossip, and the casual conversations of others affect one's relative standing and one's acceptability as a mate or as a partner in social exchange. In Euro-American society, gossiping may at times be publicly disvalued and disowned, but it remains a favorite pastime, as it no doubt is in all human societies. There is, however, something quite bizarre about much of our own society's gossip.

In every society save our own (and in recent times others, as mass entertainment spreads), one ordinarily gossips about real people, people known to oneself or at least known to those whom one knows. In our society, one very often gossips about perfect strangers, people with whom we have no common acquaintance and whom we are unlikely ever to meet. Our "knowledge" about them, moreover, may be deliberately fabricated by individuals who earn a living in this manner. Indeed, many of the people we gossip about may themselves be fabricated, that is, be entirely fictional, portrayed only by professional actors in "soap operas." We often read magazines or watch television programs exclusively devoted to the heavily fictionalized accounts of strangers, and we eagerly exchange the knowledge so gleaned when gossiping with our flesh-and-blood friends and relations.

Why should we do these strange things? First, why do we gossip at all? Second, why, in some societies, do we gossip about strangers? Let us think about these questions from an evolutionary perspective.

Evolution and the Psychology of Gossip

Gossip has to do with the exchange of information about other people. It is increasingly apparent that much of human intelligence is social intelligence, the product of selection for success in social competition: There is little doubt that we were selected for the ability to predict and influence the behavior of potential rivals for resources, present and potential allies, possible mates, and of course, close kin (Alexander, 1979; Barkow, 1983, 1989; Humphrey, 1976, 1983). Presumably, this predictive ability implies the development of elaborate internal representations of others.

Which others? It would be highly inefficient to construct elaborate internal models of everyone, so selection presumably would have favored our being discriminating. Similarly, an internal representation is not a homunculus: only some kinds of information rather than others would be incorporated. Which individuals and what kinds of information about them, then, would we have been selected to attend to? Let us rephrase this last pair of questions: which individuals are most likely to have affected the inclusive fitness of our ancestors, and what kinds of information would have had the most bearing on fitness? The short answer to "which individuals" is relatives, rivals, mates, offspring, partners in social exchange, and the very high-ranking. The short answer to "what kinds of information" is relative standing and anything likely to affect it, control over resources, sexual activities, births and deaths, current alliances/friendships and political involvements, health, and reputation about reliability as a partner in social exchange.

An evolutionary perspective suggests that we have been selected to be keenly interested in (that is, to maintain internal representations of) individuals whose relationship to us is such that, were we and they living in a Pleistocene environment, their behavior would be likely to affect our inclusive fitness. We should be especially interested in those of their activities that would have been the most likely to affect that fitness. For example, the sexual activities of a rival, a potential mate, or a relative clearly would have been fitness-relevant for our Pleistocene ancestors, and it does seem likely that, today, most of us are quite interested in gossip about the sexual activities of individuals in these categories. Thinking in terms of Pleistocene adaptation suggests that we should not find gossip about, for example, how soundly so-and-so is sleeping—a fact with relatively little fitness-relevance—nearly as interesting as gossip about with whom so-and-so is sleeping. In similar fashion, the activities of a stranger with whom we are unlikely to maintain a relationship and who is unlikely to engage in social exchange or sexual relations with a relative of ours would have had little bearing on our fitness, during the Pleistocene, and so should hold little interest for us today. Note how, by assuming that the modern practice of gossip is the result of evolved psychological mechanisms that had adaptive consequences in the Pleistocene, whatever their effects today, we begin to develop a framework and identify variables that, one hopes, eventually will lead to a formal and testable theory of gossip.

Although gossip is often an invaluable source of information, it is also unreliable. This is because selection would have favored our disseminating information in the interests, not of objective truth but of our own success in social competition (Barkow, 1989), so that we tend to derogate rivals and mask our own weaknesses (cf. Buss & Dedden 1990). We should expect evolved biases, in short, involving not just what we like to listen to but also what we like to say.

Because the fitness interests and therefore informational requirements of individuals of different age and gender are not the same, it is possible that their gossip is also not the same: but we need not exaggerate the extent of the age/gender differences suggested by a Pleistocene perspective. Our gossip influences the behavior of kin whose ages and genders are different from ours and whose behavior will affect our own inclusive fitness. Since we act as transmitters of information to and about these kin and their potential rivals and mates, selection is unlikely to have favored sharp age/gender differences in gossip. Thus, an adaptationist approach does not demand very strong age/gender differences in the content of the gossip we disseminate and find interesting, though it clearly suggests the possibility of some such differences, and hypothesizing their existence would lead to worthwhile research. Buss and Dedden (1990) have already reported, for one age-group in one society, gender differences in the content of derogation of sexual rivals.

Why Do We Gossip About Strangers?

One last question remains: why do we gossip about strangers, and even about fictitious characters? After all, we apparently evolved the psychological mechanisms underlying gossip because they helped us to exchange and manipulate information about members of our own community, real people whose activities impinged directly and indirectly upon our own fitness: Surely the health and sexual, political, and status behaviors of total and possibly nonexistent strangers with whom neither we nor our kin have any direct contact are unlikely to affect our fitness. Why then do we gossip about the strangers who inhabit our television screens and about media celebrities, and even read about them?

A possible answer is that the mass media may activate the psychological mechanisms that evolved in response to selection for the acquisition of social information.

The media may mimic the psychophysical cues that would have triggered these same mechanisms under Pleistocene conditions. As a result, strangers present only on cathode ray tubes in our living rooms, or magnified many times life-size on the screens of motion picture theatres, are mistaken for important band members by the algorithms of the evolved mechanisms of our brains. We see them in our bedrooms, we hear their voices when we dine: If this hypothesis is correct, how are we not to perceive them as our kin, our friends, perhaps even our rivals? As a result, we automatically seek information about their physical health, about changes in their relative standing, and, above all, about their sexual relationships. Treating them as friends (as partners in social exchange), we may derogate those whom we perceive as their rivals, while praising others frequently. Their behavior may leave us feeling betrayed, or sexually aroused, or politically victorious. We may be embarrassed at our strong interest in fictive people, but our knowledge of the pointlessness of this behavior is more likely to lead us to mask than to alter it (just as a knowledge of proper nutrition by itself may have little impact on eating behavior).

When the media give us constant images not of professional entertainers but of political figures, these, too, may be assimilated (and with considerably greater justification) as high-ranking members of our local community. Certainly, this has occurred with the British royal family in North America and in much of Europe. Public demand for information about such figures has created an immense industry of people who make a living seeking to discover (or invent) the most personal details of the lives of such persons, often plaguing them mercilessly (e.g., Italy's notorious *paparazzi*). A comparison of the mass media industry and that of the purveyors of "junk food" suggests itself: In modern market economies, food processors take our strong evolved preferences for salt, sugar, and fat—tastes that in the Pleistocene were indicators of scarce and valuable nutrients (Barkow, 1989)—and use them to influence our buying behavior. The mass media, it can be argued, make similar use of the evolved mechanisms that, in earlier environments, would have led to the acquisition of information about important members of our bands. In short, Pleistocene adaptations can lead to profits!

Because popular media stars did not exist in our earlier environments, there was never selection pressure in favor of our distinguishing between genuine members of our community whose actions had real effects on our lives and those of our kin and acquaintances, and the images and voices with which the entertainment industry bombards us. Media stars represent an evolutionarily unanticipated phenomenon.

Testable Hypotheses

This analysis of gossip readily yields testable hypotheses. For example, if the approach is correct then research should find the following:

1. The advertising industry and the campaign managers of politicians are correct in believing that the endorsements of media celebrities do influence purchasing and voting behavior, even when there is no "rational" reason for this influence to exist (i.e., the endorsers have no special qualifications or expertise in the fields in which they are giving advice). It would be interesting to compare buying behavior after an endorsement of a food product by a popular media figure and after one by a person with relevant credentials, such as a nutritionist. The current analysis suggests that the media figure would be the more influ-

ential. (Because many of us may experience the influence of media figures as discordant, so that we may tend to deny its existence, empirical researchers would probably want to focus on a dependent variable other than verbal response.)

2. If the evolutionary analysis presented here is valid then gossip should tend to focus on sexual activities and skill, political activities, reputation about reliability in social exchange, health and physical condition, births and deaths, and factors affecting the relative standing of others (including the control of resources).

3. If the evolutionary analysis presented here is valid then the subjects of gossip should be the relatives, rivals, etc. mentioned earlier, rather than passing strangers, media celebrities being the exception to this generalization. Though this hypothesis may seem trivial because all of us at some point have been bored by gossip about people whom we do not know, it is a specific prediction that stems directly from an evolutionary perspective and perhaps from no other.

4. The contents of gossip should differ only in detail across human societies. That Gluckman (1963) tells us that fieldworkers know that they have been accepted when local people share the gossip with them suggests that his field experience supports this hypothesis (as does my own) of the essential similarity of gossip everywhere.

While there is as yet no empirical, cross-cultural research on the content of gossip with which to confirm or deny these hypotheses, perhaps there eventually will be.

NEPOTISM AND SOCIAL STRATIFICATION

What is the relationship of social stratification to psychology and evolution? Stratification is a major organizing concept in the social sciences, where it usually refers to the division of a society into classes and castes. "A stratified society," explains the eminent political anthropologist Morton H. Fried (1967, p. 186), "is one in which members of the same sex and equivalent age do not have equal access to the basic resources that sustain life." When membership in a social stratum is ascribed by birth and mobility is considered to be in principle impossible, we usually speak of "castes" (especially when the strata are associated with particular economic activities). "Classes," on the other hand, are strata in which some mobility is at least in principle possible. Social strata, whether castes or classes, are by definition unequal in rank, wealth, and power. Anthropology, sociology, and survey research would be very different fields without the powerful and ubiquitous theoretical construct of social stratification. As for Marxism, the concept of social stratification, particularly that of class, is as central to that paradigm as the idea of "energy" is to physics.

Surprisingly, social stratification is not all that ancient a phenomenon in human history. Stratification requires a sufficiently large-scale society to produce a surplus with which to support some degree of political and economic specialization. Foraging peoples rarely have such surpluses, and our ancestors were of course foragers. Such peoples are generally organized in terms of bands, a band being "a local group composed of a small number of individuals" (Fried, 1967, 67).[3] In bands, all individuals have approximately equal access to resources and usually share an ideology of egalitarianism. This is not to say that bands or any other societies are unequivocally egalitarian—gender inequality as well as other forms of inequity may certainly exist, and the issue of how egalitarian "egalitarian" societies actually are is one of current debate in anthropology (Flanagan, 1989). Fortunately, that debate is only tangential to the noncontroversial statement that small-scale foraging band-level societies generally lack institutionalized social hierarchies, that is, they are not stratified. Thus, so long as our ancestors were organized in terms of such bands, their societies could not have been stratified.

Of course, it is possible that the megafauna of the Pleistocene, or possibly self-renewing supplies of shellfish, at times permitted the accumulation of economic surpluses and thus of relatively high population densities (much as the annual salmon run did for Northwest Coast peoples such as the Kwakiutl). Very rich, naturally occurring stands of grain in some places may have had a similar effect. Such circumstances of economic abundance could have permitted the existence of socially stratified societies even prior to the domestication of plants and animals (the Neolithic era). Thus, stratified societies could conceivably have developed at any point after the advent of modern human beings. However, since these hypothetical stratified but pre-Neolithic societies would have required unusually lush conditions coupled with the appropriate technology with which to take advantage of them, it seems reasonable to assume that they would have included only a small proportion of ancestral hominid populations, at any given point in time earlier than 10,000 to 12,000 years ago (when domestication developed in the Near East). If this reasoning is valid, then prior to the Neolithic, few (if any) members of the species *Homo sapiens sapiens* would have lived in societies sufficiently large-scale and complex to support social classes. Stratification must therefore be relatively recent. If we assume that our species is on the order of 200,000 years old, then it is likely that stratification has been typical of our societies for less than 10% of that period.

Note that social stratification itself, as a group-level phenomenon generated by the social interaction of individuals over historical time, could not have been directly selected for. The psychological traits that enable individuals to generate stratification, however, presumably are products of natural selection.[4] What, then, are the evolved psychological traits that might generate social stratification?

The Psychological Characteristics Underlying Social Stratification

There appear to be three hominid psychological traits that make it possible for us to generate, under some conditions, social stratification. These traits are (a) the pursuit of high social rank; (b) nepotism; and (c) the capacity for social exchange.

1. Seeking High Social Rank

Human beings share the general primate tendency to seek high social rank. That we do so is an implicit assumption of most of the social and behavioral science literature, where the subject is discussed under such rubrics as self-esteem, rank, status, power, prestige, and the need for achievement. Because this assumption is so uncontroversial (it is rarely even treated as a hypothesis), and since I have discussed it at length and from several perspectives in previous publications (1975a, 1975b, 1980, 1989), I will refrain from justifying it in detail in these pages. But note that typifying human beings as tending to seek higher relative standing is not the equivalent of asserting that this is the sole human motivation, that it dominates us all equally, that it has the same strength at all stages of our lives, or that we all seek it in identical ways. From the pre-school years onward, however, we are much concerned with our relative standing and generally seek to improve it in various ways and to communicate to others that our rank is higher than our rivals might concede. This concern with relative standing is the first suggested psychological root of social stratification.

2. Nepotism

The second psychological trait that may underlie social stratification is the tendency of human beings everywhere to favor their own offspring over nonoffspring, close relatives over distant, and kin in general over nonkin. Let us call this familiar but little-studied psychological characteristic by its ordinary language term, "nepotism." In his paper on kin selection theory (inclusive fitness theory), Hamilton (1964) elucidated the selection pressures that can be expected to lead to the evolution of nepotism. Although the cognitive processes governing nepotism have not been studied in a detailed way in human beings, behavioral ecologists have documented the existence of nepotism in a wide variety of other species (see Krebs and Davies [1987] for examples), and have shown that the psychological mechanisms involved have the design features one would expect given the selection pressures outlined by Hamilton.

Our own society formally disavows nepotism, so that the word often connotes "corruption," but we need not confuse the ideal with the real. While there are no doubt nonnepotistic individuals among us who disown close relatives or who are unwilling to find their children summer jobs, most of us are indeed nepotistic, often making considerable sacrifices on behalf of our children and grandchildren, our siblings and their children, which we would not make for nonkin. Despite our claims of allegiance to egalitarian values, our institutions often reflect nepotism, judging from the existence of superior school systems for the children of the wealthy, our inheritance tax laws and their loopholes, the "legacy" policies of prestigious universities, and criteria for admission to certain labor unions and medical schools.

Although Europeans and North Americans may ideally consider nepotism to be "immoral" in that it violates egalitarian codes and may make for economic/bureaucratic inefficiency, other peoples often have conflicting values. For many of the world's cultures, the notion that one should hire a perfect stranger who happens to have scored well on some civil service examination or have some kind of formal certification, rather than hire one's own brother or cousin or nephew, is the height of disloyalty and immorality.[5] For Euro-Americans to be surprised at such nepotism is ethnocentric and, some would argue, hypocritical.

3. Social Exchange and the Ability to Form Coalitions

Nepotism and the pursuit of high relative standing may be necessary but are not sufficient species traits for the generation of social stratification. Upper classes are in part political alliances, or even conspiracies, to maintain shared advantage. Such alliances are possible because of our species' ability to engage in social exchange. Trivers (1971) and Axelrod and Hamilton (1981) have discussed the selection pressures that would have shaped the psychological mechanisms for engaging in social exchange, and Cosmides and Tooby (1989, this volume) have investigated the structure of the cognitive processes that govern social exchange in humans. The maintenance of élite social strata involves relationships that go far beyond the boundaries of kinship, relationships in which nonrelatives cooperate with one another for mutually advantageous ends.

As in the case of striving for social rank, it hardly seems necessary to document that the capacity for social exchange is a human psychological trait, for this assumption is ubiquitous in the social and behavioral sciences (e.g., Blau, 1964; Homans, 1967; Sahlins, 1972; Titmuss, 1971). Members of ruling social classes collectively enjoy political power and control of economic resources at the expense of others. To do so, they must engage in numerous political and economic exchanges. The capacity for social exchange is the third and last psychological root of social stratification.

The Genesis of Social Class

Here is how the three evolved psychological roots of social stratification may interact, under certain historical circumstances, to generate élites. Individuals seek high relative standing. They engage in social exchange with others in order to achieve and maintain such standing and to transmit it to their children and other close kin. These tendencies are presumably universal. In some societies, however, there is some surplus production. High relative standing automatically would tend to involve some control over that surplus. The social exchanges through which individuals maintain their high standing now have a strong economic component. Enter nepotism. Those who achieve high standing and control over resources seek to transmit these advantages to their offspring and to other close kin. Social exchange facilitates that transmission, too, as reciprocal altruism merges with kin altruism so that members of the élite aid one another's children. Higher-ranking individuals exchange aid with people of all ranks but the low-ranking cannot afford large debts and have much less to offer than do other higher-ranking individuals and families, and when the relative rank of the participants is unequal then often so too are the exchanges. Among themselves, the higher-ranking favor exchanges that eventually cement themselves and their families (despite rivalries) into relatively coherent, self-interested élites or upper classes. These élites favor their own close kin while striving to reduce or eliminate competition from the progeny of the lower strata, producing barriers to social mobility. Thus, whenever a society achieves a relationship among its population density, environment, and technology such that surplus production of food and other goods reliably results, the psychology of our species makes it very likely that social inequality and, eventually, social stratification will soon follow. For archaeologists and other social scientists, the question arises of under precisely what conditions of environment, population density, and technology did social stratification first emerge in the various regions of the world, but essaying an answer would be well beyond the scope of this contribution.[6]

Note that until fairly large-scale societies existed, with accumulated surpluses of resources, the extent to which parents could ensure the future standing of offspring and other close kin was strictly limited. In band-level societies in which relative standing depends largely upon individual accomplishment, parents cannot guarantee that their own children, let alone other kin, will have high rank. But once we have an increase in societal scale and technology such that there is a possibility of control over some resource, or a surplus of production, then the tendency toward nepotism means that parents will strive to transmit that control or surplus to their offspring (and grandparents to grandchildren). Ordinarily, they will be unable to do so without the help of others, and frequently these "others" will not be kin. If this analysis is correct then parents will find themselves engaging in social exchanges with others, in effect organizing a political conspiracy to ensure that their children and those of their partners will also enjoy positions of power. In this view it is nepotism plus social exchange in an environment in which unequal control of resources is possible that leads to social stratification.[7]

Some Hypotheses

The argument provided here was phrased in terms of postulates and assumptions. To falsify the argument, it is necessary to treat these assumptions as hypotheses. Thus, I have in effect hypothesized that (a) human beings tend to strive for higher relative standing and this striving usually takes the form of seeking control over surplus production or over the means of production; (b) human beings everywhere tend to be nepotistic; (c) the view of social exchange algorithms presented by Cosmides and Tooby (1989, this volume) is essentially correct; and (d) both cross-culturally and historically, surplus production is associated with differences in social rank in all cases, and with social stratification in most. If any of these four hypotheses is inaccurate, then the entire argument must fall.

CONCLUSIONS: ACCOUNTING FOR CULTURE

Culture is not simply human psychology, not even evolutionary psychology, "writ large." We still need the social sciences. But psychology underlies culture and society, and biological evolution underlies psychology. By thinking about all three kinds of explanation at once we build vertically integrated theory, theory that passes the compatibility test of this book's Introduction. More than that, the vertically integrated frame helps us to ask the right questions, as when, as Tooby and Cosmides (1989) point out, we think in terms of what kind of psychology might have evolved in response to the adaptive problems of the Pleistocene, or, as does the present chapter, we ask what kind of psychology may be permitting evolutionarily unanticipated social forms.

At the beginning of this book, Tooby and Cosmides (in their chapter, "The psychological foundations of culture") demolish the standard psychological assumptions upon which most social science theories rest. *What then is to replace that standard social science model?* Following Tooby and Cosmides is Symons, who in his chapter demolishes facile efforts to account for culturally patterned behavior with a psychology-free evolutionary biology. *How, then, are we to apply an evolutionary perspective to modern culture and society?* The present chapter provides a single answer to both of those questions: We replace unexamined psychological assumptions not with pure biology but with an evolutionary psychology (or at least, for those who dislike labels, with an evolutionarily-informed psychology). As research from an evolutionary perspective accumulates the possibility of an evolutionarily informed and psychology-compatible social science grows. Perhaps it will resemble the two illustrations of which this chapter consists.

NOTES

1. I am here assuming that there has been little change in the genetic bases of our capacity for culture (and of human psychology in general) since the end of the Old Stone Age. Indirect evidence for this assumption comes from the fact that the offspring of members of human pop-

ulations with very different cultures and technologies nevertheless generally assimilate into new societies with little difficulty. At the same time, there is no evidence that any human population displays a novel "mental organ" that has evolved in response to new adaptive problems.

2. Social scientists influenced by Emile Durkheim (e.g., Radcliffe-Brown, Leslie White) would disagree strongly with the assertion that there is always a relevant psychological level of explanation underlying social behavior. Though there are historical reasons for this position, it is difficult to take seriously today because it would seem to be obvious that social action necessarily involves individual actors who behave in a manner generated by their individual psychologies. Psychological and sociological accounts are at different levels of analysis and need to be compatible with one another but are not competitors. Bock (1988) is certainly right when he declares all anthropology (and by extension, all social science) psychological. Theorists who deny this logical necessity usually ignore their psychological assumptions, which, being implicit and unexamined, often are based on folk psychology.

3. I am deliberately avoiding controversies over the precise meaning of "band" because they are not germane. The interested reader is directed to Meillassoux (1973), Woodburn (1982), and (for a general review) Myers (1988).

4. I am here taking an explicitly adaptationist stance and am therefore risking accusations of "panselectionism." The assumption that human psychology is the product of natural selection leads to testable theory and much research, while failing to make that assumption leaves one scratching one's head. See Barkow (1989, pp. 8, 29–31) for further discussion of this point. Note that I am also assuming, along with Williams (1966), that it is more profitable to focus on individual rather than on group selection. A group selectionist, however, might argue that stratification may have been directly selected for.

5. This statement is based on my unpublished field notes from research among nonliterate rural cultivators in West Africa.

6. There is at least some reason to believe that the very existence of the monumental temples and tombs of ancient civilizations is evidence of strong cooperation between kings and priests in controlling populations, suggesting social exchange between these two élite groups. See Wheatley (1971) for discussion.

7. In order to simplify the discussion, I have omitted analysis of societies with nonhereditary social rank. The usual example of this type of society is that of the New Guinea "big man," who wins control over surplus by strength of personality and by a reputation for producing much while consuming little. As he redistributes the wealth that he and his associates amass, he (and they, by reflection) rise in social rank. However, because his position depends upon his own personality, he apparently cannot transmit his rank to his children. The "big man" phenomenon may perhaps be considered a first step toward social stratification. It is not known how often "big men" are themselves the sons of big men.

REFERENCES

Alexander, R. D. (1979). *Darwinism and human affairs.* Seattle: University of Washington Press.

Axelrod, R., and Hamilton, W. D. (1981). The evolution of cooperation. *Science 211,* 1360–1366.

Barkow, J. H. (1975a). Prestige and culture: A biosocial approach. *Current Anthropology, 16,* 553–572.

Barkow, J. H. (1975b). Strategies for self-esteem and prestige in Maradi, Niger Republic. In T. R. Williams (Ed.), *Psychological anthropology* (pp. 373–388). The Hague: Mouton.

Barkow, J. H. (1980). Prestige and self-esteem: A biosocial interpretation. In D. R. Omark, F. F. Strayer, & D. G. Freedman (Eds.), *Dominance relations: An ethological view of human conflict and social interaction* (pp. 319–332). New York: Garland.

Barkow, J. H. (1981). Evolution et sexualité humaine. In C. Crépault, J. Lévy, & H. Gratton (Eds.), *Sexologie contemporaine* (pp. 103–118). Québec: Les Presses de l'Université du Québec.

Barkow, J. H. (1983). Begged questions in behavior and evolution. In G. Davey (Ed.), *Animal models of human behavior* (pp. 205–222). Chichester: Wiley.

Barkow, J. H. (1989). *Darwin, sex, and status: Biological perspectives on mind and culture.* Toronto: University of Toronto Press.

Barkow, J. H. (1991). Joinings, discontinuities, and details: Darwin, sex, and status revisited. *Behavioral and Brain Sciences. 14:*320–334.

Blau, P. M. (1964). *Exchange and power in social life.* New York: Wiley.

Bock, P. K. (1988). *Rethinking psychological anthropology: Continuity and change in the study of human action* (revision of *Continuities in psychological anthropology*). New York: Freeman.

Buss, D. M., and Dedden, L. (1990). Derogation of competitors. *Journal of Social and Personal Relationships, 7,* 395–422.

Cosmides, L., & Tooby, J. (1989). Evolutionary psychology and the generation of culture, II. Case study: A computational theory of social exchange. *Ethology and Sociobiology, 10,* 51–97.

Flanagan, J. G. (1989). Hierarchy in simple "egalitarian" societies. In B. J. Siegal, A. R. Beals, & S. A. Tyler (Eds.), *Annual review of anthropology* (Vol. 18) (pp. 245–266). Palo Alto: Annual Reviews.

Fried, M. H. (1967). *The evolution of politicial society: An essay in political anthropology.* New York: Random House.

Gluckman, M. (1963). Gossip and scandal. *Current Anthropology, 4,* 307–316.

Hamilton, W. D. (1964). The evolution of social behavior. *Journal of Theoretical Biology, 7,* 1–52.

Hinde, R. A. (1987). *Individuals, relationships and culture: Links between ethology and the social sciences.* Cambridge: Cambridge University Press.

Homans, G. C. (1967). *The nature of social science.* New York: Harcourt, Brace & World.

Humphrey, N. K. (1976). The function of intellect. In P.P.G. Bateson & R. A. Hinde (Eds.), *Growing points in ethology* (pp. 303–317). Cambridge: Cambridge University Press.

Humphrey, N. K. (1983). *Consciousness regained: Chapters in the development of mind.* Oxford: Oxford University Press.

Krebs, J., & Davies, N. (1987). *An Introduction to Behavioural Ecology* (2nd ed). Boston: Blackwell Scientific Publications.

Meillasoux, C. (1973). On the mode of production of the hunting band. In P. Alexandre (Ed.), *French perspectives in African studies.* (pp. 187–203). London: Oxford University Press.

Myers, F. R. (1988). Critical trends in the study of hunter-gatherers. In B. J. Siegal, A. R. Beals, & S. A. Tyler (Eds.), *Annual review of anthropology* (Vol. 17) (pp. 261–282). Palo Alto: Annual Reviews.

Sahlins, M. D. (1972). *Stone age economics.* Chicago: Aldine-Atherton.

Titmuss, R. M. (1971). *The gift relationship: From human blood to social policy.* New York: Pantheon Books.

Tooby, J., & Cosmides, L. (1989). Evolutionary psychology and the generation of culture, I. Theoretical considerations. *Ethology and Sociobiology, 10,* 29–49.

Trivers, R. L. (1971). The evolution of reciprocal altruism. *Quarterly Review of Biology, 46,* 35–57.

Wheatley, P. (1971). *The Pivot of the Four Quarters.* Chicago: Aldine.

Williams, G. C. (1966). *Adaptation and natural selection: A critique of some current evolutionary thought.* Princeton: Princeton University Press.

Woodburn, J. (1982). Egalitarian societies. *Man, 17,* 431–451.

AUTHOR INDEX

SUBJECT INDEX